Discrete Cash Flow Examples Illustrating Basic Time Value of Money Formulas

Example Problems (all using an interest rate of i = 10% compounded annually)

In Borrowing–Lending Terminology:	Cash Flow Diagram	Solution
A firm borrows $1,000 for 8 years. How much must it repay in a lump sum at the end of the eighth year?	P = $1,000; 0; N = 8; F = ?	$F = P(F/P, 10\%, 8)$ $= \$1{,}000(2.1436)$ $= \$2{,}143.60$
A firm desires to have $2,143.60 eight years from now. What amount should be deposited now to provide for it?	F = $2,143.60; 0; N = 8; P = ?	$P = F(P/F, 10\%, 8)$ $= \$2{,}143.60(0.4665)$ $= \$1{,}000.00$
If eight annual deposits of $187.45 each are placed in an account, how much money has accumulated immediately after the last deposit?	F = ?; 1 2 3 4 5 6 7 8; A = $187.45	$F = A(F/A, 10\%, 8)$ $= \$187.45(11.4359)$ $= \$2{,}143.60$
How much should be deposited in a fund now to provide for eight end-of-year withdrawals of $187.45 each?	A = $187.45; 1 2 3 4 5 6 7 8; P = ?	$P = A(P/A, 10\%, 8)$ $= \$187.45(5.3349)$ $= \$1{,}000.00$
What uniform annual amount should be deposited each year in order to accumulate $2,143.60 at the time of the eighth annual deposit?	F = $2,143.60; 1 2 3 4 5 6 7 8; A = ?	$A = F(A/F, 10\%, 8)$ $= \$2{,}143.60(0.0874)$ $= \$187.45$
What is the size of 8 equal annual payments to repay a loan of $1,000? The first payment is due 1 year after receiving the loan.	P = $1,000; 1 2 3 4 5 6 7 8; A = ?	$A = P(A/P, 10\%, 8)$ $= \$1{,}000(0.18745)$ $= \$187.45$

8th Edition

ENGINEERING ECONOMY

ABOUT THE AUTHORS

E. Paul DeGarmo is a Professor Emeritus of Industrial Engineering and Mechanical Engineering at the University of California, Berkeley. He was a co-author (with B. Woods) in 1942 of the first edition entitled *Introduction to Engineering Economy* and has authored or co-authored every subsequent edition.

William G. Sullivan is a Professor of Industrial Engineering at the University of Tennessee, Knoxville. He obtained a Ph.D in Industrial and Systems Engineering from the Georgia Institute of Technology and is a registered professional engineer. He has been a co-author of *Engineering Economy* since the sixth edition.

James A. Bontadelli is director of industrial engineering for the Tennessee Valley Authority and an Adjunct Professor of Industrial Engineering at the University of Tennessee, Knoxville. He earned his Ph.D from The Ohio State University, and is a registered professional engineer with more than 30 years of professional practice.

8th Edition

ENGINEERING ECONOMY

E. Paul DeGarmo
William G. Sullivan
James A. Bontadelli

MACMILLAN PUBLISHING COMPANY
New York
COLLIER MACMILLAN PUBLISHERS
London

Macmillan Publishing Company,
866 Third Avenue, New York, New York 10022

Collier Macmillan Canada, Inc.

Library of Congress Cataloging-in-Publication Data

DeGarmo, E. Paul (Ernest Paul), (date)
Engineering economy / E. Paul DeGarmo, William G. Sullivan, James
A. Bontadelli. — 8th ed.
p. cm.
Bibliography: p.
Includes index.
ISBN 0-02-328634-2
1. Engineering economy. I. Sullivan, William G., 1942-
II. Bontadelli, James A. III. Title.
TA177.4.D43 1989
658.1′55—dc19 88–25874
CIP

Printing: 1 2 3 4 5 6 7 8 Year: 0 1 2 3 4 5 6 7 8

PREFACE

A successful product or service must be technically well-conceived and must also produce results that are viewed favorably by the consuming public. That is, benefits associated with products and services must exceed their costs in order to be considered valuable. This basic principle has been extended to engineering projects for which it is also true that monetary *and* nonmonetary benefits must outweigh their implementation (acquisition and operational) costs. Accordingly, the field of engineering economy is concerned with the systematic evaluation of the benefits and costs of projects involving engineering design and analysis.

Engineering economy requires the application of technical and economic analysis, with the goal of deciding which course of action best meets technical performance criteria and uses scarce capital in a prudent manner. The domain of such decision-making activity ranges from the engineer who uses advanced computer technology to design new products, structures, systems, and services to the chief executive officer considering a major business venture that may transform the organization in the years ahead.

History of the Book

The original *Introduction to Engineering Economy*, authored by Woods and DeGarmo, appeared in 1942. The text has been in continuous classroom use for nearly 50 years, and approximately 250,000 students have learned about engineering economic analysis from the pages of this book. The eighth edition of *Engineering Economy* has built upon the rich and time-tested teaching materials of earlier editions, and its publication makes the book the second oldest on the market that deals exclusively with engineering economy.

Intended Use of the Book

This book has two primary purposes: (1) to provide a sound understanding of the principles, basic concepts, and methodology of engineering economy; and (2) to develop proficiency with these methods and with the process for making rational decisions regarding situations likely to be encountered in professional practice. Consequently, *Engineering Economy* is intended to serve as a text for classroom instruction *and* as a basic reference for use by practicing engineers in all specialty areas (e.g., chemical, civil, electrical, industrial, and mechanical).

The book is also useful to persons engaged in the management of technical activities.

As a textbook, the eighth edition is written principally for the first formal course in engineering economy. The contents of the book and the accompanying *Instructor's Manual* are organized for effective presentation and teaching of the subject matter. In addition, numerous examples and case studies dedicated to important application areas are integral to almost all chapters in the book.

A 3 credit-hour semester course should be able to cover the majority of topics in this edition, and there is sufficient depth and breadth to enable an instructor to arrange course content to suit individual needs. Moreover, by including several advanced topics in this book, a continuing second course in engineering economy can also be based on the eighth edition. The suggested text materials for a 2 credit-hour and a 3 credit-hour semester course are listed in Table P-1. Also shown in Table P-1 are advanced topics that can be utilized in a continuing second course in engineering economy.

An engineering economy course may be classified, for Accreditation Board for Engineering and Technology (ABET) purposes, as part engineering science and part engineering design. It is generally advisable to develop and teach such a course at the upper division level where the course incorporates accumulated knowledge acquired in other areas of the curriculum. The 2 and 3 credit hour semester courses shown in Table P-1 include many of the elements of engineering design for accreditation purposes; namely, iterative problem solving, open-ended exercises/case studies, creativity in formulating and evaluating feasible solutions to problems, and consideration of realistic constraints (*economic*, aesthetics, safety, etc.) in problem solving.

Changes in This Edition

The eighth edition of *Engineering Economy* features a number of notable changes from the previous edition, including the following.

1. The first three chapters in the previous edition are replaced with two new chapters that develop and present the principles and basic concepts of engineering economy in a more structured and integrated manner.
2. Chapter 3 deals with money–time relationships (previously Chapter 4) and improves on this fundamental material, which has proven over the years to be highly effective in classroom use. Chapters 4 and 5 (previously Chapters 5 and 6) have been modified to explain more clearly the basic methods for evaluating prospective cash flows associated with one or more proposed alternatives (i.e., solutions) to an engineering problem.
3. Chapters 6 and 7 treat *depreciation* and *after-tax analyses* of proposed alternatives, respectively. In these two chapters, Chapters 9 and 10 of the previous edition have been substantially rewritten to incorporate changes introduced into engineering economy studies because of the Tax Reform

TABLE P–1 Suggested Coverage of Text Material in Engineering Economy Courses

Chapter	2 Credit-Hour Semester Course (3 Credit-Hour Course on a Quarter Basis)	3 Credit-Hour Semester Course	Additional Topics, Advanced Material, or Case Studies
1	All Sections	All Sections	
2	All Sections	All Sections	
3	All Sections except 3.13, 3.16, 3.23, and 3.24	All Sections	
4	All Sections except 4.11, 4.12 and Ch. 4 appendixes	All Sections except 4.11, 4.12, and Ch. 4 appendixes	Sections 4.11 and 4.12; Ch. 4 appendixes
5	All Sections except 5.6 and 5.7	All Sections	
6	All Sections except 6.10.5 through 6.14	All Sections except 6.10.5 through 6.14	Sections 6.10.5 through 6.14
7	All Sections except 7.13 and 7.14	All Sections except 7.13 and 7.14	Sections 7.13 and 7.14
8	All Sections except 8.6	All Sections except 8.6	Section 8.6
9	Sections 9.1 and 9.2	Sections 9.1 and 9.2	Sections 9.3 through 9.8
10	Sections 10.1 through 10.6	Sections 10.1 through 10.10	Sections 10.11 through 10.15
11	Omit	Optional	All Sections
12	Optional	Sections 12.1 through 12.7	Sections 12.8 through 12.14
13	Omit	Sections 13.1 through 13.10	Sections 13.11 through 13.16
14	Omit	Optional	All Sections
	Appendixes B,C,E,H,I	All Appendixes except Appendix F	Appendix F

Act of 1986. In addition, much of the material has been reorganized for more precise and logical presentation to students.

4. Chapter 8 is a new chapter devoted entirely to developing cash flow estimates of feasible alternatives. It includes such topics as the work breakdown structure technique, cost and revenue structures, cost estimating techniques (models), and life-cycle considerations in engineering design and analyses. Also included is a comprehensive case study of developing project cash flows and analyzing their after-tax economic consequences.
5. Chapter 9 deals with treatment of inflation and price changes in before- and after-tax engineering economy studies. The topics are organized into two parts: introductory discussion and examples involving general price inflation, and advanced subjects that involve differential inflation and price changes. This represents a major improvement to the coverage in our previous edition.
6. Chapter 10 (previously Chapter 7) includes changes in dealing with uncertainty, especially in the area of sensitivity analysis. Chapter 11 is a brand-new chapter that pragmatically presents various methods for incorporating multiple-attribute decision analysis into an engineering economy study. Thus, the student will be able to handle both monetary (economic) and nonmonetary attributes in a decision problem.
7. Chapter 12 (Dealing with Replacement Problems) contains minor changes to and updating of Chapter 11 in the seventh edition. Chapter 13 (Capital Financing and Allocation) contains essential material from the previous edition, but it has significant new sections dealing with the capital investment process in industry.
8. Finally, Chapter 14 is concerned with evaluation of public projects, and it has been modified with updated material on the benefit-cost ratio method of conducting such studies.

Other Special Features

The *Intructor's Manual* is designed as a comprehensive aid in teaching the text material. For each chapter, the overview and perspective discussion provided will assist the instructor with presenting the rationale and context of the material. Suggested material for the preparation of view-graphs is also included. Full solutions of all problems at the end of each chapter are presented. In addition, representative examination questions (a "test bank") are provided to assist with this important part of course instruction.

There are *three computer programs* provided in the appendixes to this book (Appendixes C, E, and F). We have used these programs in our classes and have confirmed that they enhance the learning experience in a traditionally taught engineering economy course. They also provide the basis for a nontraditional course, such as one in which self-paced learning is the paradigm. Key features of each program are as follows:

Appendix	*Program Name*	*Features*
C	ITABLE	• Generates interest tables for discrete and continuous compounding
E	BTAX	• Calculates all equivalent worth methods, two rate of return methods, and two different payback periods for a sequence of cash flows
		• Performs sensitivity analyses of sequences of cash flows
F	ATAX	• Calculates depreciation by most methods in use since 1971
		• Calculates after-tax cash flows of projects (considering inflation, replacement policies, and sensitivity questions)
		• Determines present worth of a project whose cash flows are estimated in actual dollars

All programs have been written in BASIC to permit their portability among IBM or IBM-compatible personal computers that run on MS-DOS. Furthermore, instructions for each program and ample examples of their use are provided. These three programs are contained on a single 5 1/4" diskette (double sided, double density), and instructors who adopt *Engineering Economy* for classroom use may obtain the diskette free of charge from Macmillan Publishing Company. Duplication and distribution of copies of this diskette to students may be performed on a royalty-free basis.

Additional *comprehensive case studies* have been included in the eighth edition which provide the instructor with essential material for teaching both the first formal course and a second, more advanced course in engineering economy. These case studies demonstrate the integrated application of the principles, basic concepts, and methodology in typical real world situations that are encountered by engineers. They also serve as a bridge from the classroom to professional practice.

ACKNOWLEDGMENTS

Many friends and associates have offered valuable and constructive ideas and statements that have crystallized into numerous improvements to the eighth edition. To these many individuals, we express our sincere appreciation. Also, to our students, who have unknowingly been the proving ground for much of this text, we offer our thanks. They have reacted, without compassion on some occasions, to make us aware of good and bad methods of presentation. It is our hope that they, and those who follow them, will find this book an invaluable guide to the study and practice of engineering economy.

We wish to thank the following individuals for their helpful suggestions in the preparation of the eighth edition: Matthew S. Polk, Widener University; Richard H. Bernhard, North Carolina State University; Jorge Haddock, Rensselaer Polytechnic Institute; Donald F. Haber, University of Idaho, Moscow; Daniel L. Babcock, University of Missouri, Rolla; John Dracup, University of California, Los Angeles; A. Rashid Hasan, University of North Dakota; Paul A. Nelson, Michigan Technological University; A. T. Wallace, University of Idaho; William J. Foley, Rensselaer Polytechnic Institute; Cihan Dagli, Wichita State University; Maurice K. Kurtz, Florida Institute of Technology; Ronnie B. Catipon, Franklin University; Frederic C. Menz, Clarkson University; Philip Jacobs, University of South Carolina; W. W. Shaner, Colorado State University; Klaus E. Kroner, University of Massachusetts; and Khairy A. Tourk, Illinois Institute of Technology.

Also for the many suggestions from other colleagues, students, and practicing engineers who have used the previous editions, we express our deep appreciation. We hope they will find this edition to be even more helpful.

E. PAUL DEGARMO

WILLIAM G. SULLIVAN

JAMES A. BONTADELLI

CONTENTS

CHAPTER 1: Introduction and the Basic Principles of Engineering Economy 1

CHAPTER 2: Cost Concepts and Analysis 23

CHAPTER 3: Principles of Money-Time Relationships 62

CHAPTER 7: Evaluation of Alternatives on an After-Tax Basis 296

CHAPTER 8: Developing Cash Flows 351

CHAPTER 9: Dealing with Inflation and Price Changes 395

CHAPTER 10: Dealing with Uncertainty 445

1

INTRODUCTION AND THE BASIC PRINCIPLES OF ENGINEERING ECONOMY

The following topics are discussed in this chapter:

Why is this subject important in engineering practice?
The decision-making process
Origins of engineering economy
The relationship between engineering and management
Nonmonetary factors (attributes) and multiple objectives
Capital allocation and engineering economy
What are the principles of engineering economy?
Engineering economy, methodology, and application
Overview of the book

1.1

INTRODUCTION

Engineering economy is a discipline concerned with the systematic evaluation of the costs and benefits of proposed technical and business projects and ventures. The principles and methodology of engineering economy have wide application in engineering design and technical and general management. These principles and methodology are an integral part of the daily management and operation of private companies and corporations, regulated public utilities, government units or agencies, and nonprofit organizations. Furthermore, they are utilized to analyze alternative uses of financial resources, particularly in relation to the physical assets and operation of an organization. Last, but certainly not least, engineering economy will prove to be invaluable to you in assessing the economic merits of alternative uses of your personal funds.

The objectives of this first chapter are to introduce the subject of engineering economy; to discuss its crucial role in engineering practice; to delineate, discuss, and illustrate the basic principles of the subject; and to provide an overview of the book.

In the operation of an organization, engineers and managers use engineering economy to assist with decision making in situations such as the following:

1. Selecting between alternative designs for a component, machine, structure, system, product, or service during the engineering design process.
2. Estimating and analyzing the economic consequences of automation improvements in a factory operation.
3. Selecting among proposed projects within the annual capital budget limit established in a corporation.
4. Analyzing whether the transportation equipment in the service fleet should be replaced, and at what rate.
5. Choosing between asset lease and purchase options to support a new product line within a company.

The list could go on. However, the importance of the subject to engineers and managers in either the private or public sectors of the economy is obvious from the above examples.

Engineering economic analysis, which is used interchangeably with *engineering economy,* usually includes significant technical considerations. Conceptually, engineering economic analysis is the same as that in most other types of technical analysis. Mathematical modeling, with emphasis on the economic effects, is the primary analytical technique used to select between defined feasible alternatives. Thus, engineering economy involves technical-economic analysis with a decision-assisting objective. This is true whether the decision maker is an engineer interactively analyzing design alternatives at a computer-aided design workstation or is the chief executive officer (CEO) considering a new project.

Our development of the subject uses the *cash flow approach* to predict and analyze the economic effects of feasible alternatives related to a design or problem situation. A cash flow occurs when money is transferred from one organization or individual to another. Thus, a cash flow represents the economic effects of an alternative in terms of money spent and received.

Engineers and other personnel engaged in technical and general management will find this subject vital to the accomplishment of their assigned responsibilities and to the achievement of their individual career goals. The experience of the authors and the corroborative experience of others clearly indicate the high use and utility of engineering economy in today's workplace. An engineer who is unprepared to excel at engineering economic analysis is not properly equipped for his or her total job. You do not want to be unprepared.*

1.2 THE DECISION-MAKING PROCESS

As William T. Morris stated, "two fundamental ideas kept in mind may help considerably with a study of decision making: *by a decision we mean a conceptualization of a choice situation, whether in the form of a mental image or an explicit model*; and, *all decision making involves the simplification of reality*."† These two basic concepts imply that decision making is not a single, unique process replicated the same way by different individuals; rather, the decision-making process will vary in form and content with different people, and even with the same individual under different situations.

There are several general steps, however, that appear to occur in decision making. Even though there will be variation in their delineation by different authors, they are common to all problem solving. We will represent and discuss the decision-making process in terms of the following four steps:

1. Recognition and formulation of the problem.
2. Search for feasible alternatives.
3. Analysis.
4. Selection/decision.

Explicit recognition, and then formulation of the problem, need to take place before proceeding with the decision-making process. Emphasis in this first general

*An engineering economy textbook is one of the most used references in engineering practice.

†William T. Morris, *Decision Analysis* (Columbus, Ohio: Grid, Inc., 1977), p. 4.

step is on understanding the problem. Some information gathering usually occurs, and the boundary or extent of the problem (or situation being studied) is defined. The current status of operations and the problem area are described, and the desired goals, objectives, or other results to be achieved are delineated. Also, any special conditions or constraints are identified.

The *search for (including development of) feasible alternatives* involves creativity and innovation in developing potential solutions to the problem. The objective of this second general step is to develop a list of potential alternatives, and then to screen these alternatives and select a smaller group of feasible alternatives. The potential alternatives are screened against the desired outcomes, special conditions and constraints, and internal and external requirements that need to be met. A procedure particularly useful in accomplishing this step is discussed in Section 1.7.

The *analysis of the feasible alternatives* is accomplished in the third general step. This includes selecting the criterion (or criteria) for judging the alternatives, gathering additional information and developing relevant data, defining any remaining requirements, and structuring (or modeling) the interrelationships involved. Any alternatives that become infeasible as a result of the detailed analysis are eliminated. The result of the analysis step is measurement of the remaining feasible alternatives in terms of the criterion (or criteria) selected.

The final step involves the decision—*selection of the preferred alternative*. Also, the alternative selected needs to be adequately described and an implementation plan developed. The description of the selected alternative, plus the implementation plan and instructions, should include the desired outcomes in terms that can be subsequently monitored through performance measurement.

EXAMPLE 1-1

A special architecture-engineering (A-E) team, including a representative from the facilities management staff, has been organized in your corporation to reanalyze the size of a new office and service support building being designed, and to review the use of other similar space. The detailed design for the new building is presently being done by an internal A-E group (supplemented with contract support). The problem under review concerns the projected increase in requirements for laboratory space (26%) and for personnel space (11%). Due to the status of the detailed design effort, it is important that the team present a recommendation to management within 7 weeks on the best way to meet these additional space requirements. Make a recommendation in terms of the four steps of the decision-making process.

Solution:
The team began its effort with a detailed examination of the additional space requirements and a review of the design results to date. The following information was developed as part of step 1 of the decision-making process:

1. The foundation and structural design will permit the addition of two floors to the six presently planned for the building. Each floor has 22,000 gross square feet of space.
2. The heating, ventilation, and air conditioning (HVAC) system is planned for six floors. The detailed design of the system has not started, but the cost of a larger HVAC system would have to be included if additional floors were added.
3. The additional laboratory space (26%) and personnel space (11%) will require 31,500 gross square feet.
4. There is a time constraint. The need for the new building space is urgent.
5. Additional money for the new building, beyond that already allocated, is limited.

In step 2 of the decision-making process, the team worked as a group to develop a list of potential alternatives to meet future space needs. Then, the team (1) developed in more detail the constraints, the internal and external requirements, and the desired outcomes related to space needs; and (2) screened the potential alternatives using this explicit information. The result was the identification of three feasible alternatives:

1. Add two floors to the design of the new building.
2. Rehabilitate some existing storage space for laboratory space and rent the additional space needed for personnel and storage.
3. Rent the additional space required for laboratories and personnel.

In the analysis step (step 3), the team decided an appropriate period for the study was 15 years, and the primary decision criterion was minimizing costs. However, other criteria such as meeting the time constraint and causing minimum disruption to present operations would be considered. Then, the relevant cost data and other information needed for the analysis were developed and organized. The team leader at this point divided the team into three working groups. Each group was assigned to estimate the economic consequences in cash flow form and to organize other information relevant to the decision for one alternative. After the groups finished, the whole team reviewed the cash flow and other information for each alternative. They found, when all costs were considered, that Alternative 2 was slightly better. However, Alternatives 2 and 3 were equal regarding schedule (time constraint), and Alternative 3 was preferred regarding minimizing the disruption of current operations. Alternative 1 was considered infeasible (after the detailed analysis) due to the additional funds required immediately for the new building if two floors were added to the design and due to the longer schedule that would result.

Based on these results, the team decided to recommend Alternative 2. In step 4, they completed the documentation of their analysis and submitted the report and final recommendation to senior management. Included in the report was an implementation plan for Alternative 2. After meeting with the special team early in the seventh week, management approved Alternative 2 for implementation.

1.3 ORIGINS OF ENGINEERING ECONOMY

Cost considerations and comparisons are fundamental aspects of engineering practice. This basic point was emphasized in Section 1.1. However, the development of engineering economy methodology, which is now used in nearly all engineering work, is of relatively recent origin. This does not mean that, historically, costs have usually been overlooked in engineering decisions, but the perspective that ultimate economy is a primary concern to the engineer and the availability of sound techniques to address this concern differentiate this aspect of modern engineering practice from past practices.

A pioneer in the field was Arthur M. Wellington,* a civil engineer, who in the latter part of the nineteenth century specifically addressed the role of economic analysis in engineering projects. His particular area of interest was railroad building in the United States. This early work was followed by other contributions in which the emphasis was on techniques that depended primarily on financial and actuarial mathematics.

In 1930, Eugene Grant published the first edition of his textbook.† This was a significant milestone in the development of engineering economy as we know it today. Even though he relied on the mathematics of finance as a basic part of the methodology, he definitively established that engineering economy was not just the use of these techniques in engineering problems. He placed emphasis on developing an economic point of view in engineering, and (as he stated in the preface) "this point of view involves a realization that quite as definite a body of principles governs the economic aspects of an engineering decision as governs its physical aspects."

The development of engineering economy over a period of more than a century, similar to the development of most fields of knowledge, has incorporated concepts from a number of other disciplines. This development and assimilation process has occurred rather rapidly in the past 50 years. Accounting, finance,

*Arthur M. Wellington, *The Economic Theory of Railway Location*, second ed. (New York: John Wiley and Sons, 1887).

†Eugene L. Grant, *Principles of Engineering Economy* (New York: The Ronald Press Company, 1930).

microeconomics, and engineering—combined with modeling and analysis—provided an initial core of knowledge from which the evolutionary development proceeded.*

Over time, tax law, decision analysis, probability and statistics, computer science, operations research, engineering management, cost estimating, and other disciplines and techniques have made significant contributions to its development. The field today remains dynamic, and the development of new or improved techniques is occurring routinely in both practice and academe.

1.4 THE RELATIONSHIP BETWEEN ENGINEERING AND MANAGEMENT

Business continues to become more technical. Consequently, engineers play an increasingly important role in general management as well as a dominant role in technical management. More and more decision making in government and industry involves engineers. Often, these decisions are made primarily on the basis of the technical and economic factors involved. In an increasing number of situations, however, other factors must be included in an analysis and may prevail over these considerations. When managers are not engineers, they often call upon engineers to make technical-economic analyses that lead to recommendations upon which managerial decisions can be based. Frequently, an engineer is essentially in the position of a consultant to management and must combine technical and economic knowledge and analysis to provide sound conclusions and recommendations.

With recent developments in mathematical, statistical, and computer techniques that permit the quantitative handling of increasingly complex technical-economic problems, the engineer has an opportunity to play an even more important role in the decision-making process. Not only do engineers have the mathematical and scientific background for understanding and using such techniques, but they also have the background that enables them to recognize the practical limitations of these techniques and the effect of the lack of information that usually exists in real situations. Thus, the engineer is in a position to make or recommend the necessary compromises and adjustments that will enable a realistic, although possibly not perfect, solution to be implemented.

*For instance, the important relationship between accounting and engineering economy is described in Appendix A.

1.5

NONMONETARY FACTORS AND MULTIPLE OBJECTIVES

We stated in Section 1.4 that, increasingly, noneconomic factors (also called attributes) need to be included in an analysis and considered in the decision. Thus, the decision and subsequent success of engineering and business projects can be affected to a considerable extent by *nonmonetary factors* and *multiple objectives*. Here, we use the term nonmonetary factors (attributes) to mean those considerations vital to decision making for which a market mechanism does not exist to establish value.

For example, in an individual situation, if a person goes to a store to buy a new suit of clothes and finds that one in plain black can be obtained for $75 less than one of comparable quality in other colors, he or she probably will not make the decision solely on the basis of price. The color, fabric, cut of the clothing, and other factors may prevail in the decision. The financial well-being of the manufacturer of the garments will be affected by the desires of customers for color, style, fabric, and so on. Similarly, satisfactory decisions and recommendations regarding the feasibility of new projects or changes to the current operation must take into account all factors, monetary and nonmonetary, that will affect the undertaking, as well as the multiple objectives that often must be achieved. Some of the most common nonmonetary (also called *intangible* and *irreducible*) factors that need to be considered involve general business conditions, ethical and social values, consumer likes and dislikes, environmental impact, and governmental regulations.

All businesses operate within an economic system that functions in accordance with certain general rules. Whether a product or service is socially desirable, acceptable, or needed can easily determine the success or failure of a venture. Similarly, whether a product or service meets the existing likes and dislikes of the public, which may change with time, can be important in the success or failure of an enterprise. For example, a limited list of objectives other than profit maximization or cost minimization that can be important to an organization would include the following:

1. Meeting customer expectations consistently.
2. Maximization of employee satisfaction.
3. Maintaining flexibility to meet changing demand.
4. Maintenance of a desired public image.
5. Leveling cyclic fluctuations in production.
6. Improvement of safety in operations.
7. Reduction of pollutants.

Engineering economic analysis formally provides for those objectives or factors that can be reduced to monetary terms. These results should then be evaluated with the other factors before a final decision is made. There exist analytical methods to accomplish the integrated evaluation of monetary and nonmonetary factors in decision making, if the appropriate estimates can be made. In general, one should pursue formal analysis as far as is technically and economically feasible, and then depend on the judgment and experience of the decision maker.

Although the primary focus of this book is on the correct use of procedures to analyze economic desirability, particularly when technical considerations are involved, our treatment of engineering economy includes the consideration of nonmonetary factors and multiple objectives in decision making. This subject is discussed in more detail in Chapter 11.

1.6 CAPITAL ALLOCATION AND ENGINEERING ECONOMY

In this section, we introduce the capital allocation (investment) function of an organization and briefly describe how engineering economy relates to it. A more detailed explanation of capital sourcing and allocation is provided in Chapter 13.

An outstanding phenomenon of present-day industrialized civilizations is the extent to which engineers and managers, by using capital (money and property), are able to create wealth through activities that transform various types of resources into goods and services. Naturally, those who provide capital and thereby finance these activities are concerned that it be invested to best advantage. Likewise, those who consume (spend) capital by designing and implementing new processes, products, systems, and services are equally concerned that available capital be invested wisely. An overview of capital sourcing and capital allocation activities in a typical organization is shown in Figure 1-1.

The *financing* function of an organization is responsible for obtaining capital. Capital allocation takes place as part of the *investment* function, and it is here that engineering economy plays a significant role in initiating the capital expenditure approval process. As shown in Figure 1-1, engineering economy is heavily involved with the evaluation of mutually exclusive project alternatives for improving various facets of a company's operations. Mutually exclusive means that one alternative is recommended as best in fulfilling a given function and the remaining feasible alternatives are rejected. From throughout the organization, the "best" alternatives for making different improvements are then aggregated into a project (investment) portfolio for a particular budgeting period (e.g., a

FIGURE 1-1

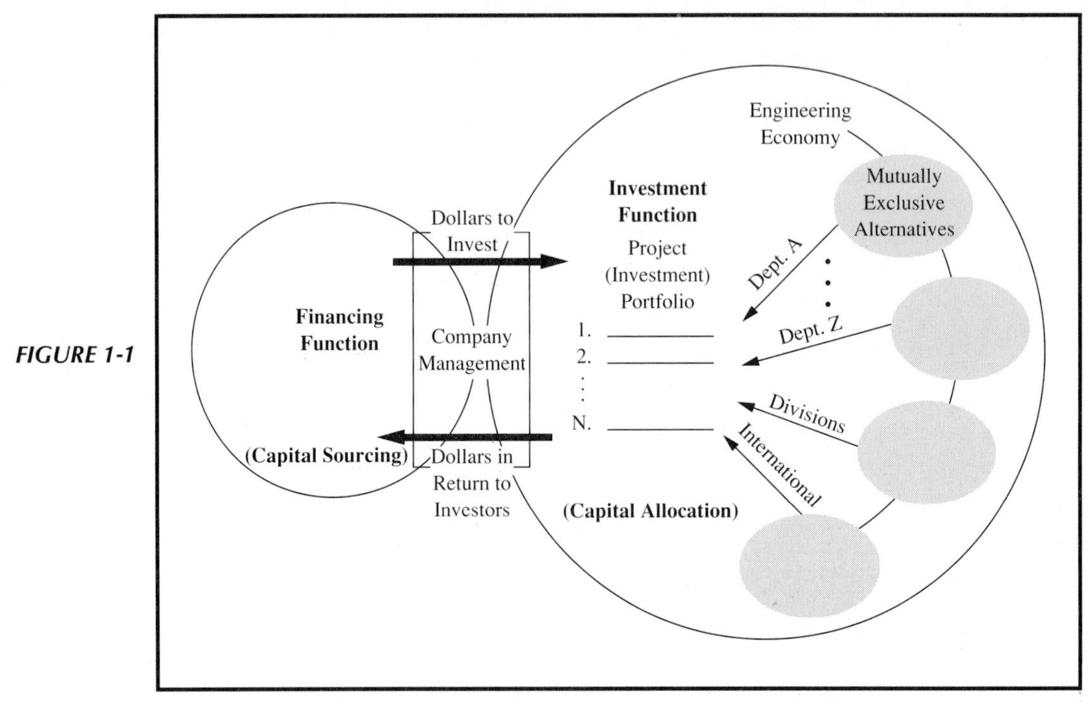

Overview of Capital Sourcing and Allocation Activities in a Typical Organization

fiscal year). This project portfolio consists of *independent* investment opportunities, meaning that a *subset* of them will be approved for funding since available capital in any organization is almost always less than the demand for it. As shown in Figure 1-1, the investment function allocates the available capital among the projects selected on a companywide basis. Management, through its activities in this function, is responsible for ensuring that a reasonable return (in dollars) is earned on these project investments so providers of capital will be motivated to furnish more when the need arises. Thus, it should be apparent why the informed practice of engineering economy as defined in Section 1.1 is an essential element in the foundation of an organization's competitive culture.

1.7 WHAT ARE THE PRINCIPLES OF ENGINEERING ECONOMY?

The development, study, and application of any methodology must begin with a basic foundation. The foundation for engineering economy is a set of principles, or fundamental concepts, that provide a sound basis for development of the methodology.* These principles must be adhered to, and in most instances are also integral to decision making. However, in engineering economic analysis experience has shown that most errors in the analysis and comparison of alternatives can be traced to some violation or lack of adherence to the following seven principles.

Principle 1: The choice (decision) is among alternatives. The feasible alternatives need to be identified and then defined for subsequent analysis. A decision involves making a choice among two or more alternatives. If there is only one alternative, no decision is required.

Developing and defining the *feasible* alternatives for evaluation is important because of the resulting impact on the quality of the decision. Engineers and managers should place a high priority on this responsibility. Creativity and innovation are essential to the process.

This effort begins with an explicit understanding of the design criteria, improvement objectives, business goals, or other results to be achieved. Then, knowledge of the existing operation and related conditions, and the ability to take a *systems viewpoint and approach,* are important in identifying and defining the feasible alternatives.

What do we mean by taking a systems viewpoint and approach in this context? In brief, the alternatives relate to an existing operation or set of conditions. Any changes, and even making no change, will affect (or interact with) the environment of the operation or conditions involved, both internal and external to the organization. Judging feasibility of an alternative, before detailed analysis, is dependent on understanding the effects each alternative will have within the operating environment as well as the potential interactions between this decision and others. Explicitly considering the impacts of each alternative within the environment of the existing operation or set of conditions is the essence of taking a systems viewpoint and approach.

Decision making in organizations often occurs under three different general situations. These are

1. An immediate problem, opportunity, or other operating need. Responding with a timely decision is a normal management function.

*The definition of the principles of engineering economy varies with different authors. We will define and discuss the foundation of the discipline in terms of seven basic principles. Examples of other delineations may be found in

1. Eugene L. Grant, W. Grant Ireson, and Richard S. Leavenworth, *Principles of Engineering Economy*, seventh ed. (New York, New York: John Wiley and Sons, 1982), pp. 4–18.
2. G. A. Fleischer, *Engineering Economy: Capital Allocation Theory* (Monterey, California: Brooks/Cole Engineering Division, 1984), pp. 5–9.
3. Report titled "Research Planning Conference for Developing a Research Framework for Engineering Economics," Gerald J. Thuesen (Editor), Georgia Institute of Technology, March 1986, pp. 10–14. The report was the result of the National Science Foundation Grant MEA-8501237.

2. Improvements or changes to an existing operation. Even though some new projects may be involved, they would normally be limited in scope.
3. New projects and ventures. These are significant extensions to the organization's present operation, or they involve the creation of a new organizational entity and operation.

An example of the first general situation above is the decision making that occurs in a firm when a grievance is filed in accordance with the provisions of the labor agreement. This is a routine action that requires a timely management response. Since this is probably a repetitive problem, the decision making in the organization should already be well defined. Even though management should be alert to the possibility of new courses of action that need to be considered, particularly when the agreement is renegotiated, an analysis and comparison of alternatives to handle this type of problem normally would not be needed.

As an illustration of the second general situation, assume that you work for a company that operates in a major metropolitan area and provides local transportation and delivery services on a contract basis to major department stores and other retail outlets. The only competitor in the area has just renewed a department store contract with prices below what your company is charging in its present contracts, and also has guaranteed reduced delivery times as part of the department store's improved service program to its customers. The management of your company is concerned about erosion of market share, particularly since some major contracts are to be renegotiated soon. Obviously, some improvements will need to be made in the operation of your company to meet the competition. Decision making related to this problem (and opportunity) is representative of the second general situation.

An example of the third situation would be a state department of transportation planning to undertake a major new 5-year program to upgrade and replace bridges in the state and federal highway network. A program of this scope should not be handled as part of normal operations.

Decision making in both the second and third situations involves the development, analysis, and comparison of alternatives. The question, then, is how do we devise the group of feasible alternatives that merit detailed analysis and comparison? The systems viewpoint and approach to do this is shown schematically in Figure 1-2.

The process begins with the creation and innovation step; that is, initially identifying all the potential alternatives that should be given further consideration. Within an organization, all personnel who can contribute to this step should be involved. The objective is to include every course of action that merits further consideration. This step is shown at the top of Figure 1-2.

Detailed analysis is not used to reduce the list of potential alternatives to the much smaller group of feasible alternatives. Instead, the engineers, managers, and other personnel in the organization with knowledge of the existing operation or set of conditions should judge the feasibility of each potential alternative. This can be accomplished using the following three steps:

FIGURE 1-2

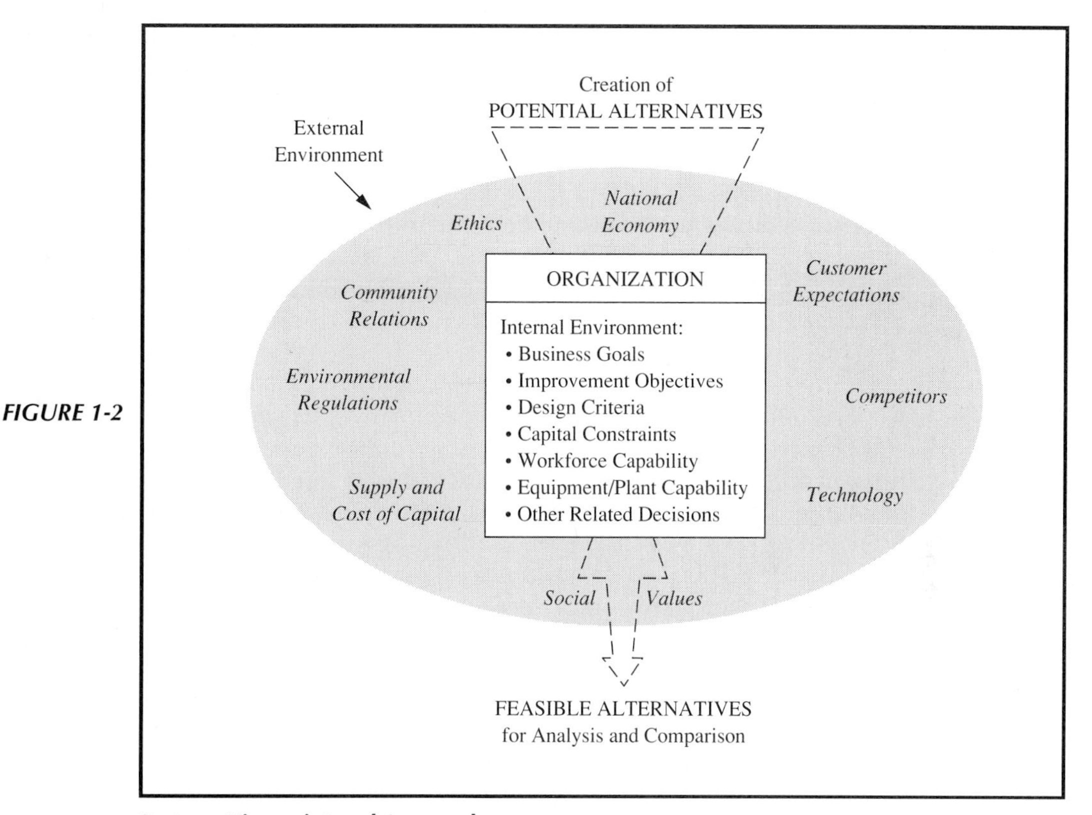

Systems Viewpoint and Approach

1. Perform basic calculations to determine whether each alternative can meet essential performance criteria.
2. Use the systems viewpoint and approach to judge how well each potential alternative will meet the internal and external requirements. Conceptually, the knowledge and experience of the appropriate people are used to consider the impacts of each alternative on the requirements of the internal and external factors to see if they appear to be met, and to determine whether the goals, objectives, and other results can be achieved. The dotted arrow in Figure 1-2 represents the second step in the process.
3. Review each of the potential alternatives that pass steps 1 and 2, and make a final judgment whether it merits a detailed analysis. The result of this final step is the small group of feasible alternatives for which a detailed analysis and comparison should yield a preferred alternative.

A feasible alternative, then, is based on judgment and a minimum of basic calculations; it can be implemented by the organization so that desired results are achieved without unacceptable consequences. As an example, pollution control equipment from Manufacturer A to treat an effluent by-product is not a feasible alternative if it will not achieve a regulated discharge standard. It would not satisfy a mandatory external requirement and would produce an unacceptable consequence. However, if the equipment from Manufacturer B will meet all these requirements and is judged capable of achieving the other results, it would be selected as a feasible alternative. Using similar reasoning, a proposed project that will cause the capital budget limit for an organization to be exceeded is not a feasible investment for that fiscal year.

One alternative that is often feasible, and too frequently overlooked, is making no change to the current operation or set of conditions (i.e., doing nothing). If you judge this option feasible, make sure it is considered in the analysis.

Principle 2: Only the differences in expected future outcomes among the alternatives are relevant to their comparison and should be considered in the decision. If all prospective outcomes of the feasible alternatives were exactly the same, there would be no basis or need for comparison. We would be indifferent among the alternatives and could make a decision using a random selection.

Obviously, it is only differences in the future outcomes of the alternatives that are important. Those outcomes common to all alternatives can be disregarded in the comparison and decision. For example, if your feasible housing alternatives were two residences with the same purchase (or rental) price, this consideration would be inconsequential to your final choice. Instead, the decision would depend on other factors such as location and annual operating and maintenance costs. This illustrates, in a simple way, Principle 2, which emphasizes the basic purpose of an engineering economic analysis, to recommend a future course of action based on the differences among feasible alternatives.

Principle 3: The prospective outcomes of the feasible alternatives, economic and other, should be consistently developed from a defined viewpoint (perspective). The perspective of the decision maker, which often is that of the owners of the firm, would normally be used. However, it is important that the viewpoint for the particular decision is defined, and then used consistently in the description, analysis, and comparison of the alternatives.

As an example, consider a public organization operating for the purpose of river basin development, including the generation and wholesale distribution of electricity from dams on the river system. A program is being planned to upgrade and increase the capacity of the power generators at two dam sites. What perspective should be used in defining the technical alternatives for the program? The "owners of the firm," in this example, would be the segment of the public that will pay the cost of the program as well as benefit from it, and that perspective should be used.

Now, let us look at an example where the viewpoint may not be that of the owners of the firm. Suppose the company in this example is a private firm, and the problem deals with operating a flexible benefit package for the employees. Also, assume the feasible alternatives for operating the plan all have the same future costs to the company. The alternatives, however, have differences from the perspective of the employees and their satisfaction is an important decision criterion. The viewpoint for this analysis, comparison, and decision should be that of the employees of the company as a group, and the feasible alternatives should be defined from their perspective.

Principle 4: Using a common unit of measurement to enumerate as many of the prospective outcomes as possible will make easier the analysis and comparison of the feasible alternatives. It is desirable to make commensurable (directly comparable) the maximum number of prospective outcomes. For economic consequences, a monetary unit such as dollars is the common measure. Also you should try to translate other outcomes, which do not initially appear to be economic, into the monetary unit. This, of course, will not be feasible with some of the outcomes. The additional effort toward this goal, however, will enhance commensurability and make easier the subsequent analysis and comparison of alternatives.

What should you do with the outcomes that are not economic; that is, the expected consequences that cannot be translated (and estimated) using the monetary unit? First, if possible, enumerate the expected future results using an appropriate unit of measurement for each outcome. If this is not feasible for one or more outcomes, describe these consequences explicitly so the information is useful to the decision maker in the comparison of the alternatives.

EXAMPLE 1-2

Consider a situation that involves the local routing of a segment of an interstate highway bypass in a typical urban area. We will assume the feasible alternatives have been reduced to two defined routes having different physical characteristics (distance, drainage requirements, bridges, etc.), environmental impacts, and other consequences. Illustrate the guidelines of Principle 4 by discussing their potential application in this situation.

Solution

The physical construction characteristics of each route can be quantified using normal engineering units of measurement, and then the construction cost can be estimated in the monetary unit from these quantities. The cost of engineering support for each alternative, and other consequences that can be routinely estimated in terms of the monetary unit, would be handled in a similar manner.

Next, we should determine which future outcomes can be estimated, with additional effort, as an economic consequence. For example, consider noise pollution. Assume the two route alternatives above have different noise levels as an undesirable consequence. The characteristics of this problem can be made explicit using technical measurement, but can it also be handled as an economic consequence? Yes: At least we should be able to facilitate comparison of the alternatives by translating this consequence into the cost of reducing the noise pollution. That is, the cost of abatement to an acceptable level for each alternative can be estimated with additional effort.

Finally, we can have another type of consequence such as an undesirable aesthetic impact on the natural environment, caused by the interstate along part of one route. Even though measurement can play only a minor role with this type of an outcome, the impact can be explicitly described in a way that will assist the decision maker. In this case, it is unlikely that a monetary unit can play any role as a common measure for comparing the two alternatives with regard to this consequence.

Principle 5: Selection of a preferred alternative (decision making) requires the use of a criterion (or criteria). The decision process should consider the outcomes enumerated in the monetary unit, and those expressed in some other unit of measurement or made explicit in a descriptive manner. The decision maker will normally select the alternative that will best serve the long-term interests of the owners of the organization.

In engineering economic analysis, the primary criterion relates to the long-term financial interests of the owners. This is based on the assumption that available capital will be allocated to provide maximum monetary return to the owners. Often, though, there are other organizational objectives you would like to achieve with your decision, and these should be considered and given weight in the selection of an alternative. Several examples were listed in Section 1.5. These nonmonetary attributes and multiple objectives become the basis for additional criteria in the decision-making process. Techniques that consider all the outcomes simultaneously within a multiple objectives framework are discussed in Chapter 11.

Principle 6: Uncertainty is inherent in projecting (or estimating) the future outcomes of the feasible alternatives and should be recognized in their analysis and comparison. The analysis of the feasible alternatives involves projecting or estimating the future consequences associated with each of them. Estimating the magnitude and the impact of future outcomes of any course of action is uncertain. Even if the alternative involves no change from current operations, the probability is high that today's estimates of future cash receipts and disbursements, for example, will not be what eventually occurs. Thus, dealing with uncertainty is an important aspect of engineering economic analysis and is the subject of Chapter 10.

Principle 7: Improved decision making results from an adaptive process; to the extent practicable, initial projected outcomes of the selected alternative and actual results achieved should be subsequently compared. A good decision-making process can result in a decision that has an undesirable outcome. Other decisions, even though relatively successful, will have results significantly different from the initial estimates of the consequences. Learning and adapting from our experience are essential, and whether in the private or public sectors of our economy, they are indicators of a good organization.

The evaluation of results versus the initial estimate of outcomes for the selected alternative is often considered impracticable or not worth the effort. Too often, no feedback to the decision-making process occurs. Organizational discipline is needed to ensure that post-evaluations of implemented decisions are routinely accomplished, and the results used to improve future analyses of alternatives and the quality of decision making. The percentage of important decisions in an organization that are not post-evaluated should be small. For example, a common mistake made in the comparison of feasible alternatives is the failure to examine adequately the impact on the decision of incremental changes in selected factors associated with the alternatives. Only post-evaluations will highlight this type of weakness in the engineering economy studies being done in an organization.

1.8 ENGINEERING ECONOMY METHODOLOGY AND APPLICATION

The methodology part of engineering economy consists of the techniques and procedures that form the technical content of the discipline. These techniques incorporate the principles discussed in Section 1.7, and are the tools applied and used in engineering practice. They are the topic of the remaining chapters of this book. The challenge at any time, in addition to the continuing development of new or improved techniques, is to apply the methodology well.

The applications of engineering economy, as we have illustrated, are diverse and found in most areas of operation in an organization. Because of this diversity in application, it would be advantageous to have a more structured way of viewing the use of engineering economy techniques in a typical organization. Various schemes could be used to form categories within which to organize applications. However, one way to think about applications is in the following two broad categories:

1. Functional Activity—planning, design, plant engineering, production, shipping, material management, transportation management, and similar activity areas of an organization or project.
2. Type of Decision—replacement (equipment and facilities), capital budget allocation, plant expansions, buy versus lease (or make versus buy), new project evaluation, and so on.

These two general categories are not mutually exclusive. The various types of decisions (Category 2) can occur in each of the functional areas (Category 1). But this simple structure, even though very general, is a practical and useful way to think about applications.

We can also form a simple analogy between the principles, methodology, and applications of engineering economy and the basic components of a building. The principles provide the foundation to the discipline upon which the methodology is developed and upon which it rests. The methodology provides the framework (or shell) of the discipline. Both the foundation (principles) and framework (methodology) provide the capability to support the functioning components; that is, the applications.

1.9 OVERVIEW OF THE BOOK

The contents of this book have been organized in a logical sequence for both teaching and applying the principles and methodology of engineering economy. In this first chapter, after a general discussion of the subject and its importance to the engineer, we presented the fundamental concepts of engineering economy in terms of seven basic principles, which were discussed and illustrated. Thus, in Chapter 1, the basic foundation of the subject has been established.

In Chapter 2, selected cost terms and other cost concepts important in engineering economy studies are presented. Also, the typical steps in an engineering economic analysis are delineated and discussed. This discussion, in terms of seven sequential steps, includes how the basic principles of engineering economy presented in Chapter 1 are incorporated and applied in practice. Thus, in Chapter 2 the foundation of the subject is further extended to support the subsequent development and application of the methodology.

Chapter 3 concentrates on the concepts of money–time relationships, specifically the development of proper techniques to consider the time value of money in manipulating the future revenues and costs associated with an alternative (course of action). Then in Chapter 4 the methods commonly used in practice to analyze the economic consequences of an alternative are developed and demonstrated. These methods, and their proper use in the analysis and comparison of alternatives, are the primary subjects of Chapter 5. Included in Chapter 5 is a discussion of the appropriate time period for an analysis as well as other topics related to the comparison of alternatives on a before-tax basis. Thus, Chapters 3, 4, and 5 together develop and discuss an essential part of the methodology needed for the remainder of the book and for performing engineering economy studies on a before-tax basis.

In Chapter 6 and Chapter 7, the additional techniques required to accomplish engineering economy studies on an after-tax basis are developed and discussed.

In the private sector, most engineering economy studies are done on an after-tax basis. Therefore, these two chapters add to the basic part of the methodology developed in Chapters 3, 4, and 5. A major portion of Chapter 6 is concerned with depreciation under the modified accelerated cost recovery system authorized by the Tax Reform Act of 1986. However, techniques applicable to assets acquired prior to the effective date of the Act are also included. Similarly, the emphasis in Chapter 7 is on performing after-tax analysis under the Tax Reform Act of 1986.

Chapter 8 considers the critical question of how to develop an estimate of the future economic consequences associated with an alternative. Developing and discussing the basic process associated with this step in an engineering economic analysis constitute a crucial part of application and practice. This topic is placed in Chapter 8, instead of earlier in the book, so that the basic methodology of analyzing and comparing the estimated revenues and costs associated with alternatives on both a before-tax and an after-tax basis could be discussed in an integrated manner. Consequently, the development of these estimates of economic impact for an alternative is given concentrated attention in a dedicated chapter.

The consideration of inflation (or deflation) and price changes is the topic of Chapter 9. This important subject needs to be understood by engineers in current practice. The concepts for handling price changes in an engineering economic analysis are discussed both comprehensively and pragmatically from an application viewpoint in this chapter.

The concern over uncertainty is a reality in engineering practice. In Chapter 10 the impact of potential variation between the estimated economic outcomes of an alternative and results that may occur is considered; that is, techniques for analyzing the consequences of uncertainty in future estimates of revenues and costs are developed and discussed.

Chapter 11 presents techniques for simultaneously dealing with economic (monetary) and nonmonetary factors (attributes) when multiple objectives are involved in the decision. More and more decisions involve the explicit consideration of nonmonetary factors in choosing a course of action.

The analysis of whether existing assets should be continued in service or whether they should be replaced with new assets to meet current and future operating needs is a frequent question within an organization. In Chapter 12, techniques for addressing this question are developed and discussed. Since the replacement of assets results in a significant demand for capital, decisions made in this area are important and require particular attention.

Replacement of assets is one source of demand for capital in an organization. In Chapter 13 the proper identification and analysis of all projects and other needs for capital within an organization, and the capital financing and capital allocation process to meet these needs, are addressed. This basic process is crucial to the welfare of an organization since it affects most operating outcomes, whether in terms of current product quality and service effectiveness or long-term capability to compete in the world marketplace.

Numerous public works projects are authorized, financed, and operated by local, state, and federal agencies. The subject of Chapter 14 is how to conduct engineering economy studies for such public projects by using the benefit–cost ratio technique.

In this chapter we have provided an introduction to the discipline of engineering economy, including a brief summary of its background and continuing development. The importance to an engineer of having a good working knowledge of the principles and methodology of the subject in today's workplace was emphasized. The engineer who is unprepared to handle the economic aspects of an engineering decision just as competently as its physical aspects is not properly equipped to perform his or her total job.

The application of the principles and methodology of engineering economy is primarily concerned with the evaluation of alternative uses of capital. Engineering economic analysis is used to assist decision making in many different situations. This is true whether the decision maker is an engineer involved in an alternative design question or a senior manager considering a new project. Thus, as business continues to become more technical, the engineering and management functions are increasingly interconnected and the importance of engineering economy to the operation of an organization increases.

The foundation for engineering economy is a set of fundamental concepts that provide a sound basis for the development and application of the methodology. We have delineated and discussed these fundamental concepts in terms of seven principles. These principles are crucial to the discipline, and your understanding of them is critical to the accomplishment of good engineering economy studies. Here are the seven principles:

1. The choice (decision) is among alternatives. The feasible alternatives need to be identified and then defined for subsequent analysis.
2. Only the differences in expected future outcomes among the alternatives are relevant to their comparison and should be considered in the decision.
3. The prospective outcomes of the feasible alternatives, economic and other, should be consistently developed from a defined viewpoint (pespective).
4. Using a common unit of measurement to enumerate as many of the prospective outcomes as possible will make easier the analysis and comparison of the feasible alternatives.

5. Selection of a preferred alternative (decision making) requires the use of a criterion (or criteria). The decision process should consider the outcomes enumerated in the monetary unit, and those expressed in some other unit of measurement or made explicit in a descriptive manner.
6. Uncertainty is inherent in projecting (or estimating) the future outcomes of the feasible alternatives and should be recognized in their analysis and comparison.
7. Improved decision making results from an adaptive process; to the extent practicable, initial projected outcomes of the selected alternative and actual results achieved should be subsequently compared.

1.10 PROBLEMS

1-1. List ten typical situations in the operation of an organization where an engineering economic analysis would significantly assist decision making. You may assume a specific type of organization (e.g., manufacturing firm, medical health center and hospital, transportation company, government agency, etc.) if it will assist in the development of your answer (state any assumptions). (1.1)*

1-2. Explain why the subject of engineering economy is important to the practicing engineer. (1.1–1.4)

1-3. Assume that your employer is a manufacturing firm that produces several different electronic consumer products. What are five nonmonetary attributes that may be important when a significant change is considered in the design of the current best-selling product? (1.5; Principle 5)

1-4. Explain the meaning of the following terms: (1.6)

a. Capital financing.

b. Capital allocation (within an organization).

c. Mutually exclusive alternatives.

1-5. Will the increased use of automation increase the importance of engineering economy studies? Why or why not?

1-6. Explain the meaning of the statement "the choice (decision) is among alternatives." (Principle 1)

*The number(s) in parentheses at the end of a problem refer to the section(s) in the chapter most closely related to the problem. In these problems, and occasionally in the problems of subsequent chapters, a fundamental principle discussed in Section 1.7 will be referred to.

1-7. Describe the outcomes that should be expected from a feasible alternative. What are the differences between potential alternatives and feasible alternatives? (Principle 1)

1-8. Define uncertainty. What are some of the basic causes of uncertainty in engineering economy studies? (Principle 6)

1-9. You have discussed with a coworker in the engineering department the importance *of explicitly defining the viewpoint (perspective) from which future outcomes of a course of action* being analyzed are to be developed. Explain what you mean by a viewpoint or perspective. (Principle 3)

1-10. Describe three situations in which the monetary differences among engineering alternatives could be less important than the nonmonetary differences among them. (Principle 2)

1-11. Two years ago you were a member of the project team that analyzed whether your company should upgrade some building, equipment, and related facilities to support the expanding operation of the company. The project team analyzed three feasible alternatives, one of which is making no changes in facilities, and the remaining two involve significant facility changes. Now, you have been selected to lead a post-evaluation team. Delineate your technical plan for comparing the estimated consequences (developed two years ago) of implementing the selected alternative with the results that have been achieved. (Principle 7)

1-12. Describe how it might be feasible in an engineering economic analysis to consider the following different situations in terms of the monetary unit: (Principle 4)

a. A piece of equipment that is being considered as a replacement for an existing item has greater reliability; that is, the mean time between failure (MTBF) during operation of the new equipment has been increased 40% in comparison with the present item.

b. A company manufactures wrought iron patio furniture for the home market. Some changes in material and metal treatment, which involve increased manufacturing costs, are being considered to significantly reduce the problem with rusting.

c. A large foundry operation has been in the same location in a metropolitan area for the past 35 years. Even though it is in compliance with current air pollution regulations, the continuing residential and commercial development of that area is causing an increasing expectation on the part of local residents for improved environmental control by the foundry. The company considers community relations to be quite important.

2

COST CONCEPTS AND ANALYSIS

The objectives of Chapter 2 are to (1) describe some of the basic cost terminology and concepts encountered throughout this book, and (2) illustrate how they should be used in engineering economic analysis and decision making. The following topics are discussed in this chapter:

Sunk and opportunity costs
Fixed, variable, and incremental costs
Recurring and nonrecurring costs
Direct, indirect, and overhead costs
Standard costs
Cash versus book costs
What is "the life cycle?"
Life-cycle costs
Breakeven analysis
The average unit cost function
Present economy studies
Accounting and engineering economy studies
What are the steps in an engineering economic analysis?

2.1
INTRODUCTION

Accomplishing engineering design to meet economic needs and to achieve competitive operations in private and public sector organizations depends on a prudent balance between what is technically feasible and what is economically acceptable. Unfortunately, there is no short-cut method to reach this balance between technical and economic feasibility. Thus, engineering economic analysis concepts and methods should be carefully applied to provide results that will help to attain an acceptable balance.

The word *cost* has meanings that vary in usage.* Since concepts are ideas generalized from particular instances or situations, the *cost concepts* used in an engineering economy study will depend on the problem or situation and on the decision to be made. As we shall demonstrate, such studies involve an integration of cost concepts with the basic principles discussed in Chapter 1. Consequently, the content of Chapter 2, which involves selected cost concepts and their use, is important to the methodology and applications covered in subsequent chapters of the book.

2.2
COST TERMINOLOGY

In this section, selected cost terms used extensively in engineering economy are defined and illustrated. Additional cost concepts, selected because of their general importance, are discussed in the next section.

2.2.1 Sunk Cost

A *sunk cost* is one that has occurred in the past and has no relevance to estimates of future costs and revenues related to an alternative course of action.

*For purposes of this book, the words *cost* and *expense* are used interchangeably. Many of the commonly used terms in engineering economy are defined in Appendix B.

Thus, it is common to all the feasible alternatives, is not part of the future (prospective) cash flows, and can be disregarded in an engineering economic analysis. We need to be able to recognize sunk costs and then handle them properly in an analysis. Specifically, we need to be alert in any situation that involves a past expenditure that cannot be recovered, or capital that has already been invested and cannot be retrieved, for the possible existence of sunk costs.

The concept of sunk cost may be illustrated by the following simple example. Suppose Joe College finds a motorcycle he likes and pays $40 as a down payment, which will be applied toward the $1,300 purchase price but which must be forfeited if he decides not to take the cycle. Over the weekend, Joe finds another motorcycle he considers equally desirable for a purchase price of $1,230. For purposes of deciding which cycle to purchase, the $40 is a sunk cost and thus would not enter into the decision except that it lowers the remaining cost of the first cycle. The decision then is between paying $1,260 ($1300 − $40) for the first motorcycle versus $1,230 for the second motorcycle.

In summary, sunk costs result from past decisions and therefore are irrelevant in the analysis and comparison of alternatives that affect the future. Even though it is sometimes emotionally difficult to do, sunk costs should be ignored except possibly to the extent that their existence assists you to better anticipate what will happen in the future.

EXAMPLE 2-1

Assume that you bought stock in a company several years ago for $10 per share. The stock performed badly, and 1 year ago it was selling for $1 per share. You kept the stock and now it is selling for $3 per share. Your concern at this point is whether to sell or keep the stock. You might reason that selling would result in a ($10 − $3)/$10 = 70% loss from the original price, or in a ($3 − $1)/$1 = 200% gain over the low selling price. Actually, at the present, the $10 per share original price is also a past capital investment, which is a sunk cost, and the $1 per share is also a past value and is irrelevant. The valid viewpoint at the present is to consider the likely future performance of your current investment in the stock (at $3 per share) compared to that of some other investment. Accordingly, you should guard against the tendency in such situations to "throw good money after bad" by allowing the past to improperly influence your view of the future.

EXAMPLE 2-2

Another classic example of a sunk cost occurs in the replacement of assets. Suppose that your firm is considering the replacement of a piece of equipment. It originally cost $50,000, is presently shown on the company records with a value of $20,000, and can be sold for an estimated $5,000. For purposes of replacement analysis, the $50,000 is a sunk cost. However, the viewpoint is sometimes taken that the sunk cost should be considered as the difference between the value shown in the company records and the present realizable selling price. According to this viewpoint, the sunk cost is $20,000 minus $5,000, or $15,000. Neither the $50,000 nor the $15,000, however, should be considered in an engineering economic analysis except for the manner in which the $15,000 may affect income taxes, which will be discussed in Chapter 7.

2.2.2 Opportunity Cost

An *opportunity cost* is incurred because of the use of limited resources such that the opportunity to use those resources to monetary advantage in an alternative use is forgone. Thus, it is the cost of the best rejected (i.e., forgone) opportunity, and is often hidden or implied.

As an example, suppose that a project involves the use of vacant warehouse space presently owned by a company. The cost for that space to the project should be the income or savings that possible alternative uses of the space may bring to the firm. In other words, the opportunity cost for the warehouse space should be the income derived from the best alternative use of the space. This may be more than or less than the average cost of that space obtained from the accounting records of the company.

As another illustration, consider a student who could earn $20,000 for working during a year and who chooses instead to go to school for a year and spend $5,000 to do so. The opportunity cost of going to school for that year is $25,000: $5,000 cash outlay and $20,000 for income forgone. (This neglects the influence of income taxes and assumes that the student has no earning capability while in school.)

EXAMPLE 2-3

The concept of an opportunity cost is often encountered when the replacement of a piece of equipment or other capital asset is being analyzed. Let us reconsider Example 2-2, where your firm is considering the replacement of an existing piece of equipment which originally cost $50,000, is presently shown on the company records with a value of $20,000, but has a present market value of only $5,000. For purposes of an engineering economic analysis of whether to replace the existing piece of equipment, the present investment in that equipment should be considered as $5,000; for by keeping the equipment, the firm is giving up the *opportunity* to obtain $5,000 from its disposal. Thus, the $5,000 immediate selling price is really the investment cost of not replacing the piece of equipment and is based on the opportunity cost concept. This is why the relevant amount in the engineering economic analysis is the $5,000 market value.

2.2.3 Fixed, Variable, and Incremental Costs

Fixed costs are those which are unaffected by changes in activity level over a feasible range of operations for the capacity or capability available. Some typical fixed costs include insurance and taxes on facilities, general administrative salaries, license fees, and interest costs on a given amount of borrowed capital.

Of course, any cost is subject to change, but fixed costs tend to remain constant over a specific range of operating conditions. When large changes in usage of resources occur, or when plant expansion or shutdown is involved, you should expect fixed costs to be affected.

Variable costs are those associated with an operation that will vary in relation to changes in the quantity of output or other measures of activity level. If you were making an engineering economic analysis of a proposed change to an existing operation, the variable costs may be the primary part of the prospective differences between the present and changed operation as long as the range of activities is not significantly changed. For example, the costs of material and labor used in a product or service are variable costs since they vary with output level.

An *incremental cost*, or an *incremental revenue*, refers to the additional cost, or revenue, that will result from increasing the output of a system by one or more units. Incremental costs are also referred to as *marginal costs* in this book. Reference is frequently made to the incremental cost (which is actually an increase in variable cost) associated with a very limited change in output or activity level. For instance, the incremental cost per mile for driving an automobile may be $0.27, but this depends on several considerations, such as total mileage driven during the year (normal operating range), age of the automobile, and so forth. Also, it is common to read of the "incremental cost of producing a barrel of oil" and the "incremental cost to the state for educating a student." As these examples would indicate, the incremental cost (or revenue) is often quite difficult to determine in practice.

EXAMPLE 2-4

In connection with surfacing a new highway, the contractor has a choice of two sites on which to set up the asphalt mixing plant equipment. A subcontractor will be paid $0.20 per cubic yard per mile for hauling the asphalt paving material from the mixing plant to the job site. Factors relating to the two site alternatives are as follows:

Cost Factor	Site A	Site B
Average hauling distance	3 miles	2 miles
Monthly rental of site	$100	$500
Cost to set up and remove equipment	$1,500	$2,500
If site B is selected, there will be an added charge of $64 per day for a flagman		

The job involves 50,000 cubic yards of mixed asphalt paving material. It is estimated that 4 months (17 weeks of 5 working days per week) will be required for the job. Compare the two sites in terms of their fixed and variable costs. Which is the better site?

Solution

The fixed and variable costs (labeled C_F and C_V respectively) for this job are listed below. Both site rental costs and the cost of the flagman at Site B would be constant for the total job, but the hauling cost would vary in total amount with the distance and thus with the total quantity of "yd^3-miles."

Cost Element	Site A		Site B	
Rent (C_F)	4 × $100 =	$ 400	4 × $500 =	$ 2,000
Setup/Removal (C_F)		1,500		2,500
Flagman (C_F)			5 × 17 × $64 =	5,440
Hauling (C_V)	50,000 × $0.20 × 3 =	30,000	50,000 × $0.20 × 2 =	20,000
Total		$31,900		$29,940

Thus, site B is less costly.

EXAMPLE 2-5

Four college students who live in the same geographical area intend to go home for Christmas vacation (a distance of 400 miles each way). One of the students has an automobile and agrees to take the other three if they will pay the cost of running the automobile for the trip. When they return from the trip, the owner presents each of them with a bill for $102.40 stating that she has kept careful records of the cost of operating the car, and that based on an annual average of 15,000 miles, the cost per mile is $0.384. The three others feel that the charge is too high and ask to see the cost figures on which it is based. The owner shows them the following list:

Cost Element	Cost per Mile
Gasoline	$0.120
Oil and lubrication	0.021
Tires	0.027
Depreciation	0.150
Insurance and taxes	0.024
Repairs	0.030
Garage	0.012
Total	$0.384

The three riders, after reflecting on the situation, form an opinion that only the costs for gasoline, oil and lubrication, tires, and repairs are a function of mileage driven (variable costs) and thus could be caused by the trip. Since these four costs total only \$0.198 per mile, and thus \$158.40 for the 800-mile trip, the share for each student would be \$158.40/3 = \$52.80. Obviously, the opposing views are substantially different. Which, if either, is correct? What are the consequences of the two different viewpoints in this matter, and what should be the decision-making criterion?

Solution

In this instance, assume that the owner of the automobile agreed to accept \$52.80 per person from the three riders, based on the variable costs that were purely incremental for the Christmas trip versus the owner's average annual mileage. That is, the \$52.80 per person is the "with a trip" cost relative to the "without" alternative.

Now, what would the situation be if the three students, because of the low cost, returned and proposed another 800-mile trip the following weekend? And what if there were several more such trips on subsequent weekends? Quite clearly, what started out to be a small marginal (and temporary) change in operating conditions—from 15,000 miles per year to 15,800 miles—soon would become a normal operating condition of 18,000 or 20,000 miles per year. On this basis it would not be valid to compute the extra cost per mile as \$0.198.

Since the normal operating range would be changed, the fixed costs would also have to be considered. A more valid incremental cost would be obtained by computing the total annual cost if the car were driven, say, 18,000 miles, then subtracting the total cost for 15,000 miles of operation, and thereby determining the cost of the 3,000 additional miles of operation. From this difference, the cost per mile for the additional mileage could be obtained. In this instance, the total cost for 15,000 miles of driving per year was 15,000 × \$0.384 = \$5,760. If the cost of 18,000 miles per year of service, due to increased depreciation, repairs, and so forth turned out to be \$6,570, it is evident that the cost of the additional 3,000 miles is \$810. Then the corresponding incremental cost per mile due to the increase in the operating range would be \$0.27. Therefore, if several weekend trips were expected to become a normal operation, the owner would be on more reasonable economic ground to quote a cost of \$0.27 per mile for even the first trip.

2.2.4 Recurring and Nonrecurring Costs

These two general cost terms are often used to describe various types of expenditures. *Recurring costs* are those that are repetitive and occur when an organization produces similar goods or services on a continuing basis. Variable costs are also recurring costs since they repeat with each unit of output. But recurring costs are not limited to variable costs. A fixed cost that is paid on a repeatable basis is a recurring cost. For example, in an organization providing architectural and engineering services, office space rental, which is a fixed cost, is also a recurring cost.

Nonrecurring costs, then, are those that are not repetitive even though the total expenditure may be cumulative over a relatively short period of time. Typically, nonrecurring costs involve developing or establishing a capability or capacity to operate. For example, the purchase cost for real estate upon which a plant will be built is a nonrecurring cost, as is the cost of constructing the plant itself.

2.2.5 Direct, Indirect, and Overhead Costs

These frequently encountered cost terms involve most of the cost elements that also fit into the previous overlapping categories of fixed and variable costs, and recurring and nonrecurring costs. *Direct costs* are those that can be reasonably measured and allocated to a specific output or work activity. The labor and material costs directly associated with a product or service, or a construction activity, are direct costs. Materials needed to make a pair of scissors typify a direct cost.

Indirect costs are those that are difficult to attribute or allocate to a specific output or work activity. Sometimes the term is used with a limited definition and refers to those cost elements closely linked to the support of operations that would involve too much effort to directly allocate to a specific output. In this usage, they are costs allocated through some selected formula (such as proportional to direct labor hours) to the outputs or work activities. For example, the cost of commonly used tools in a plant is treated as an indirect cost. On the other hand, the term is sometimes used in a broader context to mean all costs incurred for the general operation of the business that cannot be directly allocated.

The term *overhead costs* is often used to mean all expenditures that are not direct costs. These costs normally are divided into different categories that vary in usage. Two common categories are indirect costs (with the limited definition discussed above) and *general and administrative (G and A) costs*. Selling costs may be included in overhead, or defined as a separate category.

There are different methods used to allocate overhead costs among products, services, or activities. The most commonly used methods involve allocation in proportion to direct labor cost, direct labor hours, direct materials costs, the sum of direct labor and direct materials cost (referred to as *prime cost* in a manufacturing operation), or machine hours. In each of these methods it is necessary to know what the total overhead costs have been, or are estimated to be, for a time period (typically a year) to allocate them to the production (or service delivery) outputs. Also, total overhead costs are associated with a certain level of production. This is an important condition that should be remembered when dealing with unit cost data (see Section 2.3.2).

We can illustrate direct, indirect, and overhead costs using a typical project situation such as the construction of an addition to an existing plant. Through a project management system, the work would be planned, scheduled, and controlled—including cost control—by defined activities. The costs of labor and material for each activity are direct costs; that is, they are directly charged to each activity as they are used in accomplishing the work. Then there are other project costs associated with accomplishing the work that would be very difficult to allocate directly to each construction activity.

For example, consider the costs of operating a tool crib, distributing construction material, providing common and miscellaneous material, supplying compressed air to most of the site from a single source, and so on. All such indirect costs are allocated to the construction activities through a formula, probably as a function of direct labor hours. Then we have costs associated with central purchasing, payroll administration, general project management, insurance, taxes, and so on. These are general and administrative overhead costs that are allocated among the work activities by formula. Again, the allocation to the activities may be done as a function of direct labor hours, direct material costs, or through some other relationship such as the sum of the two.

2.2.6 Standard Costs

Standard costs are representative costs per unit of output that are established in advance of actual production or service delivery. They are developed from the direct labor hours, materials, and support functions (with their established costs per unit) planned for the production or delivery process. For example, a standard cost for manufacturing one unit of an automotive part such as a starter would be developed as follows:

Standard Cost Element	Sources of Data for Standard Costs
Direct labor +	Process routing sheets, standard times, standard labor rates
Direct material +	Material quantities per unit, standard unit material costs
Factory overhead costs	Total factory overhead costs allocated based on prime costs (direct labor plus direct material costs)
= Standard Cost (per unit)	

Standard costs play an important role in cost control and other management functions. Some representative uses are

1. Estimating future manufacturing or service delivery costs.
2. Measuring operating performance by comparing actual cost per unit with the standard unit cost.
3. Preparing bids on products or services requested by customers.
4. Establishing the value of work-in-process and finished inventories.

In practice, variations in actual manufacturing or service delivery costs per unit from the standard cost may be caused by several factors. For example, changes in the actual purchase costs of materials from those used to calculate the standard cost is one common source of variation. Also, variation between the actual quantities of material and labor used and the standard quantities per unit is another factor often involved in the control of operating costs. Early identification and correction of these types of problems are important in effective cost control. In addition, updating standard cost calculations on a periodic basis is required to ensure their continued usefulness.

2.2.7 Cash Cost versus Book Cost

A cost that involves payment of cash is called a *cash cost* to distinguish it from one that does not involve a cash transaction but is reflected only in the accounting system as a *noncash* cost. This noncash cost is often referred to as a *book cost*. Cash costs are estimated from the perspective established for the analysis (Principle 3, Section 1.6) and are the future expenses incurred for the alternatives being analyzed. Book costs are costs that do not involve cash payments, but rather represent the recovery of past expenditures over a fixed period of time. The most common example of book costs is a charge called depreciation for the use of assets such as plant and equipment. In engineering economic analysis, only those costs that are cash flows or potential cash flows from the defined perspective for the analysis need to be considered. Depreciation, for example, is not a cash flow and is important in an analysis only because it affects income taxes, which are cash flows. We discuss the topics of depreciation and income taxes in Chapters 6 and 7 respectively.

2.2.8 Life-Cycle Cost

In engineering practice, the term *life-cycle cost* is often encountered. This term refers to a summation of all the costs, both recurring and nonrecurring, related to a product, structure, system, or service during its life span. The *life cycle* is illustrated in Figure 2-1. It begins with identification of the economic need or want (the requirement), and ends with retirement and disposal activities. It is a time horizon that has to be defined in the context of the specific situation, whether it is a highway bridge or a jet engine for a commercial aircraft. The

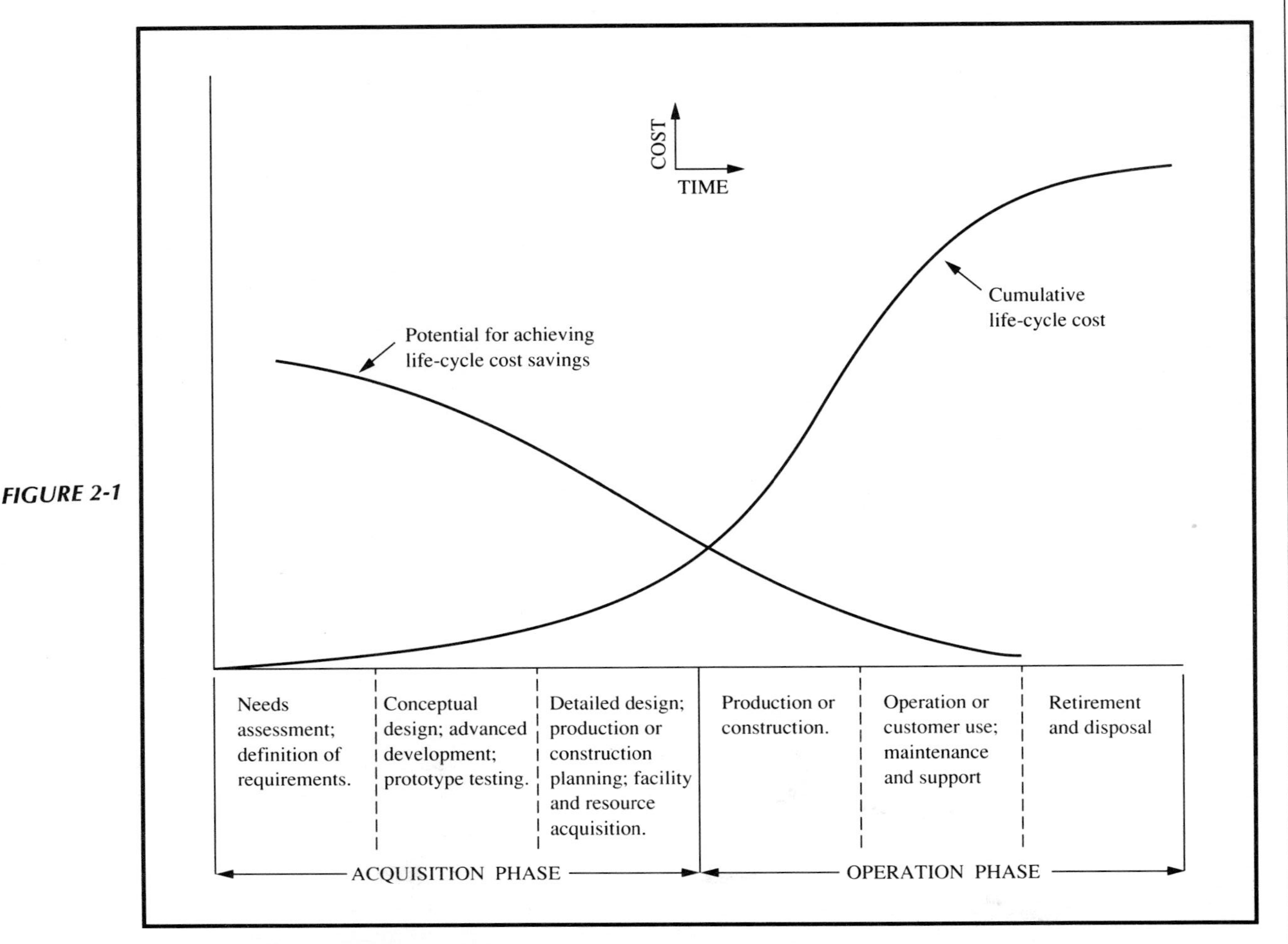

FIGURE 2-1 ***The Life-Cycle and Its Relative Cost Profiles***

end of the life cycle may be projected on a functional or an economic basis. For example, the time over which a piece of equipment or a structure is able to perform economically may be shorter than that permitted by physical capability. Changes in the design efficiency of a boiler represent this situation. The old boiler may be able to produce the steam required but not economically enough for the intended use.

The life cycle may be divided into two general time periods, the acquisition phase and the operation phase; although in some definitions the life cycle is divided into more than two phases, we will use the two-phase concept. As shown in Figure 2-1, each of these phases is further subdivided into interrelated but different activity periods that occur progressively.

The acquisition phase begins with an analysis of the economic need or want—the analysis necessary to make explicit the requirement or concept for the product, structure, system, or service. Then, with the requirement explicitly defined, the other activities in the acquisition phase can proceed in a logical sequence. The conceptual design activities translate the defined technical and operational requirements into a preferred preliminary design. Included in these activities are development of the feasible alternatives, and engineering economic analyses to assist in selection of the preferred preliminary design. Also, advanced development and prototype testing activities to support the preliminary design work occur during this period.

The next group of activities in the acquisition phase involves detailed design and the planning for production or construction. This is followed by the activities necessary to prepare, acquire, and make ready for operation the facilities and other resources needed for the production, delivery, or construction of the product, structure, system, or service involved. Again, engineering economy studies are an essential part of the design process to analyze and compare alternatives, and to assist in determining the final detailed design.

In the operation phase, the production, delivery, or construction of the end item(s) or service and their operation or customer use occurs. This phase ends with retirement from active operation or use and, often, disposal of the physical assets involved. The priorities for engineering economy studies during the operation phase are achieving efficient and effective support to operations, determining whether (and when) replacement of assets should occur, and projecting the timing of retirement and disposal activities.

The costs incurred during the acquisition phase are primarily nonrecurring except for some fixed costs that will continue in the operation phase. Costs during the operation phase are primarily recurring in nature except for those involved with the retirement and disposal activities.

In Figure 2-1, relative cost profiles are shown (as a function of the activity periods in the life cycle) for both cumulative costs and the potential for achieving cost savings. The cumulative life-cycle cost begins to increase rapidly at the end of the acquisition phase. In general, 80–90 percent of the life-cycle cost typically occurs during the operation phase. In contrast, the period of highest potential for achieving life-cycle cost savings is during the acquisition phase; that is, during the needs analysis, preliminary design, and detailed design activities. This is why engineering economic analysis must be an integral part of the design process.

Thus, the purpose of the life-cycle cost concept is to make explicit the interrelated effects of costs over the total life span for a product, structure, system, or service. The objective of the design process is to minimize the life-cycle cost, while meeting other performance requirements, by making the right trade-offs between prospective costs during the acquisition phase and those during the operation phase.

The cost elements of the life cycle that need to be considered in engineering economic analysis will vary with the situation. Because of their common use, however, several basic life-cycle cost categories are defined.

The *investment cost*, or *first cost*, is the capital required for most of the activities in the acquisition phase. This category includes many separate, detailed cost elements. For example, the costs of the initial needs analysis and feasibility study, research and development, and all the design effort and prototype testing are in this category. In simple cases, such as acquiring specific equipment, these costs may be incurred as a single expenditure. On a large, complex construction project, however, a series of expenditures over an extended period could be incurred as the various acquisition phase activities are progressively accomplished.

EXAMPLE 2-6

Consider the situation where the equipment and related support for a new CAD/CAM work station are being acquired for the engineering department that you work in. The applicable cost elements and estimated expenditures are as follows:

Cost Element	Cost
Install a leased telephone line for communication	$ 1,100/month
Lease CAD/CAM software (includes installation and debugging)	550/month
Purchase hardware (CAD/CAM workstation)	20,000
Purchase a 9600 baud modem	2,500
Purchase a high-speed printer	1,500
Purchase a four-color plotter	10,000
Shipping costs	500
Initial training (in house) to gain proficiency with CAD/CAM software	6,000

What is the investment cost of this CAD/CAM system?

Solution

The investment cost in this example is the sum of all the cost elements except the two monthly lease expenditures: specifically, the sum of the initial costs for the CAD/CAM work station, modem, printer, and plotter ($34,000); shipping cost ($500); and the initial training cost ($6,000). These cost elements result in a total investment cost of $40,500. The two cost elements that involve lease payments on a monthly basis (telephone line and CAD/CAM software) are part of the recurring costs in the operation phase.

The term *working capital* refers to the funds required for current assets (i.e., other than fixed assets such as equipment, facilities, etc.) that are needed for the start and subsequent support of operation activities. The cost elements in this category are associated initially with activities in the acquisition phase of the life cycle which then support activities during the operation phase. For example, products cannot be made or services delivered without some materials available in inventory. Functions such as maintenance cannot be supported without a minimum level of spare parts, tools, trained personnel, and other resources. Then, a controlled level of these resources must be maintained to support continuing operations. Also, some cash must be available to pay employee salaries and the other immediate expenses of operation. The amount of working capital needed will vary with the project and the product, structure, system, or service involved. However, some or all of the investment in working capital is usually recovered during disposal activities.

The *operation and maintenance cost* category contains most of the recurring cost elements associated with the operation phase of the life cycle. The direct and indirect costs of operation associated with the five primary resource areas—people, machines, materials, energy, and information—are a major part of the costs in this category. Additional costs in the overhead classification, not already included in the indirect costs of the five resource areas, are another significant part of the operation and maintenance costs. Depending on the organization's accounting system, these additional overhead costs may include a number of cost elements. As discussed in Section 2.2.5, costs in this category include certain inventory charges, insurance costs, general and administrative (G and A) costs, and a number of miscellaneous expenses that cannot be charged directly to an output.

The *disposal cost* category includes those nonrecurring costs of shutting down the operation and the retirement and disposal of assets at the end of the life cycle. The costs incurred will depend on the operating situation. Normally, costs associated with personnel, materials, transportation, one-time special activities, and overhead can be expected. These costs, in some instances, will be partially offset by receipts from the sale of assets with remaining market value. The organization may receive more in revenues from these assets than the cost of the retirement and disposal activities, but a net disposal cost often occurs.

2.3

APPLICATION OF COST CONCEPTS

We begin this section with a discussion of economic *breakeven analysis*. This basic technique has many applications. The analysis procedure is explained and some typical applications are explored.

The next topic discussed is the concept of the *average unit cost* function. This analysis technique is often complementary in use with breakeven analysis, and is quite simple and often useful when doing an engineering economic analysis.

We conclude this section with a discussion of *present economy studies*; that is, situations where the economic considerations involve short enough time periods that money–time relationships, which are the topic of Chapter 3, can be disregarded.

2.3.1 Economic Breakeven Analysis

We stated in Section 2.2.3 that costs such as insurance and taxes, G and A expenses, and research and development costs occur in virtually all organizations, remain constant over a normal range of output volume or activity level, and are referred to as *fixed* costs. In addition, there are the *variable* costs of operation that vary in relation to changes in the output volume or other measures of activity level. All costs during the operation phase of the life cycle, except the retirement and disposal costs, are in one of these two general cost classifications. The relationship of these costs to each other, and to the total cost function (with variable costs assumed to be a linear function of output volume) is shown in Figure 2-2. Thus, at any level of output volume or activity level (X) during a time period (e.g., year), that is within the normal range of operation for the available capacity,

$$C_T = C_F + C_V \qquad (2\text{-}1)$$

where C_T is the total cost and C_F and C_V denote the total fixed and total variable costs, respectively. Also, for the linear relationship assumed for the variable costs,

$$C_V = c_v(X) \qquad (2\text{-}2)$$

where c_v is the variable cost per unit of output.

The total sales revenue (S_T) during the same time period, when the selling price per output unit (s_p) is assumed to remain constant over the range of

FIGURE 2-2

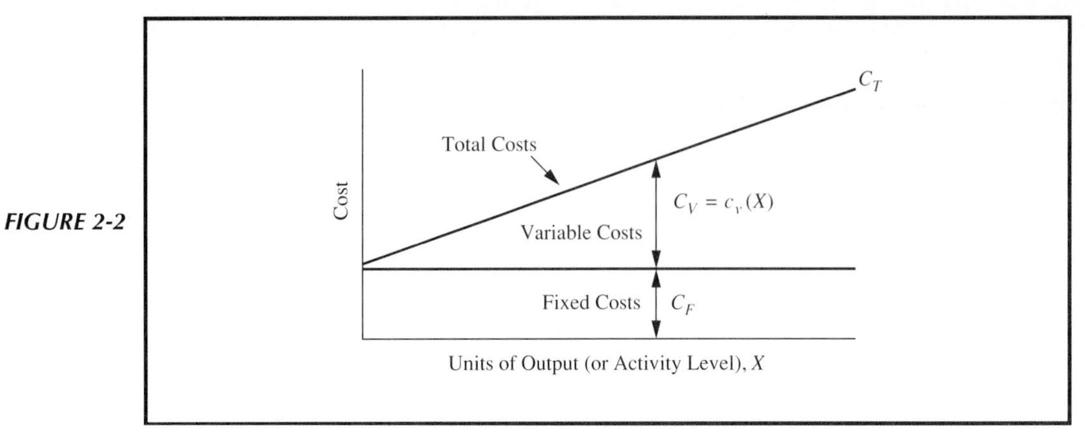

Typical Fixed, Variable, and Total Costs as a Function of Output Volume or Activity Level

demand, is a linear function equal to the selling price per unit times output volume. Thus, the total sales revenue is

$$S_T = s_p(X) \tag{2-3}$$

when demand during the time period is assumed equal to the output volume or activity level. For this situation, the relationship over a given time period between output volume or activity level and fixed, variable, and total costs and total revenue is shown in Figure 2-3.

By representing the revenue and costs of a business operation in this graphical form, it is possible to portray the estimated profit (or loss) for any output volume or activity level and any sales demand. Output or activity level for a time period can be measured in units of product or service, capacity utilization, or sales demand. Such a graphical portrayal of revenue and costs, as a function of the organization's output, is called a *breakeven chart*. It is essentially a continuous income (profit and loss) statement (Appendix A) for all feasible levels of operational activity within the available capacity. Breakeven charts are very useful for analyzing new projects, for portraying and understanding the effects of variations in fixed and variable costs on the profitability of a business operation, and for other applications. Thus, they may be used to help analyze and portray the effects of proposed changes in operation for many different situations.

In Figure 2-3, the line $C'C_F$ represents the fixed costs of production or service delivery. The line $C'C_T$ shows the variation in total variable cost with output, and since its starting point is at C' (i.e., includes fixed costs), it actually represents total production costs. The gross revenue from sales is represented by the line $0\ S_T$. Because $C'C_T$ represents the total costs of production, and $0\ S_T$ the total revenue from sales, the intersection of these two lines is called the *breakeven point*.

FIGURE 2-3

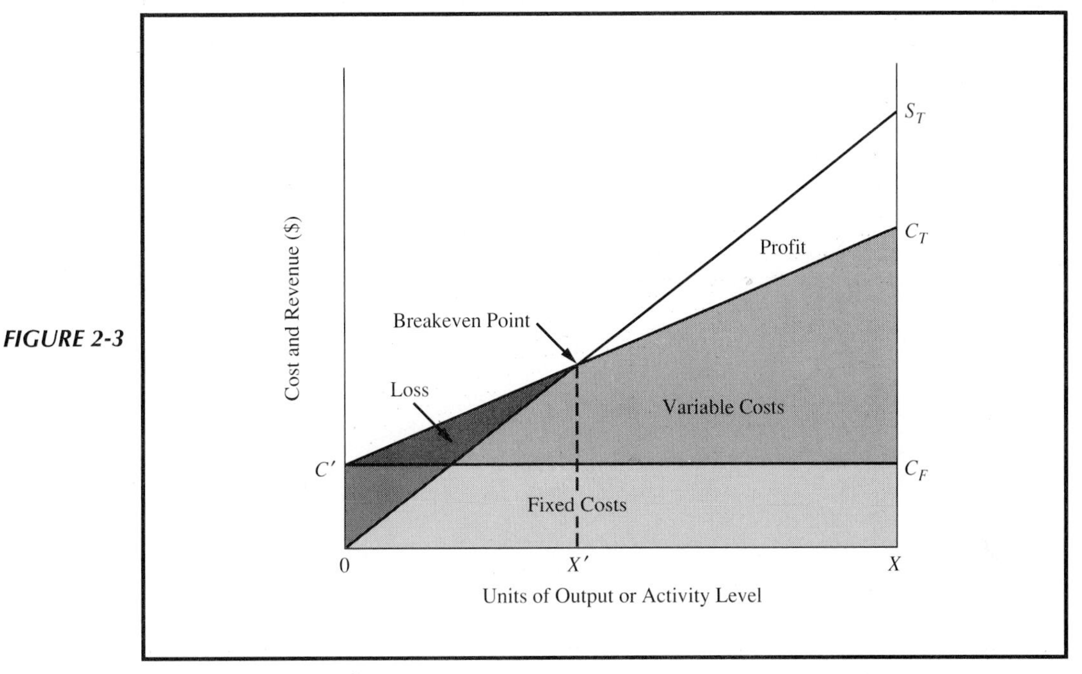

Typical Breakeven Chart

At the breakeven point (X'), under the linear relationships assumed, fixed costs equal the output (and assumed sales demand) multiplied by the difference between selling price and variable cost per unit. That is,

$$C_F = (s_p - c_v)(X') \tag{2-4}$$

To determine the breakeven point, we have

$$X' = \frac{C_F}{(s_p - c_v)} \tag{2-5}$$

Obviously, to reach a breakeven position, the selling price has to be greater than the variable cost per unit. Also, at the breakeven point the organization will neither make a profit nor incur any loss. If the output volume (and assumed sales demand) for the time period, however, is greater than X', a profit will result. Similarly, when the output volume is less than X', a loss from operation will be incurred by the organization.

EXAMPLE 2-7

As stated earlier in this section, one use of breakeven charts is to help analyze the relative effects on the breakeven point (X') of changes in the fixed and variable costs of an operation. For example, suppose the fixed cost associated with the production of a product is \$500,000 per year. Also, assume that the variable cost is \$20,000 and the selling price is \$30,000 for each percentage point of annual output capacity (and assumed sales demand). Thus, the maximum sales per year are \$3,000,000 (at 100 percent of output capacity), and we have

$$C_F = \$500,000 \text{ per year} \qquad \text{(Fixed cost)}$$
$$c_v = \$20{,}000/1\% \text{ of annual output capacity} \qquad \text{(Variable cost/unit)}$$
$$s_p = \$30{,}000/1\% \text{ of annual output capacity} \qquad \text{(Selling price/unit)}$$

Determine the breakeven point for this situation and plot the breakeven chart.

Solution:
For these conditions, the breakeven point (Equation 2-5), expressed as a percent of annual output capacity (and assumed sales demand), is

$$X' = \frac{C_F}{(s_p - c_v)} = \frac{\$500{,}000}{\$30{,}000 - \$20{,}000} = 50 \text{ percent}$$

This specific situation is illustrated in Figure 2-4. For any sales demand greater than \$1,500,000 per year (50 percent of available output capacity), the firm will make a profit. Without any changes in C_F, s_p, or c_v, it will earn a profit of \$10,000 for each 1 percent of output capacity above the present breakeven point (50 percent) used to meet sales demand. Also, it will incur a loss in operation on this product if sales demand utilizes less than 50% of available annual capacity. In general, for output volume X and variable cost per unit c_v, we have

$$\text{Profit (Loss)} = (\$30{,}000 - c_v)(X - X')$$

Market competition often creates pressure to lower the breakeven point of an operation; the lower the breakeven point, the less likely that a loss will occur during market fluctuations. Also, if the selling price remains constant, a larger profit will be achieved at any level of operation above the reduced breakeven point.

EXAMPLE 2-8

Assume that as a member of the engineering department you have been assigned the task of analyzing the sensitivity (in the situation described in Example 2-7) of X' to changes in the fixed and variable costs in your firm. With this information, your department can help set some goals and begin the development and analysis of alternative ways to reduce the breakeven point. Perform the sensitivity analysis and show the results in table form. The initial values for variable cost (c_v) and selling price (s_p) per 1 percent of

FIGURE 2-4

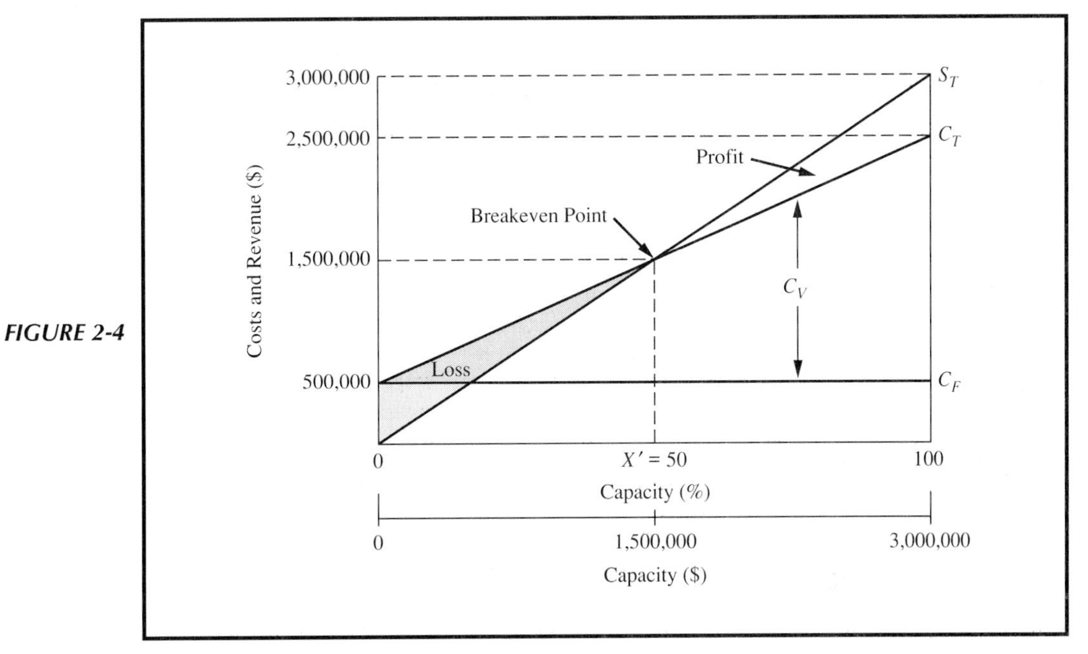

Breakeven Chart (Example 2-7)

available capacity are $20,000 and $30,000 respectively. The initial fixed cost (C_F) is $500,000.

Solution:

The results of the first four iterations of the sensitivity analysis with selected C_F and c_v values are shown in Table 2-1, along with the present situation (Iteration 0). In the first iteration, the variable cost (c_v) is reduced 10% to $18,000 per 1% of available output capacity. With this change the breakeven point becomes X' = $500,000/($30,000 − $18,000) = 41.7 (percent of capacity utilization), or equivalent to an annual sales demand of 41.7 ($30,000) = $1,251,000. Thus, a 10% reduction in the variable cost lowers X' to 41.7 for

TABLE 2–1 Breakeven Point Analysis

	Change (%)		Breakeven Point Capacity	Sales	Annual Profit ($1000) Percent of Capacity Utilization			
Iteration	C_F	c_V	(%)	($1000)	30	50	70	100
0	0	0	50.0	1,500	−200	0	200	500
1	0	−10	41.7	1,251	−140	100	340	700
2	−16.6	0	41.7	1,251	−117	83	283	583
3	−10	−15	34.6	1,038	− 60	200	460	850
4	−20	−16.6	30.0	900	0	266	533	932

a reduction of (50 − 41.7)/50 = 16.6% in the breakeven point. Also, from Table 2-1 we see that the loss that would be incurred at a 30 percent level of capacity utilization would be reduced by [−200 − (−140)]/ − 200 = 30% while profit at the 100 percent level would be increased by (700 − 500)/500 = 40%.

Iteration 2 shows that a reduction in the fixed cost (C_F) of 16.6 percent would lower X' to 41.7, the same as the 10 percent reduction in the variable cost per unit (c_v) in the first iteration. Also, iteration 4 indicates that one combination that will lower the breakeven point to 30 ($X' = 30$ percent of capacity utilization) is a 20% reduction in the fixed cost and a 16.6% reduction in the variable cost. Thus, reducing C_F to $400,000 per year (−20%), and c_v to $16,680 per 1 percent of available output capacity (−16.6%), appear to be goals that merit detailed analysis and would result in no operating loss at a 30 percent level of capacity utilization. Also, this lowering of the breakeven point can provide some flexibility to reduce the selling price when operating at any level of utilization above 30 percent (to be more price competitive), or provide higher profit if the selling price is not changed.

2.3.2 Average Unit Cost Function

Most engineering projects and business operations are designed to operate more efficiently at a certain level of capacity utilization. Deviations from this level will affect the variable cost of operation and possibly the fixed cost, and will impact the *average unit cost* of the product or service. These impacts will, in turn, affect the total profit.

Because of the importance of this cost concept in some engineering economy studies, we need to consider it in more detail. A simple way of doing so is to begin with the relationships depicted in Figure 2-2 between fixed, variable, and total costs and output volume for a specified time period. We will modify the linear relationship previously assumed between the variable cost (and total costs) and output volume to make the points discussed more explicit. Let us consider the condition where the variable cost per unit of output (c_v) is not constant when the level of output (X) is greater than a value K. This level would be determined by the Production Planning Department based on the normal operating capacity of the plant. Specifically, consider the situation where

$$c'_v = c_v(1 + \alpha)^{(X-K)/b}$$

and the total variable costs for an interval of time for output X are

$$C_V = c_v(X) \text{ when } X \leq K$$
$$C_V = c'_v(X) = c_v(1 + \alpha)^{(X-K)/b} \text{ when } K < X \leq L$$

where α, b, K, and L are constants with assigned values. In this case, the variable cost per unit at output level X, when $K < X \leq L$ is increasing at the rate α percent for each b-units of additional total output. This could result from increased labor costs due to overtime pay, increased maintenance costs, and other factors. The level $X = L$ is the upper operational limit feasible without adding an increment to the fixed cost (C_F). This situation is shown in Figure 2-5.

Given these relationships, what is the average unit cost function for this operational range of output volume? To answer this, we investigate the average cost per unit (C_U) as a function of X, the output volume. The average unit cost at any output volume is equal to the total cost of operation at that level divided by the number of output units.

$$C_U = C_T/X = \frac{(C_F + C_V)}{X} \tag{2-6}$$

And, in the situation described, we have

$$C_U = \frac{C_F}{X} + c_v \text{ when } X \leq K$$

$$C_U = \frac{C_F}{X} + c_v(1 + \alpha)^{(X-K)/b} \text{ when } K < X \leq L$$

FIGURE 2-5

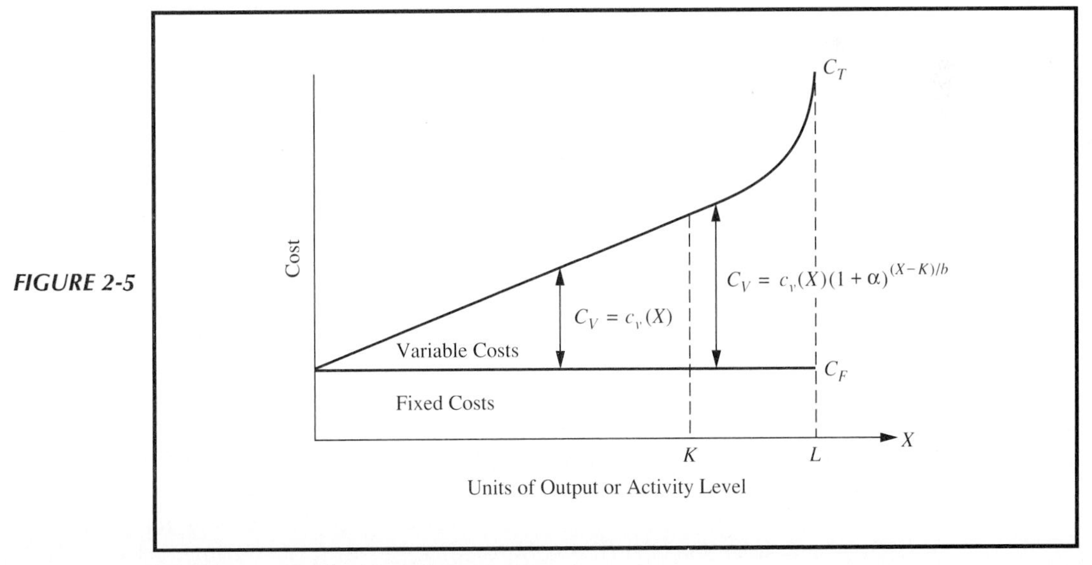

Nonlinear Variable Cost When X>K

EXAMPLE 2-9

In Example 2-8, we discussed an operation in a firm where the fixed costs were \$500,000 per year, and the variable cost and selling price were \$20,000 and \$30,000, respectively, per 1% of annual output capacity. For these conditions, the breakeven point was 50% of the annual output capacity (and assumed sales demand). We analyzed the impact on the breakeven point and the estimated annual profit of selected reductions in the fixed and variable costs.

In this example, we will modify the variable cost to the situation depicted in Figure 2-5. Specifically, we have

$$c'_v = \$20{,}000(1.05)^{(X-60)/4}$$

where $\alpha = 5$ percent, $K = 60$ (percent of capacity utilization), and $b = 4$ (percent of capacity). That is, when the output volume (X) exceeds 60 percent of operating capacity, the variable cost per unit will increase at the rate of 5 percent for each additional 4 percent of total output capacity used up to the limit (L) of 100 percent. For this situation, what is the average unit cost function?

Solution

The average unit cost function can be calculated using Equation 2-6.

$$C_U = \frac{(C_F + C_V)}{X}$$

And, for this situation we have

$$C_U = \frac{\$500{,}000}{X} + \$20{,}000 \text{ when } X \leq 60$$

$$C_U = \frac{\$500{,}000}{X} + \$20{,}000(1.05)^{(X-60)/4} \text{ when } 60 < X \leq 100$$

The results of the average unit cost calculations are shown in Table 2-2. The fixed cost component of the unit cost decreases over the range of operation as the output volume increases. The variable cost component, in this example, remains constant through $X = 60$ (percent of capacity utilization). After this point, it begins increasing with X at a rate greater than the rate of decrease in the fixed component. Thus, the minimum average cost per unit of output occurs when $X = 60$. The average unit cost function for this example is depicted in Figure 2-6.

In Table 2-2, it should be noted that at the original breakeven point ($X = 50$), $C_U = s_p = \$30,000$ (the selling price per unit of output). Also, with the nonlinear and relatively rapid increase in the unit variable cost (c'_v) when $X > 60$ percent utilization of available capacity, the total unit cost again equals

TABLE 2–2 Calculation of Average Unit Cost for Example 2-9

Percent of Capacity Utilization (X)	Average Unit Cost (C_U)		
	Fixed Cost Component	Variable Cost Component	Total
3	166,667	20,000	186,667
10	50,000	20,000	70,000
20	25,000	20,000	45,000
30	16,667	20,000	36,667
40	12,500	20,000	32,500
50	10,000	20,000	30,000
55	9,091	20,000	29,091
60	8,333	20,000	28,333
60.1	8,319	20,024	28,343
61	8,196	20,245	28,441
65	7,692	21,258	28,950
70	7,143	22,595	29,738
71.5	6,994	23,006	30,000
80	6,250	25,526	31,776
90	5,556	28,837	34,393
100	5,000	32,578	37,578

FIGURE 2-6

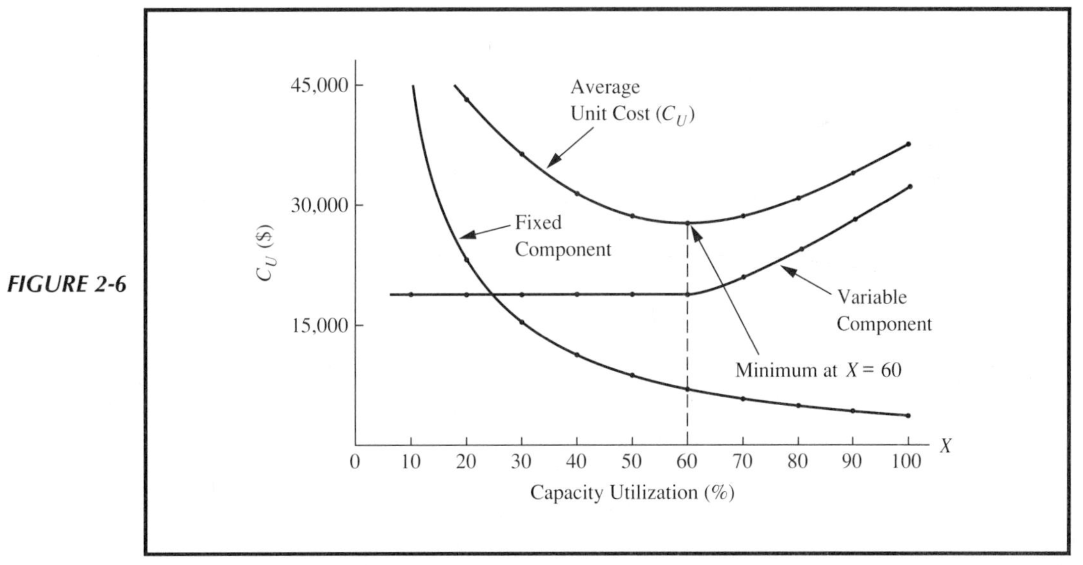

Average Unit Cost Function for Example 2-9

\$30,000 at $X = 71.5$. Therefore, in this example, an annual profit occurs only in the operating range $50 < X < 71.5$, with the maximum profit occurring when $X = 60$, the output volume at which the average unit cost is a minimum. These results are shown in Figure 2-7.

Examples 2-8 and 2-9 illustrate the potential of breakeven charts and unit cost functions in engineering economy studies. Estimation and analysis of the consequences of making changes in an operation, or of implementing a new project, can be aided by the use of these techniques.

2.3.3 Present Economy Studies

When the influence of time on money is not a significant consideration, cost analyses are usually called *present economy studies*. Typical situations involving present economy studies are

1. There is no initial investment of capital; only immediate operating costs and other factors are involved. As an example, assume you are employed by Company A and are making plans for a business trip. You can travel by commercial aircraft, which will require 3 hours elapsed travel time and the rental of a car at your destination. The other alternative is to travel by automobile, which will take 7 hours. Here the basic considerations are

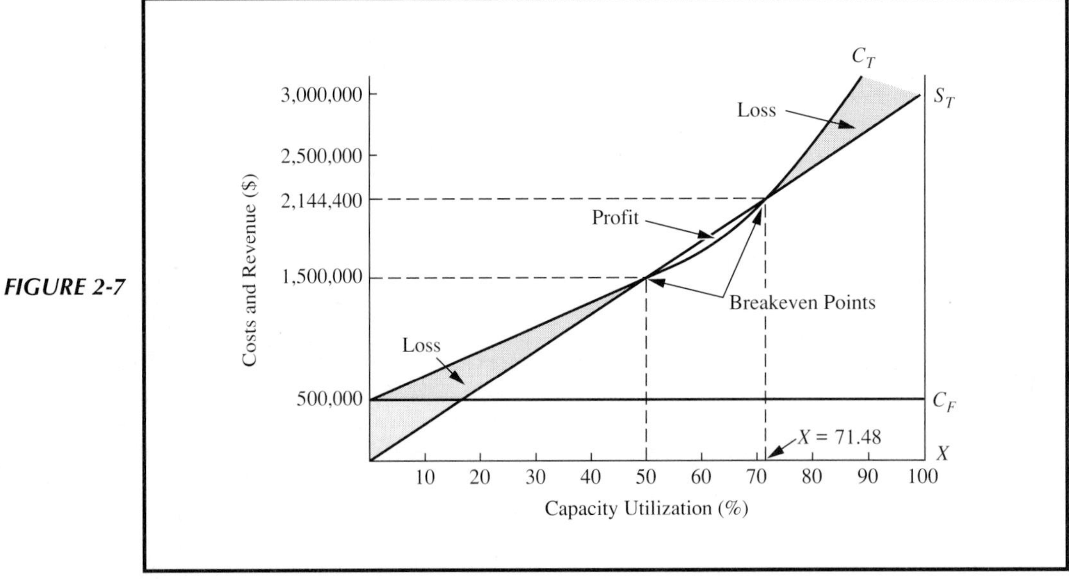

FIGURE 2-7 ***Breakeven Chart for Example 2-9***

the immediate costs, value of your time, and nonmonetary factors (e.g., fatigue).

2. There is an initial investment of capital, but after this first cost the remaining life-cycle cost is estimated to be the same, or directly proportional to the initial investment. Thus, the alternative with the lowest first cost will be the most economical over the life cycle. As an illustration, consider the construction of an interstate highway bridge overpass. Whether a longitudinally reinforced concrete slab or a precast (prestressed) concrete design is used, the maintenance and other life-cycle costs of the two designs would be proportionally the same. (However, if one or more steel design alternatives were also being considered, a present economy study probably would not be appropriate. Maintenance and other related costs would be expected to vary among the alternatives, and the cost analysis should be based on the life cycle of the structure.)
3. The differences in revenues and costs among the alternatives all occur within a limited time period (1 year or less is a general guideline), or any future differences are estimated to remain proportional to those in the first time period. This is often the case when the decision is between alternative materials in manufacturing. For example, if using low-alloy/high-yield strength steel in a particular application is estimated to give better revenue and cost results than low-carbon steel, this relative advantage would be expected to remain in future time periods.

Present economy studies tend to have certain characteristics. First, they are relatively simple in form; any complexity usually involves problems common to all cost analysis such as identifying all the relevant elements of cost, and the availability of data for estimating the prospective revenues and costs. Second, the required decision (selection of an alternative) usually is quite easily made, once the comparative revenue and cost data are organized. Third, the potential alternatives can be numerous, often involving a continuum of values for one or more critical factors that differentiate the alternatives. In this case, the problem is to determine the feasible alternatives for detailed analysis within a reasonable amount of effort and length of time.

In engineering practice, situations that give rise to present economy studies are quite common, and two typical situations are discussed below. Recognizing these situations will often save considerable analysis effort.

2.3.3.1 Total Cost in Material Selection

In a large proportion of cases, economic selection among materials cannot be based solely on the costs of the materials. Very frequently, a change in materials will affect the processing costs, and shipping costs may also be altered. A good example of this is the product illustrated in Figure 2-8. The part was produced in considerable quantities on a high-speed turret lathe, using 1112 screw-machine steel costing $0.30 per pound. A study was made to determine whether it might

FIGURE 2-8

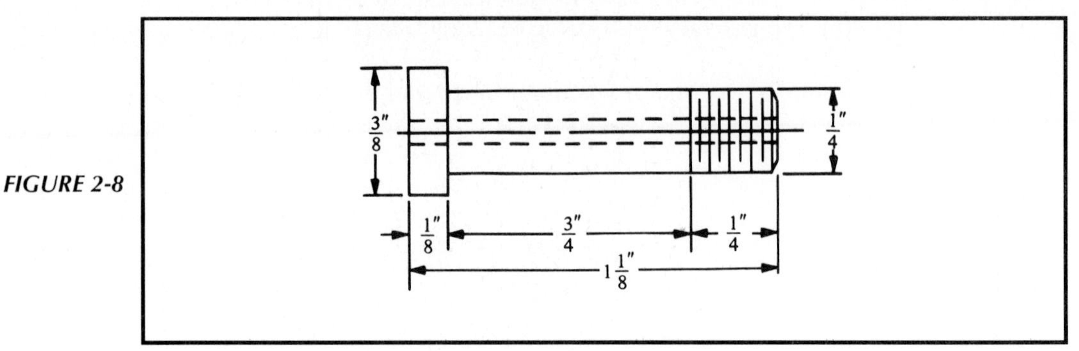

Small Screw Machine Product

be cheaper to use brass screw stock, costing $1.40 per pound. Since the weight of steel required per piece was 0.0353 pound and of brass was 0.0384 pound, the material cost per piece was $0.0106 for steel and $0.0538 for brass. However, when the manufacturing and standards departments were consulted, it was found that, although 57.1 parts per hour were being produced using steel, the output would be 102.9 parts per hour if brass were used. Inasmuch as the machine operator was paid $7.50 per hour and the overhead cost for the turret lathe was $10.00 per hour, the total-cost comparison for the two materials was as follows:

	1112 Steel			**Brass**		
Material	$0.30 × 0.0353	=	$0.0106	$1.40 × 0.0384	=	$0.0538
Labor	$7.50/57.1	=	0.1313	$7.50/102.9	=	0.0729
Overhead[a]	$10.00/57.1	=	0.1751	$10.00/102.9	=	0.0972
Total Cost per Piece			$0.3170			$0.2239
Saving per piece by use of brass = $0.3170 − $0.2239 = $0.0931						

[a]A given overhead rate applied to different alternatives without modification may be invalid for economic analyses even though useful in after-the-fact accounting allocations for whatever alternative is used. Unless explicitly stated otherwise, any given overhead rate is assumed to be applicable to all alternatives considered in the examples of this book.

Because a large number of parts were made each year, the saving of $93.10 per thousand was a substantial amount. It is also clear that costs other than the cost of material were of basic importance in the economy study.

Later history of this same product illustrates the fact that shipping costs also must often be considered in the selection between materials. After the part had been made from brass stock for several years, it was found that it was desirable to supply domestic and foreign assembly plants of the company by using air freight as a shipping medium. This led to a study of the possible use of a heat-treated aluminum alloy. This material cost $0.85 per pound and the cost of heat-treating each part, at an outside plant, was $0.018. Production studies indicated that the aluminum alloy could be machined at the same speeds as the brass stock.

The specific gravities of the brass and aluminum alloy are 8.7 and 2.75, respectively, and the raw and finished weights of the parts were as follows:

	Brass (lb)	Aluminum Alloy (lb)
Raw material	0.0384	(0.0384)(2.75/8.7) = 0.01213
Finished part	0.0150	(0.0150)(2.75/8.7) = 0.00474

Consequently, the comparative costs, including shipping at $3.00 per pound of finished part, were as follows:

	Brass	Aluminum Alloy
Material	$0.0538	$0.0103
Labor	0.0729	0.0729
Heat treatment	—	0.0180
Overhead[a]	0.0972	0.0972
Shipping	0.0450	0.0142
Total Cost per Piece	0.2689	0.2126

[a]See the footnote to the table on page 48.

A decision was made to use the aluminum alloy for those parts that were to be shipped by air to the subsidiary plants and for the parts that were consumed at the local plant. There was no advantage to using brass even when shipping costs could be omitted.

Care should be taken in making economic selections between materials to ensure that any differences in yields or resulting scrap are taken into account. Very commonly, alternative materials do not come in the same stock sizes, such as sheet sizes and bar lengths. This may considerably affect the yield obtained from a given weight of material; similarly, the resulting scrap may differ for different materials. This factor can have serious economic implications when one of the materials is considerably more costly than another. Determination of these effects is an illustration of where experience may be most helpful.

2.3.3.2 Alternative Machine Speeds

Machines frequently can be operated at different speeds, resulting in different rates of product output. However, this usually results in different frequencies of machine downtime to permit servicing or maintaining the machine, such as resharpening or adjusting tooling. Such situations lead to engineering economy studies to determine the optimum or preferred operating speed.

A simple example of this type involved the planing of lumber. Lumber put through the planer increased in value by $0.10 per board foot. When the planer was operated at a cutting speed of 5,000 feet per minute, the blades had to be sharpened after 2 hours of operation, and lumber could be planed at the rate of 1,000 board feet per hour. When the machine was operated at 6,000 feet per minute, the blades had to be sharpened after $1\frac{1}{2}$ hours of operation, and the rate of planing was 1,200 board feet per hour. Each time the blades were changed, the machine had to be shut down for 15 minutes. The blades, unsharpened, cost $50 per set and could be sharpened 10 times before having to be discarded. Sharpening cost $10 per set. The crew that operated the planer changed and reset the blades. At what speed should the planer be operated?

Because the labor cost for the crew would be the same for either speed of operation, and because there was no discernible difference in wear upon the planer, these factors did not have to be included in the study.

In problems of this type, the operating time plus the delay time due to the necessity for tool changes constitute a cycle time that determines the output from the machine. The time required for a complete cycle determines the number of cycles that can be completed in a period of available time—for example, 1 day—and a certain portion of each complete cycle is productive. The actual productive time will be the product of the productive time per cycle and the number of cycles per day.

	Value per Day
At 5,000 feet per minute	
Cycle time = 2 hours + 0.25 hour = 2.25 hours	
Cycles per day = 8 ÷ 2.25 = 3.555	
Value added by planing = 1,000 × 3.555 × 2 × $0.10 =	$711.00
Cost of resharpening blades = 3.555 × $10 = $35.55	
Cost of blades = 3.555 × $50/10 = 17.78	
Total Cost	−53.33
Net increase in value per day	$657.67
At 6,000 feet per minute	
Cycle time = 1.5 hours + 0.25 hour = 1.75 hours	
Cycles per day = 8 ÷ 1.75 = 4.57	
Value added by planing = 4.57 × 1.5 × 1,200 × $0.10 =	$822.60
Cost of resharpening blades = 4.57 × $10 = $45.70	
Cost of blades = 4.57 × $50/10 = 22.85	
Total Cost	−68.55
Net increase in value per day	$754.05

Thus it was more economical to operate at the higher speed, in spite of the more frequent sharpening of blades that was required.

It should be noted that this analysis assumes that the added production can be used. If, for example, the maximum production needed is equal to or less than that obtained by the slower machine speed (1,000 × 3.555 cycles × 2 hours = 7,110 board feet per day), then the value added would be the same for each speed, and the decision then should be based on which speed minimizes total cost.

This type of study is of great importance in connection with metal-cutting machine tool operations. Changes of cutting speeds can have a great effect on tool life. In addition, because the cost of machine tools and wage rates has increased, it is important that productivity be maintained at as high a level as possible. Under these conditions, it has frequently been found that increased cutting speeds give greater overall economy, even though the cutting-tool life is considerably less than was accepted practice in former years. This is particularly true if rapid means can be devised for changing tools when required.

2.4 ACCOUNTING AND ENGINEERING ECONOMY STUDIES

In Section 1.1, we emphasized that engineers and managers use the principles and methodology of engineering economy to assist decision making. Thus, engineering economy studies provide information upon which current decisions pertaining to the *future* operation of an organization can be based.

After a decision to invest capital in a project has been made and the money has been invested, those who supply and manage the capital want to know the financial results. Therefore, accounting procedures are established so that financial events relating to the investment can be recorded and summarized and *financial performance* determined. At the same time, through the use of proper financial information, controls can be established and utilized to aid in guiding the operation toward the desired financial goals. *General accounting* and *cost accounting* are the procedures that provide these necessary services in a business organization. Thus, accounting data are primarily concerned with *past* and *current* financial events even though such data are often used to make projections about the future.

Accounting procedures are similar to recording data in a scientific experiment. A recorder reads the pertinent gauges and meters and records all the essential data during the course of an experiment. From these data it is possible to determine the results of the experiment and to prepare a report. Similarly, an accountant records all significant financial events connected with an investment and the operation of an organization, and from these data he or she can determine what the results have been and can prepare financial reports. Just as an engineer, by taking cognizance of what is happening during the course of an experiment and making suitable corrections, can gain more information and better results from the experiment, managers must also rely on accounting reports to make corrective decisions in order to improve the current and future financial performance of the business.

General accounting is a source of much of the past financial data needed in making estimates of future financial conditions. Accounting is also a source of data for analyses that might be made regarding how well the results of a capital investment turned out compared to the results that were predicted in the engineering economic analysis.

Cost accounting, or management accounting, is a phase of accounting that is of particular importance because it is concerned principally with decision making and control in a firm. Consequently, cost accounting is the source of some of the cost data that are needed in engineering economy studies. Modern cost accounting may satisfy any or all of the following objectives:

1. Determination of the cost of products or services.
2. Provision of a rational basis for pricing goods or services.
3. Provision of a means for controlling expenditures.
4. Provision of information on which operating decisions may be based and the results evaluated.

Although the basic objectives of cost accounting are simple, the exact determination of costs usually is not. As a result, some of the procedures used are arbitrary conventions that make it possible to obtain reasonably accurate answers for most situations but may contain a considerable percentage of error in other cases.

An adequate understanding of the origins and meaning of accounting data is needed in order to properly use or not use that data in making estimates about future revenues and costs, and in comparing actual versus predicted results. However, if we understand how the accounting costs were determined, we should be able to break them down into their basic elements, and then use these cost elements to supply some of the information needed. Thus, an understanding of the basic objectives and procedures of general and cost accounting will enable us to make best use of available cost information, and to avoid needless work and serious mistakes. A brief discussion of these procedures is provided in Appendix A. Also, good references are readily available in this field.*

*For example: Evan F. Bornholtz, "Company and Cost Accounting," in *Handbook of Industrial Engineering*, ed. Gavriel Salvendy (New York: John Wiley and Sons, 1982), pp. 9.1.1–9.1.20. This is a good summary reference and overview. College textbooks on accounting provide excellent additional detail.

2.5

STEPS IN AN ENGINEERING ECONOMIC ANALYSIS

As we discussed in Chapter 1 (Section 1.1), the results of an engineering economy study are arrived at through a technical-economic analysis using a structured approach and mathematical modeling techniques. These results are then used to assist decision making. Thus, engineering economic analysis is an aid to decision making that involves the conceptualization and modeling of a choice situation and the simplification of reality, and which is not unique in its form or content.

The engineering economic analysis procedure, however, does include two general characteristics in addition to incorporating the basic principles of engineering economy.* These are (1) the choice or decision situation is made *explicit* with the form of the model used; and, (2) the procedure does have generally recognized common steps even though there will be variation in their delineation by different individuals. We represent and discuss the engineering economic analysis procedure in terms of these seven steps:

1. Recognition and formulation of the problem.
2. Development of the feasible alternatives.
3. Development of the net cash flow (and other prospective outcomes) for each feasible alternative.
4. Selection of a criterion (or criteria) for determining the preferred alternative.
5. Analysis and comparison of the feasible alternatives.
6. Selection of the preferred alternative.
7. Post-evaluation of results.

The general relationships of these steps with the principles of engineering economy, and the four general steps in decision making (Chapter 1, Section 1.2), are shown in Figure 2-9.

Before proceeding with a brief discussion of each of the seven steps, we need to recognize that their application occurs within two different situations. The first situation involves an engineering economic analysis which can be conducted as a separable study, even though it usually would be part of a larger project effort. For example, this would be the case when a change to current operations is being considered, or the feasibility of a new project is being evaluated. Such situations are very common. The second general situa-

*Reference to the principles of engineering economy (or a numbered principle) is to the seven principles of engineering economy discussed in Section 1.7 and summarized in Figure 2-9.

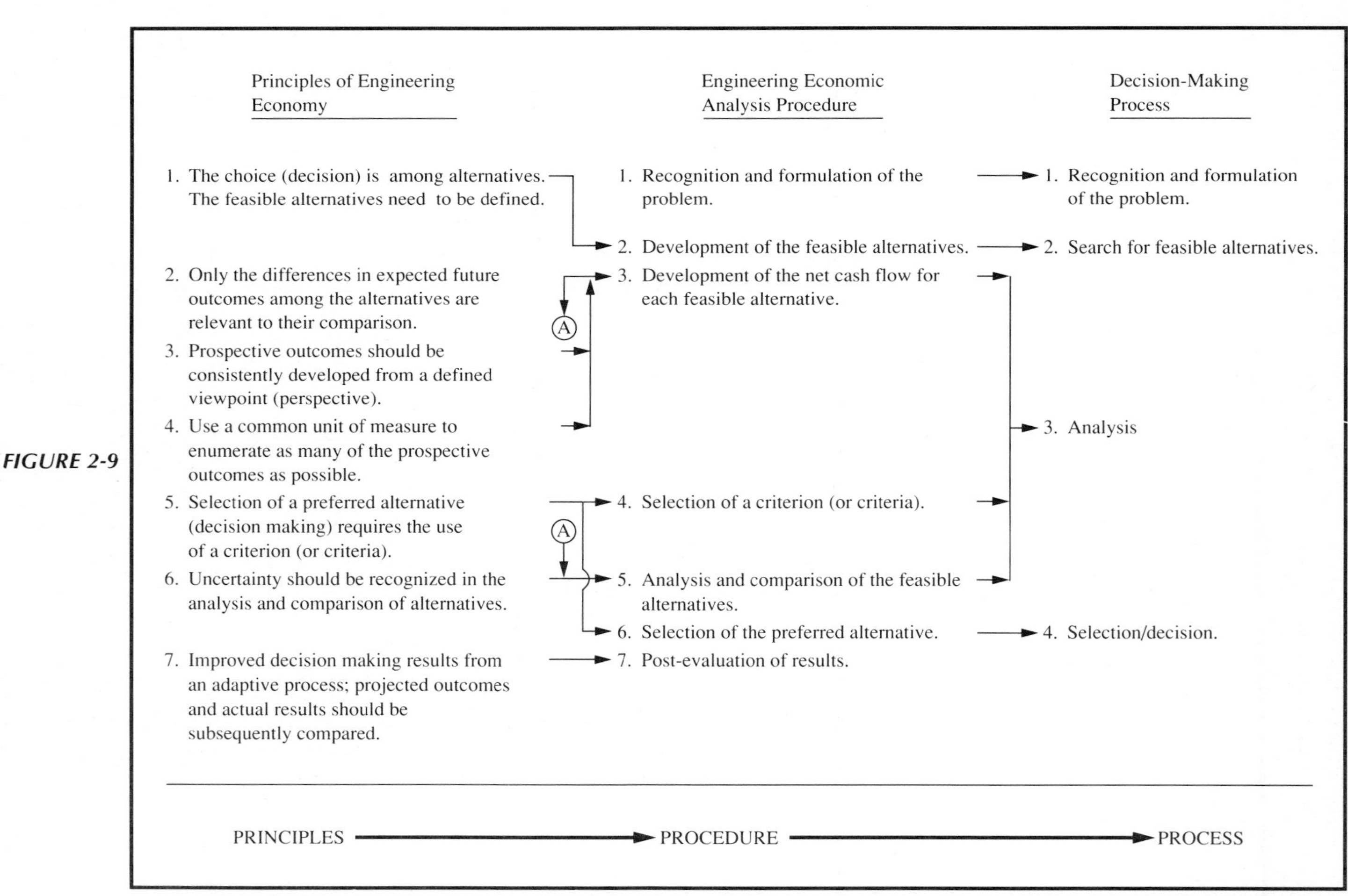

FIGURE 2-9 *The General Relationships between the Principles, the Economic Analysis Procedure and the Decision-Making Process*

tion occurs in the engineering design process, and is particularly evident with the increasing capability of the computer to support an integrated, interactive design process. Even though the same seven engineering economic analysis steps are applicable in this situation, their application is imbedded in the design process, and they occur in a more interactive mode.

Step 1: Recognition and formulation of the problem. This first step is very important since it provides the basis for the rest of the engineering economic analysis effort. Unfortunately, this step is often not done well in engineering practice. It is not adequate to just "think about" the problem. What is important is that the problem be well understood before proceeding with the rest of the analysis.

The term *problem* is used here generically. It is intended to include all reasons for doing the engineering economic analysis such as consideration of a new project, or other situations where the investment of capital is involved or affected. What you want to achieve is getting the particular situation well defined and underestood, both logically and in written form. The description of the situation should include the goals, objectives, and other results to be achieved with the selected alternative. If this is not done, the quality of the feasible alternatives developed in Step 2 could be jeopardized.

Step 2: Development of the feasible alternatives. This step incorporates Principle 1, and is accomplished using the systems viewpoint and approach discussed in Section 1.7 and illustrated in Figure 1-2. The objective of this step is to reduce the number of alternatives for subsequent detailed analysis without eliminating the best option. Otherwise, regardless of the selection made, the results will be suboptimal.

We need to emphasize again that one alternative that may be feasible, and too frequently overlooked, is making no change to the current operation or set of conditions (i.e., doing nothing). Also, there are some actions involving limited changes that should be considered as potential alternatives. For example, should an interim solution to a problem be considered, or is it best to delay a decision until a future point in time? If these types of alternatives are judged to have some merit, they should be included as potential alternatives for further consideration. Creating these and the other potential alternatives, and then deciding which ones should be included in the set of feasible alternatives, is the essence of Step 2.

Step 3: Development of the net cash flow (and other prospective outcomes) for each feasible alternative. This step incorporates the concepts of Principles 2, 3, and 4.

A modified version of Figure 1-2, as shown in Figure 2-10, will be used to illustrate the concept of the *net cash flow* associated with an alternative. Figure 2-10 represents an organization as having only one window with its external environment. Also, assume all monetary transactions of the organization flow through this window—the receipt of revenues from customers and other busi-

FIGURE 2-10

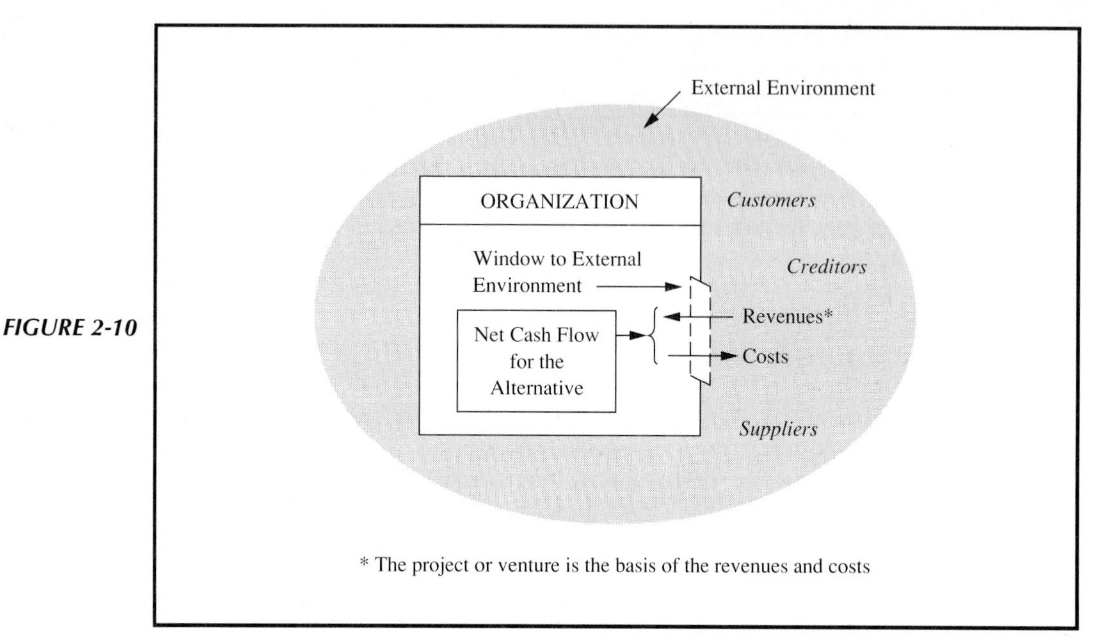

Net Cash Flow for a Proposed Feasible Alternative

ness sources; the disbursement of funds to pay suppliers, creditors, and salaries and wages of employees; and the other costs incurred in operating the organization. The key to developing the net cash flow associated with an alternative is estimating what would happen to the revenues and costs as seen at this window if the particular alternative were implemented.

In order to estimate the economic impact of an alternative on the operation of an organization, as seen at the window with its external environment, the estimating process has to incorporate three concepts:

1. Perspective (viewpoint) of the net cash flow. As stated in Principle 3, the prospective outcomes of an alternative should be consistently developed from a defined viewpoint or perspective. This viewpoint is dependent on the reason for the engineering economic analysis, but usually the perspective of the owners of the organization is used. That is, the net cash flow would be seen from their viewpoint unless determined otherwise.
2. The concept of an estimating baseline. The net cash flow for an alternative represents what is estimated to happen to future revenues and costs, as seen at the window of the organization with its external environment, from the perspective being used. Therefore, the estimated changes in revenues and costs associated with an alternative have to be relative to a baseline consistently used for all feasible alternatives. This baseline is defined and applied in either of two ways.

The first method is the *total revenue and cost approach*. That is, the "no-change" alternative is explicitly included in the set of feasible alternatives and the total revenues and costs (net cash flow) for it are estimated. Thus, when the total cost and revenue baseline approach is used, the net cash flow for the no-change alternative represents the projected revenues and costs of the current operation or situation. Similarly, the net cash flow for each of the other feasible alternatives is estimated.

The second method that can be used is the *differential* approach. Using this approach, the net cash flow for the no-change alternative is defined as zero whether or not it is explicitly defined as one of the feasible alternatives. The estimated net cash flow for each of the other feasible alternatives then represents the estimated differences (changes) in revenues and costs from the current operation or situation (no-change alternative). Whichever estimating baseline approach is used in a study, it must be consistently applied for all feasible alternatives. *A common error is to inadvertently use both baseline definitions when developing the individual net cash flow for each alternative. For example, the total cost and revenue baseline might be used for the no-change alternative, but maintenance costs in the other feasible alternatives are estimated by using differences from current operations.*

3. Common unit of measurement. As discussed in relation to Principle 4, enumerating as many of the future outcomes of the alternatives as possible in a common monetary unit such as dollars will make subsequent analysis and comparison of the feasible alternatives easier. This principle should be employed to the maximum extent possible.

In summary, the *net cash flow* for an alternative is the result of the economic impacts viewed at the organization's window, in terms of the estimated changes in revenues and costs (that is, the transfer of money from one organization to another), when the estimates are made by properly applying the three concepts discussed above. This description of a net cash flow for an alternative is fundamental to engineering economic analysis.

What is the meaning (in Step 3) of the term *other prospective outcomes?* Simply stated, these are outcomes of an alternative related to factors in the internal or external environment of the organization that cannot be quantified using the monetary unit. In Step 3, these other factors (e.g., some environmental impacts, community relations, competitive position, etc.) need to be identified and the prospective outcomes for each alternative explicitly measured or described in a consistent manner for each factor.

Step 4: Selection of a criterion (or criteria) for determining the preferred alternative. This step incorporates Principle 5. As we discussed in relation to Principle 5, the decision maker will normally select the alternative that will best serve the long-term interests of the owners of the organization, and in engineering economic analysis the financial interest is a primary criterion. However, for a specific situation the decision maker must include the relevant nonmonetary attributes and multiple objectives involved. Thus, the criteria selected must be commensurate with the particular situation.

Step 5: Analysis and comparison of the feasible alternatives. In Step 3, we discussed the development of the net cash flow for each feasible alternative. The net cash flow, in conjunction with the economic criterion selected in Step 4, provide the basis for the economic evaluation and comparison of the alternatives.

Morever, the consideration of uncertainty (Principle 6) in the estimates of future outcomes will be required, as will their comparison based only on the differences among the alternatives (Principle 2).

Step 6: Selection of the preferred alternative. When the first five steps of the engineering economic analysis have been done properly, the alternative that should be recommended to the decision maker is simply a result of the analysis effort. Thus, key elements in accomplishing this step are the soundness of the technical-economic analysis and effectiveness in communicating the result of the analysis to the decision maker. This communication includes providing assistance in considering questions important to the final decision that could not be explicitly or adequately addressed in the analysis. The criterion (or criteria) is the basis for selecting the preferred alternative.

Step 7: Post-evaluation of results. This final step implements Principle 7, and is accomplished after the results achieved from the alternative selected by the decision maker are known. Thus, Step 7 is really the follow-up step to a previous analysis that compares actual results achieved with the previously estimated outcomes. The objective is to learn how to do better analyses, and the feedback from post-implementation evaluations is important to the continuing improvement of operations in any organization. Unfortunately, like Step 1, this final step is often not done consistently or well in engineering practice; therefore, it needs particular attention to ensure feedback for use in subsequent studies.

In this chapter, we have discussed selected cost terminology and concepts important in engineering economic analysis. It is important that the meaning and use of various cost terms are understood in order to communicate effectively. If this is not achieved, misunderstanding can occur in the internal communications of an organization. Also included in this chapter is a discussion of the life-cycle cost concept and its application in engineering practice.

The other general concepts discussed in the chapter were economic breakeven analysis, the average unit cost function, and present economy studies. Each of these techniques was described and illustrated, and their potential application in engineering economic analysis demonstrated.

The use of present economy studies in noncomplex engineering decision making can provide satisfactory results and save considerable analysis effort. When an adequate engineering economic analysis can be accomplished by considering the various monetary consequences that occur in a short time period (usually one year or less), then a present economy study should be used.

The general and cost accounting records of an organization are a primary source of financial information. These records, however, are concerned with the past and current financial events of an organization (even though they are often used to make financial projections), while an engineering economic analysis is concerned with estimates of *future* outcomes. These records are often the source of some of the cost and other financial data used in engineering economy studies. An understanding of accounting data is necessary before it can be properly used, or modified for use, in engineering economy studies.

The engineering economic analysis procedure does have generally recognized steps. We discussed the analysis procedure in terms of the seven steps shown in Figure 2-9, and illustrated their relationships with the principles of engineering economy and the general steps in decision making.

The concept of the net cash flow associated with an alternative is fundamental to engineering economy studies involving complex engineering problems in which economic consequences occur over a longer time period (usually more than one year). This intermediate step in the engineering economic analysis procedure is fundamental to resolving the feasible alternatives into a common evaluation and comparison framework.

The discussion of methodology in subsequent chapters builds on the concepts of this chapter, and concentrates on analyzing the economic consequences of alternative courses of action using proper techniques.

2.6 PROBLEMS

2-1. A company in the process industry produces a chemical compound which is sold to manufacturers for use in the production of certain plastic products. The plant that produces the compound employs approximately 300 people. Develop a list of six different cost elements that would be *fixed*, and a similar list of six cost elements that would be *variable*.(2.2)

2-2. Refer to Problem 2-1 and your answer to it. (2.2)

a. Develop a table that shows the cost elements you defined and categorized as fixed and variable. Indicate which of these costs are also *recurring, nonrecurring, direct,* or *indirect*.

b. Identify one additional cost element for each of the cost categories: recurring, nonrecurring, direct, and indirect.

2-3. Within the context of the plant operation used in Problem 2-1, describe and illustrate an *incremental cost* situation.(2.2)

2-4. Discuss the relationship between indirect costs and overhead costs.(2.2)

2-5. What is the relationship between the term *cash cost* and the *cash flow concept*?(2.2)

2-6. A *breakeven point* was defined as:

$$X' = \frac{\text{fixed costs/time period}}{\text{selling price/unit} - \text{variable cost/unit}}$$

Suppose that the ABC Corporation has a production (and sales) capacity of $1,000,000 per month. Its fixed costs are $350,000 per month, and the variable costs—over a considerable range of volume—are $0.50 per dollar of sales. (2.3)

a. What is the annual breakeven point volume (X')? Develop (graph) the breakeven chart.

b. What would be the effect on X' of decreasing the variable cost per unit by 25% if the fixed costs thereby increased by 10%?

c. What would be the effect on X' if the fixed costs were decreased by 10% and the variable cost per unit increased by the same percentage?

2-7. The annual fixed costs for a plant are $100,000, and the variable costs are $140,000 at 70 percent utilization of available capacity with net sales of $280,000. What is the breakeven point in units of production if the selling price per unit is $40? (2.3)

2-8. Refer to Problem 2-6. Graph the average unit cost function for this situation as originally given for part (a). At what annual output within the present annual production (and sales) capacity of $12,000,000 does the minimum average cost per unit (C_u) occur? Why? (2.3)

2-9. An operator of a fleet of small diesel trucks has been using oil-filter cartridge A on each truck. This type of filter cartridge costs $5.00 per unit, and it has been replaced each 5,000 miles, with 1 quart of oil being added each 1,000 miles thereafter until the next oil change. A salesperson for filter cartridge B, which costs only $2.00, claims that if B is used and replaced each 2,000 miles, with the oil also being changed, no oil will have to be added between changes, and oil costing only $0.90 per quart can be used in place of the present oil, which costs $1.25 per quart. If the engines require 6 quarts of oil when the oil is changed, which type of filter would be more economical? (2.3)

2-10. A company finds that, on the average, two of its engineers, always traveling together, spend 60 hours each month in flying time on commercial airlines in making service calls to customers' plants. Also, the cost for airline tickets, airport buses, car rental, and so on is approximately $2,000 per person per month. An air charter service offers to supply a small business jet and pilot on 24-hour notice at a cost of $1,200 per month plus $125 per hour of flying time and $25 per hour for waiting time on the ground at the destination. It states that experience for similar situations has shown that using the charter service will reduce total travel time by 50%. The company estimates that the cost of car rental at destinations probably would amount to about $250 per month if the charter service is used, and the average waiting time will be about 40 hours per month. It also estimates that each engineer's time is worth $40 per hour to the company. Should the charter service be used? (2.3)

2-11. In your own words, describe the concept of the *net cash flow* for a project alternative. (2.5)

2-12. For each of the seven steps in the engineering economic analysis process, describe the activities that normally would be accomplished in that part of the analysis. (2.5)

2-13. Why does the use of accounting data in engineering economy studies need to be carefully watched? (2.4)

2-14. In the design of an automobile engine part, an engineer has a choice of either a steel casting or an aluminum alloy casting. Either material provides the same service. However, the steel casting weighs 8 ounces, compared with 5 ounces for the aluminum casting. Every pound of extra weight in the automobile has been assigned a penalty of $6 to account for increased fuel consumption during the life cycle of the car. The steel casting costs $3.20 per pound while the aluminum alloy can be cast for $7.40 per pound. Machining costs per casting are $5.00 for steel and $4.20 for aluminum. Which material should the engineer select and what is the difference in unit costs? (2.3.3)

2-15. In your own words, describe the life-cycle cost concept. Why is the potential for acheiving life-cycle cost savings greatest in the acquisition phase of the life cycle?

3

PRINCIPLES OF MONEY-TIME RELATIONSHIPS

The objective of this chapter is to provide an understanding of the *return to capital* in the form of interest (or profit), and to illustrate how basic time-value calculations are made relating to the cost of capital in engineering economy studies.

The following topics are discussed in this chapter:

Return to capital
Origins of interest
Simple interest
Compound interest
The concept of equivalence
Cash flow diagrams/tables
Interest formulas
Arithmetic sequences of cash flows
Geometric sequences of cash flows
Nominal versus effective interest rates
Interest rates that vary with time
Continuous compounding

3.1 INTRODUCTION

In Section 1.6, we introduced the concept of capital and briefly discussed its investment and allocation within an organization. In summary, capital is wealth in the form of money or property that is capable of being used to produce more wealth.

The majority of engineering economy studies involve commitment of capital for extended periods of time, so the effect of time must be considered. In this regard, it is recognized that a dollar today is worth more than a dollar one or more years from now because of the interest (or profit) it can earn. Therefore, money has a *time value*.

3.2 WHY CONSIDER RETURN TO CAPITAL?

Capital in the form of money for the people, machines, materials, energy, and other things needed in the operation of an organization may be classified into two basic categories. *Equity capital* is that owned by the individuals who have invested their money or property in a business project or venture in the hope of receiving a profit. *Debt capital*, often called *borrowed capital*, is obtained from lenders (e.g., through the sale of bonds) for investment. In return, the lenders receive interest from the borrowers.

Normally, the lenders do not receive any other benefits that may accrue from the investment of the capital borrowed. They have no ownership in the organization and do not participate as fully in the risks of the project or venture as the owners do. Thus, their fixed return on the capital loaned, in the form of interest, is more assured (has less risk) than the receipt of profit by the owners of equity capital. If the project or venture is successful, the return to the owners of equity capital (profit) can be substantially more than the interest received by lenders of capital; however, the owners could lose some or all of their money invested while the lenders still could receive all the interest owed plus repayment of the money borrowed by the firm.

There are fundamental reasons why return to capital in the form of interest and profit is an essential ingredient of engineering economy studies. First, interest and profit pay the providers of capital for forgoing its use during the time the capital is being used. The fact that the supplier can realize a return on capital acts as an *incentive* to accumulate capital by savings, thus postponing immediate consumption in favor of creating wealth in the future. Second, interest and profit are payments for the *risk* the investor takes in permitting another person, or an organization, to use his or her capital.

In typical situations, investors must decide whether the expected return on their capital is sufficient to justify buying into a proposed project or venture. If capital is invested in a project, investors would expect, as a minimum, to receive a return at least equal to the amount they have sacrificed by not using it in some other available opportunity of comparable risk. This interest or profit available from an alternative investment is called the opportunity cost of using capital in the proposed undertaking. Thus, whether borrowed capital or equity capital is involved, there is a cost for the capital employed in the sense that the project or venture must provide a return sufficient to be financially attractive to suppliers of money or property.

In summary, whenever capital is required in engineering and other business projects and ventures, it is essential that proper consideration be given to its cost. Ignoring the cost of capital (through oversight or intentionally) strips out a primary incentive that investors might have for placing their capital in a proposed project. Without such inducements, it is clear that few of us would choose to save and invest in a better future. Consequently, the productive use of capital has to correctly take this expected return into account during the period of time that the capital is committed. The remainder of this chapter deals with time value of money principles, which are vitally important to the proper evaluation of engineering projects that form the foundation of a firm's competitiveness, and hence to its very survival.

3.3 THE ORIGINS OF INTEREST

Like taxes, interest has existed from earliest recorded human history. Records reveal its existence in Babylon in 2000 B.C. In the earliest instances interest was paid in money for the use of grain or other commodities that were borrowed; it was also paid in the form of grain or other goods. Many of the existing interest practices stem from the early customs in the borrowing and repayment of grain and other crops.

History also reveals that the idea of interest became so well established that a firm of international bankers existed in 575 B.C., with home offices in Babylon. Its income was derived from the high interest rates it charged for the use of its money for financing international trade.

Throughout early recorded history, typical annual rates of interest on loans of money were in the neighborhood of 6 to 25%, although legally sanctioned rates as high as 40% were permitted in some instances. The charging of exorbitant interest rates on a loan was termed *usury*, and prohibition of usury is found in the Law of Moses.

During the Middle Ages, interest taking on loans of money was generally outlawed on scriptural grounds. In 1536 the Protestant theory of usury was established by John Calvin, and it refuted the notion that interest was unlawful. Consequently, interest taking again became viewed as an essential and legal part of doing business. Eventually, published interest tables became available to the public.

3.4 SIMPLE INTEREST

When the total interest earned or charged is directly proportional to the initial amount of the loan (principal), the interest rate, and the number of interest periods for which the principal is committed, the interest and interest rate are said to be *simple*. Simple interest is not used frequently in commercial practice in modern times.

When simple interest is applicable, the total interest, $\underline{I}$, earned or paid may be computed in the formula

$$\underline{I} = (P)(N)(i) \tag{3-1}$$

where P = principal amount lent or borrowed
N = number of interest periods (e.g., years)
i = interest rate per interest period

The total amount repaid at the end of N interest periods is $P + \underline{I}$. Thus, if \$1,000 is loaned for 3 years at a simple interest rate of 10% per annum, the interest earned will be

$$\underline{I} = \$1000 \times 0.10 \times 3 = \$300$$

The total amount owed at the end of three years would be \$1,000 + \$300 = \$1,300.

3.5 COMPOUND INTEREST

Whenever the interest charge for any interest *period* (a year, for example) is based on the remaining principal amount plus any accumulated interest charges up to the beginning of that period, the interest is said to be *compound*. The effect of compounding of interest can be shown by the following table for $1,000 loaned for three periods at an interest rate of 10% compounded per period.

Period	(1) Amount Owed at Beginning of Period	(2) = (1) × 10% Interest Charge for Period	(3) = (1) + (2) Amount Owed at End of Period
1	$1,000.00	$100.00	$1,100.00
2	$1,100.00	$110.00	$1,210.00
3	$1,210.00	$121.00	$1,331.00

As you can see, a total of $1,331.00 would be due for repayment at the end of the third period. If the length of a period is 1 year, the $1,331.00 at the end of three periods (years) can be compared with the $1,300.00 given earlier for the same problem with simple interest. A graphical comparison of simple interest and compound interest is given in Figure 3-1. The difference is due to the effect of *compounding*, which essentially is the calculation of interest on previously earned interest. This difference would be much greater for larger amounts of money, higher interest rates, or greater numbers of years. Thus, simple interest does consider the time value of money but does not involve compounding of interest. Compound interest is much more common in practice than is simple interest and is used throughout the remainder of this book.

FIGURE 3-1

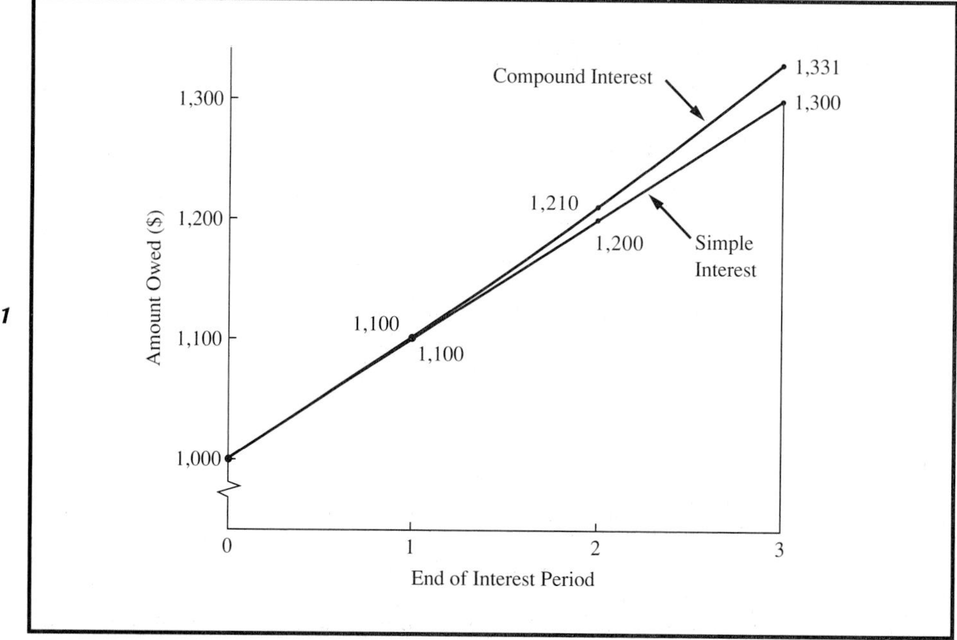

3.6

THE CONCEPT OF EQUIVALENCE

Alternatives should be compared insofar as possible when they produce similar results, serve the same purpose, or accomplish the same function. This is not always possible to achieve in some types of economy studies as we shall see later, but now our attention is directed at answering the question: How can alternatives for providing the same service or accomplishing the same function be compared when interest is involved over extended periods of time? Thus, we should consider the comparison of alternative options, or proposals, by reducing them to an *equivalent basis* that is dependent on (1) the interest rate, (2) the amounts of money involved, (3) the timing of the monetary receipts and/or disbursements, and (4) the manner in which the interest, or profit, on invested capital is repaid and the initial capital recovered.

To understand better the mechanics of interest and to expand on the notion of economic equivalence, we consider a situation in which we borrow $8,000 and agree to repay it in four years at an annual interest rate of 10%. There are many plans by which the principal of this loan (i.e., $8,000) and the interest on it can be repaid. For simplicity, we have selected four plans to demonstrate the idea of economic equivalence. Here equivalence means that all four plans are equally desirable to the borrower. Thus, any one is tradeable for one of the others. In each the interest rate is 10% and the original amount borrowed is $8,000; thus, differences among plans rest with items (3) and (4) above. The four plans are shown in Table 3-1, and it will soon be apparent that all are *equivalent* at an interest rate of 10% per year.

It can be seen in plan 1 that $2,000 of the loan principal is repaid at the end of each of the years 1 through 4. As a result, the interest we repay at the end of a particular year is affected by how much we still owe on the loan at the beginning of that year. Our end-of-year payment is just the sum of $2,000 and interest paid on the beginning-of-year amount owed.

Plan 2 indicates that none of the loan principal is repaid until the end of the fourth year. Our interest cost each year is $800, and it is repaid at the end of years 1 through 4. Since interest does not accumulate in either plan 1 or plan 2, compounding of interest is not present. Notice that $3,200 in interest is paid with plan 2, whereas only $2,000 is paid in plan 1. The reason, of course, is that we had use of the $8,000 principal for four years in plan 2 but, on average, had use of much less than $8,000 in plan 1.

Plan 3 requires that we repay equal end-of-year amounts of $2,524 each. Later in this chapter (Section 3.10) we will show how the $2,524 per year is computed. For our purposes here, the student should observe that the four end-of-year payments in plan 3 completely repay the $8,000 loan principal with interest at 10%.

TABLE 3–1 Four Plans for Repayment of $8,000 in 4 Years with Interest at 10%

(1) Year	(2) Amount Owed at Beginning of Year	(3) = 10% × (2) Interest Accrued for Year	(4) = (2) + (3) Total Money Owed at End of Year	(5) Principal Payment	(6) = (3) + (5) Total End-of-Year Payment
Plan 1: At End of Each Year Pay $2,000 Principal Plus Interest Due					
1	$8,000	$ 800	$8,800	$2,000	$ 2,800
2	6,000	600	6,600	2,000	2,600
3	4,000	400	4,400	2,000	2,400
4	2,000	200	2,200	2,000	2,200
	20,000 $-yr	$2,000 (total interest)		$8,000	$10,000 (total amount repaid)
Plan 2: Pay Interest Due at End of Each Year and Principle at End of Four Years					
1	$8,000	$ 800	$8,800	$ 0	$ 800
2	8,000	800	8,800	0	800
3	8,000	800	8,800	0	800
4	8,000	800	8,800	8,000	8,800
	32,000 $-yr	$3,200 (total interest)		$8,000	$11,200 (total amount repaid)
Plan 3: Pay In 4 Equal End-of-Year Payments					
1	$8,000	$ 800	$8,800	$1,724	$ 2,524
2	6,276	628	6,904	1,896	2,524
3	4,380	438	4,818	2,086	2,524
4	2,294	230	2,524	2,294	2,524
	20,960 $-yr	$2,096 (total interest)		$8,000	$10,096 (total amount repaid)
Plan 4: Pay Principle and Interest in One Payment at End of Four Years					
1	$ 8,000	$ 800	$ 8,800	$ 0	$ 0
2	8,800	880	9,680	0	0
3	9,600	968	10,648	0	0
4	10,648	1,065	11,713	8,000	11,713
	37,130 $-yr	$3,713 (total interest)		$8,000	$11,713 (total amount repaid)

Finally, plan 4 shows that no interest and no principal are repaid for the first three years of the loan period. Then at the end of the fourth year, the original loan principal plus accumulated interest for the four years is repaid in a single lump-sum amount of $11,712.80 (rounded in Table 3-1 to $11,713). Both plans 3 and 4 involve compound interest. The total amount of interest repaid in plan 4 is the highest of all plans considered. Not only was the principal repayment in plan 4 deferred until the end of year 4, but we also deferred all interest payment until that time. If annual interest rates rise above 10%, can you see that plan 4 causes bankers to turn gray-haired rather quickly?

This brings us back to the notion of economic equivalence. If interest rates remain constant at 10% for the plans shown in Table 3-1, all four plans are equivalent. This assumes one can freely borrow and lend at the 10% rate. Hence, we would be indifferent about whether the principal is repaid early in the loan's life (e.g., plans 1 and 3) or repaid at the end of year 4 (e.g., plans 2 and 4). *Economic equivalence is established, in general, when we are indifferent between a future payment, or series of future payments, and a present sum of money.*

To see *why* the four plans in Table 3-1 are equivalent at 10%, we could plot the amount owed at the beginning of each year (column 2) versus the year. The area under the resultant curve represents the *dollar-years* that the money is owed. For example, the dollar-years for plan 1 equals 20,000, which is obtained from this graph:

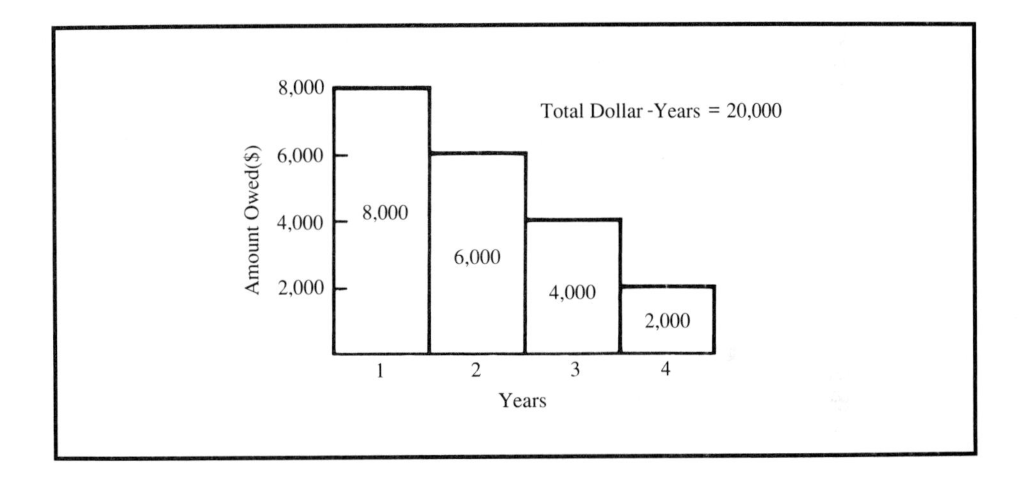

When total dollar-years are calculated for each plan and divided into total interest paid over the four years (the sum of column 3), one discovers that the ratio is constant:

Plan	Area Under Curve (dollar-years)	Total Interest Paid	Ratio of Total Interest to Dollar-Years
1	$20,000	$2,000	0.10
2	32,000	3,200	0.10
3	20,960	2,096	0.10
4	37,130	3,713	0.10

Because the ratio is constant at 0.10 for all plans, we can deduce that all repayment methods considered in Table 3-1 are equivalent even though each involves a different total end-of-year payment in column 6. Dissimilar dollar-years of borrowing, by itself, does not necessarily allow one to conclude that different loan repayment plans *are* equivalent or *are not* equivalent. In summary, equivalence is established when total interest paid, divided by dollar-years of borrowing, is a constant ratio among financing plans.

One last important point to emphasize is that the loan repayment plans of Table 3–1 are equivalent only at an interest rate of 10%. If these plans are evaluated with methods presented later in this chapter at interest rates other than 10%, one plan can be identified that is superior to the other three. For instance, when $8,000 has been lent at 10% interest and subsequently the cost of borrowed money increases to 15%, the *lender* would prefer plan 1 in order to recover his or her funds quickly so that they might be reinvested at higher interest rates.

3.7 NOTATION AND CASH FLOW DIAGRAMS/TABLES

The following notation is utilized for compound interest calculations:

i = effective interest rate per interest period
N = number of compounding periods
P = present sum of money; the *equivalent* worth of one or more cash flows at a reference point in time called the present
F = future sum of money; the *equivalent* worth of one or more cash flows at a reference point in time called the future
A = end-of-period cash flows (or *equivalent* end-of-period values) in a uniform series continuing for a specified number of periods, starting at the end of the first period

These and other commonly used symbols and terminology that appear throughout this book are defined in Appendix B.

The use of cash flow (time) diagrams and/or tables is strongly recommended for situations in which the analyst needs to clarify or visualize what is involved when flows of money occur at various times. In summary, a cash flow is the difference between total cash inflows (receipts) and cash outflows (expenditures) for a specified period of time (e.g. one year). As discussed in Chapter 2 (Section 2.5), cash flows are important in engineering economy because they form the

FIGURE 3-2

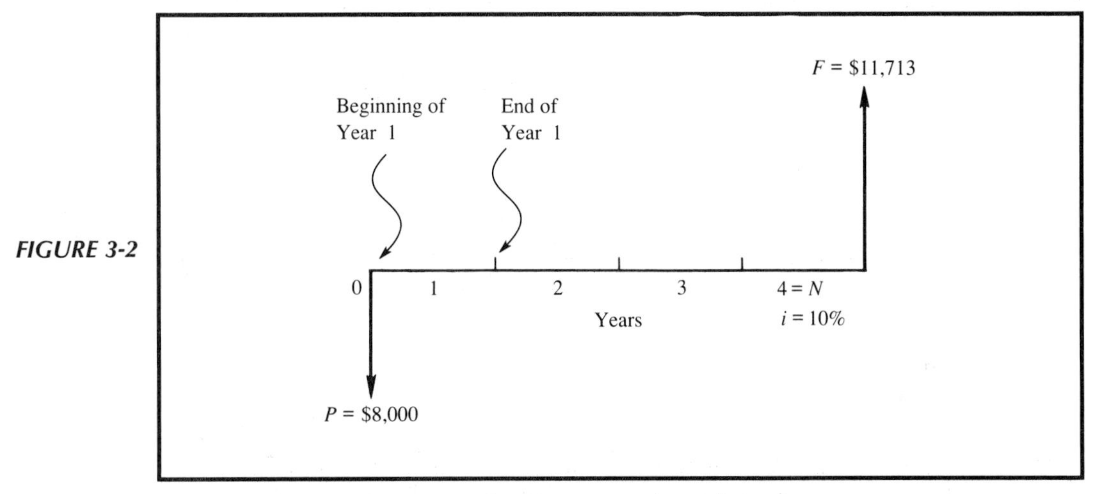

Cash Flow Diagram for Plan 4 of Table 3-1 (Lender's Viewpoint)

basis for evaluating alternatives. Indeed, the usefulness of the cash flow diagram for economic analysis problems is analogous to the use of the free-body diagram for mechanics problems.

Figure 3-2 shows a cash flow diagram for plan 4 of Table 3-1, and Figure 3-3 depicts the net cash flow of plan 3. These two figures also illustrate the definition of the above symbols and their placement on a cash flow diagram. Notice that all cash flows have been placed at the end of the year to correspond with the convention used in Table 3-1.

The cash flow diagram employs several conventions:

1. The horizontal line is a *time scale* with progression of time moving from left to right. The period (or year) labels are applied to intervals of time rather than points on the time scale. Note, for example, that the end of period

FIGURE 3-3

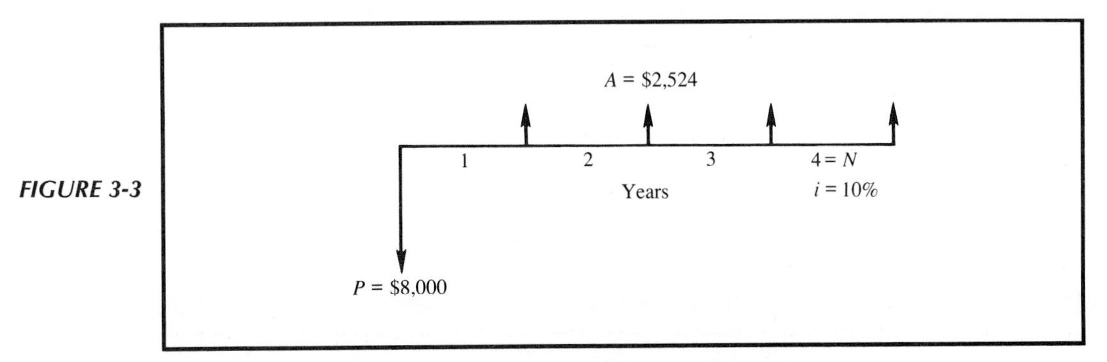

Cash Flow Diagram for Plan 3 of Table 3-1 (Lender's Viewpoint)

2 is coincident with the beginning of period 3. Only if specific dates are employed should the points in time rather than intervals be labeled.

2. The arrows signify cash flows. If a distinction needs to be made, downward arrows represent expenses (negative cash flows or cash outflows) and upward arrows represent receipts (positive cash flows or cash inflows).
3. The cash flow diagram is dependent on point of view. For example, the situations shown in Figures 3-2 and 3-3 were based on cash flow as seen by the lender. If the direction of all arrows had been reversed, the problem would be diagrammed from the borrower's viewpoint.

EXAMPLE 3-1

Before evaluating the economic merits of a proposed investment, the XYZ Corporation insists that its engineers develop a cash flow diagram of the proposal. An investment of $10,000 can be made that will produce uniform annual revenue of $5,310 for five years and then have a positive salvage value of $2,000 at the end of year 5. Annual expenses will be $3,000 at the end of each year for operating and maintaining the project. Draw a cash flow diagram for the five-year life of the project.

Solution

As shown in Figure 3-4, the initial investment of $10,000 and annual expenses of $3,000 are cash outflows, while annual revenues and the salvage value are cash inflows.

FIGURE 3-4

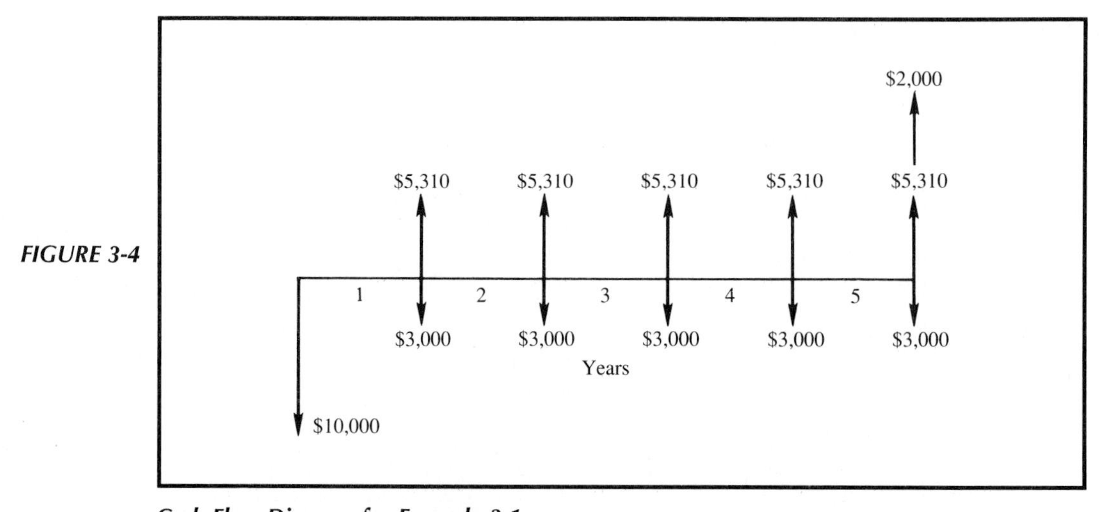

Cash Flow Diagram for Example 3-1

Notice that the beginning of a given year is the end of the preceding year. For example, the beginning of year 2 is the end of year 1.

End of Year	Net Cash Flow
0	−$10,000
1	2,310
2	2,310
3	2,310
4	2,310
5	4,310

When monetary returns each year to owners of the $10,000 capital are considered, the economic attractiveness of the project can be determined. This general subject involves different methods of compounding interest (or forgone profits) and also different conventions and assumptions concerning the timing of cash flows.

Example 3-2 presents a situation in which cash flows are represented in tabular form to facilitate the analysis of engineering plans/designs.

EXAMPLE 3-2

In the renovation of a small office building, two feasible alternatives for upgrading the heating, ventilating, and air conditioning (HVAC) system have been identified:

Alternative A: *Rebuild (overhaul) the Existing HVAC system*

- Equipment, labor, and materials to upgrade $18,000
- Annual cost of electricity 32,000
- Annual maintenance 2,400

Alternative B: *Install a new HVAC system that utilizes existing ductwork*

- Equipment, labor, and materials to install $60,000
- Annual cost of electricity 9,000
- Annual maintenance 16,000
- Replacement of a major component 4 years hence ... 9,400

At the end of eight years, the estimated salvage value for alternative A is $2,000, and for alternative B it is $8,000. Assume both alternatives will provide comparable service (comfort) over an eight-year time period, and assume that the major component replaced in alternative B will have no salvage value at the end of year 8. (a) In a cash flow table, use end-of-year convention and tabulate the cash flows for both alternatives. (b) Determine the annual cash flow difference between the alternatives ($B - A$). (c) Compute the cumulative difference through the end of year 8. (The cumulative difference is the sum of differences, $B - A$, from year 0 through year 8.)

Solution

The cash flow table for this example is given as Table 3-2. Based on these results, several points can be made: (1) "doing nothing" is not an option—either A or B must be selected; (2) even though positive and negative cash flows are included in the table, on balance we are investigating two "cost-only" alternatives; (3) a decision between the two alternatives can be made just as easily on the *difference* in cash flows (i.e., on the avoidable difference) as it can on the stand-alone net cash flows for alternatives A and B; (4) alternative B has cash flows identical to those of alternative A *except for* the differences shown in the table, so if the avoidable difference can "pay its own way," alternative B is the recommended choice; (5) cash flow changes caused by inflation or other suspected influences could have easily been inserted into the table and included in the analysis; and (6) it takes 6 *years* for the extra

TABLE 3–2 Cash Flow Table for Example 3-2

End-of-Year	Alternative *A* Net Cash Flow	Alternative *B* Net Cash Flow	Difference ($B - A$)	Cumulative Difference
0(now)	−$ 18,000	−$ 60,000	−$42,000	−$42,000
1	− 34,400	− 25,000	9,400	− 32,600
2	− 34,400	− 25,000	9,400	− 23,200
3	− 34,400	− 25,000	9,400	− 13,800
4	− 34,400	− 34,400	0	− 13,800
5	− 34,400	− 25,000	9,400	− 4,400
6	− 34,400	− 25,000	9,400	5,000
7	− 34,400	− 25,000	9,400	14,400
8	− 34,400 + 2,000	− 25,000 + 8,000	15,400	29,800
Total	−$291,200	−$261,400		

\$42,000 investment in alternative *B* to generate sufficient cumulative savings in annual expenses to justify the higher investment (this ignores the time value of money).

Later in this chapter, we shall consider the time value of money in arriving at recommendations concerning choices between alternatives. It should be apparent that a cash flow table serves to clarify the timing of cash flows, the assumptions that are being made, and the data that are available.

The remainder of Chapter 3 deals with the development and illustration of equivalence (time value of money) principles for assessing the economic attractiveness of investments such as those proposed in Examples 3-1 and 3-2.

3.8 INTEREST FORMULAS FOR DISCRETE COMPOUNDING AND DISCRETE CASH FLOWS

Table 3-3 provides a summary of the six most common discrete compound interest factors, and it utilizes notation of the preceding section. These factors are derived and explained by example problems in the following sections. The formulas are for *discrete compounding*, which means that the interest is compounded at the end of each finite-length period, such as a month or a year. Furthermore, the formulas also assume discrete (i.e., lump-sum) cash flows spaced at the end of equal time intervals on a cash flow diagram. Discrete compound interest factors are given in Appendix C, where the assumption is made that i remains constant during the N compounding periods.

TABLE 3–3 Discrete Compounding Interest Factors and Symbols[a]

To Find:	Given:	Factor by Which to Multiply "Given"[a]	Factor Name	Factor Functional Symbol[b]
For single cash flows:				
F	P	$(1 + i)^N$	Single payment compound amount	$(F/P, i\%, N)$
P	F	$\frac{1}{(1 + i)^N}$	Single payment present worth	$(P/F, i\%, N)$

[a] i, effective interest rate per interest period; N, number of interest periods; A, uniform series amount (occurs at the end of each interest period); F, future worth; P, present worth.

[b] The functional symbol system is used throughout this book.

TABLE 3–3 ***Continued***

To Find:	Given:	Factor by Which to Multiply "Given"[a]	Factor Name	Factor Functional Symbol[b]
For uniform series (annuities):				
F	A	$\frac{(1+i)^N - 1}{i}$	Uniform series compound amount	$(F/A, i\%, N)$
P	A	$\frac{(1+i)^N - 1}{i(1+i)^N}$	Uniform series present worth	$(P/A, i\%, N)$
A	F	$\frac{i}{(1+i)^N - 1}$	Sinking fund	$(A/F, i\%, N)$
A	P	$\frac{i(1+i)^N}{(1+i)^N - 1}$	Capital recovery	$(A/P, i\%, N)$

[a] *i*, effective interest rate per interest period; *N*, number of interest periods; *A*, uniform series amount (occurs at the end of each interest period); *F*, future worth; *P*, present worth.

[b] The functional symbol system is used throughout this book.

3.9 INTEREST FORMULAS RELATING PRESENT AND FUTURE WORTHS OF SINGLE CASH FLOWS

Figure 3-5 shows a cash flow diagram involving a present single sum, P, and a future single sum, F, separated by N periods with interest at $i\%$ per period. Throughout this chapter a dashed arrow, such as that shown in Figure 3-5, indicates the quantity to be determined. Two formulas relating a given P and its unknown equivalent F are provided below.

FIGURE 3-5

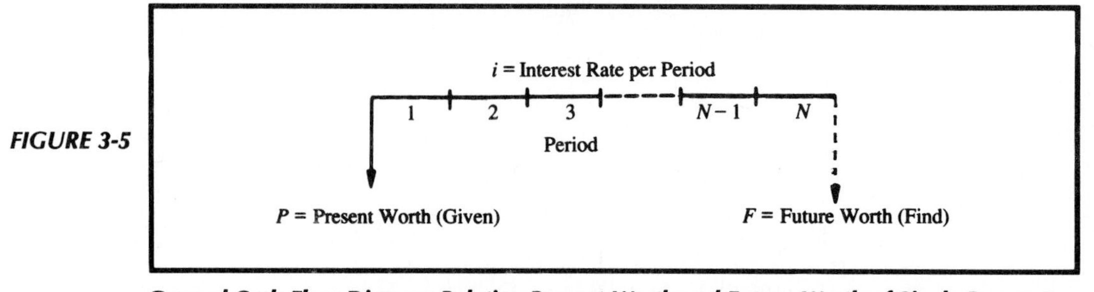

General Cash Flow Diagram Relating Present Worth and Future Worth of Single Payments

3.9.1 Finding F When Given P

If an amount of P dollars exists at a point in time and $i\%$ is the interest (profit or growth) rate per period, the amount will grow to a future amount of $P + Pi = P(1 + i)$ by the end of one period; by the end of two periods, the amount will grow to $P(1 + i)(1 + i) = P(1 + i)^2$; by the end of three periods, the amount will grow to $P(1 + i)^2(1 + i) = P(1 + i)^3$; and by the end of N periods the amount will grow to

$$F = P(1 + i)^N \tag{3-2}$$

EXAMPLE 3-3

Suppose you borrow \$8,000 now, with the promise to repay the loan principal plus accumulated interest in four years at $i = 10\%$ per year. How much would you owe at the end of four years?

Solution

Year	Amount Owed at Start of Year	Interest Owed for Each Year	Amount Owed at End of Year	Total-End-of-Year Payment
1	$P = \$8{,}000$	$iP = \$800$	$P(1 + i) = \$8{,}800$	0
2	$P(1 + i) = \$8{,}800$	$iP(1 + i) = \$880$	$P(1 + i)^2 = \$9{,}680$	0
3	$P(1 + i)^2 = \$9{,}680$	$iP(1 + i)^2 = \$968$	$P(1 + i)^3 = \$10{,}648$	0
4	$P(1 + i)^3 = \$10{,}648$	$iP(1 + i)^3 = \$1{,}065$	$P(1 + i)^4 = \$11{,}713$	$F = \$11{,}713$

In general, we see that $F = P(1 + i)^N$, and the total amount to be repaid is \$11,713. This further illustrates plan 4 in Table 3-1 in terms of notation that we shall be using throughout this book. The compounding effect in this example is apparent in Figure 3-6.

The quantity $(1 + i)^N$ is commonly called the *single payment compound amount factor*. Numerical values for this factor are given in the second column from the left in tables of Appendix C, for a wide range of values of i and N. In this book we shall use the functional symbol $(F/P, i\%, N)$ for $(1 + i)^N$. Hence Equation 3-2 can be expressed as

$$F = P(F/P, i\%, N) \tag{3-3}$$

where the factor in parentheses is read "find F given P at $i\%$ interest per period for N interest periods." Note that the sequence of F and P in F/P is the same as in the initial part of Equation 3-3 where the unknown quantity, F, is placed

FIGURE 3-6

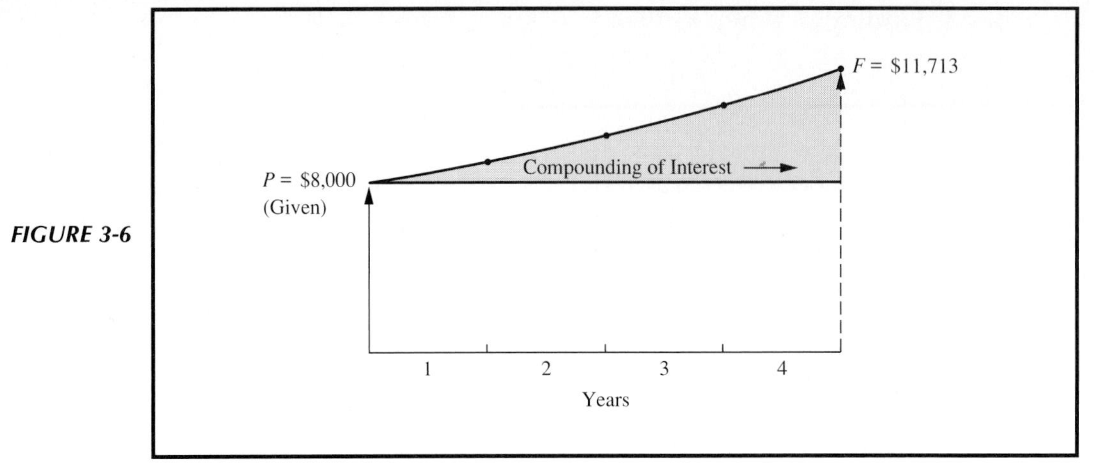

Illustration of Compounding Effect

on the left-hand side of the equation. This sequencing of letters is true of all functional symbols used in this book, and makes them easy to remember.

Another example of "finding F when given P," together with a cash flow diagram and solution, is given in Table 3-4. Note in Table 3-4 that for each of the six common discrete compound interest circumstances covered, two problem statements are given—(a) in borrowing-lending terminology, and (b) in equivalence terminology—but they both represent the same cash flow situation. Indeed, there are generally many ways in which a given cash flow situation can be expressed.

In general, a good way to interpret a relationship such as Equation 3-3 is that the calculated amount, F, at the point in time at which it occurs, *is equivalent to* (i.e., can be traded for) the known value, P, at the point in time at which it occurs, for the given interest or profit rate, i.

3.9.2 Finding *P* When Given *F*

From Equation 3-2, $F = P(1 + i)^N$. Solving this for P gives the relationship

$$P = F\left(\frac{1}{1 + i}\right)^N = F(1 + i)^{-N} \qquad (3\text{-}4)$$

The quantity $(1 + i)^{-N}$ is called the *single payment present worth factor*. Numerical values for this factor are given in the third column of the tables in Appendix C for a wide range of values of i and N. We shall use the functional symbol (P/F, $i\%$, N) for this factor. Hence

$$P = F(P/F, i\%, N) \tag{3-5}$$

EXAMPLE 3-4

An investor has an option to purchase a tract of land that will be worth \$10,000 in six years. If the value of the land increases at 8% each year, how much should the investor be willing to pay now for this property?

Solution

The purchase price can be determined from Equation 3-5 and Table C-12 in Appendix C as follows:

$$P = \$10{,}000(P/F, 8\%, 6)$$
$$P = \$10{,}000(0.6302)$$
$$= \$6{,}302$$

The discounting effect in this example is illustrated in Figure 3-7.

Another example of this type of problem, together with a cash flow diagram and solution, is given in Table 3-4.

FIGURE 3-7

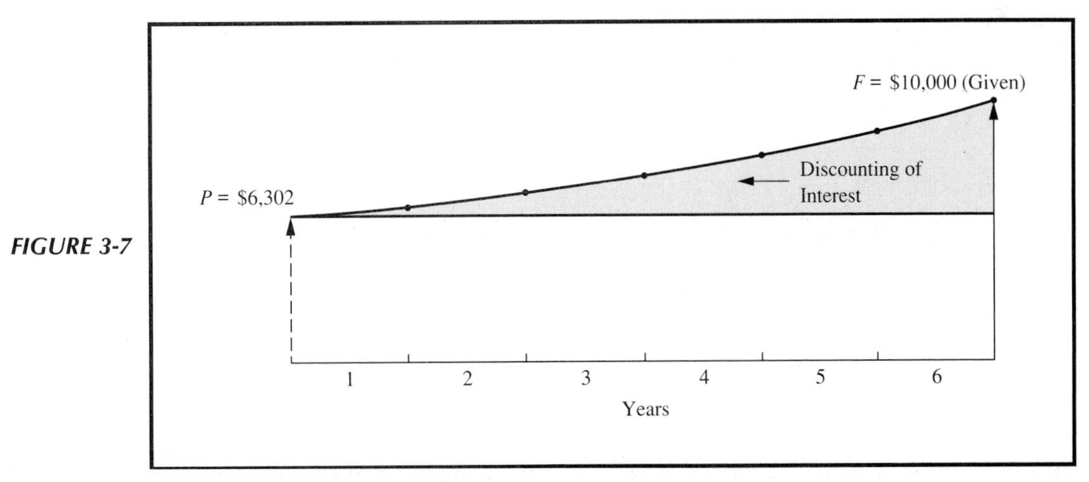

Illustration of Discounting Effect

TABLE 3–4 Discrete Cash Flow Examples Illustrating Equivalence

Example Problems (all using an interest rate of $i = 10\%$ compounded annually)

To Find:	Given:	(a) In Borrowing–Lending Terminology:	(b) In Equivalence Terminology:	Cash Flow Diagram[a]	Solution
For single cash flows:					
F	P	A firm borrows \$1,000 for 8 years. How much must it repay in a lump sum at the end of the eighth year?	What is the equivalent worth at the end of 8 years of \$1,000 at the beginning of those 8 years?	$P = \$1,000$; 0; $N = 8$; $F = ?$	$F = P(F/P, 10\%, 8)$ $= \$1,000(2.1436)$ $= \$2,143.60$
P	F	A firm desires to have \$2,143.60 eight years from now. What amount should be deposited now to provide for it?	What is the equivalent present worth of \$2,143.60 received 8 years from now?	$F = \$2,143.60$; 0; $N = 8$; $P = ?$	$P = F(P/F, 10\%, 8)$ $= \$2,143.60(0.4665)$ $= \$1,000.00$
For uniform series:					
F	A	If eight annual deposits of \$187.45 each are placed in an account, how much money has accumulated immediately after the last deposit?	What amount at the end of the eighth year is equivalent to eight end-of-year payments of \$187.45 each?	$F = ?$; 1 2 3 4 5 6 7 8; $A = \$187.45$	$F = A(F/A, 10\%, 8)$ $= \$187.45(11.4359)$ $= \$2,143.60$

[a]The cash flow diagram represents the example as stated in borrowing-lending terminology .

TABLE 3–4 *Continued*

Example Problems (all using an interest rate of $i = 10\%$ compounded annually)

To Find:	Given:	(a) In Borrowing–Lending Terminology:	(b) In Equivalence Terminology:	Cash Flow Diagram	Solution
P	A	How much should be deposited in a fund now to provide for eight end-of-year withdrawals of $187.45 each?	What is the equivalent present worth of eight end-of-year payments of $187.45 each?	$A = \$187.45$; 1 2 3 4 5 6 7 8; $P = ?$	$P = A(P/A, 10\%, 8)$ $= \$187.45(5.3349)$ $= \$1,000.00$
A	F	What uniform annual amount should be deposited each year in order to accumulate $2,143.60 at the time of the eighth annual deposit?	What uniform payment at the end of eight successive years is equivalent to $2,143.60 at the end of the eighth year?	$F = \$2,143.60$; 1 2 3 4 5 6 7 8; $A = ?$	$A = F(A/F, 10\%, 8)$ $= \$2,143.60(0.0874)$ $= \$187.45$
A	P	What is the size of 8 equal annual payments to repay a loan of $1,000? The first payment is due 1 year after receiving the loan.	What uniform payment at the end of 8 successive years is equivalent to $1,000 at the beginning of the first year?	$P = \$1,000$; 1 2 3 4 5 6 7 8; $A = ?$	$A = P(A/P, 10\%, 8)$ $= \$1,000(0.18745)$ $= \$187.45$

3.10

INTEREST FORMULAS RELATING A UNIFORM SERIES (ANNUITY) TO ITS PRESENT AND FUTURE WORTHS

Figure 3-8 shows a general cash flow diagram involving a series of uniform receipts, each of amount A, occurring at the end of each period for N periods with interest at $i\%$ per period. Such a uniform series is often called an *annuity*. It should be noted that the formulas and tables below are derived such that A occurs at the *end* of each period, and thus:

1. P (present worth) occurs one interest period before the first A (uniform payment).
2. F (future worth) occurs at the same time as the last A, and N periods after P.
3. A (annual worth) occurs at the end of periods 1 through N, inclusive.

The timing relationship for P, A, and F can be observed in Figure 3-8. Four formulas relating A to F and P are developed below.

3.10.1 Finding F When Given A

If a cash flow in the amount of A dollars occurs at the end of each period for N periods and $i\%$ is the interest (profit or growth) rate per period, the future

FIGURE 3-8

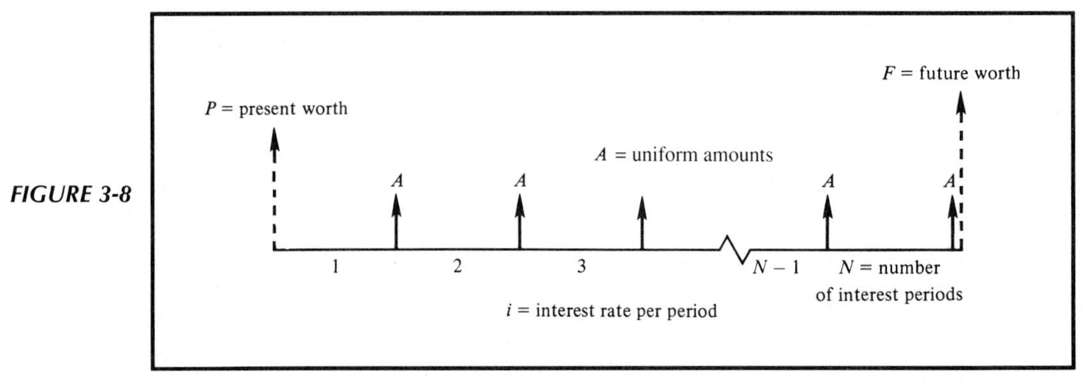

General Cash Flow Diagram Relating Uniform Series (Ordinary Annuity) to its Present Worth and Future Worth

worth, F, at the end of the Nth period is obtained by summing the future worths of each of the cash flows. Thus,

$$\begin{aligned} F &= A(F/P, i\%, N-1) + A(F/P, i\%, N-2) + A(F/P, i\%, N-3) + \cdots \\ &\quad + A(F/P, i\%, 1) + A(F/P, i\%, 0) \\ &= A[(1+i)^{N-1} + (1+i)^{N-2} + (1+i)^{N-3} + \cdots + (1+i)^1 + (1+i)^0] \end{aligned}$$

The bracketed terms comprise a geometric sequence having a common ratio of $(1+i)^{-1}$. Recall that the sum of the first N terms (S_N) of a geometric sequence is

$$S_N = \frac{a_1 - ba_N}{1-b} \qquad (b \neq 1)$$

where a_1 is the first term in the sequence, a_N the last term, and b the common ratio. If we let $b = (1+i)^{-1}$, $a_1 = (1+i)^{N-1}$, and $a_N = (1+i)^0$, then

$$F = A\left[\frac{(1+i)^{N-1} - \frac{1}{(1+i)}}{1 - \frac{1}{(1+i)}}\right]$$

which reduces to

$$F = A\left[\frac{(1+i)^N - 1}{i}\right] \tag{3-6}$$

The quantity $\{[(1+i)^N - 1]/i\}$ is called the *uniform series compound amount factor*. Numerical values for this factor are given in the fourth column of the tables in Appendix C for a wide range of values of i and N. We shall use the functional symbol $(F/A, i\%, N)$ for this factor. Hence Equation 3-6 can be expressed as

$$F = A(F/A, i\%, N) \tag{3-7}$$

Examples of this type of problem are provided below and in Table 3-4.

EXAMPLE 3-5

A woman desires to have \$100,000 in her retirement savings plan after working for 25 years. She will accomplish this by depositing A dollars each year in a savings account that earns 6% per year. How much must she save each year?

Solution

The annual deposit required to accumulate \$100,000 at 6% annual interest is

$$\begin{aligned} A &= \$100{,}000(A/F, 6\%, 25) \\ &= \$100{,}000(0.0182) \\ &= \$1{,}820.00 \end{aligned}$$

3.10.2 Finding P When Given A

From Equation 3-2, $F = P(1 + i)^N$. Substituting for F in Equation 3-6, one determines that

$$P(1 + i)^N = A\left[\frac{(1 + i)^N - 1}{i}\right]$$

Dividing both sides by $(1 + i)^N$,

$$P = A\left[\frac{(1 + i)^N - 1}{i(1 + i)^N}\right] \tag{3-8}$$

Thus, Equation 3-8 is the relation for finding the equivalent present worth (as of the beginning of the first period) of a uniform series of end-of-period cash flows of amount A for N periods. The quantity in brackets is called the *uniform series present worth factor*. Numerical values for this factor are given in the fifth column of the tables in Appendix C for a wide range of values of i and N. We shall use the functional symbol $(P/A, i\%, N)$ for this factor. Hence

$$P = A(P/A, i\%, N) \tag{3-9}$$

EXAMPLE 3-6

If a certain machine undergoes a major overhaul, its output can be increased by 20%—which translates into extra cash flow of \$20,000 at the end of each year for five years. If $i = 15\%$ per year, how much can we afford to invest to overhaul this machine?

Solution

The increase in cash flow is \$20,000 per year, and it continues for five years at 15% annual interest. The upper limit on what we can afford to spend is

$$\begin{aligned} P &= \$20{,}000(P/A, 15\%, 5) \\ &= \$20{,}000(3.3522) \\ &= \$67{,}044 \end{aligned}$$

Another example of this type of problem, together with a cash flow diagram and solution, is given in Table 3-4.

3.10.3 Finding A When Given F

Taking Equation 3-6 and solving for A, one finds that

$$A = F\left[\frac{i}{(1+i)^N - 1}\right] \tag{3-10}$$

Thus, Equation 3-10 is the relation for finding the amount, A, of a uniform series of cash flows occurring at the end of each of N interest periods that would be equivalent to (have the same value as) its future worth, F, occurring at the end of the last period. The quantity in brackets is called the *sinking fund factor*. Numerical values for this factor are given in the sixth column of the tables in Appendix C for a wide range of values of i and N. We shall use the functional symbol (A/F, $i\%$, N) for this factor. Hence

$$A = F(A/F, i\%, N) \tag{3-11}$$

An example of this type of problem, together with a cash flow diagram and solution, is given in Table 3-4.

3.10.4 Finding A When Given P

Taking Equation 3-8 and solving for A, one finds that

$$A = P\left[\frac{i(1+i)^N}{(1+i)^N - 1}\right] \tag{3-12}$$

Thus, Equation 3-12 is the relation for finding the amount, A, of a uniform series of cash flows occurring at the end of each of N interest periods that would be equivalent to, or could be traded for, the present worth, P, occurring at the beginning of the first period. The quantity in brackets is called the *capital recovery factor*.* Numerical values for this factor are given in the seventh column of the tables in Appendix C for a wide range of values of i and N. We shall use the functional symbol (A/P, $i\%$, N) for this factor. Hence

$$A = P(A/P, i\%, N) \tag{3-13}$$

An example that utilizes the equivalence between a present lump-sum amount and a series of equal uniform annual amounts starting at the end of year 1 and continuing through year four is provided in Table 3-1 as plan 3.

*The capital recovery factor is more conveniently expressed as $i/[1 - (1+i)^{-N}]$ for computation with a hand-held calculator.

Equation 3-13 yields the equivalent value of A that repays the \$8,000 loan plus 10% interest per year over four years:

$$A = \$8,000(A/P, 10\%, 4) = \$8,000(0.3155) = \$2,524$$

The entries in columns 3 and 5 of plan 3 in Table 3-1 can now be better understood. Interest owed at the end of year 1 equals \$8,000(0.10), and therefore the principal repaid out of the total end-of-year payment of \$2,524 is the difference, \$1,724. At the beginning of year 2, the amount of principal owed is \$8,000 − \$1,724 = \$6,276. Interest owed at the end of year 2 is \$6,276(0.10) ≅ \$628, and the principal repaid at that time is \$2,524 − \$628 = \$1,896. The remaining entries in plan 3 are obtained by performing these calculations for years 3 and 4.

A graphical summary of plan 3 is given in Figure 3-9. Here it can be seen that 10% interest is being paid on the beginning-of-year amount owed and that year-end payments of \$2,524 which consist of interest and principal bring the amount owed to \$0 at the end of the fourth year. (The exact value of A is \$2,523.77 and produces an exact value of \$0 at the end of four years.) It is important to note that all the uniform series interest factors in Table 3-3 involve the same concept as the one illustrated in Figure 3-9.

Another example of a problem where we desire to compute an equivalent value for A, from a given value of P and a known interest rate and number of compounding periods, is given in Table 3-4.

For an annual interest rate of 10%, the reader should now be convinced from Table 3-4 that \$1,000 at the beginning of year 1 is equivalent to the \$187.45 at the end of years 1–8, which is then equivalent to \$2,143.60 at the end of year 8.

FIGURE 3-9

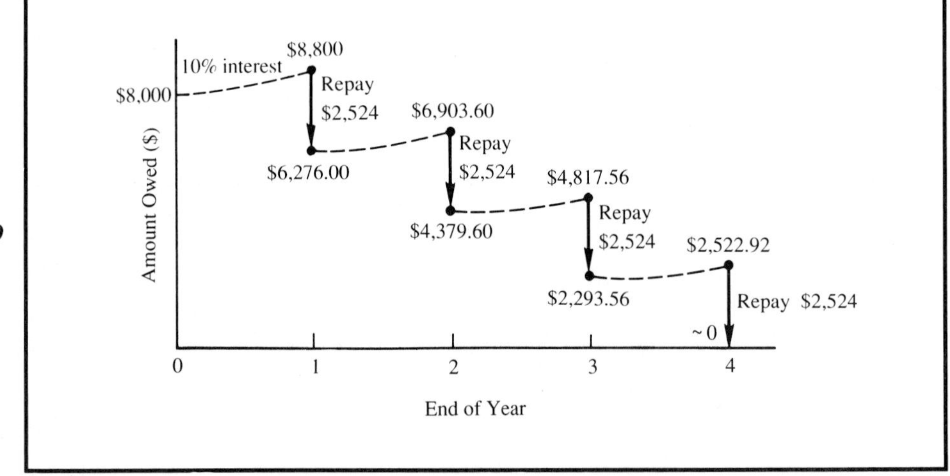

Relationship of Cash Flows for Plan 3 of Table 3-1 to Repayment of the \$8,000 Loan Principal

3.11

DEFERRED ANNUITIES (UNIFORM SERIES)

All annuities (uniform series) discussed to this point involved the first cash flow being made at the end of the first period, and they are called *ordinary annuities*. If the cash flow does not begin until some later date, the annuity is known as a *deferred annuity*. If the annuity is deferred J periods ($J < N$), the situation is as portrayed in Figure 3-10, in which the entire framed ordinary annuity has been moved forward from "time present," or "time 0," by J periods. It must be remembered that in an annuity deferred for J periods the first payment is made at the end of period ($J + 1$), assuming that all periods involved are equal in length.

The present worth at the end of period J of an annuity with cash flows of amount A is, from Equation 3-9, $A(P/A, i\%, N - J)$. The present worth of the single amount $A(P/A, i\%, N - J)$ as of time 0 will then be

$$A(P/A, i\%, N - J)(P/F, i\%, J)$$

EXAMPLE 3-7

To illustrate the discussion above, suppose that a father, on the day his son is born, wishes to determine what lump amount would have to be paid into an account bearing interest at 12% compounded annually to provide payments of $2,000 on each of the son's 18th, 19th, 20th, and 21st birthdays.

Solution

The problem is represented in Figure 3-11. One should first recognize that an ordinary annuity of four payments of $2,000 each is involved, and that the

FIGURE 3-10

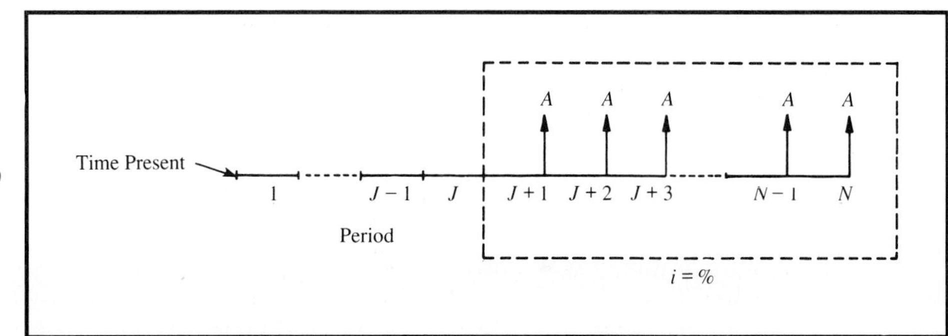

General Cash Flow Representation of a Deferred Annuity (Uniform Series)

FIGURE 3-11

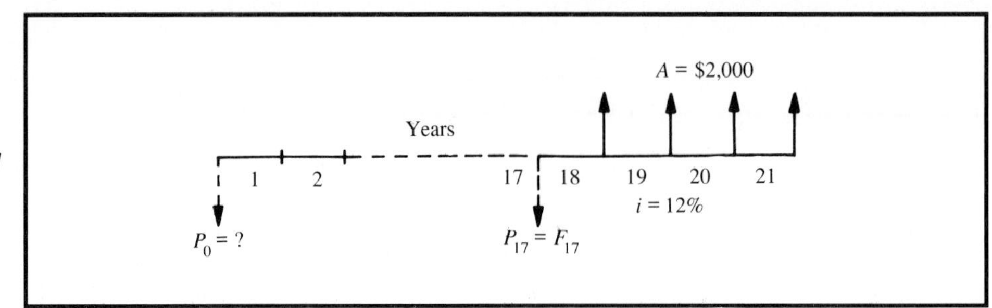

Cash Flow Diagram of the Deferred Annuity Problem in Example 3-7

present worth of this annuity occurs at the 17th birthday when a $(P/A, i\%, N - J)$ factor is utilized. In this problem, $N = 21$ and $J = 17$. It is often helpful to use a subscript with P or F to denote the point in time. Hence

$$P_{17} = A(P/A, 12\%, 4) = \$2,000(3.0373) = \$6,074.60$$

Note the dashed arrow in Figure 3-11, denoting P_{17}. Now that P_{17} is known, the next step is to calculate P_0. With respect to P_0, P_{17} is a future worth, and hence it could also be denoted F_{17}. Money at a given point in time, such as end of period 17, is the same regardless of whether it is called a present worth or a future worth. Hence

$$P_o = F_{17}(P/F, 12\%, 17) = \$6,074.60(0.1456) = \$884.46$$

which is the amount that the father would have to deposit.

EXAMPLE 3-8

As an addition to the problem in Example 3-7, suppose that it is desired to determine the equivalent worth of the four $2,000 payments as of the son's 24th birthday. Physically, this could mean that the four payments never were withdrawn or possibly that the son took them and immediately redeposited them in an account also earning interest at 12% compounded annually. Using our subscript system, we desire to calculate F_{24} as shown in Figure 3-12.

Solution
One way to work this is to calculate

$$F_{21} = A(F/A, 12\%, 4) = \$2,000(4.7793) = \$9,558.60$$

To determine F_{24}, F_{21} becomes P_{21}, and

$$F_{24} = P_{21}(F/P, 12\%, 3) = \$9,558.60(1.4049) = \$13,428.88$$

FIGURE 3-12

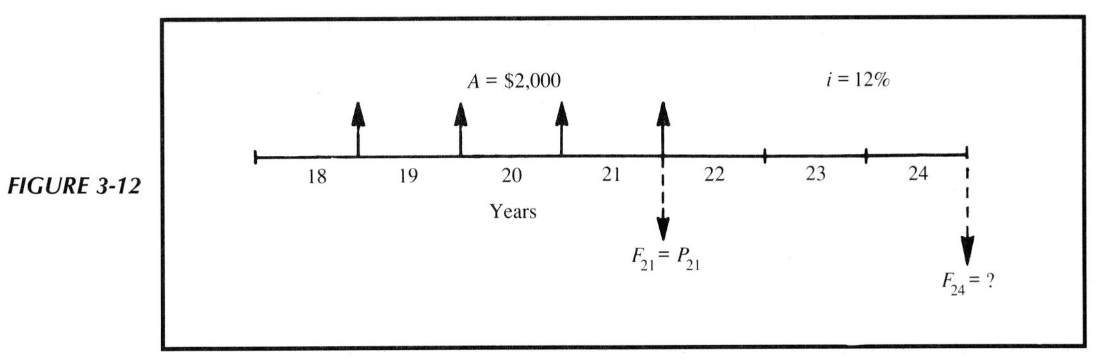

Cash Flow Diagram for the Deferred Annuity Problem in Example 3-8

Another, quicker way to work the problem is to recognize that the $P_{17} =$ \$6,074.60 and $P_0 =$ \$884.46 are each equivalent to the four \$2,000 payments. Hence one can find F_{24} directly, given P_{17} or P_0. Using P_0, we obtain

$$F_{24} = P_0(F/P, 12\%, 24) = \$884.46(15.1786) = \$13{,}424.86$$

which checks closely with the previous answer. The two numbers differ by \$4.02, which can be attributed to round-off error in the interest factors.

3.12 UNIFORM SERIES WITH BEGINNING-OF-PERIOD CASH FLOWS

It should be noted that up to this point all the interest formulas and corresponding tabled values for uniform series have assumed *end-of-period* cash flows. These same tables can be used for cases in which beginning-of-period cash flows exist merely by remembering that:

1. P (present worth) occurs one interest period before the first A (uniform series amount).
2. F (future worth) occurs at same time as the last A, and N periods after P.

On a beginning-of-period cash flow diagram, all cash flows that occur during a time period are placed at the point designated as the beginning of the period.

FIGURE 3-13

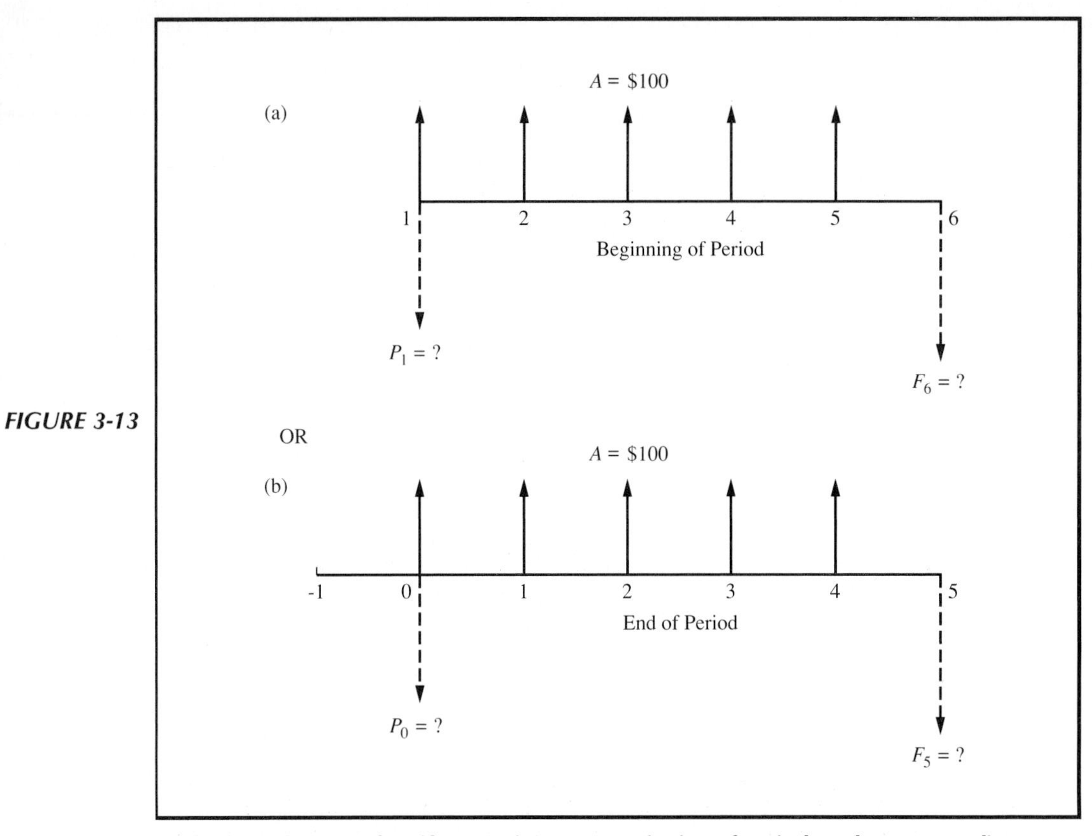

Cash Flow Diagram of Uniform Series at (a) Beginning of Period, and Corresponding Diagram at (b) End of Period

EXAMPLE 3-9

Figure 3-13a is a cash flow diagram depicting a uniform series of five beginning-of-period cash flows of $100 each. Thus, the first cash flow is at the beginning of the first period, and the fifth is at the beginning of the fifth period. (The corresponding end-of-period cash flow diagram is shown in Figure 3-13b.) If the interest rate is 10%, it is desired to find (a) the present worth of the uniform series at the beginning of the first period, and (b) the future worth at the *end* of the fifth period.

Solution

(a) There are several ways to work these types of problems. To find the present worth, P_0, in Figure 3-13b, one way is first to calculate

$$P_{-1} = A(P/A, 10\%, 5) = \$100(3.7908) = \$379.08$$

This P is at time -1 because that is one period before the first A cash flow. Also note that the interest factor is for five periods because there were five cash flows. Next, the problem is to find P_0 given P_{-1}. In this case P_0 becomes a future worth and could be denoted F_0. Hence

$$P_0 = F_0 = P_{-1}(F/P, 10\%, 1) = \$379.08(1.100) = \$416.99$$

Alternatively, one could directly utilize the Appendix C interest factors to determine the present worth in Figure 3-13a in this manner:

$$\begin{aligned} P_1 &= \$100 + \$100(P/A, 10\%, 4) \\ &= 100 + 100(3.1699) \\ &= \$416.99 \end{aligned}$$

(b) to find F_6 in Figure 3-13a, the first logical step is to calculate

$$F_5 = A(F/A, 10\%, 5) = \$100(6.1051) = \$610.51$$

Note that the F is at time 5 because that is at the same time as the last A cash flow. Also note that the interest factor is again for five periods, corresponding to the number of cash flows. Next, the problem is to find F_6 given F_5. In this case F_5 becomes a present worth and could be denoted P_5. Hence

$$F_6 = P_5(F/P, 10\%, 1) = \$610.51(1.10) = \$671.56$$

A convenient way to determine F_5 in Figure 3-13b would be to start with \$379.08 as of time -1, or \$416.99 as of time 0, and to calculate the future worth at time 5.

3.13 UNIFORM SERIES WITH MIDDLE-OF-PERIOD CASH FLOWS

Rather than using end-of-period or beginning-of-period cash flow conventions, some companies and individuals prefer to assume that discrete cash flows occur at the middle of each time period. Fortunately, the $(P/A, i, N)$, $(F/A, i, N)$, and $(P/F, i, N)$ interest factors of Appendix C can be utilized in the case of middle-of-period cash flows simply by *multiplying* the Appendix C interest factor by a "half-period correction" factor. When i is the effective interest rate per period, the half-period correction factor (HPC) to apply to interest factors in Appendix C is

$$\text{HPC} = \sqrt{1 + i}$$

When the $(A/F, i, N)$, $(A/P, i, N)$, and $(F/P, i, N)$ factors are involved, the Appendix C values are *divided* by the half-period correction factor to obtain midperiod interest factors.

EXAMPLE 3-10

Figure 3-14 is a cash flow diagram of an annuity consisting of four middle-of-year amounts of \$200. If i/year = 10%, determine (a) the equivalent present worth of the uniform series at the beginning of the first year, and (b) the equivalent future worth at the end of the fourth year.

Solution

(a) The half-period correction factor is $\sqrt{1.10} = 1.04881$. Therefore, the equivalent value of P_0, utilizing the $(P/A, 10\%, 4)$ factor from Appendix C, can be determined in two steps:

$$\begin{aligned} P_{-1/2} &= \$200(P/A, 10\%, 4) \\ &= \$200(3.1699) = \$633.98 \\ P_0 &= P_{-1/2}(\text{HPC}) = \$664.92 \end{aligned}$$

(b) The equivalent value of F_4 is determined with a $(F/A, 10\%, 4)$ factor from Appendix C and a $\sqrt{1.10}$ HPC:

$$\begin{aligned} F_4 &= \$200(F/A, 10\%, 4)(1.04881) \\ &= \$200(4.4641)(1.04881) \\ &= \$973.50 \end{aligned}$$

EXAMPLE 3-11

If \$1,500 is to be received at the end of five years when i per year is 15%, what middle-of-year uniform amount would be equivalent to this future amount? The cash flow diagram is shown in Figure 3-15.

FIGURE 3-14

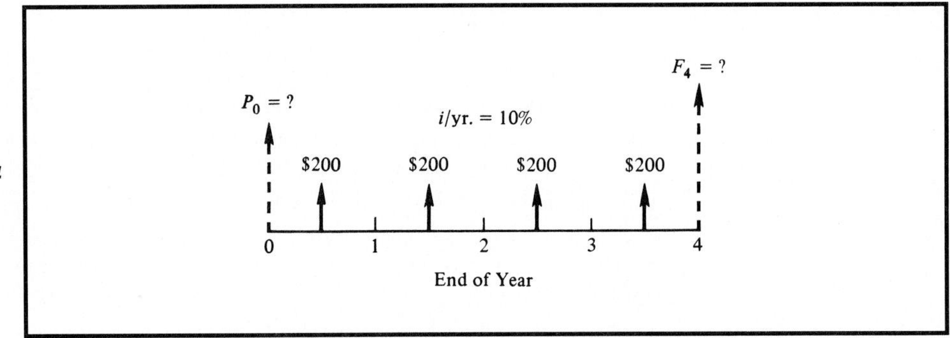

Cash Flow Diagram of a Uniform Series of Mid-Year Amounts in Example 3-10

FIGURE 3-15

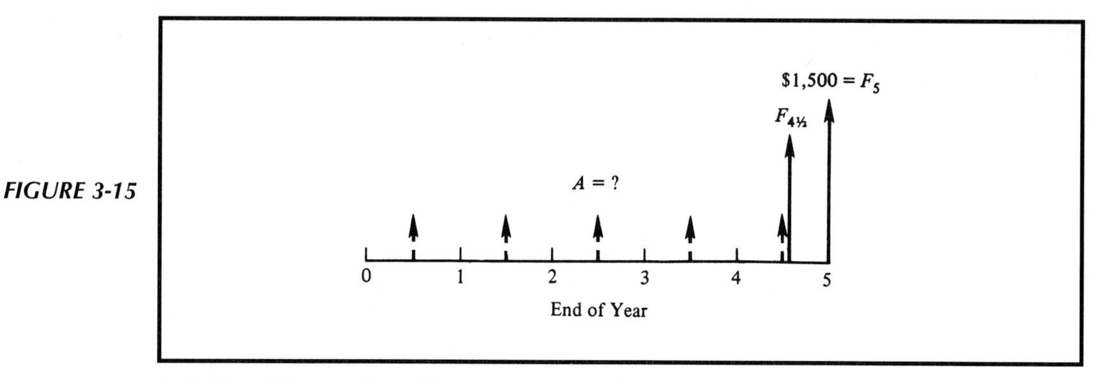

Cash Flow Diagram for Mid-Year Amounts of Example 3-11

Solution
The half-period correction is $\sqrt{1.15} = 1.0724$, so the following relationships, utilizing the appropriate Appendix C factor, can be used to solve for the middle-of-year equivalent A:

$$F_{4\frac{1}{2}}(\text{HPC}) = F_5 \qquad \text{or} \qquad F_{4\frac{1}{2}} = F_5\left(\frac{1}{\text{HPC}}\right)$$

$$A = F_{4\frac{1}{2}}(A/F, 15\%, 5)$$

$$= \frac{\$1,500}{1.0724}(0.1483) = \$207.43$$

3.14

EQUIVALENT PRESENT WORTH, FUTURE WORTH, AND ANNUAL WORTH

The reader should now be comfortable with equivalence problems that involve discrete compounding of interest and discrete cash flows. All compounding of interest takes place once per time period (e.g., a year) and to this point cash flows also occur once per time period. This section provides two examples involving equivalence calculations based on the commonly used end-of-year cash flow convention.

EXAMPLE 3-12

Figure 3-16 depicts an example problem with a series of year-end cash flows extending over eight years. The amounts are $100 for the first year, $200 for

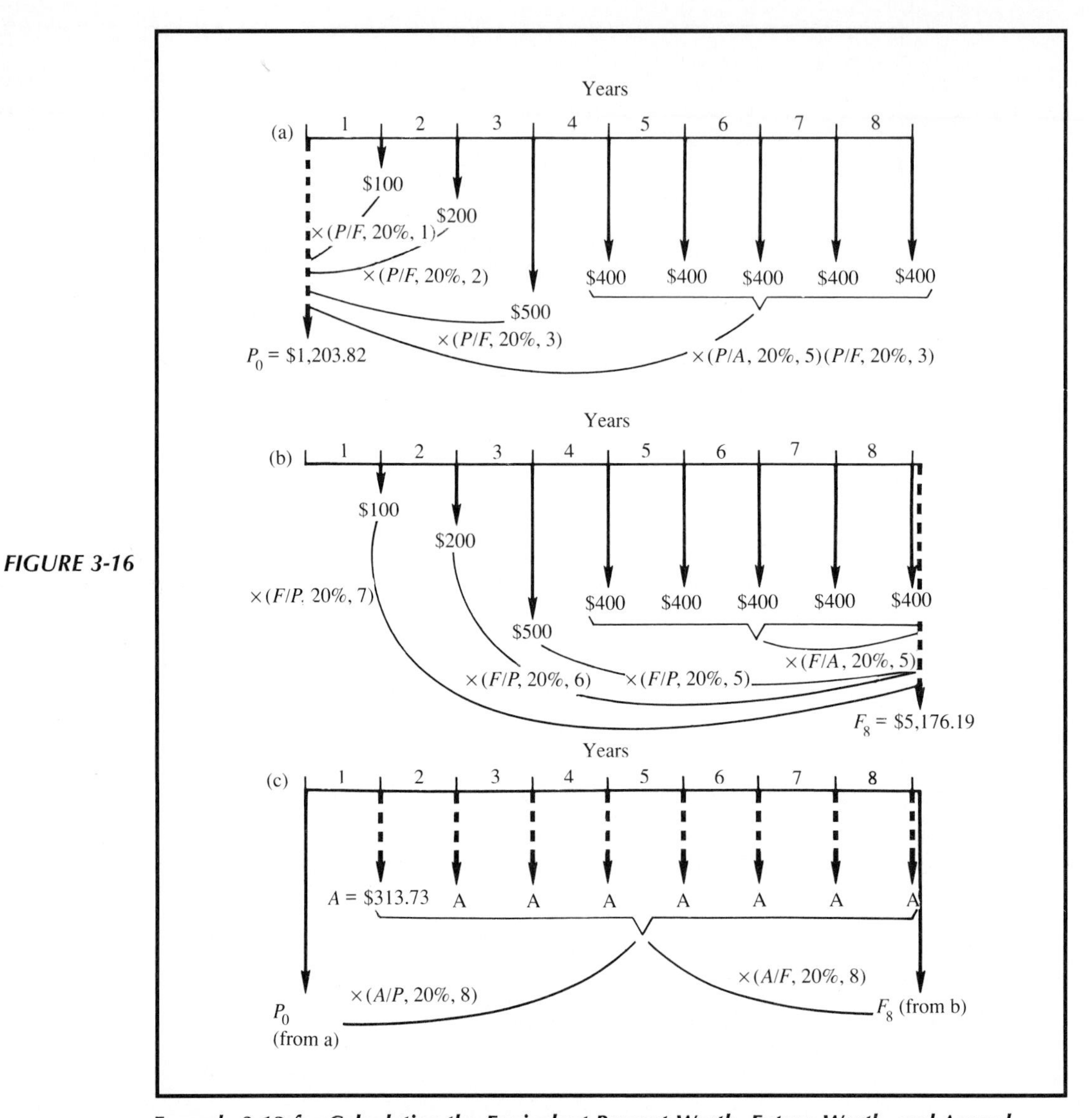

FIGURE 3-16

Example 3-12 for Calculating the Equivalent Present Worth, Future Worth, and Annual Worth

the second year, $500 for the third year, and $400 for each year from the fourth through the eighth. These could represent something like the expected maintenance expenditures for a certain piece of equipment or payments into a fund. Note that the payments are shown at the end of each year, which is a standard assumption for this book and for economic analyses in general unless one has information to the contrary. It is desired to find the equivalent

(a) present worth, (b) future worth, and (c) annual worth of these cash flows if the annual interest rate is 20%.

Solution

(a) To find the equivalent present worth, P_0, one needs to sum the worth of all payments as of the beginning of the first year (time 0). The required movements of money through time are shown graphically in Figure 3-16a.

$$\begin{aligned} P_0 = \; & F_1(P/F, 20\%, 1) && = \$100(0.8333) && = \$\ \ 83.33 \\ & + F_2(P/F, 20\%, 2) && + \$200(0.6944) && + \ 138.88 \\ & + F_3(P/F, 20\%, 3) && + \$500(0.5787) && + \ 289.35 \\ & + A(P/A, 20\%, 5) && + \$400(2.9906) && \\ & \quad \times (P/F, 20\%, 3) && \quad \times (0.5787) && + \ 692.26 \\ & && && \$1{,}203.82 \end{aligned}$$

(b) To find the equivalent future worth, F_8, one can sum the worth of all payments as of the end of the eighth year (time 8). Figure 3-16b indicates these movements of money through time. However, since the equivalent present worth is already known to be $1,203.82, one can calculate directly

$$F_8 = P_0(F/P, 20\%, 8) = \$1{,}203.82(4.2998) = \$5{,}176.19$$

(c) The equivalent annual worth of the irregular series can be calculated directly from either P_0 or F_8, as follows:

$$A = P_0(A/P, 20\%, 8) = \$1.203.82(0.2606) = \$313.73$$

or

$$A = F_8(A/F, 20\%, 8) = \$5{,}176.19(0.0606) = \$313.73$$

The computation of A from P_0 and F_8 is shown in Figure 3-16c. Thus, one finds that the irregular series of payments shown in Figure 3-16 is equivalent to $1,203.82 at time 0, $5,176.19 at time 8, or a uniform series of $313.73 at the end of each eight years.

EXAMPLE 3-13

Transform the cash flows on the left-hand side of Figure 3-17 to their equivalent cash flows on the right-hand side. That is, take the left-hand quantities as givens and determine the unknown value of Q in terms of H in Figure 3-17. The interest rate is 10% per year.

FIGURE 3-17

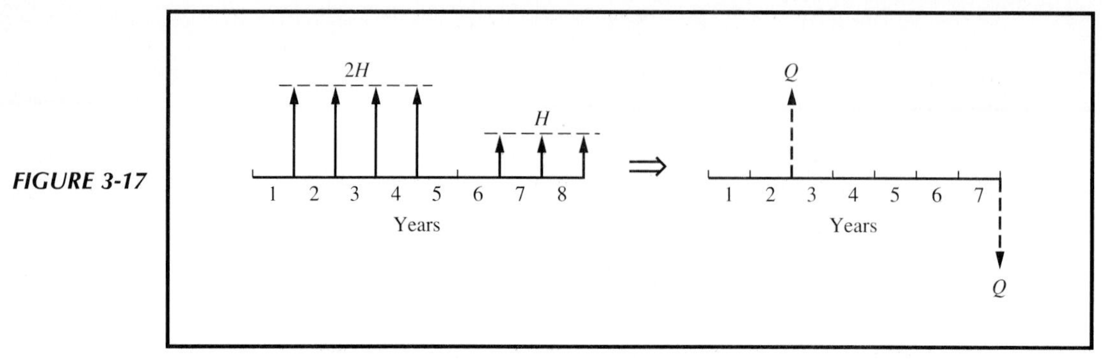

Cash Flow Diagrams for Example 3-13

Solution

If all cash flows on the left are discounted to year 0, we have $P_0 = 2H(P/A, 10\%, 4) + H(P/A, 10\%, 3)(P/F, 10\%, 5) = 7.8839H$. When cash flows on the right are also discounted to year 0, we can solve for Q in terms of H. (Notice that Q at the end of year (EOY) 2 is positive, Q at EOY 7 is negative, and the two Q values must be equal in amount.)

$$7.8839H = Q(P/F, 10\%, 2) - Q(P/F, 10\%, 7)$$

or

$$Q = 25.172H$$

3.15 INTEREST FORMULAS RELATING A UNIFORM GRADIENT OF CASH FLOWS TO ITS ANNUAL AND PRESENT WORTHS

Some economic analysis problems involve receipts or expenses that are projected to increase or decrease by a uniform *amount* each period, thus constituting an arithmetic sequence of cash flows. For example, maintenance and repair expenses on specific equipment may increase by a relatively constant amount each period. This situation can be modeled with a uniform gradient as demonstrated below.

Figure 3-18 is a cash flow diagram of a sequence of end-of-period cash flows increasing by a constant amount, G, each period. The G is known as the

FIGURE 3-18

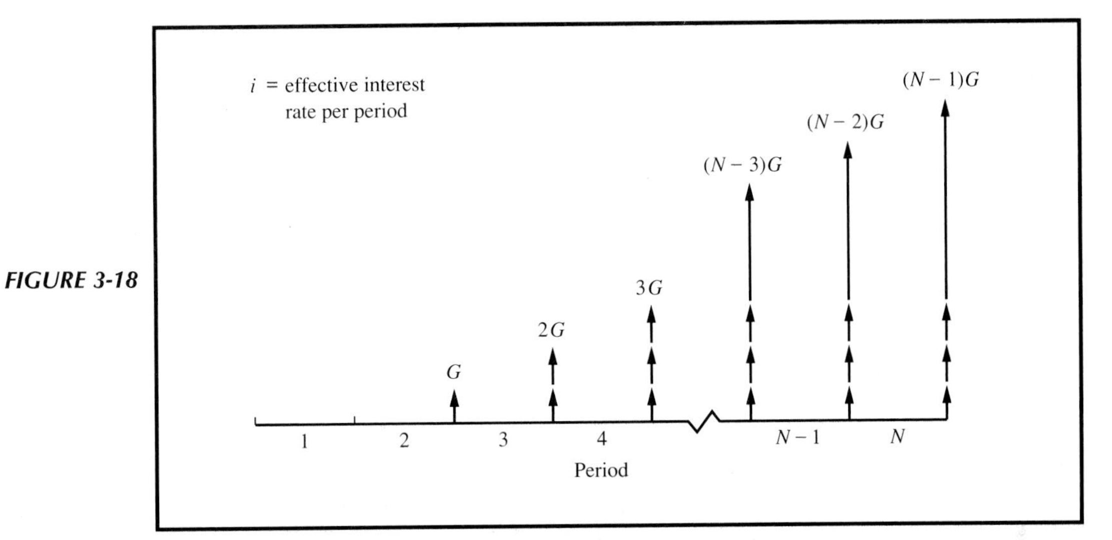

Cash Flow Diagram for a Uniform Gradient Increasing by G Dollars per Period

uniform gradient amount. Note that the timing of cash flows on which the derived formulas and tabled values are based is as follows:

End of Period	Payments
1	0
2	G
3	$2G$
.	.
.	.
.	.
$N-1$	$(N-2)G$
N	$(N-1)G$

3.15.1 Finding F When Given G

The future worth, F, of the arithmetic sequence of cash flows shown in Figure 3-18 is

$$F = G(F/A, i\%, N-1) + G(F/A, i\%, N-2) + \cdots \\ + G(F/A, i\%, 2) + G(F/A, i\%, 1)$$

or

$$F = G\left[\frac{(1+i)^{N-1}-1}{i} + \frac{(1+i)^{N-2}-1}{i} + \cdots + \frac{(1+i)^{2}-1}{i} + \frac{(1+i)^{1}-1}{i}\right]$$

$$= \frac{G}{i}\left[(1+i)^{N-1} + (1+i)^{N-2} + \cdots + (1+i)^{2} + (1+i)^{1} + 1\right] - \frac{NG}{i}$$

$$= \frac{G}{i}\left[\sum_{k=0}^{N-1}(1+i)^{k}\right] - \frac{NG}{i}$$

$$F = \frac{G}{i}(F/A, i\%, N) - \frac{NG}{i} \qquad (3\text{-}14)$$

3.15.2 Finding A When Given G

From Equation 3-14, it is easy to develop an expression for A as follows:

$$A = F(A/F, i, N)$$

$$= \left[\frac{G}{i}(F/A, i, N) - \frac{NG}{i}\right](A/F, i, N)$$

$$= \frac{G}{i} - \frac{NG}{i}(A/F, i, N)$$

$$= \frac{G}{i} - \frac{NG}{i}\left[\frac{i}{(1+i)^{N}-1}\right]$$

$$A = G\left[\frac{1}{i} - \frac{N}{(1+i)^{N}-1}\right] \qquad (3\text{-}15)$$

The term in braces in Equation 3-15 is called the *gradient to uniform series conversion factor*. Numerical values for this factor are given in Table C-25 of Appendix C for a wide range of i and N values. We shall use the functional symbol $(A/G, i\%, N)$ for this factor. Thus

$$A = G(A/G, i\%, N) \qquad (3\text{-}16)$$

3.15.3 Finding P When Given G

We may now utilize Equation 3-15 to establish the equivalence between P and G:

$$
\begin{aligned}
P &= A(P/A, i\%, N) \\
&= G\left[\frac{1}{i} - \frac{N}{(1+i)^N - 1}\right]\left[\frac{(1+i)^N - 1}{i(1+i)^N}\right] \\
&= G\left[\frac{(1+i)^N - 1 - Ni}{i^2(1+i)^N}\right] \\
&= G\left\{\frac{1}{i}\left[\frac{(1+i)^N - 1}{i(1+i)^N} - \frac{N}{(1+i)^N}\right]\right\}
\end{aligned} \tag{3-17}
$$

The term in braces in Equation 3-17 is called the *gradient to present worth conversion factor*. It can also be expressed as $(1/i)[(P/A, i\%, N) - N(P/F, i\%, N)]$. Numerical values for this factor are given in Table C-24 of Appendix C for a wide assortment of i and N values. We shall use the functional symbol $(P/G, i\%, N)$ for this factor. Hence

$$P = G(P/G, i\%, N) \tag{3-18}$$

3.15.4 Computations Using G

Be sure to notice that the direct use of gradient conversion factors applies when there is no cash flow at the end of period 1, as in Example 3-14. There may be an A amount at the end of period 1, but it is treated separately, as illustrated in Examples 3-15 and 3-16. A major advantage of using gradient conversion factors (i.e., computational time savings) is realized when N becomes large.

EXAMPLE 3-14

As an example of the straightforward use of the gradient conversion factors, suppose that certain end-of-year cash flows are expected to be \$1,000 for the *second* year, \$2,000 for the third year, and \$3,000 for the fourth year, and that if interest is 15% per year, it is desired to find the equivalent (a) present worth at the beginning of the first year, and (b) uniform annual worth at the end of each of the four years.

Solution

It can be observed that this schedule of cash flows fits the model of the arithmetic gradient formulas with $G = \$1,000$ and $N = 4$ (see Figure 3-18). Note there is no cash flow at the end of the first period.

(a) The present worth can be calculated as

$$P_0 = G(P/G, 15\%, 4) = \$1,000(3.79) = \$3,790$$

(b) The uniform annual worth can be calculated from Equation 3-16 as

$$A = G(A/G, 15\%, 4) = \$1{,}000(1.3263) = \$1{,}326.30$$

Of course, once the present worth is known, the uniform annual worth can be calculated as

$$A = P_0(A/P, 15\%, 4) = \$3{,}790(0.3503) = \$1{,}326.30$$

EXAMPLE 3-15

As a further example of the use of arithmetic gradient formulas, suppose that one has payments as follows:

End of Year	Payment
1	$5,000
2	6,000
3	7,000
4	8,000

and that one wishes to calculate their equivalent present worth at $i = 15\%$ using arithmetic gradient interest formulas.

Solution

The schedule of payments is depicted in the top diagram of Figure 3-19. The bottom two diagrams of Figure 3-19 show how the original schedule can be broken into two separate sets of payments, a uniform series of $5,000 payments plus an arithmetic gradient of $1,000 that fits the general gradient model for which factors are tabled. The summed present worths of these two separate sets of payments equal the present worth of the original problem. Thus, using the symbols shown in Figure 3-19, we have

$$\begin{aligned} P_{0T} &= P_{0A} + P_{0G} \\ &= A(P/A, 15\%, 4) + G(P/G, 15\%, 4) \\ &= \$5{,}000(2.8550) + \$1{,}000(3.79) = \$14{,}275 + 3{,}790 = \$18{,}065 \end{aligned}$$

The uniform annual worth of the original payments could be calculated with the aid of Equation 3-16 as follows:

$$\begin{aligned} A_T &= A + A_G \\ &= \$5{,}000 + \$1{,}000(A/G, 15\%, 4) = \$6{,}326.30 \end{aligned}$$

FIGURE 3-19

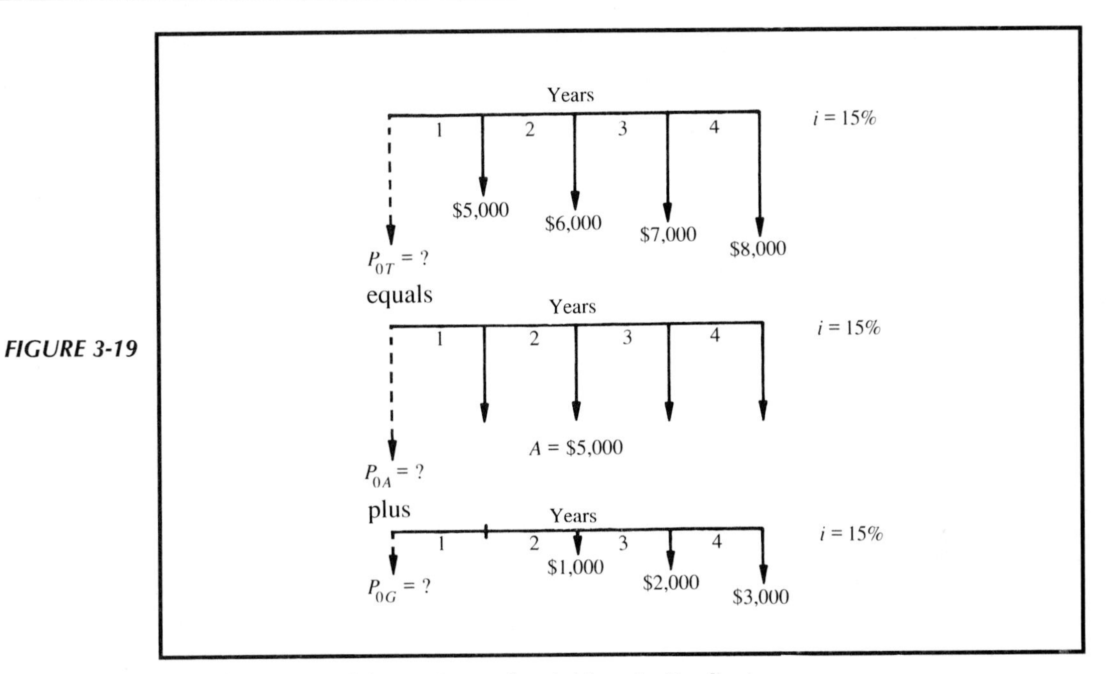

Example 3-15 Involving an Increasing Arithmetic Gradient

A_T is equivalent to P_{0T} because $6,326.30(P/A, 15%, 4) = $18,061, which is the same value obtained above (subject to round-off error).

EXAMPLE 3-16

For another example of the use of arithmetic gradient formulas, suppose that one has payments which are timed in exact reverse of the payments depicted in Example 3-15. The top diagram of Figure 3-20 shows these payments as follows:

End of Year	Payment
1	$8,000
2	7,000
3	6,000
4	5,000

Calculate the equivalent present worth at $i = 15\%$ using arithmetic gradient interest factors.

FIGURE 3-20

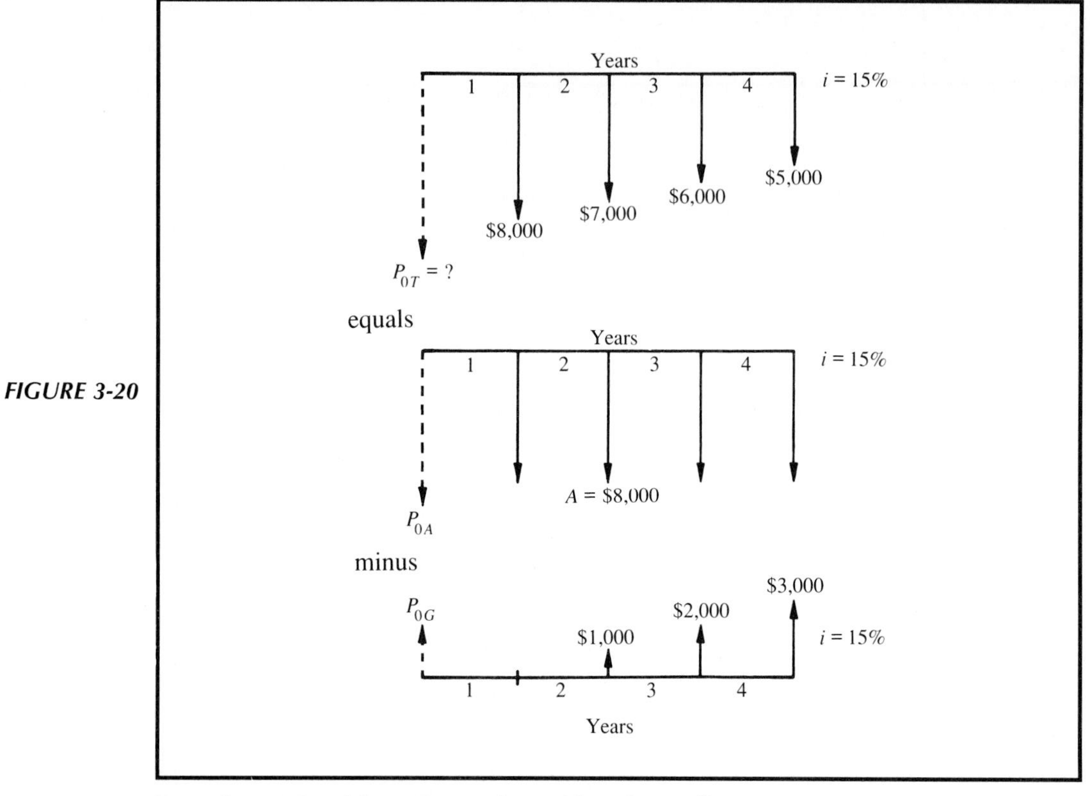

Example 3-16 Involving a Decreasing Arithmetic Gradient

Solution
The bottom two diagrams of Figure 3-20 show how these payments can be broken into two separate sets of payments. It must be remembered that the arithmetic gradient formulas and tables provided are for increasing gradients only. Hence one must subtract an *increasing* gradient of payments that *did not* occur. Thus

$$\begin{aligned} P_{0T} &= P_{0A} - P_{0G} \\ &= A(P/A, 15\%, 4) - G(P/G, 15\%, 4) \\ &= \$8{,}000(2.8550) - \$1{,}000(3.79) \\ &= \$22{,}840 - \$3{,}790 = \$19{,}050 \end{aligned}$$

Again, the uniform annual worth of the original decreasing series of payments could be calculated by the same rationale, or by finding A, given that the equivalent present worth is \$19,050:

$$A = \$19{,}050(A/P, 15\%, 4) = \$6{,}672.55$$

Note from Examples 3-15 and 3-16 that the present worth of $18,065 for an increasing arithmetic gradient series of payments is different from the present worth of $19,050 for an arithmetic gradient of payments of like amounts but reversed timing. This difference would be even greater for higher interest rates and gradient payments and exemplifies the marked effect of timing of cash flows on equivalent worths.

3.16 INTEREST FORMULAS RELATING A GEOMETRIC SEQUENCE OF CASH FLOWS TO ITS PRESENT AND ANNUAL WORTHS

Some economic equivalence problems involve projected cash flow patterns that are increasing at a constant *rate*, $\bar{f}$, each period. A fixed amount of a commodity that inflates in price at a constant rate each year is a typical situation that can be modeled with a geometric sequence of cash flows. The resultant end-of-period cash flow pattern is referred to as a geometric gradient series and has the general appearance shown in Figure 3-21. Notice that *the initial cash flow in this series, A_1, occurs at the end of period 1* and that $A_k = (A_{k-1})(1 + \bar{f}), 2 \leq k \leq N$. The Nth term in this geometric sequence is $A_N = A_1(1 + \bar{f})^{N-1}$, and the common ratio throughout the sequence is $(A_k - A_{k-1})/A_{k-1} = \bar{f}$.

Each term in Figure 3-21 could be discounted, or compounded, at interest rate i per period to obtain a value of P or F, respectively. However, this becomes quite tedious for large N, so it is convenient to have a single equation instead.

To develop a compact expression for P at interest rate i per period for the cash flows of Figure 3-21, consider the following summation:

$$P = \sum_{k=1}^{N} A_k(1 + i)^{-k} = \sum_{k=1}^{N} A_1(1 + \bar{f})^{k-1}(1 + i)^{-k}$$

or

$$P = \frac{A_1}{1 + \bar{f}} \sum_{k=1}^{N} \left(\frac{1 + \bar{f}}{1 + i}\right)^k \qquad (3\text{-}19)$$

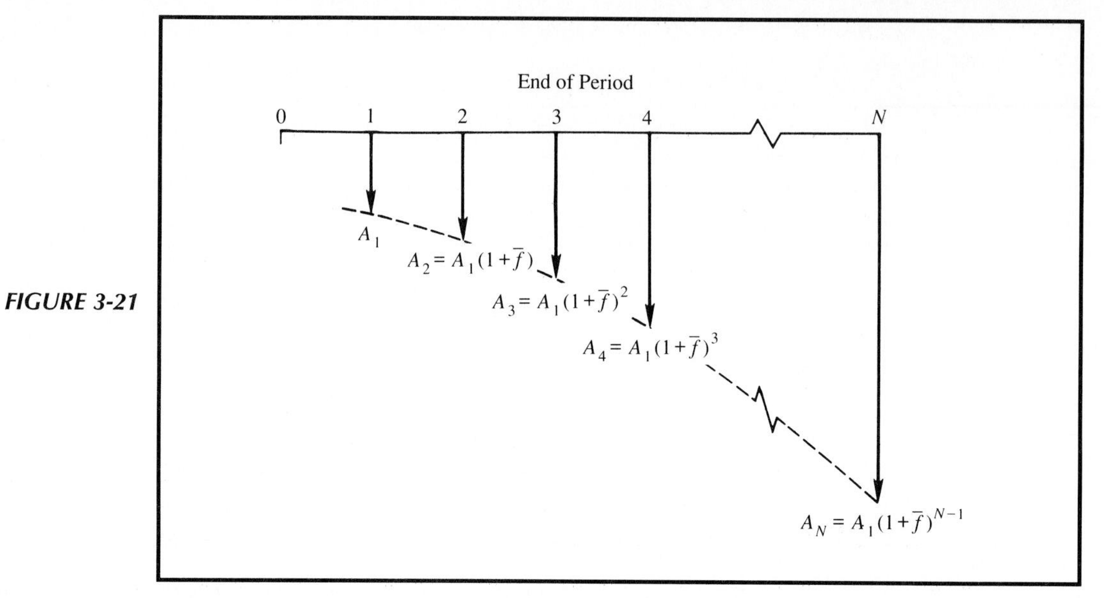

FIGURE 3-21 ***Cash Flow Diagram for a Geometric Sequence of Cash Flows Increasing at a Constant Rate of $\bar{f}$ per Period***

When $i \neq \bar{f}$, we can simplify Equation 3-19 by defining a "convenience rate," i_{CR}, as follows:

$$i_{CR} = \frac{1 + i}{1 + \bar{f}} - 1$$

The convenience rate can also be written as $i_{CR} = (i - \bar{f})/(1 + \bar{f})$. In the situation where $i \neq \bar{f}$, Equation 3-19 can thus be rewritten as

$$\begin{aligned} P &= \frac{A_1}{1 + \bar{f}} \sum_{k=1}^{N} \left(\frac{1 + i}{1 + \bar{f}} \right)^{-k} \\ &= \frac{A_1}{1 + \bar{f}} \sum_{k=1}^{N} (1 + i_{CR})^{-k} \\ &= \frac{A_1}{1 + \bar{f}} (P/A, i_{CR}\%, N)^{\dagger} \end{aligned} \tag{3-20}$$

†When $\bar{f}$ exceeds i, i_{CR} is negative and the above summation is valid only when N is finite.

Equation 3-20 makes use of the fact that

$$(P/A, i_{CR}\%, N) = \sum_{k=1}^{N}(1 + i_{CR})^{-k} = \sum_{k=1}^{N}(P/F, i_{CR}\%, k)$$

When $i_{CR} = \bar{f}$, Equation 3-20 reduces to

$$P = \frac{A_1}{1 + \bar{f}}(P/A, 0\%, N) = \frac{NA_1}{1 + \bar{f}} \tag{3-21}$$

The interested reader can verify Equation 3-21 by applying L'Hôpital's Rule to the $(P/A, i_{CR}\%, N)$ factor in Equation 3-20 and taking the limit as $i_{CR} \rightarrow 0$.

Values of i_{CR} used in connection with Equation 3-20 are typically not included in tables in Appendix C. Because i_{CR} is usually a noninteger interest rate, resorting to the definition of a $(P/A, i_{CR}\%, N)$ factor (see Table 3-3) and substituting terms into it is a satisfactory way to obtain values of these interest factors.

The end-of-period uniform annual equivalent, A, of a geometric gradient series can be determined from Equation 3-20 (or Equation 3-21) as follows:

$$A = P(A/P, i\%, N) \tag{3-22}$$

The year 0 "base" of this annuity, which increases at a constant rate of $\bar{f}\%$ per period, is A_0 and equals

$$A_0 = P(A/P, i_{CR}\%, N) \tag{3-23}$$

The difference between A and A_0 can be seen in Figure 3-22. Finally, the future equivalent of this geometric gradient series is simply

$$F = P(F/P, i\%, N) \tag{3-24}$$

Additional discussion of geometric sequences of cash flows is provided in Chapter 9 (Section 9.4), which deals with inflation and price changes.

EXAMPLE 3-17

Consider the end-of-year geometric sequence of cash flows in Figure 3-23 and determine the P, A, A_0, and F equivalent values. The rate of increase is 20% per year after the first year, and the annual interest rate is 25%.

FIGURE 3-22

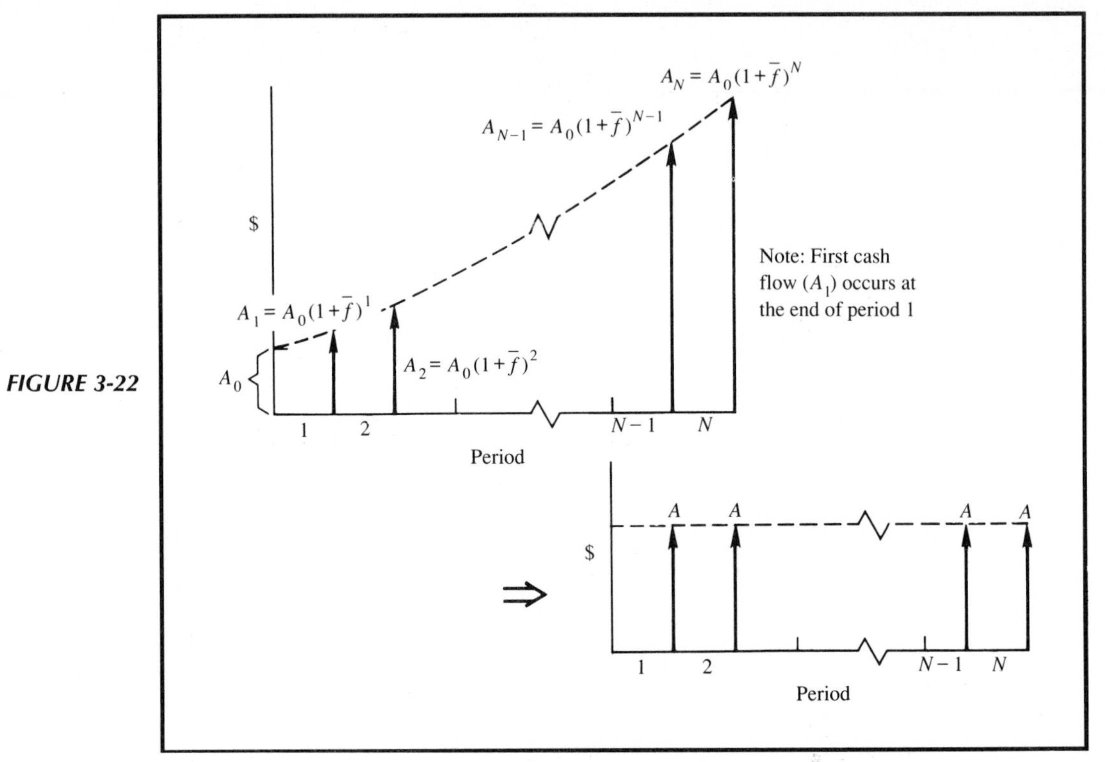

Graphical Interpretation of A and A_0 Terms in a Geometric Gradient Series when $f > 0$

FIGURE 3-23

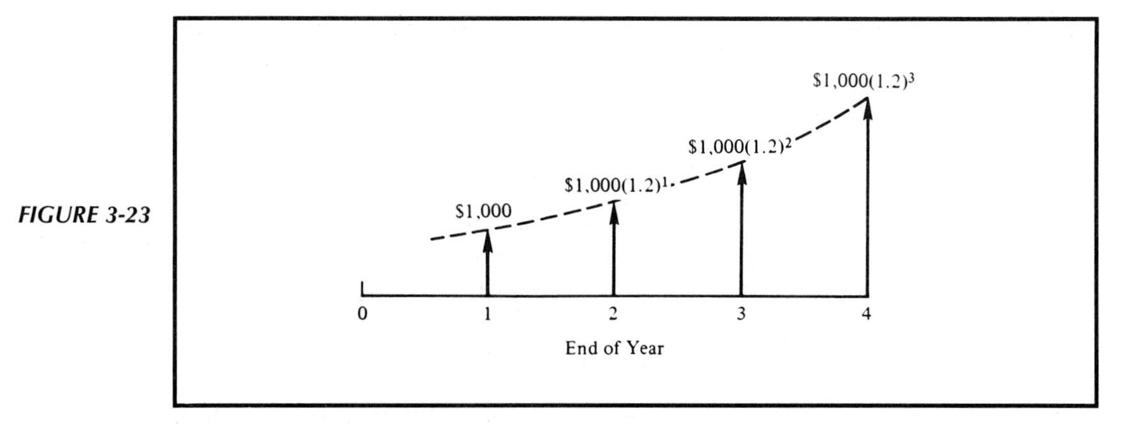

Cash Flow Diagram for Example 3-17

Solution

$$P = \frac{\$1{,}000}{1.2}\left(P/A, \frac{25\% - 20\%}{1.20}, 4\right) = \$833.33(P/A, 4.167\%, 4)$$

$$= \$833.33\left[\frac{(1.04167)^4 - 1}{0.04167(1.04167)^4}\right]$$

$$= \$833.33(3.6157) = \$3{,}013.08$$

$$A = \$3{,}013.08(A/P, 25\%, 4) = \$1{,}275.86$$

$$A_0 = \$3{,}013.08(A/P, 4.167\%, 4)$$

$$= \$3{,}103.08\left[\frac{0.04167(1.04167)^4}{(1.04167)^4 - 1}\right] = \$833.34$$

$$F = \$3{,}013.08(F/P, 25\%, 4) = \$7{,}356.15$$

EXAMPLE 3-18

A heat pump is being considered as a replacement for an existing electric resistance furnace. Based on winter heating requirements, it has been estimated that \$600 per year will be saved through reduced electricity bills. These savings have been computed at *present* electricity rates. If electricity price is expected to increase at an average rate of 14% per year into the foreseeable future, how much can we justify spending now for the heat pump if the interest rate is 12% per year? Assume that the life of the heat pump is 15 years and that its salvage value at that time is negligible.

Solution

Present savings are \$600 per year, so savings by the end of year 1 will be \$600(1.14) = \$684. With end-of-year cash flow convention, the upper bound on expenditure that can be justified now for the purchase of this heat pump is

$$P = \frac{\$684}{1.14}\left(P/A, \frac{12\% - 14\%}{1.14}, 15\right)$$

$$P = \$600(P/A, -1.75\%, 15)$$

$$P = \$600\left[\frac{(0.9825)^{15} - 1}{-0.0175(0.9825)^{15}}\right]$$

$$= \$600(17.326)$$

$$= \$10{,}395$$

3.17

NOMINAL AND EFFECTIVE INTEREST RATES

Very often, the interest period, or time between successive compounding, is something less than one year. It has become customary to quote interest rates on an annual basis, followed by the compounding period if different from one year in length. For example, if the interest rate is 6% per interest period and the interest period is six months, it is customary to speak of this rate as "12% compounded semiannually." The basic annual rate of interest is known as the *nominal rate*, 12% in this case. A nominal interest rate is represented by r. The actual annual rate on the principal is not 12% but something greater, because of the compounding that occurs twice during a year.

Thus, the frequency at which a nominal interest rate is compounded each year can have a pronounced effect on the dollar amount of interest earned. For instance, consider $1,000 to be invested for three years at a nominal rate of 12% compounded semiannually. The interest earned during the first year would be as follows:

First six months:

$$Pi = \$1,000 \times 0.06 = \$60$$

Total principal and interest at beginning of the second period:

$$P + Pi = \$1,000 + \$60 = \$1,060$$

Interest earned during second six months:

$$\$1,060 \times 0.06 = \$63.60$$

Total interest earned during year:

$$\$60.00 + \$63.60 = \$123.60$$

Effective annual interest rate:

$$\frac{\$123.60}{\$1,000} \times 100 = 12.36\%$$

If this process is repeated for years 2 and 3, the *accumulated* (compounded) *amount of interest* can be plotted as in Figure 3-24. Suppose the same $1,000 had been invested at 12% compounded *monthly*. The accumulated interest over three

FIGURE 3-24

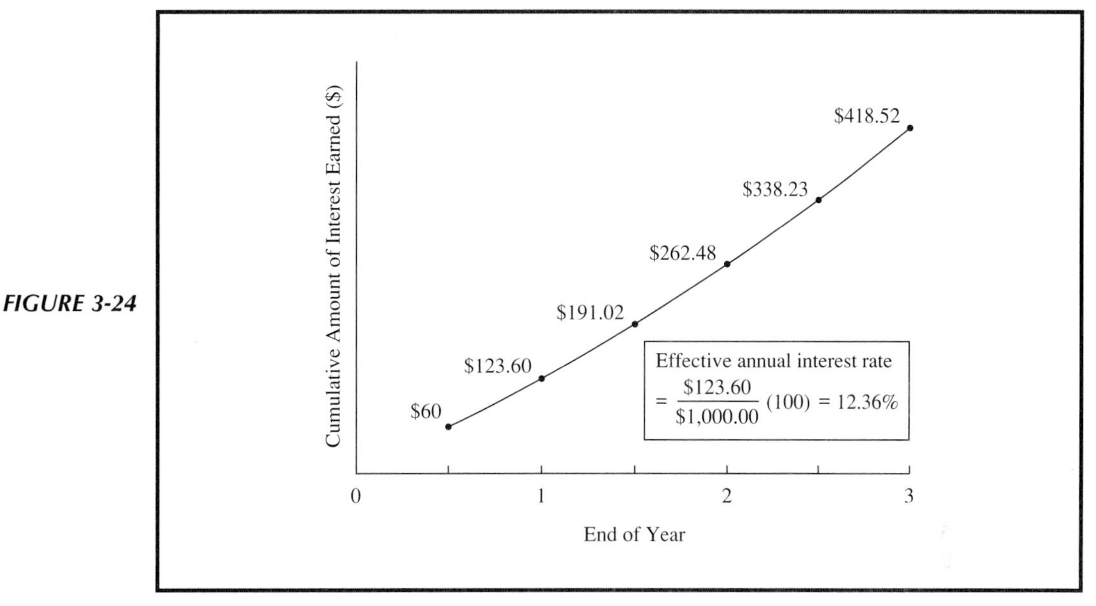

$1,000 Compounded at a Semiannual Frequency (r = 12%)

years that results from semiannual and monthly compounding is shown in Figure 3-25.

The actual or exact rate of interest earned on the principal during one year is known as the *effective rate*. It should be noted that effective interest rates always are expressed on an annual basis, unless specifically stated otherwise. In this text the effective interest rate per year is customarily designated by i and the nominal interest rate per year by r. In engineering economy studies in which compounding is annual, $i = r$. The relationship between effective interest, i, and nominal interest, r, is

$$i = (1 + r/M)^M - 1 \tag{3-25}$$

$$= (F/P, r/M, M) - 1$$

where M is the number of compounding periods per year. It is now clear from Equation 3-25 why $i > r$ when $M > 1$.

The effective rate of interest is useful for describing the compounding effect of interest earned on interest within one year. Table 3-5 shows effective rates for various nominal rates and compounding periods. The federal "truth in lending" law requires a statement regarding the annual percentage rate (APR) being charged in contracts involving borrowed money. The APR is a nominal interest rate.

As a point of interest, the reader will now realize that in Examples 3-10 and 3-11 an assumption was made regarding an interest rate per six months which was compounded (M = 2) to an effective interest rate per year:

$$i = (1 + r/6\text{ mo.})^2,\text{ or } 1 + r/6\text{ mo.} = \sqrt{1 + i} = \text{HPC}$$

FIGURE 3-25

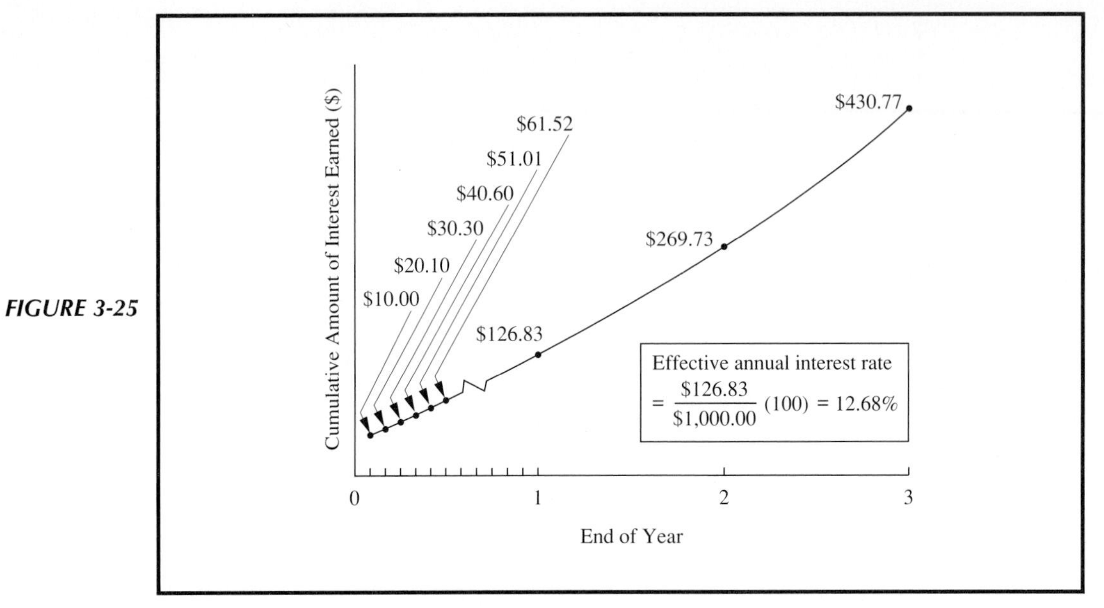

$1,000 Compounded at a Monthly Frequency (r = 12%)

EXAMPLE 3-19

A credit card company charges an interest rate of 1.375% per month on the unpaid balance of all accounts. The annual interest rate, they claim, is 12(1.375%) = 16.5%. What is the effective rate of interest per year being charged by the company?

Solution

Interest tables in Appendix C are based on time periods that may be annual, quarterly, monthly, and so on. Because we have no 1.375% tables (or 16.5%

TABLE 3–5 Effective Interest Rates for Various Nominal Rates and Compounding Frequencies

Compounding Frequency	Number of Compounding Periods per Year, M	Effective Rate (%) for Nominal Rate of 6%	8%	10%	12%	15%	24%
Annually	1	6.00	8.00	10.00	12.00	15.00	24.00
Semiannually	2	6.09	8.16	10.25	12.36	15.56	25.44
Quarterly	4	6.14	8.24	10.38	12.55	15.87	26.25
Bimonthly	6	6.15	8.27	10.43	12.62	15.97	26.53
Monthly	12	6.17	8.30	10.47	12.68	16.08	26.82
Daily	365	6.18	8.33	10.52	12.75	16.18	27.11

tables), Equation 3-25 must be used to compute the effective rate of interest in this example:

$$i = \left(1 + \frac{0.165}{12}\right)^{12} - 1$$
$$= 0.1781, \text{ or } 17.81\%/\text{year}$$

Note that $r = 12(1.375\%) = 16.5\%$, which is the APR.

3.18 INTEREST PROBLEMS WITH COMPOUNDING MORE OFTEN THAN ONCE PER YEAR

3.18.1 Single Amounts

If a nominal interest rate is quoted and the number of compounding periods per year and number of years are known, any problem involving future, annual or present worths can be calculated by straightforward use of Equations 3-3 and 3-25, respectively.

EXAMPLE 3-20

Suppose a $100 lump-sum amount is invested for 10 years at 6% compounded quarterly. How much is it worth at the end of the tenth year?

Solution

There are four compounding periods per year, or a total of $4 \times 10 = 40$ periods. The interest rate per interest period is $6\%/4 = 1.5\%$. When the values are used in Equation 3-3, one finds that

$$F = P(F/P, 1.5\%, 40) = \$100.00(1.814) = \$181.40$$

Alternatively, the effective interest rate from Equation 3-25 is 6.14%. Therefore, $F = \$100.00(1.0614)^{10} = \$181.40$.

3.18.2 Uniform Series and Gradient Series

When there is more than one compounded interest period per year, the formulas and tables for uniform series and gradient series can be used as long as there is a cash flow at the end of each interest period, as shown in Figures 3-8 and 3-18 for uniform series and gradient series, respectively.

EXAMPLE 3-21

Suppose that one has a beginning indebtedness of \$10,000 which is to be repaid by equal end-of-month installments for five years with interest at 12% compounded monthly. What is the amount of each payment?

Solution
The number of installment payments is $5 \times 12 = 60$, and the interest rate per month is $12\%/12 = 1\%$. When these values are used in Equation 3-13, one finds that

$$A = P(A/P, 1\%, 60) = \$10{,}000(0.0222) = \$222$$

EXAMPLE 3-22

Certain operating expenditures are expected to be 0 at the end of the first six months, \$1,000 at the end of the second six months, and to increase by \$1,000 at the end of each six-month period thereafter for a total of four years. It is desired to find the equivalent uniform payment at the end of each of the eight six-month periods if nominal interest is 20% compounded semiannually.

Solution
A cash flow diagram is shown in Figure 3-26, and the solution is

$$A = G(A/G, 10\%, 8) = \$1,000(3.0045) = \$3,004.50$$

Notice that $r/6$ months $= 10\%$.

FIGURE 3-26

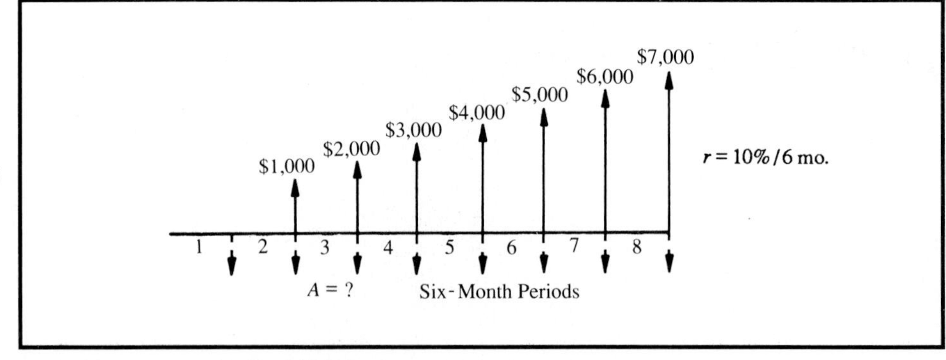

Arithmetic Gradient with Compounding More Often than Once per Year in Example 3-22

3.19

INTEREST PROBLEMS WITH UNIFORM CASH FLOWS LESS OFTEN THAN COMPOUNDING PERIODS

In general, if i is the effective interest rate per interest period and there is a uniform cash flow, X, at the *end* of each kth interest period ($k > 1$), then the equivalent payment, A, at the end of each interest period is

$$A = X(A/F, i\%, k) \tag{3-26}$$

By similar reasoning, if i is the effective interest rate per interest period and there is a uniform payment, X, at the *beginning* of each kth interest period, then the equivalent payment, A, at the end of each interest period is

$$A = X(A/P, i\%, k) \tag{3-27}$$

EXAMPLE 3-23

Suppose that there exists a series of 10 end-of-year payments of $1,000 each and it is desired to compute the equivalent worth of those payments as of the end of the tenth year if interest is 12% compounded quarterly. The problem is depicted in Figure 3-27.

Solution

Interest is 12%/4 = 3% per quarter, but the uniform series cash flows are not at the end of each quarter. Hence, one must make special adaptations to fit the interest formulas to the tables provided. To solve this type of problem, an equivalent cash flow must be computed for the time interval that corresponds to the stated compounding frequency, or an effective interest rate must be determined for the interval of time separating cash flows.

One useful adaptation procedure is to take the number of compounding periods over which a cash flow occurs and convert the cash flow into its equivalent uniform end-of-period series. The upper cash flow diagram in Figure 3-28 shows this approach applied to the first year (four interest periods) in the example of Figure 3-27. The uniform end-of-quarter payment, equivalent to $1,000 at the end of the year with interest at 3% per quarter, can be calculated by using Equation 3-26:

$$A = F(A/F, 3\%, 4) = \$1{,}000(0.2390) = \$239$$

FIGURE 3-27

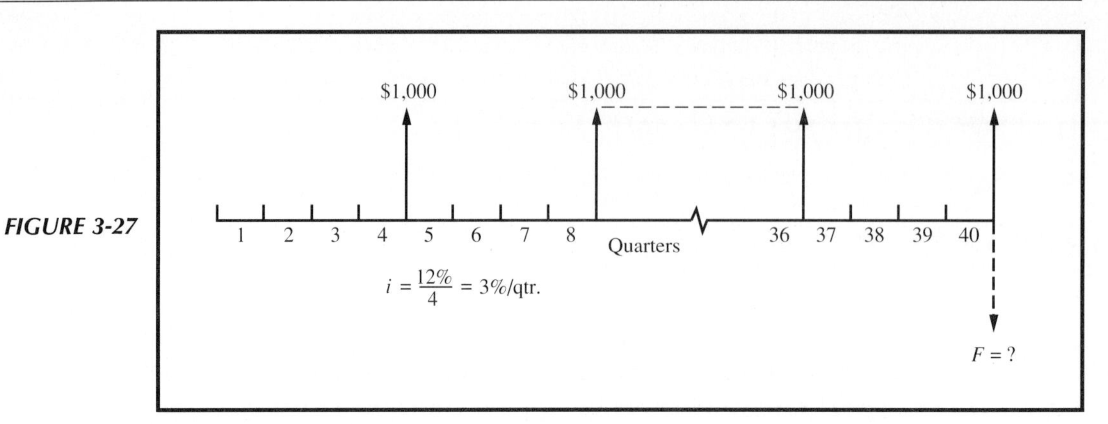

Uniform Series with Cash Flows Less Often than Compounding Periods in Example 3-23

Thus, \$239 at the end of each quarter is equivalent to \$1,000 at the end of each year. This is true not only for the first year but also for each of the 10 years under consideration. Hence the original series of 10 end-of-year payments of \$1,000 each can be converted to a problem involving 40 end-of-quarter payments of \$239 each, as shown in the lower cash flow diagram of Figure 3-28.

The equivalent worth at the end of the tenth year (40th quarter) may then be computed as

$$F = A(F/A, 3\%, 40) = \$239(75.4012) = \$18,021$$

Another adaptation procedure for handling uniform series with cash flows less often than compounding periods is to find the exact interest rate for each

FIGURE 3-28

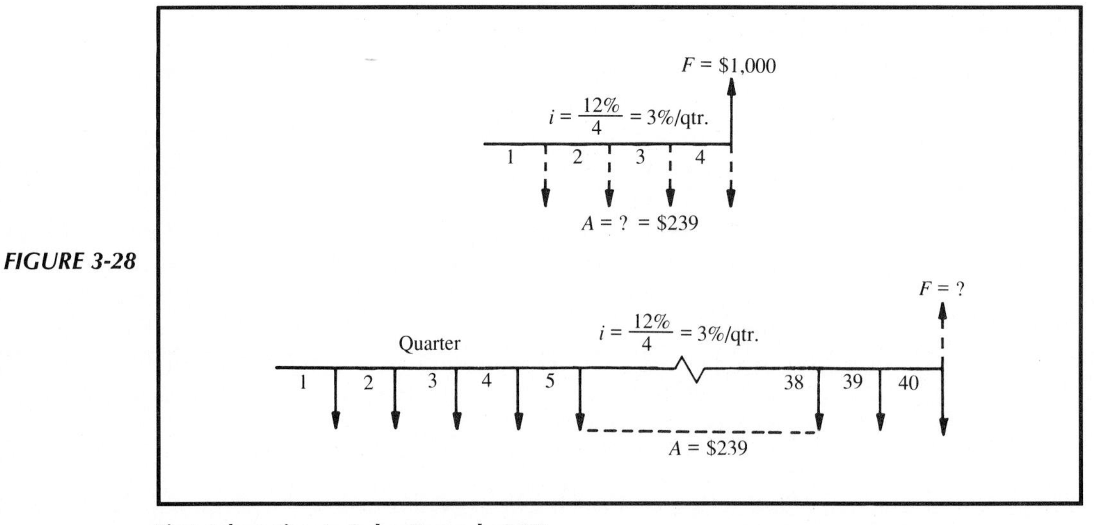

First Adaptation to Solve Example 3-23

time period separating cash flows and then to straightforwardly apply the interest formulas and tables for that exact interest rate. For Example 3-23, interest is 3% per quarter and payments occur each year. Hence, the interest rate to be found is the exact rate each year, or the *effective rate* per year. The effective rate per year that corresponds to 3% per quarter (12% nominal) can be found from Equation 3-25:

$$\left(1 + \frac{0.12}{4}\right)^4 - 1 = (F/P, 3\%, 4) - 1 = 0.1255$$

Hence, the original problem in Figure 3-27 can now be expressed as shown in Figure 3-29. The future worth of this series can then be found as

$$F = A(F/A, 12.55\%, 10) = \$1{,}000(F/A, 12.55\%, 10) = \$18{,}022$$

Because interest factors are not commonly tabled for $i = 12.55\%$, one must compute the $(F/A, 12.55\%, 10)$ factor by substituting $i = 0.1255$ and $N = 10$ into its algebraic equivalent, $[(1 + i)^N - 1]/i$.

Substitution into the algebraic equivalent will give the exact answer (same as the \$18,021 by the first adaptation method in Figure 3-28) except for any round-off error). Linear interpolation for the factor may also be utilized as a good approximation in most problems.

Yet another adaptation for problems involving uniform series cash flows that occur less often than compounding periods is to treat each cash flow as a single sum. This is usually unsatisfactory because of the number of computations involved, but it is well to recognize the possibility. Thus, the solution to Example 3-23 can be computed by recognizing that the first cash flow is compounded 36 interest periods at 3% per period, the second cash flow is

FIGURE 3-29

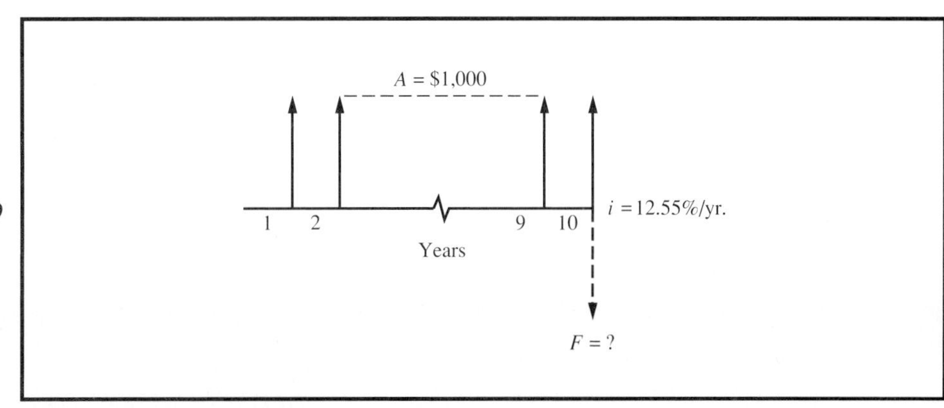

Second Adaptation to Solve Example 3-23

compounded 32 interest periods, and so on. Hence the problem can be solved as

$$F = \$1{,}000[(F/P, 3\%, 36) + (F/P, 3\%, 32) + \cdots + (F/P, 3\%, 4) + (F/P, 3\%, 0)]$$

$$= \$1{,}000(2.8983 + 2.5751 + \cdots + 1.1255 + 1.00)$$

The exact answer of \$18,021 would be obtained from this calculation if there were no interpolation or round-off error.

3.20

INTEREST PROBLEMS WITH UNIFORM CASH FLOWS OCCURRING MORE OFTEN THAN COMPOUNDING PERIODS

When cash flows occur more often than compounding frequency, a rather unrealistic situation develops as illustrated in Example 3-24.

EXAMPLE 3-24

Suppose that an individual insists on making weekly payments of \$50 into a savings account upon which nominal interest of 10% is compounded quarterly. What is the equivalent lump-sum future amount in the account at the end of two years? There are several assumptions that might be made in working this problem, and two of them are shown below.

Solution

Assumption 1: There is no interest computed on any amount except that which is in the account by the end of each quarter. In this case the amount deposited in each quarter totals \$50(13) = \$650. Because the nominal interest per quarter is 2.5%, the compound amount at the end of two years is

$$F = \$650(F/A, 2.5\%, 8) = \$5{,}678.48$$

Assumption 2: There is an appropriate nominal interest rate per week, r_{52}, that compounds to $i = 2.5\%$ each quarter. This rate is

$$(1 + r_{52})^{13} - 1 = 0.025 \qquad \text{or} \qquad r_{52} = 0.0019$$

Therefore, the future amount at the end of two years with weekly compounding at r_{52} per week is

$$F = \$50(F/A, 0.19\%, 104)$$
$$= \$5{,}743.33$$

Of the two assumptions, probably the first is the more realistic and is recommended for solving this type of problem.

3.21 INTEREST RATES THAT VARY WITH TIME

When the interest rate on a loan can vary with, for example, the Federal Reserve Board's discount rate, it is necessary to take this into account when determining the future worth of the loan. It is becoming commonplace to see interest-rate "escalation riders" on some types of loans. Example 3-25 demonstrates how this situation is treated.

EXAMPLE 3-25

A person has made an arrangement to borrow $1,000 now and another $1,000 two years hence. The entire obligation is to be repaid at the end of 4 years. If the projected interest rates in years 1, 2, 3, and 4 are 10%, 12%, 12%, and 14%, respectively, how much will be repaid as a lump-sum amount at the end of four years?

Solution

This problem can be solved by compounding the amount owed at the beginning of each year by the interest rate that applies to each individual year and repeating this process over the four years to obtain the total future worth:

$$F_1 = \$1{,}000(F/P, 10\%, 1) = \$1{,}100$$
$$F_2 = \$1{,}100(F/P, 12\%, 1) = \$1{,}232$$
$$F_3 = (\$1{,}232 + \$1{,}000)(F/P, 12\%, 1) = \$2{,}500$$
$$F_4 = \$2{,}500(F/P, 14\%, 1) = \$2{,}850$$

To obtain the present worth of a series of future amounts, a procedure similar to that above would be utilized with a sequence of single-payment present worth factors. In general, the present worth of a cash flow occurring at the end of period N can be computed with Equation 3-28, where i_k is the interest rate for the kth period:

$$P = \frac{F_N}{\prod_{k=1}^{N}(1 + i_k)} \tag{3-28}$$

For instance, if $F_4 = \$1,000$ and $i_1 = 10\%$, $i_2 = 12\%$, $i_3 = 13\%$, and $i_4 = 10\%$,

$$\begin{aligned} P &= \$1{,}000(P/F, 10\%, 1)(P/F, 12\%, 1)(P/F, 13\%, 1)(P/F, 10\%, 1) \\ &= \$1{,}000[(0.9091)(0.8929)(0.8850)(0.9091)] = \$653 \end{aligned}$$

3.22 INTEREST FACTOR RELATIONSHIPS

The following relationships among the six basic discrete compounding interest factors should be recognized:

$$(P/F, i\%, N) = \frac{1}{(F/P, i\%, N)} \tag{3-29}$$

$$(A/P, i\%, N) = \frac{1}{(P/A, i\%, N)} \tag{3-30}$$

$$(A/F, i\%, N) = \frac{1}{(F/A, i\%, N)} \tag{3-31}$$

$$(F/A, i\%, N) = (P/A, i\%, N)(F/P, i\%, N) \tag{3-32}$$

$$(P/A, i\%, N) = \sum_{k=1}^{N}(P/F, i\%, k) \tag{3-33}$$

$$(F/A, i\%, N) = \sum_{k=0}^{N-1}(F/P, i\%, k) \tag{3-34}$$

$$(A/F, i\%, N) = (A/P, i\%, N) - i \tag{3-35}$$

These same relationships exist among the corresponding continuous compounding interest factors discussed in the following sections.

3.23

INTEREST FORMULAS FOR CONTINUOUS COMPOUNDING AND DISCRETE CASH FLOWS

In most business transactions and economy studies, interest is compounded at the end of discrete periods of time and, as has been discussed previously, cash flows are assumed to occur in discrete amounts at the beginning, middle, or end of such periods. *This practice will be used throughout the remaining chapters of this book.* However, it is evident that in most enterprises cash is flowing in and out in an almost continuous stream. Because cash, whenever available, can usually be used profitably, this situation creates opportunities for very frequent compounding of the interest earned. So that this condition can be dealt with when continuously compounded interest rates are available, the concepts of continuous compounding and continuous cash flow are sometimes used in economy studies. Actually, the effects of these procedures compared to discrete compounding are rather small in most cases.

Continuous compounding assumes that cash flows occur at discrete intervals (e.g., once per year), but that compounding is continuous throughout the interval. For example, with a nominal rate of interest per year of r, if the interest is compounded M times per year, at the end of one year one unit of principal will amount to $[1 + (r/M)]^M$. If $M/r = p$, the foregoing expression becomes

$$\left[1 + \frac{1}{p}\right]^{rp} = \left[\left(1 + \frac{1}{p}\right)^{p}\right]^{r} \tag{3-36}$$

Since

$$\lim_{p\to\infty}\left(1 + \frac{1}{p}\right)^{p} = e^1 = 2.71828\ldots,$$

Equation 3-36 can be written as e^r. Consequently, the *continuously compounded compound amount factor (single cash flow)* at $r\%$ nominal interest for N years is e^{rN}. Using our functional notation, we express this as

$$(F/P,\underline{r}\%,N) = e^{rN} \tag{3-37}$$

Note that the symbol $\underline{r}$ is directly comparable to that used for discrete compounding and discrete cash flows ($i\%$) except that $\underline{r}\%$ is used to denote the nominal rate *and* the use of continuous compounding.

Since e^{rN} for continuous compounding corresponds to $(1 + i)^N$ for discrete compounding, e^r is equal to $(1 + i)$. Hence, we may correctly conclude:

$$i = e^r - 1 \tag{3-38}$$

By use of this relationship, the corresponding values of (P/F), (F/A), and (P/A) for continuous compounding may be obtained from Equations 3-4, 3-7, and 3-9, respectively, by substitution of $e^r - 1$ for i in these equations. Thus, for continuous compounding and discrete cash flows,

$$(P/F,\underline{r}\%,N) = \frac{1}{e^{rN}} = e^{-rN} \tag{3-39}$$

$$(F/A,\underline{r}\%,N) = \frac{e^{rN} - 1}{e^r - 1} \tag{3-40}$$

$$(P/A,\underline{r}\%,N) = \frac{1 - e^{-rN}}{e^r - 1} = \frac{e^{rN} - 1}{e^{rN}(e^r - 1)} \tag{3-41}$$

Values for $(A/P, \underline{r}\%, N)$, and $(A/F, \underline{r}\%, N)$ may be derived through their inverse relationships to $(P/A, \underline{r}\%, N)$ and $(F/A, \underline{r}\%, N)$, respectively. All the continuous compounding, discrete cash flow interest factors and their uses are summarized in Table 3-6.

Because continuous compounding is used rather infrequently in this text, detailed values for $(A/F, \underline{r}\%, N)$ and $(A/P, \underline{r}\%, N)$ are not given in Appendix D. However, the tables in Appendix D do provide values of $(F/P, \underline{r}\%, N)$, $(P/F, \underline{r}\%, N)$, $(F/A, \underline{r}\%, N)$, and $(P/A, \underline{r}\%, N)$ for a limited number of interest rates.

It is important to note that tables of interest and annuity factors for continuous compounding are tabulated in terms of nominal annual rates of interest.

EXAMPLE 3-26

Suppose that one has a present amount of $1,000 and it is desired to determine what equivalent uniform end-of-year payments could be obtained from it for 10 years if interest is 20% compounded continuously ($M = \infty$).

Solution

Here we utilize this formulation:

$$A = P(A/P,\underline{r}\%,N)$$

Since the (A/P) factor is not tabled for continuous compounding, we substitute its inverse (P/A), which is tabled in Appendix D. Thus

$$A = P \times \frac{1}{(P/A,\underline{20}\%,10)} = \$1{,}000 \times \frac{1}{3.9054} = \$256$$

TABLE 3–6 Continuous Compounding and Discrete Cash Flows—Interest Factors and Symbols[a]

To Find:	Given:	Factor by Which to Multiply "Given"	Factor Name	Factor Functional Symbol
For single cash flows:				
F	P	e^{rN}	Continuous compounding compound amount (single cash flow)	$(F/P, \underline{r}\%, N)$
P	F	e^{-rN}	Continuous compounding present worth (single cash flow)	$(P/F, \underline{r}\%, N)$
For uniform series (annuities):				
F	A	$\dfrac{e^{rN}-1}{e^r-1}$	Continuous compounding compound amount (uniform series)	$(F/A, \underline{r}\%, N)$
P	A	$\dfrac{e^{rN}-1}{e^{rN}(e^r-1)}$	Continuous compounding present worth (uniform series)	$(P/A, \underline{r}\%, N)$
A	F	$\dfrac{e^r-1}{e^{rN}-1}$	Continuous compounding sinking fund	$(A/F, \underline{r}\%, N)$
A	P	$\dfrac{e^{rN}(e^r-1)}{e^{rN}-1}$	Continuous compounding capital recovery	$(A/P, \underline{r}\%, N)$

[a] $\underline{r}$, nominal annual interest rate, compounded continuously; N, number of periods (years); A, uniform series amount (occurs at end of each year); F, future worth; P, present worth.

It is interesting to note that the answer to the same problem, except with discrete annual compounding ($M = 1$), is

$$A = P(A/P, 20\%, 10)$$
$$= \$1{,}000(0.2385) = \$239$$

EXAMPLE 3-27

A person needs $12,000 immediately as a down payment on a new home. Suppose that he can borrow this money from his insurance company. He will be required to repay the loan in equal payments, made every six months over the next eight years. The nominal interest rate being charged is 7% compounded continuously. What is the amount of each payment?

Solution

The nominal interest rate per six months is 3.5%. Thus, A each 6 months is \$12,000(A/P, $\underline{r}$ = 3.5%, 16). By substituting terms in Equation 3-41 and then using its inverse which is tabled in Appendix D, we determine the value of A per 6 months to be \$997, that is, A = \$12,000/(P/A, $\underline{r}$ = 3.5%, 16).

3.24 INTEREST FORMULAS FOR CONTINUOUS COMPOUNDING AND CONTINUOUS CASH FLOWS

Continuous flow of funds means a series of cash flows occurring at infinitesimally short intervals of time; this corresponds to an annuity having an infinite number of short periods. This formulation could apply to companies having receipts and expenses that occur frequently during each working day. In such cases the interest normally is compounded continuously. If the nominal interest rate per year is $\underline{r}$ and there are p payments per year, which amount to a total of one unit per year, then, by use of Equation 3-8, for one year the present worth at the beginning of the year is

$$P = \frac{1}{p}\left\{\frac{[1 + (r/p)]^p - 1}{r/p[1 + (r/p)]^p}\right\} = \frac{[1 + (r/p)]^p - 1}{r[1 + (r/p)]^p} \tag{3-42}$$

The limit of $[1 + (r/p)]^p$ as p approaches infinity is e^r. By letting the present worth of one unit per year, flowing continuously and with continuous compounding of interest, be called the *continuous compounding present worth factor (continuous, uniform cash flow over one period)*, one finds

$$(P/\bar{A},\underline{r}\%, 1) = \frac{e^r - 1}{re^r} \tag{3-43}$$

where $\bar{A}$ is the amount flowing uniformly and continuously over one year (here \$1).

For $\bar{A}$ flowing each year over N years, as depicted in Figure 3-30,

$$(P/\bar{A},\underline{r}\%, N) = \frac{e^{rN} - 1}{re^{rN}} \tag{3-44}$$

which is the *continuous compounding present worth factor (continuous, uniform cash flows)*.

FIGURE 3-30

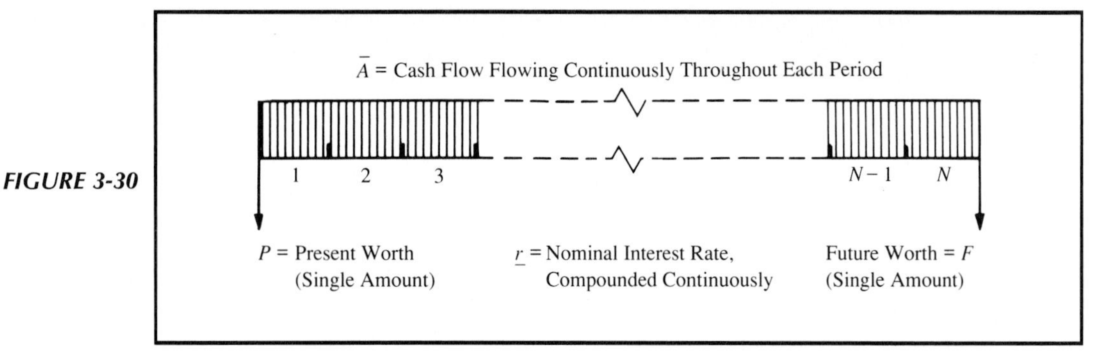

General Cash Flow Diagram for Continous Compounding, Continous Cash Flows

Equation 3-43 can also be written

$$(P/\bar{A},\underline{r}\%,1) = e^{-r}\left[\frac{e^r - 1}{r}\right] = (P/F,\underline{r}\%,1)\left[\frac{e^r - 1}{r}\right]$$

Because the present worth of \$1 per year, flowing continuously with continuous compounding of interest, is $(P/F, \underline{r}\%, 1)(e^r - 1)/r$, it follows that $(e^r - 1)/r$ must also be the compound amount of \$1 per year, flowing continuously with continuous compounding of interest. Consequently, the *continuous compounding compound amount factor (continuous, uniform cash flow over one year)* is

$$(F/\bar{A},\underline{r}\%,1) = \frac{e^r - 1}{r} \tag{3-45}$$

For N years,

$$(F/\bar{A},\underline{r}\%,N) = \frac{e^{rN} - 1}{r} \tag{3-46}$$

Equation 3-46 can also be developed by integration in this manner:

$$F = \bar{A}\int_0^N e^{rt}dt = \bar{A}\left(\frac{1}{r}\right)\int_0^N re^{rt}dt$$

or

$$F = \frac{\bar{A}}{r}(e^{rt})\Big|_0^N = \bar{A}\left[\frac{e^{rN} - 1}{r}\right]$$

This is the *continuous compounding compound amount factor (continuous uniform cash flows for N years).*

TABLE 3–7 Continuous Compounding Continuous Uniform Cash Flows—Interest Factors and Symbols[a]

To Find:	Given:	Factor by Which to Multiply "Given"	Factor Name	Factor Functional Symbol
F	$\bar{A}$	$\frac{e^{rN}-1}{r}$	Continuous compounding compound amount (continuous, uniform cash flows)	$(F/\bar{A}, \underline{r}\%, N)$
P	$\bar{A}$	$\frac{e^{rN}-1}{re^{rN}}$	Continuous compounding present worth (continuous, uniform cash flows)	$(P/\bar{A}, \underline{r}\%, N)$
$\bar{A}$	F	$\frac{r}{e^{rN}-1}$	Continuous compounding sinking-fund (continuous, uniform cash flows)	$(\bar{A}/F, \underline{r}\%, N)$
$\bar{A}$	P	$\frac{re^{rN}}{e^{rN}-1}$	Continuous compounding capital recovery (continuous, uniform cash flows)	$(\bar{A}/P, \underline{r}\%, N)$

[a] $\underline{r}$, nominal annual interest rate, compounded continuously; N, number of periods (years); $\bar{A}$, amount of money flowing continuously and uniformly during each period; F, future worth; P, present worth.

Values of $(P/\bar{A}, \underline{r}\%, N)$ and $(F/\bar{A}, \underline{r}\%, N)$ are given in the tables in Appendix D for various interest rates. Values for $(\bar{A}/P, \underline{r}\%, N)$ and $(\bar{A}/F, \underline{r}\%, N)$ can be readily obtained through their inverse relationship to $(P/\bar{A}, \underline{r}\%, N)$ and $(F/\bar{A}, \underline{r}\%, N)$, respectively. A summary of these factors and their use is given in Table 3-7.

EXAMPLE 3-28

What will be the future worth at the end of five years of a uniform, continuous cash flow, at the rate of \$500 per year for five years, with interest compounded continuously at the nominal annual rate of 8%?

Solution

$$F = \bar{A}(F/\bar{A}, \underline{8}\%, 5) = \$500 \times 6.1478 = \$3{,}074$$

It is interesting to note that if this cash flow had been in year-end amounts of \$500 with discrete annual compounding of $i = 8\%$, the future worth would have been

$$F = A(F/A, 8\%, 5) = \$500 \times 5.8666 = \$2{,}933$$

If the year-end payments had occurred with 8% nominal interest compounded continuously, the future worth would then have been

$$F = A(F/A,\underline{8}\%,5) = \$500 \times 5.9052 = \$2,953$$

It is clear that for a given A amount and continuous compounding of a given nominal interest rate, continuous funds flow produces the largest-valued future worth.

EXAMPLE 3-29

What is the future worth of $10,000 per year that flows continuously for 8 1/2 years if the nominal interest rate is 10% per year? Continuous compounding is utilized.

Solution

There are seventeen 6-month periods in 8 1/2 years and the $\underline{r}$ per six months is 5%. The $\bar{A}$ every six months is $5,000, so $F = \$5,000(F/\bar{A},\underline{5}\%,17) = \$133,964.50$.

This formulation is utilized to enable us to find an interest factor having an integer-valued N. The same answer could have been obtained by resorting to the definition of the $(F/\bar{A},\underline{r}\%,N)$ factor given in Table 3-7 with $N = 8.5$ years:

$$F = \$10,000\left[\frac{e^{0.10(8.5)} - 1}{0.10}\right]$$
$$= \$133,964.50$$

Chapter 3 has presented the fundamental time value of money relationships which are utilized throughout the remainder of this book. These relationships among cash flows at different points in time allow us to perform economic equivalence calculations for alternative designs, machines, systems, and so on, so that capital can be wisely invested. The general process for allocating capital and the role that engineering economy plays in this process was discussed in Chapter 1 (Section 1.6).

The emphasis in Chapter 3 is on the notion of economic equivalence, whether the relevant cash flows and interest rates are discrete or continuous. Students should feel comfortable with the material in this chapter before embarking on their journey through subsequent chapters.

3.25 PROBLEMS

The number in parentheses ( ) that follows each problem indicates the section from which the problem is taken. (Refer to Appendix I of this book for answers to even-numbered problems.)

3-1. What lump-sum amount of interest will be paid on a $10,000 loan that was made on August 1, 1988, and repaid on November 1, 1994, with ordinary simple interest at 12% per year? (3.4)

3-2. An automobile dealer advertises the availability of "simple" interest of 0.5% per month on an automobile loan as follows:

$$\begin{aligned} \text{Amount to be financed} &= \$10{,}000 \\ \text{Time period to repay loan} &= 24 \text{ months} \\ \text{Monthly amount to be repaid} &= \frac{\$10{,}000}{24} + 0.005(\$10{,}000) \\ &= \$466.67 \end{aligned}$$

Does this financing plan really involve simple interest? Explain your answer. (3.4)

3-3. How much interest is *payable each year* on a loan of $2,000 if the interest rate is 10% per year when half of the loan principal will be repaid as a lump sum at the end of three years and the other half will be repaid in one lump-sum amount at the end of six years? How much interest will be paid over the six-year period? (3.4)

3-4. In Problem 3-3, if the interest had not been paid each year but had been allowed to compound, how much interest would be due to the lender as a lump sum at the end of the sixth year? How much extra interest is being paid here (as compared to Problem 3-3) and what is the reason for the difference? (3.5)

3-5. Draw a cash flow diagram for $10,500 being loaned out at an interest rate of 15% per annum over a period of six years. How much simple interest would be repaid as a lump-sum amount at the end of the sixth year? (3.6, 3.7)

3-6. A future amount, F, is equivalent to $1,500 now when eight years separates the amounts and the annual interest is 12%. What is the value of F? (3.9)

3-7. A present obligation of $12,000 is to be repaid in equal uniform annual amounts, each of which includes repayment of the debt (principal) and interest on the debt, over a period of six years. If the interest rate per year is 10%, what is the amount of the annual repayment? (3.10)

3-8. Suppose that the $12,000 in Problem 3-7 is to be repaid at a rate of $2,000 per year plus the interest that is owed and based on the beginning-of-year unpaid principal. Compute the total amount of interest repaid in this situation and compare it with that of Problem 3-7. Why are the two amounts different? (3.6)

3-9. A person desires to accumulate $2,500 over a period of seven years so that a cash payment can be made for a new roof on a summer cottage. To have this amount when it is needed, annual payments will be made to a savings account that earns 8% annual interest per year. How much must each annual payment be? Draw a cash flow diagram. (3.10)

3-10. Mrs. Green has just purchased a new car for $12,000. She makes a down payment of 30% of the negotiated price and then makes payments of $303.68 for each month thereafter for 36 months. Furthermore, she believes the car can be sold for $3,500 at the end of three years. Draw a cash flow diagram of this situation from Mrs. Green's viewpoint. (3.7)

3-11. If $25,000 is deposited now into a savings account that earns 12% per year, what uniform annual amount could be withdrawn at the end of each year for 10 years so that nothing would be left in the account after the tenth withdrawal? (3.10)

3-12. It is estimated that a certain piece of equipment can save $6,000 per year in labor and materials costs. The equipment has an expected life of five years and no salvage value. If the company must earn a 20% rate of return on such investments, how much could be justified now for the purchase of this piece of equipment? Draw a cash flow diagram. (3.10)

3-13. Suppose that installation of Low-Loss thermal windows in your area is expected to save $150 a year on your home heating bill for the next 18 years. If you can earn 8% a year on other investments, how much could you afford to spend now for these windows? (3.10)

3-14. A proposed product modification to avoid production difficulties will require an immediate expenditure of $14,000 to modify certain dies. What annual savings must be realized to recover this expenditure in four years with interest at 10%? (3.10)

3-15. What must be the prospective saving in money six years hence to justify a present investment of $2,250? Use interest at 12%, compounded annually. (3.10)

3-16. Suppose that $10,000 is borrowed now at 15% interest per annum. A partial repayment of $3,000 is made four years from now. The amount that will remain to be paid then is most nearly:

a. $7,000 **b.** $8,050 **c.** $8,500
d. $13,000 **e.** $14,490 (3.9)

3-17. A machine costs $20,000 and has an estimated life of eight years and a scrap value of $2,000. If it is assumed that the interest rate is 8%, compounded annually, the uniform annual amount that must be set aside at the end of each of the eight years to replace the machine is most nearly:

a. $1,692 **b.** $2,170 **c.** $2,250
d. $1,880 **e.** $3,480 (3.10)

3-18. Suppose you have $10,000 cash today and can invest it at 10% interest each year. How many years will it take you to become a millionaire? (3.9)

3-19. Determine the present equivalent and annual equivalent value of the following cash flow pattern when i = 8% per year. (3.14)

End of Year	**0**	**1**	**2**	**3**	**4**	**5**	**6**	**7**
Amount ($)	−1,500	+500	+500	+500	+400	+300	+200	+100

3-20. Maintenance costs for a new bridge with an expected 50-year life are estimated to be $1,000 each year for the first 5 years, followed by a $10,000 expenditure in the fifteenth year and a $10,000 expenditure in year 30. If $i = 10\%$ per year, what is the equivalent uniform annual cost over the entire 50-year period? (3.14)

3-21. Equal end-of-year payments of $263.80 each are being made on a $1,000 loan at 10% effective interest per year.

a. How many payments are required to repay the entire loan?

b. Immediately after the second payment, what lump-sum amount would completely pay off the loan? (3.10)

3-22. Suppose $400 is deposited at the beginning of each year into a bank account that pays interest annually ($i = 10\%$). If 12 payments are made into the account, how much would be accumulated in this fund by the end of the twelfth year? (3.10, 3.12)

3-23. John Q. wants his estate to be worth $65,000 at the end of 10 years. His net worth is now zero. He can accumulate the desired $65,000 by depositing $3,887 at the end of each year for the next 10 years. At what interest rate must his deposits be invested? Give an answer to the nearest tenth of a percent. (3.10)

3-24. Suppose that the parents of a young child decide to make annual payments into a savings account, with the first payment being made on the child's fifth birthday and the last payment being made on the fifteenth birthday. Then starting on the child's eighteenth birthday, the withdrawals shown below will be made. If the effective annual interest rate is 10% during this period of time, what is the amount of the annuity in years 5 through 15? (3.10, 3.15)

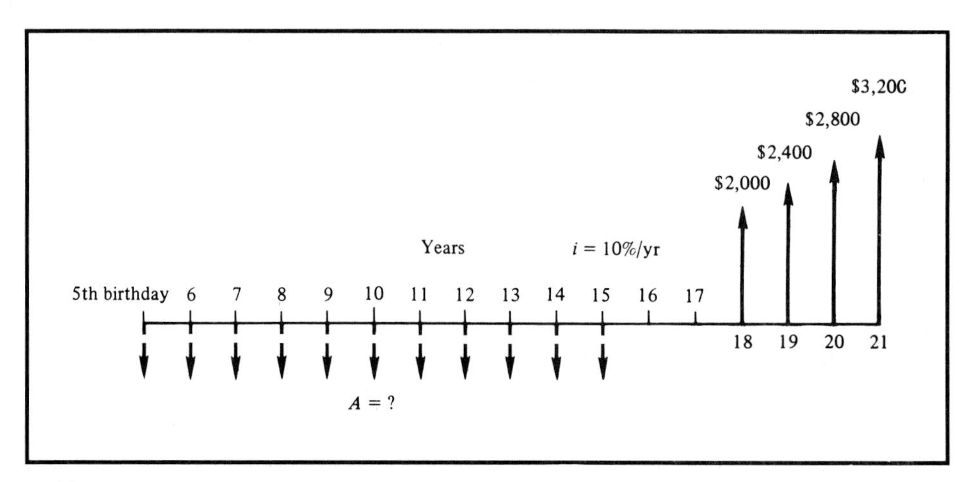

Problem 3-24

3-25. What lump sum of money must be deposited into a bank account at the present time so that $500 per month can be withdrawn for five years, with the first withdrawal scheduled for six years from today? The interest rate is 1% per month. (*Hint:* Monthly withdrawals begin at the end of month 72.) (3.11)

3-26. Calculate the future worth at the end of 1991, at 8% compounded annually, of this savings account: (3.10, 3.15)

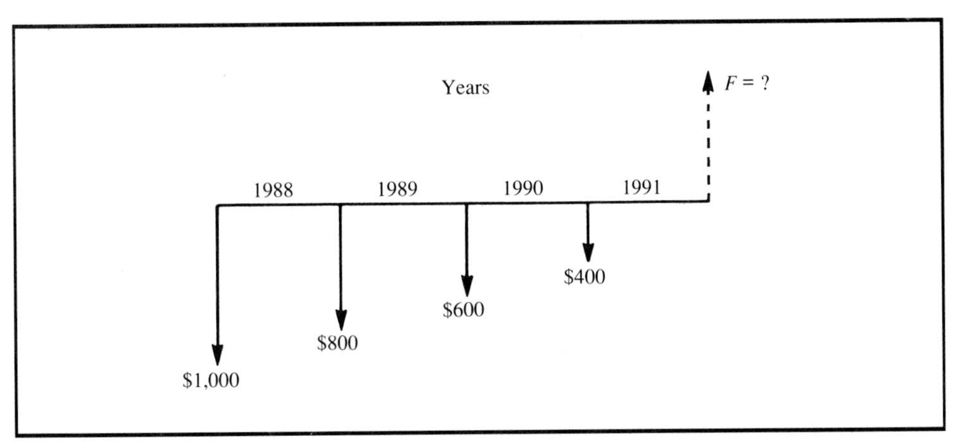

Problem 3-26

3-27. Convert the following cash flow pattern to a uniform series of end-of-year costs over a seven-year period. Let $i = 12\%$. (3.14)

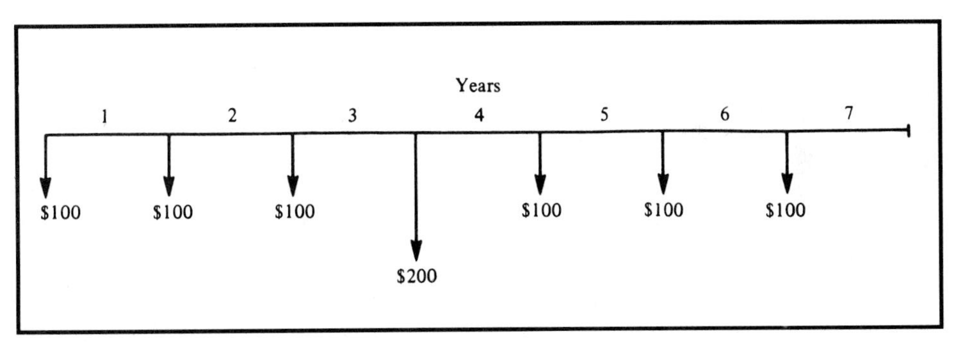

Problem 3-27

3-28. Repairs to an existing warehouse can be made immediately for $10,000. If instead the repair work is postponed for four years, the estimated cost will be $25,000. The useful life of the warehouse is not affected by when the repairs are made. Assume there are no extra costs incurred by postponing the work. If the company requires a 20% return on its investment in this situation, when should the repairs be made? (3.9)

3-29. A certain fluidized-bed combustion vessel has a first cost of $100,000, a life of 10 years, and negligible salvage value. Annual costs of materials, maintenance, and electric power for the vessel are expected to total $8,000. A major relining of the combustion vessel will occur during the fifth year at a cost of $20,000; during this year the vessel will *not* be in service. If the interest rate is 15% per year, what is the lump-sum equivalent cost of this project at the present time ($t = 0$)? Assume that a beginning-of-year cash flow convention is being utilized. (3.12, 3.14)

3-30. A certain government agency utilizes midperiod cash flow convention in its engineering economy studies. Project R-127 has this estimated cash flow pattern:

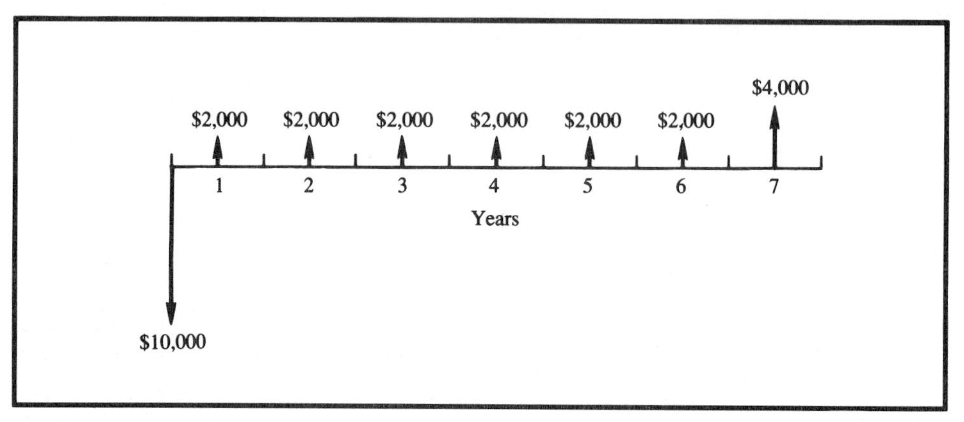

Problem 3-30

If $i = 10\%$ per year, answer the following questions.

a. What is the *present equivalent* value of this project at the beginning of year 1?
b. What is the *annual equivalent* value of the project over the seven-year period?
c. What is the *future equivalent* value at the end of year 7? (3.13)

3-31. An expenditure of $20,000 is made to modify a materials-handling system in a small job shop. This modification will result in first-year savings of $2,000, second-year savings of $4,000, and savings of $5,000 per year thereafter. How many years must the system last if a 25% per year return on investment is required? The system is tailor-made for this job shop and has no salvage value at any time. Use mid-year cash flow convention to solve this problem. (3.13)

3-32. A woman arranges to repay a $1,000 bank loan in 10 equal payments at a 10% effective annual interest rate. Immediately *after* her third payment she borrows another $500, also at 10%. When she borrows the $500, she talks the banker into letting her repay the remaining debt of the first loan and the entire amount of the second loan in 12 equal annual payments. The first of these 12 payments would be made one year after she receives the $500. Compute the amount of each of the 12 payments. (3.14)

3-33. You purchase special equipment that reduces defects by $10,000 per year on item A. This item is sold on contract for the next five years. After the contract expires, the special equipment will save approximately $2,000 per year for five more years. You assume that the machine has no salvage value at the end of 10 years. How much can you afford to pay for this equipment now, if you require a 25% return on your investment? All cash flows are end-of-year amounts. (3.14)

3-34. Suppose you have an opportunity to invest in a fund that pays 12% interest compounded annually. Today, you invest $10,000 into this fund. Three years later (end of year 3), you borrow $5,000 from a local bank at 10% effective annual interest and invest it in the fund. Two years later (EOY 5), you withdraw enough money from the fund to repay the bank loan and all interest due on it. Three years from this withdrawal (EOY 8) you start taking $2,000 per year out of the fund. After five more years (EOY 12), you have withdrawn your original $10,000. The amount remaining in the fund is earned interest. *How much remains?* Hint: Draw a cash flow diagram. (3.14)

3-35. Find the uniform annual amount that is equivalent to a gradient series in which the first year's payment is $500, the second year's payment is $600, the third payment is $700, and so on, and there is a total of 10 payments. The annual interest rate is 8%. (3.15)

3-36. Suppose that annual income from a rental is expected to start at $1,200 per year and decrease at a uniform rate of about $40 each year for the 15-year expected life of the property. The investment cost is $7,000 and i is 10% per year. Is this a good investment? (3.15)

3-37. For a repayment schedule that starts at the end of the year 3 at $Z and proceeds for years 4 through 10 as $2Z, $3Z, . . . , what is the value of Z if the principal of this loan is $10,000 and the interest rate is 10% compounded annually? (3.15)

3-38. The heat loss through the exterior walls of a certain poultry processing plant is estimated to cost the owner $3,000 next year. A salesman from Superfiber Insulation, Inc., has told you, the plant engineer, that he can reduce the heat loss by 80% with the installation of $15,000 worth of Superfiber now. If the cost of heat loss rises by $200 per year (gradient) after the next year and the owner plans to keep the present building for 10 more years, what would you recommend if the cost of money is 12% per year? (3.15)

3-39. Find the equivalent of Q in this cash flow diagram: (3.14, 3.15)

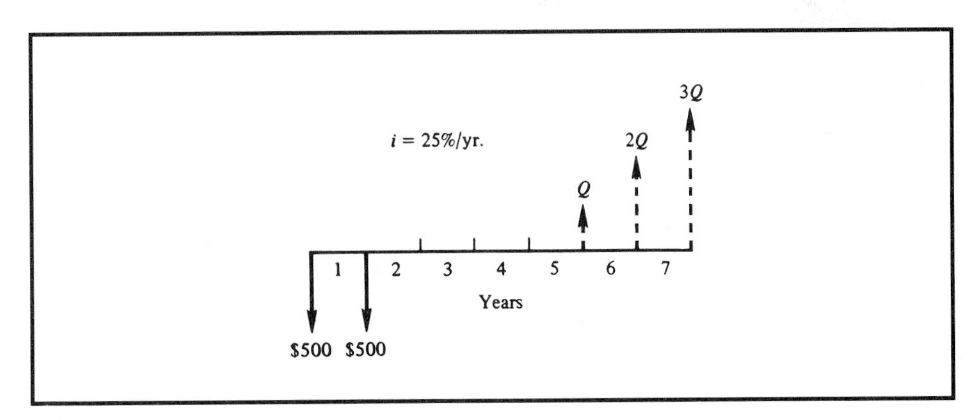

Problem 3-39

3-40. Solve for the value of G below so that the left-hand cash flow diagram is equivalent to the one on the right. Let $i = 10\%$ per year. (3.14, 3.15)

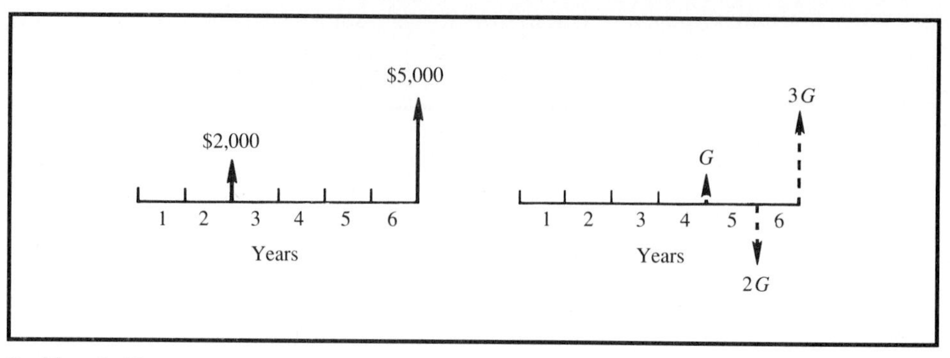

Problem 3-40

3-41. What value of N comes closest to making the left cash flow diagram equivalent to the one on the right-hand side? Let $i = 15\%$ per year. (3.14, 3.15)

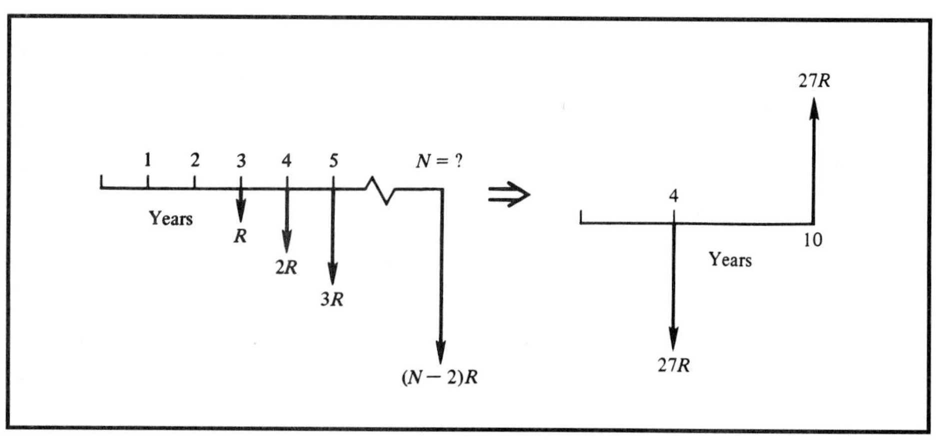

Problem 3-41

3-42. Find the value of B on the left-hand diagram that makes the two cash flow diagrams below equivalent at $i = 10\%$/year. (3.14, 3.15)

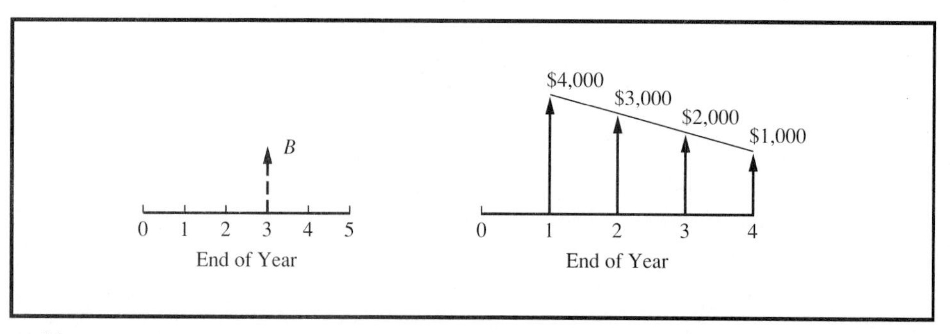

Problem 3-42

3-43. Determine the value of W on the right-hand diagram that makes the two cash flow diagrams shown below equivalent when i = 15%/year. (3.14, 3.15)

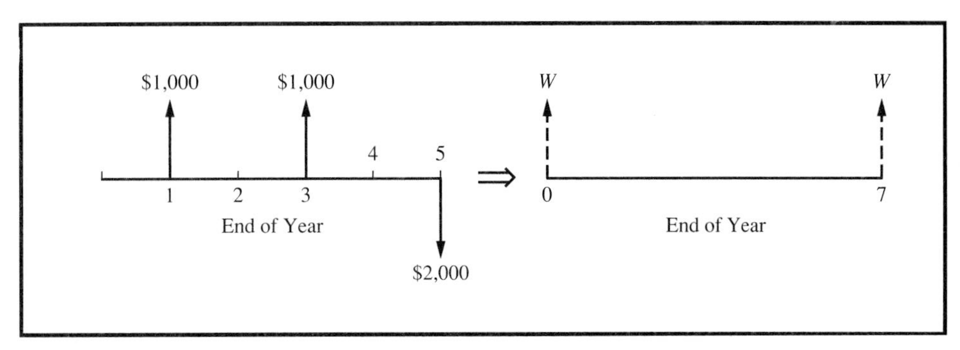

Problem 3-43

3-44. A geometric gradient that increases at $\bar{f}$ = 6% per year for 15 years is shown below. The annual interest rate is 12%. What is the present equivalent value of this gradient? (3.16)

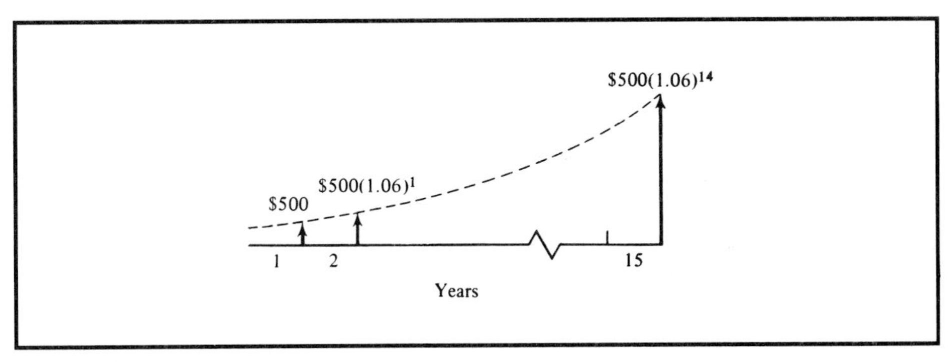

Problem 3-44

3-45. An electronic device is available that will reduce this year's labor costs by $10,000. The equipment is expected to last for eight years. If labor inflates at an average rate of 7% per year and the interest rate is 15% per year, answer these questions:

a. What is the maximum amount that we could justify spending for the device?

b. What is the uniform annual equivalent value (A) of labor costs over the eight-year period?

c. What annual "year 0" amount (A_0) that inflates at 7% per year is equivalent to the answer in part (a)? (3.16)

3-46. You are the manager of a large crude oil refinery. As part of the refining process, a certain heat exchanger (operated at high temperatures and with abrasive material flowing through it) must be replaced every year. The replacement and downtime cost in the first year is $75,000. It is expected to increase due to inflation at a rate of 8% per year for five years, at which time this particular heat exchanger will no longer be needed. If the company's cost of capital is 18% per year, how much could you afford to spend for a higher-quality heat exchanger so that these annual replacement and downtime costs could be eliminated? (3.16)

3-47. Solve for X such that the cash receipt at year 0 is equivalent to the cash outflows in years 1 through 6. (3.16)

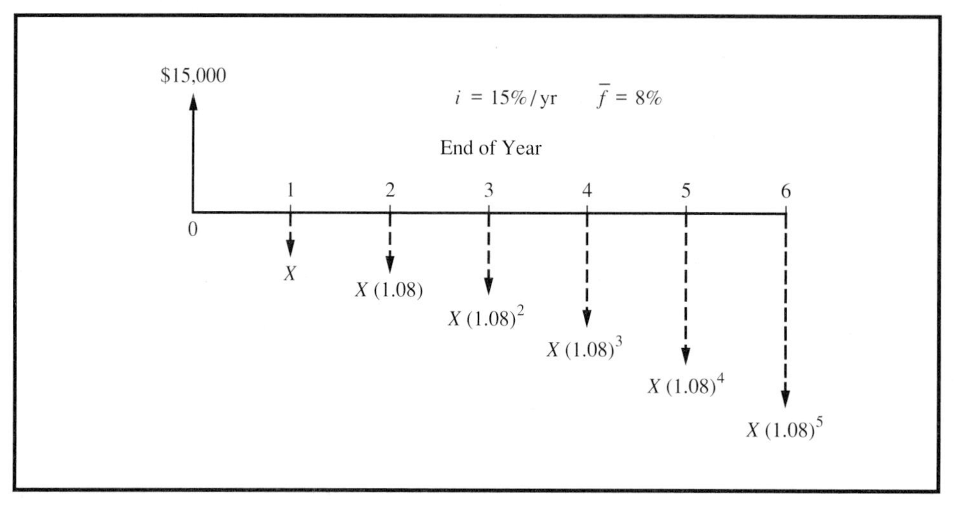

Problem 3-47

3-48. Compute the effective annual interest rate in each of these situations:

a. 10% nominal interest, compounded semiannually.
b. 10% compounded quarterly.
c. 10% compounded continuously.
d. 10% compounded weekly. (3.17)

3-49.

a. A certain savings and loan association advertises that it pays 8% interest, compounded quarterly. What is the *effective* interest rate per annum? If you deposit $5,000 now and plan to withdraw it in three years, how much would your account be worth at that time? (3.17)

b. If instead you decide to deposit $800 every year for three years, how much could be withdrawn at the end of the third year? Suppose that, instead, you deposit $400 every six months for three years, now what would the accumulated amount be? (3.17, 3.19)

3-50. You have just learned that the ABC Corporation has a bond that costs $350 now and eight years later pays a lump-sum amount of $1,000. The cash flow diagram looks like this:

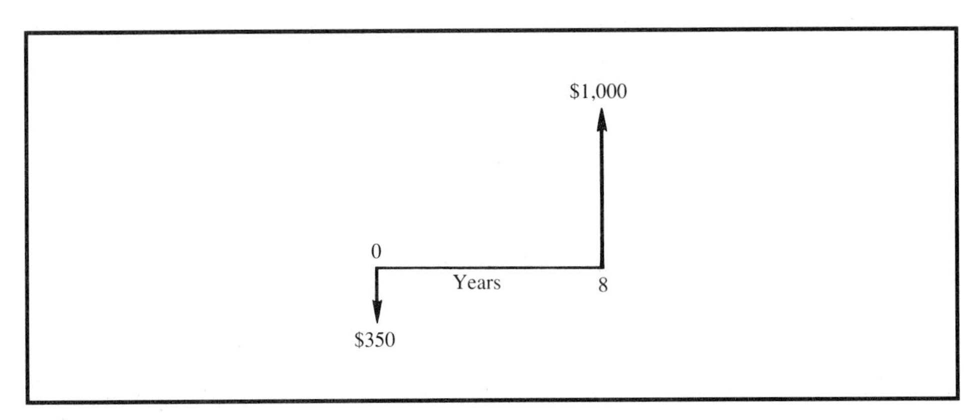

Problem 3-50

If you bought one of these bonds, what interest rate would you earn on your investment? Calculate your answer to the nearest 0.1 of 1%. (3.9)

3-51. Exactly 140 years ago, your great-grandmother deposited $200 in a large New York bank as part of a savings plan. She forgot all about the deposit during her lifetime. Two years ago the bank notified your family that the account was worth $191,516. What was the effective annual interest rate in this situation? (3.9)

3-52. As often quoted, the prepaid premium of insurance policies covering natural disasters (floods, etc.) during a three-year period is 2.4 times the cost for a single year of coverage. What rate of interest is being earned on the extra investment if a three-year policy is purchased now rather than three one-year policies at the beginning of each of the years? (3.10, 3.12)

3-53. Suppose you have just borrowed $7,500 at 10% nominal interest compounded quarterly. What is the total lump-sum, compounded amount to be paid by you at the end of a six-year loan period? (3.18)

3-54. How many deposits of $100 each must a person make at the end of each month if she desires to accumulate $3,350 for a new home entertainment center? Her savings account pays 9% nominal interest compounded monthly. (3.18)

3-55. You have used your credit card to purchase automobile tires for $340. Unable to make payments for seven months, you then write a letter of apology and enclose a check to pay your bill in full. The credit card company's nominal interest rate is 18% compounded monthly. For how much should you write the check? (3.18)

3-56. How long does it take a given amount of money to triple itself if the money is invested at a nominal rate of 15%, compounded monthly? (3.18)

3-57. A local bank offers a "Vacation Made Easy" plan as follows. Each participant in the plan deposits an amount of money, A, at the end of each week for 50 weeks with no interest being paid by the bank. Then the bank makes the 51st and 52nd payments and returns to each participant a grand total of 52 A at the end of week 52 that is used to pay for the vacation. What is the true interest rate per year being earned *by participants* in this plan? Assume that an opportunity exists for weekly compounding if participants elect another investment plan elsewhere. (3.19)

3-58. Determine the present equivalent value of $400 paid every three months over a period of seven years in each of these situations:

a. The interest rate is 12%, compounded annually.
b. The interest rate is 12%, compounded quarterly.
c. The interest rate is 12%, compounded continuously. (3.19)

3-59.

a. What extra semiannual expenditure for five years would be justified for the maintenance of a machine in order to avoid an overhaul costing $3,000 at the end of five years. Assume nominal interest at 8%, compounded semiannually. (3.19)
b. What is the annual equivalent worth of $125,000 now, when 18% nominal interest per year is compounded monthly? Let $N = 15$ years. (3.19)

3-60.

a. What equal monthly payments will repay an original loan of $1,000 in six months at a nominal rate of 6% compounded monthly? What is the effective annual rate?
b. For part (a), what is the effective quarterly rate? (3.17)

3-61. If the nominal interest rate is 10% and compounding is semiannual, what is the present worth of the following receipts? (3.15, 3.17)

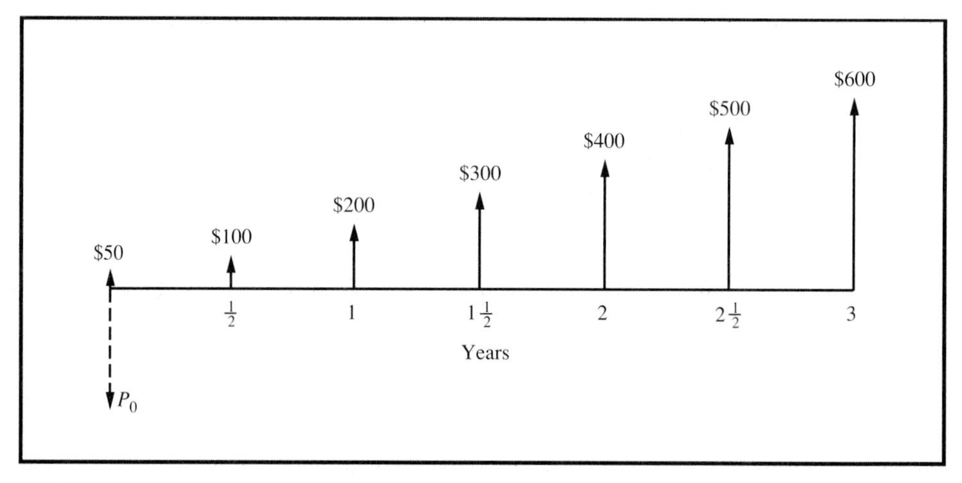

Problem 3-61

3-62. Find the value of A that is equivalent to the gradient shown below if the nominal interest rate is 12% compounded monthly. (3.15, 3.19)

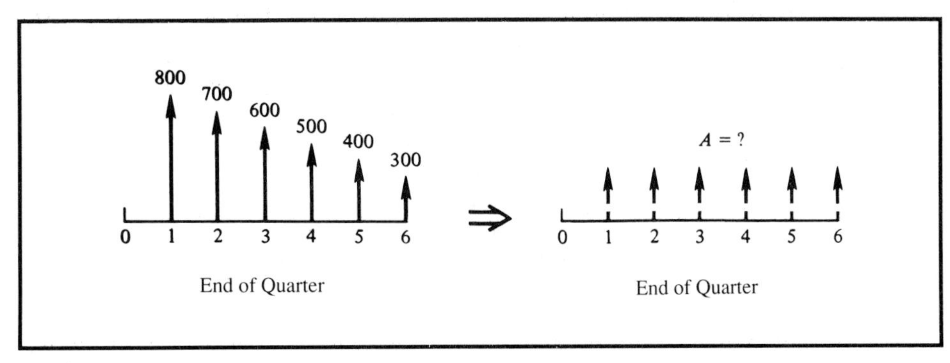

Problem 3-62

3-63. Rework problem 3-54 when the savings account earns 9% nominal interest compounded semiannually. (3.20)

3-64. Suppose that you have a money market certificate earning an average annual rate of interest, which varies over time as follows:

Year k	1	2	3	4	5
i_k	14%	12%	10%	10%	12%

If you invest $5,000 in this certificate at the beginning of year 1 and do not add or withdraw any money for five years, what is the value of the certificate at the end of the fifth year? (3.21)

3-65. Determine the present equivalent value of this cash flow diagram when the annual interest rate (i_k) varies as indicated. (3.21)

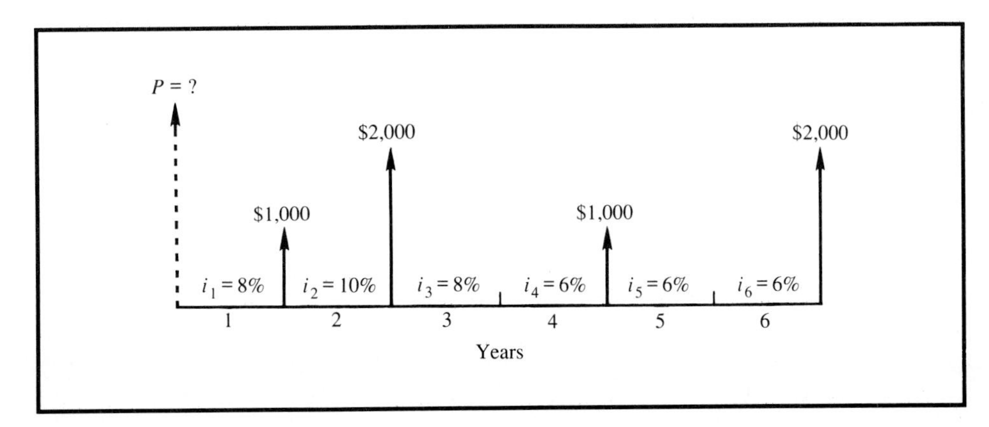

Problem 3-65

3-66. If a nominal interest rate of 8% is compounded continuously, determine the unknown quantity in each of the following situations.

a. What uniform end-of-year amount for 10 years is equivalent to $8,000 at the end of year 10?

b. What is the present equivalent value of $1,000 per year for 12 years?

c. What is the future worth at the end of the sixth year of $243 payments every six months during the six years? The first payment occurs six months from the present and the last occurs at the end of the sixth year.

d. Find the equivalent lump-sum amount at the end of year nine when $P_0 = \$1,000$ and a nominal interest rate of 8% is compounded continuously. (3.23)

3-67. Find the value of the unknown quantity Z in the diagram below when $\underline{r} = 10\%$ continuously compounded (3.23)

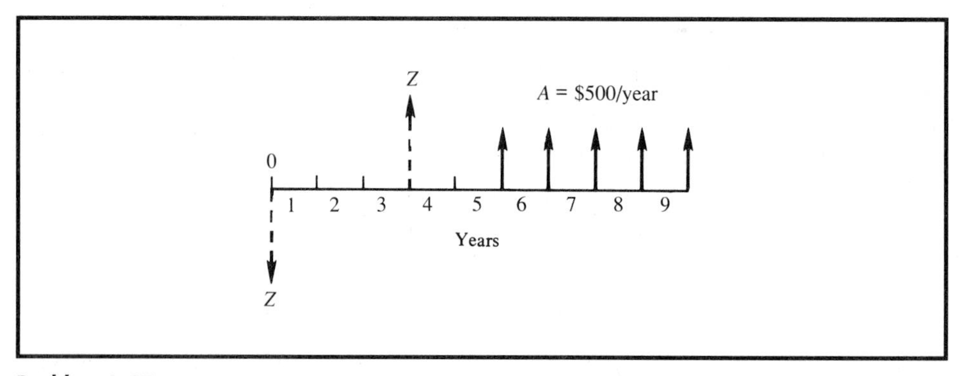

Problem 3-67

3-68. A man deposited $2,000 in a savings account when his son was born. The nominal interest rate was 8% per year, compounded continuously. On the son's 18th birthday, the accumulated sum is withdrawn from the account. How much would this accumulated amount be? (3.23)

3-69. Find the value of P in this cash flow diagram: (3.23)

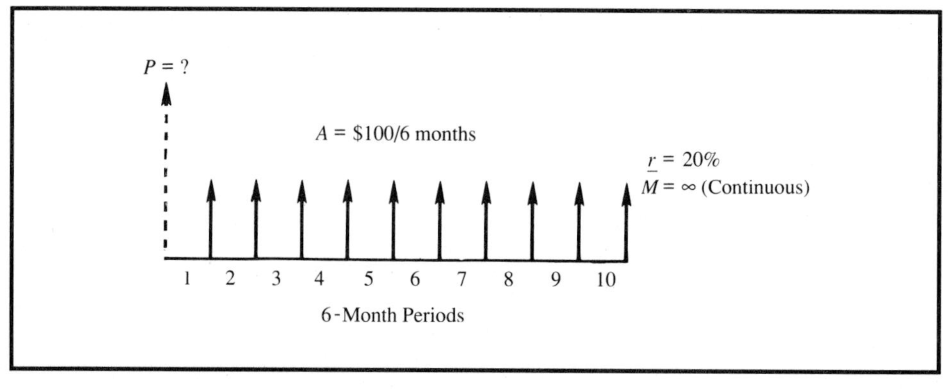

Problem 3-69

3-70. A person needs $12,000 immediately as a down payment on a new home. Suppose that he can borrow this money from his company credit union. He will be required to repay the loan in equal payments, *made every six months* over the next eight years. The annual interest rate being charged is 10% compounded continuously. What is the amount of each payment? (3.23)

3-71.

a. What is the present worth of a uniform series of annual payments of $3,500 each for five years if the interest rate, compounded continuously, is 10%.

b. An amount of $7,000 is invested in a certificate of deposit (CD) and will be worth $16,000 in nine years. What is the continuously compounded nominal interest rate for this CD? (3.23)

3-72.

a. Many persons prepare for retirement by making monthly contributions to a savings program. Suppose that $100 is set aside each month and invested in a savings account that pays 12% interest each year, compounded continuously. Determine the accumulated savings in this account at the end of 30 years.

b. In part (a), suppose that an annuity will be withdrawn from savings that have been accumulated at the end of year 30. The annuity will extend from the end of year 31 to the end of year 40. What is the value of this annuity if the interest rate and compounding frequency in part (a) do not change? (3.23)

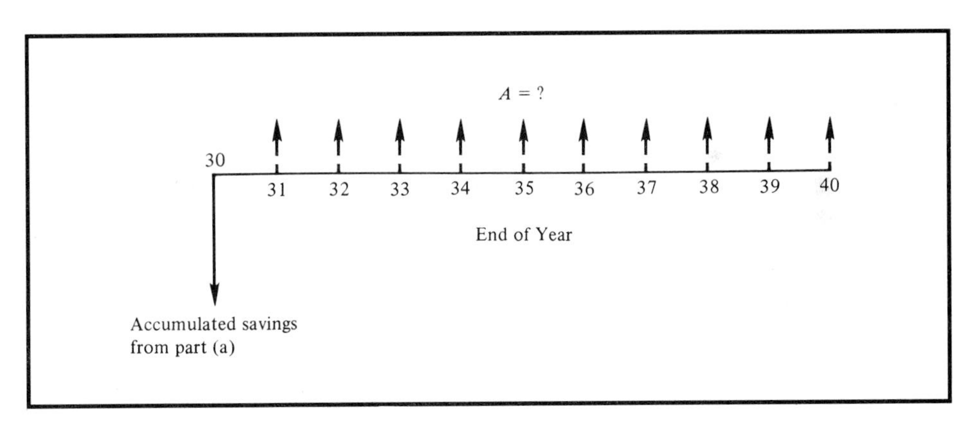

Problem 3-72 (b)

3-73.

a. What is the future worth of a continuous funds flow amounting to $10,500 per year when $\underline{r} = 15\%$, $M = \infty$, and $N = 12$ years?

b. If the nominal interest rate is 10% per year, continuously compounded, what is the future value of $10,000 per year flowing continuously for 8.5 years?

c. Let $\overline{A} = \$7,859$ per year with $\underline{r} = 15\%$, $M = \infty$. How many years will it take to have $1 million in this account? (3.24)

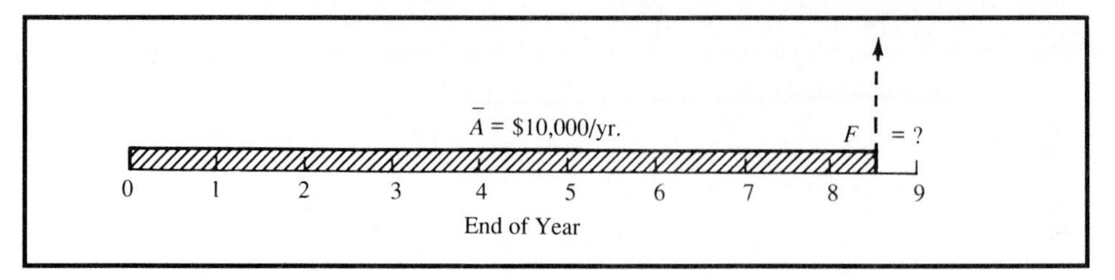

Problem 3-73(b)

3-74. For how many years must an investment of \$63,000 provide a continuous flow of funds at the rate of \$16,000 per year so that a nominal interest rate of 10%, continuously compounded, will be earned? (3.24)

3-75. What is the present value of the following continuous funds flow situations?

a. \$1,000,000 per year for four years at 10% compounded continuously.

b. \$6,000 per year for 10 years at 8% compounded annually.

c. \$500 per quarter for 6.75 years at 12% compounded continuously. (3.24)

3-76. What is the difference in present equivalent worths for the two cash flow diagrams shown below? (3.24)

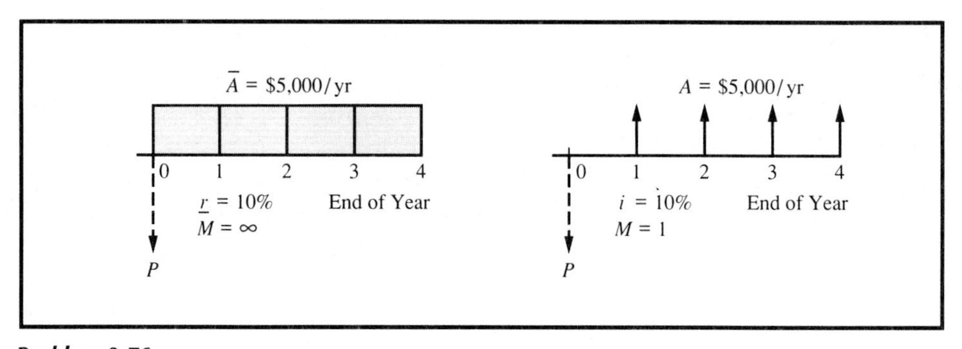

Problem 3-76

4

APPLICATIONS OF MONEY-TIME RELATIONSHIPS

The two primary objectives of this chapter are (1) to illustrate the mechanics of several basic methods for making engineering economy studies considering the time value of money, and (2) to describe briefly the underlying assumptions and interrelationships of these methods.

The principal topics discussed in Chapter 4 are as follows:

Five basic methods for assessing economic worth:
  a. present worth
  b. annual worth
  c. future worth
  d. internal rate of return
  e. external rate of return

A computer program for selected methods
Payback period
Industrial case studies

4.1
INTRODUCTION

All engineering economy studies of capital projects should consider the return that a given project will or should produce. A basic question this book addresses is whether a proposed capital investment and its associated expenditures can be recovered by revenues (or savings) over time *in addition to* a return on the capital that is sufficiently attractive in view of risks involved and potential alternative uses. The interest and money–time relationships of Chapter 3 emerge as essential ingredients in answering this question, and they are applied to many different types of problems in this chapter.

Because patterns of capital investment, revenue (or savings) cash flows, and cost cash flows can be quite different in various projects, there is no single method for performing engineering economic analyses that is ideal for all cases. Consequently, several methods are commonly used in practice. All produce equally satisfactory results and lead to the same decision in cases where the inherent assumptions of each method are applicable.

All engineering economy study methods described in this chapter are for before-tax studies. However, it is recognized that income taxes are an ever-present and somewhat unpleasant fact of life; most corporations pay out a sizeable portion of their gross profits in the form of income taxes, both to state and federal governments. Chapter 7 and parts of subsequent chapters show how to make after-tax studies.

4.2
FIVE BASIC METHODS

In this chapter we concentrate on the correct use of five basic methods for assessing the economic merits of a single cash flow profile resulting from a proposed problem solution. Later in Chapter 5 multiple (two or more) problem solutions, or alternatives, are considered. The methods discussed in Chapter 4 are classified as follows:

Equivalent worth

1. Present worth (PW)
2. Annual worth (AW)
3. Future worth (FW)

Rate of return

1. Internal rate of return (IRR)
2. External rate of return (ERR)

The first three methods convert all cash flows into equivalent worths at some point or points in time using an interest rate (before taxes) equal to the minimum attractive rate of return (MARR). Establishing the MARR as the interest rate to be used for discounting purposes within an organization is usually a policy issue resolved by management based on a number of considerations. These considerations include (1) the number of potential projects (investment opportunities), their purpose, and their financial attractiveness; (2) the availability, source, and cost of capital funds; and (3) the perceived risks associated with the investment opportunities. Also, the type of organization involved will affect the selection. In general, the MARRs within government organizations are less than those used by public utilities, which in turn are less than the MARRs normally established within a private sector industry.

The MARR should be chosen to maximize the economic well-being of an organization. However, in practice the MARR is a decision criterion that is provided to the analyst for purposes of evaluating the economic merits of alternative courses of action. How explicitly and adeptly an organization specifies its MARR is seldom clear. Determination of the MARR is a critical policy issue that affects the strategic welfare of any enterprise, and this topic is discussed in more detail in Chapter 13 (Section 13.9).

The last two methods listed are different ways to calculate an annual rate of profit or savings resulting from an investment so that a rate of return can, in turn, be compared against the MARR. The "payback period" is a method that typically ignores the time value of money, and it is also briefly discussed. Another popular method is the benefit–cost technique. It will be discussed in connection with public sector engineering economy studies in Chapter 14.

Unless otherwise specified, end-of-period cash flow convention and discrete compounding of interest are utilized throughout this and subsequent chapters. A planning horizon, or study (analysis) period,[*] *of N compounding periods (usually years) is used to evaluate prospective investments throughout the remainder of the book.*

4.3 THE PRESENT WORTH METHOD

The present worth (PW) method is based on the concept of equivalent worth of all cash flows relative to some base or beginning point in time called the

[*] The terms study period and analysis period are used interchangeably in this book. The concept of an analysis or study period is discussed in more detail in Chapter 5.

present. That is, all cash inflows and outflows are discounted to the base point at an interest rate that is generally the MARR.

The present worth of an alternative is a measure of how much money will have to be put aside now to provide for one or more future expenditures. It is assumed that such cash placed in reserve earns interest at a rate equal to a firm's MARR.

To find the PW of a series of cash receipts and/or expenses, it is necessary to discount future amounts to the present by using an interest rate for the appropriate study period (years, for example) in the following manner.

$$PW = F_0(1 + i)^0 + F_1(1 + i)^{-1} + F_2(1 + i)^{-2} + \cdots + F_k(1 + i)^{-k} + \cdots + F_N(1 + i)^{-N} \tag{4-1}$$

where i = effective interest rate, or MARR, per compounding period
k = index for each compounding period ($0 \leq k \leq N$)
F_k = future cash flow at the end of period k
N = number of compounding periods in the planning horizon

The relationship given in Equation 4-1 is based on the assumption of a constant interest rate throughout the life of a particular project. If the interest rate is assumed to change, the present worth must be computed in two or more steps as illustrated in Chapter 3.

The higher the interest rate and the further into the future a cash flow occurs, the lower is its present worth. This is shown graphically in Figure 4-1. As long as the present worth, PW, (i.e., present equivalent of cash inflows minus cash outflows) is greater than or equal to zero, the project is economically justified; otherwise, it is not acceptable.

If receipts or savings are not known so that only cash outflows (expenses) are relevant, the method is characterized by negative-valued present worth, and then it is often referred to as the present worth-cost (PW–C) method.

EXAMPLE 4-1

An investment of $10,000 can be made in a project that will produce a uniform annual revenue of $5,310 for 5 years and then have a salvage value of $2,000. Annual expenses will be $3,000 each year for operation and maintenance. The company is willing to accept any project that will earn an annual return of 10% or more, before income taxes, on all invested capital. Show whether this is a desirable investment by using the present worth method.

FIGURE 4-1

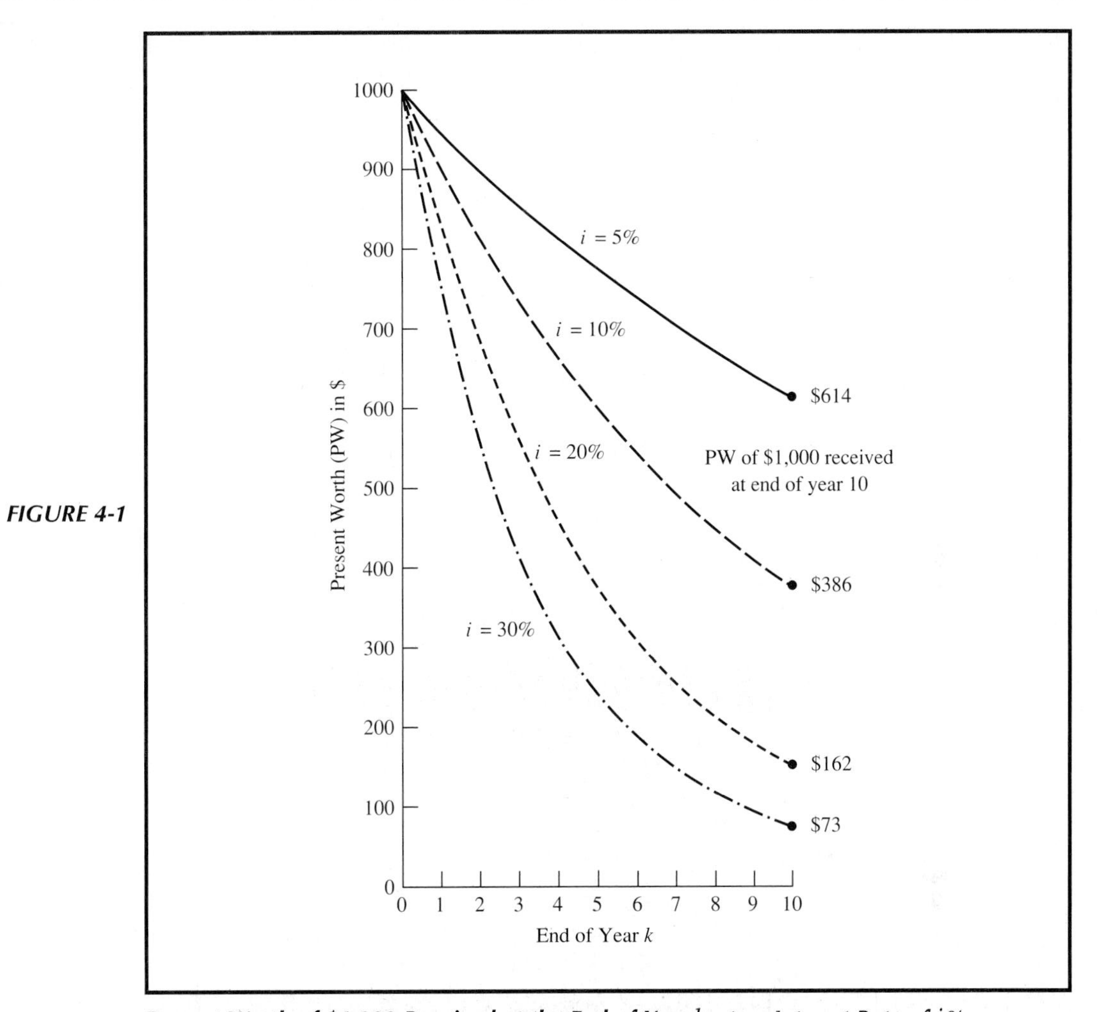

Present Worth of $1,000 Received at the End of Year k at an Interest Rate of i%

Solution

	Present Worth	
	Outflows	**Inflows**
Annual revenue: $5,310(P/A, 10%, 5)		$20,125
Salvage value: $2,000(P/F, 10%, 5)		1,245
Investment	−$10,000	
Annual expenses: $3,000(P/A, 10%, 5)	− 11,370	
Total	−$21,370	$21,370
Total PW		$ 0

Since total PW = $0, the project is shown to be marginally acceptable.

EXAMPLE 4-2

A piece of new equipment has been proposed by engineers to increase the productivity of a certain manual welding operation. The initial investment (first cost) is \$25,000, and the equipment will have a salvage value of \$5,000 at the end of a study period of 5 years. Increased productivity attributable to the equipment will amount to \$8,000 per year after extra operating costs have been subtracted from the value of the additional production. Thus, it is obvious that the "do nothing" alternative has been considered in estimating these cash flows. A cash flow diagram for this equipment is given in Figure 4-2. If the firm's minimum attractive rate of return (before income taxes) is 20% per year, is this proposal a sound one? Use the present worth method.

Solution

$$\text{Total PW} = \text{PW of cash receipts} - \text{PW of cash outlays}$$

or

$$\begin{aligned}\text{Total PW} &= \$8{,}000(P/A, 20\%, 5) + \$5{,}000(P/F, 20\%, 5) - \$25{,}000\\ &= \$934.29\end{aligned}$$

Because PW > 0, this equipment is economically justified.

FIGURE 4-2

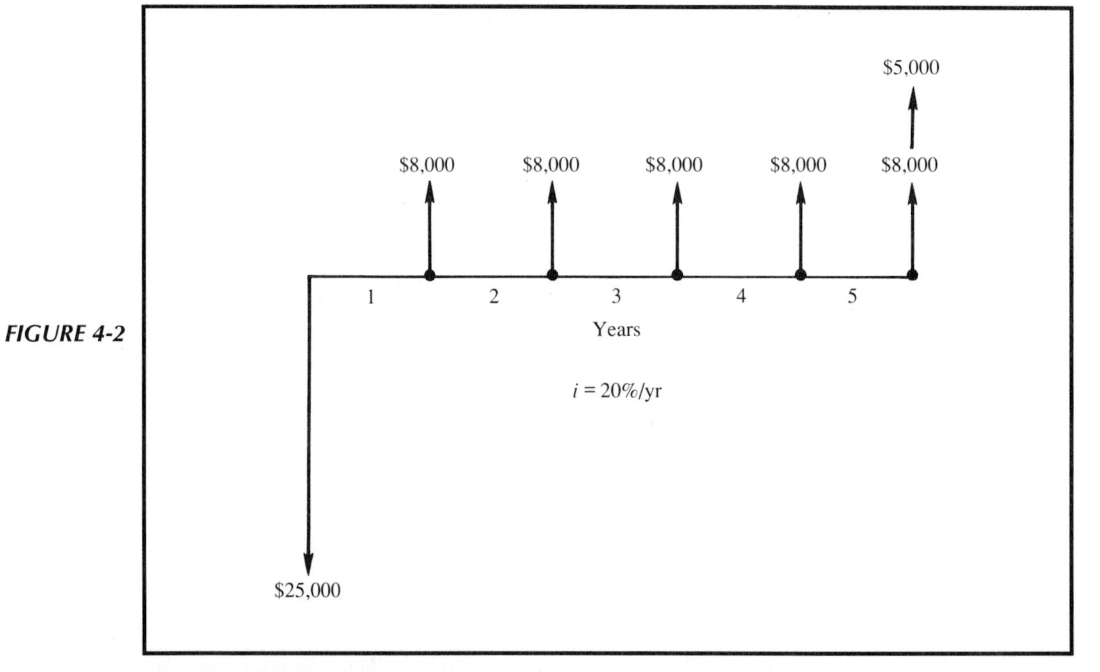

Cash Flow Diagram for Example 4-2

TABLE 4–1 Cumulative PW Calculations for Example 4–2

End of Year k	(A) Cash Flow	(B) Present Worth of Cash Flow at i = 20%	(C) Cumulative PW at i = 20% through Year k	(D) Cum. PW at i = 0% through Year k
0	−$25,000	−$25,000	−$25,000	−$25,000
1	8,000	6,667	− 18,333	− 17,000
2	8,000	5,556	− 12,777	− 9,000
3	8,000	4,630	− 8,147	− 1,000
4	8,000	3,858	− 4,289	7,000
5	13,000	5,223	934	20,000

Based on Example 4-2, Table 4-1 can be utilized to plot the cumulative present worth of cash flows through year k. The graphs of cumulative present worth shown in Figure 4-3 are plotted from columns (C) and (D) of Table 4-1.

The minimum attractive rate of return in Example 4-2 (and in other examples throughout this chapter) is to be interpreted as an effective interest rate (i).

FIGURE 4-3

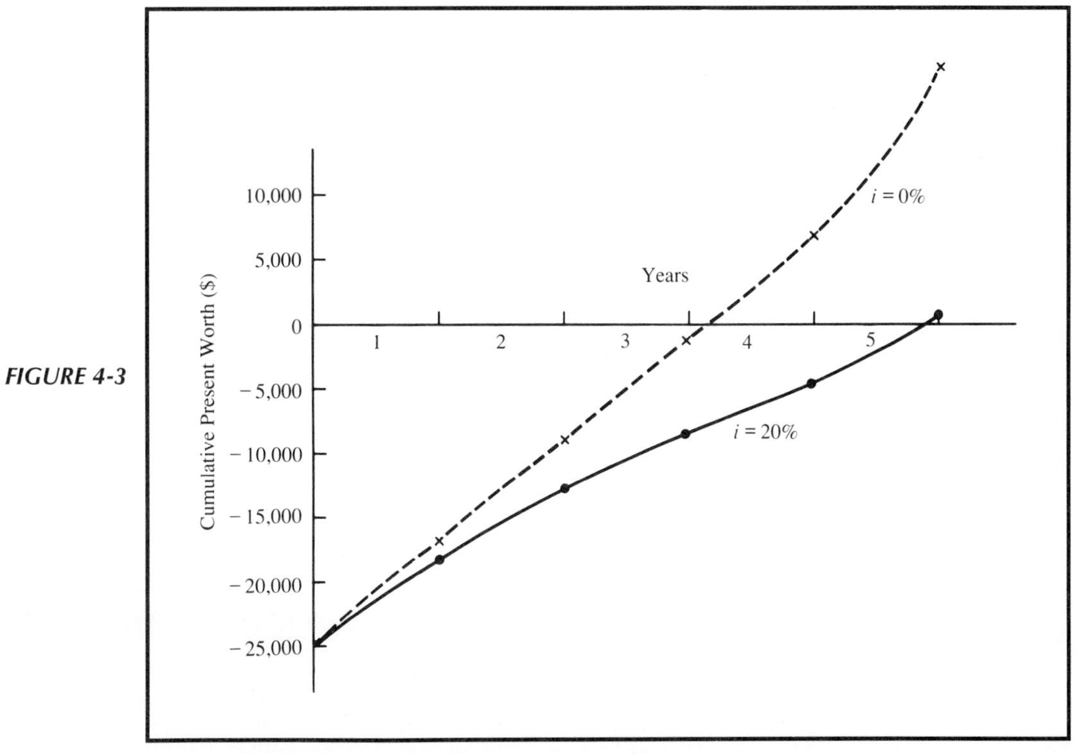

Graph of Cumulative Present Worth for Example 4-2

Here $i = 20\%$ per year. Cash flows are discrete, end-of-year amounts. If *continuous compounding* had been specified for a nominal interest rate (r) of 20%, the present worth would have been calculated by using Appendix D interest factors:

$$\begin{aligned} \text{PW} &= -\$25{,}000 + \$8{,}000(P/A, \underline{r} = 20\%, 5) \\ &\quad + \$5{,}000(P/F, \underline{r} = 20\%, 5) \\ &= -\$25{,}000 + \$8{,}000(2.8551) + \$5{,}000(0.3679) \\ &= -\$319.60 \end{aligned}$$

Consequently, with continuous compounding the equipment would not be economically justifiable. The reason is that the higher effective interest rate ($e^{0.20} - 1 = 0.2214$) reduces the present worth of future positive cash flows but does not affect the present worth of the capital invested at the beginning of year 1.

A popular application of the present worth method is to establish value (current selling price) for bonds and other types of securities that provide fixed interest income. Example 4-3 illustrates the *bond valuation* problem.

EXAMPLE 4-3

A certain U.S. Treasury bond that matures in 8 years has a face value of \$10,000. This means that the bondholder will receive \$10,000 cash when the bond's maturity date is reached. The bond stipulates a fixed interest rate of 8% per year, but interest payments are made to the bondholder every 3 months and amount to 2% of the face value.

A prospective buyer of this bond would like to earn 10% nominal interest on his investment because interest rates in the economy have risen since the bond was issued. How much should this buyer be willing to pay for the bond?

Solution

To establish the value of this bond in view of stated conditions, the present worth of future cash flows during the next 8 years (the study period) must be evaluated. Interest payments are quarterly, as shown in Figure 4-4, so the present worth is computed at $10\%/4 = 2.5\%$ per quarter for the remaining $8(4) = 32$ quarters of the bond's life:

$$\begin{aligned} P &= \frac{\$200}{\text{qtr}}\left(P/A, \frac{2.5\%}{\text{qtr}}, 32 \text{ qtrs}\right) + \$10{,}000\left(P/F, \frac{2.5\%}{\text{qtr}}, 32 \text{ qtrs}\right) \\ &= \$4{,}369.84 + \$4{,}537.71 \\ &= \$8{,}907.55 \end{aligned}$$

Thus, the buyer should pay no more than \$8,907.55 when 10% nominal interest is desired.

FIGURE 4-4

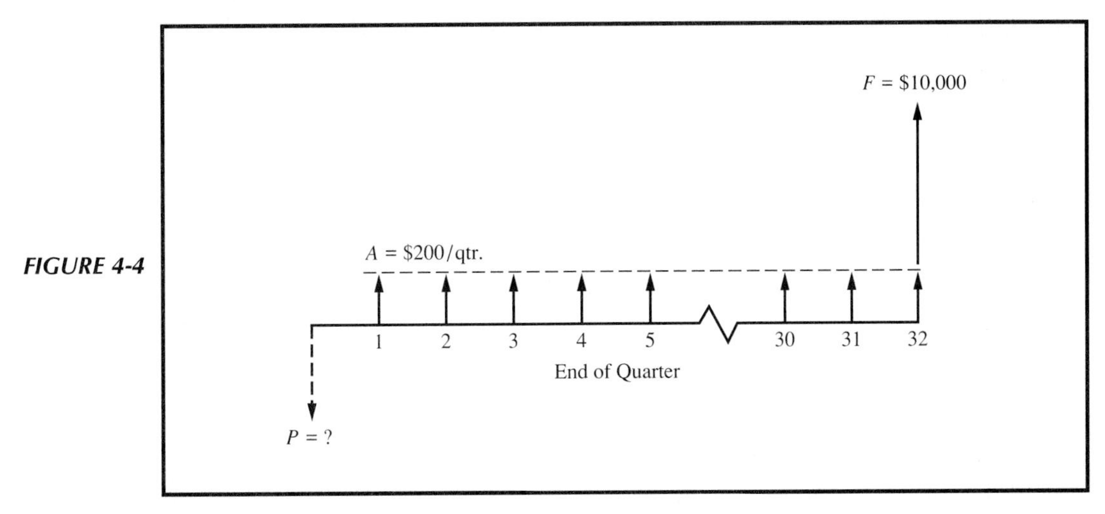

Cash Flow Diagram for Example 4-3

When the life of an investment or the study (analysis) period is infinite, a special case of the present worth method is the *capitalized worth* method, which will be discussed in Chapter 5.

4.4 THE ANNUAL WORTH METHOD

The *annual worth* (AW) of a project is a uniform annual series of dollar amounts, for a stated study period, that is *equivalent* to the cash inflows (receipts or savings) and/or cash outflows (expenses) under consideration. In other words, the annual worth of a project is its annual equivalent receipts (or savings) ($\underline{R}$) minus annual equivalent expenses ($\underline{E}$), less its annual equivalent capital recovery (CR) amount which is defined below. An annual equivalent value of $\underline{R}$, $\underline{E}$, and CR is computed as the MARR. The study period is denoted by N, which is usually years. In equation form the annual worth is

$$\text{AW} = \underline{R} - \underline{E} - \text{CR} \tag{4-2}$$

As long as the annual worth is greater than or equal to zero, the project is economically attractive; otherwise, it is not. An annual worth of zero means that an annual return equal to the MARR has been earned.

The capital recovery cost (CR) for a project is the equivalent uniform annual cost of the capital invested. It is an annual amount that covers the following two items:

1. Depreciation (loss in value of the asset).
2. Interest on invested capital (minimum attractive rate of return).

As an example, consider a machine or other asset that will cost $10,000, last 5 years, and have a salvage value of $2,000. Further, the interest on invested capital, i, is 10%.

It can be shown that no matter which method of calculating an asset's loss in value over time is used, the equivalent annual cost of capital recovery is the same. For example, if straight line depreciation* is used, the equivalent annual cost of capital recovery is calculated to be $2,310, as shown in Table 4-2.

There are several convenient formulas by which capital recovery cost may be calculated to obtain the same answer as above. Probably the easiest formula to understand involves finding the annual equivalent of the initial investment and then subtracting the annual equivalent of the salvage value. Thus

$$\text{CR} = I(A/P, i\%, N) - S(A/F, i\%, N) \tag{4-3}$$

where I = initial investment for the project†
S = salvage (residual) value at end of study period
N = project study period

TABLE 4–2 Calculation of Equivalent Annual Capital Recovery Cost Assuming Straight Line Depreciation

Year	Value of Investment at Beginning of Year[a]	Straight Line Depreciation	Interest on Beginning-of-Year Investment at i = 10%	Capital Recovery Cost Each Year	Present Worth of Capital Recovery Cost at i = 10%
1	$10,000	$1,600	$1,000	$2,600	$2,600(P/F, 10%, 1) = $2,364
2	8,400	1,600	840	2,440	2,440(P/F, 10%, 2) = 2,016
3	6,800	1,600	680	2,280	2,280(P/F, 10%, 3) = 1,713
4	5,200	1,600	520	2,120	2,120(P/F, 10%, 4) = 1,448
5	3,600	1,600	360	1,960	1,960(P/F, 10%, 5) = 1,217
					$8,758

Total CR cost = $8,758 (A/P, 10%, 5) = $2,310

[a]This is also referred to later as the beginning-of-year unrecovered investment.

*Explanations of various methods of depreciation are given in Chapter 6.

†In some cases the investment will be spread over several periods. In such situations, I is the PW of all investment amounts.

When Equation 4-3 is applied to the example in Table 4-2, the CR amount is

$$\begin{aligned} CR &= \$10{,}000(A/P, 10\%, 5) - \$2{,}000(A/F, 10\%, 5) \\ &= \$10{,}000(0.2638) - 2{,}000(0.1638) = \$2{,}310 \end{aligned}$$

Another way to calculate the CR cost is to add the annual sinking fund depreciation charge (or deposit) to the interest on the original investment (sometimes called minimum required profit). Thus

$$CR = (I - S)(A/F, i\%, N) + I(i\%) \tag{4-4}$$

When Equation 4-4 is applied to the example in Table 4-2, the CR amount is

$$\begin{aligned} CR &= (\$10{,}000 - \$2{,}000)(A/F, 10\%, 5) + \$10{,}000(10\%) \\ &= \$8{,}000(0.1638) + \$10{,}000(0.10) = \$2{,}310 \end{aligned}$$

Yet another very popular way to calculate the CR cost is to add the equivalent annual cost of the depreciable portion of the investment and the interest on the nondepreciable portion (salvage value). Thus

$$CR = (I - S)(A/P, i\%, N) + S(i\%) \tag{4-5}$$

Applied to the same example as above,

$$\begin{aligned} CR &= (\$10{,}000 - \$2{,}000)(A/P, 10\%, 5) + \$2{,}000(10\%) \\ &= \$8{,}000(0.2638) + \$2{,}000(0.10) = \$2{,}310 \end{aligned}$$

EXAMPLE 4-4

Considering the same project as in Example 4-1, show whether it is justified using the AW method.

Solution

	Annual Worth	
	Outflows	**Inflows**
Annual revenue:		\$5,310
Annual expenses	−\$3,000	
CR cost[a] = (\$10,000 − \$2,000)(*A/P*, 10%, 5) + \$2,000(10%)	− 2,310	
Total	−\$5,310	\$5,310
Total AW		\$ 0

[a]Uses Equation 4-5.

Since total AW = \$0, the project earns exactly 10% and is thus minimally acceptable. Of course, nonmonetary considerations would probably sway the decision one way or another.

EXAMPLE 4-5

By using the annual worth method, determine whether the equipment described in Example 4-2 should be recommended.

Solution
The AW method applied to Example 4-2 yields the following:

$$\begin{aligned}\text{AW}(20\%) &= \overbrace{\$8000}^{\underline{R}-\underline{E}} - \overbrace{[\$25{,}000(A/P, 20\%, 5) - \$5000(A/F, 20\%, 5)]}^{\text{CR amount (Eq. 4-3)}} \\ &= \$8000 - (\$8359.49 - \$671.90) \\ &= \$312.41\end{aligned}$$

Because its AW is positive, the equipment more than pays for itself over a period of 5 years while earning a 20% return per year on the unrecovered investment. In fact, the annual equivalent "surplus" is $312.41, which means that the equipment provided more than a 20% return on beginning-of-year unrecovered investment. This piece of equipment should be recommended as an attractive investment opportunity.

EXAMPLE 4-6

An investment company is considering building a 25-unit apartment complex in a growing town. Because of the long-term growth potential of the town, it is felt that the company could average 90% of full rent for the whole complex each year. If the following items are reasonably accurate predictions, what is the minimum monthly rent that should be charged if a 12% MARR is desired? Use the annual worth method.

Land investment	$50,000
Building investment	$225,000
Study period, N	20 years
Rent per unit per month	?
Upkeep per unit per month	$35
Property taxes and insurance per year	10% of total investment

Solution
The procedure for solving this problem is first to determine the equivalent annual worth of all costs at the MARR of 12% per year. To earn exactly 12% on this project, the annual rental income, adjusted for 90% occupancy, must equal the AW of costs:

$$\begin{aligned}\text{Initial investment} &= \$50{,}000 + \$225{,}000 = \$275{,}000\\ \text{Taxes and insurance/year} &= 0.1(\$275{,}000) = \$27{,}500\\ \text{Upkeep/year} &= \$35(12 \times 25)(0.9) = \$9{,}450\\ \text{Capital recovery cost/year} &= \$275{,}000(A/P, 12\%, 20)\\ &\quad -\$50{,}000(A/F, 12\%, 20) = \$36{,}123\end{aligned}$$

(We assume that investment in land is recovered at the end of year 20 and that annual upkeep is directly proportional to the occupancy rate.)

$$\text{Equivalent annual worth of costs} = \$27{,}500 + \$9{,}450 + \$36{,}123 = \$73{,}073$$

Minimum *annual* rental required = \$73,073 and the monthly rental amount, $\hat{R}$, is

$$\hat{R} = \frac{\$73{,}073}{(12 \times 25)(0.9)} = \$270.64$$

The annual worth method is sometimes called the annual cost (AC) method when only costs are involved. The goal (when the do-nothing alternative is not an option) is to select the least negative AC when several alternatives are being compared.

Many decision makers prefer the annual worth method because it is relatively easy to interpret when one is accustomed to working with annual income statements and cash flow summaries.

4.5 THE FUTURE WORTH METHOD

Because a primary objective of all time value of money methods is to maximize the future wealth of the owners of a firm, the future worth (FW) criterion has become increasingly popular in recent years. With this method, the future worth of an alternative can be calculated in view of the MARR and compared with the do-nothing option. If FW $\geq$ 0, the alternative would be recommended.

The future worth method is exactly comparable to the present worth method except that all cash inflows and outflows are compounded forward to a reference point in time called the future.

EXAMPLE 4-7

Evaluate the future worth of the equipment described in Example 4-2. Show the relationship among FW, PW, and AW for this example.

Solution

$$FW = -\$25{,}000(F/P, 20\%, 5) + \$8{,}000(F/A, 20\%, 5) + \$5{,}000$$
$$= \$2{,}324.80 > 0$$

Again, the equipment is shown to be a good investment. The FW is a larger number compared to the corresponding AW and PW, and it is equivalent:

$$PW = \$2{,}324.80(P/F, 20\%, 5) = \$934.29$$
$$AW = \$2{,}324.80(A/F, 20\%, 5) = \$312.41$$

These results were obtained in Examples 4-2 and 4-5, respectively.

To this point the PW, AW, and FW methods have utilized a known and constant MARR over the study period. Each method produces a measure of merit expressed in dollars and is equivalent to the other two. Another group of methods is now considered that produces an interest rate as its measure of merit. The two methods are internal rate of return (IRR) and external rate of return (ERR).

4.6 THE INTERNAL RATE OF RETURN METHOD

The internal rate of return (IRR) method is the most widely used rate of return method for performing engineering economic analyses. It commonly is called by several other names, such as investor's method, discounted cash flow method, and profitability index.

This method solves for the interest rate that equates the equivalent worth of an alternative's cash inflows (receipts or savings) to the equivalent worth of cash outflows (expenditures, including investments). Equivalent worth may be computed with any of the three methods discussed above. The resultant interest rate is termed the "internal rate of return" (IRR). For a single alternative, the IRR is not positive unless both receipts and expenses are present in the cash flow pattern, and the sum of receipts exceeds the sum of all cash outflows.

By using a present worth formulation, the IRR is the $i'\%$* at which

$$\sum_{k=0}^{N} R_k(P/F, i'\%, k) = \sum_{k=0}^{N} E_k(P/F, i'\%, k) \qquad (4\text{-}6)$$

*i' is often used in place of i to mean the interest rate that is to be determined.

where R_k = net receipts or savings for the kth year
E_k = net expenditures including investments for the kth year
N = project life (or study period)

Once i' has been calculated, it is then compared with the MARR to assess whether the alternative in question is acceptable. If $i' \geq$ MARR the alternative is acceptable; otherwise, it is not.

A popular variation of Equation 4-6 for computing the IRR for an alternative is to determine the i' at which its *net* present worth is zero. In equation form, the IRR is the value of i' at which

$$\sum_{k=0}^{N} R_k(P/F, i'\%, k) - \sum_{k=0}^{N} E_k(P/F, i'\%, k) = 0 \tag{4-7}$$

For an alternative with a single investment at the present time ($k = 0$) followed by a series of positive cash flows over N, a graph of net present worth versus the interest rate typically has the general form shown in Figure 4-5. The point at which PW = 0 in Figure 4-5 defines i', which is the project's internal rate of return.

The value of i' can also be determined as the interest rate at which FW = 0 or AW = 0. For example, by setting net FW equal to zero, Equation 4-8 would result:

$$\sum_{k=0}^{N} R_k(F/P, i'\%, N-k) - \sum_{k=0}^{N} E_k(F/P, i'\%, N-k) = 0 \tag{4-8}$$

Another way to interpret the IRR is through an unrecovered investment balance diagram. Figure 4-6 shows how much of the original investment in an

FIGURE 4-5

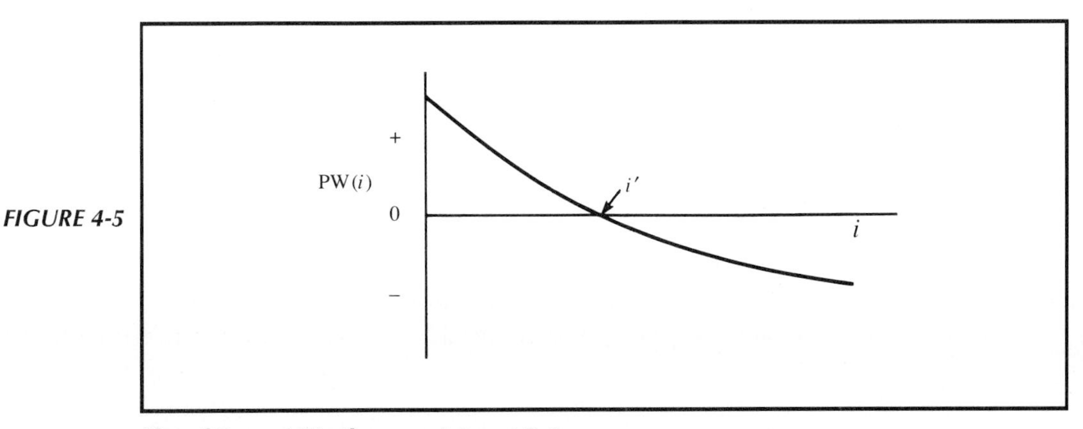

Plot of Present Worth versus Interest Rate

FIGURE 4-6

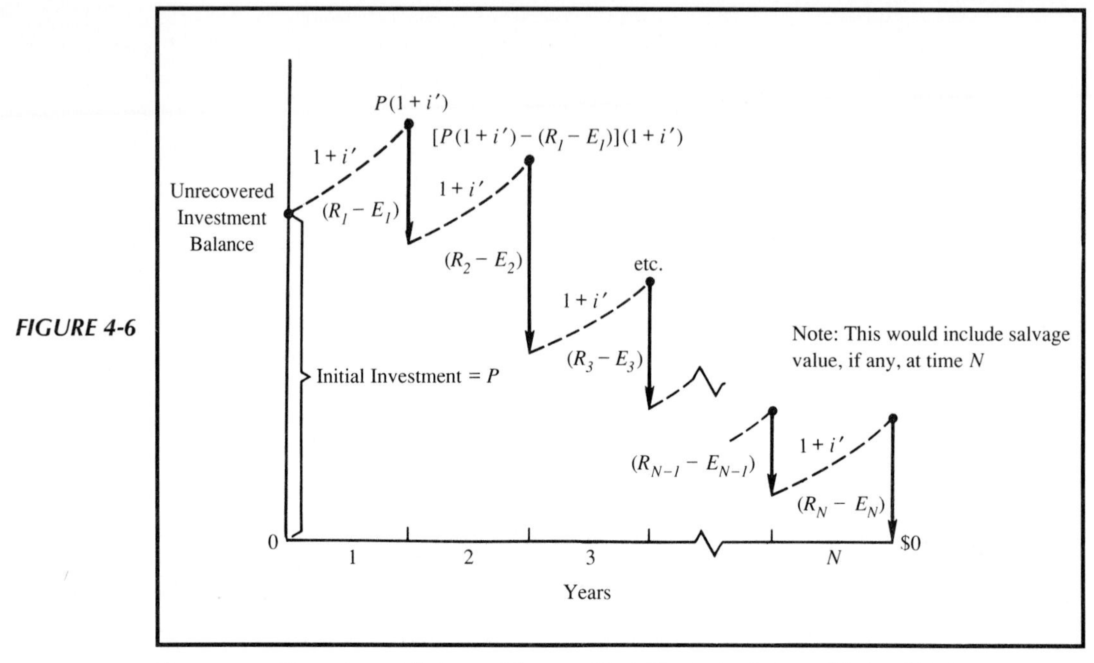

Unrecovered Investment Balance Diagram That Illustrates Internal Rate of Return

alternative is still to be recovered as a function of time. The downward arrows in Figure 4-6 represent annual returns, $(R_k - E_k)$ for $1 \leq k \leq N$, against the unrecovered investment and the dashed lines indicate the opportunity cost of interest, or profit, on the beginning-of-year investment balance. The IRR is that value of i' in Figure 4-6 that causes the unrecovered investment balance to exactly equal 0 at the end of the study period (year N) and thus represents the *internal* earning rate of a project. It is important to notice that i' is calculated on the *beginning-of-year unrecovered* investment through the life of a project rather than on the total initial investment.

The method of solving Equations 4-6 to 4-8 normally involves trial-and-error calculations until the $i'\%$ is converged upon or can be interpolated. Example 4-8 presents a typical solution using the common convention of "+" signs for cash inflows and "−" signs for cash outflows.

EXAMPLE 4-8

An investment of $10,000 can be made in a project that will produce a uniform annual revenue of $5,310 for 5 years and then have a salvage value of $2,000. Annual expenses will be $3,000 each year for operation and maintenance costs. The company is willing to accept any project that will earn at least 10% per year, before income taxes, on all invested capital. Determine whether it is acceptable by using the IRR method.

Solution

By writing an equation for net present worth and setting it equal to zero, we can determine the IRR.

$$0 = -\$10{,}000 + (\$5{,}310 - \$3{,}000)(P/A, i'\%, 5) + \$2{,}000(P/F, i'\%, 5);\ i'\% = ?$$

If we did not already know the answer from Example 4-1 ($i' = 10\%$), we would probably try a relatively low i', such as 5%, and a relatively high i', such as 25%.

$$\begin{aligned} \text{At } i' = 5\%: &\ -\$10{,}000 + \$2{,}310(4.3295) \\ &+ \$2{,}000(0.7835) = +\$1{,}568 \\ \text{At } i' = 25\%: &\ -\$10{,}000 + \$2{,}310(2.6893) \\ &+ \$2{,}000(0.3277) = -\$3{,}132 \end{aligned}$$

Since we have both a positive and a negative PW, the answer is bracketed. Linear interpolation can be used to find an approximation of the unknown $i'\%$ as shown in Figure 4-7. (The dashed curve in Figure 4-7 is what we are linearly approximating.) The answer, $i'\%$, can be obtained graphically to be approximately 12.0%, which is the interest rate at which the PW = \$0.

Linear interpolation for the answer, $i'\%$, can be accomplished by using the similar triangles dashed in Figure 4-7.

$$\frac{25\% - 5\%}{\$1{,}568 - (-\$3{,}132)} = \frac{i'\% - 5\%}{\$1{,}568 - \$0}$$

FIGURE 4-7

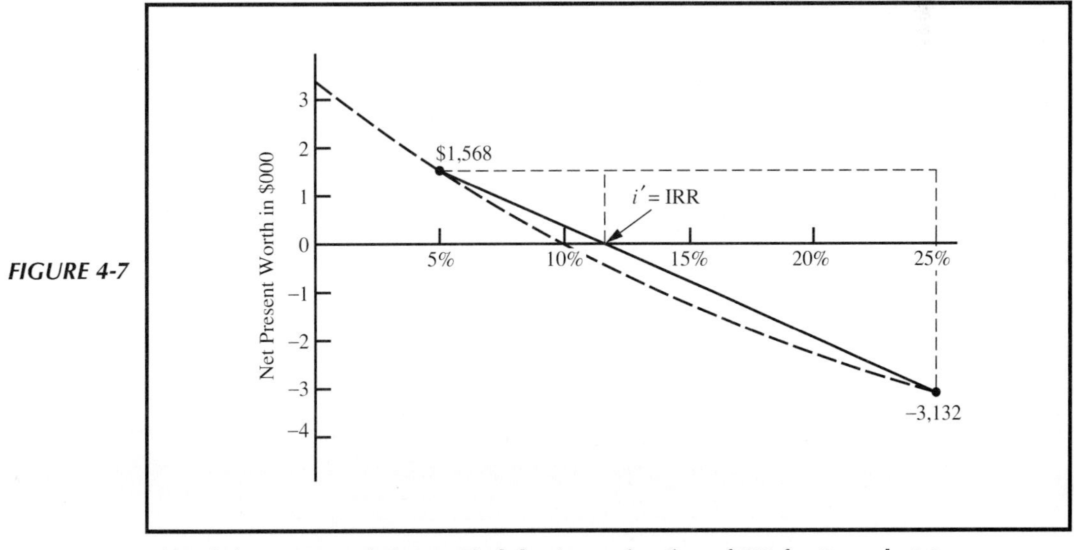

Use of Linear Interpolation to Find the Approximation of IRR for Example 4-8

or

$$i'\% = 5\% + \frac{\$1{,}568}{\$1{,}568 - (-\$3{,}132)}(25\% - 5\%)$$
$$= 5\% + 6.7\% = 11.7\%$$

The approximate solution above illustrates the trial-and-error process, together with linear interpolation. The error in this answer is due to nonlinearity of the PW function and would be less if the range of interest rates used in the interpolation had been smaller.

From the results of Examples 4-1 and 4-4, we already know that the project is minimally acceptable and that $i' =$ MARR $= 10\%$. We can confirm this by trying $i = 10\%$ in the PW equation, as follows:

$$\text{At } i = 10\%: -\$10{,}000 + (\$5{,}310 - \$3{,}000)(P/A, 10\%, 5) + \$2{,}000(P/F, 10\%, 5) = 0$$

EXAMPLE 4-9

A piece of new equipment has been proposed by engineers to increase the productivity of a certain manual welding operation. The initial investment (first cost) is \$25,000, and the equipment will have a salvage value of \$5,000 at the end of its expected life of 5 years. Increased productivity attributable to the equipment will amount to \$8,000/year after extra operating costs have been subtracted from the value of the additional production. A cash flow diagram for this equipment is given in Figure 4-2. Evaluate the internal rate of return of the proposed equipment. Is the investment a good one? Recall that the MARR is 20%.

Solution

By utilizing Equation 4-7, the following expression is obtained:

$$\$8{,}000(P/A, i'\%, 5) + \$5{,}000(P/F, i'\%, 5) - \$25{,}000 = 0 \qquad (4\text{-}9)$$

To initiate solving Equation 4-9 by trial and error, Table 4-3 is helpful.

TABLE 4–3 Computation of Selected PW (*i*) in Example 4–9

i'	PW(i'%)
0.00	\$8,000(5) + \$5,000(1) − \$25,000 = \$20,000
0.10	8,000(3.7908) + 5,000(0.6209) − 25,000 = 8,430.90
0.20	8,000(2.9906) + 5,000(0.4019) − 25,000 = 934.30
0.25	8,000(2.6893) + 5,000(0.3277) − 25,000 = −1,847.10

FIGURE 4-8

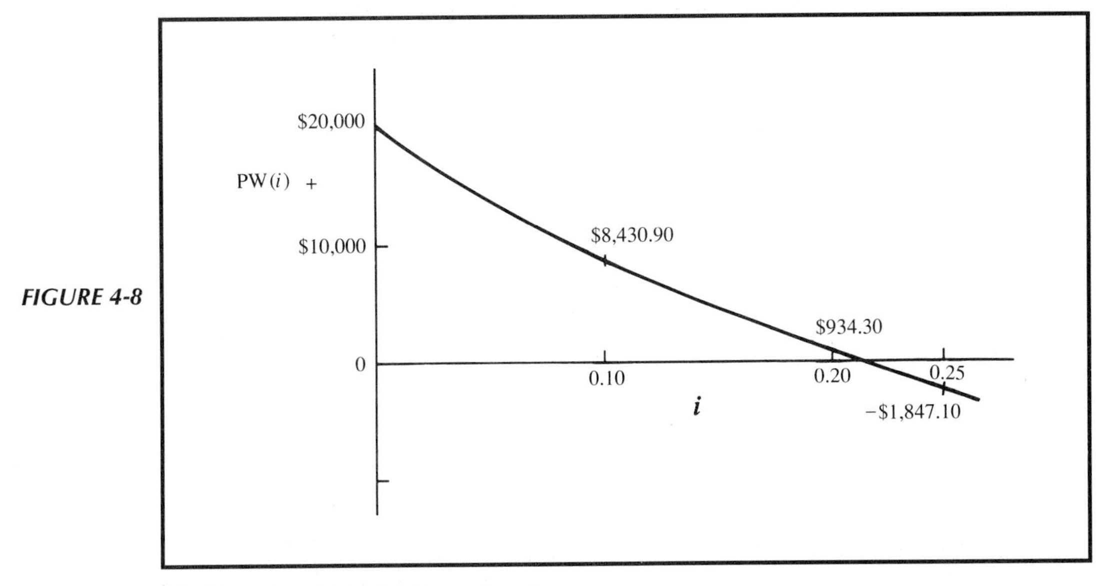

PW Plotted against i for Example 4-9

The PW computations in Table 4-3 are illustrated in Figure 4-8.

By inspection, the value of i' where PW $= 0$ is about 22%. For most applications an i' value of 22% is accurate enough since our major concern is whether i' exceeds the MARR. A more nearly exact value of i' can be determined by directly solving Equation 4-9 with repeated trial-and-error calculations ($i' = 21.577\%$) or by using the computer program in Appendix E. Clearly, this is economically attractive because $21.577\% > 20\%$.

A final point needs to be illustrated for Example 4-9. The unrecovered investment balance diagram is provided in Figure 4-9, and the reader should notice that $i' = 21.577\%$ is a rate of return calculated on the beginning-of-year unrecovered investment. The internal rate of return is *not* an average return each year based on the total investment of $25,000.

A rather common application of the internal rate of return method is in so-called *installment financing* types of problems. These problems are associated with financing arrangements for purchasing merchandise "on time." An interest, or finance, charge is typically based on the total amount of money borrowed and is paid by the borrower at the beginning of the loan instead of on the unpaid loan balance as illustrated by Figure 4-9. Such a finance charge is, of course, not in accord with the true definition of interest. To determine the true interest rate being charged in such cases, the IRR method is frequently employed. Examples 4-10 and 4-11 are representative installment financing problems.

FIGURE 4-9

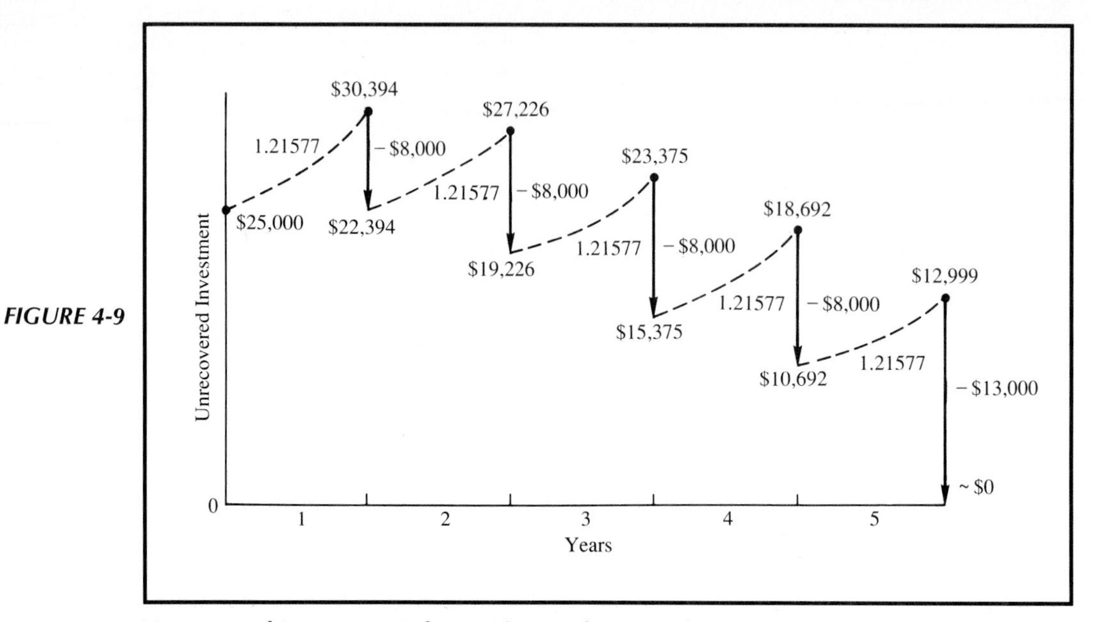

Unrecovered Investment Balance Diagram for Example 4-9

EXAMPLE 4-10

The Fly-by-Night finance company advertises a "bargain 6% plan" for financing the purchase of automobiles. To the amount of the loan being financed, 6% is added for each year money is owed. This total is then divided by the number of months over which the payments are to be made, and the result is the amount of the monthly payments. For example, a woman purchases a $10,000 automobile under this plan and makes an initial cash payment of $2,500. She wishes to pay the balance in 24 monthly payments.

	Purchase price	= $10,000
−	Initial payment	= 2,500
=	Balance due, (P)	= 7,500
+	6% finance charge = 0.06 × 2 years × $7,500	= 900
=	Total to be paid	= 8,400
∴	Monthly payments = $8,400/24	= $ 350

What rate of interest does she actually pay?

Solution

Because there are to be 24 payments of $350 each, made at the end of each month, these constitute an annuity at some unknown rate of interest, $i'\%$, that should be computed only upon the unpaid balance. Therefore,

$$P = A(P/A, i'\%, N)$$

$$\$7{,}500 = \$350(P/A, i'\%, N)$$

$$(P/A, i'\%, N) = \frac{\$7{,}500}{\$350} = 21.43$$

By examination of the interest tables for P/A factors at $N = 24$ that come closest to 21.43, one finds that (P/A, 3/4%, 24) = 21.8891 and (P/A, 1%, 24) = 21.2434.

Linear interpolation gives

$$i'\% = 1\% - \frac{21.43 - 21.2434}{21.8891 - 21.2434}(1\% - 3/4\%) = 0.93\%$$

Since payments are monthly, 0.93% is the interest rate being charged per month. The nominal rate paid on the borrowed money is 0.93%(12) = 11.16% compounded monthly. This corresponds to an effective annual interest rate of $[(1 + 0.0093)^{12} - 1] \times 100 \cong 12\%$. Notice that prospective reinvestment of $350 each month by the borrower at 0.93% is assumed. What appeared at first to be a real bargain turns out to involve effective annual interest at recent typical rates for automobile loans.

EXAMPLE 4-11

A small company needs to borrow $160,000. The local (and only) banker makes this statement: "We can loan you $160,000 at a very favorable rate of 12% per year for a 5-year loan. However, to secure this loan you must agree to establish a checking account (with no interest) in which the *minimum* average balance is $32,000. In addition, your interest payments are due at the end of each year and the principal will be repaid in a lump-sum amount at the end of year 5." What is the true effective interest rate being charged?

Solution

The cash flow diagram from the banker's viewpoint appears in Figure 4-10. The interest rate (IRR) that establishes equivalence of positive and negative cash flows can easily be computed:

$$P_0 = F_5(P/F, i'\%, 5) + A(P/A, i'\%, 5)$$

$$\$128{,}000 = \$128{,}000(P/F, i'\%, 5) + \$19{,}200(P/A, i'\%, 5)$$

If we try $i' = 15\%$, we discover that $128,000 = $128,000. Therefore, the true effective interest rate is 15%.

FIGURE 4-10

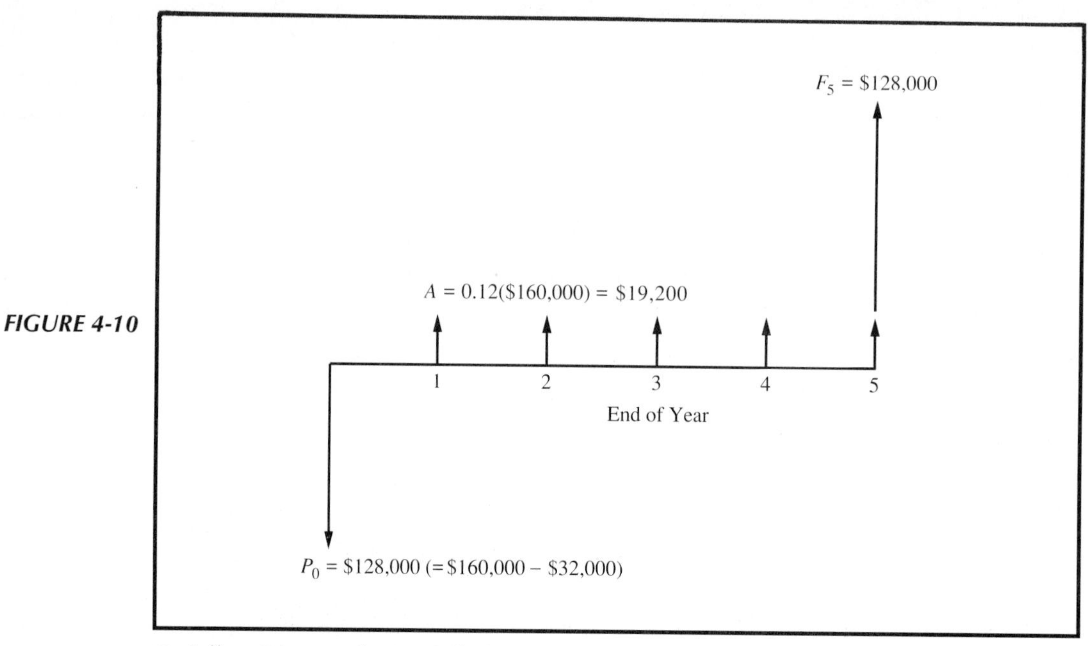

Cash flow Diagram for Example 4-11

4.6.1 Difficulties Associated with the IRR Method

The PW, AW, and FW methods utilize the assumption that net receipts less expenses (positive recovered funds) each time period are reinvested at the MARR during the study period, N. However, the IRR method is based on the assumption that recovered funds, if not consumed in each time period, are reinvested at $i'\%$ rather than at the MARR. This latter assumption may not mirror reality in some problems, thus making IRR an unacceptable method for analyzing engineering alternatives.* This can usually be remedied by the ERR method, which is described later.

Other difficulties with the IRR method include its computational intractability and the occurrence of multiple internal rates of return in some types of problems. A computational aid for determining the internal rate of return for a single alternative having both positive and negative cash flows or for cash flow differences between two alternatives is given in Appendix 4-A. A procedure

*See H. Bierman and S. Smidt, *The Capital Budgeting Decision: Economic Analysis of Investment Projects*, New York: Macmillan Publishing Company, 1984. The term *internal* rate of return means that the value of this measure depends only on the cash flows from an investment and not on any assumptions about reinvestment rates: "One does not need to know the reinvestment rates to compute the internal rate of return. However, one may need to know the reinvestment rates to compare alternatives." (p. 60).

for dealing with multiple rates of return is discussed and demonstrated in Appendix 4-B. Generally speaking, multiple rates are not too meaningful for decision-making purposes and another method of evaluation must be utilized.

Another possible drawback to the IRR method is that it must be carefully applied and interpreted in the analysis of two or more alternatives when only one of them is to be selected (i.e., mutually exclusive alternatives). This is discussed further in Chapter 5. The key advantage of the method is its widespread acceptance by industry, where various types of rates of return and ratios are routinely used in making project selections.

4.6.2 Selecting Trial Rates of Return When Using the IRR Method

Users of the IRR method are often perplexed as to how to select an initial (and perhaps subsequent) rate of return to reduce the number of trials required to obtain an acceptably accurate answer. The following is an intuitive approach when calculations are performed manually.

An approximate rate of return can be initially obtained by ignoring the time value of money and determining the average annual profit as a percentage of the average investment. For the project in Example 4-1, this can be done as follows:

Cash inflow:	
Annual receipts: $5,310 × 5	$26,550
Salvage value	2,000
Total	$28,550
Cash outflow:	
Annual disbursements: $3,000 × 5	−$15,000
Investment	− 10,000
Net Cash Inflow (Profit)	$ 3,550

$$\text{Average profit per year} = \frac{\$3{,}550}{5} = \$710$$

$$\text{Average investment} = \frac{\$10{,}000 + \$2{,}000}{2} = \$6{,}000$$

$$\frac{\text{Average profit per year}}{\text{Average investment}} = \frac{\$710}{\$6{,}000} = 11.8\%$$

Thus, one might start with a first trial rate of return of 12%. Substituting this into Equation 4-7 for finding the IRR gives a negative PW. It should be kept in mind that a lower interest rate will result in higher-valued present worth factors for both single sums and uniform series; also, if the positive terms need to become larger so that the present worth will be ≥ 0, the next trial rate should be lower.

It is recommended that the second trial rate be sufficiently different from the first trial rate so that the answer will be bracketed between a positive and a negative PW. One can then interpolate to find the IRR as demonstrated in Example 4-8.

4.7 THE EXTERNAL RATE OF RETURN METHOD

The reinvestment assumption of the IRR method noted in Section 4.6.1 may not be valid in an engineering economy study. For instance, if a firm's MARR is 20% per year and the IRR for a project is 42.4%, it may not be possible for the firm to reinvest net cash proceeds from the project at much more than 20%. This situation, coupled with the computational demands and possible multiple interest rates associated with the IRR method, has given rise to other rate of return methods, such as the external rate of return (ERR) method, that can remedy some of these weaknesses.

The ERR method takes into account the external interest rate (ϵ) at which net cash flows generated (or required) by a project over its life can be reinvested (or borrowed) outside the firm. If this external reinvestment rate happens to equal the project's IRR, then the ERR method produces results identical to those of the IRR method.

In general, all cash outflows are discounted to period 0 (the present) at ϵ% per compounding period while all cash inflows are compounded to period N at ϵ%. The external rate of return is then the interest rate that establishes equivalence between the two quantities. In equation form, the ERR is the i'% at which

$$\sum_{k=0}^{N} E_k(P/F, \epsilon\%, k)(F/P, i'\%, N) = \sum_{k=0}^{N} R_k(F/P, \epsilon\%, N-k) \qquad (4\text{-}10)$$

where R_k = excess of receipts over expenditures in period k
E_k = excess of expenditures over receipts in period k
N = project life or number of periods for the study
ϵ = external reinvestment rate per period

Graphically, we have the following:

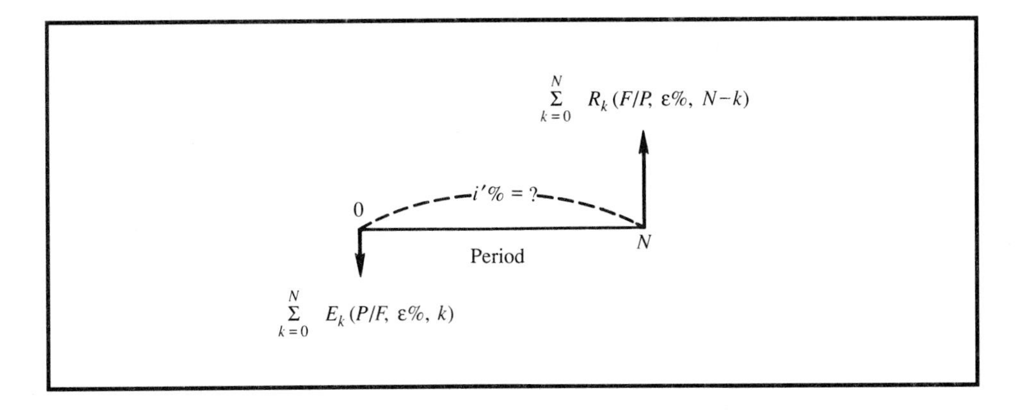

A project is acceptable when the $i'\%$ of the ERR method is greater than or equal to the firm's MARR.

The external rate of return method has two basic advantages over the IRR method:

1. It usually can be solved for directly rather than by trial and error.
2. It is not subject to the possibility of multiple rates of return. (*Note:* The multiple rate of return problem with the IRR method is discussed in Appendix 4-B.)

EXAMPLE 4-12

Referring to Example 4-9, suppose that ϵ = MARR = 20% per year. What is the alternative's external rate of return, and is the alternative acceptable?

Solution

By utilizing Equation 4-10, we have this relationship to solve for i':

$$\$25{,}000(F/P, i'\%, 5) = \$8{,}000(F/A, 20\%, 5) + \$5{,}000$$

$$(F/P, i'\%, 5) = \frac{\$64{,}532.80}{\$25{,}000} = 2.5813$$

$$i' = 20.88\%$$

Because $i' >$ MARR, the alternative is barely justified.

EXAMPLE 4-13

When ϵ = 15% and MARR = 20%, determine whether the project whose cash flow diagram appears in Figure 4-11 is acceptable.

FIGURE 4-11

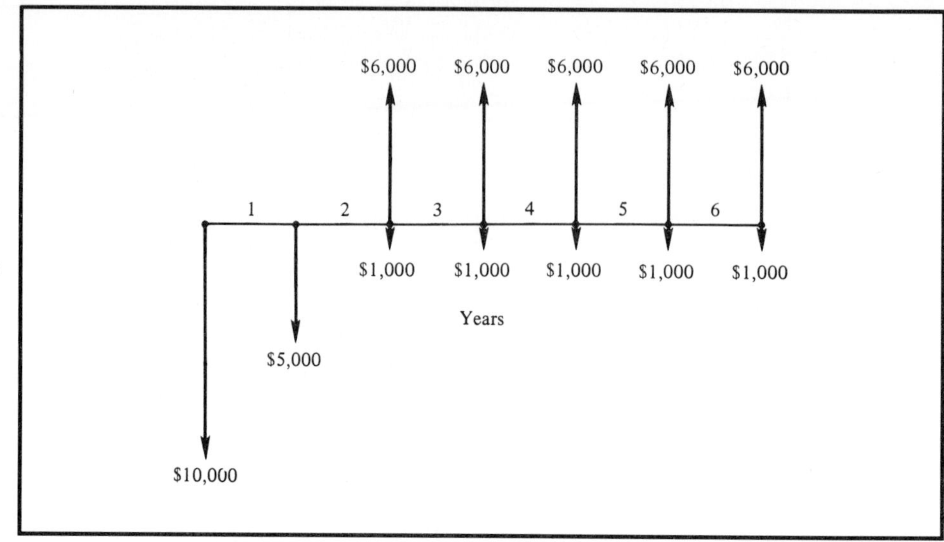

Cash Flow Diagram for Example 4-13

Solution

$$E_0 = \$10,000 \; (k = 0)$$
$$E_1 = \$5,000 \; (k = 1)$$
$$R_k = \$5,000 \text{ for } k = 2, 3, \ldots, 6$$
$$[\$10{,}000 + \$5{,}000(P/F, 15\%, 1)](F/P, i'\%, 6) = \$5{,}000(F/A, 15\%, 5); i' = 15.3\%$$

The i' is less than MARR = 20%; therefore, this project would be unacceptable according to the ERR method.

EXAMPLE 4-14

John Doe purchased for $8,900 a U.S. Treasury Bond having a face value of $10,000. The bond pays nominal interest of 8% on a quarterly basis (i.e., $200 every 3 months), and it matures in 8 years. John is very conservative and has thus decided to deposit his interest earnings from the bond in a bank account that pays 6% nominal interest compounded quarterly.

(a) What is the external rate of return on the bond?
(b) What is the internal rate of return on the bond?
(c) Why is there a difference in answers to (a) and (b)?

Solution

(a) The ERR can be computed with Equation 4-10 when $\epsilon = 1.5\%$ per quarter and $N = 32$ quarters:

$$\$8{,}900(F/P, i'\%, 32) = \$200(F/A, 1.5\%, 32) + \$10{,}000$$
$$(F/P, i'\%, 32) = 2.038, \text{ or } (1 + i')^{32} = 2.038$$

Thus,

$$i' = 2.25\%/\text{quarter}$$
$$= [(1.0225)^4 - 1] \times 100 = 9.31\%/\text{year}$$

(b) The IRR is determined with Equation 4-6 as follows:

$$\$200(P/A, i'\%, 32) + \$10{,}000(P/F, i'\%, 32) = \$8{,}900$$

By trial and error,

$$i' = 2.50\%/\text{quarter}$$
$$= [(1.025)^4 - 1] \times 100 = 10.38\%/\text{year}$$

(c) The difference between answers arises because cash proceeds from the bond are reinvested at 6% nominal interest in (a), and they are assumed to be reinvested at 10% nominal interest in (b) which does not represent John's true situation.

4.8 A COMPUTER PROGRAM FOR SELECTED METHODS

To assist with the solution of engineering economy problems by methods discussed in this and subsequent chapters, a computer program called BTAX is provided in Appendix E. The program is written in BASIC and is suitable for most personal computers. The results of running this computer program for Examples 4-2, 4-5, 4-7, 4-9, and 4-12 are shown below. Other capabilities of BTAX are described in Appendix E.

```
1988 REVISION - INCORPORATES RATE OF RETURN
METHODS
DO YOU WANT TO SEE INSTRUCTIONS FOR DATA
INPUT (Y/N) ? Y

** INSTRUCTIONS FOR ENTERING CASH FLOW DATA **
```

```
THIS PROGRAM GIVES THE USER THE OPTION OF
CALCULATING ANY OF THE FOLLOWING INDICATORS :
INTERNAL RATE OF RETURN, PRESENT WORTH, FUTURE
WORTH, ANNUAL WORTH, SIMPLE PAYBACK, DISCOUNTED
PAYBACK PERIOD, AND EXTERNAL RATE OF RETURN.

END OF PERIOD CASH FLOWS ARE ENTERED IN THE
FOLLOWING MANNER :

                  C, J1, J2

WHERE C = CASH FLOW FOR EACH YEAR FROM THE YEAR
J1 THROUGH J2.

ALL CALCULATIONS ARE BASED ON YEAR 0 BEING THE
PRESENT PERIOD (IF NEGATIVE YEARS ARE ENTERED,
THE PRESENT WORTH IS BASED ON PERIOD 0 THAT WAS
ENTERED BY THE USER.)

THE NUMBER OF DIFFERENT CASH FLOW VALUES ASKED
FOR IN THE PROGRAM IS THE NUMBER OF DIFFERENT
CASH FLOW SEQUENCES THAT YOU WILL ENTER, FOR
EXAMPLE :

CASH FLOW        PERIOD
---------        ------

10000            0
    0            1
10000            2
10000            3
10000            4

FOR THE ABOVE CASH FLOWS, THE NUMBER OF
DIFFERENT CASH FLOW SEQUENCES IS 3 AND THE
CASH FLOWS SHOULD BE ENTERED AS FOLLOWS :

                  10000,0,0
                      0,1,1
                  10000,2,4

ALL CASH FLOWS THAT OCCUR AT THE END OF A
PARTICULAR TIME PERIOD MUST BE ENTERED AS A NET
```

```
AMOUNT. FOR EXAMPLE, A -10000 AT THE END OF
YEAR 2 AND A +2000 AT THE END OF YEAR 2 WOULD BE
ENTERED AS -8000,2,2.

A NET CASH FLOW AMOUNT IS ENTERED IN THIS MANNER
FOR EACH TIME PERIOD IN THE LIFE OF THE
ALTERNATIVE BEING CONSIDERED.

e.g. A PRODUCT HAVING A $10000 EXPENDITURE AT
END OF YEAR 0 FOLLOWED BY SAVINGS OF $3000/YEAR
AT END OF YEAR 1 TO 6 AND A FINAL SALVAGE
VALUE OF $2000 AT END OF YEAR 6 WOULD BE
ENTERED AS FOLLOWS :

                    -10000,0,0
                      3000,1,5
                      5000,6,6

THE USER IS LIMITED TO END-OF-PERIOD CASH FLOW
CONVENTION BUT HAS THE OPTION OF SELECTING
DISCRETE OR CONTINUOUS COMPOUNDING. MULTIPLE
INTERNAL RATES OF RETURN CANNOT BE IDENTIFIED.

PLEASE ANSWER THE FOLLOWING QUESTIONS:
```

Example 4-2

```
HOW MANY DIFFERENT CASH FLOW VALUES ARE THERE ? 3
ENTER CASH FLOW, FIRST PERIOD, LAST PERIOD
(SEPARATE BY COMMA)
? -25000,0,0
? 8000,1,4
? 13000,5,5

YEAR     CASH FLOW

0        -25,000.00
1          8,000.00
2          8,000.00
3          8,000.00
4          8,000.00
5         13,000.00

WANT TO MAKE ANY CORRECTIONS (Y/N) ? N
```

```
SELECT A METHOD FROM THE FOLLOWING:
(ENTER PW/FW/AW/IRR/ERR/SPP/DPP/TABLE/ ?/END):PW
ENTER INTEREST RATE AS DECIMAL (e.g. ENTER 10%
                                        AS .1) = .2
DO YOU WANT CONTINUOUS (CON) OR DISCRETE (DIS)
                                    COMPOUNDING? DIS
→ PW = 934.2832
```

Example 4-5

```
WANT ANOTHER RUN (Y/N) ? Y
ENTER NEW/OLD/?/CHANGE ? OLD
(ENTER PW/FW/AW/IRR/ERR/SPP/DPP/TABLE/?/END):AW
ENTER INTEREST RATE AS DECIMAL (e.g. ENTER 10%
                                        AS .1) = .2
DO YOU WANT CONTINUOUS (CON) OR DISCRETE (DIS)
                                   COMPOUNDING ? DIS
→ AW = 312.405
```

Example 4-7

```
WANT ANOTHER RUN (Y/N) ? Y
ENTER NEW/OLD/?/CHANGE ? OLD
(ENTER PW/FW/AW/IRR/ERR/SPP/DPP/TABLE/?/END):FW
ENTER INTEREST RATE AS DECIMAL (e.g. ENTER 10%
                                        AS .1) = .2
DO YOU WANT CONTINUOUS (CON) OR DISCRETE (DIS)
                                   COMPOUNDING ? DIS
→ FW = 2324.793
```

Example 4-9

```
WANT ANOTHER RUN (Y/N) ? Y
ENTER NEW/OLD/?/CHANGE ? OLD
(ENTER PW/FW/AW/IRR/ERR/SPP/DPP/TABLE/ ?/END):IRR
DO YOU WANT DISCRETE (DIS) OR CONTINUOUS (CON)
                                 INTEREST RATE ? DIS
→ IRR IS BETWEEN 21.577 AND 21.578 %
```

Example 4-12

```
WANT ANOTHER RUN (Y/N) ? Y
ENTER NEW/OLD/?/CHANGE ? OLD
(ENTER PW/FW/AW/IRR/ERR/SPP/DPP/TABLE/ ?/END):ERR
ENTER EXTERNAL REINVESTMENT RATE E AS DECIMAL? .2
DO YOU WANT DISCRETE (DIS) OR CONTINUOUS (CON)
                                  INTEREST RATE? DIS
→ ERR IS BETWEEN 20.883 AND 20.88401 %
```

```
WANT ANOTHER RUN (Y/N) ? N
```

4.9

SUMMARY COMPARISON OF ENGINEERING ECONOMY STUDY METHODS

The reader may well have wondered why five different methods have been presented in this chapter when any one of them will give a valid answer for the reinvestment assumption inherent in that method. The best answer is that preferences differ among analysts, decision makers, and organizations. It must always be remembered that results of an engineering economy study should be communicated to decision makers in terms they can readily understand. Additionally, one method may be easier to use than another method because of the particular cash flow patterns involved.

Usually equivalence methods such as AW, PW, and FW are easier to use computationally than rate of return methods. However, many decision makers prefer to analyze study results in terms of rates of return. Business enterprises generally adopt one, or at most two, analysis techniques for broad categories of alternatives to be analyzed.

The three equivalence methods are directly related so that one amount can be obtained from the other by the following relations:

$$AW = PW(A/P, i\%, N) = FW(A/F, i\%, N) \quad (4\text{-}11)$$

$$PW = AW(P/A, i\%, N) = FW(P/F, i\%, N) \quad (4\text{-}12)$$

$$FW = AW(F/A, i\%, N) = PW(F/P, i\%, N) \quad (4\text{-}13)$$

All these methods involve the built-in assumption that funds can be reinvested at $i\%$ per compounding period, which is normally the MARR. This same assumption is often utilized in the ERR method. However, it is possible to consider that funds are reinvested at an external rate, $\epsilon \neq$ MARR. The IRR method, on the other hand, has the built-in assumption that all funds are reinvested at the particular IRR rate computed for the project generating those funds.*

In general, the numerical answer for a given project will be close for each of the rate of return methods. If the reinvestment rate is the same as the rate of return calculated, then IRR = ERR. This and other conditions relating answers by the two rate of return methods are summarized in the following table.

*This is explained further in H. Bierman and S. Smidt, *The Capital Budgeting Decision: Economic Analysis of Investment Projects*, New York: Macmillan, 1984, pp. 57–60.

Reinvestment Rate	ERR
= IRR	= IRR
> IRR	> IRR
< IRR	< IRR

4.10 THE PAYBACK (PAYOUT) PERIOD METHOD

All methods presented thus far reflect the *profitability* of a proposed alternative for a study period of N. The payback method, which is often called the simple payout method, mainly indicates a project's *liquidity* rather than its profitability. Historically, the payback method has been used as a measure of a project's riskiness since liquidity deals with how fast an investment can be recovered. A low-valued payback period is considered desirable. Quite simply, the payback method calculates the number of years required for positive cash flows to equal the total investment, I. Hence the simple payback period is the *smallest* value of θ ($\theta \leq$ N) for which this relationship is satisfied under end-of-year cash flow convention:

$$\sum_{k=1}^{\theta}(R_k - E_k) - I \geq 0 \tag{4-14}$$

The *simple* payback period, θ, ignores the time value of money and all cash flows that occur after θ. If this method is applied to the investment project in Example 4-8, the number of years for payout can be calculated to be

$$\sum_{k=1}^{4.33}(\$5{,}310 - \$3{,}000) - \$10{,}000 > 0$$

When $\theta = N$ (the last time period in the planning horizon), the salvage value is included in the determination of a payback period. As can be seen from Equation 4-14, the payback period does not indicate anything about project desirability except the speed with which the investment will be recovered. The payback period can produce misleading results, and it is definitely not recommended except as supplemental information in conjunction with one of the five methods previously discussed.

Sometimes the *discounted* payback period, $\theta'(\theta' \leq N)$, is calculated so that the time value of money is considered:

$$\sum_{k=1}^{\theta'} (R_k - E_k)(P/F, i\%, k) - I \geq 0 \tag{4-15}$$

where $i\%$ is the minimum attractive rate of return, I is the investment made at the present time ($k = 0$), and θ' is the smallest value that satisfies Equation 4-15. This variation of the simple payback period produces the *breakeven life* of a project in view of the time value of money. However, neither payback period calculation includes cash flows occurring after θ (or θ'). Another serious defect is that they do not take into consideration the economic life of physical assets. Thus, these methods will be misleading if one alternative that has a longer (less desirable) payout period than another produces a higher rate of return (or net present worth) on the invested capital.

The use of the payout period for making investment decisions should be avoided except as a measure of how quickly invested capital will be recovered, which is an indicator of project risk. For instance, the breakeven life of the investment described in Table 4-1 is 5 years (in column C), which corresponds to the discounted payback period at $i = 20\%$. Payback periods of 3 years or less are often desired in U.S. industry, so under this criterion the equipment in Example 4-2 would be regarded as too risky even though its profitability is acceptable.

EXAMPLE 4-15

Considering the alternative presented in Example 4-9, calculate the simple payback period and the discounted payback period. In view of a minimum acceptable payback period of 5 years, is this alternative economically attractive?

Solution

Payback period:

$$\left[\sum_{k=1}^{\theta=3.13} (\$8,000)_k\right] - \$25,000 \simeq 0$$

Thus $\theta = 3.13$ years and the alternative is acceptable. (With end-of-year cash flow convention, this value of θ would be rounded up to 4 years.) Notice that the $5,000 salvage value was not considered in Equation 4-14 to determine the simple payback period.

Discounted payback period:

$$\left[\sum_{k=1}^{\theta'=5} (\$8{,}000)_k(P/F, 20\%, k)\right] + \$5{,}000(P/F, 20\%, 5) - \$25{,}000 > 0$$

or

$$\$8{,}000(P/A, 20\%, 5) + \$5{,}000(P/F, 20\%, 5) - \$25{,}000 > 0$$

at the project's expected life of $N = 5$ years. The alternative is minimally acceptable with the discounted payback method since $\theta' = 5$ years. In this case, the salvage value was considered in determining the value of θ'. Observe that when $\theta' = 4$, Equation 4-15 is not satisfied.

4.11 AN EXAMPLE OF A PROPOSED INVESTMENT TO REDUCE COSTS

EXAMPLE 4-16

A manufacturer of jewelry is contemplating the installation of a system that will recover a larger portion of the fine particles of gold and platinum that result from the various manufacturing operations. At the present time a little over $45,000 worth of these metals is being lost per year, and it is anticipated that, because of the growth of the company, this amount will increase by $5,000 each year for the next 10 years (the study period). The proposed system, involving a network of exhaust ducts and separators, will recover approximately two-thirds of the gold and platinum that otherwise would be lost. The complete installation would cost $140,000. The best estimates for the operating costs of the system, obtained from operations of similar systems, are $10,000 per year for operating expense, $1,800 per year for maintenance and repairs, and 2% of the first cost annually for property taxes and insurance. The company would require the investment to be recovered with interest in 10 years. The MARR of the company, before taxes, is 15%. Should the recovery system be installed?

Solution

Such an investment is made to reduce some of the operating expenses, in this case the cost of the material used. Thus, the saving (income) to be obtained by making an investment is almost entirely within the control of the investors. The company knows exactly what expenses have been. If the efficiency of the proposed equipment is known, the only factors that should affect the saving are the variation of production, operation, and maintenance expenses of the proposed equipment. In most cases of this type, these items are known or may be predicted quite accurately. The company would have a good idea of how its volume might vary. Operation and maintenance expenses can usually be estimated accurately, especially if historical data are available on the proposed equipment.

Using the data given, we find that the present worth and internal rate of return calculations would be as shown in Table 4-4. In deciding whether or not the calculated internal rate of return of 16.8% is sufficient to justify investment, each factor that might contribute to the risk must be examined so that a measure of the total risk may be obtained. In this case it appears that the factors are quite well controlled or known. There is little reason to believe that much more risk would be involved than is present in all the normal operations of the company. Thus, the company can use its own experience as a basis of comparison. If the company is sound and its business quite stable, a return of 16.8% coupled with a simple payback period of 6 years should be satisfactory.

It may be seen that when capital is invested in a going concern in order to bring about reduction in costs, the risk is usually easier to assess and is often much less than when entirely new enterprises are involved. As a result, the rate of return required for such investments is often lower.

EXAMPLE 4-16A

Using the cash flow data of Example 4-16, determine the (**a**) IRR, (**b**) PW at 15%, (**c**) simple payback period, and (**d**) discounted payback period at 15% with the computer program (BTAX) in Appendix E.

Solution

```
PLEASE ANSWER THE FOLLOWING QUESTIONS:

HOW MANY DIFFERENT CASH FLOW VALUES ARE THERE? 11
ENTER CASH FLOW, FIRST PERIOD, LAST PERIOD
(SEPARATE BY COMMA)
? -140000,0,0
? 18733,1,1
? 22067,2,2
? 25400,3,3
? 28733,4,4
```

TABLE 4–4 Tabular Determination of Present Worth and Internal Rate of Return[a] for the Proposed System in Example 4–16

Year End, N	Investment	Recovery	Costs	Net Cash Flow	$(P/F, 15\%, N)$	Present Worth at $i = 15\%$	$(P/F, 20\%, N)$	Present Worth at $i = 20\%$
0	−$140,000	—	—	−$140,000	1.000	−$140,000	1.000	−$140,000
1	—	$33,333	−$14,600	18,733	0.8696	16,290	0.8333	15,610
2	—	36,667	− 14,600	22,067	0.7561	16,685	0.6944	15,323
3	—	40,000	− 14,600	25,400	0.6575	16,701	0.5787	14,699
4	—	43,333	− 14,600	28,733	0.5718	16,430	0.4823	13,858
5	—	46,667	− 14,600	32,067	0.4972	15,944	0.4019	12,888
6	—	50,000	− 14,600	35,400	0.4323	15,303	0.3349	11,855
7	—	53,333	− 14,600	38,733	0.3759	14,560	0.2791	10,810
8	—	56,667	− 14,600	42,067	0.3269	13,752	0.2326	9,785
9	—	60,000	− 14,600	45,400	0.2843	12,907	0.1938	8,799
10	—	63,333	− 14,600	48,733	0.2472	12,047	0.1615	7,870
Total				$197,333		$ 10,619		−$ 18,503

[a] IRR ≅ 15% + [$10,619/(10,619 + 18,503)] × (20% − 15%) ≅ 16.8%.

```
? 32067,5,5
? 35400,6,6
? 38733,7,7
? 42067,8,8
? 45400,9,9
? 48733,10,10

WANT TO MAKE ANY CORRECTIONS ON CASH FLOW
                                        (Y/N) ? N
```

```
    SELECT A METHOD FROM THE FOLLOWING:
(a) (ENTER PW/FW/AW/IRR/ERR/SPP/DPP/TABLE/
                                        ?/END):IRR
    DO YOU WANT DISCRETE (DIS) OR CONTINUOUS (CON)
                              INTEREST RATE? DIS
    → IRR IS BETWEEN 16.64 AND 16.641 %
```

```
    WANT ANOTHER RUN (Y/N) ? Y
    ENTER NEW/OLD/?/CHANGE ? OLD
(b) (ENTER PW/FW/AW/IRR/ERR/SPP/DPP/TABLE/
                                        ?/END):PW
    ENTER INTEREST RATE AS DECIMAL (e.g. ENTER 10%
                                  AS .1) = .15
    DO YOU WANT CONTINUOUS (CON) OR DISCRETE (DIS)
                                COMPOUNDING? DIS
    → PW = 10616.36
```

```
    WANT ANOTHER RUN (Y/N) ? Y
    ENTER NEW/OLD/?/CHANGE ? OLD
(c) (ENTER PW/FW/AW/IRR/ERR/SPP/DPP/TABLE/
                                        ?/END):SPP
    → SIMPLE PAYBACK PERIOD = YEAR 6
```

```
    WANT ANOTHER RUN (Y/N) ? Y
    ENTER NEW/OLD/?/CHANGE ? OLD
(d) (ENTER PW/FW/AW/IRR/ERR/SPP/DPP/TABLE/
                                        ?/END):DPP
    ENTER INTEREST RATE AS DECIMAL (e.g. ENTER 10%
                                  AS .1) = .15
    DO YOU WANT CONTINUOUS (CON) OR DISCRETE (DIS)
                                COMPOUNDING? DIS
    → DISCOUNTED PAYBACK PERIOD = YEAR 10
```

```
    WANT ANOTHER RUN (Y/N) ? N
```

4.12

A CASE STUDY OF A LARGE INDUSTRIAL INVESTMENT OPPORTUNITY

The following case study illustrates the before-tax cash flow analysis of a typical industrial plant expansion problem. It involves many types of cash flows and numerous considerations that can be readily expanded upon in later chapters. Particular attention is directed to variable, fixed, and sunk costs that are present in addition to expenditures that are *capitalized* (e.g., investments) versus those such as materials costs that are *expensed*.

EXAMPLE 4-17

I. Problem Statement

Product X-21, a food preservative, is manufactured in a facility with a nominal capacity of 180,000 pounds per year. The ultimate capacity of the facility can be increased, at some loss in cost efficiency, by process modifications and the use of extensive amounts of overtime, to 195,000 pounds per year. With the present marketing strategy, sales are expected to level off at about 190,000 pounds per year in 1990 and remain at that volume. Therefore, the existing capacity will be sufficient under the present marketing strategy.

However, an opportunity has become available to enter a new market for an X-21 product modified slightly to give it better warm weather stability. A German firm has a trade secret on the modifying process but is willing to enter into a nondisclosure agreement for a one-time fee of $75,000. A proposal has been submitted to management to expand the existing X-21 plant to permit the company to enter this new market. *Management has requested an economic analysis of this proposal and a recommendation.* Your assignment is to provide this recommendation.

If the buyer chooses the biweekly mortgage, a total savings of $113,590 will result over the life of the loan (a 48.5% reduction in total interest cost of the conventional monthly mortgage). Keep in mind that under either arrangement, approximately the same amount ($860) is being paid each month to the mortgage holder. The interest savings above become less dramatic when interest rates are lower, but biweekly mortgages still represent a source of considerable savings for fixed-rate loans that involve the same interest rate for monthly or biweekly mortgages.

II. General Background Information

A. Demand schedule: The demand schedule for X-21 is as follows:

Year	Total Demand (lb)	Incremental Sales from This Project (lb)
1989	175,000	0
1990	190,000	0
1991	220,000	30,000
1992	250,000	60,000
1993–2005	260,000	70,000

Both the modified and regular grades of X-21 will be sold at the same price of \$39.50 per pound. All demand over 190,000 pounds per year is attributable to the new market.

B. Existing manufacturing facility: The existing facility consists of two production lines, each with an ultimate capacity of 97,500 pounds per year. A new line of the same capacity can be added in an existing building, which because of federal restrictions and its unique layout, cannot be used for any other purpose. That is, no cost need be allocated for the building space. The new line can be used for both regular X-21 and the modified X-21.

C. Expansion plan: The total installed cost of the new line is estimated at \$430,000, assuming project operation by January 1, 1991. Of this total, engineering expenditures of \$40,000 would be required in 1989 and all other capital costs would be incurred in 1990.

D. Project life: The company considers all projects of this type and degree of risk to have a 15-year life. Thus, commercial operation of the new production line would begin in early 1991 and terminate at the end of year 2005.

E. Inflation: The analysis will ignore the effects of inflation (in Chapter 9, this topic is considered).

F. Federal and state income taxes: The analysis will ignore all income taxes (income taxes are considered in Chapter 7).

G. Interest rate and financing: The minimum acceptable before-tax rate of return for projects with this degree of risk is 20% per year (effective). (The financing decision for large projects of this type is discussed in Chapter 13.)

H. Other profitable company operations: These will offset any short-term negative cash flows.

I. Expenditures: Those occurring throughout a year will be assumed for cash flow purposes to occur at the end of the year.

III. Cash Flow Data

A. Capitalizable project expenditures: A total of \$430,000 will be required to install the new production line. In 1989, \$40,000 is needed, and the remainder of \$390,000 would be expected in 1990. (Capitalizable expenditures are those expenditures that must be depreciated using the methods discussed in Chapter 6.)

B. Project expenditures chargeable to operations: Certain expenditures will result from the dismantlement and rearrangement of special equipment that will be required to fit the new line into the existing building. Estimated before-tax cost for this work is $48,600.

C. Related projects and work orders: Various types of auxiliary equipment will have to be replaced at 5-year intervals. Therefore, equipment replacement expenditures of $120,000 are planned for 1995 and 2000.

D. Net change in working capital: All firms require capital for day-to-day operations, including funds to support prepaid expenses, inventories, and bank balances. Such *working capital* has associated with it an opportunity cost that in many cases is a substantial item in an engineering economy study. Working capital caused by a project is normally treated as an asset whose first cost and salvage value are equal. Thus, the annual opportunity cost (i) of working capital in year k, WC_k, is $i\,WC_k$.

The change in working capital should be roughly proportional to the change in sales. Working capital in 1988 was approximately $200,000 when sales were 164,000 pounds. The following *net changes in working capital* are computed using this proportionality:

Year	Net Change from Previous Year
1989	$13,400
1990	18,300
1991	36,600
1992	36,600
1993	12,200
1994–2005	0

Only changes in 1991 and later are pertinent to the analysis.

E. Prepaid know-how: A single payment of $75,000 to the German firm is required before startup. This payment will be made at the end of 1990.

F. Pretax cash flows from commercial operation (incremental due to the new line):

1. ***Fixed costs:*** Fixed costs for supervisory salaries, general plant overhead, insurance, and property taxes will increase by $750,000 per year if this project is implemented. This increase will remain constant over the 15-year life of the project.
2. ***Variable costs:*** These costs consist primarily of raw materials costs, direct labor, utilities, and the variable portion of maintenance costs. Variable costs are expected to be $9.18 per pound in 1991. As operating experience is gained, yields should increase, resulting in variable costs of $8.83 per pound in 1992 and $7.56 per pound in 1993 and thereafter.

3. ***General and administrative (G and A) costs:*** G and A costs are budgeted for $900,000 in 1991 when the product is first introduced. After 1991, G and A costs are budgeted for $1,100,000 per year. This amounts to about 40% of sales revenue after sales have leveled out.

The following table summarizes the annual before-tax cash flows in years 1991-2005 attributable to this proposed project.

Year	(1) Fixed Cost (thousands)	(2) Variable Cost (thousands)	(3) G and A Costs (thousands)	Sales Revenues (thousands)	Sales Revenue Less Costs in Columns 1–3 (thousands)
1991	$750	$275.4	$ 900	$1,185	−$740.4
1992	750	529.8	1,100	2,370	− 9.8
1993–2005	750	529.2	1,100	2,765	385.8

G. Introductory costs: Introductory costs will be relatively small. The only significant introductory cost is training personnel in modifications to the process and technical assistance during the startup. The total introductory cost in 1990 is estimated at $8,500.

H. Sales Revenues: During 1991, sales revenues are expected to be $39.50/pound × 30,000 pounds/year = $1,185,000. They level out to $39.50 × 70,000 pounds/yr = $2,765,000 from 1993 through 2005.

IV. Analysis Results and Recommendations

A cash flow worksheet for this example is given in Table 4-5. From the before-tax cash flows in column 9, the project's internal rate of return can be calculated based on Equation 4-7:

$$0 = -\$40,000 - \$522,100(P/F, i'\%, 1) - \$777,000(P/F, i'\%, 2) - \cdots + \$471,200(P/F, i'\%, 16)$$

By trial and error, $i' = 18.3\% <$ MARR. Because this determination is quite tedious, it is suggested that the computer program in Appendix E be employed to verify the project's IRR.

The before-tax MARR is 20% and the present worth at the end of 1989 is −$87,137. This may also be confirmed by utilizing BTAX in Appendix E. Thus, the IRR and PW measures of profitability indicate that the company should *not* enter the new market for a modified X-21 product. Even if these criteria had signaled a favorable project (e.g., IRR ≥ 20%), a detailed analysis of uncertainty and nonmonetary considerations should be performed before a decision is made. These topics are addressed in Chapters 10 and 11, respectively.

TABLE 4–5 Cash Flow Worksheet for Example 4–17

	(1)	(2)	(3)	(4)	(5)	(6)	(7)	(8)	(9)
End of Year	Capitalizable Project Expenditures	Expenditures Chargeable to Operations	Related Projects and Work Orders	Net Change in Working Capital	Prepaid Know-How	Total Investment Cash Flow: Cols. 1 + 2 + 3 + 4 + 5	Before-Tax Cash Flows During Commercial Operation	Introductory Costs	Before-Tax Cash Flows for the Entire Project
1989	−$ 40,000					−$ 40,000			−$ 40,000
1990	− 390,000	−$48,600			−$75,000	− 513,600		−$8,500	− 522,100
1991				−$36,600		− 36,600	− 740,400		− 777,000
1992				− 36,600		− 36,600	− 9,800		− 46,400
1993				− 12,200		− 12,200	385,800		$373,600
1994							385,800		$385,800
1995			−$120,000			− 120,000	385,800		$265,800
1996							385,800		$385,800
1997							385,800		$385,800
1998							385,800		$385,800
1999							385,800		$385,800
2000			− 120,000			− 120,000	385,800		$265,800
2001							385,800		$385,800
2002							385,800		$385,800
2003							385,800		$385,800
2004							385,800		$385,800
2005				+ 85,400[a]		+ 85,400[a]	385,800		$471,200

[a]This is return of working capital at the end of the project.

4.13

MONTHLY VERSUS BIWEEKLY MORTGAGES

In some areas of the country, biweekly mortgages are available and offer substantial savings in interest payments relative to conventional monthly mortgages. Example 4-18 examines both types of mortgages and verifies that big savings in interest are possible with a biweekly mortgage.

EXAMPLE 4-18

With a \$75,000, 30-year mortgage on a new home, a certain buyer is presented with the possibility of making mortgage payments every 2 weeks (26 times a year) for 17.5 years or making payments once each month (12 times a year) for 30 years. Each bimonthly payment is about half the monthly payment at an annual percentage rate, of APR, of 13.5%.

$$\text{Monthly Payment} = \$75{,}000(A/P, \frac{13.5\%}{12}\text{ per month, 360 payments}) = \$859$$

$$\begin{aligned}\text{Biweekly Payment} &= \$75{,}000(A/P, \frac{13.5\%}{26}\text{ every 2 weeks, 455 payments})\\ &= \$430\end{aligned}$$

With biweekly payments, note that 13.5%/26 = 0.51923% is charged every 2 weeks on the outstanding (unpaid) loan principal, and there are 17.5 years × 26 payments/yr = 455 payments of \$430 each.

How much interest is saved by this buyer if he elects to make biweekly mortgage payments?

Solution

The total cost of interest over the life of the biweekly mortgage is far less than that of the conventional monthly mortgage repayment arrangement because the buyer is repaying the loan principal more rapidly. For the \$75,000 loan in question, total interest payments under each type of mortgage are as follows:

	Monthly Payments $\frac{\$859}{\text{month}} \times$ 360 payments =	Biweekly Payments $\frac{\$430}{\text{biweekly}} \times$ 455 payments
Total Payments	\$309,240	\$195,650
Less Initial Loan	\$ 75,000	\$ 75,000
Total Interest Payments	\$234,240	\$120,650

If the buyer chooses the biweekly mortgage, a total savings of $113,590 will result over the life of the loan (a 48.5% reduction in total interest cost of the conventional monthly mortgage). Keep in mind that under either arrangement, approximately the same amount ($860) is being paid each month to the mortgage holder. The savings above become less dramatic when interest rates are lower, but biweekly mortgages still represent a source of considerable savings for fixed-rate loans that involve the same interest rate for monthly or biweekly mortgages.

Throughout this chapter we have examined these five basic methods for evaluating the financial *profitability* of a single project: present worth, annual worth, future worth, internal rate of return, and external rate of return. Two supplemental methods for assessing a project's *liquidity* were also presented: the simple payback period and the discounted payback period. Computational procedures, assumptions, and acceptance criteria for all methods were discussed and illustrated with examples.

4.14 PROBLEMS

Unless stated otherwise, discrete compounding of interest and end-of-period cash flows should be assumed in all problem exercises in the remainder of the book. The number in parentheses at the end of a problem refers to the chapter section most closely related to the problem.

4-1. Uncle Wilbur's trout ranch is now for sale for $40,000. Annual property taxes, maintenance, supplies, and so on are estimated to continue to be $3,000 per year. Revenues from the ranch are expected to be $10,000 next year and then decline by $500 per year thereafter through the tenth year. If you bought the ranch, you would plan to keep it only 5 years and at that time sell it for the value of the land, which is $15,000. If your desired annual rate of return is 12%, should you become a trout rancher? Use the present worth method. (4.3)

4-2. A company is considering constructing a plant to manufacture a proposed new product. The land costs $300,000, the building costs $600,000, the equipment costs $250,000, and $100,000 working capital is required. It is expected that the product will result in sales of $750,000 per year for 10 years, at which time the land can be sold for $400,000, the building for $350,000, the equipment for $50,000, and all of the working capital recovered. The annual out-of-pocket expenses for labor, materials, and all other items are estimated to total $475,000. If the company requires a minimum return of 25% on projects of comparable risk, determine if it should invest in the new product line. Use the present worth method. (4.3)

4-3

a. Determine the PW and AW of the following proposal when the MARR is 15%. (4.3, 4.4)

	Proposal A
First cost	$10,000
Expected life	5 years
Salvage value[a]	−$ 1,000
Annual receipts	8,000
Annual expenses	4,000

[a] A negative salvage value means there is a net cost to dispose of an asset.

b. Determine the IRR for proposal A. Is it acceptable? (4.6)

4-4.

a. Evaluate machine XYZ on the basis of the present worth method when the minimum attractive rate of return is 12%. Pertinent cost data are as follows: (4.3)

	Machine XYZ
First cost	$13,000
Useful life	15 years
Salvage value	$ 3,000
Annual operating costs	100
Overhaul—end of fifth year	200
Overhaul—end of tenth year	550

b. Determine the capital recovery cost of machine XYZ by all three formulas presented in the text. (4.4)

4-5.

a. A certain service can be performed satisfactorily by process *R*, which has a first cost of $8,000, an estimated life of 10 years, no salvage value, and annual net receipts (revenues − expenses) of $2,400. Assuming a minimum attractive rate of return of 18% before income taxes, find the annual worth of this process and specify whether you would recommend it.

b. A compressor that costs $2,500 has a 5-year useful life and a salvage value of $1,000 after 5 years. At nominal interest of 10%, compounded quarterly, what is the *annual* capital recovery cost of the compressor? (4.4)

4-6. You purchased a building five years ago for $100,000. Its annual maintenance cost has been $5,000 per year. At the end of 3 years, you have spent $9,000 on roof repairs. At the end of five years, you sell the building for $120,000. During the period of ownership, you rented the building for $10,000 per year paid at the *beginning* of each year. Use the annual worth method to evaluate this investment when your MARR is 12% per year. (4.4)

4-7. A friend of yours just purchased a $10,000 bond that was discounted to sell for $7,500. It is a 6% bond with interest payable annually that matures in 7 years. In exactly 3 years, your friend plans to sell the bond at a price that gives the buyer a 12% interest rate compounded annually. For how much will your friend sell the bond? (4.3)

4-8. On January 1, 1990 your brother bought a used car for $8,200 and he agreed to make a down payment of $1,500 and repay the balance in 36 equal payments with the first payment due February 1. The interest rate is 13.8%, compounded monthly. During the summer your brother made enough money so that he decided to repay the entire balance due on the car as of September 1. How much did he repay on September 1? (4.3)

4-9. Find the internal rate of return (IRR) in each of these situations: (4.6)

a.

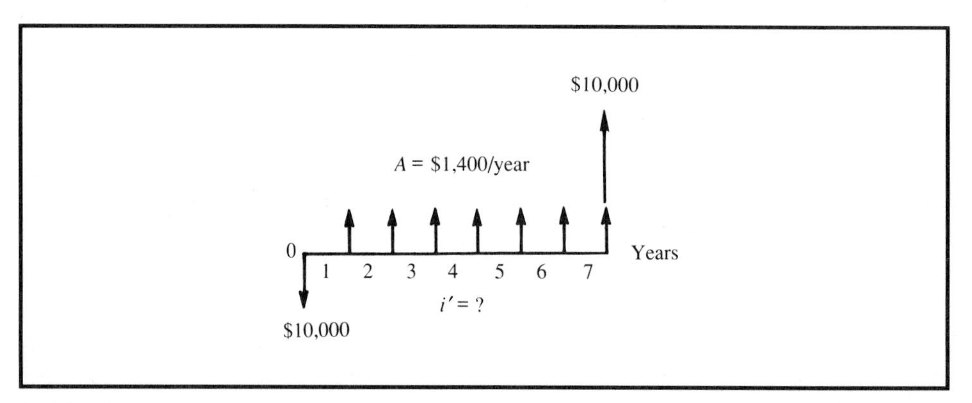

Problem 4-9a

b.

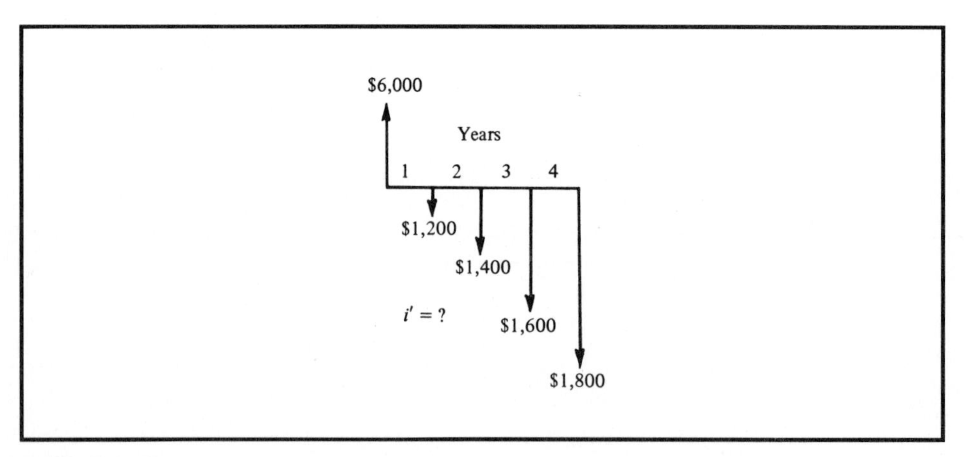

Problem 4-9b

c. You purchased a used car for $4,200. After you make a $1,000 down payment on the car, the salesperson looks in her *Interest Calculations Made Simple* handbook and announces: "The monthly payments will be $160 for the next 24 months and the first payment is due 1 month from now." (Draw a cash flow diagram.)

d.

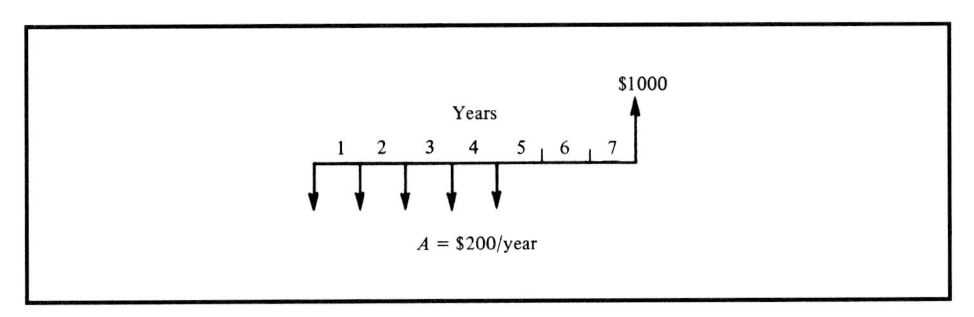

Problem 4-9d

4-10. Rework part (c) of Problem 4-9 by using the ERR method when ϵ = MARR = 8% per year. (4.7)

4-11. Plot the present worth of part (a) of Problem 4-9 as a function of the interest rate. The MARR is equal to 8%. Now plot the future worth of part (a) versus i. What similarities do you see? (4.6)

4-12. Draw an unrecovered investment balance diagram for part (a) of Problem 4-9 using the IRR determined in that problem. (4.6)

4-13. A small company purchased now for $23,000 will lose $1,200 each year the first 4 years. An additional $8,000 invested in the company during the fourth year will result in a profit of $5,500 each year from the fifth year through the fifteenth year. At the end of 15 years the company can be sold for $33,000.

a. Determine the internal rate of return (IRR). (4.6)
b. Determine the ERR when ϵ = 8%. (4.7)
c. Calculate the future worth if MARR = 12%. (4.5)

4-14.

a. Monthly amounts of $200 each are deposited into an account that earns 12% nominal interest, compounded quarterly. After 48 deposits of $200 each, what is the *future equivalent* worth of the account? State your assumptions. (4.5)
b. A "Christmas Plan" requires deposits of $10 per week for 52 weeks each year. The stated nominal interest rate is 20% compounded semiannually. What is the *present equivalent worth* of the plan (beginning of week 1)? (4.3)

4-15. A $20,000 ordinary life insurance policy for a 22-year-old female can be obtained for annual premiums of approximately $250. This type of policy (ordinary life) would pay a death benefit of $20,000 in exchange for annual premiums of $250 that are paid during the lifetime of the insured person. If the average life expectancy of a 22-year-old female is 77 years, what interest rate establishes equivalence between cash outflows and inflows for this type of insurance policy? Assume that all premiums are paid on a beginning-of-year basis and that the last premium is paid on the female's 76th birthday. (4.6)

4-16. On January 1, 1980, a government bond was purchased for $9,400. The face value of the bond is $10,000 and 8% nominal interest is paid on the face value four times each year. Thus, every 3 months the bondholder receives $200 as a cash payment. When the bond matures on January 1, 1990, $10,000 is paid to the bondholder. What is the effective rate of interest being earned in this situation? (4.6)

4-17. An individual approaches the Loan Shark Agency for $1,000 to be repaid in 24 monthly installments. The agency advertises an interest rate of 1 1/2% per month. They proceed to calculate his monthly payment in the following manner:

Amount requested	$1,000
Credit investigation	25
Credit risk insurance	5
Total	$1,030

$$\text{Interest: } (\$1030)(24)(0.015) = \$371$$
$$\text{Total owed: } \$1,030 + \$371 = \$1,401$$
$$\text{Payment: } \frac{\$1,401}{24} = \$58.50$$

What effective annual interest rate is the individual paying? (4.6)

4-18. In a certain foreign country, a man who wanted to borrow $10,000 for a 1-year period was informed he would only have to pay $2,000 in interest (i.e., a 20% interest rate). The lender stated that the total owed to him, $12,000, would be repaid at the rate of $1,000 per month for 12 months. What was the true effective interest being charged in this transaction? Why is it greater than the apparent rate of 20%? (4.6)

4-19. Suppose that you borrow $500 from the Easy Credit Company with the agreement to repay it over a 3-year period. Their stated interest rate is 9% per year. They show you the following items in determining the monthly payment:

Principal	$500
Total interest: 0.09(3 years)($500)	135
Loan application fee	15

They ask you to pay the interest immediately, so you leave with $365 in your pocket. Your monthly payment is calculated as follows:

$$\frac{\$500 + \$15}{36} = \$14.30/\text{month}$$

What is the effective interest rate per year? (4.6)

4-20. A zero-coupon certificate involves payment of a fixed sum of money now with future lump-sum withdrawal of an accumulated amount. Earned interest is not paid out periodically, but instead compounds to become the major component of the accumulated amount paid when the zero-coupon certificate matures. Consider a certain zero-coupon certificate that was issued on March 25, 1988, and matures on January 30, 2005. A person who purchases a certificate for $13,500 will receive a check for $54,000 when the certificate matures. What is the interest rate (yield) that will be earned on this certificate? (4.6)

4-21. List the advantages and disadvantages of each of the five basic methods for performing engineering economy studies (PW, AW, FW, IRR, ERR). (4.3–4.7)

4-22. The Anirup Food Processing Company is presently using an outdated method for filling 25-pound sacks of dry dog food. To compensate for weighing inaccuracies inherent to this packaging method, the process engineer at the plant has estimated that each sack is overfilled by 1/8 pound on the average. A modern method of packaging is now available that would eliminate overfilling (and underfilling). The production quota for the plant is 300,000 sacks per year for the next 6 years, and a pound of dog food costs this plant $0.15 to produce. The present system has no salvage value and will last another 4 years, and the modern method has an estimated life of 4 years with a salvage value equal to 10% of its investment cost, I. It is also estimated that the present packaging operation will cost $2,100 per year more to maintain than the modern method. If the minimum attractive rate of return is 12% for this company, what amount, I, could be justified for the purchase of the modern packaging method? (*Hint:* Find I at which the present worth of savings from the modern method equals zero.) (4.3)

4-23. Evaluate the acceptability of the following project with all methods discussed in Chapter 4. Let MARR $=$ 15%, $\epsilon =$ 12%, maximum acceptable θ $=$ 5 years, and maximum acceptable θ' $=$ 6 years.

***Project*: R137-A**
***Title*: Syn-Tree Fabrication**
***Description*: Establish a production facility to manufacture synthetic palm trees for sale to resort areas in Alaska**
***Cash Flow Estimates*:**

Year	Amount (thousands)
0	−$1,500
1	200
2	400
3	450
4	450
5	600
6	900
7	1,100

4-24. Your boss has just presented you with a summary of projected costs and annual receipts (before taxes) for a new product line. He asks you to calculate the before-tax internal rate of return for this investment opportunity. What would you present to your boss and how would you explain the results of your analysis? (It is widely known that the boss likes to see graphs of present worth versus interest rate for this type of problem.) (4.6)

End of Year	Net Cash Flow
0	− $450,000
1	− 42,500
2	+ 92,800
3	+ 386,000
4	+ 614,600
5	− $202,200

4-25. Rework Problem 4-24 with the ERR method and an external reinvestment rate of 12%. Why are answers to the two problems different? (4.7)

4-26. In order to enter the market to produce a new toy for children, a manufacturer will have to make an immediate investment of $60,000 and additional investments of $5,000 at the end of 1 year and $3,000 more at the end of 2 years. Competing toys now are being produced by two large manufacturers. From a fairly extensive study of the market, it is believed that sufficient sales can be achieved to produce year-end, before-tax, net cash flows as follows:

Year	Cash Flow	Year	Cash Flow
1	−$10,000	6	$26,000
2	5,000	7	26,000
3	5,000	8	20,000
4	20,000	9	15,000
5	21,000	10	7,000

In addition, while it is believed that after 10 years the demand for the toy will no longer be sufficient to justify production, it is estimated that the physical assets would have a scrap value of about $8,000. If capital is worth not less than 12% before taxes, would you recommend undertaking the project? Make a recommendation with each of these methods: (a) future worth, and (b) discounted payback with a maximum acceptable payback of 5 years. (4.5, 4.10)

4-27. Your firm is currently paying $250 a month to a commercial garbage collection agency to haul waste paper to the city dump. The paper could be sold as waste paper if it were baled and strapped. A paper baler is available at the following conditions:

Purchase price = $6,500
Labor to operate baler = $3,500/year
Strapping material = $300/year
Life of baler = 30 years
Salvage value = $500
MARR = 10%/year

If it is estimated that 500 bales would be produced per year, what would the selling price per bale to a wastepaper dealer have to be to make this project acceptable? (4.4)

4-28. Joe Roe is considering establishing a company to produce impellers for water pumps. An investment of $100,000 will be required for the plant and equipment, and $15,000 will be required for working capital. It is expected that the property will last for 15 years, at which time only the working capital part of the investment can be recovered. It is estimated that sales will be $200,000 per year, and that operating expenses will be as follows:

Materials	$40,000 per year
Labor	$70,000 per year
Overhead	$10,000 + 10% of sales per year
Selling expense	$5,000 per year

Joe has a regular job paying $30,000 per year, but he will keep that job even if he establishes this company. If Joe expects to earn at least 15% on his capital, should this investment be made? Use the annual worth method. (4.4)

4-29. The equipment in one department of a company is operating at only 75% of capacity because of the fact that the painting booths are overloaded. By building and equipping a new painting shed on adjoining land that is owned by the company, at a total cost of $14,000 the output could be stepped up to 100% (4,000 units per year). It is estimated that the useful life of the new paint shed and equipment would be 10 years. The old paint booths could be utilized for 10 more years. To operate at full capacity would require employing 2 more machinists and 2 painters at monthly salaries of $1,200 each. Indirect labor costs would be 10% of direct labor costs. Operation and maintenance costs per month on the new equipment are estimated to be $50 per 1,000 units processed. The cost for materials is $50 per unit. Annual expenses for taxes and insurance on the new facilities would be 8% of the first cost. The sales department assures management that the full output of 4,000 units per year can be sold at the present selling price of $200 per unit. Capital earns an average of 20% before income taxes. What would you recommend? (4.4)

4-30. A machine that is not equipped with a brake "coasts" 30 seconds after the power is turned off upon completion of each workpiece, thus preventing removal of the work from the machine. The time per piece, exclusive of this stopping time, is 2 minutes. The machine is used to produce 40,000 pieces per year. The operator receives $6.50 per hour and the machine overhead rate is $4.00 per hour. How much could the company afford to pay for a brake that would reduce the stopping time to 3 seconds, if it would have a life of 5 years? Assume zero salvage value, capital worth 10%, and that repairs and maintenance on the brake would total not over $250 per year. (4.3)

4-31. A meat-packing company is considering producing a new product, the entire domestic supply of which is now manufactured by three large companies. A new plant would be required that would cost $200,000 and be built on land that now is owned by the company but not used. The plant would have an expected life of 20 years. Annual costs for labor would be $60,000 and for material, $110,000. These would provide for an annual output of 600,000 pounds. This would constitute 25% of the present domestic consumption and would be 80% of the plant capacity. Advertising and other overhead expenses would amount to $50,000 per year. Taxes and insurance would total 3% of the value of the plant. The product is a very stable one and at present is selling for 55 cents per pound. Over a period of years the price has varied by plus or minus 15%.

a. Would you recommend that the company go ahead on this project? Capital is available and earns not less than 15%.

b. What is the minimum selling price for the product that would justify the investment? (4.4)

4-32. The ABC Company recently started producing a new product, in addition to two others it has been producing for several years. One of the major parts of this new product now is being purchased from another company at a cost of $7.00 per unit. One of the officers of the ABC Company believes it would be advisable to purchase the required equipment, at a cost of $7,000, that would permit making this component. This equipment would have a capacity of 7,000 units per year and a useful life of at least 5 years. It could be installed in the existing plant provided a small storage shed were built at a cost of $2,000 to make the necessary floor space available. Material costs would be $1.10 per unit, and direct labor costs $2.40 per unit. Incremental overhead is 50% of direct labor cost. There would also have to be an added annual charge of 2% of the first cost of the storage shed to cover taxes and insurance. The company now is purchasing 4,000 of the parts per year.

a. If capital is worth 15%, should the part be purchased or made in the plant?

b. What volume would be required to justify purchasing the equipment? Use the annual worth method. (4.4)

4-33. A company has the opportunity to take over a redevelopment project in an industrial area of a city. No immediate investment is required, but it must raze the existing buildings over a 4-year period and at the end of the fourth year invest $2,400,000 for new construction. It will collect all revenues and pay all costs for a period of 10 years, at which time the entire project, and properties thereon, will revert to the city. The net cash flow is estimated to be as follows:

Year End	Net Cash Flow
1	$ 500,000
2	300,000
3	100,000
4	− 2,400,000
5	150,000
6	200,000
7	250,000
8	300,000
9	350,000
10	400,000

Tabulate present worth versus the interest rate and determine whether multiple IRRs exist. If so, use the ERR method to determine a rate of return when $\epsilon = 8\%$. (4.6, 4.7, App. 4B)

4-34. The prospective exploration for oil in the outer continental shelf by a small, independent drilling company has produced a rather curious pattern of cash flows, as follows:

End of Year	Net Cash Flow
0	−$ 520,000
1–10	+ 200,000
10	− 1,500,000

The $1,500,000 expense at the end of year 10 will be incurred by the company in dismantling the drilling rig.

a. Over the 10-year period, plot present worth versus the interest rate (i) in an attempt to discover whether multiple rates of return exist. (4.6)

b. Based on the projected net cash flows and results in part (a), what would you recommend regarding the pursuit of this project? Customarily, the company expects to earn at least 20% on invested capital before taxes. (App. 4B))

4-35. A certain project has net receipts equaling $1,000 now, has costs of $5,000 at the end of the first year, and then earns $6,000 at the end of the second year.

a. Show that multiple rates of return exist for this problem when using the IRR method (i' = 100%, 200%). (App. 4B)

b. If an external reinvestment rate of 10% is available, what is the ERR for this project? (4.7)

Appendix 4-A: Aids for Calculation of the Internal Rate of Return

The computations to determine the internal rate of return, IRR, can be rather laborious, particularly if the periodic cash flows do not follow some pattern to which tabled interest factors can be readily applied.

Figure 4-A-1 is a form that can be used to simplify the calculations if the analyst elects *not* to use the computer program in Appendix E. Net cash flows for each year of a project's life are entered in the column headed "CASH FLOW (0% Int. Rate)." These are the present worths at 0%. Each of these cash flows is then multiplied by the factor in the adjacent subcolumn labeled "Factor" [which are (P/F, i%, N) factors] and the result entered in the next subcolumn, headed "Present Worth." These are the present worths at a 10% interest rate. The calculations are repeated for the 20%, 40%, and 60% interest rates as needed and the columns added to obtain the total present worth of expenses (A) and the total present worth of receipts (B) for each trial interest rate. At each interest rate the total present worth of expenditures is divided by the total present worth of receipts and the result entered in the space designated "Ratio A/B."

The internal rate of return sought is the interest rate at which A equals B, or at which "ratio A/B" equals unity. If one of the interest rates used does not result in the unity ratio,

FIGURE 4-A-1

PROJECT ____________ Analyst ____________ Date ____________

	TIMING		CASH FLOW (0% Int. Rate)	10% Int. Rate		20% Int. Rate		40% Int. Rate		60% Int. Rate	
	Year	Period	Expenses	Factor	Present Worth	Factor	Present Worth	Factor	Present Worth	Factor	Present Worth
BEFORE		5th Yr		1.464		2.073		3.842		6.560	
		4th		1.333		1.728		2.744		4.100	
	1987	3rd	$60,000	1.210	$72,000	1.440	$86,400	1.960		2.560	
	1988	2nd	5,000	1.100	5,500	1.200	6,000	1.400		1.600	
	1989	1st	13,000	1.000	13,000	1.000	13,000	1.000		1.000	
	TOTALS (A)		78,000		90,500		105,400				
	Cal. Year	Period	Receipts	Factor	Present Worth	Factor	Present Worth	Factor	Present Worth	Factor	Present Worth
AFTER ZERO TIME	90	1st Yr	$18,000	0.909	$16,400	0.833	$15,100	0.714		0.624	
	91	2nd	22,000	.826	18,200	.694	15,300	.510		.300	
	92	3rd	23,000	.751	17,400	.579	13,400	.364		.244	
	93	4th	22,000	.683	15,100	.482	10,600	.260		.152	
	94	5th	18,000	.621	11,200	.402	7,200	.186		.095	
	95	6th	11,000	.565	6,200	.335	3,700	.133		.059	
	96	7th	5,000	.513	2,560	.279	1,400	.095		.037	
	97	8th	15,000	.466	7,000	.233	3,300	.067		.023	
		9th		.424		.194		.048		.015	
		10th		.385		.162		.035		.009	
		11th		.351		.135		.024		.006	
		12th		.318		.112		.018		.004	
		13th		.290		.094		.013		.002	
		14th		.263		.078		.009		.001	
		15th		.239		.065		.006			
		16th		.217		.054		.005			
		17th		.197		.045		.003			
		18th		.180		.038		.002			
		19th		.164		.031		.002			
		20th		.149		.026		.001			
		21st		.135		.022					
		22nd		.123		.018					
		23rd		.112		.015					
		24th		.102		.013					
		25th		.092		.011					
	TOTALS (B)		$134,000		$94,060		$70,000				
	RATIO A/B		0.58		0.96		1.51				

Tabular Determination of the Internal Rate of Return for a Proposed New Venture

we hope that the unity ratio will be bracketed so that the answer can be interpolated. Figure 4-A-2 is provided to help with interpolation.

To illustrate the use of these aids, a sample problem and the associated computations are shown in Table 4-A-1. For the sample problem, an initial investment of $60,000 is required, and an additional investment of $20,000 will be required at the end of the second year. The project would be terminated at the end of 10 years, at which time it is expected that a recovery of $20,000 will be obtained from the salvage value of the assets. The revenues and expenses are estimated to be as shown for the various years.

The net cash flow amounts in the right-hand column of Table 4-A-1 can be transferred to the "CASH FLOW (0% Int. Rate)" column of Figure 4-A-1. For computational simplicity, the "zero time" is taken to be the year of the last negative cash flow, which is the end of year 1989. Computations using Figure 4-A-1 result in ratios of present worth of expenses to present worth of receipts (called "Ratio A/B") of 0.58 for 0% interest, 0.96 for 10% interest, and 1.51 for 20% interest. Hence the interest rate (IRR), at which the ratio is 1.00, is bracketed between 10% and 20%. Linear interpolation gives the answer:

$$\text{IRR} = i' = 10\% + \frac{1.00 - 0.96}{1.51 - 0.96}(20\% - 10\%) = 10.7\%$$

Figure 4-A-2 is a chart for graphical interpolation. To obtain the answer for the example in Table 4-A-1 and Figure 4-A-1, note that the three plotted points form a curvilinear relationship. Such graphical interpolation, if performed on large, accurately scaled paper, can result in greater accuracy than linear interpolation.

Appendix 4-B: The Multiple Rate of Return Problem With the IRR Method

Whenever the IRR method is used and the cash flows reverse sign (from net outflow to net inflow or the opposite) more than once over the period of study, one should be alert to the rather remote possibility that either no interest rate or multiple interest rates may exist. Actually, the maximum number of possible rates of return in the $(-1, \infty)$ interval for any given project is equal to the number of cash flow reversals over time. As an example, consider the following project for which the IRR is desired.

EXAMPLE 4-B-1

Year	Net Cash Flow
0	$ 500
1	− 1,000
2	0
3	250
4	250
5	250

FIGURE 4-A-2

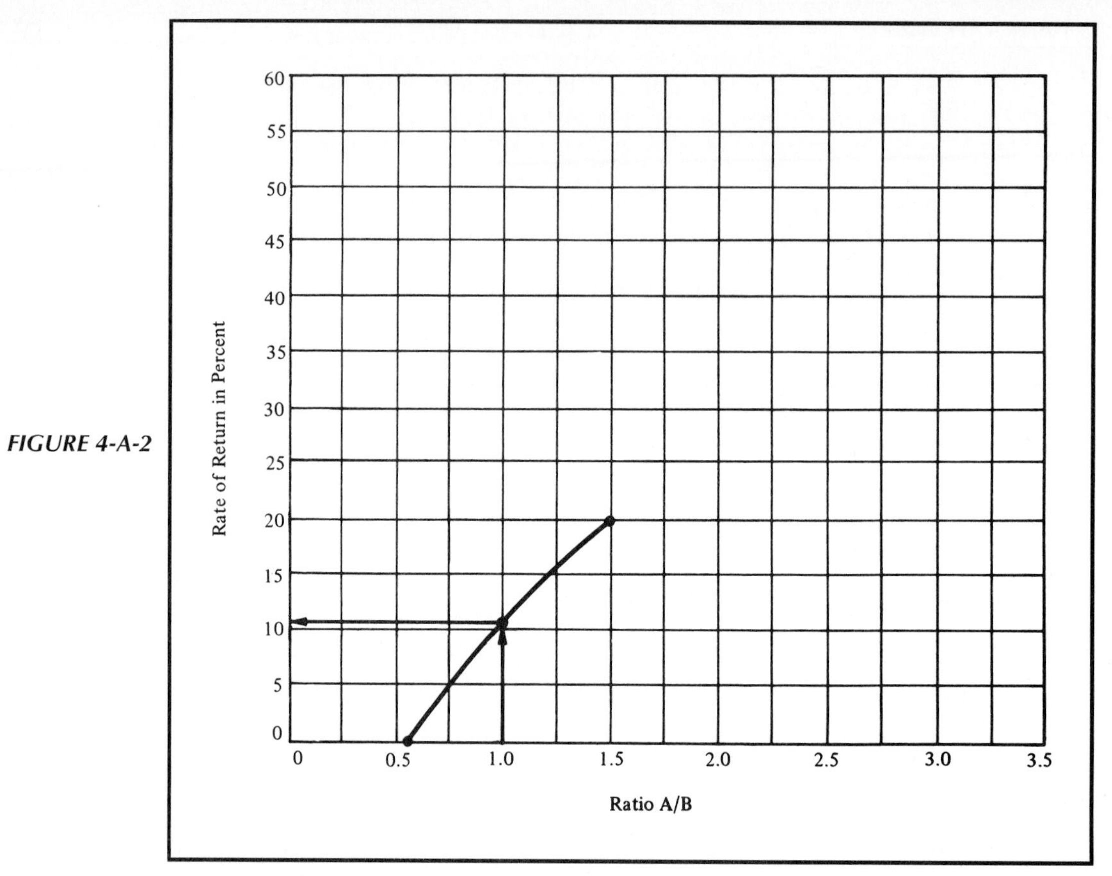

Rate of Return Interpolation Chart with Entries for the Example Project of Figure 4-A-1

TABLE 4–A–1 Investments, Revenues, and Costs for a Proposed New Venture

End of Year	(A) Investment (−) or Recovery (+) of Capital	(B) Receipts (or Savings) (+)	(C) Expenses (−)	(D) = (A) + (B) + (C) (Transferred to Figure 4-A-1) Net Cash Flow (+) or (−)
1987	−$60,000	—	—	−$60,000
1988	—	$20,000	$25,000	− 5,000
1989	− 20,000	35,000	25,000	− 13,000
1990	—	48,000	30,000	18,000
1991	—	52,000	30,000	22,000
1992	—	55,000	32,000	23,000
1993	—	55,000	33,000	22,000
1994	—	50,000	32,000	18,000
1995	—	35,000	24,000	11,000
1996	—	25,000	20,000	5,000
1997	+ 20,000	10,000	15,000	15,000

Solution

		PW at 35%		PW at 63%	
Year	Net Cash Flow	Factor	Amount	Factor	Amount
0	\$ 500	1.3500	\$ 675	1.6300	\$ 815
1	− 1,000	1.0000	− 1,000	1.0000	− 1,000
2	0				
3	250	0.5487	137	0.3764	94
4	250	0.4064	102	0.2309	58
5	250	0.3011	75	0.1417	35
PW			Σ = 11		Σ = 2

Thus, the present worth of the net cash flows equals zero for interest rates of about 35% and 63%. Whenever multiple answers such as these exist, it is likely that none is correct.

An effective way to overcome this difficulty and obtain a plausible answer is to manipulate cash flows as little as necessary so that there is only one reversal of the cash flows over time. This can be done by using the MARR or an external reinvestment rate to manipulate the funds, and then solving for the rate of return by using Equation 4-6 or 4-7. For example, if the minimum attractive rate of return is 10%, the +\$500 at year 0 can be compounded to year 1 to be \$500(F/P, 10%, 1) = +\$550. This, added to the −\$1,000 at year 1, equals −\$450. The −\$450, together with the remaining cash flows, which are all positive, now fit the condition that there be only one reversal in the cash flows over time. The return on invested capital at which the present worth of the cash flows equals 0 can now be shown to be 19%, per the following table:

		PW at 19%	
Year	Cash Flows	Factor	Amount
1	−\$450	1.0000	−\$450
3	250	0.7062	177
4	250	0.5934	148
5	250	0.4987	125
PW			Σ = 0

It should be noted that whenever a manipulation of cash flows such as the above is done, the calculated rate of return will vary according to what cash flows are manipulated and at what interest rate.

Probably the most straightforward way to overcome this multiple rate of return problem in general is to use the ERR method. For Example 4-B-1, the ERR equals 12.4% when the reinvestment rate (ϵ) is 10%:

$$\$1{,}000(P/F, 10\%, 1)(F/P, i'\%, 5) = \$500(F/P, 10\%, 5) + \$250(F/A, 10\%, 3)$$

$$(F/P, i', 5) = 1.632$$

$$i' = 0.124\ (12.4\%)$$

This differs from the 19% return in the table above because the ERR method makes the assumption that all positive cash flows are reinvested at 10%. The 19% return on invested capital assumes that funds can be reinvested at 19%.

EXAMPLE 4-B-2

Use the ERR method to analyze the cash flow pattern shown below. The internal rate of return is indeterminant (none exists) so that the IRR is not a workable procedure. The external reinvestment rate is 12% and MARR equals 15%.

Year	Cash Flows
0	\$5000
1	− 7000
2	2000
3	2000

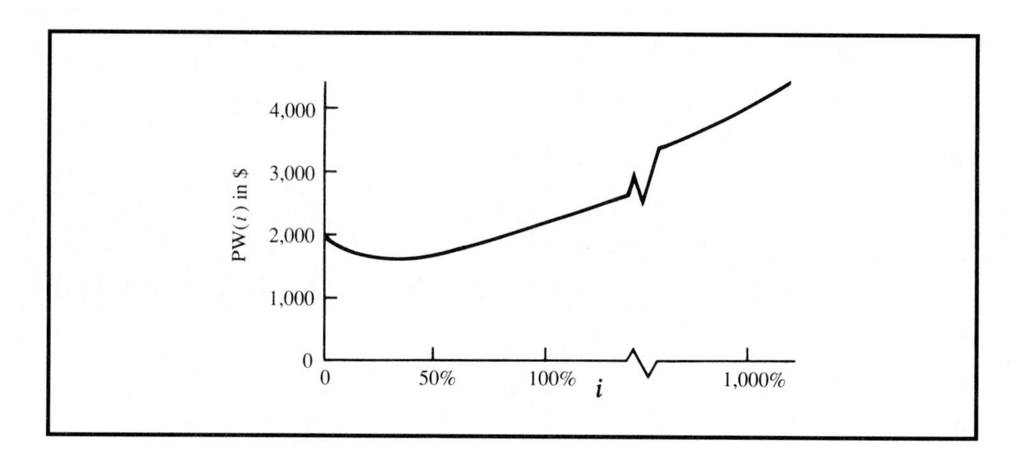

Solution

The ERR method provides these results:

$$\begin{aligned} \$7{,}000(P/F, 12\%, 1)(F/P, i'\%, 3) &= \$5{,}000(F/P, 12\%, 3) \\ &\quad + \$2{,}000(F/P, 12\%, 1) + \$2{,}000 \\ (F/P, i', 3) &= 1.802 \\ i' &= 21.7\% \end{aligned}$$

Thus, the ERR is greater than the MARR. Apparently, the project having this cash flow pattern would be acceptable. The PW at 15% is equal to \$1,740.36, which confirms the acceptability of this project.

5

COMPARING ALTERNATIVES

The primary objective of Chapter 5 is to develop and demonstrate methodology for the economic analysis and comparison of mutually exclusive alternatives.

The following topics are discussed in this chapter:

Basic philosophy for comparing mutually exclusive alternatives
Equivalent worth methods
What analysis (study) period should be used?
Rate of return methods
The capitalized worth method
Mutually exclusive combinations of projects
Comparisons among independent projects

5.1
INTRODUCTION

Most engineering projects and business ventures can be accomplished by more than one feasible method or alternative. When the selection of one of these alternatives excludes the consideration of any other alternative, the alternatives are called *mutually exclusive*. Typically, the feasible alternatives being considered require the investment of different amounts of capital, and their annual revenues and costs may vary. Sometimes the alternatives may have different useful lives. Because different levels of investment normally produce varying economic outcomes, we must perform an engineering economy study to determine which one of the mutually exclusive alternatives is preferred, and how much capital should be invested.

A seven-step procedure for accomplishing engineering economy studies was discussed in Section 2.5. In this chapter, we address Step 5 (analysis and comparison of the feasible alternatives) and Step 6 (selection of the preferred alternative) of the analysis procedure. For now, we focus on comparing mutually exclusive alternatives on the basis of economic considerations alone.

Five basic methods of analyzing cash flows were discussed in Chapter 4, and they are applicable to the analyses in this chapter. These methods provide a *basis for economic comparison* of feasible alternatives for an engineering project or other business undertaking. When correctly applied using the MARR as an economic decision criterion, these methods result in the correct selection of a preferred alternative from a set of mutually exclusive feasible alternatives.

5.2
BASIC PHILOSOPHY FOR COMPARING ALTERNATIVES

A decision that involves selection among two or more mutually exclusive alternatives must incorporate the fundamental purpose of capital investment; namely, to obtain at least the minimum attractive rate of return (MARR) for each dollar invested. In practice, there are usually a limited number of feasible alternatives to consider, so that the return associated with each increment of investment (rather than each unit of capital) must be evaluated. The problem of deciding which mutually exclusive alternative should be selected is made easier if we adopt a simple policy: *The feasible alternative that requires the minimum investment of capital and produces satisfactory functional results will be chosen unless the incre-*

mental capital associated with an alternative having a larger investment can be justified with respect to its incremental savings (or benefits).

Under this policy we consider the alternative that requires the least investment of capital, and which has a return equal to or greater than the MARR, to be the *base alternative*. Sometimes the base alternative is to "do nothing," or maintain the status quo. The investment of additional amounts of capital over the base alternative usually results in increased capacity, increased quality, increased revenues, decreased operating expenses, or increased life. Before additional money is invested, it must be shown that each avoidable increment of capital can pay its own way relative to other available investment opportunities. In summary, if the extra benefits obtained by investing additional capital are better than could be obtained from investment of the same capital elsewhere, the investment probably should be made. If this is not the case, we obviously would not invest more than the minimum amount required, including the possibility of nothing at all.

When mutually exclusive alternatives are compared using equivalent worth methods such as present worth, the above policy is implemented by selecting the course of action that maximizes the equivalent worth computed at the minimum attractive rate of return. This is demonstrated in Section 5.3. If company policy requires that the internal rate of return method or the external rate of return method be used to decide among competing alternatives, an analysis procedure is required whereby each increment of capital must justify itself by producing a rate of return that meets or exceeds the MARR. We shall illustrate this procedure in Section 5.4 and show that the selected alternative matches the results of analyzing mutually exclusive alternatives with an equivalent worth method.

5.3 EQUIVALENT WORTH METHODS

In Chapter 4, we learned that equivalent worth methods (PW, AW, and FW) convert all relevant cash flows into equivalent present, annual, or future amounts at the MARR. If a single project is under consideration, it is acceptable (earns at least the MARR) if the equivalent worth of the best feasible alternative for doing it is greater than or equal to 0; otherwise, it is not acceptable. These methods all assume that recovered funds (net cash inflows) can be reinvested at the MARR.

If two or more mutually exclusive alternatives are being compared and receipts or savings (cash inflows) as well as costs (cash outflows) are known, the feasible alternative that has the highest net equivalent worth should be selected as long as that equivalent worth is greater than or equal to zero. If only costs are known or are being considered when project revenues for all alternatives are the same, that feasible alternative which has the least negative equivalent worth of costs should be selected. The examples that follow in this section illustrate two general situations: (1) all the feasible alternatives under consideration have identical useful lives, and (2) the alternatives have different useful lives. In each case, the analyst must determine the analysis (study) period that best reflects actual conditions particular to the capital investment situation being evaluated and provides a sound basis for decision making.

5.3.1 Identical Useful Lives

When the mutually exclusive alternatives for a project or proposed venture have the same useful lives which match the analysis period, selection of the best feasible alternative is straightforward. All equivalent worth methods may be applied to this situation when (1) estimated revenues and costs vary among alternatives as in Example 5-1, and (2) estimated costs only are relevant to the comparison and differ among alternatives as in Example 5-2.

EXAMPLE 5-1

Three mutually exclusive investment alternatives for implementing an office automation plan in an engineering design firm are being considered. *Investment alternatives are those with an initial capital investment that produces positive cash flows from increased revenue, savings through reduced costs, or both.* The study period is 10 years, and the useful lives of all three feasible alternatives are also 10 years. Residual (salvage) values of all alternatives are assumed to be zero at the end of 10 years. If the firm's MARR is 10% per year, which alternative should be selected in view of the following estimates?

	Alternative		
	A	B	C
Investment (first) cost	\$390,000	\$920,000	\$660,000
Net annual receipts less expenses	69,000	167,000	133,500

Solution of Example 5-1 by the PW Method

$$PW_A(10\%) = -\$390{,}000 + \$69{,}000(P/A, 10\%, 10) = \$33{,}975$$
$$PW_B(10\%) = -\$920{,}000 + \$167{,}000(P/A, 10\%, 10) = \$106{,}143$$
$$PW_C(10\%) = -\$660{,}000 + \$133{,}500(P/A, 10\%, 10) = \$160{,}300$$

Based on the PW method, alternative C would be selected. The order of preference is C > B > A, where C > B means "C is preferred to B."

Solution of Example 5-1 by the AW Method

$$AW_A(10\%) = [PW_A(10\%)](A/P, 10\%, 10) = \$5{,}529$$
$$AW_B(10\%) = [PW_B(10\%)](A/P, 10\%, 10) = \$17{,}274$$
$$AW_C(10\%) = [PW_C(10\%)](A/P, 10\%, 10) = \$26{,}088$$

Here the annual worth is computed from the present worth, and alternative C is again chosen.

Solution of Example 5-1 by the FW Method

$$FW_A(10\%) = [PW_A(10\%)](F/P, 10\%, 10) = \$88{,}123$$
$$FW_B(10\%) = [PW_B(10\%)](F/P, 10\%, 10) = \$275{,}307$$
$$FW_C(10\%) = [PW_C(10\%)](F/P, 10\%, 10) = \$415{,}776$$

Future worth is also a multiple of present worth as shown, and again the choice is alternative C. For all three equivalent worth methods in this example, notice that C > B > A.

EXAMPLE 5-2

A company is going to install a new plastic-molding press. Four different presses are available. *You should observe that this is a set of four mutually exclusive cost alternatives; that is, there is an initial investment followed by negative cash flows.* The essential differences in initial investment and operating costs are shown below.

	Press			
	A	B	C	D
Investment (installed)	$6,000	$7,600	$12,400	$13,000
Useful life (years)	5	5	5	5
Annual operation and maintenance costs:				
Power	680	680	1,200	1,260
Labor	6,600	6,000	4,200	3,700
Maintenance	400	450	650	500
Property taxes and insurance	120	152	248	260
Total Annual Costs	$7,800	$7,282	$ 6,298	$ 5,720

Each press will produce the same number of units. However, because of different degrees of mechanization, some require different amounts and grades of labor and have different operation and maintenance costs. None is expected to have a salvage value and the study period is 5 years. Any capital invested is expected to earn at least 10% before taxes. Which press (feasible alternative) should be chosen?

TABLE 5–1 Comparison of Four Molding Presses Using PW Method (Example 5–2)

	Press			
	A	B	C	D
Present worth of				
Investment	\$ 6,000	\$ 7,600	\$12,400	\$13,000
Costs:				
(total annual costs)				
× (P/A, 10%, 5)	29,568	27,605	23,874	21,683
Total PW (Costs)	\$35,568	\$35,205	\$36,274	\$34,683

Solution of Example 5-2 by the PW Method

When alternatives for which revenues are considered equal are compared using the present worth method, that method is sometimes descriptively called the present worth–cost method. The alternative that has the minimum (least negative) PW is judged to be the most desirable. Table 5-1 shows the analysis by the PW method. The economic criterion is to choose that alternative with the minimum PW of costs, which is press D. The order of preference among the feasible alternatives in *decreasing* order is press D, press B, press A, and press C. This rank ordering is identical for all methods considered when they are correctly applied.

Solution of Example 5-2 by the AW Method

Table 5-2 shows the analysis by the annual worth method. When only costs are present, this method is sometimes called the annual cost method. The economic criterion is to choose that alternative with the minimum AW of costs which is press D.

Solution of Example 5-2 using the FW Method

Table 5-3 shows the analysis by the FW method, which may be referred to as the future worth-cost method when only costs are involved. Once again, press D is shown to be the least costly, and the order of desirability of the four alternatives is the same as for the PW and AW methods.

TABLE 5–2 Comparison of Four Molding Presses Using AW Method (Example 5–2)

	Press			
	A	B	C	D
Annual costs:				
Operation and maintenance costs	\$7,800	\$7,282	\$6,298	\$5,720
Capital recovery				
= (investment) × (A/P, 10%, 5)	1,583	2,005	3,271	3,429
Total AW (Costs)	\$9,383	\$9,287	\$9,569	\$9,149

TABLE 5–3 Comparison of Four Molding Presses Using FW Method (Example 5–2)

	Press			
	A	B	C	D
Future worth of				
Investment:				
(investment) × (F/P, 10%, 5)	$ 9,663	$12,240	$19,970	$20,937
Costs:				
(total annual costs)				
× (F/A, 10%, 5)	47,620	44,457	38,450	34,921
Total FW (Costs)	$57,283	$56,697	$58,420	$55,858

5.3.2 Different Useful Lives

In many cases, alternatives that require larger investments of capital than others will have higher annual revenues (or lower costs when revenues are equal) as well as longer useful lives. Unequal lives among feasible alternatives somewhat complicate their analysis and comparison. To make engineering economy studies in such cases, we must adopt some procedure that will put the feasible alternatives on a comparable basis. Two types of assumptions regarding the analysis period are employed: (1) the repeatability assumption, and (2) the coterminated assumption.

The *repeatability assumption* involves two main conditions:

1. The analysis period over which the alternatives are being compared is either indefinitely long or equal to a common multiple of the lives of the alternatives.
2. The economic consequences that are estimated to happen in an alternative's initial life span will happen also in all succeeding life spans (replacements), if any, for each feasible alternative.

The repeatability assumption has limited use in engineering practice because actual situations seldom meet both conditions.

The *coterminated assumption* involves the use of a finite analysis period for all feasible alternatives. This planning horizon may be the period of needed service or any selected length of time such as (1) the life of the shorter-lived alternative, (2) the life of the longer-lived alternative, (3) less than the shortest alternative life, (4) greater than the longest alternative life, or (5) in between the shortest and longest lives. The key point is that the selected analysis (study) period is appropriate for the decision situation under investigation.

In Examples 5-3 and 5-4, equivalent worth methods are utilized to evaluate mutually exclusive alternatives when estimated annual revenues are different from one alternative to the next. The repeatability assumption is illustrated in

Example 5-3, and the coterminated assumption is shown in Example 5-4. In Examples 5-5 and 5-6, mutually exclusive cost alternatives are analyzed with the repeatability and coterminated assumptions, respectively.

EXAMPLE 5-3

The following data have been estimated for two feasible investments, A and B, for which revenues as well as costs are known and which have *different* lives. If the minimum attractive rate of return is 10%, show which feasible alternative is more desirable by using equivalent worth methods. Use the repeatability assumption.

	A	B
Investment (first) cost	\$3,500	\$5,000
Annual revenue	1,900	2,500
Annual cost	645	1,383
Useful life (years)	4	8
Salvage value at end of useful life	0	0

Solution of Example 5-3 Using the PW Method
The expected lives of the alternatives are 4 and 8 years, respectively. The lowest common multiple is 8 years, which will be taken to be the length of the analysis period. Table 5-4 shows that alternative B has a greater net PW, so it is the better choice. Notice that the two alternatives are explicitly evaluated over an 8-year analysis (study) period.

Solution of Example 5-3 Using the AW Method
Since alternative B has a higher net AW (see Table 5-5), it is again shown to be the better economic choice. It is important to observe that annual worths

TABLE 5–4 Example 5–3 (PW Method)

	A	B
Annual revenue:		
\$1,900(*P*/*A*, 10%, 8)	\$10,136	
\$2,500(*P*/*A*, 10%, 8)		\$13,337
Total PW of Revenue	\$10,136	\$13,337
Annual cost:		
\$645(*P*/*A*, 10%, 8)	3,441	
\$1,383(*P*/*A*, 10%, 8)		7,378
Original investment	3,500	5,000
First replacement: \$3,500(*P*/*F*, 10%, 4)	2,390	
Total PW of Costs	\$ 9,331	\$12,378
PW of revenue − PW of costs	\$805	\$959

TABLE 5–5 Example 5–3 (AW Method)

	A	B
Annual revenue	$1,900	$2,500
Annual cost:		
Expenses	645	1,383
CR cost:		
$3,500(*A/P*, 10%, 4)	1,104	
$5,000(*A/P*, 10%, 8)		937
Total Annual Equivalent Expenses	$1,749	$2,320
AW of revenue − AW of costs	$151	$180

are compared over different lives (4 and 8 years) but that the analysis period is implicitly assumed to be 8 years. The annual worth of each alternative can also be quickly computed as follows:

$$\begin{aligned} AW_A(10\%) &= [PW_A(10\%)](A/P, 10\%, 8) \\ &= \$805(0.1874) \\ &= \$151 \\ AW_B(10\%) &= [PW_B(10\%)](A/P, 10\%, 8) \\ &= \$959(0.1874) \\ &= \$180 \end{aligned}$$

EXAMPLE 5-3A

Use BTAX in Appendix E to perform the calculations in Example 5-3.

Alternative A

```
HOW MANY DIFFERENT CASH FLOW VALUES ARE THERE? 2
ENTER CASH FLOW, FIRST PERIOD, LAST PERIOD
                                   (SEPARATE BY COMMA)
? -3500,0,0
? 1255,1,4
YEAR    CASH FLOW
0       -3,500.00
1        1,255.00
2        1,255.00
3        1,255.00
4        1,255.00

WANT TO MAKE ANY CORRECTIONS ON CASH FLOW (Y/N)?
                                                 N
SELECT A METHOD FROM THE FOLLOWING:
(ENTER PW/FW/AW/IRR/ERR/SPP/DPP/TABLE/?/END):AW
ENTER INTEREST RATE AS DECIMAL (e.g. ENTER 10%
                                     AS .1) = .1
```

```
DO YOU WANT CONTINUOUS (CON) OR DISCRETE (DIS)
                                    COMPOUNDING ? DIS
→ AW = 150.8522
WANT ANOTHER RUN (Y/N)?Y

HOW MANY DIFFERENT CASH FLOW VALUES ARE
                                          THERE ? 1
ENTER CASH FLOW, FIRST PERIOD, LAST PERIOD
                                  (SEPARATE BY COMMA)
? 150.85,1,8

YEAR    CASH FLOW
1       150.85
2       150.85
3       150.85
4       150.85
5       150.85
6       150.85
7       150.85
8       150.85

WANT TO MAKE ANY CORRECTIONS ON CASH FLOW (Y/N)?
                                                   N
SELECT A METHOD FROM THE FOLLOWING:
(ENTER PW/FW/AW/IRR/ERR/SPP/DPP/TABLE/?/END):PW
ENTER INTEREST RATE AS DECIMAL (e.g. ENTER 10%
                                         AS .1) = .1
DO YOU WANT CONTINUOUS (CON) OR DISCRETE (DIS)
                                    COMPOUNDING ? DIS
→ PW = 804.7736
```

Alternative B

```
HOW MANY DIFFERENT CASH FLOW PERIODS ARE
                                          THERE? 2
ENTER CASH FLOW, FIRST PERIOD, LAST PERIOD
                                  (SEPARATE BY COMMA)
? -5000,0,0
? 1117,1,8

YEAR    CASH FLOW
0       -5,000.00
1        1,117.00
2        1,117.00
3        1,117.00
4        1,117.00
5        1,117.00
```

```
6          1,117.00
7          1,117.00
8          1,117.00

WANT TO MAKE ANY CORRECTIONS ON CASH FLOW (Y/N)?
                                                N
SELECT A METHOD FROM THE FOLLOWING:
(ENTER PW/FW/AW/IRR/ERR/SPP/DPP/TABLE/?/END): PW
ENTER INTEREST RATE AS DECIMAL (e.g. ENTER 10%
                                   AS .1) = .1
DO YOU WANT CONTINUOUS (CON) OR DISCRETE (DIS)
                             COMPOUNDING ? DIS
→ PW = 959.1121

WANT ANOTHER RUN (Y/N) ? y
ENTER NEW/OLD/?/CHANGE ? old
(ENTER PW/FW/AW/IRR/ERR/SPP/DPP/TABLE/?/END): aw
ENTER INTEREST RATE AS DECIMAL (e.g. ENTER 10%
                                   AS .1) = .1
DO YOU WANT CONTINUOUS (CON) OR DISCRETE (DIS)
                             COMPOUNDING ? dis
→ aw = 179.7798
```

EXAMPLE 5-4

Suppose that Example 5-3 is modified such that the expected period of required service from Alternative A or B is only 4 years. (Perhaps the company has a firm contract to produce a manufactured good for exactly 4 years.) A choice must be made between A and B in view of the coterminated life of 4 years and a MARR = 10%. Which feasible alternative is more desirable?

Solution of Example 5-4 Using the AW Method

A key question here concerns the estimated residual, or salvage, value of Alternative B at the end of year 4 (S_4). A variety of approaches can be used to make this estimate. When there is no information to the contrary, it is convenient to assume the following for the alternative whose useful life has been truncated (Alternative B):

$$S_4 = \text{present worth of remaining capital recovery cost amounts} + \text{present worth of salvage value at end of year 8}$$

Because the salvage value at the end of year 8 is zero, S_4 can be computed in accordance with the definition above:

$$S_4 = [\$5{,}000(A/P, 10\%, 8)](P/A, 10\%, 4) = \$937.22(3.1699) = \$2{,}971$$

TABLE 5–6 Example 5–4 (AW Method)

	A	B
Annual revenue	\$1,900	\$2,500
Annual cost:		
Expenses	645	1,383
CR cost:		
\$3,500(*A*/*P*, 10%, 4)	1,104	
\$5,000(*A*/*P*, 10%, 4) − \$2,971(*A*/*F*, 10%, 4)		937
Total Annual Cost	\$1,749	\$2,320
AW of revenue − AW of costs	\$ 151	\$ 180

Calculations of the AW for Alternatives A and B are summarized in Table 5-6.

Alternative B is the better choice. With the assumption above concerning the coterminated salvage value of B, the annual worths in Example 5-4 are identical to those in Example 5-3! This, of course, would not be true if a different salvage value estimate had been used for Alternative B. However, the *analysis procedure* would be the same. For example, if the salvage value of Alternative B had been estimated to be one-half of the initial investment (i.e. \$2,500), the AW of B would be \$2,500 − \$1,383 − [\$5,000(*A*/*P*, 10%, 4) − \$2,500(*A*/*F*, 10%, 4)] = \$78. Alternative A then becomes the better choice.

Now we consider two examples that involve only costs. Example 5-5 illustrates the repeatability assumption, and Example 5-6 shows how to deal with a selected study period (cotermination assumption) which is shorter than the useful life of the longer-lived alternative.

EXAMPLE 5-5

A selection is to be made between two structural designs. Because revenues do not exist (or can be assumed to be equal), only negative cash flows (costs) are shown, as follows:

	Structure M	Structure N
Investment (first) cost	\$12,000	\$40,000
Salvage value at end of life	0	10,000
Annual expenses	2,200	1,000
Useful life (years)	10	25

Using the repeatability assumption, determine which structure is better if the MARR is 15%.

Solution of Example 5-5 by the PW Method

A basic principle in comparing mutually exclusive alternatives is that all alternatives should be compared over the *same length of time* (study period). With the repeatability assumption, this is most conveniently chosen to be the lowest common multiple of the lives for the alternatives. In this case 50 years of service can be provided by structure M and replacing it four times—a total of five structures. The same length of service can be obtained by structure N and replacing it once—a total of two such structures. Calculations are shown in Table 5-7. Thus, the PW of costs for structure M is lower, indicating that M is better.

Solution of Example 5-5 Using the AW Method

With the repeatability assumption, the equivalent annual cash flow of each life span in the common multiple of years is identical for a particular alternative. This allows structures M and N to be compared with minimal effort, as shown in Table 5-8. Recall that in Section 4.4 we defined the AC method to be synonymous with the AW method when only costs are involved. Since $4,591 is less than $7,141, structure M is again shown to be the better economic choice. Notice that with the AW method, repeatability permits alternatives to be compared over their different life spans, thus reducing the amount of computational effort required.

If the period of needed service is less than a common multiple of the lives, that should be reflected in the analysis period selected for an engineering economy study. A cotermination point that may be appropriate for the situation is the life of the shortest-lived alternative. Example 5-6 illustrates this particular coterminated assumption.

EXAMPLE 5-6

Suppose that we are faced with the same two feasible alternatives as in Example 5-5, as follows:

	Structure M	Structure N
Investment (first) cost	$12,000	$40,000
Salvage value at end of life	0	10,000
Annual expenses	2,200	1,000
Useful life (years)	10	25
MARR	15%	15%

It is desired to compare the economics of the alternatives by terminating the study period at the end of 10 years and assuming a residual value for structure N at that time of, say, $25,000. Table 5-9 shows that structure M still has the

TABLE 5–7 Example 5–5 (PW Method)

Present Worth	Structure M
Original investment	−$12,000
First replacement: $12,000(*P*/*F*, 15%, 10)	− 2,966
Second replacement: $12,000(*P*/*F*, 15%, 20)	− 733
Third replacement: $12,000(*P*/*F*, 15%, 30)	− 181
Fourth replacement: $12,000(*P*/*F*, 15%, 40)	− 44
Annual expenses: $2,200(*P*/*A*, 15%, 50)	− 14,653
Total PW	−$30,577

Present Worth	Structure N
Original investment	−$40,000
First replacement: ($40,000−$10,000)(*P*/*F*, 15%, 25)	− 911
Annual expenses: $1,000(*P*/*A*, 15%, 50)	− 6,661
Salvage of last replacement: $10,000(*P*/*F*, 15%, 50)	9
Total PW	−$47,563

TABLE 5–8 Example 5–5 (AW Method Using Repeated Life)

	Structure M	Structure N
Annual cost:		
Expenses	$2,200	$1,000
CR cost:		
$12,000(*A*/*P*, 15%, 10)	2,391	
($40,000 − $10,000)(*A*/*P*, 15%, 25)		
+ $10,000(15%)		6,141
Total AW(Cost)	$4,591	$7,141

TABLE 5–9 Example 5–6 (AW Method Using Coterminated Life)

	Structure M	Structure N
Annual cost:		
Expenses	$2,200	$1,000
CR cost:		
$12,000(*A*/*P*, 15%, 10)	2,391	
($40,000 − $25,000)(*A*/*P*, 15%, 10)		
+ $25,000(15%)		6,739
Total AW(Cost)	$4,591	$7,739

lower annual cost. (This result does not have to agree with the decision for Example 5-5 using the repeatability assumption.)

It should be expected that different assumptions for determining the salvage value of the longer-lived alternative *could* result in different decisions. This did not happen in Example 5-6 because the choice is insensitive to estimates of residual value for structure N. In fact, a salvage value of $40,000 for structure N at the end of year 10 would not reverse the preference for structure M!

Of course, in many situations it is not reasonable to expect cash flows to remain unchanged for life spans after the first. Inflation and price changes could cause subsequent life spans to experience different cash flows. In such situations the PW or FW method is usually easiest to use for comparison of alternatives after estimating the magnitude and timing of expected cash flows over a common (identical) analysis period.

5.4 RATE OF RETURN METHODS

The philosophy and policy discussed in Section 5.2 is especially relevant when using rate of return methods to evaluate mutually exclusive alternatives. The best feasible alternative produces satisfactory functional results and requires the minimum investment of capital, unless a larger investment can be justified with respect to the incremental savings (benefits) and costs it produces. This policy can be reduced to three principles when applying rate of return methods:

1. Each increment of capital must justify itself by producing a sufficient rate of return on that increment.
2. Compare a higher investment alternative against a lower investment alternative only when the latter is acceptable.
3. Select the feasible alternative that requires the largest investment of capital as long as the incremental investment is justified by savings that earn at least the MARR.

This section is organized the same as Section 5.3. Thus, the rate of return methods discussed in Chapter 4 are first applied to mutually exclusive alternatives having identical useful lives and a common analysis (study) period. Then, the rate of return methods are applied to situations in which lives of the alternatives are not identical and do not match the analysis period. In both cases, feasible alternatives with different net revenues less costs, as well as those with costs only, will be considered. You should be careful to observe that the rate of return methods, when correctly applied to mutually exclusive alternatives, will produce the same preferred choice as do the equivalent worth methods.

5.4.1 Identical Useful Lives

It is to be expected that many investment situations will result in different revenues being produced by two or more feasible alternatives. In general, a firm should be willing to invest additional amounts of capital as long as each increment is justified by a sufficient return (at least equal to the MARR). This economic criterion of choice is explicitly applied when using a rate of return method and implicitly exists when using an equivalent worth method. This selection procedure is further illustrated in Example 5-7 for a mutually exclusive set of feasible alternatives.

EXAMPLE 5-7

Suppose that we have been requested to evaluate these six mutually exclusive feasible alternatives for a project (arranged in ascending order of initial investment) using the IRR method. The useful life of each alternative is 10 years, salvage values equal zero, and the MARR is 10%. Also, net annual revenues less costs vary among all alternatives. If the study period is 10 years, which alternative should be chosen?

	Alternative					
	A	B	C	D	E	F
Investment (first) cost	$900	$1,500	$2,500	$4,000	$5,000	$7,000
Annual revenues less costs	150	276	400	925	1,125	1,425

Solution of Example 5-7 Using the IRR Method
For each of the feasible alternatives, the IRR can be computed by determining the interest rate at which the PW equals zero. We can also solve for the IRR by determining the interest rate at which the FW or AW is equal to zero (using the AW is illustrated below for alternative A):

$$0 = -\$900(A/P, i'\%, 10) + \$150; \; i'\% = ?$$

By trial and error, we determine that $i'\% = 10.6\%$. In the same manner, IRRs of all the feasible alternatives are computed and summarized:

	A	B	C	D	E	F
IRR on Total Investment	10.6%	13.0%	9.6%	19.1%	18.3%	15.6%

At this point, *only alternative C is unacceptable* because its IRR is less than the MARR = 10%. Also, Alternative A is the base alternative from which to begin the incremental analysis procedure since it is the feasible alternative with the lowest initial investment whose IRR (10.6%) is equal to or greater than the MARR (10%).

In Example 5-7, it is not necessarily correct to select the alternative that maximizes the IRR on total investment. That is to say, alternative D may not be the best choice since maximization of IRR on total investment does not guarantee maximization of equivalent worth. From Section 5.3 we learned that the alternative with the largest PW, AW, and FW is the proper choice when the overall measure of effectiveness is maximization of an organization's future wealth to its owners.

To make the correct choice when using the IRR method, we must examine each increment of capital to see if it will pay its own way. The analysis is performed in Table 5-10, where the symbol Δ is used to mean *incremental* or *change in*, and the letters on the ends of an arrow designate the feasible alternatives for which the increment is considered. Table 5-10 provides the analysis of the five remaining feasible alternatives, and the IRRs on incremental cash flows are again computed using the AW method.

From Table 5-10, it is apparent that alternative E would be chosen because it requires the largest investment for which the last increment of investment capital is justified. That is, we desire to invest additional increments of the $7,000 presumably available for this project as long as each avoidable increment can earn 10% or better.

It was assumed in Example 5-7 (and all other examples involving mutually exclusive alternatives, unless noted to the contrary) that available capital for a project *not* committed to one of the feasible alternatives is invested in some other project or opportunity where it will earn a return equal to the MARR. Therefore, in this case the $2,000 left over by selecting Alternative E instead of F can earn 10% elsewhere, which is more than we could obtain by investing it in F. All equivalent worth methods of Chapter 4 made use of this assumption (i.e., other investment opportunities exist at the MARR).

TABLE 5–10 Comparison of Five Acceptable Feasible Alternatives Using the IRR Method (Example 5–7)

Increment considered	A	A→B	B→D	D→E	E→F
Δ Investment (first) cost	$900	$600	$2,500	$1,000	$2,000
Δ Annual revenues less costs	$150	$126	$649	$200	$300
IRR on Δ investment	10.6%	16.4%	22.6%	15.1%	8.1%
Is increment justified?	Yes	Yes	Yes	Yes	No

Three errors commonly made in this type of analysis are to choose the feasible alternative (1) with the highest overall IRR on total investment, or (2) with the highest IRR on an incremental investment, or (3) with the largest investment that has an IRR greater than or equal to the MARR. None of these criteria are generally correct. For instance, in Example 5-7 one might erroneously choose Alternative D rather than Alternative E because the IRR for the increment from B to D is 22.6% and from D to E only 15.1% (error #2 above). A more obvious error, perhaps, is the temptation to maximize the IRR on total investment and select Alternative D (error #1). The third error would be committed by selecting Alternative F for the reason that it has the largest total investment with an IRR greater than the MARR (15.6% > 10%).

EXAMPLE 5-8

This example illustrates the IRR method applied to the cost alternatives evaluated previously in Example 5-2. The study period is 5 years, and the useful lives of all four presses are also 5 years. For convenience, relevant information is repeated below. The MARR is 10% per year. Using the IRR method, which press should be chosen?

	Press			
	A	B	C	D
Investment (first) cost	$6,000	$7,600	$12,400	$13,000
Useful life (years)	5	5	5	5
Annual operation and maintenance costs:				
Power	$ 680	$ 680	$ 1,200	$ 1,260
Labor	6,600	6,000	4,200	3,700
Maintenance	400	450	650	500
Property taxes and insurance	120	152	248	260
Total Annual Costs	$7,800	$7,282	$ 6,298	$ 5,720

Solution of Example 5-8 Using the IRR Method

For feasible cost alternatives, the IRR on total investment is not defined because there are no revenues. Therefore, the base alternative is the one with the least initial investment, and we proceed directly with an analysis of incremental cash flows with the understanding that the base alternative (rather than "do nothing") will be selected only if each increment of capital

TABLE 5–11 Comparison of Four Molding Presses Using IRR Method (Example 5–8)

	Press			
	A	B	C	D
Investment (first) cost	\$6,000	\$7,600	\$12,400	\$13,000
Annual cost	\$7,800	\$7,282	\$ 6,298	\$ 5,720
Useful life (years)	5	5	5	5
Increment considered:		A→B	B→C	B→D
Δ Investment cost		\$1,600	\$4,800	\$5,400
Δ Annual savings		\$ 518	\$ 984	\$1,562
IRR on Δ investment		18.6%	0.8%	13.8%
Is increment justified?		Yes	No	Yes

cannot produce a return equal to or greater than the MARR through cost reductions (savings).

Table 5-11 provides a tabulation of the incremental cash flows and a calculated IRR for each *increment* of investment considered. As in Table 5-10, the symbol Δ is used to mean *incremental,* or *change in,* and the letters on the end of arrows indicate the feasible alternatives for which the increment is considered. Note that the incremental cash flows between the cost alternatives are in fact investment alternatives.

The first increment subject to analysis is the \$7,600 − \$6,000 = \$1,600 extra investment required for press B compared to press A. For this increment of investment, annual costs are reduced by \$7,800 − \$7,282 = \$518. The IRR on the incremental investment is the interest rate at which the present worth of the incremental cash flow is equal to zero. Thus,

$$0 = -\$1{,}600 + \$518(P/A, i'\%, 5);\ i' = ?$$

and the value of i' (by trial and error) which satisfies this relationship is $i' = 18.6\%$. Since 18.6%> 10% (the MARR), the increment A → B is justified.

The next increment subject to analysis is the \$12,400 − \$7,600 = \$4,800 extra investment required for press C compared to press B. This increment results in annual cost reduction of \$7,282 − \$6,298 = \$984. The IRR on increment B → C can then be determined by finding the i' at which

$$0 = -\$4{,}800 + \$984(P/A, i'\%, 5); i' = ?$$

Thus i' = IRR can be found to be approximately 0.8%. [*Note:* $(P/A, 0\%, 5) =$ 5.00.] Since 0.8%< 10%, the increment B → C is *not* justified, and we can say that press C itself is not justified.

The next increment that should be analyzed is B → D, not C → D, because press C has already been shown to be unacceptable and can no longer be a valid basis for comparison with other alternatives.* For B → D, the incremental investment is \$13,000 − \$7,600 = \$5,400 and the incremental saving in annual expenses is \$7,282 − \$5,720 = \$1,562. The IRR on increment B → D can then be determined by finding the i' at which

$$0 = -\$5{,}400 + \$1{,}562(P/A, i'\%, 5);\ i'\% = ?$$

The value of i' can be found to be 13.8%. Since 13.8% > 10%, the increment B → D is justified.

Based on the preceding analysis, press D would be the choice because it is the highest investment for which each increment of investment capital is justified. In choosing press D, one is actually accepting the first \$6,000 investment (base alternative) as necessary without justification, the \$1,600 increment A → B earning 18.6% and the \$5,400 increment B → D earning 13.8%. The IRR on the \$7,000 increment, A → D, is 14.8%. Figure 5-1 depicts these results for all separable increments that comprise the total investment of \$13,000 in press D.

The underlying rationale for the type of analysis in Example 5-8 is that the firm wants to invest capital if and only if it is necessary or will earn at least the MARR, which in this case is 10%. Thus, the firm supposedly has opportunities to invest elsewhere any capital not used for one of the presses in comparable risk projects where at least 10% per year can be earned.

FIGURE 5-1

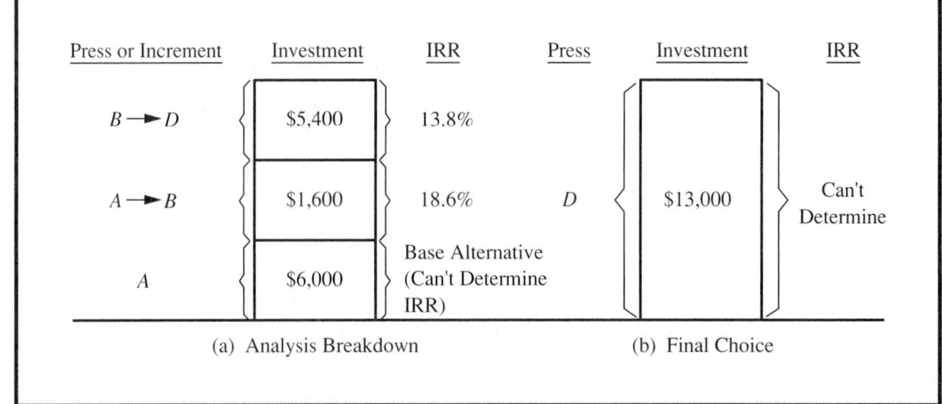

Representation of Increments and IRR on Increments Considered in Justifying Press D in Example 5-8. (a) Analysis Breakdown, and (b) Final Choice

*If the IRR were computed for increment C → D, it would be in excess of 90%. This means that press D is *very* attractive compared to press C, but since press C is not justified, this is not a valid basis for comparison.

EXAMPLE 5-9

In an automotive parts plant, a study team is analyzing an improvement project to increase the productivity of a flexible manufacturing center. The estimated cash flows for the three feasible alternatives being analyzed are shown in Table 5-12. The analysis period being used is 6 years, and the MARR for capital investments at the plant is 20%. Using the external rate of return (ERR) method, which alternative should be selected (ϵ = MARR).

Solution of Example 5-9 Using the ERR Method

The rationale and criteria for using the ERR method to compare feasible alternatives are the same as for the IRR method. The only difference is in the calculation methodology. However, in this example we also demonstrate a solution table which explicitly shows both the total cash flows for the feasible alternatives and the subsequent incremental cash flows. This format (shown in Table 5-12) is preferred to the format used in Table 5-11 when the cash flows for the alternatives (and consequently the incremental cash flows) are nonuniform after time $k = 0$.

Table 5-12 provides a tabulation of the calculation and acceptability of each additional increment of investment considered. Since these three feasible alternatives are a mutually exclusive set of investment alternatives, the first incremental cash flow analyzed is the difference between doing nothing ($0) and implementing Alternative A. For this increment of investment (Alt. A) the present worth of the first cost and the annual costs (at $i = \epsilon\%$) is just

TABLE 5–12 Comparison of Three Feasible Alternatives Using the ERR Method (Example 5–9)

End of	Alternative Cash Flows			Incremental Cash Flows		
Period	A	B	C	A[a]	A→B	A→C
0	−$640,000	−$680,000	−$755,000	−$640,000	−$ 40,000	−$115,000
1	262,000	− 40,000	205,000	262,000	− 302,000	− 57,000
2	290,000	392,000	406,000	290,000	102,000	116,000
3	302,000	380,000	400,000	302,000	78,000	98,000
4	310,000	380,000	390,000	310,000	70,000	80,000
5	310,000	380,000	390,000	310,000	70,000	80,000
6	260,000	380,000	324,000	260,000	120,000	64,000
			Incremental Analysis:			
			Δ PW of First and Annual Costs	− 640,000	− 291,657	− 162,498
			Δ FW of Annual Revenue	2,853,535	651,091	685,082
			ERR (on Δ PW costs)	28.3%	14.3%	27.1%
			Is Increment Justified?	Yes	No	Yes

[a] The incremental cash flow between making no change ($0) and implementing Alternative A.

the −$640,000 investment (first) cost. Therefore, the ERR on this increment of investment is (using Equation 4-10):

$$\begin{aligned} \$640{,}000(F/P, i'\%, 6) &= \$262{,}000(F/P, 20\%, 5) + \cdots + \$260{,}000 \\ &= \$2{,}853{,}535 \\ (F/P, i'\%, 6) = (1 + i'\%)^6 &= \$2{,}853{,}535/\$640{,}000 = 4.4586 \\ (1 + i'\%) &= (4.4586)^{1/6} = 1.2829 \\ i'\% &= 0.2829, \text{ or ERR} = 28.3\% \end{aligned}$$

Using a MARR = 20%, this increment of investment is justified and Alternative A is an acceptable base alternative. By using similar calculations, the increment A → B, earning 14.3%, is not justified and increment A → C, earning 27.1%, is justified. Therefore, Alternative C is the preferred alternative for the improvement project.

EXAMPLE 5-9A

Use the computer program in Appendix E (BTAX) to determine the ERR of incremental cash flows in Example 5-9.

Solution

Alt. A

```
HOW MANY DIFFERENT CASH FLOW VALUES ARE
                                      THERE ? 7
ENTER CASH FLOW, FIRST PERIOD, LAST PERIOD
                               (SEPARATE BY COMMA)
? -640000,0,0
? 262000,1,1
? 290000,2,2
? 302000,3,3
? 310000,4,4
? 310000,5,5
? 260000,6,6

YEAR    CASH FLOW
0       -640,000.00
1        262,000.00
2        290,000.00
3        302,000.00
4        310,000.00
5        310,000.00
6        260,000.00

WANT TO MAKE ANY CORRECTIONS TO CASH FLOW (Y/N)?
                                                N
```

```
SELECT A METHOD FROM THE FOLLOWING:
(ENTER PW/FW/AW/IRR/ERR/SPP/DPP/TABLE/
                                        ?/END):ERR
ENTER EXTERNAL REINVESTMENT RATE E AS DECIMAL? .2
DO YOU WANT DISCRETE (DIS) OR CONTINUOUS (CON)
                                 INTEREST RATE ? DIS
→ ERR IS BETWEEN 28.29199 AND 28.29299 %
```

```
      WANT ANOTHER RUN (Y/N) ? Y
      ENTER NEW/OLD/?/CHANGE ? NEW
A → B HOW MANY DIFFERENT CASH FLOW VALUES ARE THERE ? 7
      YEAR   CASH FLOW
      0        -40,000.00
      1       -302,000.00
      2        102,000.00
      3         78,000.00
      4         70,000.00
      5         70,000.00
      6        120,000.00

      WANT TO MAKE ANY CORRECTIONS ON CASH FLOW (Y/N)?
                                                    N
      SELECT A METHOD FROM THE FOLLOWING:
      (ENTER PW/FW/AW/IRR/ERR/SPP/DPP/TABLE/
                                          ?/END):ERR
      ENTER EXTERNAL REINVESTMENT RATE E AS DECIMAL?
                                                   .2
      DO YOU WANT DISCRETE (DIS) OR CONTINUOUS (CON)
                                   INTEREST RATE ? DIS
      → ERR IS BETWEEN 14.32 AND 14.321 %
```

```
      WANT ANOTHER RUN (Y/N) ? Y
      ENTER NEW/OLD/?/CHANGE ? NEW
A → C HOW MANY DIFFERENT CASH FLOW VALUES ARE THERE ?
                                                    7

      YEAR   CASH FLOW
      0       -115,000.00
      1        -57,000.00
      2        116,000.00
      3         98,000.00
      4         80,000.00
      5         80,000.00
      6         64,000.00
```

```
WANT TO MAKE ANY CORRECTIONS ON CASH FLOW (Y/N)?
                                               N
SELECT A METHOD FROM THE FOLLOWING:
(ENTER PW/FW/AW/IRR/ERR/SPP/DPP/TABLE/?/END):ERR
ENTER EXTERNAL REINVESTMENT RATE E AS DECIMAL?
                                              .2
DO YOU WANT DISCRETE (DIS) OR CONTINUOUS (CON)
                              INTEREST RATE? DIS
→ ERR IS BETWEEN 27.1 AND 27.101 %

WANT ANOTHER RUN (Y/N) ? N
```

5.4.2 Different Useful Lives

When the useful lives of mutually exclusive alternatives are not the same, the repeatability assumption may be used in their comparison if the selected analysis period corresponds to a common multiple of the useful lives. If the analysis period is shorter (or longer) than the common multiple of lives, the cotermination assumption is appropriate. Again, the underlying principle is to compare mutually exclusive alternatives on an equivalent basis that includes a realistic and identical analysis period.

To illustrate the IRR method when useful lives are different, we return to Example 5-3 and reanalyze it under two assumed conditions: (1) the repeatability assumption is realistic and the analysis period is 8 years, and (2) the coterminated assumption must be utilized because the decision situation requires an analysis period of 6 years.

EXAMPLE 5-10

Two feasible alternatives for a project (considered previously in Example 5-3) have the following data:

	A	B
Investment (first) cost	\$3,500	\$5,000
Annual revenue	1,900	2,500
Annual cost	645	1,383
Salvage value at end of useful life	0	0
Useful life (years)	4	8

As you can see, net annual revenues less costs are different for the alternatives, which also have different useful lives. If the MARR is 10%, show which feasible alternative is better by using the IRR method when (a) the study period is 8 years and repeatability is a realistic assumption, and (b) the study period is 6 years, at which time salvage values of A and B are expected to be $1,500 and $1,000, respectively.

Solution

(a) Because the repeatability assumption is specified, the least common multiple of years that includes the analysis period is 8 years. With this in mind, the first step is to compute the IRR on the total investment in Alternatives A and B. Recall that the MARR is 10%. This computation is shown below by setting the PW equal to zero over an 8-year period.

Alternative A:

$$0 = -\$3{,}500 + \$1{,}255(P/A, i'_A\%, 8) - \$3{,}500(P/F, i'_A\%, 4); \quad i'_A\% = ?$$

By trial and error, $i'_A = 16.2\%$ and Alternative A is acceptable.

Alternative B:

$$0 = -\$5{,}000 + \$1{,}117(P/A, i'_B\%, 8); \quad i'_B\% = ?$$

By trial and error, $i'_B = 15.1\%$ and Alternative B is also acceptable.

If, at this point, the analyst made a choice based on maximizing the IRR on total investment, Alternative A would be recommended. However, in our discussion in Section 5.4.1 (Example 5-7), it was emphasized that a correct choice among alternatives may *not* result from selecting the alternative having *the greatest rate of return on total investment*.

To obtain the correct choice, we must examine the incremental investment and net annual receipts less costs associated with Alternative B relative to Alternative A. The reader will observe that Alternative B has the same cash flows as A, *except* for the avoidable increment of investment that must be justified in order to make B the preferred feasible alternative, and the differences in revenues and costs. The IRR at which the PW of the incremental cash flows equals zero is determined as follows:

Alternative A → Alternative B:

$$0 = -\$1{,}500 - \$138(P/A, i'\%, 8) + \$3{,}500(P/F, i'\%, 4); \quad i'_{A \to B} = ?$$

By trial and error, the IRR on the incremental cash flow (A → B) is 12.7%. Because 12.7% is greater than the MARR of 10%, Alternative B should be chosen. (It is left to the reader to show that there are no multiple rates of return in this example.)

Alternatively, the IRR on the net incremental cash flow could be calculated by finding the $i'\%$ at which the AWs of the two alternatives are equal:

$$-\$3{,}500(A/P, i'\%, 4) + \$1{,}900 - \$645 = -\$5{,}000(A/P, i'\%, 8) + \$2{,}500 - \$1{,}383$$

Again, it can be determined that $i'_{A \to B} = 12.7\%$ causes both alternatives to have equal annual worths. Since this is greater than 10% (the MARR) the avoidable increment of investment in Alternative B is justified.

The reason for ranking errors that can occur when selections among mutually exclusive alternatives are based on maximization of IRR on total investment can be seen in Figure 5-2. When the MARR lies to the left of $IRR_{A \to B}$, an incorrect choice will be made by selecting an alternative that maximizes rate of return. This is because the IRR method assumes reinvestment of cash flows at the calculated rate of return (16.2% and 15.1%, respectively, for Alternatives A and B in this case). But the PW method assumes reinvestment at 10%. In this example, reinvestment can occur at a rate as high as 12.7% without causing a reversal in preference.

From Figure 5-2 it can be seen that $PW_B > PW_A$ at the MARR of 10% even though $IRR_A > IRR_B$. This ranking inconsistency is avoided by examining $IRR_{A \to B}$, which correctly leads to the selection of Alternative B.

FIGURE 5-2

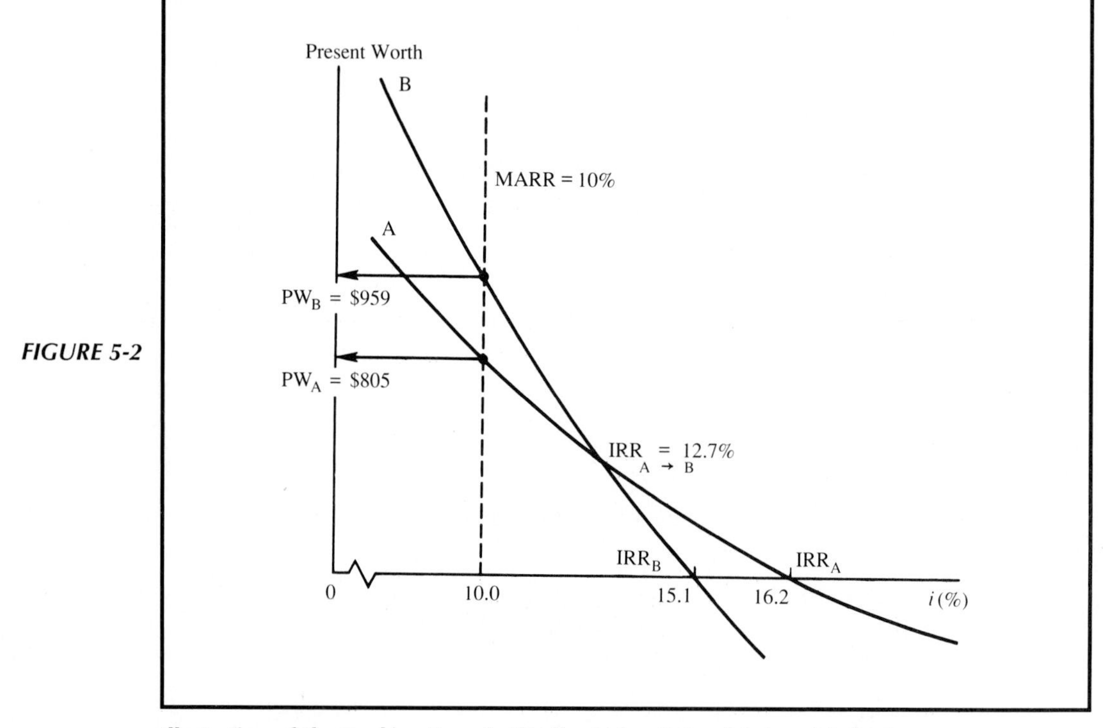

Illustration of the Ranking Error in Studies Using Rate of Return Methods

EXAMPLE 5-10A

Determine the internal rate of return on the incremental investment (A → B) in part a of Example 5-10 using the computer program in Appendix E (BTAX).

Solution

Calculation of IRR on A → B (Data Already Entered)

```
SELECT A METHOD FROM THE FOLLOWING:
(ENTER PW/FW/AW/IRR/ERR/SPP/DPP/TABLE/
                                   ?/END):table

YEAR    CASH FLOW
0       -1,500.00
1         -138.00
2         -138.00
3         -138.00
4        3,362.00
5         -138.00
6         -138.00
7         -138.00
8         -138.00

WANT TO MAKE ANY CORRECTIONS ON CASH FLOW (Y/N)?
                                               N
SELECT A METHOD FROM THE FOLLOWING:
(ENTER PW/FW/AW/IRR/ERR/SPP/DPP/TABLE/
                                  ?/END): IRR
PLEASE WAIT...
DO YOU WANT DISCRETE (DIS) OR CONTINUOUS (CON)
                              INTEREST RATE? DIS
→ IRR IS BETWEEN 12.708 AND 12.709 %
```

(b) If we coterminate the analysis period at the end of 6 years, the cash flow diagrams for Alternative A, Alternative B, and their incremental difference (A → B) are shown in Figure 5-3.

The IRR of Alternative A is 13.9%, for Alternative B, 12.6% and IRR on the increment A → B, 10.1%. These values were determined by solving for the $i'\%$ at which the PW of each cash flow equals zero. Thus, both alternatives are acceptable and, again, Alternative B is selected because its incremental cash flow produces a return greater than 10% (the MARR). Once again, notice that the feasible alternative having the larger IRR on total investment is not the choice.

FIGURE 5-3

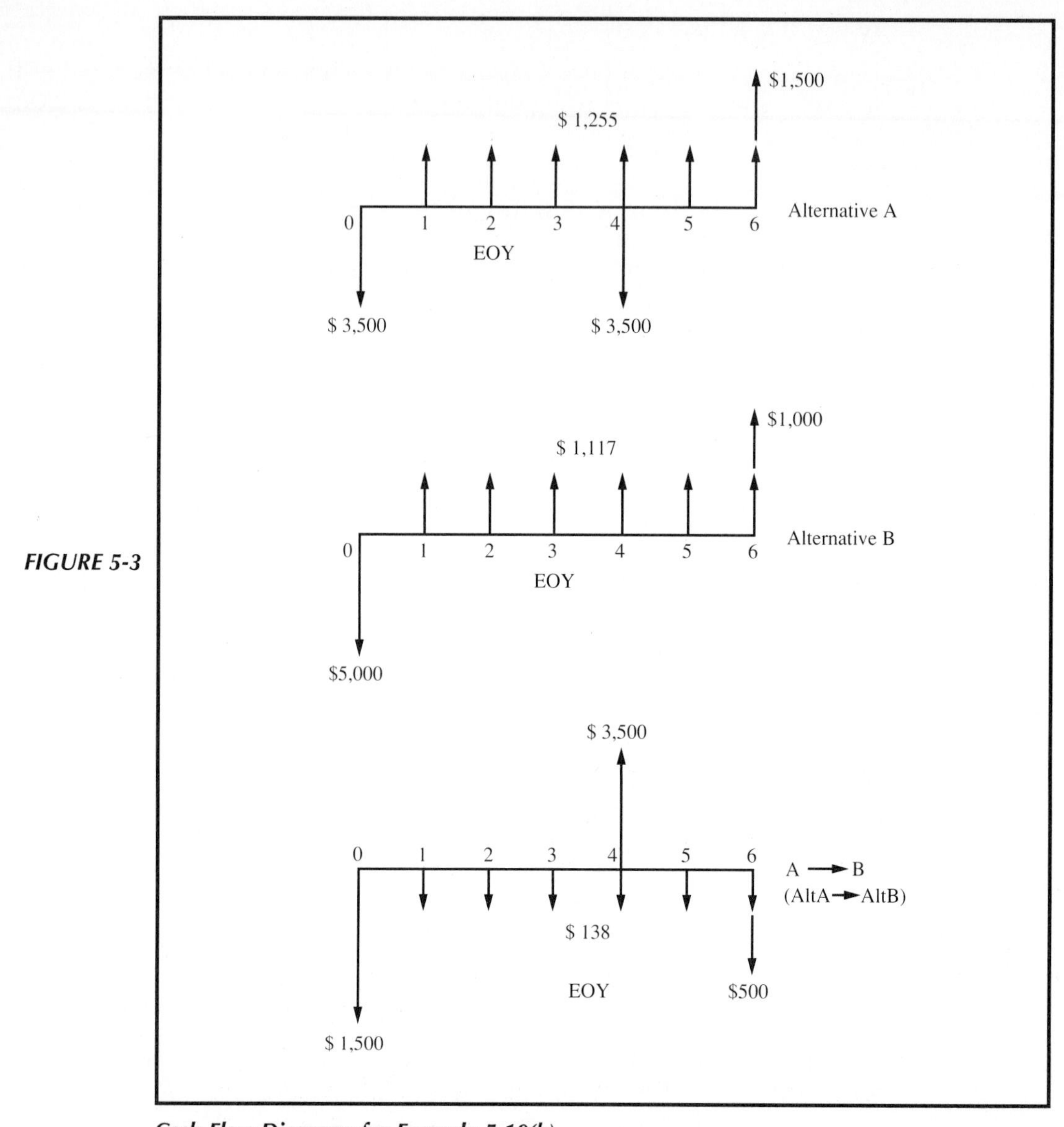

Cash Flow Diagrams for Example 5-10(b)

By this point in the chapter, two key observations are clear concerning the comparison of mutually exclusive alternatives: (1) equivalent worth methods are computationally less cumbersome to use and their results fairly easy to interpret, and (2) rate of return methods, if utilized, may not produce correct choices if the analyst or the manager insists on maximizing return on total investment. That is, incremental analysis must be used with rate of return methods to ensure that the best feasible alternative is selected. To emphasize these points further, consider the assignment given to Cynthia Jones in Example 5-11.

EXAMPLE 5-11

The owner of a downtown parking lot has retained a prestigious architect-engineering firm to determine whether it would be financially attractive to construct an office building on the site now being used for parking. Cynthia Jones, a newly hired civil engineer, has been requested to perform the analysis and offer a recommendation. The data she has assembled on the feasible alternatives developed by the project team are summarized below:

Mutually Exclusive Alternative	Total Investment (including land)	Total Net Annual Income from Property
A. Keep existing parking lot	$ 200,000	$ 22,000
B. Construct one-story building	4,000,000	600,000
C. Construct two-story building	5,550,000	720,000
D. Construct three-story building	7,500,000	960,000

(a) The analysis period selected is 15 years. For each alternative, the property has an estimated residual value at the end of 15 years that is *equal* to the total investment shown. The owner of the parking lot understands the IRR method, but Cynthia's boss has always insisted on a present worth analysis. Therefore, she decides to make the analysis by using both methods. If the MARR equals 10% per year, which alternative should Cynthia recommend?

(b) Repeat (a), except now assume that all alternatives have an estimated residual value of 50% of the total investment shown.

Solution

(a) The present worths of each alternative are computed as follows:

$$PW_A(10\%) = -\$200{,}000 + \$22{,}000(P/A, 10\%, 15) + \$200{,}000(P/F, 10\%, 15)$$
$$= \$15{,}210$$
$$PW_B(10\%) = -\$4{,}000{,}000 + \$600{,}000(P/A, 10\%, 15) + \$4{,}000{,}000(P/F, 10\%, 15)$$
$$= \$1{,}521{,}260$$
$$PW_C(10\%) = -\$5{,}550{,}000 + \$720{,}000(P/A, 10\%, 15) + \$5{,}550{,}000(P/F, 10\%, 15)$$
$$= \$1{,}255{,}060$$
$$PW_D(10\%) = -\$7{,}500{,}000 + \$960{,}000(P/A, 10\%, 15) + \$7{,}500{,}000(P/F, 10\%, 15)$$
$$= \$1{,}597{,}360$$

To maximize present worth, Cynthia would recommend to her boss the construction of a three-story office building (Alternative D).

The internal rate of return method is now applied to the analysis of this project:

	Mutually Exclusive Alternative A	B	C	D
Total investment	\$200,000	\$4,000,000	\$5,550,000	\$7,500,000
Net annual income	22,000	600,000	720,000	960,000
Residual value	200,000	4,000,000	5,550,000	7,500,000
IRR[a]	11%	15%	13%	12.8%

[a] When initial investment equals residual (salvage) value, the IRR = net annual income ÷ initial investment.

Cynthia observes that IRR on total investment is maximized for alternative B, but she knows this is not necessarily the correct choice. She notes that all alternatives have IRRs that exceed MARR = 10%; therefore, Alternative A (which is the feasible alternative with the lowest initial investment cost whose IRR ≥ MARR) is an acceptable base alternative from which to begin the analysis of the incremental cash flows. The next step (Table 5-13) is to examine the IRRs on these incremental cash flows after ranking the alternatives from lowest total investment to the highest.

Based on this analysis, Cynthia would recommend that the owner of the parking lot seriously consider constructing a three-story office building (Alternative D).

(b) When the residual value of the property is estimated to be one-half of the total investment, each alternative's present worth can be quickly recomputed:

$$\begin{aligned} PW_A(10\%) &= -\$200{,}000 + \$22{,}000(P/A, 10\%, 15) + \$100{,}000(P/F, 10\%, 15) \\ &= -\$8{,}730 \end{aligned}$$

TABLE 5–13 Example 5–11 with Residual Value Equal Total Investment (IRR Method)

	Incremental Analysis of Alternatives A[a]	A→B	B→C	B→D
Additional investment (cost)	\$200,000	\$3,800,000	\$1,550,000	\$3,500,000
Additional annual income	22,000	578,000	120,000	360,000
Additional residual value	200,000	3,800,000	1,550,000	3,500,000
Incremental IRR	11%	15.2%	7.7%	10.3%
Decision	Accept keeping parking lot.	Accept one-story bldg.,reject parking lot	Keep one-story bldg., reject two-story bldg.	Accept three-story bldg., reject one-story bldg.

[a] The incremental cash flow between making no change (\$0) and implementing Alternative A.

Similarly,

$$\begin{aligned}PW_B(10\%) &= \$1{,}042{,}460\\ PW_C(10\%) &= \$590{,}730\\ PW_D(10\%) &= \$699{,}610\end{aligned}$$

Under this assumption for the residual value, the one-story building (Alternative B) would be recommended.

Again, the internal rate of return method requires more time and computation effort:

	Mutually Exclusive Alternative A	B	C	D
Total investment	\$200,000	\$4,000,000	\$5,550,000	\$7,500,000
Net annual income	22,000	600,000	720,000	960,000
Residual value	100,000	2,000,000	2,775,000	3,750,000
IRR[a]	9.3%	13.8%	11.6%	11.4%

[a] For example, the IRR of Alternative A is computed as follows: $0 = -\$200{,}000 + \$22{,}000\,(P/A, i'\%, 15) + \$100{,}000\,(P/F, i'\%, 15)$; $i'\% = ?$. With the Appendix E computer program, $i' = 9.3\%$.

Alternative A is now unacceptable (9.3% < 10%) and cannot serve as the base alternative from which to proceed with the incremental analysis procedure. However, Alternative B is acceptable and has the lowest initial investment of the remaining three feasible alternatives, so the incremental analysis would proceed as shown in Table 5-14.

Finally, Cynthia concludes that the one-story building is best by using the IRR method. At this point, she tells her boss: "If I ever have to repeat this sort of analysis involving mutually exclusive alternatives, I'm going to insist on using an equivalent worth method such as present worth!" The boss agrees and comments later to an associate that Cynthia is on her way to the top in this organization.

TABLE 5–14 Example 5–11 with Residual Value Equal One-Half Total Investment (IRR Method)

	Incremental Analysis of Alternatives B[b]	B→C	B→D
Additional investment (cost)	\$4,000,000	\$1,550,000	\$3,500,000
Additional annual income	600,000	120,000	360,000
Additional residual value	2,000,000	775,000	1,750,000
Incremental IRR[a]	13.8%	5.5%	8.5%
Decision	Accept one-story bldg., reject "no change"	Keep one-story bldg., reject two-story bldg.	Keep one-story bldg., reject three-story bldg.

[a] For instance, the IRR of B→C is determined as follows: $0 = -\$1{,}550{,}000 + \$120{,}000(P/A, i'\%, 15) + \$775{,}000(P/F, i'\%, 15)$; $i' = 5.5\%$.

[b] The incremental cash flow between making no change (\$0) and implementing Alternative B.

5.5

COMPARISON OF ALTERNATIVES USING THE CAPITALIZED WORTH METHOD

One special variation of the present worth method mentioned in Chapter 4 involves the determination of the worth of all receipts and/or expenses over an infinite length of time. This is known as the *capitalized worth* (CW) method. If expenses only are considered, results obtained by this method can be more appropriately expressed as *capitalized cost*. This is a convenient basis for comparing mutually exclusive alternatives when the period of needed service is indefinitely long or when the common multiple of the lives is very long and the repeatability assumption is applicable.

The capitalized worth of a perpetual series of end-of-period uniform payments A, with interest at $i\%$ per period, is $A(P/A, i\%, \infty)$. From the interest formulas, it can be seen that $(P/A, i\%, N) \rightarrow 1/i$ as N becomes very large. Thus, capitalized worth $= A/i$ for such a series, as can also be seen from the relation

$$P = A(P/A, i\%, \infty) = A\left[\lim_{N\to\infty} \frac{(1+i)^N - 1}{i(1+i)^N}\right] = A\left(\frac{1}{i}\right)$$

The P is often referred to as the *capitalized worth* of A (and denoted CW).

The annual worth of a series of payments of amount $\$X$ at the end of each kth period with interest at $i\%$ per period is $\$X(A/F, i\%, k)$. The capitalized worth of such a series can thus be calculated as $\$X(A/F, i\%, k)/i$.

EXAMPLE 5-12

Suppose a firm wishes to endow an advanced manufacturing processes laboratory at a prestigious university. The endowment principal will earn interest that averages 8% per annum, which will be sufficient to cover all expenditures incurred in the establishment and maintenance of the laboratory for an indefinitely long period of time (forever). Cash requirements of the laboratory are estimated to be \$100,000 now; \$30,000 per year over an indefinitely long future; and \$20,000 at the end of every fourth year (forever) for equipment replacement.

(a) For this particular problem, what study period (N) is, practically speaking, defined to be "forever"?

(b) What amount of endowment principal (none of which will be spent) is required to earn enough interest to support the cash requirements of this laboratory forever?

Solution

(a) A practical definition (approximation) of "forever" (infinity) is dependent on the interest rate. By examining the $(P/A, i\%, N)$ factor as N increases, we observe that this factor approaches a value of $1/i$. For $i = 8\% (1/i = 12.5000)$, you will notice that the $(P/A, 8\%, N)$ factor equals 12.4943 when $N = 100$. Therefore, $N = 100$ is essentially forever (∞) when $i = 8\%$. As the interest rate gets larger, the approximation of forever drops dramatically. For instance, when $i = 20\% (1/i = 5.0000)$, forever can be approximated using about 40 years; the $(P/A, 20\%, N)$ factor equals 4.9966 when $N = 40$.

(b) The capitalized worth of the cash requirements is synonymous with the endowment principal needed to support the laboratory forever. By using the relationship that CW = (equivalent annual cost)/i, we can compute the amount of the endowment:

$$\begin{aligned}\text{CW} &= \frac{\$100{,}000(A/P, 8\%, \infty) + 30{,}000 + 20{,}000(A/F, 8\%, 4)}{0.08} \\ &= \frac{\$8{,}000 + 30{,}000 + 4{,}438}{0.08} \\ &= \$530{,}475\end{aligned}$$

where the factor value for $(A/P, 8\%, \infty)$ is given in Table C-12 (Appendix C) as equal to 0.08000.

Another way of considering the amount of endowment principal needed in this example is to reason that we first have to establish the facility (\$100,000), and then have enough left in the endowment fund to earn a return that will meet the annual maintenance costs (\$30,000) and the periodic replacement of equipment needs (\$20,000 at the end of each fourth year). Using this logic, we have

$$\begin{aligned}\text{CW} &= \$100{,}000 + \frac{30{,}000 + 20{,}000(A/F, 8\%, 4)}{0.08} \\ &= \$100{,}000 + \frac{(30{,}000 + 4{,}438)}{0.08} \\ &= \$530{,}475\end{aligned}$$

which, of course, is the same CW amount as our previous calculation.

EXAMPLE 5-13

Compare structures M and N given in Example 5-5 by the capitalized worth (cost) method. Thus (as shown in Table 5-15), structure M is the better alternative, which is, of course, consistent with the results for Example 5-5 by the other methods of comparison. As an aside, since we have previously

TABLE 5–15 Example 5–13 (CW Method)

	Capitalized Worth	
	Structure M	Structure N
First cost:	\$12,000	\$40,000
Replacements:		
\$12,000(A/F, 15%, 10)/0.15	3,940	
[(\$40,000 − \$10,000)(A/F, 15%, 25)]/0.15		940
Annual expenses:		
\$2,200/0.15	14,667	
\$1,000/0.15		6,667
Total Capitalized Worth (Cost)	\$30,607	\$47,607

calculated the equivalent annual worth (AW) for these two alternatives, the easiest way to have determined the capitalized worth was to use the relation

$$\text{CW} = \frac{\text{AW}}{i}$$

$$\text{For structure M: } -\$4{,}591/0.15 = -\$30{,}607$$
$$\text{For structure N: } -\$7{,}141/0.15 = -\$47{,}607$$

5.6 DEFINING MUTUALLY EXCLUSIVE INVESTMENT ALTERNATIVES IN TERMS OF COMBINATIONS OF PROJECTS

It is helpful to categorize investment opportunities (projects) in three major groups as follows:

1. ***Mutually exclusive:*** at most one project out of the group can be chosen.
2. ***Independent:*** the choice of a project is independent of the choice of any other project in the group so that all or none of the projects may be selected, or some number in between.
3. ***Contingent:*** the choice of a project is conditional on the choice of one or more other projects.

It is common for decision makers to be faced with sets of mutually exclusive, independent, and/or contingent investment projects. For example, a contractor might be considering investing in a dump truck, and/or a power shovel, and/or an office building. For each of these types of investments (projects), there may be two or more mutually exclusive alternatives (i.e., brands of dump trucks, types of power shovels, and designs of office buildings). While the choice of an office building is probably independent of that of either dump trucks or power shovels, the choice of any type of power shovel may be contingent (conditional) on the decision to purchase a dump truck.

A general approach, then, requires that all investment projects be listed and that all the feasible combinations of projects be enumerated. *Such combinations of projects will then be mutually exclusive*. Each combination of projects is mutually exclusive since each is unique and the acceptance of one combination of investment projects precludes the acceptance of any of the other combinations. The net cash flow of each combination is determined simply by adding, period by period, the cash flows of each project included in the mutually exclusive combination being considered.

For example, suppose that we have three projects: A, B, and C. If the projects themselves are all mutually exclusive, then the four possible mutually exclusive combinations are shown in binary form in Table 5-16. If, by chance, the firm feels that one of the projects must be chosen (i.e., it is not permissible to turn down all alternatives), then mutually exclusive combination 1 would be eliminated from consideration.

If the three projects are independent, there are eight mutually exclusive combinations, as shown in Table 5-17.

To illustrate one of the many possible instances of contingent projects, suppose that A is contingent on the acceptance of both B and C, and that C is contingent on the acceptance of B. Now there are four mutually exclusive combinations: (1) do nothing, (2) B only, (3) B and C, and (4) A, B, and C.

Suppose that one is considering two independent sets of mutually exclusive projects. That is, projects A1 and A2 are mutually exclusive, as are B1 and B2. However, the selection of any project from the set of projects A1 and A2 is independent of the selection of any project from the set of projects B1 and B2. *Independent* means that the choice of a project from the A set does not affect the choice from the B set. For example, the decision problem may be to select

TABLE 5–16 Combinations of Three Mutually Exclusive Projects[a]

Mutually Exclusive Combination	Project X_A	X_B	X_C	Explanation
1	0	0	0	Accept none
2	1	0	0	Accept A
3	0	1	0	Accept B
4	0	0	1	Accept C

[a] For each investment project there is a binary variable X_j that will have the value 0 or 1 indicating that project j is rejected (0), or accepted (1). Each row of binary numbers represents an investment alternative in terms of a combination of projects (mutually exclusive combination). This convention is used throughout this book.

TABLE 5–17 Mutually Exclusive Combinations of Three Independent Projects

Mutually Exclusive Combination	Project			Explanation
	X_A	X_B	X_C	
1	0	0	0	Accept none
2	1	0	0	Accept A
3	0	1	0	Accept B
4	0	0	1	Accept C
5	1	1	0	Accept A and B
6	1	0	1	Accept A and C
7	0	1	1	Accept B and C
8	1	1	1	Accept A, B, and C

TABLE 5–18 Mutually Exclusive Combinations for Two Independent Sets of Mutually Exclusive Projects

Mutually Exclusive Combination	Project			
	X_{A1}	X_{A2}	X_{B1}	X_{B2}
1	0	0	0	0
2	1	0	0	0
3	0	1	0	0
4	0	0	1	0
5	0	0	0	1
6	1	0	1	0
7	1	0	0	1
8	0	1	1	0
9	0	1	0	1

at most one dump truck out of two models being considered, and to select at most one office building out of two designs being considered. Table 5-18 shows all mutually exclusive combinations for this situation.

Example 5-14 provides an illustration of how to enumerate the mutually exclusive project combinations (investment alternatives) from independent sets of projects and then select an optimal set (*portfolio*) under an available capital constraint.

EXAMPLE 5-14

The following are proposed projects, their interrelationships, and respective cash flows for the coming budgeting period. Some of the projects are mutually exclusive as noted below, and B1 and B2 are independent of C1 and C2. Also, certain projects are dependent of others that may be included in the final portfolio. Using the PW method and MARR = 10%, determine what

TABLE 5–19 Project Cash Flows and Present Worths (Example 5–14)

	Cash Flow ($000s) for End of Year					
Project	0	1	2	3	4	PW ($000s) at MARR = 10%
B1	−50	20	20	20	20	13.4
B2	−30	12	12	12	12	8.0
C1	−14	4	4	4	4	−1.3
C2	−15	5	5	5	5	0.9
D	−10	6	6	6	6	9.0

combination of projects is best if the capital to be invested is (a) unlimited, and (b) limited to $48,000.

Project B1 }
Project B2 } mutually exclusive

Project C1 }
Project C2 } mutually exclusive and dependent (contingent) on the acceptance of B2

Project D contingent on the acceptance of C1

Solution

The PW for each project by itself is shown in the right-hand column of Table 5-19. As a sample calculation, the PW for project B1 is

$$-\$50{,}000 + \$20{,}000(P/A, 10\%, 4) = \$13{,}400$$

The mutually exclusive project combinations are shown in Table 5-20.

The combined cash flows and the PW for each mutually exclusive combination are shown in Table 5-21.

TABLE 5–20 Mutually Exclusive Project Combinations (Example 5–14)

Mutually Exclusive Combination	Project				
	B1	B2	C1	C2	D
1	0	0	0	0	0
2	1	0	0	0	0
3	0	1	0	0	0
4	0	1	1	0	0
5	0	1	0	1	0
6	0	1	1	0	1

TABLE 5–21 Combined Project Cash Flows and Present Worths (Example 5–14)

Mutually Exclusive Combination	Cash Flow ($000s) for End of Year 0	1	2	3	4	Invested Capital ($000s)	PW ($000s) at MARR = 10%
1	0	0	0	0	0	0	0
2	−50	20	20	20	20	50	13.4
3	−30	12	12	12	12	30	8.0
4	−44	16	16	16	16	44	6.7
5	−45	17	17	17	17	45	9.0
6	−54	22	22	22	22	54	16.0

Examination of the right-hand column reveals that mutually exclusive combination 6 has the highest PW if capital available (in year 0) is unlimited, as specified in part (a). If, however, capital available is limited to $48,000, as specified in part (b), mutually exclusive combinations 2 and 6 are both not feasible. Of the remaining mutually exclusive combinations, 5 is best, which means that a portfolio consisting of projects B2 and C2 would be selected for a net PW = $9,000.

For problems that involve a relatively small number of projects, the general technique just presented for arranging various types of projects into mutually exclusive combinations is computationally practical. However, for larger numbers of projects, the number of mutually exclusive combinations becomes quite large, and a computer program should be used to perform the calculations.

5.7 COMPARISONS AMONG INDEPENDENT PROJECTS

All previous examples in this chapter have involved the analysis of mutually exclusive alternatives related to a project, or in Example 5-14, a mutually exclusive set of investment alternatives in terms of combinations of projects. However, any of the five basic engineering economy study methods can also be used for the comparison of independent projects. That is, the choice of one project does not affect the choice of any other, and any number of projects may be chosen as long as sufficient capital is available.

EXAMPLE 5-15

Given the following independent projects, determine which should be chosen using the annual worth method. The minimum required rate of return is 10%, and there is no limitation on total investment funds available.

Project	Investment (First) Cost, *I*	Life, *N*(yrs)	Salvage Value, *S*	Net Annual Cash Flow, *A*
X	$10,000	5	$10,000	$2,300
Y	12,000	5	0	2,800
Z	15,000	5	0	4,067

Solution

As shown in Table 5-22, projects X and Z, having positive annual worths, would be satisfactory for investment, but project Y would not be satisfactory. The same indication of satisfactory projects and the unsatisfactory project would be obtained using the other equivalent worth or rate of return methods.

In many problems involving selections among independent projects, different revenues (or savings) and useful lives are present. Because these projects are typically nonrepeating, it is usually assumed that positive cash flows of shorter-lived projects are reinvested at the MARR over a period of time corresponding to the life of the longest-lived project. The following example illustrates this assumption.

EXAMPLE 5-16

A large corporation is considering the funding of several independent, non-repeating projects for enlarging freshwater harbors in three areas of the country. Its available capital this year for such projects is $200 million, and

TABLE 5–22 Example 5–15 (AW Method)

Project	(1) Net Annual Cash Flow, *A*	(2) Capital Recovery Cost: $(I-S) \times (A/P, 10\%, 5) + S(10\%)$	$(3)=(1)-(2)$ Annual Worth
X	$2,300	$1,000	$1,300
Y	2,800	3,166	− 366
Z	4,067	3,957	110

the firm's MARR is 10%. In view of the data below, which project(s), if any, should be funded?

Project	Investment (First), Cost *I*	Net Annual Benefits, *A*	Useful Life, *N*
A	$93,000,000	$13,000,000	15 years
B	55,000,000	9,500,000	10 years
C	71,000,000	10,400,000	30 years

Solution

With assumed reinvestment of cash inflows at 10% interest, the present worth of each project can be computed over its useful life and utilized in the selection process. If each independent project is acceptable, we must next determine which combination of projects maximizes present worth without exceeding $200 million.

Again, all three projects (Table 5-23) cannot be chosen without violating the limitation on available funds. Projects A and C should be recommended because this combination maximizes PW (= $32,918,944) with a total investment of $164 million. The leftover funds ($200 − $164 = $36 million) would presumably be invested elsewhere at the MARR of 10%.

TABLE 5–23 Example 5–16 (PW Method)

Project	(1) Investment (First) Cost, *I*	(2) PW of Benefits $A(P/A, 10\%, N)$	(3) = (2) − (1) Present Worth
A	$93,000,000	$98,879,034	$ 5,879,034
B	55,000,000	58,373,388	$ 3,373,388
C	71,000,000	98,039,910	$27,039,910

Chapter 5 has built on the previous chapters in which the principles and applications of money–time relationships were developed. Specifically, Chapter 5 introduced several difficulties associated with selecting the best alternative from a mutually exclusive set of feasible candidates when using time value of money concepts. Moreover, alternatives with unequal lives, various types of dependencies, cost-only versus different revenues and costs, and funding constraints were considered in deciding how to maximize the productivity of invested capital based on the MARR. In light of these complications, we learned that choosing the alternative with the largest equivalent worth (or least negative in the case of cost alternatives) using the MARR would produce this desired result.

If a rate of return method is being utilized to analyze and rank (in preferential order) mutually exclusive alternatives, each avoidable increment of additional capital must earn at least the MARR to ensure that the best alternative is chosen. Examples were provided to illustrate correct computational procedures for avoiding the ranking inconsistency that sometimes occurs when equivalent worth and rate of return methods are applied to the same set of mutually exclusive alternatives. Also, we considered projects with perpetual lives. In this situation, the capitalized worth method of economic evaluation was applied. The chapter concluded by demonstrating the evaluation of independent and/or contingent projects using these same methods.

5.8 PROBLEMS

5-1. Four mutually exclusive alternatives are being evaluated, and their costs and revenues are itemized below.

a. If the MARR is 12%/year, use the present worth method to determine which alternatives are *economically acceptable*.

b. If the total capital available is $200,000, which alternative should be *selected*? (5.3)

	Mutually Exclusive Alternative			
	I	II	III	IV
Investment (first) cost	$100,000	$152,000	$184,000	$220,000
Net annual receipts less costs	15,200	31,900	35,900	41,500
Salvage value	10,000	0	15,000	20,000
Useful life (years)	10	10	10	10

5-2. In the design of a new facility, the following mutually exclusive alternatives are under consideration:

	Alternative 1	Alternative 2	Alternative 3
Investment (first) cost	$28,000	$16,000	$23,500
Net cash flow/year	5,500	3,300	4,800
Salvage value	1,500	0	500
Useful life (years)	10	10	10

Assume that the interest rate (MARR) is 15%. Use the following to choose the best of these three feasible alternatives:

a. AW method (5.3)
b. IRR method (5.4)

5-3. The Consolidated Oil Company must install antipollution equipment in a new refinery to meet federal clean air legislation. Four types of equipment (feasible alternatives) are being considered, which will have investment and annual operating costs as follows:

	Equipment Type			
	A	B	C	D
Investment (first) cost	$600,000	$760,000	$1,240,000	$1,600,000
Annual Expenses:				
Power	68,000	68,000	120,000	126,000
Labor	40,000	45,000	65,000	50,000
Maintenance	660,000	600,000	420,000	370,000
Taxes and insurance	12,000	15,000	25,000	28,000

Assuming a useful life of 10 years for each type of equipment, no salvage value, and that the company wants a before-tax minimum return of 15% on its capital, calculate the present worth of each alternative to determine which one should be purchased. (5.3)

5-4. Work problem 5-3 using: (5.4)

a. the IRR method
b. the ERR method when ϵ (the reinvestment rate) equals 15%.

5-5. A large sailboat is being built in a coastal town. In specifying air conditioning equipment and insulation for the living quarters of the vessel, engineers have found various combinations of compressors and insulation thickness to be feasible. A small air conditioner compressor causes the need for more insulation but has slightly lower operating and maintenance costs compared to a larger compressor. Three combinations (feasible alternatives) are being considered that have the following investment and annual costs.

	Combination		
	A	B	C
First cost of compressor	$7,000	$ 4,200	$ 5,500
First cost of insulation	7,000	14,000	10,000
Out-of-pocket costs/year	950	300	600

Each system is expected to have a 20-year life, with no salvage value. The before-tax minimum attractive rate of return is 20%. Compare the alternatives using the future worth method. (5.3)

5-6. Consider the following two mutually exclusive alternatives related to an improvement project and recommend which one (if either) should be implemented. Use the present worth method. (5.3)

	Machine	
	A	B
Investment (first) cost	$20,000	$30,000
Salvage value	4,000	0
Annual receipts	10,000	14,000
Annual costs	4,400	8,600
Useful life (years)	5	10
Minimum attractive rate of return = 15%		
Analysis period = 10 years, assuming "repeatability"		

5-7. A small branch office of a major retailing firm is planning to purchase a small microprocessor. Three vendors have supplied cost, useful life, and salvage data (shown below). The minimum attractive rate of return is 12%.

	Machine		
	A	B	C
Investment (first) cost	$800	$1,400	$900
Annual service contract cost	180	150	170
Salvage value	200	600	100
Useful life (years)	5	8	5

If the expected annual saving in labor is $500 regardless of which machine is purchased, what recommendation would you make to the boss? Assume that the coterminated life assumption is valid and that after 5 years (selected study period) the market value of machine B is expected to be $825. (5.3)

5-8. A construction company is going to purchase several light-duty trucks. Its MARR before taxes is 18%. It is considering two makes, and the following relevant data are available (5.3)

	Wiltsbilt	Big Mack
Investment (first) cost	$10,000	$15,000
Salvage value at end of life	2,000	3,000
Annual out-of-pocket costs	4,000	3,000
Useful life (estimated by manufacturer)	3 years	5 years

a. Which type of truck should be selected when the repeatability assumption is appropriate?
b. Which type of truck would you recommend if the selected analysis period is 3 years (coterminated assumption) and it is estimated that a Big Mack truck will have a salvage value of $5,600 at that time?

5-9. A real estate operator has a 30-year lease on a plot of land. Estimates of the annual costs and revenues of various types of structures on the piece of land are as follows.

	First Cost of Structure	Revenues Less Costs
Apartment house	$300,000	$69,000/year
Theater	200,000	40,000/year
Department store	250,000	55,000/year
Office building	400,000	76,000/year

Each structure is expected to have a salvage value equal to 20% of its first cost. If the investor requires a minimum attractive rate of return of at least 12% before taxes on all investments, which structure (if any) should be selected? Use the annual worth method. (5.3)

5-10. A certain service can be performed satisfactorily either by process R or process S. Process R has a first cost of $8,000, an estimated service life of 10 years, no salvage value, and annual net receipts (revenues − costs) of $2,400. The corresponding figures for process S are $18,000, 20 years, salvage value equal to 20% of first cost, and $4,000. Assuming a minimum attractive rate of return of 15% before income taxes, find the future worth of each process and specify which you would recommend. Use the repeatability assumption. (5.3)

5-11. Compare the AW (equivalent uniform annual costs) of the following two electric pumps if the minimum acceptable rate of return (MARR) is 10%. The study period for the pump will be exactly 5 years, because the government contract for which the pump is needed will continue only for the next 5 years. What assumptions did you have to make in your analysis? (5.3)

	Circle D Pump	Qwik Sump Pump
Investment (first) cost	$4,000	$7,000
Annual expenses	800	900
Salvage value	0	1,000
Useful life (years)	3	7

5-12. In the Rawhide Co., Inc. (leather products manufacturer), decisions regarding approval of proposals for plant investment are based upon a stipulated minimum attractive rate of return of 20% before income taxes. The following five packaging devices were compared assuming a 10-year life and zero salvage value for each. Which one (if any) should be selected? Make any additional calculations you think are needed to make a comparison using the IRR method. (5.4)

	Packaging Equipment				
	A	B	C	D	E
Investment (first) cost	$38,000	$50,000	$55,000	$60,000	$70,000
Net annual return	11,000	14,100	16,300	16,800	19,200
Rate of return (IRR)	26.1%	25.2%	26.9%	25.0%	24.3%

5-13. Consider the three mutually exclusive alternatives below and use the internal rate of return method to make a selection. The feasible alternative chosen must provide service for a 10-year period. The MARR is 12%. State all assumptions you make. (5.4)

	A	B	C
Investment (first) cost	$2,000	$8,000	$20,000
Revenues less costs	$ 600/year	$2,220/year	$ 3,600/year
Salvage value	0	0	0
Project life (years)	5	5	10

5-14. Two high-speed backhoes are being considered by the Apex Construction Company to replace a present piece of equipment. The equipment will be needed for only 3 years. Cost and other data for the proposed backhoes are as follows:

	Backhoe M	Backhoe N
Purchase price	$50,000	$100,000
Annual savings	25,000	60,000
Estimated salvage value at end of year 3	20,000	10,000
Useful life of backhoe (years)	3	5

Over what *range* of values of the MARR is alternative N preferred to alternative M? (5.4)

5-15. A comparison is being made between the mutually exclusive alternatives M2 and N2 in view of these estimated net cash flows:

	Alternative	
End of Year	M2	N2
0	−$1,500	−$12,000
1	575	4,400
2	575	4,400
3	575	4,400
4	575	4,400
5	0	4,400

A coterminated life of 4 years is used in the analysis, and the MARR is 15%. What recommendation should be made with (a) the present worth method, and (b) the internal rate of return method? Should the recommendations be the same? (5.3, 5.4)

5-16. Work Problem 5-9 by using the ERR method when the reinvestment rate is 12%. Calculate the present worth of each alternative at 12%. Why do all methods lead to the same choice? (5.3, 5.4)

5-17. A manufacturing company is considering purchasing a 10-horsepower (hp) electric motor which it estimates will run an average of 6 hours per day for 250 days per year. Past experience indicates that (1) its annual cost for taxes and insurance averages 2.5% of first cost, (2) it must make 10% on invested capital before income taxes are considered, and (3) it must recover capital invested in machinery within 5 years. Two motors are offered to the company. Motor A costs $340 and has a guaranteed efficiency of 85% at the indicated operating load. Motor B costs $290 and has a guaranteed efficiency of 80% at the same operating load. Electric energy costs the company 2.3 cents per kilowatt-hour (kWh), and 1 hp = 0.746 kW.

Use the internal rate of return method to choose the better electric motor. Then compare the two motors using the present worth method. (5.3, 5.4)

5-18. A recent engineering graduate has encountered a puzzling situation in her analysis of three mutually exclusive alternatives related to a project. Her cash flow data and results with three different analysis methods are shown below when the MARR is 8%. Comment on her results and rework the analysis as you would have done it. Which economic measure of merit is correct in this situation? (5.3, 5.4)

		Cash Flows for		
End of Year		Alt. I	Alt. II	Alt. III
0		−$100,000	−$100,000	−$100,000
1		0	110,000	0
2		155,000	0	0
3		0	0	120,000
4		0	0	0
5		− 21,000	10,000	50,000
Measure of merit:	PW	$18,595	$8,658	$29,289
	IRR	19.5%	15.6%	16.2%
	Simple payback	2 years	1 year	3 years

5-19. Estimates for a proposed development are as follows. Plan A has a first cost of $50,000, a life of 25 years, a $5,000 salvage value, and annual maintenance costs of $1,200. Plan B has a first cost of $90,000, a life of 50 years, no salvage value, and annual maintenance costs of $6,000 for the first 15 years, and $1,000 per year for years 16 through 50. Assuming interest at 10%, compare the two plans by use of the capitalized worth method. (5.5)

5-20. In the design of a certain system, two mutually exclusive alternatives are under consideration. These feasible alternatives are as follows.

	Plan A	Plan B
Investment (first) cost	$50,000	$120,000
Salvage value	10,000	20,000
Annual costs	9,000	5,000
Useful life (years)	20	50

If *perpetual service life* is assumed, which of these alternatives do you recommend? The MARR is 10%. (5.5)

5-21. Use the *capitalized worth method* to determine which mutually exclusive bridge design to recommend. The minimum attractive rate of return is 15%. (5.5)

	Bridge Design A	Bridge Design B
Investment (first) cost	$274,000	$326,000
Annual cost of upkeep	10,000	8,000
Interim upgrade cost	50,000 (every sixth year)	42,000 (every seventh year)
Salvage value	0	0
Useful life (years)	83	92

5-22. Determine the capitalized worth in Problem 5-11, and compute the percentage difference in present worth and capitalized worth for the Circle D pump. Explain why this difference exists. (5.3, 5.5)

5-23. Which of the following independent projects would you recommend if no more than $30,000 is available for investment? The MARR is 20%. (5.6)

End of Year	Project 1	Project 2	Project 3
0	−$12,000	−$10,000	−$15,000
1	5,000	5,000	6,000
2	5,000	5,000	6,000
3	5,000	3,000	6,000
4	5,000	4,000	6,000
Rate of return (IRR)	24.1%	27.2%	21.9%

5-24. An organization is reviewing the following four independent capital investment projects for improving its building and grounds. The firm's budget next year for making such improvements is limited to $550,000. (a) List all mutually exclusive combinations ("portfolios") of projects and determine the investment requirements of each. (b) Which combinations are feasible? (5.6)

Project	Capital Requirement
1	$100,000
2	250,000
3	225,000
4	170,000

5-25. Three independent projects (A, B, and C) are under consideration, and no more than $150,000 can be spent now to implement any combination of them. Project D is dependent on the acceptance of Project A. If the MARR is 15% per year, which feasible combination would you recommend? (5.6)

	Cash Flow ($000) for End of Year			
Project	**0**	**1**	**2**	**3**
A	−100	40	40	60
B	−120	25	50	85
C	− 30	6	19	11
D	− 20	10	10	5

5-26. The Upstart Corporation is trying to decide between two industrial cranes. Crane A and Crane B below are mutually exclusive, and one of them must be chosen immediately.

	Crane	
	A	**B**
Investment (first) cost	$250,000	$370,000
Operating cost/year	22,000	8,000
Salvage value (end of useful life)	100,000	125,000
Useful life (years)	15	18
PW over useful life	−$366,353	−$408,923

Crane A has an extension boom that is optional. It would cost an extra $25,000 but would *save* an estimated $5,000 per year in operating cost. Crane B comes equipped with an extension boom, but an optional auger attachment can be purchased for $10,000. The auger would save $8,000 per year in drilling setup costs. The before-tax MARR is 15% at Upstart. Carefully state your assumptions and make a recommendation regarding which alternative to select. The availability of funds is not a limiting factor in this situation. (5.6)

5-27. Four projects are under consideration by your company. Projects A and C are mutually exclusive, projects B and D are mutually exclusive and cannot be implemented unless project A *or* C has been selected. No more than $140,000 can be spent at time 0. The before-tax MARR is 15%. The estimated cash flows are as follows:

	Project			
End of Year	A	B	C	D
0	−$100,000	−$20,000	−$120,000	−$30,000
1	40,000	6,000	25,000	6,000
2	40,000	10,000	50,000	10,000
3	60,000	10,000	85,000	19,000

Form all mutually exclusive combinations of these projects (investment alternatives) in view of the specified relationships and determine which one should be selected. (5.6)

5-28. Your company has $20,000 in "surplus" funds which it wishes to invest in new revenue-producing projects. There have been *three* independent sets of mutually exclusive projects developed. The useful life of each is 5 years and all salvage values are zero. You have been asked to perform an internal rate of return analysis to select the best combination of projects. If the MARR is 12%, which combination of projects would you recommend? (5.6)

	Project	First Cost	Net Annual Benefits
Mutually exclusive	A1	$ 5,000	$1,500
	A2	7,000	1,800
Mutually exclusive	B1	12,000	2,000
	B2	18,000	4,000
Mutually exclusive	C1	14,000	4,000
	C2	18,000	4,500

5-29. A firm is considering the development of several new products. The products under consideration are listed below; products in each group are mutually exclusive.

Group	Product	Development Cost	Annual Net Cash Income
A	A1	$ 500,000	$ 90,000
	A2	650,000	110,000
	A3	700,000	115,000
B	B1	600,000	105,000
	B2	675,000	112,000
C	C1	800,000	150,000
	C2	1,000,000	175,000

At most one product from each group will be selected. The firm has a minimum attractive rate of return of 10% and a budget limitation on development costs of $2,100,000. The life of all products is assumed to be 10 years, with no salvage value. (5.6)

a. List all feasible mutually exclusive combinations (investment alternatives).

b. Using the present worth method, determine which alternative should be selected.

5-30. Three independent projects are being considered:

	Project		
	X	Y	Z
Investment (first) cost	$100	$150	$200
Uniform annual savings	16.28	22.02	40.26
Useful life (years)	10	15	8
Computed IRR over the useful life	10%	12%	12%

The before-tax MARR is 10%, so all projects appear to be acceptable. At the end of their useful lives, projects X and Z will be replaced with other projects that have a 10% internal rate of return. Which projects should be chosen if investment funds are limited to $250? (5.7)

6

DEPRECIATION AND DEPLETION

Depreciation and depletion are accounting concepts that establish annual deductions against before-tax income so that the effects of use and time on property value can be reflected in a firm's financial statements. A fairly elementary treatment of depreciation and depletion is provided in Chapter 6, and the major topics covered are

What is depreciation?
Definitions of value
Purposes of depreciation
Actual depreciation revealed by time
Types of depreciation
Requirements of a depreciation method
What can be depreciated?
The Tax Reform Act of 1986
Depletion
Accounting for depreciation funds

6.1
INTRODUCTION

This chapter discusses various methods of depreciation that must be utilized in performing the after-tax engineering economy studies we will deal with in Chapter 7. We also demonstrate how depletion affects after-tax analyses. There have been three distinct periods of income tax legislation which have dramatically altered the depreciation practices of U.S businesses. They are (1) pre-1981, (2) January 1, 1981 through July 31, 1986, and (3) August 1, 1986 until the present (the Tax Recovery Act of 1986). Most of Chapter 6 concerns the third time period, but because current practices are built upon the other two it is necessary to discuss briefly their provisions as well. Appendix 6-A discusses the primary depreciation methods that were utilized prior to 1981.

6.2
WHAT IS DEPRECIATION?

Depreciation is the decrease in value of physical properties with the passage of time. More specifically, depreciation is an accounting concept that establishes an annual deduction against before-tax income such that the effect of use and time on an asset's value can be reflected in a firm's financial statements. Although the fact that depreciation does occur is easily established and recognized, the determination of its magnitude in advance is not easy. The actual amount of depreciation can never be established until the asset is retired from service. Because depreciation is a *noncash cost* that affects income taxes, we must learn to consider it properly when making after-tax engineering economy studies in Chapter 7.

From a business viewpoint, a physical asset has value because one expects to receive future monetary benefits through its possession and use. These benefits are in the form of future cash flows resulting from (1) the use of the asset to produce salable goods or services, or (2) the ultimate sale of the asset. It is because of these anticipated cash flows that an asset has commercial value. Depreciation, then, represents an estimate of decrease in an asset's value because its ability to produce future cash flows will, most likely, decrease over time.

6.3

DEFINITIONS OF VALUE

Because depreciation is defined as a decrease in *value* of an asset, it is necessary to give some consideration to the meaning of value. Unfortunately, we discover that there are several meanings attached to it. Probably the best definition of value, in a commercial sense, is that it is the present worth of all the future profits that are to be received through ownership of a particular property. This undoubtedly excellent definition is, however, difficult to apply in actual practice, inasmuch as we can seldom determine profits far in advance. Thus, several other measures of value are commonly used, some of which are approximations of the foregoing definition.

The most commonly encountered measure of value is *market value*. This is what will be paid by a willing buyer to a willing seller for a property where each has equal advantage and is under no compulsion to buy or sell. Buyers are willing to pay the market price because they believe it approximates the present value of what they will receive through ownership, including some rate of interest or profit. For new properties the cost on the open market is used as the original value.

Next to market value, probably the most important kind of value is *use value*. This is what the property is worth to the owner as an operating unit. A property may be worth more to the person who possesses it and has it in operation than it would be to someone else who, if he or she purchased it, might have to spend additional funds to move it and get it into operation. Use value is, of course, very closely related to our original definition of value. It is difficult to determine for the same reasons.

A third type of value is known as *fair value*. This usually is determined by a disinterested party in order to establish a price that is fair to both seller and buyer.

Book value is the worth of a property as shown on the accounting records of a company. It is ordinarily taken to mean the original cost of the property less the amounts that have been charged as depreciation expense. It thus represents the amount of capital that remains invested in the property and must be recovered in the future through the depreciation accounting process. It should be remembered, however, that because companies may use various depreciation accounting methods that produce different results, book value may have little or no relationship to the actual or market value of the property involved.

Salvage, or *residual, value* is the price that can be obtained from the sale of the property after it has been used. A positive salvage value implies that the property has further utility. It is affected by several factors. The reason of the present owner for selling may influence the salvage value. If the owner is selling because there is very little commercial need for the property, this will affect the residual value; change of ownership will probably not increase the commercial utility of the article. Salvage value will also be affected by the present cost of reproducing the property; price levels may either increase or decrease the residual value. A third factor that may affect salvage value is the location of the property. This is particularly true in the case of structures that must be moved in order to be of further use.

It may be seen that the various definitions of value vary considerably. Although people normally possess property so that they may receive benefits from it, some of the benefits frequently are not in the form of money. This fact further complicates the setting of value in monetary terms in order to place an ordinary commercial value on property.

6.4 OTHER DEFINITIONS

Because this chapter makes use of numerous terms that are not generally included in an engineering student's vocabulary, an abbreviated set of definitions is presented here. This list is intended to supplement definitions provided in Section 6.3.

Accelerated Cost Recovery System (ACRS) A system of depreciation that is mandatory for most tangible assets placed in service from January 1, 1981, through July 31, 1986.

adjusted (cost) basis Changes to the original cost basis of a property caused by various types of improvements or casualty losses. If the cost basis is adjusted, the depreciation deduction may have to be changed, depending on the reason for the adjustment and on the method of depreciation used.

Asset Depreciation Range (ADR) guideline period The midpoint of the asset depreciation range, in years, which is also the midpoint life for use in the ACRS.

amortization The distribution of the cost of certain types of property, usually intangible, over a prescribed period of time. Such distributions are usually made in equal installments and apply to such items as franchises, pollution control facilities, and starting a business.

basis (or "cost basis") The cost of acquiring an asset (purchase price) plus normal costs of making the asset serviceable. This is the amount from which depreciation is deducted.

class life The life that would apply for depreciation purposes to an item after 1980 based on the asset depreciation range (ADR) system and other factors.

depletion The gradual lessening in value of a mineral deposit or a tract of timber by its removal for use in a business activity. A depletion allowance is made for such "wasting" assets.

expensing The recovery of all or part of the cost of an asset through deductions from income in the current tax year.

intangible property Property that cannot be seen or touched, such as patents, contracts, franchises, copyrights, and licenses.

Modified Accelerated Cost Recovery System (MACRS) A system of depreciation established under the Tax Reform Act of 1986 which applies to all tangible property placed in service after 1986 (or after July 31, 1986, by choice).

personal property Property that can be moved from one location to another, such as a machine tool or piece of office furniture, and is not classified as real property.

property An asset—such as a building, computer, or patent.

real property Property that is land, and generally anything tangible that is erected on, growing on, or attached to land. A building is considered to be real property, for example.

recovery percentage A percentage (or rate, expressed as a decimal) for each year of the ACRS or MACRS recovery period that is utilized to compute a depreciation deduction.

recovery period The period of time over which the unadjusted basis of items in a particular class of property is recovered. Under MACRS, tangible property that is not real property is placed into one of six recovery periods (3, 5, 7, 10, 15, or 20 years).

tangible personal property Property that can be seen or touched—such as automobiles, machinery, or equipment—and is not real property.

tax life The period of time over which depreciation deductions are used to offset taxable income in determining federal income taxes. Tax life is often shorter than useful life.

unadjusted (cost) basis Usually the initial cost of a property, less any amount considered to be an expense in making the asset serviceable.

useful life Generally synonymous with ADR guideline period, but useful life may also be the period of time an asset is kept in productive use in a trade or business.

6.5 PURPOSES OF DEPRECIATION

Because property decreases in value, it is desirable to consider the effect that this depreciation has on engineering projects. Primarily, it is necessary to consider depreciation for two reasons:

1. To provide for the recovery of capital that has been invested in physical property.
2. To enable the cost of depreciation to be charged to the cost of producing products or services that result from the use of the property. Depreciation cost is deductible in computing profits on which income taxes are paid.

To understand these purposes, consider the following example.

EXAMPLE 6-1

Mr. Smith invested $3,000 in a machine for making a special type of concrete building tile as an avocation. He found that with his own labor in operating the machine he could produce 500 tiles per day. Working 300 days per year, he could make 150,000 tiles. He was able to sell the tiles for $50 per thousand. The necessary materials and power cost $20 per thousand tiles.

At the end of the first year, he had sold 150,000 tiles and computed his total profit, at the rate of $30 per thousand, to be $4,500. This continued for 2 more years, at which time the machine was worn out and would not operate satisfactorily. To continue in business, he would have to purchase a new machine.

During the 3-year period, believing he was actually making a profit of $4,500 per year, he had spent all of his supposed profits on other interests. He suddenly found that he no longer had his original $3,000 of capital, his machine was worn out, and he had no money with which to purchase a new one. What error had Mr. Smith made in his reasoning and accounting?

Solution

Analysis of the situation reveals that Mr. Smith had not recognized that depreciation was occurring, and he had made no provision for recovering the capital invested in the tile machine. The machine, which was valued at $3,000 when purchased, had decreased in value until it was worthless.

Through this depreciation, $3,000 of capital had been used in making tiles. Depreciation was just as much a cost of producing the tiles as was the cost of the material and power. However, depreciation differs from these other costs in that it always is paid or committed in advance. *Thus it is essential that depreciation be considered so that the capital used to prepay this cost may be recovered.* Failure to do this will always result in the depletion of capital.

Because capital must be maintained, it is necessary that the recovery be made by charging the depreciation that has taken place to the cost of producing whatever has been produced. Thus, in the case of the tile machine, production of 450,000 tiles "consumed" the machine. We might say that each thousand tiles produced decreased the value of the machine $3,000/450 = $6.67. Therefore, $6.67 should be charged as the cost of depreciation for making each thousand tiles. Adding this cost of depreciation to $20 (the cost of materials and power) gives the true cost of producing 1,000 tiles. With the true cost known, the actual profit can be determined. At the same time, with depreciation charged as a cost, a means for recovery of capital is provided.

6.6 ACTUAL DEPRECIATION REVEALED BY TIME

Depreciation differs from other costs in several respects. *First,* although its actual magnitude cannot be determined until the asset is retired from service, it always is paid or committed in advance. Thus, when we purchase an asset, we are prepaying all the future depreciation cost. *Second*, throughout the life of the asset we can only estimate what the annual or periodic depreciation cost is. Consequently, we must estimate the depreciation cost in engineering economy studies. Obviously, such estimates will not be entirely accurate, but this should not be too disturbing inasmuch as the same is true of virtually all other cost items in an engineering economy study.

A *third* difference is the fact that while much usually can be done to control the ordinary out-of-pocket costs, such as labor and material costs, relatively little can be done to control depreciation cost once an asset has been acquired, except, perhaps, through maintenance expenditures. Further, many of the factors that affect depreciation costs are external to the person or organization that owns the asset. If future conditions change and the demand for a product decreases, there may be a decline in the amount of material used, and probably a decrease in the profits. However, the depreciation cost, having been prepaid, may continue as before, and the result may be a loss of capital through failure to recover what has been prepaid.

6.7 TYPES OF DEPRECIATION

Depreciation, or the decrease in value of an asset, has several causes, some of which are very difficult to predict or anticipate. Decreases in value with the passage of time may be broadly classified as follows:

1. Normal depreciation: (a) physical, (b) functional.
2. Depreciation due to changes in price level.

Physical depreciation is due to lessening of the physical ability of a property to produce results. Its common causes are wear and deterioration. These cause operation and maintenance costs to increase and output to decrease. As a result, the profits may decrease. Physical depreciation is mainly a function of time and use.

Functional depreciation, often called obsolescence, is more difficult to determine than physical depreciation. It is the decrease in value that is due to the lessening in the demand for the function that the property was designed to render. This lessening may be brought about in many ways: Styles change, population centers shift, more efficient machines are produced, or markets are saturated. Also, increased demand may mean that an existing machine is no longer able to produce the required volume; this inadequacy is another cause of functional depreciation.

Depreciation due to changes in price levels is almost impossible to predict and is seldom accounted for in engineering economy studies. When price levels rise during inflationary periods, even if all the capital invested at the time of original purchase has been recovered, this recovered capital will not be sufficient to provide an identical replacement. Although there has been a recovery of the invested capital, the capital has decreased in value. Thus it is the capital, not the property, that has depreciated. Inflating annual depreciation to compensate for this phenomenon is not permitted in determining profits for income tax purposes.

6.8 REQUIREMENTS OF A DEPRECIATION METHOD

From the standpoint of management, a depreciation method should

1. Provide for the recovery of invested capital as rapidly as is consistent with the economic facts involved.
2. Not be too complex.
3. Ensure that the book value will be reasonably close to the market value at any time.
4. Be accepted by the Internal Revenue Service (IRS), if the method is also to be used for determining federal income taxes.

These requirements are somewhat contradictory and are not easily met. As a result, numerous methods for computing depreciation have been devised. Each is based on some hypothesis regarding loss of an asset's value versus time (or usage) and is an attempt to solve the complex depreciation problem in a reasonably simple and satisfactory manner. Because there are contradictory factors involved, and because future and unknown factors exist, it can be expected that perfection will not be achieved through the use of any depreciation method.

For engineering economic analysis purposes the requirements of a depreciation method are somewhat different. Obviously, it should provide for the recovery of capital and the proper assignment of depreciation cost over the estimated life of the asset. Property used in connection with the production of income is depreciated when its estimated life is greater than 1 year. The amount of depreciation claimed in a given year is influenced by an asset's value, estimated life, salvage value (if any), date in service, and the method of calculating depreciation. But, equally important, a depreciation method should account properly for the flow of capital funds that are recovered, and which thereby reduce the amount of capital remaining invested in a project. These recovered funds thus are available to the firm for other use or investment. Finally, the method used must permit the proper evaluation of the profitability of an investment being considered in an engineering economy study.

6.9

WHAT CAN BE DEPRECIATED?*

Depreciable property is property used in a business to produce income. The cost of depreciable property can be deducted from business income for income tax purposes over a future period of time. Many different kinds of property can be depreciated (e.g., machinery, buildings, vehicles, patents, copyrights, furniture, and equipment).

Depreciable property may be classified as *tangible* or *intangible*. Tangible property is any property that can be seen or touched. Intangible property is property, such as a copyright or franchise, that is not tangible. Additionally, depreciable property may be classified as *real* or *personal*. Personal property is any property, such as machinery or equipment, that is not real estate. Real property is land and generally anything that is erected on, growing on, or attached to land. However, land itself is never depreciable.

Property is depreciable if it meets these requirements:

1. It must be used in business or held for the production of income.
2. It must have a determinable life, and that life must be longer than one year.
3. It must be something that wears out, decays, gets used up, becomes obsolete, or loses value from natural causes.

In general, if property does not meet all three of these conditions, it is not depreciable.

6.10

THE TAX REFORM ACT OF 1986 AND ITS DEPRECIATION PROVISIONS

In the fall of 1986, Congress passed one of the most extensive income tax reforms in the history of the United States. This landmark legislation is the Tax Reform Act of 1986 (TRA 86), and the changes it made to income tax law are so far-reaching that every individual and corporate taxpayer in America will

*Various methods of depreciation approved by the Internal Revenue Service (IRS) are fully described in IRS Publication 534, *Depreciation*, which is published by the U.S. Government Printing Office. Portions of Chapter 6 are adapted from this reference.

be affected for years to come. TRA 86 became effective January 1, 1987, except for certain depreciation practices that were retroactive to August 1, 1986.

The Tax Reform Act of 1986 is intended to be "revenue neutral" over the 1987–1991 budget period. This means that aggregate federal income tax liabilities are not expected to change due to tax reform. According to the Congressional Joint Committee on Taxation, TRA 86 reduces individual income taxes by $122 billion, but matches this reduction by a nearly identical increase ($120 billion) in corporate income taxes. Thus, based on these estimates, total federal income tax revenues will be virtually unchanged compared to prior law; TRA 86 merely shifts more of the income tax burden from individuals to corporations.

What is the economic goal of those who sponsored and supported the Tax Reform Act of 1986? The primary objective of the legislation is to improve the efficiency and operation of the U.S. economy by relying more on market-oriented forces and less on tax considerations as the impetus behind economic decision making. TRA 86 recognizes that income taxes exert a fundamental influence on decisions that people make in their individual and business lives to work, save, invest, or take risk. By broadening the tax base and by reducing tax rates at both the individual and corporate levels, TRA 86 is intended to reduce the distorting impact of taxes and allow these decisions to be made more on the basis of other economic considerations and less on the basis of taxes. With market, rather than tax, considerations guiding economic behavior, the output of the economy should be (according to this view) more attuned to market forces.

6.10.1 Accelerated Cost Recovery System

The Tax Reform Act of 1986 revamps the Accelerated Cost Recovery System (ACRS) enacted in 1981, by which depreciation deductions for income tax purposes are determined. One of the more important provisions adapted from the Economic Recovery Tax Act (ERTA) of 1981 concerns ACRS. Hence, the original ACRS is first discussed, followed by a description of the modified ACRS established by the Tax Reform Act of 1986.

ACRS is mandatory for most tangible depreciable assets *placed in service* from January 1, 1981, through July 31, 1986. Previous methods had required estimates of useful life and salvage value. Under ACRS, however, salvage value and useful life estimates are not directly utilized in calculating depreciation.

ACRS allows a business to recover the cost *basis* of recovery property over a *recovery period*. The cost (*unadjusted*) basis is normally the cash purchase price of a property plus the cost of making the asset serviceable.

EXAMPLE 6-2

In 1984, your firm purchased a used machine for $10,500 for use in producing income. An additional $1,000 was spent to recondition the machine. What is the unadjusted basis?

Solution

The unadjusted basis in the machine is $11,500 (total costs of $10,500 plus $1,000).

A recovery period under ACRS is determined by the class life of the property: generally 3 years, 5 years, 10 years, or 15 years. The class life of a particular asset is obtained from IRS tables, as will be illustrated shortly. A recovery percentage for each year of the property's recovery period is obtained from Table 6-1 to compute ACRS depreciation deductions. The deduction is calculated as follows:

$$d_k(p) = r_k(p) \cdot B \tag{6-1}$$

where $d_k(p)$ = depreciation deduction in year k for recovery property class p

$r_k(p)$ = ACRS percentage (as a decimal) for year k in recovery property class p

B = unadjusted basis of the recovery property

ACRS cannot be used for property placed in service before 1981. Also, ACRS cannot be used for intangible depreciable property. Recovery property is tangible depreciable property placed in service after 1980, and it generally includes

TABLE 6–1 ACRS Percentages for Property Placed in Service from January 1, 1981, through July 31, 1986

	Applicable Percentage (r_k)			
Recovery Year (k)	3-Year Property	5-Year Property	10-Year Property	15-Year Property
1	25	15	8	5
2	38	22	14	10
3	37	21	12	9
4		21	10	8
5		21	10	7
6			10	7
7			9	6
8			9	6
9			9	6
10			9	6
11				6
12				6
13				6
14				6
15				6

Source: *Depreciation,* Internal Revenue Service Publication 534, U.S. Government Printing Office, December, 1987 (rev.).

new or used property acquired after 1980 for use in trade or business or to be held for the production of income. Property acquired and used for any purposes before 1981 is not recovery property.

ACRS, like the other methods, does not allow deductions for the cost of property, such as land, which has no determinable life. Furthermore, salvage value is treated as being equal to zero under ACRS. Do not reduce the cost basis by any salvage value when calculating deductions under ACRS.

EXAMPLE 6-3

An asset, purchased in early 1985, had a cost basis of $7,500 and was classified as 5-year ACRS property. The anticipated useful life was 7 years, and its salvage value at that time was estimated to be $1,000. (a) What was the depreciation deduction in 1986 (year 2)? (b) If the asset is sold at the end of 1988 for $2,000, what is the difference between the selling price and book value?

Solution

(a) From Table 6-1, the appropriate ACRS percentage allowance for the second year of ownership (r_2) is 22%. Thus, the depreciation deduction is $7,500(0.22) = $1,650, based on Equation 6-1. Notice that useful life and estimated salvage value do not affect the ACRS depreciation deduction calculation. (b) The book value of the asset is its initial cost basis less cumulative depreciation through the end of 1988:

$$\$7{,}500 - \$7{,}500(0.15 + 0.22 + 0.21 + 0.21) = \$1{,}575$$

Thus, the difference between selling price and book value is $2,000 − $1,575 = $425.

Under ACRS, each item of recovery property is assigned to a property class. These classes establish the recovery periods over which the unadjusted basis of items in a class are recovered. We shall concern ourselves with 3-, 5-, 10-, and 15-year property only and describe each further below.

A primary factor in determining the ACRS recovery period is the property's *asset depreciation range (ADR) guideline period*. These guideline periods were developed prior to 1981, and an abbreviated listing of ADR guideline periods is provided in Table 6-2.

For 3-, 5-, 10-, and 15-year property, the full first-year percentage applies no matter when in the tax year the property was placed in service. By knowing

TABLE 6–2 Selected Asset Depreciation Range Guideline Periods[a]

Description of Depreciable Assets	Guideline Period (years)
Transportation	
Automobiles, taxis	3
Buses	9
General-purpose trucks:	
Light	4
Heavy	6
Air transport (commercial)	12
Petroleum	
Exploration and drilling assets	14
Refining and marketing assets	16
Manufacturing	
Sugar and sugar products	18
Tobacco and tobacco products	15
Carpets and apparel	9
Lumber, wood products, and furniture	10
Chemicals and allied products	9.5
Cement	20
Fabricated metal products	12
Electronic components	6
Rubber products	14
Communication	
Telephone	
Central-office buildings	45
Distribution poles, cables, etc.	34
Radio and television broadcasting	6
Electric utility	
Hydraulic plant	50
Nuclear	20
Transmission and distribution	30
Services	
Office furniture and equipment	10
Computers and peripheral equipment	6
Recreation—bowling alleys, theaters, etc.	10

[a] *Source: Depreciation,* Internal Revenue Service Publication 534, U.S. Government Printing Office, December 1981 (rev.).

the ADR guideline period from Table 6-2, it is generally possible to determine a property's ACRS class life, as illustrated in Table 6-3.

Observe that the ACRS percentages in Table 6-1 were applicable to depreciable property placed in service during 1981–1986. A realistic situation involving ACRS provisions is presented in Example 6-4. Modified ACRS percentages were established by the Tax Reform Act of 1986 and are discussed in the following section.

TABLE 6–3 ACRS Class Lives for Tangible Personal Property, Based on ADR Guideline Period[a]

ACRS Class Life	ADR Guideline Period	Examples of Property
3-year	4 years or less	Automobiles, taxis, and light duty trucks; also property used in connection with research and experimentation.
5-year	Greater than 4 years but less than or equal to 18.5 years	Most kinds of production machinery and office furniture, computers, copiers, and general-purpose tools such as drills.
10-year	Greater than 18.5 years but less than or equal to 25 years	Public utility equipment
15-year	More than 25 years	Public utility property

[a] *Source: Depreciation,* Internal Revenue Service Publication 534, U.S. Government Printing Office, December 1987 (rev.).

EXAMPLE 6-4

In May of 1984 your company traded in a computer, used in its business, which had a book value at that time of $25,000. A new, faster computer system having a fair market value of $400,000 was acquired. Because the vendor accepted the older computer as a trade-in, a deal was agreed to whereby your company would pay $325,000 cash for the new computer system.

(a) What is the ACRS class life of the new computer?

(b) How much depreciation can be deducted each year based on this class life?

Solution

(a) The new computer has an ADR guideline period of 6 years (see Table 6-2). Hence, its ACRS recovery period is 5 years (see Table 6-3).

(b) The cost (unadjusted) basis for this property is $350,000, which is the sum of the $325,000 cash price of the computer and the $25,000 that was allowed on the trade-in (in this case the trade-in was treated as a nontaxable transaction).

ACRS percentages that apply to the $350,000 cost basis are found in Table 6-1. An allowance (half-year) is built into the year 1 percentage, so it does not matter that the computer was purchased in May of 1984 instead of, say,

November of 1984. The depreciation deductions for 1984 through 1988 can be computed with Equation 6-1 and the 5-year ACRS percentages as follows:

Property	Date Placed in Service	Cost Basis	ADR Guideline Period	ACRS Recovery Period
Computer System	May 1984	$350,000	6 years	5 years
	Year	**Depreciation Deductions**		
	1984	15% × 350,000 = $ 52,500		
	1985	22% × 350,000 = 77,000		
	1986	21% × 350,000 = 73,500		
	1987	21% × 350,000 = 73,500		
	1988	21% × 350,000 = 73,500		
		Total $350,000		

6.10.2 Modified ACRS

The modified ACRS (MACRS) method created by the Tax Reform Act of 1986 is now the principal means for computing depreciation. The main differences between the prior and the new law are the class lives of assets and the methods for recovering their costs. The new system continues to rely on the pre-1981 Asset Depreciation Range (ADR) guideline periods as the primary basis for classifying assets (see Table 6-2). Under MACRS, assets are assigned to one of six classes of depreciable personal property or to one of two classes of real property. The cost basis of assets in the more short-lived 3-, 5-, 7- and 10-year property classes is recovered at a rate based on the 200% declining balance depreciation method. A switch to the straight-line method is permitted when needed to optimize the deductions. Costs of assets in the 15- and 20-year classes are recovered using the 150% declining balance method, again switching to the straight-line method at the optimal time. *The declining balance and straight-line methods of depreciation are discussed in a later section, as is determination of the optimal switchover time.* The straight-line method must be used for all real estate. A summary of MACRS class lives, permissible depreciation methods, and special rules included in the Tax Reform Act of 1986 is provided in Table 6-4.

Because MACRS as established by the Tax Reform Act of 1986 does not provide statutory cost-recovery allowances, averaging conventions are needed. Generally, a half-year convention applies to first-year and last-year allowances for depreciable personal property, and a mid-month convention applies to real property. Under a special rule, when more than 40% of asset additions in any year are placed in service in the last quarter of the taxable year, a mid-quarter convention is used to compute the cost-recovery allowances for all additions during the year.

TABLE 6–4 MACRS Class Lives and Permissible Methods for Calculating Depreciation Rates[a]

MACRS Class Life and Depreciation Method	ADR Guideline Period	Special Rules
3-year, 200% declining balance	4 years or less	Includes some race horses. Excludes cars and light trucks.
5-year, 200% declining balance	More than 4 years to less than 10	Includes cars and light trucks, semiconductor manufacturing equipment, qualified technological equipment, computer-based central office switching equipment, some renewable and biomass power facilities, and research and development property.
7-year, 200% declining balance	10 years to less than 16	Includes single-purpose agricultural and horticultural structures and railroad track. Includes property with no ADR midpoint.
10-year, 200% declining balance	16 years to less than 20	None.
15-year, 150% declining balance	20 years to less than 25	Includes sewage treatment plants, telephone distribution plants, and equipment for two-way voice and data communication.
20-year, 150% declining balance	25 years or more	Excludes real property with ADR midpoint of 27.5 years or more. Includes municipal sewers.
27.5-year, straight-line	N/A	Residential rental property.
31.5 year, straight-line	N/A	Nonresidential real property.

[a]*Source:* Arthur Andersen and Company, *Tax Reform 1986: Analysis and Planning,* Chicago, 1986, p. 112. Reproduced with permission of Arthur Andersen & Co.

6.10.2.1 Half-Year Convention

Under 1981 ACRS rules, a half-year convention was built into the statutory allowance in the year the asset was placed in service, but the cost of 3-year property, for example, was still recovered within 3 tax years. Starting in 1987 under MACRS, the computational mechanics of the half-year convention for personal property could be viewed as effectively adding another tax year onto the cost-recovery period. For example, the cost of 3-year property is recovered by using the half-year convention for the first year, then the full allowance for the second and third years is deducted, and any remaining balance is deducted for the fourth year. The recovery period is still 3 years, but the deductions are spread over 4 tax years.

Under MACRS, the half-year convention treats all property placed in service, or disposed of, during a tax year as placed in service, or disposed of, at the midpoint of that tax year. A half-year of depreciation is allowable for the first year during which property is placed in service, *regardless* of when the property is actually placed in service during the tax year. For each of the remaining years of the recovery period, a full year of depreciation is taken. If property is held for the entire recovery period, a half-year of depreciation (i.e., the remaining book value) is allowable for the year following the end of the recovery period. If property is disposed of before the end of the recovery period, only a half-year of depreciation is allowable for the year of disposal (i.e. sale).

6.10.2.2 Mid-quarter Convention

If during any tax year the aggregate cost bases of all property placed in service during the last 3 months of that tax year exceed 40% of the aggregate bases of all property placed in service during the year, a mid-quarter convention is utilized for calculating all depreciation deductions. The basis of either residential rental property or nonresidential real property cannot be included in the aggregate bases of the property when making this calculation. Under a mid-quarter convention, all property placed in service, or disposed of, during any quarter of a tax year is treated as placed in service, or sold, at the midpoint of the quarter.

To determine the MACRS deduction for property subject to the mid-quarter convention, first calculate depreciation for the *full* tax year and then multiply by the following percentages for the quarter of the tax year the property is placed in service.

Quarter of tax year	*Percentage*
First	87.5%
Second	62.5%
Third	37.5%
Fourth	12.5%

6.10.2.3 MACRS Examples

Table 6-5 provides a summary of the MACRS rates that result from the conditions set forth in Table 6-4 and the half-year averaging convention discussed in Section 6.10.2.1. Examples of the MACRS method using Table 6-5 in this section are followed by a discussion of historical depreciation methods (including the declining balance and straight-line methods) that are the basis for computing the rates in Table 6-5. Next, the 5-year modified ACRS rates from Table 6-5 are derived for the student in Section 6.10.5. An overview of the computation of depreciation deductions prescribed by the Tax Reform Act of 1986 is given in Figure 6-1.

EXAMPLE 6-5

A firm purchased a new piece of semiconductor manufacturing equipment in July of 1988. The cost basis for the equipment is $100,000. Determine (a) the

TABLE 6–5 Modified ACRS Rates per Tax Reform Act of 1986

	Class Life					
Year	3-year[a]	5-year[a]	7-year[a]	10-year[a]	15-year[b]	20-year[b]
1	0.3333	0.2000	0.1429	0.1000	0.0500	0.0375
2	0.4445	0.3200	0.2449	0.1800	0.0950	0.0722
3	0.1481	0.1920	0.1749	0.1440	0.0855	0.0668
4	0.0741	0.1152	0.1249	0.1152	0.0770	0.0618
5		0.1152	0.0893	0.0922	0.0693	0.0571
6		0.0576	0.0892	0.0737	0.0623	0.0528
7			0.0893	0.0655	0.0590	0.0489
8			0.0446	0.0655	0.0590	0.0452
9				0.0656	0.0591	0.0447
10				0.0655	0.0590	0.0447
11				0.0328	0.0591	0.0446
12					0.0590	0.0446
13					0.0591	0.0446
14					0.0590	0.0446
15					0.0591	0.0446
16					0.0295	0.0446
17						0.0446
18						0.0446
19						0.0446
20						0.0446
21						0.0223

[a] These rates are determined by applying the 200% declining balance method to the appropriate class life with the half-year convention applied to the first and last years. Rates for each class life must sum to 1.0000.

[b] These rates are determined with the 150% declining balance method instead of the 200% declining balance method and are rounded off to 4 significant digits.

Source: Depreciation, Internal Revenue Service Publication 534, U.S. Government Printing Office, December, 1987 (rev.).

FIGURE 6-1

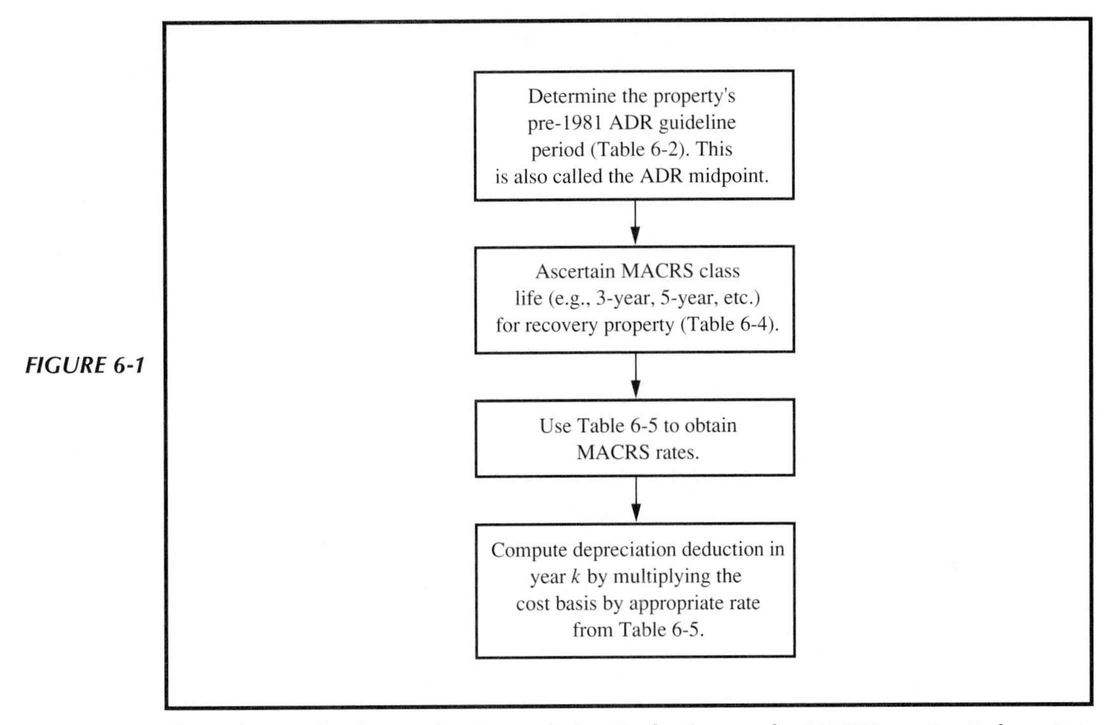

Flow Diagram for Computing Depreciation Deductions under MACRS per Tax Reform Act of 1986

depreciation charge permissible in the fourth year (1991), (b) the book value at the end of 1989, and (c) the cumulative depreciation through 1990.

Solution

From Table 6-2, it may be seen that semiconductor (electronic) manufacturing equipment has an ADR guideline period of 6 years, and from Table 6-4 it has a 5-year MACRS class life. The rates that apply are given in Table 6-5.

(a) The depreciation deduction, or capital recovery allowance, that is allowable in 1991 (d_4) is 0.1152 (\$100,000) = \$11,520.

(b) The book value at the end of 1989 (BV_2) is the cost basis less depreciation charges in years 1 and 2:

$$BV_2 = \$100{,}000 - \$100{,}000(0.20 + 0.32) = \$48{,}000$$

(c) Accumulated depreciation through 1990, D_3^*, is the sum of depreciation in 1988, 1989, and 1990, or

$$\begin{aligned} D_3^* &= d_1 + d_2 + d_3 \\ &= \$100,000(0.20 + 0.32 + 0.192) \\ &= \$71,200 \end{aligned}$$

A computer program called ATAX, which is described in Appendix F, is designed to compute depreciation and perform after-tax economic evaluations. ATAX is used in Example 6-5A to illustrate several depreciation calculations.

EXAMPLE 6-5A

Repeat the calculations in parts a, b, and c of Example 6-5 by using the ATAX computer program in Appendix F.

```
MAIN MENU

1. Asset placed in service on, or before December
   31, 1980.
   Applicable methods: Straight Line (SL)
                       Sum of Years Digits (SYD)
                       Declining Balance (DB)

2. Asset placed in service between January 1,
   1981 and July 31, 1986.
   Applicable method: Accelerated Cost Recovery
                      System (ACRS)

3. Asset placed in service between August 1,
   1986 to present.
   Applicable method: Modified ACRS (MACRS)
0. Exit to system

   Enter option: 3

Menu 3: MACRS

1. Calculate Depreciation of the asset.
2. Evaluate an Investment.
0. Return to last menu.
   Enter option: 1

3.1 MACRS Method
Enter the Useful Life of the Asset in Years (N): 6
Enter the Investment (I): 100000
Enter the Class Life: 5
```

```
You entered:
N = 6
I = 100000
Class = 5
Is this correct (y/n) ?: y

MACRS Depreciation.
```

Yr	EOY Book Value	Depreciation	MACRS	Cumulative
1	80000.00	20000.00	0.2000	20000.00
2	48000.00	32000.00	0.3200	52000.00
3	28800.00	19200.00	0.1920	71200.00
4	17280.00	11520.00	0.1152	82720.00
5	5760.00	11520.00	0.1152	94240.00
6	0.00	5760.00	0.0576	100000.00

```
I = 100000   N = 6   MACRS CLASS = 5

Do you want to print the table (y/n)? y
```

EXAMPLE 6-6

During 1989, your firm purchases a machine for $4,000, which is placed in service in January; office furniture for $1,000, which is placed in service in September; and a computer for $5,000, which is placed in service in October. A calendar year is used as your firm's tax year. The firm uses MACRS depreciation for the machine, the office furniture, and the computer. The aggregate cost basis of all property placed in service in 1989 is $10,000. Since the basis of the computer ($5,000), which is placed in service during the last 3 months of the tax year, exceeds 40% of the aggregate basis of all property ($10,000) placed in service during 1989, the mid-quarter convention must be utilized for all property. The machine placed in service in January is 5-year property, the office furniture is 7-year property, and the computer is 5-year property. All are depreciated using MACRS with the mid-quarter convention. (a) What depreciation deduction is permissible in 1989? (b) What depreciation deduction is permissible in 1990 for the property that was placed in service during 1989?

Solution

(a) The depreciation for the furniture is determined by first dividing 1 by 7 to get 14.29%. Since 7-year property is depreciated using the 200% declining balance method (see Table 6-4), multiply 14.29% by 2 to get the 200% declining balance rate of 28.6%. Then multiply this rate by $1,000 (basis) to get the depreciation of \$286 for a full year.* The furniture was placed in service in the third quarter of the tax year; therefore, multiply \$286 by 37.5% (mid-quarter percentage for the third quarter) to get a deduction for the furniture of \$107 (d_1) for 1989.

The depreciation for the machine and the computer is calculated by first dividing 1 by 5 to get 20%. Since 5-year property is depreciated using the 200% declining balance method, multiply 20% by 2 to get the 200% declining balance rate of 40%. Then multiply the basis of the machine (\$4,000) and the basis of the computer (\$5,000) by 40% to get depreciation deductions of \$1,600 and \$2,000, respectively, for a full year. The machine was placed in service in the first quarter, so the \$1,600 is multiplied by 0.875 to obtain a first-year depreciation (d_1) of \$1,400. The computer was placed in service in the fourth quarter of the tax year; therefore, multiply the \$2,000 by 12.5% (mid-quarter percentage for the fourth quarter) to get a deduction for the computer of \$250 for 1989.

(b) From (a), we can determine the book value at the *end* of 1989 to be

$$\begin{aligned}\text{Machine: } & \$4{,}000 - \$1{,}400 = \$2{,}600\\ \text{Furniture: } & \$1{,}000 - \$107 = \$893\\ \text{Computer: } & \$5{,}000 - \$250 = \$4{,}750\end{aligned}$$

The 1990 depreciation deductions (d_2) for these assets can then be found by multiplying the book value by the 200% declining rate:

$$\begin{aligned}\text{Machine } & \$2{,}600(2/5) = \$1{,}040\\ \text{Furniture } & \$893(2/7) = \$255\\ \text{Computer } & \$4{,}750(2/5) = \$1{,}900\end{aligned}$$

6.10.3 The Alternate MACRS Method

The Tax Reform Act of 1986 provides for the use of an alternate method. Election to adopt the alternate MACRS method for a class of property applies to all property in that class that is placed in service during the tax year. The decision to use the alternate MACRS method, once made, is irrevocable for the assets involved.

A company may use the alternate MACRS method for most property. Under this method, depreciation is calculated using the straight-line method of depreciation with no salvage value. Regarding automobiles, light general purpose trucks, and any computer or peripheral equipment, the straight-line method is applied over a 5-year recovery period with a half-year convention. For most other property, the straight-line method is applied with a half-year convention over the MACRS class life of the property.

*200% declining balance depreciation is discussed in Section 6.10.4.2.

6.10.4 Methods Used to Develop Modified ACRS Rates

This section describes the straight-line and declining balance methods of computing depreciation and illustrates the optimal switchover from an accelerated declining balance method to the straight-line method. Recall that the salvage value at the end of ACRS and MACRS class life is zero for all rates in Tables 6-1 and 6-5. In the following section (6.10.5), we derive the five-year MACRS rates which utilize optimal switchover and half-year convention as mandated by the Tax Reform Act of 1986.

6.10.4.1 The Straight-Line Method

The straight-line method of computing depreciation assumes that the loss in value is directly proportional to the age of the asset. This straight-line relationship gives rise to the name of the method. The following definitions are used in the equations below. If we define:

N = depreciable life of the asset in years
B = unadjusted cost basis
d_k = annual depreciation deduction in the kth year ($1 \le k \le N$)
BV_k = book value at the end of year k
S = salvage value at the end of the depreciable life of the asset
D_k^* = cumulative depreciation through year k

then

$$d_k = \frac{B - S}{N} \tag{6-2}$$

$$D_k^* = \frac{k(B - S)}{N} \tag{6-3}$$

$$BV_k = B - D_k^* \tag{6-4}$$

This method of computing depreciation is widely used. It is simple and gives a uniform annual depreciation charge. Its proponents argue that inasmuch as other costs, as well as the depreciable life, must be estimated, there is little reason for attempting to use a more complex formula.

6.10.4.2 The Declining Balance Method

In the declining balance method, sometimes called the *constant percentage method* or the Matheson formula, it is assumed that the annual cost of depreciation is a fixed percentage of the book value at the *beginning* of the year. The ratio of the depreciation in any one year to the book value at the beginning of the year is constant throughout the life of the asset and is designated by R ($0 \le R \le 1$). In this method $R = 2/N$ when, for example, 200% declining balance is being used, and N equals the depreciable life of an asset. The following relationships hold true for the declining balance method:

$$d_1 = B(R) \tag{6-5}$$

$$d_k = B(1 - R)^{k-1}(R) \tag{6-6}$$

$$D_k^* = B[1 - (1 - R)^k] \tag{6-7}$$

$$BV_k = B(1 - R)^k \tag{6-8}$$

$$BV_N = B(1 - R)^N \tag{6-9}$$

Even though the declining balance procedure is rather simple to apply, it has two weaknesses. The annual cost of depreciation is different each year and, from a calculation viewpoint, this is inconvenient. Also, with this formula an asset can never depreciate to zero value because the salvage value (S) is not utilized in Equations 6-5 to 6-9. This is not a serious difficulty since a switch from the declining balance method to any slower method of depreciation (such as straight-line) can be made so that a zero (or some other) book value in year N is reached. With regard to Table 6-4, values of R that are utilized in developing the MACRS rates are $2/N$ (200% declining balance) and $1.5/N$ (150% declining balance).

Proponents of the declining balance method assert that its results more nearly parallel the actual market value of an asset than do those obtained by the straight-line method. This undoubtedly is true in the case of such things as automobiles, where new models and style changes are large factors in the establishment of the market value. However, it may not be true of many industrial and commercial structures and some equipment. The following example illustrates the straight-line and declining balance depreciation methods.

EXAMPLE 6-7

A new electric saw for cutting lumber in a furniture manufacturing plant has a cost basis of $4,000 and a 10-year ADR guideline period. It was placed in service on January 1, 1980. The estimated salvage value of the saw is zero at the end of 10 years. What will be the depreciation deduction during the sixth year, the book value at the end of the sixth year, and the cumulative depreciation cost through the sixth year?

Solution by the Straight-Line Method

Applying Equations 6-2, 6-3, and 6-4 to Example 6-7, we obtain

$$d_6 = \frac{\$4,000 - 0}{10} = \$400$$

$$D_6^* = \frac{6(\$4,000 - 0)}{10} = \$2,400$$

$$BV_6 = \$4,000 - \frac{6(\$4,000 - 0)}{10} = \$1,600$$

Solution by the 200 Percent Declining Balance Method

Applying the declining balance relationships in Equations 6-6, 6-7, and 6-8 to Example 6-7, we obtain

$$R = 2/N = 0.2$$
$$d_6 = \$4,000(1 - 0.2)^5(0.2) = \$262.14$$
$$D_6^* = \$4,000[1 - (1 - 0.2)^6] = \$2,951.42$$
$$BV_6 = \$4,000(1 - 0.2)^6 = \$1,048.58$$

Solution by the 150 Percent Declining Balance Method

In this situation $R = 1.5/N = 0.15$, and we obtain these results:

$$d_6 = \$4,000(1 - 0.15)^5(0.15) = \$266.22$$
$$D_6^* = \$4,000[1 - (1 - 0.15)^6] = \$2,491.60$$
$$BV_6 = \$4,000(1 - 0.15)^6 = \$1,508.60$$

EXAMPLE 6-7A

Determine d_6, D_6^*, and BV_6 with ATAX (Appendix F) for each of the three depreciation methods illustrated in Example 6-7.

```
Menu 1: SL, SYD, DB

1. Calculate Depreciation of the asset.
2. Evaluate an Investment.
0. Return to last menu.

Enter option :1
1.1.1 Straight Line Method
Enter the useful life of the asset in years (N):
                                              10
Enter the investment (I): 4000
Enter salvage value at the end of year 10 (S): 0

You entered:
N = 10
I = 4000
S = 0
```

```
Is this correct (y/n) ?: y

Straight Line Depreciation.
```

Yr	EOY Book Value	Depreciation	Cumulative
1	3600.00	400.00	400.00
2	3200.00	400.00	800.00
3	2800.00	400.00	1200.00
4	2400.00	400.00	1600.00
5	2000.00	400.00	2000.00
6	1600.00	400.00	2400.00
7	1200.00	400.00	2800.00
8	800.00	400.00	3200.00
9	400.00	400.00	3600.00
10	0.00	400.00	4000.00

```
I = 4000  N = 10  S = 0
```

```
1.1.3 Declining Balance Method
Enter the useful life of the asset in years (N):
                                              10
Enter the investment (I): 4000
Enter the Declining Balance Rate, e.g. 1.5, 2.0
                                        (R):2.0
You entered:
N = 10
I = 4000
R = 2.0

Is this correct (y/n) ?y

Declining Balance Depreciation.
```

Yr	EOY Book Value	Depreciation	Cumulative
1	3200.00	800.00	800.00
2	2560.00	640.00	1440.00
3	2048.00	512.00	1952.00
4	1638.40	409.60	2361.60
5	1310.72	327.68	2689.28
6	1048.58	262.14	2951.42
7	838.86	209.72	3161.14
8	671.09	167.77	3328.91
9	536.87	134.22	3463.13
10	429.50	107.37	3570.50

```
I = 4000  N = 10  R = 2
```

```
1.1.3 Declining Balance Method

Enter the useful life of the asset in years (N):
                                              10
Enter the investment (I): 4000
Enter the Declining Balance Rate, e.g. 1.5, 2.0
                                        (R): 1.5

You entered:
N = 10
I = 4000
R = 1.5

Is this correct (Y/N)? Y

Declining Balance Depreciation.
```

Yr	EOY Book Value	Depreciation	Cumulative
1	3400.00	600.00	600.00
2	2890.00	510.00	1110.00
3	2456.50	433.50	1543.50
4	2088.02	368.48	1911.97
5	1774.82	313.20	2225.18
6	1508.60	266.22	2491.40
7	1282.31	226.29	2717.69
8	1089.96	192.35	2910.04
9	926.47	163.49	3073.53
10	787.50	138.97	3212.50

```
I = 4000   N = 10   R = 1.5
```

6.10.4.3 Declining Balance with Switchover to Straight Line

Because the declining balance method never reaches a book value of zero, it is permissible to switch from this method to the straight-line method so that an asset's BV_N will be zero (or some other desired salvage amount).

Table 6-6 illustrates a switchover from double declining balance depreciation to straight-line depreciation for Example 6-7. The switchover occurs in the year where a larger depreciation amount is obtained from the straight-line method. From Table 6-6, it is apparent that d_6 = \$262.14. The book value at the end of year 6 (BV_6) is \$1,048.58. Additionally, the reader will observe that BV_{10} is \$4,000 − \$3,570.50 = \$429.50 without switchover to the straight-line method in Table 6-6. With switchover, BV_{10} equals 0. It is clear that this asset's d_k, D_k^*, and BV_k in years 7 through 10 are established from the straight-line method which

TABLE 6–6 Switchover from the 200% Declining Balance Method to the Straight-Line Method

		Depreciation Method		
Year, k	(1) Beginning-of-Year Book Value[a]	(2) 200% Declining Balance Method[b]	(3) Straight-Line Method[c]	(4) Depreciation Amount Selected[d]
1	$4,000.00	$ 800.00	> $400.00	$ 800.00
2	3,200.00	640.00	> 355.56	640.00
3	2,560.00	512.00	> 320.00	512.00
4	2,048.00	409.60	> 292.57	409.60
5	1,638.40	327.68	> 273.07	327.68
6	1,310.72	262.14	= 262.14	262.14 (switch)
7	1,048.58	209.72	< 262.14	262.14
8	786.44	167.77	< 262.14	262.14
9	524.30	134.22	< 262.14	262.14
10	262.16	107.37	< 262.14	262.14
		$3,570.50		$4,000.00

[a] Column 1 for year k less column 4 for year k equals the entry in column 1 for year $k + 1$.
[b] 20% ($= 2/N$) of column 1.
[c] Column 1 divided by remaining years from beginning of year through the tenth year.
[d] Select the larger amount in column 2 or column 3.

permits the full cost basis to be depreciated over the 10-year ADR guideline period.

EXAMPLE 6-8

A large manufacturer of sheet metal products in the Midwest purchased a new, modern computer-controlled flexible manufacturing system in October of 1991 for $3.0 million. Because this company would not be profitable until the new technology had been in place for several years, it elected to utilize alternate MACRS in computing its depreciation deductions. Thus, the company could "slow down" its depreciation allowances in hopes of postponing its income tax advantages until it became a profitable concern. What depreciation allowances can be claimed for the new system?

Solution

From Table 6-2, the ADR guideline period for a manufacturer of fabricated metal products is 12 years. This flexible manufacturing system would normally be depreciated using 7-year modified ACRS rates (Table 6-4). However, under the alternate MACRS method, the straight-line method with no salvage value is applied to the 7-year recovery period using half-year convention. Consequently, depreciation in year 1 (1991) would be

$$\frac{1}{2}\left(\frac{\$3,000,000}{7}\right) = \$214,286$$

Depreciation deductions in years 2–7 (1992–1997) would be $428,571 each year, and depreciation in year 8 (1998) would be $214,288. Notice how the half-year convention extends depreciation deductions over 8 years.

6.10.5 Development of MACRS Rates for Class Life of Five Years

It is now possible to make use of the 200% declining balance method and the straight-line method to illustrate how the MACRS rates shown in Table 6-5 were developed in accordance with the Tax Reform Act of 1986.

As mentioned previously, the MACRS recovery periods and their associated rates are based on pre-1981 ADR guideline periods. Furthermore, the MACRS rates are determined by switching from a 200% or 150% declining balance method to the straight-line method in order to optimize depreciation deductions. Finally, a half-year convention applies to first- and last-year allowances for depreciable personal property, and a mid-month convention applies to real property. (In this chapter we have confined our attention to personal property.)

Generally speaking, the half-year convention is used to simplify the prorating of first-year depreciation because it assumes that property is placed in service at the midpoint of the first year and removed from use halfway into service year $N + 1$. For example, there is no need to prorate 8/12 of an asset's first-year depreciation when the item is placed in service on May 1. If the asset has a class life of 5 years, some residual depreciation will be claimed in year 6; for a class life of N years, depreciation must be taken over $N + 1$ years. Special provisions may apply in some cases such that the mid-quarter convention previously discussed may be necessary.

All of these conditions are now utilized to develop MACRS rates for a class life of 5 years. Table 6-7 shows the calculation of each year's depreciation by the 200% declining balance method and straight-line method so the larger depreciation allowance can be selected. Notice the half-year convention affects the year 1 and year 6 depreciation rates. Also note that the switchover to straight-line depreciation occurs during year 5 and that zero salvage value is built into the rates.

EXAMPLE 6-9

In late 1986, the ABC company purchased for $10,000 a depreciable asset having a 10-year ADR guideline period and a 7-year MACRS class life. The company elected to use MACRS for the property, and the cost basis of the property was $10,000. What are the depreciation deductions in each of the 8 years for which depreciation can be claimed (1986–1993)?

Solution

Since this is 7-year property, first divide 1 by 7 to get the basic rate of 1/7, or 14.29%. Multiply 14.29% by 2 to determine the 200% declining balance rate

TABLE 6–7 Development of MACRS Rates for 5-Year Personal Property

Beginning of Year	(A) Depreciation with 200% Declining Balance Method[a]	(B) Depreciation with Straight Line Method[b]	(C) = Max{ A,B} MACRS Rate
1	$0.4(B)(0.5) = 0.2B$[c]	$\frac{B}{5.5}(0.5) = 0.091B$	$0.2B$
2	$0.4(B - 0.2B) = 0.32B$	$\frac{1}{4.5}(B - 0.2B) = 0.178B$	$0.32B$
3	$0.4(B - 0.2B - 0.32B)$ $= 0.192B$	$\frac{1}{3.5}(B - 0.2B - 0.32B)$ $= 0.137B$	$0.192B$
4	$0.4(B - 0.2B - 0.32B$ $- 0.192B) = 0.1152B$	$\frac{1}{2.5}(B - 0.2B - 0.32B$ $- 0.192B) = 0.1152B$	$0.1152B$
5	$0.4(B - 0.2B - 0.32B - 0.192B$ $- 0.192B - 0.1152B)$ $= 0.0691B$	$\frac{1}{1.5}(B - 0.2B - 0.32B$ $-0.192B - 0.1152B)$ $= 0.1152B)$	$0.1152B$[d]
6	Not applicable	$0.1728B - 0.1152B$ $= 0.0576B$	$0.0576B$

[a] $R = 2/N = 0.4$, and half-year convention applies to year 1 depreciation.

[b] Salvage value is zero at end of class life, and half-year convention applies to year 1 depreciation.

[c] B denotes the unadjusted basis of recovery property (see Section 6.10.1) and typically remains constant over the asset's class life.

[d] Switch from 200% declining balance to straight line method occurs in year 5.

of 28.58%. Then multiply \$10,000 by 28.58% to get \$2,858, which is adjusted for half-year convention to arrive at depreciation for 1986 of \$1,429.

For 1987, the depreciation deduction will be figured by subtracting \$1,429 from \$10,000 to get the book value of the property (\$8,571). Multiply \$8,571 by 28.58% to determine the depreciation deduction of \$2,450 for 1987. For 1988, following the same procedure, multiply (\$8,571 − \$2,450 =) \$6,121 by 28.58% to get a depreciation deduction of \$1,749. For 1989, multiply (\$6,121 − \$1,749 =) \$4,372 by 28.58% for a deduction of \$1,249. The remaining basis at the beginning of 1990 will be \$3,122 and the declining balance deduction would be \$3,122(0.2858) = \$892.

By switching to the straight-line method in 1991, the deduction would be \$892, which is \$2,230 divided by the 2.5 remaining years in the depreciation period, for both 1991 and 1992. For 1993 (final year) it is possible to deduct \$447 (a half-year depreciation). The total depreciation that will have been claimed for this property is its original cost of \$10,000.

EXAMPLE 6-9A

Use the ATAX computer program to verify the calculations in Example 6-9.

Solution

```
3.1 MACRS Method
Enter the Useful Life of the Asset in Years (N):
                                              10

Enter the Investment (I): 10000
Enter the MACRS Class Life: 7

You entered:
N = 10
I = 10000
Class = 7
Is this correct (y/n) ?: y

MACRS Depreciation.
```

Yr		EOY Book Value	Depreciation[a]	MACRS	Cumulative
1	(1986)	8571.00	1429.00	0.1429	1429.00
2		6122.00	2449.00	0.2449	3878.00
3		4373.00	1749.00	0.1749	5627.00
4		3124.00	1249.00	0.1249	6876.00
5		2231.00	893.00	0.0893	7769.00
6		1338.00	893.00	0.0893	8662.00
7		446.00	892.00	0.0892	9554.00
8	(1993)	0.00	446.00	0.0446	10000.00
9		0.00	0.00	0.0000	10000.00
10		0.00	0.00	0.0000	10000.00

```
I = 10000   N = 10   MACRS CLASS = 7
```

[a]Minor differences of $2 or less exist between the solutions of Example 6-9 and the ATAX program because of round-off error in the manual calculations.

6.11 MULTIPLE ASSET DEPRECIATION ACCOUNTING

To this point we have considered depreciation on individual assets. Very frequently, for accounting purposes, identical or even somewhat dissimilar assets may be considered as a group in regard to depreciation. This group procedure has some advantages.

Although it may be known that the *average* life of assets in a group will be N years, it is recognized that some items will last less than N years, and some will last longer than N years. If N years is assumed to be the depreciable life of an individual asset, and if it should be retired from service in less than N years at a salvage value less than the book value, there would be a loss of capital. On the other hand, if it should be used for more than N years and then sold for any amount, a gain of capital would be realized. By using *multiple asset* accounting, adjustments for such losses or gains are avoided since such a "dispersion effect" in the lives is recognized when the average life for the group is adopted.

It is fairly common practice, for tax purposes, to use item accounts for buildings, structures, and high-value equipment and to use group accounts for general equipment. In the following portions of this book we shall assume item depreciation, since this will provide some degree of simplicity and will highlight the effects of income taxes relative to gains and losses that may occur.

6.12 DEPLETION

When natural resources are being consumed in producing products or services, the term *depletion* is used to indicate the decrease in value of the resource base that has occurred. The term is commonly used in connection with mining properties, oil and gas wells, timber lands, and so on. In any given parcel of mineral property, for example, there is a definite quantity of ore, oil, or gas available. As some of the resource is extracted and sold, the reserve decreases and the value of the property normally diminishes.

However, there is a difference in the manner in which the amounts recovered through depletion and depreciation must be handled. In the case of depreciation, the property involved usually may be replaced with similar property when it has become fully depreciated. In the case of depletion of mineral or other natural resources, such replacement usually is not possible. Once the gold has been removed from a mine, or the oil from an oil well, it cannot be replaced. Thus, in a manufacturing or other business where depreciation occurs, the principle of maintenance of capital is practiced, and the amounts charged for depreciation expense are reinvested in new equipment so that the business may continue in operation indefinitely. On the other hand, in the case of a mining or other mineral industry, the amounts charged as depletion cannot be used to replace the sold natural resource, and the company, in effect, may sell itself out of business, bit by bit, as it carries out its normal operations. Such companies frequently pay out to the owners each year the amounts recovered as depletion. Thus, the annual payment to the owners is made up of two parts: (1) the profit that has been earned, and (2) a portion of the owner's capital that is being returned, marked as depletion. In such cases, if the natural resource were eventually completely consumed, the company would be out of business, and the stockholder would hold stock that was theoretically worthless but would have received back all of his or her invested capital.

In the actual operation of many natural resource businesses, the depletion funds may be used to acquire new properties, such as new mines and oil-producing properties, and thus give continuity to the enterprise.

There are two ways to compute depletion allowances: (1) the cost method and (2) the percentage method. The cost method applies to all types of property subject to depletion and is the more widely used method. Under the cost method a *depletion unit* is determined by dividing the adjusted cost basis of a property by the number of units remaining to be mined or harvested. (Units may be feet of timber, tons of ore, etc.) The deduction (depletion allowance) for a given tax year is then calculated as the product of number of units *sold* during the year and the depletion unit, in dollars.

In practice, depletion may also be based on a percentage of the year's income in accordance with IRS regulations. Depletion allowances on mines and other natural deposits, including geothermal deposits, may be computed as a percentage of gross income, provided that the amount charged for depletion does not exceed 50% of the *net income* before deduction of the depletion allowance. The percentage method can be used for most types of metal mines, geothermal deposits and coal mines, but not for timber. Generally, the use of percentage depletion for oil and gas is not allowed except for certain domestic oil and gas production. Typical percentage depletion allowances* in effect in 1988 were

Sulphur and uranium; domestically mined lead, zinc, nickel, and asbestos	22%
Gold, silver, copper, iron ore, and oil shale from U.S. deposits; geothermal wells in the United States	15%
Coal, lignite, and sodium chloride	10%
Clay, gravel, sand, and stone	5%

It is possible that the total amount charged for depletion over the life of a property under this procedure may be far more than the original cost. When the percentage method applies to a property, depletion allowances must be calculated by both the cost method and the percentage method. The larger

*Depletion allowances are established by the IRS and may be revised with new federal income tax legislation.

allowance may be taken and used to reduce the basis of the property for purposes of refiguring the depletion unit as necessary. Figure 6-2 provides the logic for determining whether percentage or cost depletion is allowable in a given tax year.

Example 6-10 illustrates the cost method of determining a depletion allowance.

EXAMPLE 6-10

The WGS Zinc Company bought an ore-bearing parcel of land in January, 1989, for $2,000,000. The recoverable reserves in the mine were estimated to be 500,000 tons.

(a) If 75,000 tons of ore were mined during the year and 50,000 tons were sold, what was the depletion allowance for 1989?

(b) Suppose at the end of 1989 reserves were re-evaluated and found to be only 400,000 tons. If 50,000 additional tons are sold in 1990, what is the depletion allowance for 1990?

Solution

(a) The depletion unit is $2,000,000/500,000 tons = $4.00 per ton. A depletion allowance, based on units sold, for 1989 is 50,000 tons ($4.00/ton) = $200,000.

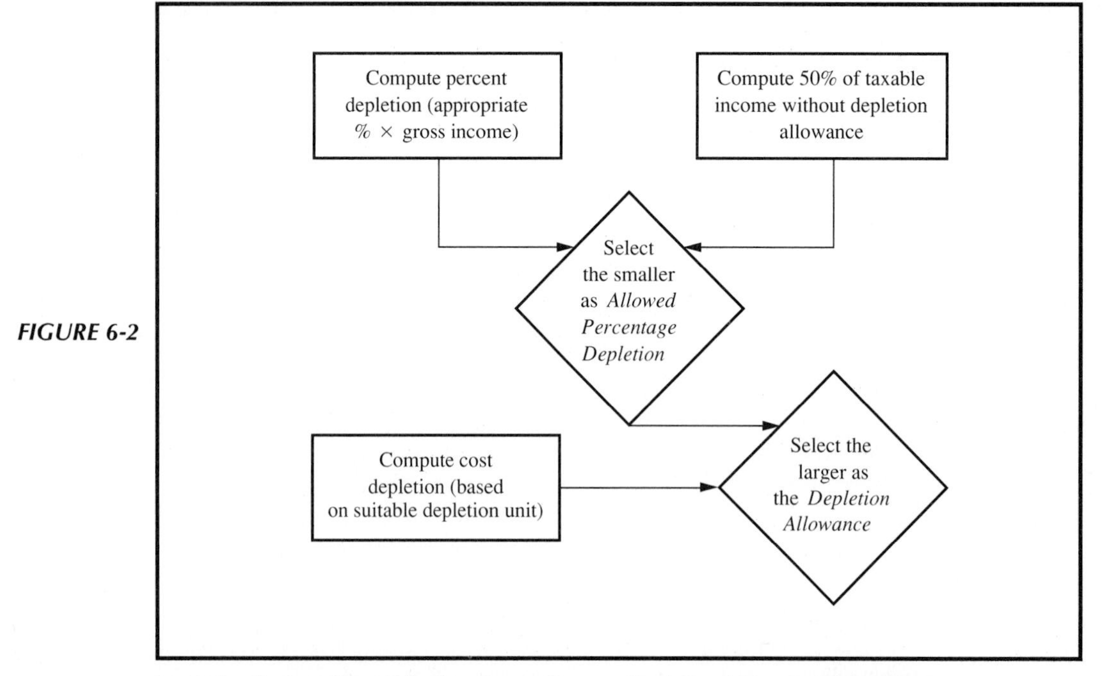

FIGURE 6-2

Logic for Determining Whether Percentage or Cost Depletion is Allowable

(b) The adjusted basis at the beginning of 1990 would be \$2,000,000 − \$200,000 = \$1,800,000. The depletion unit would be \$1,800,000/400,000 tons = \$4.50/ton, and a depletion allowance of 50,000 tons (\$4.50/ton) = \$225,000 is permissible in 1990.

6.13 ACCOUNTING FOR DEPRECIATION FUNDS*

The procedure by which accounting is made for the flow of depreciation funds and their subsequent reinvestment in the business often somewhat mystifies engineers who have not had any training in accounting. Quite probably, terminology that is applied to some of the accounts involved has added to the confusion.

The basic procedure can be illustrated by the simplified balance sheets shown in Figure 6-3. These relate to a man, John Doe, who starts out with an investment of \$3,000 in cash; this condition is portrayed in balance sheet (a) of Figure 6-3.

FIGURE 6–3 Balance Sheets, Showing the Manner in Which Cash is Converted into Depreciable Assets and Recovered through Depreciation Accounting

Balance Sheet (a)

Assets			*Liabilities and Ownership*	
Cash		\$3,000	John Doe, Ownership	\$3,000
	Total	\$3,000	Total	\$3,000

Balance Sheet (b)

Assets			*Liabilities and Ownership*	
Cash		\$ 0	John Doe, Ownership	\$3,000
Truck		3,000		
	Total	\$3,000	Total	\$3,000

Balance Sheet (c)

Assets			*Liabilities and Ownership*	
Cash and other assets		\$1,000	John Doe, Ownership	\$3,000
Truck	\$3,000			
Less amount charged for depreciation to date	1,000	2,000		
	Total	\$3,000	Total	\$3,000

*Reading of Appendix A is a prerequisite to this section.

Mr. Doe then uses his $3,000 to purchase a truck. The relationship of his assets, liabilities, and ownership for this condition is shown in balance sheet (b). At the end of a year he computes the depreciation on his truck to be $1,000, assuming a 3-year life and zero salvage value and using straight-line depreciation. He therefore sets aside, out of revenue, this $1,000 and reinvests it in the business, perhaps keeping some of it in the form of cash but using some of it for the purchase of other needed assets. This new financial state is portrayed in balance sheet (c), which at the same time shows the original value of the truck and also its depreciated value. In this general manner, the flow of depreciation money into the company and the fact that it is retained and used in the form of either cash or other assets, are recorded. Unfortunately, the item labeled "Less amount charged for depreciation to date" in Figure 6-3 is called by a variety of names, including *reserve for depreciation,* a term that is very often confusing.

6.14 DETERMINATION OF PROPERTY LIFE FROM MORTALITY DATA

As has been pointed out in this chapter, one of the most difficult problems in connection with estimating depreciation costs is the determination of the life that a property may be expected to have. With the Modified Accelerated Cost Recovery System, the IRS has simplified the question of selecting a depreciable life by limiting the number of class lives to eight categories. However, there are situations in which the *probable life* of an asset is desired—independent of IRS guidelines. For certain properties and under certain conditions, well-established statistical techniques can be used. However, it must be pointed out that in order to use these techniques, it is necessary (1) to wait until substantially all of a group of identical properties have been retired from service, or (2) to have in service a substantial number of identical properties of all possible ages. Also, where rapid technological progress is occurring, or functional depreciation is a primary determining factor, these techniques will be of little help. Thus, these procedures are of definite, but limited, assistance in solving the problem of economic life in economy studies.

Just as human beings are born, live, grow old, and die, physical properties are produced, put into service, render service, and are removed from use. The same procedures that are used to determine the probable mortality, average life, and life expectancy for human beings may also be used for determining and estimating the life that may be expected from physical property, *provided that certain conditions exist*. These basic conditions are as follows:

1. A sufficient number of basically identical units must be involved so that averages may be used.
2. The service conditions must be the same for future units as for those from which the mortality data were obtained.
3. The study period must be long enough to assure valid data.

Table 6-8, for example, gives mortality data for a certain type of electric lamp. What one desires is hours of service from each lamp. The result for the group is therefore measurable as lamp-hours of service, the multiple (product) of the number of lamps and the average number of hours the group provided.

In terms of the survivor curve in Figure 6-4, this product is the area under the curve. For a given age, the area under the portion of the survivor curve to the right of the ordinate at that age gives the service in lamp-hours provided by the lamps surviving at the age selected. Hence, this gives a rule by which one may obtain the average life, beyond a given age, of all units reaching the age. This is the *expectancy* at the given life, which is obtained by dividing the area to the right of the ordinate at the given life by the value of that ordinate. The sum of this value, which is the expectancy, and the life at which it is obtained, is called the *probable life*. In Figure 6-4, the probable life curve is to the right of the survivor curve.

TABLE 6–8 Mortality Data for Electric Lamps

Life (Hours)	Lamps in Service (Number Surviving)	Remaining Life Expectancy	Probable Life
1,999.5	0	0	1,999.5
1,899.5	755	50.0	1,949.5
1,799.5	1,897	89.8	1,889.3
1,699.5	4,237	112.5	1,812.0
1,599.5	7,559	141.1	1,740.6
1,499.5	11,334	177.2	1,676.7
1,399.5	15,637	214.9	1,614.4
1,299.5	20,620	250.5	1,550.0
1,199.5	26,131	287.4	1,486.9
1,099.5	31,969	325.8	1,425.3
999.5	37,857	367.5	1,367.0
899.5	43,745	411.2	1,310.7
799.5	49,483	457.7	1,257.2
699.5	54,994	506.8	1,206.3
599.5	59,977	560.6	1,160.1
499.5	64,280	619.7	1,119.2
399.5	68,055	682.5	1,082.0
299.5	71,377	748.5	1,048.0
199.5	73,717	823.1	1,022.6
99.5	74,859	909.8	1,009.3
−0.5	75,614	1,000.2	999.7

The computation of expectancy and probable life for the electric lamp mortality case can be performed by the area method. The survivor curve is treated as a broken line joining points at the class interval boundaries in Table 6-8. Beginning at the right end of the curve, this gives the ordinate 0 at life 1,999.5 hours; the ordinate 755, or 1.0%, at life 1,899.5 hours, and so on, back to 75,614 or 100.0% at life − 0.5 hours (zero life). (See Table 6-8.) These values are for the survivor curve. For the probable life curve, sum the areas beginning at the right, and divide the sum of the partial areas back to a given life by the value of the ordinate at that life. For life 699.5 hours, divide the sum of the partial areas to the right by the number of lamps in service at 699.5 hours, 54,994, to obtain the expectancy, 506.8 hours. The average total life of lamps in service at 699.5 hours is therefore 699.5 + 506.8 = 1,206.3 hours.

FIGURE 6-4

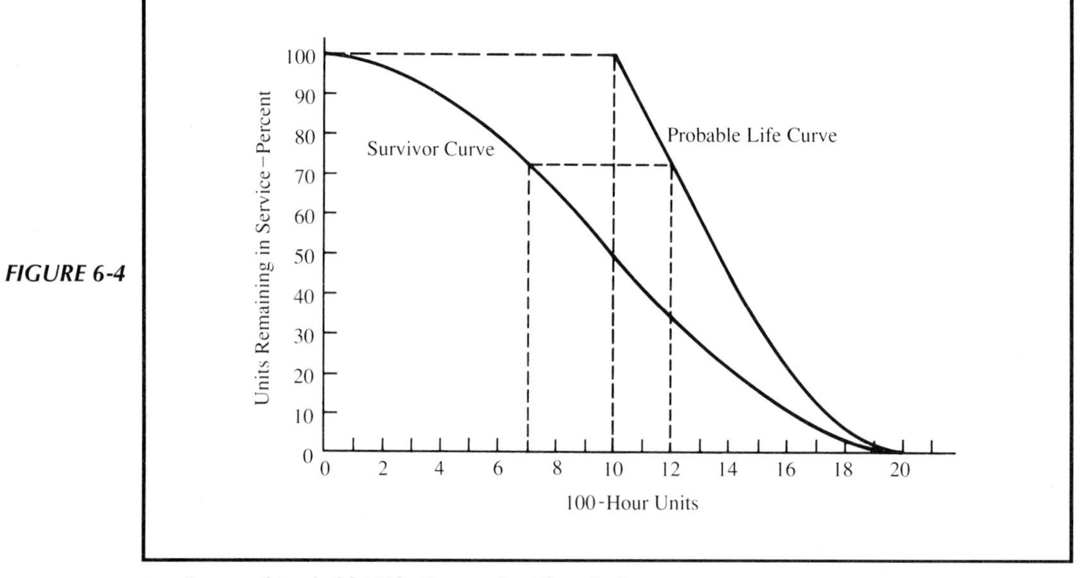

Survivor and Probable Life Curves for Electric Lamps

As of early 1988, the provisions of the Tax Reform Act of 1986 are completely phased in regarding the determination of depreciation deductions for most qualifying assets purchased after July 31, 1986. During the life of the eighth edition of *Engineering Economy*, modifications to this Act are almost certain to occur. The practitioner is advised to review IRS Publication 534, *Depreciation* (latest revision), or some other authoritative source of current federal income tax provisions affecting depreciation and depletion. Students are encouraged to remember the *process* by which depreciation and depletion allowances are generally computed as well as the *role* that depreciation and depletion play in determining after-tax cash flows and after-tax profitability, discussed in Chapter 7.

6.15 PROBLEMS

The number in parentheses () that follows each problem references the section from which the problem is taken.

6-1. Depreciation may be one of the least understood but most important aspects of engineering economic analyses. Explain why this is likely to be true. (6.2)

6-2. Define depreciation. What are the two main purposes of depreciation? (6.2, 6.5)

6-3. The accounting need for depreciation arises for several reasons. Describe what these reasons are. How are they related to definitions of the value of an asset? (6.3, 6.6)

6-4. How is the cost of depreciation different from other production or service expenses such as labor, material, and overhead? (6.6)

6-5. What conditions must a property satisfy to be considered "depreciable"? (6.9)

6-6. What are the basic requirements of a depreciation method? (6.8)

6-7. Explain the difference between obsolescence and physical depreciation. (6.7)

6-8. What are the effects of changes in price levels on capital recovery for purposes of asset replacement? (6.7)

6-9. What is the difference between real property and personal property? (6.9)

6-10. The basic intent of the Tax Reform Act of 1986 was to accomplish what? Why was this intent so significant? (6.10)

6-11. Explain the Accelerated Cost Recovery System. How does it differ from the service output method of depreciation that is discussed in Appendix 6-A? (6.10, App. 6-A)

6-12. What is the difference between an ADR guideline period and an ACRS recovery period? (6.10.1)

6-13. Explain how the cost basis of ACRS recovery property is determined. (6.10.1)

6-14. Can ACRS percentages be computed for a depreciable property's useful life? Explain your answer. (6.10.1)

6-15. How does modified ACRS (MACRS) differ from its predecessor, ACRS? (6.10.2)

6-16. When does the mid-quarter convention apply in MACRS? (6.10.2)

6-17. Why would a business elect to use the alternate MACRS method rather than the MACRS rates in Table 6-5? (6.10.3)

6-18. Derive the 3-year MACRS rates when half-year convention is mandatory. To check your answer, refer to Table 6-5 for the correct rates. (6.10.5)

6-19. Develop a logic flow chart for determining the optimal year to switch from 200% declining balance to straight-line depreciation. (6.10.4)

6-20. A company purchased a machine for $15,000. It paid shipping costs of $1,000 and nonrecurring installation costs amounting to $1,200. At the end of 3 years, the company had no further use for the machine, so it spent $500 for having the machine dismantled and was able to sell the machine for $1,500.

a. What is the total investment cost for this machine?

b. The company had depreciated the machine on a straight-line basis, using an estimated life of 5 years and $1,000 salvage value. By what amount did the recovered depreciation fail to cover the actual depreciation? (6.10.1)

6-21. Refer to Example 6-4. Rework this example assuming the computer is placed in service on August 15, 1989. (6.10.2)

6-22. An asset for drilling that was purchased by a petroleum refinery in 1988 has a first cost of $60,000 and an estimated salvage value of $12,000. The ADR guideline period for useful life is taken from IRS Publication 534 (Table 6-2), and the MACRS recovery period is 7 years. Compute the depreciation amount in the third year and the book value at the end of the fifth year of life by each of these methods:

a. The straight-line method over useful (ADR) life. (6.10.4)

b. The sum-of-the-years'-digits method over useful (ADR) life. (App. 6-A)

c. The 200% declining balance method (with $R = 2/N$) over useful (ADR) life. (6.10.4)

d. The MACRS method, using rates from Table 6-5. (6.10.2)

6-23. By each of the methods indicated below, calculate the book value of a highpost binding machine at the end of 4 years if the item originally cost $1,800 and had an estimated salvage value of $400. The ADR guideline period is 8 years and the ACRS recovery period is 5 years.

a. The machine was purchased in 1982 and ACRS percentages in Table 6-1 are to be used. (6.10.1)

b. The machine was purchased in 1989 and MACRS rates in Table 6-5 are applicable. (6.10.2)

c. Alternate MACRS is to be utilized over the ACRS recovery period of 5 years with half-year convention. (6.10.3)

6-24. An optical scanning machine was purchased in 1984 for $150,000. It is to be used for reproducing blueprints of engineering drawings, and its ADR guideline period is 10 years. The estimated market value of this machine at the end of 10 years is $30,000.

a. What is the ACRS class life of the machine? (6.10.1)

b. Based on your answer to (a), what was the depreciation deduction in 1987? (6.10.1)

c. What was the book value at the beginning of 1986? (6.10.1)

6-25. A piece of earthmoving equipment was purchased in 1989 by the Jones Construction Company. The cost basis was $300,000 and the equipment's ADR guideline period is 6 years.

a. Determine the MACRS depreciation deductions for this property. (6.10.2)
b. If the earthmoving equipment had been purchased in 1985, what would its ACRS depreciation deductions have been? (6.10.1)
c. Compute the difference in present worth of the two sets of depreciation deductions in (a) and (b) if $i = 12\%$ per year.

6-26. An asset was purchased in 1983 (6 years ago) for $6,400. At that time its ACRS class life and salvage value were estimated to be 5 years and $1,000, respectively. The ADR guideline period is 10 years. If the asset is sold now for $1,500, what is the difference between its market value of $1,500 and its present book value if depreciation has been by
a. The straight-line method (based on ADR guideline period)? (6.10.4)
b. The sum-of-the-years'-digits method (based on ADR guideline period)? (App. 6-A)
c. The ACRS method? (6.10.1)

6-27. During the first quarter of 1988, a pharmaceutical company purchased a mixing tank that had a retail (fair market) price of $120,000. It replaced an older, smaller mixing tank that had a book value of $15,000 in the second quarter of 1988. Because a special promotion was underway, the old tank was used as a trade-in for the new one, and the cash price (including delivery and installation) was set at $99,500. The ADR guideline period for the new mixing tank was 9.5 years.
a. Under MACRS, what is the depreciation deduction in 1990? (6.10.2)
b. Under MACRS, what is the book value at the end of 1991? (6.10.2)
c. If 200% declining balance depreciation had been applied to this problem, what would be the cumulative depreciation through the end of 1991? (6.10.4)

6-28. A company purchased a machine for $30,000 and depreciated it using the 1988 modified ACRS rates for a 5-year class life. Show that the annual costs of depreciation plus interest at 12% per year on the beginning-of-year book value are equal to the uniform year-end capital recovery cost (refer to Table 4-1). (6.10.2)

6-29. A piece of equipment that cost $5,000 was found to have a trade-in value of $4,000 at the end of the first year, $3,200 at the end of the second year, $2,560 at the end of the third year, $2,048 at the end of the fourth year, and $1,638 at the end of the fifth year. (6.6)
a. Determine the actual depreciation (difference in trade-in values) that occurred during each year.
b. Using a 10% interest rate and based on (a), compute the interest on the remaining investment each year as was done in Table 4-1.
c. Determine the sum of depreciation and interest on the unrecovered investment for each year.
d. Compare the results from (c) with the uniform year-end amount that represents capital recovery with interest and show that they are equivalent.

6-30. An asset purchased in the first quarter of 1988 for $24,000 has an ADR guideline period of 19 years and an estimated market value of $4,000 at that time. In this instance the aggregate bases of all property placed in service during the last 3 months of the tax year (1988) exceed 40% of the aggregate bases of *all* property placed in service in 1988. **(a)** Use the appropriate convention to determine the depreciation deduction for this particular asset. **(b)** Compare results in (a) with the depreciation that could have been claimed if the 40% restriction had not been invoked. (6.10.2)

6-31. A marble quarry is estimated to contain 900,000 tons of stone, and the ZARD mining company just purchased this quarry for $1,800,000. If 100,000 tons of marble can be sold each year and the average selling price per ton is $8.60, calculate the first year's depletion allowance for **(a)** the cost depletion method and **(b)** percentage depletion at 5% per year. ZARD's net income, before deduction of a depletion allowance, is $350,000. (6.12)

6-32. A gas well in Oklahoma has reserves of 2,000,000 mcf in the ground. The initial cost basis was $800,000, and during the first year of operation a depletion allowance of $280,000 was taken. At the beginning of the second year of operation, the reserves were re-estimated to be 1,400,000 mcf. What is the value of the depletion unit with the cost method? (6.12)

6-33. The mortality statistics on a certain telephone cable are as follows:

Years in Use	Percent Displaced Each Year	Years in Use	Percent Displaced Each Year
1	0.4	15	7.3
2	0.8	16	7.1
3	1.5	17	6.5
4	1.9	18	5.4
5	2.1	19	5.0
6	2.9	20	4.1
7	3.4	21	3.4
8	4.0	22	2.6
9	4.5	23	2.4
10	5.1	24	1.9
11	5.3	25	1.4
12	5.9	26	1.1
13	6.5	27	0.5
14	7.0		

Assume a given installation consists of 100 miles of cable that costs $10,000 per mile.

a. If 12% of the installation is 15 years old, find the expectancy and present value of this portion of the installation, using sinking fund depreciation with interest at 10%.

b. If the probable life of all the cable is taken arbitrarily to be 18 years, what is the present value of the 12% that is 15 years old? (Use sinking fund depreciation.) (6.14; App. 6-A)

6-34. A machine costing $4,000 is estimated to be usable for 400 units and then have no salvage value.

a. What would be the depreciation charge for a year in which 150 units were produced? (App. 6-A)

b. What would be the total depreciation charged after 600 units were produced? (App. 6-A)

Appendix 6-A: Pre-ACRS Methods of Depreciation

Before ACRS was enacted, other methods were used to calculate depreciation. For property placed in service before 1981, these methods must be utilized. Furthermore, these methods cannot be used for property that qualifies for ACRS and MACRS.

There are many different methods of figuring depreciation that are acceptable. Any method that is reasonable and acceptable to the Internal Revenue Service may be used if applied consistently. The conditions at the end of the tax year during which a firm depreciates a piece of property determine whether or not the method is reasonable.

Two common pre-ACRS methods of depreciation have already been discussed in Chapter 6: (1) straight-line method and (2) declining balance method. Other methods, to be described in Appendix 6-A, are (1) sum-of-the-years' digits, (2) sinking fund, and (3) service output.

To calculate depreciation using these methods, three things must be determined about the property to be depreciated. They are (1) its cost basis, (2) its useful life, and (3) its estimated salvage value at the end of its useful life. The amount of the deduction in any year also depends on which method of depreciation is chosen.

These methods differ from ACRS and MACRS in the following ways:

1. Instead of taking depreciation deductions over a specified recovery period, they were taken over the useful life of the property.
2. Instead of figuring the deduction for the first year using percentages in the tables or the half-year or mid-quarter convention, the analyst computed the deduction that would be allowed for a full year and then prorated it for the part of the year that the asset was actually in service. A firm could depreciate property only for the part of the year it was in service.
3. Instead of figuring deductions using the full amount of the property's depreciable cost basis, a firm had to take salvage value into account when calculating depreciation.

Some of the major aspects of pre-1981 depreciation calculations are described below:

Basis. To deduct the proper amount of depreciation each year, first determine the cost basis in the property to be depreciated. The cost basis used for figuring depreciation is the same as the cost basis that would be used for figuring the gain on a sale, and the original cost basis is usually the purchase price.

Adjusted basis. Events will often change the cost basis of a piece of property. This changed basis is called the adjusted basis. Some events, such as improvements, increase basis. Events such as deductible casualty losses and depreciation decrease basis.

Useful life. The useful life of a piece of property is an estimate of how long it can be expected to be of use in a business capacity. It is the length of time over which yearly depreciation deductions will be made from the cost basis in the property. It is not how long the property will last, but how long it will continue to be useful.

Many things affect the useful life of property, such as (1) frequency of use, (2) age when acquired, (3) repair policy, and (4) environmental conditions. The useful life can also be affected by technological improvements, reasonably foreseeable economic changes, and other causes. All these factors should be considered before deciding upon a useful life for the property.

The useful life of the same type of property varies from user to user. A company used the general experience of the industry until it was able to determine useful life of a particular property.

Determining salvage value. Salvage value is the estimated value of property at the end of its useful life. It is what a firm expects to get for the property if sold after there is no longer any productive use for the item. The salvage value of a piece of property must be estimated when it is first acquired.

Salvage value is affected by both how the property is used and how long it is kept in service. If it is policy to dispose of property that is still in good operating condition, the salvage value may be relatively large. However, if policy is to use property until it is no longer usable, its salvage value may be its junk value.

Net salvage. Net salvage is the salvage value of a piece of property minus what it costs to remove it when disposed of. A firm could choose either salvage value or net salvage when it calculated depreciation. If the cost to remove the property was more than the estimated salvage value, then net salvage for income tax purposes was zero. That is, salvage value could never be less than zero.

Ten percent rule. If a company acquired personal property that has a useful life of 3 years of more, it could use a figure for a salvage value that was less than the actual estimate. The company then subtracted from its estimate of salvage value an amount equal to 10% of the basis in the property. If salvage value was less than 10% of basis, salvage value was ignored when depreciation is calculated.

Restrictions. For the first two-thirds of the useful life of depreciable property, there was a limit on how much depreciation could be deducted. A firm was not permitted to depreciate property faster than it would have been depreciated under the declining balance method. At the end of each tax year, the total amount of depreciation could not be more than the total amount of depreciation a company would have been allowed to deduct using the declining balance method.

Sum-of-the-Years'-Digits Method (SYD). To compute the depreciation deduction by the SYD method, the digits corresponding to the number of each permissible year of life are first listed in reverse order. The sum of these digits is then determined.* The depreciation factor for any year is the number from the reverse-ordered listing for that year divided by the sum of the digits. For example, for a property having a life of 5 years, SYD depreciation factors are as follows:

Year	Number of the Year in Reverse Order (digits)	SYD Depreciation Factor
1	5	5/15
2	4	4/15
3	3	3/15
4	2	2/15
5	1	1/15
Sum of the digits	= 15	

The depreciation for any year is the product of the SYD depreciation factor for that year and the difference between the cost basis (B) and the salvage value (S). The general expression for the annual cost of depreciation for any year k, when the total life is N, is

$$d_k = (B - S)\left[\frac{2(N - k + 1)}{N(N + 1)}\right] \tag{6-A-1}$$

*The sum of digits for a life N equals $1 + 2 + \cdots + N = \text{SYD} = N(N + 1)/2$.

The book value at the end of year k is

$$BV_k = B - \left[\frac{2(B-S)}{N}\right]k + \left[\frac{(B-S)}{N(N+1)}\right]k(k+1) \qquad \text{(6-A-2)}$$

and the cumulative depreciation through the kth year is simply

$$D_k^* = B - BV_k \qquad \text{(6-A-3)}$$

When Equations 6-A-1 to 6-A-3 are applied to the data of Example 6-7, the results are

$$\text{Sum of the years' digits (SYD)} = 10(11)/2 = 55$$

$$\text{Depreciation factor for the sixth year} = 5/55$$

$$d_6 = \$4{,}000\left[\frac{2(10-6+1)}{10(11)}\right] = \$363.64$$

$$BV_6 = \$4{,}000 - \left[\frac{2(4{,}000)}{10}\right]6 + \left[\frac{4{,}000}{10(11)}\right]6(7) = \$727.27$$

$$D_6^* = \$4{,}000 - \$727.27 = \$3{,}272.73$$

Book values for the four "historical" methods that have been applied to the data of Example 6-7 are plotted in Figure 6-A-1. In this diagram it should be noted that the

FIGURE 6-A-1

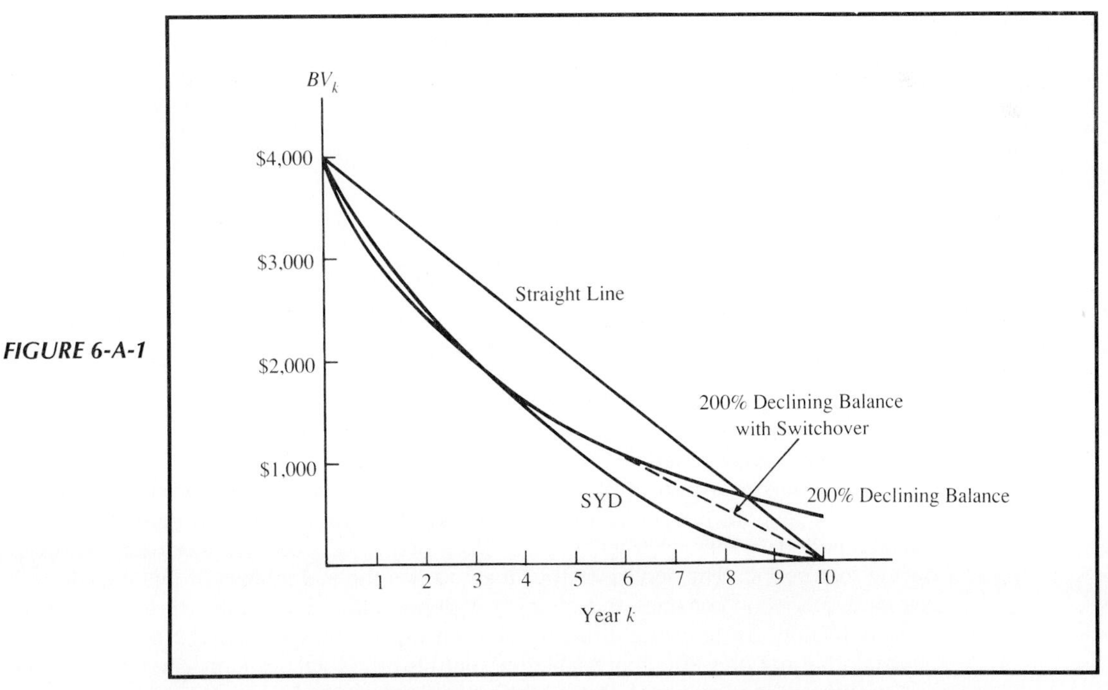

Comparison of Book Values Obtained by Various Depreciation Methods in Example 6-7

SYD method, like the declining balance method, provides for very rapid (accelerated) depreciation during the early years of life. Further, the SYD method enables properties to be depreciated to zero value and is easier to use than the declining balance method. This method also tends to reduce chances that the book value of an asset will exceed actual, or resale, value at any time. However, use of the SYD, or any other accelerated depreciation method, in effect reduces the computed profits of a corporation during the early years of asset life and thus reduces income taxes in those early years.

The Sinking Fund Method. The sinking fund method assumes that a sinking fund is established in which funds will accumulate for replacement purposes. The total depreciation that has taken place up to any given time is assumed to be equal to the accumulated value of the sinking fund (including interest earned) at that time.

With this method, if the estimated life, salvage value, and interest rate on the sinking fund are known, a uniform yearly deposit can be computed. The cost of depreciation for any year is the sum of this deposit and accumulated interest for that year. For an interest rate of $i\%$, these equations are used in connection with the sinking fund method of depreciation:

$$d = (B - S)(A/F, i\%, N) \tag{6-A-4}$$

$$d_k = d(F/P, i\%, k - 1) \tag{6-A-5}$$

$$D_k^* = (B - S)(A/F, i\%, N)(F/A, i\%, k) \tag{6-A-6}$$

$$BV_k = B - [(B - S)(A/F, i\%, N)(F/A, i\%, k)] \tag{6-A-7}$$

$$D_k^* = B - BV_k \tag{6-A-8}$$

Applying these equations to the data of Example 6-7 produces the following results when $i = 8\%$:

$$\begin{aligned} d &= (\$4{,}000 - \$0)(A/F, 8\%, 10) \\ &= (\$4{,}000)(0.0690) = \$276.12 \end{aligned}$$

$$d_6 = (\$276.12)(F/P, 8\%, 5) = \$405.71$$

$$\begin{aligned} BV_6 &= \$4{,}000 - \$4{,}000(A/F, 8\%, 10)(F/A, 8\%, 6) \\ &= \$4{,}000 - \$2{,}025.60 = \$1{,}974.40 \end{aligned}$$

$$D_6^* = \$4{,}000 - \$1{,}974.40 = \$2{,}025.60$$

The Service Output Method. Some companies attempt to compute the depreciation of various assets based on their output. When equipment is purchased, an estimate is made of the amount of service an asset will render during its useful life. Depreciation for any period is then charged on the basis of the service that has been rendered during that period. The service output method has the advantages of making the unit cost of depreciation constant and giving low depreciation deductions during periods of low production. Some of the difficulty in applying this method may be understood by realizing that not only the depreciable life, but also the total amount of service that the equipment will render during this period, must be estimated.

The *machine-hour method* of depreciation is a variation of the service output method. It is applied here to Example 6-7. Suppose it is expected that the electric saw will be used a total of 10,000 hours over a period of 10 years and then will have no salvage value. In the sixth year of operation, the estimated usage is 800 hours and the cumulative usage by the end of year 6 should be about 6,400 hours. If depreciation is based on hours of use, the quantities d_6, D_6^*, and BV_6 for the saw are as computed

$$\text{depreciation/hour} = \frac{\$4{,}000}{10{,}000 \text{ hours}} = \$0.40/\text{hour}$$

$$d_6 = 800 \text{ hours } (\$0.40/\text{hr}) = \$320$$

$$D_6^* = 6{,}400 \text{ hours } (\$0.40/\text{hr}) = \$2{,}560$$

$$BV_6 = \$4{,}000 - \$2{,}560 = \$1{,}440$$

7

EVALUATION OF ALTERNATIVES ON AN AFTER-TAX BASIS

The objectives of this chapter are (1) to explain the differences between before-tax and after-tax economic analyses, and (2) to illustrate some important features of the Tax Reform Act of 1986 as they relate to after-tax studies. The following topics are covered in Chapter 7:

Distinctions among different types of taxes
What are the differences between before-tax and after-tax studies?
Taxable income of business firms
Taxable income of individuals
Effective income tax rates
Ordinary income (and losses)
What are capital gains and losses?
Investment tax credits
A procedure for making after-tax engineering economy studies
Illustrations of computing after-tax cash flows
After-tax comparisons of cost alternatives
After-tax evaluations of specific financing arrangements
The after-tax effect of depletion allowances

7.1 INTRODUCTION

Most organizations consider the effect of income taxes on the financial results of a proposed project or venture because income taxes usually represent a significant cash outflow that cannot be ignored. Depreciation deductions, as computed in Chapter 6, affect the taxable income that an organization generates; thus, we are now in a position to describe how income tax liabilities (or credits) and after-tax cash flows are determined in practice. Selected provisions of the Tax Reform Act of 1986 (TRA 86) are discussed, followed by a number of examples that demonstrate the use of a general format for conducting after-tax evaluations.

7.2 DISTINCTIONS AMONG DIFFERENT TYPES OF TAXES

Before discussing the consequences of income taxes in engineering economy studies, we need to distinguish between income taxes and several other types of taxes.

1. *Income taxes* are assessed as a function of gross revenues minus certain allowable deductions and exemptions. They are levied by the federal, most state, and occasionally municipal governments.
2. *Property taxes* are assessed as a function of the "value" of real estate, business, and personal property. Hence, they are independent of the income or profit of an individual or a business. They are levied by municipal, county, and/or state governments.
3. *Sales taxes* are assessed on the basis of purchases of goods and/or services, and are thus independent of gross income or profits. They are normally levied by state, municipal, or county governments.
4. *Excise taxes* are federal taxes assessed as a function of the sale of certain goods or services often considered nonnecessities, and are hence independent of the income or profit of an individual or a business. Although they are usually charged to the manufacturer or original provider of the goods or services, the cost is passed on to the consumer.

In this chapter we shall be concerned only with income tax considerations. Income taxes resulting from the profitable operation of a firm normally are taken into account in evaluating engineering projects and business ventures. The reason is quite simple; income taxes associated with a proposed project may represent a major cash outflow that should be considered together with other cash inflows and outflows in assessing the overall economic attractiveness of that project. There are many other taxes not directly associated with the income-producing capability of a new project (e.g., property taxes and excise taxes), but they are usually negligible when compared with federal and state income taxes. When these other types of taxes are included in engineering economy studies, they are normally deducted from gross revenues as any other operating expense would be. For instance, an annual property tax computed as 4% of the investment cost would be subtracted from revenues in determining the before-tax cash flows that we considered in Chapters 4 and 5.

7.3 DIFFERENCES BETWEEN BEFORE-TAX AND AFTER-TAX STUDIES

Up to this point, there has been no consideration of income taxes in our discussion of engineering economy studies—that is, only before-tax studies have been conducted—for two reasons. First, in the types of studies considered thus far, income taxes often do not have any major effect on the decisions that would be made; therefore, the accuracy of the study would not be improved by their inclusion. Second, by not complicating the studies with income tax effects, we could place primary emphasis on basic engineering economy principles and methodology. However, there is a wide variety of individual and industrial capital investment problems in which income taxes do affect the choice among alternatives, and after-tax studies must consequently be made.

Some of the basic differences between before-tax and after-tax studies can be explained in terms of Figure 7-1, which shows the flow of capital within a typical firm. This diagram illustrates how funds are generated so that the firm can continue as a going concern from year to year. New debt and equity capital are initially obtained from external sources so that investments in buildings, land, and equipment can be made and supported with sufficient amounts of working capital to meet payrolls, purchase materials, and so forth. As gross revenues are produced through the sale of goods and/or services, they are reduced by operating expenses in arriving at operating income. Then operating income has depreciation and interest on debt capital subtracted from it to yield net income

FIGURE 7-1

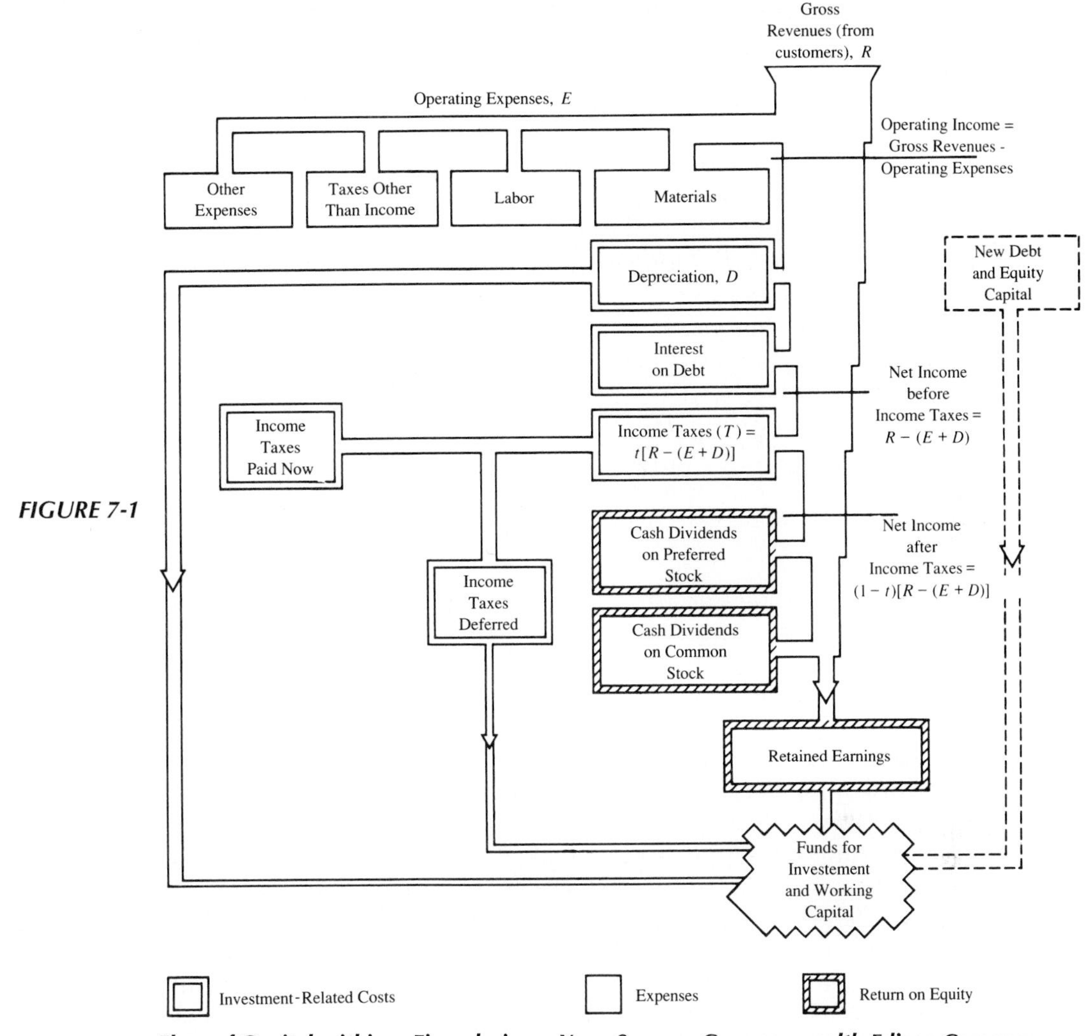

Flow of Capital within a Firm during a Year. **Source:** ***Commonwealth Edison Company, Chicago, IL***

before income taxes, which is often referred to as *profit before income taxes.* The symbols used in Figure 7-1 are defined as follows:

$R =$ gross revenues from customers.

$E =$ operating expenses (these are current and tend to be proportional to the extent of business activity and thus can be controlled to a degree) plus interest paid for the use of borrowed capital (which is a financial cost due to the use of borrowed money; this cost does not exist if 100% equity capital is employed by a firm).

D = depreciation; this is the recognition of cost due to loss in value of assets, such as property, buildings, or equipment; as was explained in Chapter 6, the annual magnitude of this cost can only be estimated.

T = income taxes; these are costs to an organization that are obviously not ordinary costs because they depend on profits remaining after expenses are paid or accounted for.

t = effective income tax rate used for computing income taxes.

Two classes of profit are of concern in engineering economy studies. The first is profit before income taxes; the other is profit after income taxes. We can formalize the relationship between net income before taxes (*NIBT*) and net income after taxes (*NIAT*):

$$NIBT = R - (E + D) \tag{7-1}$$

$$NIAT = NIBT - T \tag{7-2}$$

$$= R - (E + D) - t[R - (E + D)]$$

$$= (1 - t)[R - (E + D)] \tag{7-3}$$

Because the interest paid for the use of borrowed capital is included as an expense in Equations 7-1 to 7-3, the profits indicated in Figure 7-1 belong to the owners of equity capital (i.e., stockholders). This figure indicates that profits after income taxes (NIAT) are distributed as cash dividends on preferred stock and common stock, with the remainder going to retained earnings. Once the firm is profitable, funds for further investment and increased working capital are accumulated from retained earnings, depreciation reserves, deferred income taxes, and new equity-debt capital as shown.

Investment capital is thus *transformed* into goods and services that the company hopes will result in after-tax profits. However, after-tax profits in Figure 7-1 are usually different in amount from the after-tax cash flows that are produced by a project. If depreciation deductions are added back to net income after taxes, a project's after-tax cash flow (ATCF) in a particular year can be determined by Equation 7-4:

$$\text{ATCF} = \text{NIAT} + D \tag{7-4}$$

Engineering economy studies consider the after-tax cash flows that a project produces in evaluating the financial performance of the investment. The ATCF represents the amounts of money that a project or venture contributes to (or drains from) the treasury of a firm and is regarded as a better indicator of economic benefits than is net income after taxes. The remainder of this chapter is devoted to explaining how a project's after-tax cash flow is obtained in practice.

Another difference between before-tax and after-tax studies is related to the interest rate (i.e., the MARR) that is used in performing time value of money calculations. An approximation of the before-tax MARR requirement, which includes the effect of income taxes, for studies involving only before-tax cash flows can be obtained from the following relationship:

$$\text{(before-tax MARR)}[(1 - \text{ effective income tax rate})] \cong \text{after tax MARR}$$

Thus,

$$\text{before-tax MARR} = \frac{\text{after-tax MARR}}{1 - \text{(effective income tax rate)}} \tag{7-5}$$

An expression for determining the effective income tax rate is given in Section 7.6.

In practice, it is usually desirable to make after-tax analyses in any income-tax–paying organization. After-tax analyses can be performed by exactly the same methods (PW, IRR, etc.) as before-tax analyses. The only difference is that after-tax cash flows must be used in place of before-tax cash flows, and the calculation of a measure of merit is based on an *after-tax* minimum attractive rate of return.

The mystery behind the sometimes complex computation of income taxes is reduced when one recognizes that income taxes paid are just another type of disbursement, while income taxes saved (through business deductions, expenses, or direct tax credits) are identical to other kinds of reduced expenses (e.g., savings).

The basic concepts underlying federal and state income tax regulations that apply to most economic analyses of capital investments generally can be understood and applied without difficulty. This chapter is not intended to be a comprehensive treatment of federal tax law. Rather, we utilize in this chapter some of the more important provisions of the Tax Reform Act of 1986, followed by illustrations of a general procedure for computing after-tax cash flows and conducting after-tax analyses.

7.4 TAXABLE INCOME OF BUSINESS FIRMS

At the end of each tax year, a corporation must calculate its net (or taxable) income or loss. Several steps are involved in this process, beginning with the calculation of *gross income*. Gross income represents the gross profits from operations (revenues from sales minus the cost of goods sold) plus income from dividends, interest, rent, royalties, and gains (or losses) on the exchange of capital assets. The corporation may deduct from gross income all ordinary and necessary operating expenses to conduct the business *except capital expenditures*.

Deductions for depreciation are permitted each tax period as a means of consistently and systematically recovering capital. Consequently, allowable expenses and depreciation deductions may be used to determine taxable income as shown in Equation 7-6:

$$\text{Taxable income} = \text{gross income} - \text{all expenses except capital expenditures} - \text{depreciation deductions} \tag{7-6}$$

As we discussed in Section 7.3, this taxable income is often referred to as *net income before taxes*, and when income taxes are subtracted from it, the remainder is called the *net income after taxes*. There are two types of income for tax computation purposes: ordinary income (and losses) and capital gains (and losses). These types of income will be explained in Section 7.7 and Section 7.8, respectively.

EXAMPLE 7-1

In 1990 a company generates $1,500,000 of gross income and incurs operating expenses of $800,000. Interest payments on borrowed capital amount to $48,000. The total depreciation deductions in 1990 equal $114,000. What is the taxable income of this firm?

Solution

Based on Equation 7-6, this company's taxable income in 1990 would be

$$\$1{,}500{,}000 - \$800{,}000 - \$48{,}000 - \$114{,}000 = \$538{,}000$$

7.5 TAXABLE INCOME OF INDIVIDUALS

An individual's total income is essentially what is called *adjusted gross income* for federal income tax purposes. Adjusted gross income consists of wages, salaries, and/or tips—plus other income such as interest, royalties, and pensions—less adjustments to income for moving expenses, alimony payments, and so on. From this amount individuals may subtract personal exemptions and allowable deductions to determine their *taxable income*. In 1989 personal exemptions are provided at the rate of $2,000 for the taxpayer and each dependent, and after 1989 this exemption will be indexed to inflation. Allowable deductions can also be claimed, within some limits, for such items as large medical costs, state and local taxes, interest on borrowed money, charitable contributions, and casualty losses. A standard deduction may be taken by people who do not itemize their allowable deductions.

Taxpayers who *do not* itemize deductions are permitted to reduce their adjusted gross income by a specified standard amount. These standard deductions are indexed for inflation, beginning in 1989, based on the following 1988 standard deductions categorized by filing status:

Filing Status	*1988 Standard Deduction*
Married/Joint Return	\$5,000
Head of Household	\$4,400
Single	\$3,000
Married/Separate Return	\$2,500

Thus, taxable income of individual taxpayers is determined with Equation 7-7:

$$\text{Taxable income} = \text{adjusted gross income} - \text{personal exemption deductions} - \text{other allowable deductions} \quad (7\text{-}7)$$

EXAMPLE 7-2

Jayne Doe has an adjusted gross income of \$60,000 in 1989, resulting from her salary and interest on savings accounts. She files her federal tax return as a *single* taxpayer and elects to take the standard deduction which has been indexed for inflation at 6% above the 1988 standard deduction. What is Jayne's taxable income in 1989?

Solution

In 1989 the personal exemption deduction that Jayne can claim is \$2,000. Furthermore, she can take the standard deduction which amounts to \$3,000 (1.06) = \$3,180. Therefore, from Equation 7-7, Jayne's taxable income is \$60,000 − \$2,000 − \$3,180 = \$54,820.

7.6 CALCULATION OF EFFECTIVE INCOME TAX RATES

Income taxes are levied on both personal and corporate incomes. Although the regulations of most of the states with income taxes have the same basic features as the federal regulations, there is great variation in the tax rates. State income taxes are in most cases much less than federal taxes and often can be closely approximated as a constant percentage of federal taxes. Therefore, no attempt will be made to discuss state income taxes. An understanding of the applicable federal income tax regulations usually will enable the analyst to apply the proper procedures if state income taxes must also be considered.

To illustrate the calculation of an effective (combined federal and state) income tax rate for a corporation, suppose that the federal income tax rate is 34% and the state income tax rate is 6%. Further assume the common case in which taxable income is computed the same way for both taxes except that state taxes are deductible from taxable income for federal tax purposes but federal taxes are not deductible from taxable income for state tax purposes. The general expression for the effective income tax rate (t) is

$$t = \text{federal rate} + \text{state rate} - (\text{federal rate})(\text{state rate}) \tag{7-8}$$

In this example the effective income tax rate would be

$$t = 0.34 + 0.06 - 0.34(0.06) = 0.3796, \text{ or about } 38\%$$

It is the effective income tax rate on increments of taxable income that is of importance in engineering economy studies.

7.7 ORDINARY INCOME (AND LOSSES)

Ordinary income is the net income that results from the regular business operations (such as the sale of products or services) performed by a corporation or individual. For federal income tax purposes, virtually all ordinary income adds to taxable income and is subject to a graduated rate scale (higher rates for higher income).

7.7.1 Corporate Provisions

The federal income tax rates resulting from the Tax Reform Act of 1986 are given in Table 7-1. For example, suppose that a firm in 1989 has a gross income of \$5,270,000, expenses (excluding capital) of \$2,927,500, and depreciation of \$1,874,300. Its taxable income and federal income tax would be determined with Equation 7-6 and Table 7-1 as follows:

TABLE 7–1 Corporate Federal Income Tax Rates in Effect in 1989

Taxable Income	Tax Rate
Less than $50,001	15%
$50,001–$75,000	25%
$75,001–$100,000	34%
$100,001–$335,000	39%[a]
Greater than $335,000	34%

[a] The 39% rate serves to counteract lower rates on taxable income under $75,001 and creates a flat 34% rate for all taxable income above $335,000.

taxable income	= gross income − expenses − depreciation	
	= $5,270,000 − $2,927,500 − $1,874,300	
	= $468,200	
income tax	= 15% of first $50,000	$7,500
	+ 25% of next $25,000	6,250
	+ 34% of next $25,000	8,500
	+ 39% of next $235,000	91,650
	+ 34% of remaining $133,200	45,288
	Total	$159,188

The total income tax liability in this case is $159,188. The average income tax rate ($159,188 / $468,200 = 0.34) is identical to a flat rate of 34% on taxable income. This will always be true when a firm's taxable income is greater than $335,000. Hence, the federal income tax rate on taxable income over $335,000 is clearly 34%. Because engineering economy studies are concerned with incremental differences among alternatives, we shall be using a 34% *incremental* federal income tax rate, as shown in Figure 7-2, for most studies concerning corporations with large taxable incomes.

EXAMPLE 7-3

A corporation is expecting an annual taxable income of $45,000. It is considering investing an additional $100,000, which is expected to create an added annual net cash flow (receipts minus expenses) of $35,000 and an added annual depreciation charge of $20,000. What is the corporation's federal tax liability based on rates in effect in 1989 (a) without the added investment, and (b) with the added investment?

Solution

(a) *Income Taxes/Year*	*Rate*	*Amount*
On first $45,000	15%	$6,750
	Total	$6,750

FIGURE 7-2

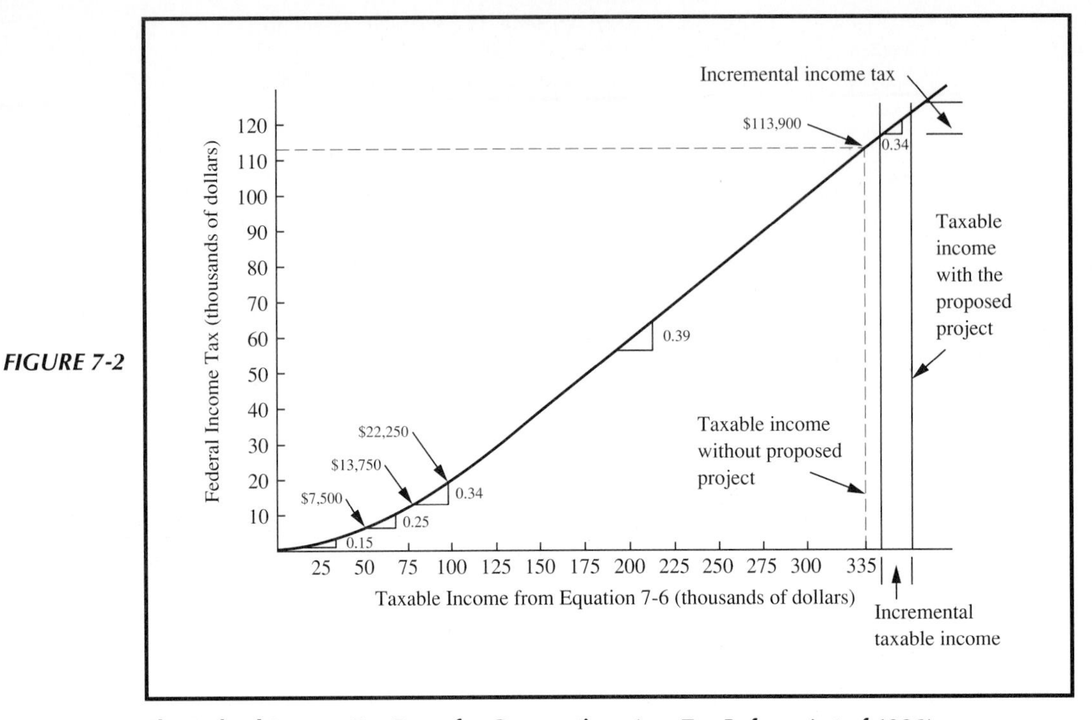

The Federal Income Tax Rates for Corporations (per Tax Reform Act of 1986)

(b) Taxable income

Before added investment		$45,000
+ added net cash flow		35,000
− depreciation		(20,000)
	Total	$60,000

Income Taxes on $60,000	*Rate*	*Amount*
On first $50,000	15%	$7,500
On next $10,000	25%	2,500
	Total	$10,000

The increased income tax liability from the investment is $3,250.

As an added note, the determination of change in tax liability can usually be determined more readily by an incremental approach. For instance, this example involved changing the taxable income from $45,000 to $60,000 as a result of the new investment. Thus, the change in income taxes per year (for 1989) could be calculated as:

First $50,000 − $45,000 = $5,000 at 15%	=	$ 750
Next $60,000 − $50,000 = $10,000 at 25%	=	2,500
	Total	$3,250

The average federal income tax rate on the additional $35,000 − $20,000 = $15,000 of taxable income is calculated as ($3,250 / $15,000) = 0.2167, or 21.67%.

In addition to lowering the maximum rate on corporate taxable income from 46% to 34%, the Tax Reform Act of 1986 created a new *alternative minimum tax* (AMT) system that is intended to ensure that any corporation with economic income pays a minimum amount of federal income tax. Now, corporations must compute their income tax liability as illustrated in this section, and many must also compute their AMT according to a rather complex set of rules that is beyond the scope of our discussion. Corporations now pay the *maximum* of income tax resulting from the system using rates in Table 7-1 or from the AMT system. It is generally acknowledged that the AMT is the most far-reaching and complex business tax provision in the Tax Reform Act of 1986.

7.7.2 Individual Taxpayer Provisions

Concerning individual taxpayers, an income tax liability is determined according to a so-called regular system based on rates shown in Table 7-2. Individual tax rates are now a maximum of 28%, compared to the previous maximum of 50%. Moreover, an alternative minimum tax (AMT) is also computed by taxpayers having very high incomes. As in the case of corporate AMT, discussion of the individual AMT is too complex for our introductory treatment of income taxes.

What is now different about individual federal taxation is the extent to which the Tax Reform Act of 1986 segregates individual taxpayers into two distinct categories. One group of taxpayers has experienced very little change, other than the effect of reduced tax rates. This group is made up of individuals whose income sources are conventional—salaries, wages, dividends, and interest from

TABLE 7–2 Individual Federal Income Tax Rates on Taxable Income in 1989

	Taxable Income for Filing Status:			
Tax Rate	**Married/ Joint Return**	**Head of Household**	**Single**	**Married/ Separate Returns**
15%	First $29,750	First $23,900	First $17,850	First $14,875
28	$29,751–$71,900	$23,901–$61,650	$17,851–$43,150	$14,876–$35,950
33[a]	$71,901–$149,250	$61,651–$123,790	$43,151–$89,560	$35,951–$113,300
28	Over $149,250	Over $123,790	Over $89,560	Over $113,300

[a] The 33% rate creates a flat 28% rate for all taxable income above $149,250.

such traditional sources as savings accounts, money market funds, certificates of deposit, and savings bonds.

The other group of taxpayers has entered uncharted waters. This group is composed of sophisticated investors, usually with diversified portfolios and often with tax-favored investments. From a tax viewpoint, the diversified investor has discovered that the economics of investing and the evaluation of investment strategies have changed significantly.

EXAMPLE 7-4

John and Mary Smith file a joint federal income tax return on their adjusted gross income of $44,750. They have two young children living at home, so a total of four personal exemptions can be claimed. The Smiths also have other allowable *itemized* deductions that will reduce adjusted gross income by $4,000. How much money in federal income taxes do they owe to the U.S. Treasury in 1989?

Solution

Each personal exemption in 1989 is worth $2,000; therefore, the Smiths can claim $8,000 in personal exemptions when computing their taxable income. They can also claim a standard deduction of $5,000, which exceeds their $4,000 itemized deduction.*

$$\text{Taxable income} = \$44{,}750 - \$8{,}000 - \$5{,}000 = \$31{,}750$$

From Table 7-2 the married/joint return rate on the first $29,750 of taxable income is 15% and taxable income in excess of $29,750 is taxed at a 28% rate. Thus, the Smith's federal income tax liability in 1989 is:

$$\text{Income Tax Owed} = 0.15(\$29{,}750) + 0.28(\$31{,}750 - \$29{,}750) = \$5{,}022.50$$

7.8 CAPITAL GAINS (AND LOSSES)

When a capital asset† is disposed of for more (or less) than its book value, the resulting capital gain (or capital loss) historically had been taxed (or saved

*This assumes inflation in 1989 is nil (see Section 7.5).

†Capital assets include all property owned by a taxpayer except (a) property held mainly for sale to customers, (b) most accounts or notes receivable, (c) depreciable property utilized to carry out the production process, (d) real property used by the business, and (e) copyrights and certain types of short-term, non-interest-bearing state/federal notes. Stock owned in another company is a familiar example of a capital asset.

taxes) at a rate which was different from that for ordinary income. However, TRA 86 eliminated the differential between income taxes on capital gains and on ordinary income. Now capital gains are taxed at a maximum rate of 34%; capital losses create tax savings in profitable firms amounting to at most 34% of the loss. Furthermore, there is no longer a distinction between long-term and short-term capital gains and losses.

In equation form the determination of capital gains and losses is straightforward:

$$\text{capital gain(loss)} = \text{net selling or disposal price} - \text{book value} \tag{7-9}$$

or

$$\text{capital gain(loss)} = \text{net selling or disposal price} - \text{original first cost} + \text{accumulated depreciation deductions} \tag{7-10}$$

In many cases, personal property used in a trade or business is sold for an amount that is greater than its book value at the time of disposal. Such gains are technically *not* capital gains, but instead are referred to as *depreciation recapture.* TRA 86 specifies that this difference between disposal price (i.e., market value) of a depreciable asset and its book value at the time of disposal is to be taxed as ordinary income. Hence, gains are taxed at a maximum rate of 34%. When an asset's selling price is less than its book value, the resulting loss on disposal creates an income tax credit that is at most 34% of the loss.

EXAMPLE 7-5

A large, profitable company has just made two transactions in which a gain and a loss have been recorded. The first transaction involved the sale of long-term bonds having a face value of $100,000. Because interest rates in the economy are now substantially higher than those in force when the bond was purchased 5 years ago, the company was able to obtain only $95,800 from the sale of these bonds. The second transaction was the sale of a piece of earthmoving equipment for $23,100. This depreciable asset has a current book value of $18,000. (a) Classify each transaction and determine its federal income tax consequence. (b) What is the net income tax liability (or credit)?

Solution

(a) The first transaction involves a capital asset, and the *capital loss* is $100,000 − $95,800 = $4,200. The resultant tax *credit* is ($4,200)(0.34) = $1,428. The second transaction is an illustration of *depreciation recapture* because the selling price is higher than the book value of the equipment. Here a *gain on disposal* is $23,100 − $18,000 = $5,100 and the income tax *liability* is $5,100(0.34) = $1,734.

(b) The net consequence of these transactions is a tax liability of $1,734 − $1,428 = $306. This figure could also be obtained by offsetting the capital loss against the gain from depreciation recapture and taxing the difference:

$$\text{Taxes owed} = 0.34[\$5{,}100(\text{gain}) - \$4{,}200(\text{loss})] = \$306$$

7.9 INVESTMENT TAX CREDIT

A special provision of the federal income tax law, *when in force*, is the investment tax credit (ITC). Originally enacted in 1962, it permits businesses to subtract from their overall tax liability a stated percentage of the basis in qualifying property purchased during a particular tax year. If the ITC is 10% of the value of a qualifying property, for instance, the net impact is to reduce the after-tax cost of the asset to 90% of its invoiced cost to a company. Qualifying property is depreciable, tangible property used in a trade or business.

The Tax Reform Act of 1986 *repealed the investment tax credit* in the interest of broadening the corporate tax base and lowering maximum income tax rates on ordinary income from 46% in 1986 to 34% in 1988 and beyond. This reduction in rates on ordinary income was made possible, in part, by elimination of the ITC. Because the ITC has been repealed and reinstated several times since 1962, it is highly likely that Congress will restore the credit in some form in the future. A simple example is provided here to illustrate the mechanics of the investment tax credit.

EXAMPLE 7-6

A firm purchased a computerized machine tool for $250,000. Suppose Congress has restored a 10% investment tax credit that applies to domestically manufactured machine tools. (a) What is the after-tax cost of acquiring this equipment? (b) What is the MACRS depreciation deduction in the first year of ownership if this property's class life is 7 years?

Solution

(a) The 10% investment tax credit is $25,000, assuming the entire investment is qualifying property. This financial incentive applies dollar-for-dollar against tax liabilities that the firm incurs, and it effectively reduces the machine tool's first cost (after taxes) to $225,000.

(b) The basis in the machine tool is unaffected by the 10% ITC. This represents another financial incentive of investment tax credits. Hence, the first-year MACRS depreciation deduction is 0.1429 ($250,000) = $35,725. (Refer to Table 6-5 for MACRS rates.)

Sections 7.4 through 7.9 provide only a minimal description of some of the main provisions of the Tax Reform Act of 1986 that are important to engineering economy studies. It is by no means complete, but it is intended to establish a basis for illustrating after-tax (i.e., after income tax) economic analyses. In general, the analyst should either search out specific provisions of the federal and/or state income tax law affecting projects being studied or seek information from persons qualified in income tax law.*

The remainder of the chapter illustrates various types of after-tax problems by using a tabular form for computing after-tax cash flows.

7.10

GENERAL PROCEDURE FOR MAKING AFTER-TAX ECONOMIC ANALYSES

After-tax economic analyses can be performed by exactly the same methods as before-tax analyses. The only difference is that after-tax cash flows are used in place of before-tax cash flows by including expenses (or savings) due to income taxes and then making equivalent worth calculations using an after-tax minimum attractive rate of return. The tax rates and governing regulations may be complex and subject to changes, but once those rates and regulations have been translated into their effect on after-tax cash flows, the remainder of the after-tax analysis is relatively straightforward.

To formalize the procedure described in previous sections for determining the net income before taxes, net income after income taxes, and after-tax cash flow of the incremental project shown in Figure 7-2, the following notation and equations are re-stated. For any given year k of the study period, $k = 0, 1, 2, \ldots, N$, let

R_k = revenues from the project; this is the positive cash flow from the project during period k
E_k = cash outflows during year k for deductible expenses and interest
D_k = sum of all noncash, or book, costs during year k, such as depreciation or depletion
t = effective income tax rate on ordinary income (federal, state, and other)
T_k = income taxes paid during year k
ATCF_k = after-tax cash flow from the project during year k

*Some applicable publications are

1. *Tax Guide for Small Business*, U.S. Internal Revenue Service Publication 334, published annually.
2. *Your Federal Income Tax*, U.S. Internal Revenue Service Publication 17, published annually.
3. J.K. Lasser, *Your Income Tax* (New York: Simon and Schuster, published annually).

Because the net income before taxes (i.e., taxable income) is $(R_k - E_k - D_k)$, the *ordinary income tax liability* when $R_k > (E_k + D_k)$ is computed with Equation 7-11:

$$T_k = -t(R_k - E_k - D_k) \tag{7-11}$$

The *net income after taxes* (NIAT) is then simply taxable income (i.e, net income before taxes) minus the tax liability amount determined by Equation 7-11:

$$\text{NIAT}_k = \underbrace{(R_k - E_k - D_k)}_{\text{taxable income}} - \underbrace{t(R_k - E_k - D_k)}_{\text{taxable income}}$$

or

$$\text{NIAT}_k = (R_k - E_k - D_k)(1 - t) \tag{7-12}$$

The *after-tax cash flow* associated with a project equals the net income after taxes plus noncash items such as depreciation:

$$\begin{aligned} \text{ATCF}_k &= \text{NIAT}_k + D_k \\ &= (R_k - E_k - D_k)(1 - t) + D_k \\ &= (1 - t)(R_k - E_k) + tD_k \end{aligned} \tag{7-13}$$

In many economic analyses of engineering and business projects, after-tax cash flows in year k are computed in terms of BTCF_k, or year k before-tax cash flows:

$$\text{BTCF}_k = R_k - E_k \tag{7-14}$$

Thus,

$$\begin{aligned} \text{ATCF}_k &= \text{BTCF}_k + T_k \\ &= (R_k - E_k) - t(R_k - E_k - D_k) \\ &= (1 - t)(R_k - E_k) + tD_k \end{aligned} \tag{7-15}$$

Tabular headings to facilitate the computation of after-tax cash flows with Equations 7-11 to 7-15 are as follows:

Year	(A) Before-Tax Cash Flow	(B) Depre-ciation	(C) = (A) − (B) Taxable Income	(D) = −t(C) Cash Flow for Income Taxes	(E) = (A) + (D) After-Tax Cash Flow
k	$R_k - E_k$	D_k	$R_k - E_k - D_k$	$-t(R_k - E_k - D_k)$	$(1 - t)(R_k - E_k) + tD_k$

Column A consists of the same information used in before-tax analyses; namely, the cash revenues or savings less the deductible expenses. Column B contains depreciation that can be claimed for tax purposes. Column C is the taxable income, or amount subject to income taxes. Column D contains the income taxes paid (or saved). Finally, column E shows the after-tax cash flows to be used directly in after-tax economic analyses just as the before-tax cash flows in column A are used in before-tax economic analyses.

A summary of the process of determining *net income after taxes* and the *after-tax cash flow* during each year of an N-year study period is provided in Figure 7-3. "Net income after taxes" is well understood in many companies, and it can be easily obtained from Figure 7-3 for purposes of making presentations to upper- level management. This format is used extensively throughout the remainder of Chapter 7, and it provides a convenient way to organize data in after-tax studies.

The column headings of Figure 7-3 indicate the arithmetic operations for computing columns C, D, and E when $k = 1, 2, \ldots, N$. When $k = 0$ and $k = N$, capital expenditures are usually involved and their tax treatment (if any) is illustrated in examples that follow. It is intended that the table be used with the conventions of + for cash inflow or savings and − for cash outflow or opportunity forgone.

7.11 ILLUSTRATION OF COMPUTATIONS OF AFTER-TAX CASH FLOWS

The following problems (Examples 7-7 and 7-8) illustrate the computation of after-tax cash flows as well as many common situations that affect income taxes. All problems include the assumption that income tax expenses (or savings) occur at the same time (year) as the income or expense that gives rise to the taxes. For purposes of comparing the effects of various situations, the after-tax *internal rate of return* and *present worth* are computed for each example. One can observe from the results of Examples 7-7 and 7-8 that the faster (i.e., earlier) the depreciation deduction, the more favorable the after-tax internal rate of return and present worth will become.

In examples that follow, the student should carefully observe the dates at which various types of property are placed in service. For instance, if an asset was placed in service prior to January 1, 1981, it must be depreciated by methods included in the appendix to Chapter 6. Most examples utilize the MACRS depreciation regulations and income tax rates that were established by the Tax Reform Act of 1986.

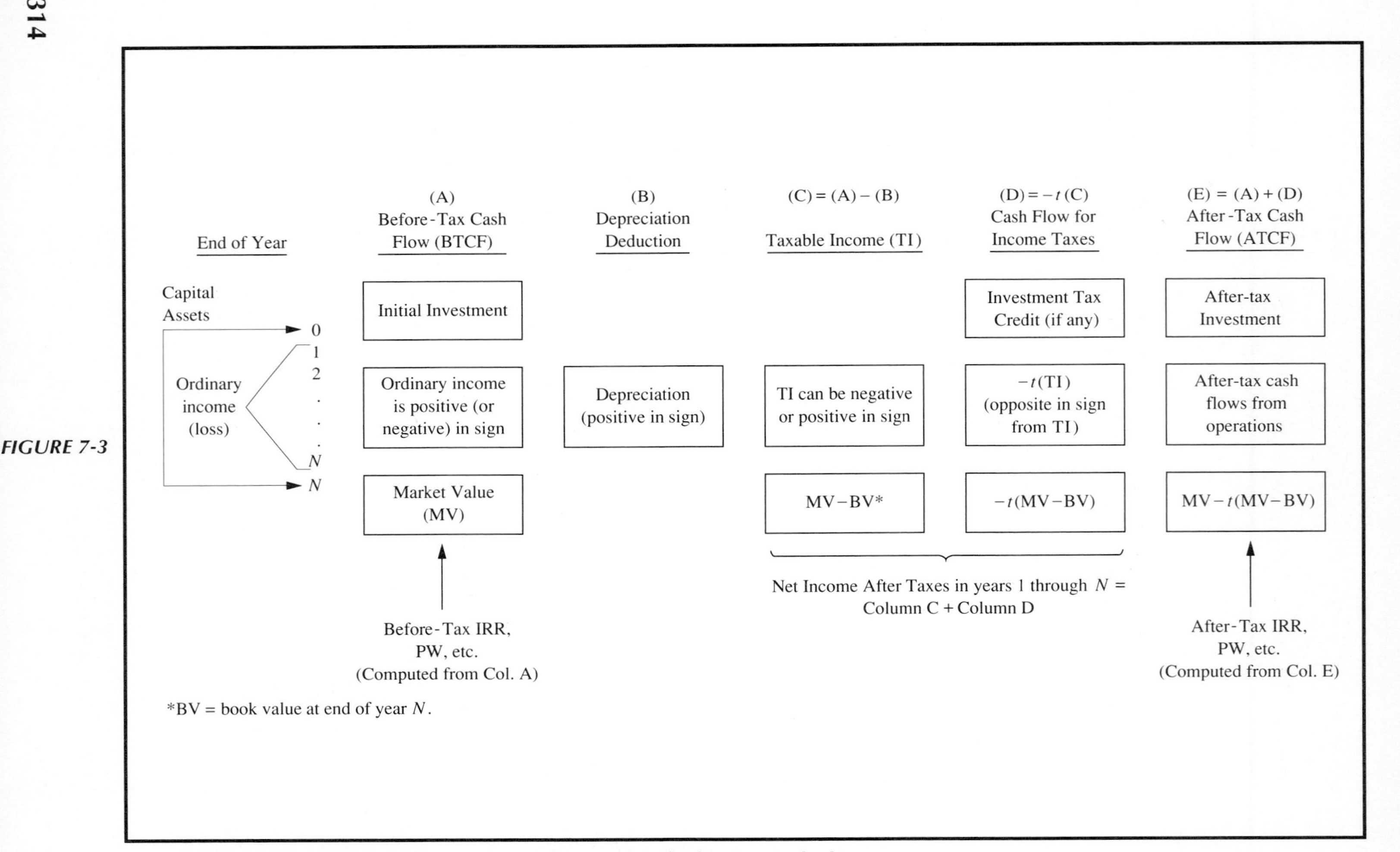

FIGURE 7-3 *General Format for Determining Net Income After Taxes and After-Tax Cash Flow*

EXAMPLE 7-7

Certain new machinery placed in service in 1989 is estimated to cost \$180,000. It is expected to *reduce* net annual operating expenses by \$36,000 per year for 10 years and to have a \$30,000 market value at the end of the tenth year. (a) Develop the before-tax and after-tax cash flows, and (b) calculate the before-tax and after-tax internal rate of return using MACRS depreciation with a federal income tax rate of 34% plus a state income tax rate of 6%. State income taxes are deductible from federal income taxes. This machinery has a MACRS class life of 5 years. (c) Calculate the after-tax present worth when the after-tax MARR = 10%. In this example the study period is 10 years, but the tax life is 6 years (which includes the carryover effect of the half-year convention).

Solution

(a) Table 7-3 applies the format illustrated in Figure 7-3 to calculate the before-tax and after-tax cash flows for this example. In column D the effective income tax rate is very close to 0.38 (from Equation 7-8) based on information provided above.

(b) The before-tax internal rate of return is computed from column A:

$$0 = -\$180{,}000 + \$36{,}000(P/A, i'\%, 10) + \$30{,}000(P/F, i'\%, 10)$$

By trial and error, $i' = 16.1\%$.

The entry in the last year is shown to be \$30,000 since the machinery will have this estimated market value. However, the asset was depreciated to zero with the MACRS method. Therefore, when the machine is sold at the end of the year 10, there will be \$30,000 of *recaptured depreciation* which is taxed at the effective income tax rate of 38%. This tax entry is shown in column D.

By trial and error, the after-tax internal rate of return for Example 7-7 is found to be 12.4%.

(c) When MARR = 10% is inserted into the PW equation at the bottom of Table 7-3, it can be determined that the after-tax present worth of this investment is \$17,209.

EXAMPLE 7-7A

Determine the after-tax PW of the machinery described in Example 7-7, using the computer program in Appendix F.

Solution

```
ATAX: After Tax Investment Evaluation

Main Menu

1. Asset placed in service on, or before
   December 31, 1980.
```

TABLE 7–3 After-Tax Cash Flow Analysis of Example 7-7

	(A)	(B) Depreciation Deduction					(C) = (A) − (B)	(D) = −0.38(C)	(E) = (A) + (D)
End of Year, k	Before-Tax Cash Flow	Basis	×	MACRS Rate	=	Deduction	Taxable Income	Cash Flow for Income Taxes	After-Tax Cash Flow
0	−$180,000	—		—		—			−$180,000
1	36,000	$180,000	×	0.2000	=	$36,000	0	0	36,000
2	36,000	$180,000	×	0.3200	=	57,600	− 21,600	+ 8,208	44,208
3	36,000	$180,000	×	0.1920	=	34,560	1,440	− 547	35,453
4	36,000	$180,000	×	0.1152	=	20,736	15,264	− 5,800	30,200
5	36,000	$180,000	×	0.1152	=	20,736	15,264	− 5,800	30,200
6	36,000	$180,000	×	0.0576	=	10,368	25,632	− 9,740	26,260
7–10	36,000	0				0	36,000	− 13,680	22,320
10	30,000						30,000[a]	− 11,400[b]	18,600
								TOTAL	$130,201

[a] Depreciation recapture = market value − book value = $30,000 − 0 = $30,000.
[b] Tax on depreciation recapture = $30,000(0.38) = $11,400.

After-Tax IRR: Set PW of column E = 0 and solve for i' in the following equation.

$$0 = -\$180{,}000 + 36{,}000\,(P/F, i', 1) + 44{,}208(P/F, i', 2) + 35{,}453(P/F, i', 3) + 30{,}200(P/F, i', 4) + 30{,}200(P/F, i', 5) + 26{,}260(P/F, i', 6) + 22{,}320(P/A, i', 4)(P/F, i', 6) + 18{,}600(P/F, i', 10);\ \text{IRR} = 12.4\%$$

```
   Applicable methods: Straight Line (SL)
                       Sum of Years Digits (SYD)
                       Declining Balance (DB)

2. Asset placed in service between January 1,
   1981 and July 31, 1986.
   Applicable method: Accelerated Cost Recovery
                      System (ACRS)

3. Asset placed in service between August 1, 1986
   to present.
   Applicable method: Modified ACRS (MACRS)

0. Exit to system

Enter option: 3
```

```
Menu 3: MACRS

1. Calculate Depreciation of the asset.
2. Evaluate an Investment.
0. Return to last menu.

Enter option :2
```

```
INVESTMENT

Cost of depreciable investment : 180000
Cost of non-depreciable investment : 0
Investment Tax Credit percentage (decimal): 0.0

Often the Useful Life of the investment is not
identical to the Study Period. Please check.

Useful Life of the investment : 10
Study Period : 10

Do you want to consider inflation/escalation
                                        [Y/N]? N
```

```
REVENUES AND EXPENSES

The amount in DOLLARS of the items below can be
CONSTANT during all, or a part, of the Useful
Life, or can be LUMPY
```

```
1. annual revenues
2. annual cost of utilities
3. annual cost of labor
4. annual cost of materials
5. annual lease cost

Enter C if CONSTANT/NONE else L if LUMPY

1c 2c 3c 4c 5c

Enter beginning (B) and ending (E) year in which
constant items apply, as pairs B,E:
Item 1 : 1,10
Item 2 : 1,10
Item 3 : 1,10
Item 4 : 1,10
Item 5 : 1,10

REVENUES AND EXPENSES

Annual revenues in dollars : 36000
Annual cost of utilities in dollars : 0
Annual cost of labor in dollars : 0
Annual cost of materials in dollars : 0
Annual lease cost in dollars : 0

TAXES and MARR

Incremental Income Tax Rate (decimal) : 0.38
Tax Rate on Capital Gain or Loss (decimal) : 0.38
The MARR for After-Tax Analysis (decimal) : 0.10

3.2 MACRS Depreciation

Enter MACRS Class Life of Investment from
                                   [3 5 7 10 15 20]
The Useful Life is 10 yrs.
Class Life: 5
Please enter:
Salvage Value for depreciation purposes and
Selling Price at time of disposal.

In MACRS assumed Salvage Value equals 0.

Selling Price of year 10 : 30000
```

YR	BTCF	DEPR	TAXABLE INCOME	INCOME TAXES	ATCF	PRESENT WORTH
0	-180000.00	0.00	0.00	0.00	-180000.00	-180000.00
1	36000.00	-36000.00	0.00	0.00	36000.00	32727.27
2	36000.00	-57600.00	-21600.00	8208.00	44208.00	36535.54
3	36000.00	-34560.00	1440.00	-547.20	35452.80	26636.21
4	36000.00	-20736.00	15264.00	-5800.32	30199.68	20626.79
5	36000.00	-20736.00	15264.00	-5800.32	30199.68	18751.63
6	36000.00	-10368.00	25632.00	-9740.16	26259.84	14823.00
7	36000.00	0.00	36000.00	-13680.00	22320.00	11453.69
8	36000.00	0.00	36000.00	-13680.00	22320.00	10412.44
9	36000.00	0.00	36000.00	-13680.00	22320.00	9465.86
10	66000.00	0.00	66000.00	-25080.00	40920.00	15766.43
				Present Worth:		17208.86

```
You have a Winner
The Present Worth of the Investment is Positive.
Do you want to print this table [Y/N]?Y

OPTIONS TO CONTINUE
1. Evaluate same Investment with different
                                      timings.
2. Evaluate another Investment.
3. Return to previous menu.

Enter option: 2
```

If the machinery in Example 7-7 had been classified as 10-year MACRS property instead of 5-year property, depreciation deductions would be "slowed down" in the early years of the study period and shifted into later years as shown in Table 7-4. Compared to Table 7-3, entries in columns (C), (D), and (E) in Table 7-4 are less favorable in the sense that a fair amount of after-tax cash flow is deferred until later years, producing a lower after-tax IRR and PW. For instance, the present worth is reduced from $17,209 in Table 7-3 to $9,061 in Table 7-4 (a 47% decrease). The only difference between Table 7-3 and Table 7-4 is the *timing of the after-tax cash flow*, which is a function of the timing and magnitude of the depreciation deductions. In fact, the curious reader can confirm that the sums of entries in columns (A) through (E) are identical (except for round-off differences) in both tables—timing of cash flows does, of course, make a difference!

It is also of interest to note the impact that the investment tax credit (ITC) would have on the after-tax IRR. If a 10% ITC could have been claimed for this investment, it would have reduced other income taxes payable by the firm, thereby reducing the out-of-pocket investment by $18,000 to $162,000. As a result, the after-tax internal rate of return is increased to 15.3% (the deprecia-

TABLE 7–4 Reworked Example 7-7 with a 10-Year MACRS Class Life

	(A)	(B) Depreciation Deduction					(C) = (A) − (B)	(D) = −0.38(C)	(E) = (A) + (D)
Year, k	Before-Tax Cash Flow	Basis	×	MACRS Rate	=	Deduction	Taxable Income	Cash Flow for Income Taxes	After-Tax Cash Flow
0	−$180,000	—		—		—			−$180,000
1	36,000	$180,000	×	0.1000	=	$18,000	$18,000	−$ 6,840	29,160
2	36,000	180,000	×	0.1800	=	32,400	3,600	− 1,368	34,632
3	36,000	180,000	×	0.1440	=	25,920	10,080	− 3,830	32,170
4	36,000	180,000	×	0.1152	=	20,736	15,264	− 5,800	30,200
5	36,000	180,000	×	0.0922	=	16,596	19,404	− 7,374	28,626
6	36,000	180,000	×	0.0737	=	13,266	22,734	− 8,639	27,361
7	36,000	180,000	×	0.0656	=	11,808	24,192	− 9,193	26,807
8	36,000	180,000	×	0.0655	=	11,790	24,210	− 9,200	26,800
9	36,000	180,000	×	0.0655	=	11,790	24,210	− 9,200	26,800
10	36,000	180,000	×	0.0655	=	11,790	24,210	− 9,200	26,800
10	30,000						30,000	− 11,400	18,600
11	0	180,000	×	0.0328	=	5,904	−5,904	+ 2,240	2,240
								TOTAL	$130,196

PW(10%) = $9,061; IRR = 11.2%

tion schedule is assumed not to change). The affected portion of the cash flow table would be as follows:

Year, k	Before-Tax Cash Flow	Depreciation Deduction	Taxable Income	Cash Flow for Income Taxes	After-Tax Cash Flow
0	−$180,000			+$18,000	−$162,000
1	36,000	$36,000	0	0	36,000
2	36,000	57,000	−$21,600	+ 8,202	44,208
.	.	.	.	.	.
.	.	.	.	.	.
.	.	.	.	.	.

The investment tax credit is normally treated as a "year 0" cash flow because of quarterly income tax payments required of corporations. The end of the first quarter is as long as a firm would wait to claim an ITC, and 3 months is obviously closer to the beginning of the first year than it is to the end of the year.

EXAMPLE 7-8

The Ajax Semiconductor Company is attempting to evaluate the profitability of adding another integrated circuit production line to its present operations. The company would need to purchase 2 or more acres of land for $275,000 (total). The facility would cost $60,000,000, have no market value at the end of 5 years and could be depreciated with the MACRS method using a class life of 5 years. An increment of working capital would be required, and its estimated amount is $10,000,000. Gross income is expected to increase by $30,000,000 per year for 5 years, and operating expenses are estimated to be $8,000,000 per year for 5 years. It is expected that depreciation will be claimed over 6 years even though the facility will be closed down after 5 years. The firm's effective income tax rate is 40%, and no investment tax credit applies. (a) Set up a table and determine the after-tax cash flow for this project. (b) What is the net income after taxes in year 3? (c) Is the investment worthwhile when the after-tax MARR is 12%?

Solution

(a) The format recommended in Figure 7-3 is followed in Table 7-5 to obtain after-tax cash flows in years 0–6. Acquisition of land as well as additional working capital are treated as nondepreciable investments whose market values at the end of year 5 are estimated to equal their first costs. (In economic evaluations, it is customary to assume that land and working capital do not inflate in value during the study period.) Also observe that the last MACRS depreciation deduction is taken at the end of year 6 due to the half-year convention. Thus, it is assumed that the tax life (6 years) exceeds the study period.

TABLE 7–5 After-Tax Analysis of Example 7-8

Year, k	(A) Before-Tax Cash Flow	(B) Depreciation Deduction	(C) = (A) − (B) Taxable Income	(D) = −0.4(C) Cash Flow for Income Taxes	(E) = (A) + (D) After-Tax Cash Flow
0	−\$60,000,000 − 10,000,000 − 275,000				−\$70,275,000
1	22,000,000	12,000,000	10,000,000	−4,000,000	18,000,000
2	22,000,000	19,200,000	2,800,000	−1,120,000	20,880,000
3	22,000,000	11,520,000	10,480,000	−4,192,000	17,808,000
4	22,000,000	6,912,000	15,088,000	−6,035,200	15,964,800
5	22,000,000	6,912,000	15,088,000	−6,035,200	15,964,800
5	10,275,000[a]				10,275,000
6	0	3,456,000	− 3,456,000	1,382,400	1,382,400

[a]Market value of working capital and land.

By using Equation 7-13, one is able to compute after-tax cash flow in year 3 (for example) to be

$$\begin{aligned} ATCF_3 &= (\$22,000,000 - \$11,520,000)(1 - 0.40) + \$11,520,000 \\ &= \$17,808,000 \end{aligned}$$

(b) The net income after taxes in year 3 can be determined with Equation 7-12:

$$NIAT_3 = (\$22,000,000 - \$11,520,000)(1 - 0.40) = \$6,288,000$$

It can also be obtained directly from Table 7-5 by adding the year 3 entries from columns C and D: \$10,480,000 − \$4,192,000 = \$6,288,000.
(c) The after-tax IRR is obtained from entries in column E of Table 7-5, and with the Appendix E computer program the IRR is found to be 11.6%. The after-tax PW equals \$852,672 at MARR = 12%. Based on economic considerations, this integrated circuit production line should be recommended since it appears to be quite attractive.

EXAMPLE 7-8A

(a) Use ATAX in Appendix F to perform the calculations shown in Example 7-8 when the tax life is 6 years but the study period is 5 years.
(b) Suppose the depreciable property in Example 7-8 (\$60,000,000) will be disposed of for \$0 at the end of year 5, and a loss on disposal of \$6,912,000 will be claimed at the end of year 5. By referring to section 6.10.2.1, you will recall that when property is disposed of *before* the end of the MACRS recovery period, only a half-year of depreciation is allowable in the year of disposal. In our case, \$3,456,000 can be claimed as depreciation in year 5, and the book value is \$6,912,900 at the end of year 5. Because the selling price (market value) is zero, the loss on disposal equals our book value of \$6,912,000. As seen from Figure 7-3, a tax credit of 0.40 (\$6,912,000) = \$2,764,800 is created at the end of year 5.

Use ATAX in Appendix F to determine the PW of these situations at a MARR = 12%, and comment on whether a company should opt for the procedure followed in part (a) when tax life is greater than the study period or for the procedure used in part (b).

Solution

(a) The data entry prompts of ATAX are indicated below. Notice that the tax life is 6 years. The final table of ATCFs is also shown (all entries are expressed in thousands of dollars), and the after-tax PW equals $852,672 as in the Example 7-8 solution.

```
INVESTMENT

Cost of depreciable investment : 60000
Cost of non-depreciable investment : 10275
Investment Tax Credit percentage (decimal) : 0.0

Often the Useful Life of the investment is not
identical to the Study Period. Please check.

Useful Life of the investment : 6
Year when non-depreciable investment is
                                    recovered : 5

Do you want to consider inflation/escalation
                                        [Y/N]?N
```

```
REVENUES AND EXPENSES

The amount in DOLLARS of the items below can be
CONSTANT during all, or a part, of the Useful
Life, or can be LUMPY
1. annual revenues
2. annual cost of utilities
3. annual cost of labor
4. annual cost of materials
5. annual lease cost

Enter C if CONSTANT/NONE else L if LUMPY

1C 2C 3C 4C 5C

Enter beginning (B) and ending (E) year in which
constant items apply, as pairs B,E:
Item 1 :1,5
Item 2 :1,5
Item 3 :1,5
Item 4 :1,5
Item 5 :1,5
```

```
REVENUES AND EXPENSES

Annual revenues in dollars :22000
Annual cost of utilities in dollars :0
Annual cost of labor in dollars :0
Annual cost of materials in dollars :0
Annual lease cost in dollars :0

TAXES and MARR

Incremental Income Tax Rate (decimal) : 0.40
Tax Rate on Capital Gain or Loss (decimal) : 0.40
The MARR for After-Tax Analysis (decimal) : 0.12

3.2 MACRS Depreciation

Enter MACRS Class Life of Investment from
                                    [3 5 7 10 15 20]
The Useful Life is 6 yrs.
Class Life: 5
Please enter:
Salvage Value for depreciation purposes and
Selling Price at time of disposal.

In MACRS assumed Salvage Value equals 0.

Selling price for year 6 : 0
```

YR	BTCF	DEPR	TAXABLE INCOME	INCOME TAXES	ATCF	PRESENT WORTH
0	-70275.00	0.00	0.00	0.00	-70275.00	-70275.00
1	22000.00	-12000.00	10000.00	-4000.00	18000.00	16071.43
2	22000.00	-19200.00	2800.00	-1120.00	20880.00	16645.41
3	22000.00	-11520.00	10480.00	-4192.00	17808.00	12675.38
4	22000.00	- 6912.00	15088.00	-6035.20	15964.80	10145.92
5	32275.00	- 6912.00	15088.00	-6035.20	26239.80	14889.17
6	0.00	- 3456.00	-3456.00	1382.40	1382.40	700.37
					Present Worth:	852.67

```
You Have a Winner
The Present Worth of the Investment is Positive.
Do you want to print this table [Y/N]?Y
```

(b) When Example 7-8 is repeated using ATAX for a loss on disposal of $6,912,000 and a tax life of 5 years, the following results are produced.

```
INVESTMENT

Cost of depreciable investment : 60000
Cost of non-depreciable investment : 10275
Investment Tax Credit percentage (decimal) : 0.0
Often the Useful Life of the investment is not
identical to the Study Period. Please check.

Useful Life of the investment : 5
Study Period : 5
Year when non-depreciable investment is
                                   recovered :5

Do you want to consider inflation/escalation
                                      [Y/N]? N
```

```
REVENUES AND EXPENSES

The amount in DOLLARS of the items below can be
CONSTANT during all, or a part, of the Useful
Life, or can be LUMPY
1. annual revenues
2. annual cost of utilities
3. annual cost of labor
4. annual cost of materials
5. annual lease cost

Enter C if CONSTANT/NONE else L if LUMPY

1C 2C 3C 4C 5C

Enter beginning (B) and ending (E) year in which
constant items apply, as pairs B,E:
Item 1 :1,5
Item 2 :1,5
Item 3 :1,5
Item 4 :1,5
Item 5 :1,5
```

```
REVENUES AND EXPENSES

Annual revenues in dollars :22000
Annual cost of utilities in dollars :0
Annual cost of labor in dollars :0
```

```
Annual cost of materials in dollars :0
Annual lease cost in dollars : 0

TAXES and MARR

Incremental Income Tax Rate (decimal) : 0.40
Tax Rate on Capital Gain or Loss (decimal) : 0.40
The MARR for After-Tax Analysis (decimal) : 0.12

3.2 MACRS Depreciation

Enter MACRS Class Life of Investment from
                                [3 5 7 10 15 20]
The Useful Life is 5 yrs.
Class Life: 5

ADJUSTING USEFUL LIFE

The Tax Life of 6 is greater than the Useful
                                Life of 5 years.

This program would proceed with the Useful Life
provided. Depreciation after Useful Life is not
recovered and the difference between Book Value
and Selling Price at the end of the Useful Life
is taxed as a capital gain or loss.

Do you want to re-enter the Useful Life [Y/N]?N

Please enter:
Salvage Value for depreciation purposes and
Selling Price at time of disposal.

In MACRS assumed Salvage Value equals 0.

Selling price for year 5 : 0
```

YR	BTCF	DEPR	TAXABLE INCOME	INCOME TAXES	ATCF	PRESENT WORTH
0	−70275.00	0.00	0.00	0.00	−70275.00	−70275.00
1	22000.00	−12000.00	10000.00	−4000.00	18000.00	16071.43
2	22000.00	−19200.00	2800.00	−1120.00	20880.00	16645.41
3	22000.00	−11520.00	10480.00	−4192.00	17808.00	12675.38
4	22000.00	− 6912.00	15088.00	−6035.20	15964.80	10145.92
5	32275.00	− 3456.00	11632.00	−4652.80	27622.20	15673.58
				Present Worth:		936.72

```
You Have a Winner
The Present Worth of the Investment is Positive.
Do you want to print this table [Y/N]?Y
```

The PW has increased by $84,047 because the tax credit from the loss on disposal of $6,912,000 is taken in year 5. Most companies will choose this procedure if they can elect to do so because after-tax profitability is greater.*

7.12 ILLUSTRATION OF AFTER-TAX ECONOMIC COMPARISONS OF COST-ONLY ALTERNATIVES

In this section the comparison of mutually exclusive alternatives involving costs only is described.

EXAMPLE 7-9

An engineering consulting firm can purchase a fully configured computer-aided design (CAD) workstation for $20,000. It is estimated that the useful life of the workstation is 7 years, and its market value in 7 years should be $2,000. Operating expenses are estimated to be $40 per 8-hour work day, and maintenance will be performed under contract for $8,000 per year. The MACRS class life is 5 years. Furthermore, the effective income tax rate is 40% and the investment tax credit is zero.

As an alternative, sufficient computer time can be leased from a time-share company at a cost of $20,000 per year. If the after-tax MARR is 10%, how many work days per year must the workstation be needed in order to justify *leasing* it?

Solution

This example involves an after-tax evaluation of purchasing depreciable property versus leasing it. We are to determine how heavily the workstation must be utilized so that the lease option is a good economic choice. A key assumption is that the cost of engineering design time is unaffected by whether the workstation is purchased or leased. Variable operations expenses associated

*Many of the problems at the end of this chapter (e.g., 7-9, 7-17, 7-20, 7-26) require an assumption to be made regarding how MACRS depreciation in year $N + 1$ is treated for an N-year study period. A conservative approach is to follow the procedure used in Example 7-8 where the tax life exceeds the study period.

with ownership result from the purchase of supplies, utilities, and so on. Hardware and software maintenance cost is fixed at $8,000 per year. It is further assumed that the maximum number of working days per year is 250.

Determination of after-tax cash flow for the lease option is relatively straightforward and is not affected by how much the workstation is utilized:

$$\text{After-tax cost of the lease} = \$20{,}000(1 - 0.40) = \$12{,}000$$

After-tax cash flows for the purchase option involve expenses that are fixed (not a function of equipment utilization) in addition to expenses that vary with equipment usage. If we let X equal the number of working days per year that the equipment is utilized, the variable cost per year of operating the workstation is $\$40X$. The after-tax analysis of the purchase alternative is shown in Table 7-6.

The after-tax annual cost of purchasing the workstation is

$$\begin{aligned}\$20{,}000(A/P, 10\%, 7) &+ 24X + [\$3{,}200(P/F, 10\%, 1) + \cdots \\ &+ \$4{,}800(P/F, 10\%, 7)](A/P, 10\%, 7) \\ &- \$1{,}200(A/F, 10\%, 7) = \$24X + \$7{,}511\end{aligned}$$

To solve for X, we equate the after-tax annual costs of both alternatives:

$$\$12{,}000 = \$24X + \$7{,}511$$

Thus $X = 187$ days/year. Therefore, if the firm expects to utilize the CAD workstation in its business *more than* 187 days per year, the equipment should be leased. The graphical summary of Example 7-9 shown in Figure 7-4 provides rationale for this recommendation. The importance of the workstation's estimated utilization, in workdays per year, is now apparent.

TABLE 7–6 After-Tax Analysis of Purchase Alternative

End of Year, k	(A) Before-Tax Cash Flow	(B) Depreciation Deduction[a]	(C) = (A) − (B) Taxable Income	(D) = −t(C) Cash Flow for Income Taxes	(E) = (A) + (D) After-Tax Cash Flow
0	−$20,000				−$20,000
1	−40X − 8,000	$4,000	−40X − 12,000	16X + 4,800	−24X − 3,200
2	−40X − 8,000	6,400	−40X − 14,400	16X + 5,760	−24X − 2,240
3	−40X − 8,000	3,840	−40X − 11,840	16X + 4,736	−24X − 3,264
4	−40X − 8,000	2,304	−40X − 10,304	16X + 4,122	−24X − 3,878
5	−40X − 8,000	2,304	−40X − 10,304	16X + 4,122	−24X − 3,878
6	−40X − 8,000	1,152	−40X − 9,152	16X + 3,661	−24X − 4,339
7	−40X − 8,000	0	−40X − 8,000	16X + 3,200	−24X − 4,800
7	2,000		2,000	−800	1,200

[a]Depreciation deduction$_k$ = $20,000 × (MACRS rate)$_k$. Refer to Table 6–5.

FIGURE 7-4

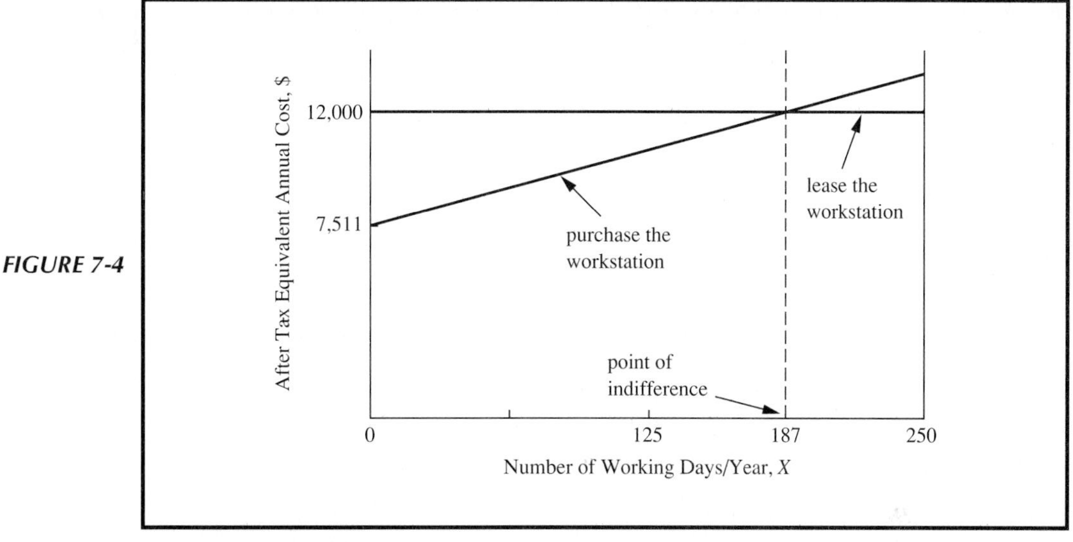

Summary of Example 7-9

EXAMPLE 7-10

It is desired to compare the economics of two industrial material handling systems. The pertinent data are as follows:

	System X	System Y
First cost	$80,000	$200,000
Useful life (also depreciation life)	20 years	40 years
Annual before-tax cash disbursement	$18,000	$ 6,000
Salvage value at end of life	20,000	40,000

Set up a table to compute the after-tax cash flows based on straight-line depreciation using the lives and salvage values given and a 30% effective income tax rate to cover both state and federal income taxes. With an after-tax MARR of 10%, show which alternative is best by the (a) internal rate of return (IRR) method and (b) present worth (PW) method. This is a small firm that is profitable in its overall operation. Assume that the "do nothing" alternative is not an option.

Before applying the various economy study methods, it should be recognized that the solutions shown below for each method employ the repeatability assumption for comparisons of alternatives having different economic lives as discussed in Chapter 5.:

1. The period of needed service over which the alternatives are being compared is either indefinitely long or a length of time equal to a common multiple of the lives.
2. What is estimated to happen in the first life span will happen in all succeeding life spans, if any.

It is also assumed that system X must be depreciated over 20 years and system Y must be depreciated over 40 years.

Table 7-7 shows the calculation of after-tax cash flows to be used in the following solutions. Notice that *negative* taxable income and *positive* income taxes (i.e., reduction in taxes owed) are possible because the firm is profitable in its overall activity.

Solution by IRR Method

The IRR cannot be found for either alternative alone, but one can use this method to determine if the incremental investment is justified. Since the economic lives differ, the easiest way to deal with this situation is to find the $i'\%$ at which the equivalent annual costs (AC) of the two alternatives are equal (or the difference in the equivalent annual costs for the two alternatives is zero). Thus,

$$AC_X = \$80{,}000(A/P, i'\%, 20) - \$20{,}000(A/F, i'\%, 20) + \$11{,}700$$

$$AC_Y = \$200{,}000(A/P, i'\%, 40) - \$40{,}000(A/F, i'\%, 40) + \$3{,}000$$

TABLE 7–7 Determination of After-Tax Cash Flows For Example 7-10

Year	(A) Before-Tax Cash Flow	(B) Depreciation Deduction	(C) = (A) − (B) Taxable Income	(D) = −0.3(C) Cash Flow for Income Taxes	(E) = (A) + (D) After-Tax Cash Flow
System X					
0	−$ 80,000				−$ 80,000
1–20	− 18,000	$3,000[a]	−$21,000	+$6,300	− 11,700
20	+ 20,000				+ 20,000
System Y					
0	− 200,000				− 200,000
1–40	− 6,000	4,000[b]	− 10,000	+ 3,000	− 3,000
40	+ 40,000				+ 40,000

[a] Annual depreciation for system X = ($80,000 − $20,000)/20 = $3,000.

[b] Annual depreciation for system Y = ($200,000 − $40,000)/40 = $4,000.

The manually calculated results are as follows:

	$i' = 5\%$	$i' = 10\%$
AC_x:	\$17,515	\$20,748
AC_y:	14,325	23,362
$AC_x - AC_y$:	\$ 3,190	−\$ 2,614

Thus,

$$i'\% = \text{IRR} = 5\% + \frac{\$3{,}190}{\$3{,}190 + \$2{,}614}(10\% - 5\%)$$
$$= 5\% + 2.8\% = 7.8\%$$

Since 7.8% < 10% (MARR), the incremental investment is not justified, and system X is the indicated choice.

Solution by the PW Method
Using the lowest common multiple of lives = 40 years as the study period, we obtain the following:

	System X	System Y
Annual expenses:		
\$11,700(*P*/*A*, 10%, 40)	−\$114,415	
\$3,000(*P*/*A*, 10%, 40)		−\$ 29,337
Original investment	− 80,000	− 200,000
First replacement: [(−\$80,000 + \$20,000) × (*P*/*F*, 10%, 20)]	− 8,919	
Salvage after 40 years		
20,000(*P*/*F*, 10%, 40)	+ 442	
40,000(P/F, 10%, 40)		+ 884
Total PW	−\$202,892	−\$228,453

Since PW_X is less negative than PW_Y, system X is again the indicated choice.

EXAMPLE 7-11

It is desired to compare the four presses in Example 5-1 on an after-tax basis by using the AW method. Assume an effective income tax rate of 50%, that 1984 ACRS percentages are used to compute depreciation deductions with a 5-year recovery period, and that the after-tax MARR is 10%. The essential data are shown in the following table, and a tabular computation of the after-tax annual cash flows for each press is shown in Table 7-8.

	Press			
	A	B	C	D
Investment	−$6,000	−$7,600	−$12,400	−$13,000
Salvage value	0	0	0	0
Annual amounts, years 1–5:				
before-tax cash flow	− 7,800	− 7,282	− 6,298	− 5,720
Useful life (years)	5	5	5	5

Solution

Based on the after-tax cash flows calculated in Table 7-8, the equivalent after-tax annual worth of press A at a MARR of 10% is

$$\begin{aligned}\mathrm{AW}_A(10\%) &= -\$6{,}000(A/P, 10\%, 5)\\ &\quad -[\$3{,}450(P/F, 10\%, 1) + \$3{,}240(P/F, 10\%, 2)\\ &\quad +\$3{,}270(P/F, 10\%, 3) + \$3{,}270(P/F, 10\%, 4)\\ &\quad +\$3{,}270(P/F, 10\%, 5)](A/P, 10\%, 5)\\ &= -\$4{,}889\end{aligned}$$

TABLE 7–8 Data and After-Tax Annual Cash Flow Computations for Comparison of the Four Molding Presses in Example 7-11 (Original Data Given in Example 5-1)

Press	Year	Before-Tax Cash Flow	Depreciation Deduction	Taxable Income	Income Taxes	After-Tax Cash Flow
A	0	−$ 6,000	0	0	0	−$ 6,000
	1	− 7,800	$ 900	−$8,700	4,350	− 3,450
	2	− 7,800	1,320	− 9,120	4,560	− 3,240
	3	− 7,800	1,260	− 9,060	4,530	− 3,270
	4	− 7,800	1,260	− 9,060	4,530	− 3,270
	5	− 7,800	1,260	− 9,060	4,530	− 3,270
B	0	− 7,600	0	0	0	− 7,600
	1	− 7,282	1,140	− 8,422	4,211	− 3,071
	2	− 7,282	1,672	− 8,954	4,477	− 2,805
	3	− 7,282	1,596	− 8,878	4,439	− 2,843
	4	− 7,282	1,596	− 8,878	4,439	− 2,843
	5	− 7,282	1,596	− 8,878	4,439	− 2,843
C	0	− 12,400	0	0	0	− 12,400
	1	− 6,298	1,860	− 8,158	4,079	− 2,219
	2	− 6,298	2,728	− 9,026	4,513	− 1,785
	3	− 6,298	2,608	− 8,902	4,451	− 1,847
	4	− 6,298	2,608	− 8,902	4,451	− 1,847
	5	− 6,298	2,608	− 8,902	4,451	− 1,847
D	0	− 13,000	0	0	0	− 13,000
	1	− 5,720	1,950	− 7,670	3,835	− 1,885
	2	− 5,720	2,860	− 8,580	4,290	− 1,430
	3	− 5,720	2,730	− 8,450	4,225	− 1,495
	4	− 5,720	2,730	− 8,450	4,225	− 1,495
	5	− 5,720	2,730	− 8,450	4,225	− 1,495

Similarly, the equivalent annual worths of presses B, C, and D can be determined:

$$AW_B(10\%) = -\$4{,}894$$
$$AW_C(10\%) = -\$5{,}194$$
$$AW_D(10\%) = -\$5{,}004$$

Hence, press A is the recommended choice, followed closely by press B. In the before-tax study of Example 5-1, press D was the choice. This demonstrates clearly that income tax considerations can change the recommendations made in before-tax engineering economy studies.

In cost-only alternatives such as those above, the *negative taxable income* and resultant *positive income taxes* should be interpreted as offsets against positive taxable income and expenses for income taxes, respectively, that arise in other profitable areas of activity in a firm.

Some of the consequences of the Tax Reform Act of 1986 may be seen in cost-only problems when Example 7-11 is reworked with MACRS depreciation over a 5-year class life and a lower effective income tax of 40%. Instead of an updated Table 7-8 to illustrate these changes, a summary of after-tax AW at 10% interest for 1981–1986 conditions, contrasted with present (1989) conditions, is provided below. The after-tax AWs become more negative in 1989 primarily because reduced income tax rates increase after-tax expenses to the firm!

		(A) ACRS (1981–1986) $t = 0.50$	**(B) MACRS (1989) $t = 0.40$**	**% Change $\left(\frac{A-B}{A} \times 100\right)$**
After-Tax	A	−\$4,889	−\$5,771	18.0% more expensive
Annual Worth	B	− 4,894	− 5,751	17.5% more expensive
at i = 10%	C	− 5,194	− 6,034	16.2% more expensive
	D	− 5,004	− 5,796	15.8% more expensive

7.13 AFTER-TAX EVALUATIONS OF SPECIFIC FINANCING ARRANGEMENTS

Engineering economy problems encountered in previous chapters have not distinguished between the types of funds (i.e., equity versus debt capital) that compose the pool of capital available for investment in engineering and business projects. Engineers and other technical personnel generally do not concern themselves with how these funds have been sourced because their mission is to ascertain the best (most profitable) uses of the capital, given that it is available (see Figure 1-1). Consequently, this book has dealt with problems from the standpoint of a firm's *overall* pool of investment capital rather than a subset of it. Hence, the interest rate (i.e., the MARR) used for time-value-of-money calculations typically reflects the cost of both borrowed capital and equity capital.

In this section we shall demonstrate how to analyze a capital investment problem from the viewpoint of the *equity investor* (i.e., the stockholder). This viewpoint may be desired when a particular undertaking is quite risky, and equity investors would like an explicit assessment of the project's profitability in terms of *their* investment. Another motivation for looking at an investment from an equity viewpoint stems from the basic aim of private enterprise, which is to *maximize the future wealth* of the owners of the firm. To meet this objective, a project's after-tax cash flows must be evaluated from the viewpoint of a firm's stockholders using an interest rate equal to their opportunity cost of capital. This can be done simply by adding a column for "Loan and Interest Cash Flow" in Figure 7-3, as is illustrated in the following example.

EXAMPLE 7-12

This example is fashioned after Example 7-7. Additional information is given concerning capital that is borrowed to purchase the machinery, and it is desired to determine the after-tax PW of equity capital. The problem is restated with additional information as follows.

Certain new machinery is estimated to cost $180,000 installed. It is expected to reduce net annual operating expenses by $36,000 per year (not including interest) for 10 years and to have $30,000 salvage value at the end of the tenth year. Fifty thousand dollars of the investment is to be borrowed at 10% interest payable annually, with *all* the principal to be repaid at the end of the tenth year. To simplify the analysis, the machinery is depreciated with the straight-line method over 10 years using a $30,000 salvage value.

Determine the after-tax equity cash flow and calculate the after-tax PW of equity capital. The opportunity cost of equity capital has been set at 18% per year.

Solution

Table 7-9 shows the recommended tabular format with the column headings indicating how the calculations are made. Notice that interest on borrowed capital is deductible as a business expense and hence reduces before-tax

TABLE 7–9 Computations to Determine the After-Tax Equity Cash Flow for Example 7-12

Year	(A) Before-Tax Cash Flow	(B) Depreciation Deduction	(C) Loan and Interest Cash Flow	(D) = (A) − (B) + Interest Portion of (C) Taxable Income	E = −0.38(D) Cash Flow for Income Taxes	(F) = (A) + (C) + (E) After-Tax Equity Cash Flow
0	− $180,000		+$50,000			−$130,000
1–10	+ 36,000	$15,000	− 5,000	+$16,000	−$6,080	+ 24,920
10	+ 30,000[a]		− 50,000			− 20,000

[a] It is assumed that the market value of the machinery in year 10 equals its book value. Thus, no income taxes are paid at the end of the study period.

savings to \$31,000. From the results of the right-hand column, the after-tax PW of equity investment can be calculated:

$$\begin{aligned} PW &= -\$130{,}000 + \$24{,}920(P/A, 18\%, 10) \\ &\quad - \$20{,}000(P/F, 18\%, 10) \\ &= -\$21{,}832 \end{aligned}$$

Because the PW is negative, this investment appears to be a poor one based on after-tax equity cash flows.

EXAMPLE 7-13

A firm is considering purchasing an asset for \$10,000, with half of this amount coming from retained earnings (equity) and the other half being borrowed for 3 years at an effective interest rate of 12% per year. Uniform annual payments, consisting of interest and loan principal, will be utilized to repay the \$5,000 loan. Straight-line depreciation over a 3-year period can be claimed, and the asset's salvage value at this time is expected to be \$4,000. Before-tax net income from the asset is estimated to be \$5,000 per year, which does not include the cost of borrowed money. Finally, the firm's effective income tax rate is 50%.

If the opportunity cost of capital to equity investors is 20% per year after taxes, is this a profitable investment from the stockholder's viewpoint?

Solution

The amount of the uniform loan repayment is \$5,000(A/P,12%,3) = \$2,081.75. As described in Chapter 4, this capital recovery amount consists of repayment of borrowed money (principal) and interest on the unpaid principal at the beginning of each year. Because interest is a deductible business expense but repayment of loan principal is not, a schedule of annual loan principal and interest must be developed as follows (to the nearest dollar):

End of Year	Interest	Principal
1	\$5,000(0.12) = \$600	\$2,082 − \$600 = \$1,482
2	(5,000 − 1,482)(0.12) = 422	2,082 − 422 = 1,660
3	(5,000 − 1,482 − 1,660)(0.12) = 223	2,082 − 223 = 1,859

The after-tax equity cash flow for the asset can now be determined as indicated in the heading of Table 7-9:

	(A)	(B)	(C) Loan Cash Flow:		(D)	(E)	(F)
Year	Before-Tax Cash flow	Deprecia-tion	Interest[a]	Principal[b]	Taxable Income	Cash Flow for Income Taxes	After-tax Equity Cash Flow
0	−$10,000			+$5,000			−$5,000
1	5,000	$2,000	−$600	−$1,482	$2,400	−$1,200	1,718[c]
2	5,000	2,000	− 422	− 1,660	2,578	− 1,289	1,629
3	5,000	2,000	− 223	− 1,859	2,777	− 1,389	1,529
3	4,000						4,000

[a] Interest is deductible from column A when computing taxable income (column D).
[b] Principal is an after-tax cost deducted from column F.
[c] $5,000 − $600 − $1,200 − $1,482 = $1,718.

The present worth of column F at 20% is $763. Because the PW is greater than zero, this investment is marginally attractive.

After-tax evaluations based on equity investment frequently utilize an annual schedule of loan interest and principal such as the one illustrated in Example 7-13. To formalize the development of such a schedule, the following discussion and equations are offered.

Consider a loan repayment scheme (e.g., a home mortgage) involving equal uniform *monthly* payments of $A = I_0(A/P, i_b\%, N)$, where typically i_b is a nominal loan interest rate per month. N is the number of months over which the loan is repaid, and I_0 is the initial lump-sum amount of the loan at time 0. In after-tax problems that include different debt-equity mixes in the financing of a project, the *interest* on a loan must be broken out of A because it is a deductible business expense whereas the remainder of A is repayment of principal (i.e., equity) and is not deductible in calculating a firm's taxable income. Graphically, the relationship between loan equity and loan interest has this general appearance:

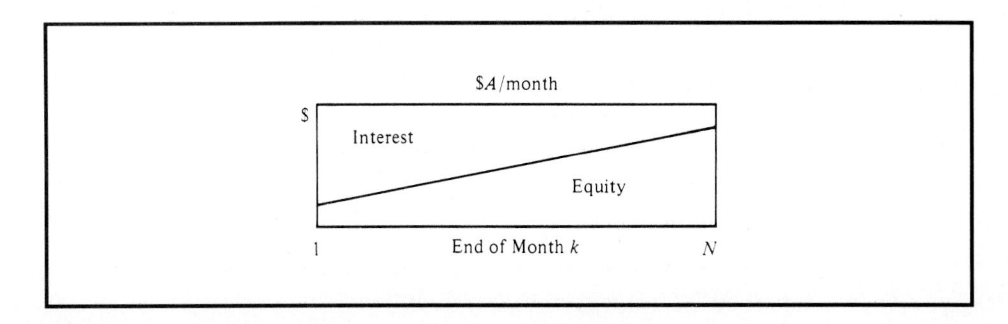

As a function of month $k (1 \leq k \leq N)$, it is desired to be able to determine quickly the interest in month k, I^*_k, and the equity in month k, E^*_k. It will always be true that $I^*_k + E^*_k = A$. The interest repaid during month 1 is $i_b I_0$, and $E^*_1 = A - i_b I_0$. In general, the interest repaid in any month is the amount of loan principal, or equity, owed at the beginning of that month multiplied by i_b. The loan principal owed at the beginning of month k, or the end of month $k - 1$, is I_{k-1} and can be easily determined by:

$$I_{k-1} = A(P/A, i_b\%, N - k + 1) \tag{7-16}$$

Thus

$$I^*_k = I_{k-1}(i_b) \tag{7-17}$$

and

$$E^*_k = A - I^*_k \tag{7-18}$$

For example, suppose that an asset is purchased for \$5,000 and a 20% down payment is made. The remaining \$4,000 is financed at $1\frac{1}{2}\%$ per month (on the unpaid balance) over 24 months. How much interest and equity are repaid during the eighteenth month?

$$A = \$4{,}000(A/P, 1\tfrac{1}{2}\%, 24) = \$199.70$$

$$\begin{aligned} I_{17} &= \$199.70(P/A, 1\tfrac{1}{2}\%, 24 - 18 + 1) = \$1{,}317.66 \text{ (from Equation 7-16)} \\ I^*_{18} &= \$1{,}317.66(0.015) = \$19.76 \text{ (from Equation 7-17)} \\ E^*_{18} &= \$199.70 - \$19.76 = \$179.94 \text{ (from Equation 7-18)} \end{aligned}$$

In many problems the present worth of before-tax and/or after-tax interest cost of a debt-financed project must be calculated over the entire loan period. The appropriate discount rate is usually a firm's MARR expressed over the same interval as i_b. Each I^*_k could be computed as shown above and the present worth then determined as follows.

$$\text{Before-tax cost of interest } (i_{BT})\text{: } \text{PW}(i_{BT}) = \sum_{k=1}^{N} I^*_k(P/F, i_{BT}\%, k) \tag{7-19}$$

$$\text{After-tax cost of interest } (i_{AT})\text{: } \text{PW}(i_{AT}) = \left[\sum_{k=1}^{N} I^*_k(P/F, i_{AT}\%, k)\right](1 - t) \tag{7-20}$$

where t is the effective income tax rate. To save time with computations, Equation 7-20 can be reduced as follows:

$$\text{PW}(i_{AT}) = (1 - t)A\left[(P/A, i_{AT}\%, N) - \frac{(P/F, i_b\%, N) - (P/F, i_{AT}\%, N)}{i_{AT} - i_b}\right] \tag{7-21}$$

Referring to the loan example above, what is the present worth of after-tax interest cost if $t = 0.46$ and $i_{AT} = 2\%$ per month? By using Equation 7-21, this result is obtained:

$$
\begin{aligned}
\text{PW}(i_{AT}) &= (1 - 0.46)(\$199.70) \\
&\quad \times \left[(P/A, 2\%, 24) - \frac{(P/F, 1.5\%, 24) - (P/F, 2\%, 24)}{0.02 - 0.015} \right] \\
&= (1 - 0.46)(\$199.70) \left[18.9139 - \frac{0.6995 - 0.6217}{0.005} \right] \\
&= \$361.68
\end{aligned}
$$

7.14 THE AFTER-TAX EFFECT OF DEPLETION ALLOWANCES

Income from investment in certain natural resources is subject to depletion allowances before income taxes are computed. Under certain conditions, notably where the taxpayer is in a relatively high income tax bracket, depletion provisions in the tax law can provide considerable economic advantages.

As an example, consider the case of a profitable corporation that has a net taxable income of \$600,000. In 1989 it spends \$400,000 to drill and develop a geothermal well that has an estimated reservoir of 10,000,000 gallons of water. Hot water is produced and sold at \$0.20 per gallon in accordance with the schedule shown in column 2 of Table 7-10 to produce the gross income shown in column 3. Column 4 shows the net cash flow after production costs have been deducted.

The depletion allowance that can be deducted in a given year may be based on a fixed *percentage* of the gross income (15% for geothermal wells), provided that the deduction *does not exceed 50% of the net income before such deduction* (column 5). Depletion, computed in this manner, is shown in column 7 of Table 7-10. Another method is to base depletion on the estimated investment cost of the product. In this case the estimated 10,000,000 gallons of water in the well cost \$400,000. Such *cost depletion* may, if desired, be charged at the rate of \$0.04 per gallon shown in column 6 of the same table.

The taxable income resulting from the most favorable application of depletion allowances (by the cost method or the percentage method) is shown in column 8. The advantage of percentage depletion allowances stems from the fact that the total depletion that can be claimed often exceeds the depreciable capital investment. However, this advantage does not show itself in this particular situation because of the relatively small fraction of reservoir capacity that is sold in years 1–5. In fact, the cost depletion allowance in column 6 is consistently

TABLE 7–10 Capital Recovery Provided by a Geothermal Well, with Cost and Percentage Depletion Allowances Used in Computation of Income Taxes

(1) Year	(2) Gallons of Water Sold	(3) Gross Income (Cash Flow)	(4) Net Income Before Taxes	(5) 50% of Net Income	(6) Cost Depletion at $0.04 per Gallon[a]	(7) Depletion Allowance at 15% of Gross Income	(8) [= (4) − either (6) or (7)] Taxable Income[b]	(9) [= −0.40(8)] Income Taxes	(10) [= (4) + (9)] Net Cash Flow After Taxes
1	700,000	$140,000	+$80,000	$40,000	$28,000	$21,000	$52,000	−$20,800	$59,200
2	600,000	120,000	+ 70,000	35,000	24,000	18,000	46,000	− 18,400	51,600
3	450,000	90,000	+ 48,000	24,000	18,000	13,500	30,000	− 12,000	36,000
4	200,000	40,000	+ 24,000	12,000	8,000	6,000	16,000	− 6,400	17,600
5	50,000	10,000	+ 2,500	1,250	2,000	1,500	500	− 200	2,300

[a] Cost depletion summary: Yr.1 $\frac{\$700,000}{10,000,000}(\$400,000) = \$28,000$

Yr.2 $\frac{600,000(\$400,000 - \$28,000)}{10,000,000 - 700,000} = \$24,000$

Yr.3 $\frac{450,000(\$372,000 - \$24,000)}{9,300,000 - 600,000} = \$18,000$

Yr.4 $\frac{200,000(\$348,000 - \$18,000)}{8,700,000 - 450,000} = \$8,000$

Yr.5 $\frac{50,000(\$330,000 - \$8,000)}{8,250,000 - 200,000} = \$2,000$

[b] In computing taxable income, the larger allowance in column 6 or column 7 is chosen as long as percentage depletion does not exceed 50% of column 4. If cost depletion exceeds percentage depletion, then cost depletion must be used during that particular year.

better (higher) than the fixed percentage allowance in column 7. When cost depletion exceeds percentage depletion, it must be used in computing taxable income, given that the basis of the property has not been exhausted. It is also noteworthy that cost depletion is not limited to 50% of net income shown in column 5, which is determined before depletion deductions are considered. (Recall that Figure 6-2 summarized the procedure for determining allowable depletion.)

Inasmuch as the firm has a net taxable income of $600,000 even before returns from the geothermal well are considered, we shall assume that the incremental tax rate is 40% in 1989, thus giving the income tax shown in column 9 of Table 7-10. Column 10 shows the net after-tax cash flow provided to the investor for years 1–5 of the well's operation. The remaining 8,000,000 gallons of hot water would presumably be sold over the subsequent 10–15 years of the well's operation.

Chapters 6 and 7 have presented important aspects of the Tax Reform Act of 1986 relating to depreciation, depletion and income taxes. It is essential to understand these topics so that correct after-tax engineering economy evaluations of proposed projects and ventures may be conducted. Depreciation and income taxes are also integral parts of subsequent chapters in this book. For instance, Chapter 9 examines the interrelationship between inflation and income taxes, and Chapter 10 describes how before- and after-tax sensitivity studies are performed. Various models for evaluating before- and after-tax replacement decisions are presented in Chapter 12.

In Chapter 7 many concepts regarding current federal income tax laws were described. For example, topics such as taxable income, effective income tax rates, taxation of ordinary income and capital gains, and investment tax credits were explained. A general format for pulling together and organizing all these apparently diverse subjects was presented in Figure 7-3. This format offers the student or practicing engineer a means of collecting on one worksheet information that is required for determining after-tax cash flows, and properly evaluating the after-tax financial results of a proposed capital investment. Figure 7-3 was then employed in numerous examples. The student's challenge is now to use this worksheet in organizing data presented in problem exercises at the end of this chapter and to answer questions regarding the after-tax desirability of the proposed undertaking(s).

One last observation must be made. The BASIC computer program included in Appendix F (called ATAX) is capable of calculating depreciation deductions by most methods explained in Chapter 6.* It can also generate the format in Figure 7-3 by which relevant income tax information is utilized to determine the after-tax cash flows and the present worth of a project. It should come in handy when doublechecking manual solutions to problem exercises at the end of Chapter 7.

*An instructor may obtain a $5\frac{1}{4}''$ floppy diskette containing all BASIC programs in the appendixes to this book simply by requesting it from the College Division of the Macmillan Publishing Company.

7.15 PROBLEMS

No investment tax credit should be used in any of the following problems (unless one is stated). One hundred percent equity financing is to be assumed in all problems except 7-27 through 7-30. Further assume end-of-year cash flows and discrete (annual) compounding of interest.

7-1. A firm's total gross revenues in 1989 were \$11,240,000. The sum of all expenditures (excluding capital expenditures) was \$5,890,000. Depreciation for the entire year came to a total of \$3,415,000.

a. Compute the federal income taxes payable in 1989.
b. What was this firm's average income tax rate in 1989?
c. What was this firm's net income after taxes?
d. Determine the after-tax cash flow for the firm in 1989. (7.3, 7.7)

7-2. Consider a firm that in 1989 had a taxable income of \$90,000 and total gross revenues of \$220,000. By referring to income tax tables in Section 7.7, answer these questions:

a. What amount of federal income taxes was paid in 1989?
b. What was the net income (after federal income taxes) in 1989?
c. What was the total amount of deductible expenses (e.g., materials, labor, fuel, interest) and depreciation claimed in 1989? (7.3, 7.4, 7.7)

7-3. For the Surefire Automatic Casting Company, gross revenues in 1989 amounted to \$7,800,000. Operating expenses were \$4,900,000 and depreciation deductions were \$1,200,000. There was no interest on borrowed money.

a. How much federal income tax was paid in 1989?
b. What was the net income after taxes in 1989?
c. What was the firm's after tax cash flow in 1989? (7.3, 7.4, 7.5)

7-4. If the incremental federal income tax rate is 34% and the incremental state income tax rate is 9%, what is the effective combined income tax rate (t) when state income taxes are deductible from federal income taxes? (7.6)

7-5. A corporation has estimated that its taxable income will be $57,000 in 1990. It has the opportunity to invest in a business venture that is expected to add $8,000 to next year's taxable income. How much in federal taxes will be owed *with* and *without* the proposed venture? (7.7)

7-6. Bob Brown had a taxable income for 1989 of $30,000 derived from his salary and interest from investments. For the same year, Joe Jones had a taxable income of $27,000, composed of $12,000 from salary and a capital gain of $15,000. Both men are married.

a. Which man had the greater after-tax income?

b. What effective tax rate did Joe pay on the $15,000 capital gain? (7.7)

7-7. Greta Goodtone, who is married, is in the 28% incremental income tax bracket. Her business manager convinced her to purchase a farm for $75,000, with the idea of improving it over a 4-year period and then selling it for a capital gain. Greta bought the farm, and she immediately spent an additional $15,000 for structural improvements. During the next 4 years the annual operating costs of the farm exceeded the revenues by $15,000 per year. However, Greta was able to deduct these annual losses on her annual income tax returns and also to claim $5,000 each year as depreciation expense. At the end of the fourth year she sold the farm for $150,000.

a. What was the before-tax internal rate of return on the investment in the farm?

b. What was the after-tax internal rate of return if the whole capital gain was taxed at 28%? (7.7, 7.8)

7-8. a. Amanda Plumrose bought some common stock 10 years ago for $5,000. She decides to sell the stock now for a capital gain of $3,000. Furthermore, the taxable income from her job is $28,000. At the end of the year how much in taxes will she owe the federal government? (7.7, 7.8)

b. A certain taxpayer has a taxable income that is subject to an incremental rate of 28%. He has the chance either to invest in tax-free municipal bonds yielding 8% or to lend money at interest to a company. If he lends to the firm, what interest rate is necessary to yield him the same after-tax return as the municipal bonds? (7.7)

7-9. In 1988 a corporation built a warehouse at a cost of $100,000, estimating that it would have a useful (ADR) life of 10 years and a market value of $120,000 after 5 years. MACRS depreciation deductions are to be taken over the appropriate class life. If this corporation is in the 40% effective income tax bracket, what will be the after-tax gain from the sale of this warehouse at the end of 5 years? Depreciation recapture that occurs at the time of disposal will be taxed at 40%. No discounting calculations are necessary. (7.7, 7.8)

7-10. Your brother purchased a sports car 5 years ago for $20,000. Because he is a physician and can depreciate the car for business purposes (straight line method over 5 years), its present book value is zero and he is able to sell it now for $20,000. What interest rate did your brother earn on this investment after income taxes are considered if his effective income tax rate is 28%? (7.8, 7.10)

7-11. An individual who was, and is, in the 28% income bracket, purchased some nondepreciable property for $20,000. Two weeks later he had an opportunity to sell it for $23,000. He held the property for 4 years and then sold it for $35,200. In the meantime he paid annual property taxes of $600. His objective in keeping the property was to achieve a 10% return on his capital, after taxes. Did he achieve his objective? (7.8, 7.10)

7-12. A centerless grinder can be purchased new for $18,000. It will have an 8-year useful life and no salvage value. Reductions in operating costs (savings) from the machine will be $8,000 for the first 4 years and $3,000 the last 4 years. Depreciation will be by the MACRS method using a class life of 5 years. A used grinder can be bought for $8,000 and will have no scrap or salvage value in 8 years. It will save a constant $3,000 per year over the 8-year period and will be depreciated $1,000 per year for 8 years. The effective income tax rate is 40%. Determine the after-tax present worth on the *incremental* investment required by the new centerless grinder. Let the MARR be 12%, after taxes. (7.10)

7-13. The management of a hospital is considering the installation of an automatic telephone switchboard, which would replace a manual switchboard and eliminate the attendant operator's position. The class of service provided by the new equipment is estimated to be at least equal to the present method of operation. A total of five operators is needed to provide telephone service (three shifts per day, 365 days per year) and each operator earns $16,000 per year. Company-paid benefits and overhead are 25% of wages. Money costs 8% after income taxes. The combined federal and state income tax rate is 40%. Annual property taxes and maintenance are 2.5% and 4% of investment, respectively. Depreciation is 15-year straight line. Disregarding inflation, how large an investment in the new equipment can be economically justified through after-tax savings obtained by eliminating the present equipment and labor costs? The existing equipment has zero salvage value. (7.10)

7-14. The owners of a small TV repair shop are planning to invest in some new circuit testing equipment. The details of the proposed investment are as follows:

First cost = $5,000
Salvage value = $0
Extra revenue = $2,000/year
Extra expenses = $800/year
Expected life = 5 years (also equal to the MACRS recovery period)
Effective income tax rate = 15%

a. If MACRS depreciation is used, calculate the present worth of after-tax cash flows when the MARR (after taxes) is 12%. Should the equipment be purchased?

b. Use the alternate MACRS depreciation method over the MACRS recovery period to calculate after-tax cash flows, and recommend whether the new equipment should be purchased if the after-tax MARR is 12%. Include the half-year correction when determining depreciation. (7.1)

7-15. A firm must decide between two systems, A and B. Their effective income tax rate is 40% and MACRS depreciation is used. If the after-tax desired return on investment is 10%, which system should be chosen? State your assumptions. (7.11)

	System A	*System B*
Initial cost	$100,000	$200,000
Class life	5 years	5 years
Useful (ADR) life	7 years	6 years
Market value at end of useful life	$ 30,000	$ 50,000
Annual revenues less expenses over useful life	20,000	40,000

7-16. You have a piece of equipment with a present book value of $192,000. Next year's depreciation will be $96,000. You can sell the equipment now for $80,000 or you can sell it one year from now for the same amount. If you do not sell it now, you will definitely sell it next year. How much before-tax cash flow must the equipment produce over the next year (assume that it all comes at the end of the year) to justify keeping the equipment for one more year? Assume an after-tax MARR of 15%. Depreciation recapture, if any, is taxed at an effective income tax rate of 40%. (7.11)

7-17. A certain piece of real estate has a first cost of $50,000. If this building is purchased in 1989, it is believed that the property will be held for 10 years and then sold for an estimated $30,000. The estimated receipts from rental are $10,000 a year throughout the 10 years. Estimated annual upkeep costs are $3,000 the first year and will increase by $300 each year to $5,700 in the tenth year. In addition, it is estimated that there will be a single outlay of $2,000 for maintenance overhaul at the end of the fifth year. Assume MACRS depreciation with a 20-year class life and an effective income tax rate of 40%. Assume that the $2,000 overhaul cost can be treated for tax purposes as a current expense in the fifth year. If an investor has a 10% after tax MARR, should he purchase the rental property? Show all calculations. (7.11)

7-18. Currently, a firm has annual operating revenues of $190,000, cost of sales of $50,000, and depreciation charges are running at $40,000 annually. A new project is proposed that will raise revenues by $30,000 and increase cost of sales by $10,000. If this new project necessitates a total capital cost of $50,000 which can be depreciated to zero salvage value at the end of its 6-year life, what is the external rate of return after federal income taxes are paid? Assume that MACRS depreciation is used with a class life of 5 years. Recovered funds from the project will be invested outside the firm at an after-tax rate of 10%. (7.11)

7-19. A company must purchase a particular asset for $10,000. The MACRS class life of the asset is 5 years, and it will have a salvage value of zero in computing depreciation deductions. However, management of this company believes the asset will have an actual market value of $2,000 at the end of its 5-year life. The annual operating and maintenance costs are $2,000 the first year and increase by $200 per year thereafter. With MARR = 12% after taxes and an effective income tax rate of 40%, calculate the after-tax equivalent annual cost of this asset. Assume that the company is profitable in its other activities, and that the recaptured depreciation is taxed at 40%. (7.12)

7-20. Your firm can purchase a machine for $12,000 to replace a rented machine. The rented machine costs $4,000 per year. The machine that you are considering would have a life of 8 years and a $5,000 salvage value at the end of its life. By how much could annual operating expenses increase and still provide a return of 10% after taxes? The firm is in a 40% income tax bracket, and revenues produced with either machine are identical. Assume alternate MACRS (straight line with half-year convention) is utilized to recover the investment in the machine and the recovery period is 5 years. (7.12)

7-21. Your boss asked you to determine whether the company can justify the purchase of a special piece of earthmoving equipment. The data for the problem are as follows:

First cost of equipment	$20,000
Depreciation	straight line with $2,000 salvage value, 10-year life
Property taxes and insurance	$500/year
Maintenance cost	$400 the first year, increasing $40 per year thereafter
Operating savings (due to increased productivity)	$7,000 the first year, decreasing $300 per year thereafter

The boss expects there to be no salvage value. Assume that the boss is correct, and make an after-tax study for the equipment when the effective income tax rate is 40%, after-tax MARR = 10%, and depreciation recapture is taxed at 40%. (7.11)

7-22. An injection molding machine can be purchased and installed for $90,000. It has a 7-year MACRS class life but will be kept in service for 8 years. The salvage value for calculating depreciation is zero, but it is believed that $10,000 can be obtained when the machine is disposed of at the end of year 8. The net annual "value added" (i.e., receipts less expenses) that can be attributed to this machine is constant over 8 years and amounts to $15,000. An effective income tax rate of 40% is used by the company and the after-tax MARR equals 15%. (7.6, 7.10, 7.11)

a. What is the approximate value of the company's before-tax MARR?
b. Determine the MACRS depreciation amounts in years 1 through 8.
c. What is the taxable income at the end of year 8 that is related to capital-investment?
d. Set up a table and calculate the after-tax cash flows for this machine.
e. Should a recommendation be made to purchase the machine?

7-23. Your company has purchased equipment (for $50,000) that will reduce materials and labor costs by $14,000 each year for N years. After N years there will be no further need for the machine, and since the machine is specially designed it will have no salvage value at any time. However, the IRS has ruled that you must depreciate the equipment on a straight-line basis with a tax life of 5 years. If the effective income tax rate is 40%, what is the minimum number of years your firm could operate the equipment to earn 10% after taxes on its investment? (7.11)

7-24. a. Suppose that you have just completed the mechanical design of a high-speed automated palletizer that has an investment cost of $3,000,000. The existing palletizer is quite old and has no salvage value. The new palletizer can be depreciated with the MACRS method using a class life of 5 years, and the equipment's market value at the end of 5 years is estimated to be $300,000. The effective income tax rate is 40%. One million pallets will be handled by the palletizer each year during the 7-year expected project life.

What net annual savings per pallet (i.e., total savings less expenses) will have to be generated by the palletizer to justify this purchase in view of an after-tax MARR of 15%? (7.11)

b. Referring to the situation described in part (a), suppose that you have estimated that these incremental savings and costs will be realized after the automated palletizer is installed:

1. Twenty-five operators will no longer be needed. Each operator earns an average of $15,000 per year in direct wages, and company fringe benefits are 30% of direct wages.
2. Property taxes and insurance amounting to 5% of the palletizer's installed cost will have to be paid over and above those paid for the present system.
3. Maintenance costs relative to the present system will increase by $15,000 per year.
4. Orders will be filled more efficiently because of the automated palletizer (less waste, quicker response). These savings are estimated at $100,000 per year.

The effective income tax rate is 40%, and the equipment's life, depreciation method, after-tax MARR, and so on, are the same as those in part (a). If the existing system has no salvage value, what is the maximum amount of money that can be invested in the automated palletizer (in view of the savings and costs itemized above) so that a 15% after-tax return on this investment is realized? (7.11)

7-25. Your company has to obtain some new production equipment for the next 6 years, and leasing is being considered. You have been directed to accomplish an after-tax study of the leasing approach. The pertinent information for the study is as follows:

Lease costs: First year, $80,000; second year, $60,000; third through sixth years, $50,000 per year. Assume that a 6-year contract has been offered by the lessor that fixes these costs over the 6-year period. Other costs (not covered in contract) are $4,000 per year, and the effective income tax rate is 40%.

a. Develop the annual after-tax cash flows for the leasing alternative.
b. If the MARR after taxes is 8%, what is the equivalent annual cost for the leasing alternative? (7.12)

7-26. Your company expects the demand for a new product to be strong over the next 5 years. Thus, additional investment in new production equipment is being considered. The market value of the equipment at the end of 5 years is estimated to be 20% of initial cost, and the estimated before-tax cash flow for the 5-year period is as follows:

Year k	*BTCF*
0 (investment)	−$200,000
1 through 5	60,000
5 (market value)	40,000

The effective corporate income tax rate is 40%; the book value of the equipment at the end of the 5-year MACRS class life is $0.

a. What is the after-tax internal rate of return (IRR)?
b. What is the after-tax external rate of return (ERR) assuming the external reinvestment rate is 12%? (7.11)

7-27. Suppose that a machine costing $11,000 can be financed entirely by borrowed funds *or* by equity capital. With borrowed funds, the loan is to be repaid at the rate of $2,000 at the end of each year for the first four years and $3,000 at the end of the fifth year. Interest charges each year are computed at 10% of the unpaid, beginning-of-year balance of the loan. Depreciation is calculated on a straight-line basis, the depreciable life is 5 years, and the estimated salvage value for depreciation purposes is $1,000. The expected before-tax cash flow attributable to the machine before deducting annual interest charges and operating costs is $10,000, and the effective income tax rate is 50%. Operating costs will amount to $3,000 per year.

a. Determine the after-tax cash flows for both financing plans, and compute the present worth of each at a MARR of 15%.

b. Compute the IRR on after-tax equity cash flow for both financing plans.

c. When borrowed funds are used to finance a project, what problems can arise when the IRR method is utilized to compare alternative financing plans? (7.13)

7-28. A firm is considering the introduction of a new product in 1991. The marketing department has estimated that the product can be sold over a period of 5 years at a price of $7.00 per unit. Sales are estimated to be 10,000 units the first year and will increase by 2,000 units each year thereafter. Manufacturing equipment necessary to produce the item will cost $200,000. It is estimated that this equipment can be sold for $50,000 at the end of year 5. MACRS depreciation with a 5-year class life will be used. The equipment will be financed by borrowing $160,000, and this loan principal is to be repaid in five equal end-of-year payments of $32,000 each. Interest at 10% per year will be paid on the outstanding loan principal at the beginning of each year. The $40,000 balance will be financed from equity funds. Operating and maintenance costs (not including taxes) will be $50,000 the first year and decrease by $3,000 each year thereafter. The firm's effective tax rate is 40%, and its minimum attractive rate of return is 15%, after taxes. Calculate the present worth of the after-tax equity cash flow assuming the corporation is profitable in its other activities. (7.13)

7-29. Suppose that you borrow $10,000 at 15% per year ($i_b$) and agree to repay this amount in three equal end-of-year payments. What is your after-tax cost of interest in each of the 3 years, and what is the present worth of the after-tax interest expense? Let $t = 0.32$ and the after-tax rate of return (i_{AT}) be 18% per year. (7.13)

7-30. Rework Example 7-12 when the $50,000 worth of borrowed funds, at 10% interest, is to be repaid in equal end-of-year amounts such that no principal remains after the tenth payment is made. Be careful to note that interest and repaid principal will vary in amounts from year to year in this situation. (7.13)

7-31. Refer to Figure 6-2 and the depletion example in Table 7-10. Suppose in years 6–10 of this well's operation that hot water can be sold for $0.22 per gallon, a constant 1,000,000 gallons per year can be sold, and the depletion allowance is 22%. The firm's expected net income, before any depletion allowance has been deducted, is $80,000 per year (in column 4 of Table 7-10). If the effective income tax rate remains at 40%, what is the net cash flow after taxes in years 6–10? (7.14)

7-32. A large mineral deposit in Wyoming is estimated to contain 1,000,000 tons of a mineral whose percentage depletion allowance is 22%. A mining company has made an initial investment of $40,000,000 to recover this ore, and the market price for the ore is \$175/ton. The company's after-tax MARR is 12% and its effective income tax rate is 40%. It is anticipated that the ore will be sold at the rate of 100,000 tons per year and that operating expenses, exclusive of depletion deductions, will be approximately \$9,000,000 per year.

a. Determine the after-tax cash flow for this mining venture when percentage depletion (or cost depletion, if appropriate) is used.

b. Determine the present worth of after-tax cash flow in part (a). (7.14)

7-33. Complete the following table which lists capital recovery provided by a mining operation, with cost percentage depletion allowances used in computation of income taxes, (Refer to Figure 6-2.) Initial reserves are 200,000 units. (7.14)

(1) Year	(2) Units Sold	(3) Gross Income (Cash Flow)	(4) Net Income Before Taxes	(5) 50% of Net Income	(6) Cost Depletion at $4.00 per Unit	(7) Depletion Allowance at 22% of gross Income	(8) [= (4) − either (6) or (7)] Taxable Income	(9) [= −0.40(8)] Income Tax	(10) [= (4) + (9)] Net Cash Flow After Taxes
1	70,000	$1,400,000	+$800,000	$400,000	?	?	?	?	?
2	60,000	1,200,000	+$700,000	350,000	?	?	?	?	?
3	45,000	900,000	+$480,000	240,000	?	?	?	?	?
4	20,000	400,000	+$240,000	120,000	?	?	?	?	?
5	5,000	85,000	+$ 25,000	12,500	?	?	?	?	?

8

DEVELOPING CASH FLOWS

The objectives of this chapter are to discuss an integrated approach for developing cash flows for the feasible alternatives being analyzed in a study, and to delineate and illustrate selected techniques that will assist in accomplishing such estimates. The following topics are discussed in this chapter:

Components of an integrated approach
What is a work breakdown structure (WBS)?
The cost and revenue structure
Estimating techniques (models)
Estimating cash flows for a typical small project
Developing cash flows (a case study)

8.1
INTRODUCTION

In Chapter 2 (Section 2.5), we discussed the engineering economic analysis procedure in terms of seven steps which are relisted here:

1. Recognition and formulation of the problem.
2. Development of the feasible alternatives.
3. Development of the net cash flow (and other prospective outcomes) for each feasible alternative.
4. Selection of a criterion (or criteria) for determining the preferred alternative.
5. Analysis and comparison of the feasible alternatives.
6. Selection of the preferred alternative.
7. Post-evaluation of results.

In Chapters 3 through 7, the methodology needed to accomplish Steps 4, 5, and 6 was developed and demonstrated. In this chapter, we return to Step 3 above.

Because engineering economy studies deal with outcomes of decisions made at the present time that extend into the future, estimating the future cash flows for the feasible alternatives is a critical step in the analysis procedure. A decision based on the analysis is economically sound only to the extent that these cost and revenue estimates are representative of what subsequently will occur.

Figure 8-1 provides a perspective of Step 3 in terms of its activities and highlights its pivotal role in the engineering economic analysis procedure. In Step 1, the need for doing an analysis is identified and decided upon. Also, the specific situation (improvement opportunity, new project or venture, etc.) is explicitly defined; the desired outcomes in terms of goals, objectives, and other results are developed; and any special conditions and constraints that need to be met are delineated.

The activities in Step 2 result in the selection and description of the feasible alternatives to be analyzed in the engineering economy study. First, the requirements internal and external to the organization that need to be met are described, and the list of potential alternatives is developed. Then, each potential alternative is screened (using the systems viewpoint and approach discussed in Section 1.7) against these requirements, any special conditions that were identified in Step 1, and the apparent capability of achieving the desired outcomes. The potential alternatives that pass this explicit screening become the feasible alternatives.

Thus, Step 3 begins with *feasible alternatives* to be analyzed in the study *already selected*, and the differences between them *already highlighted*. Also, other

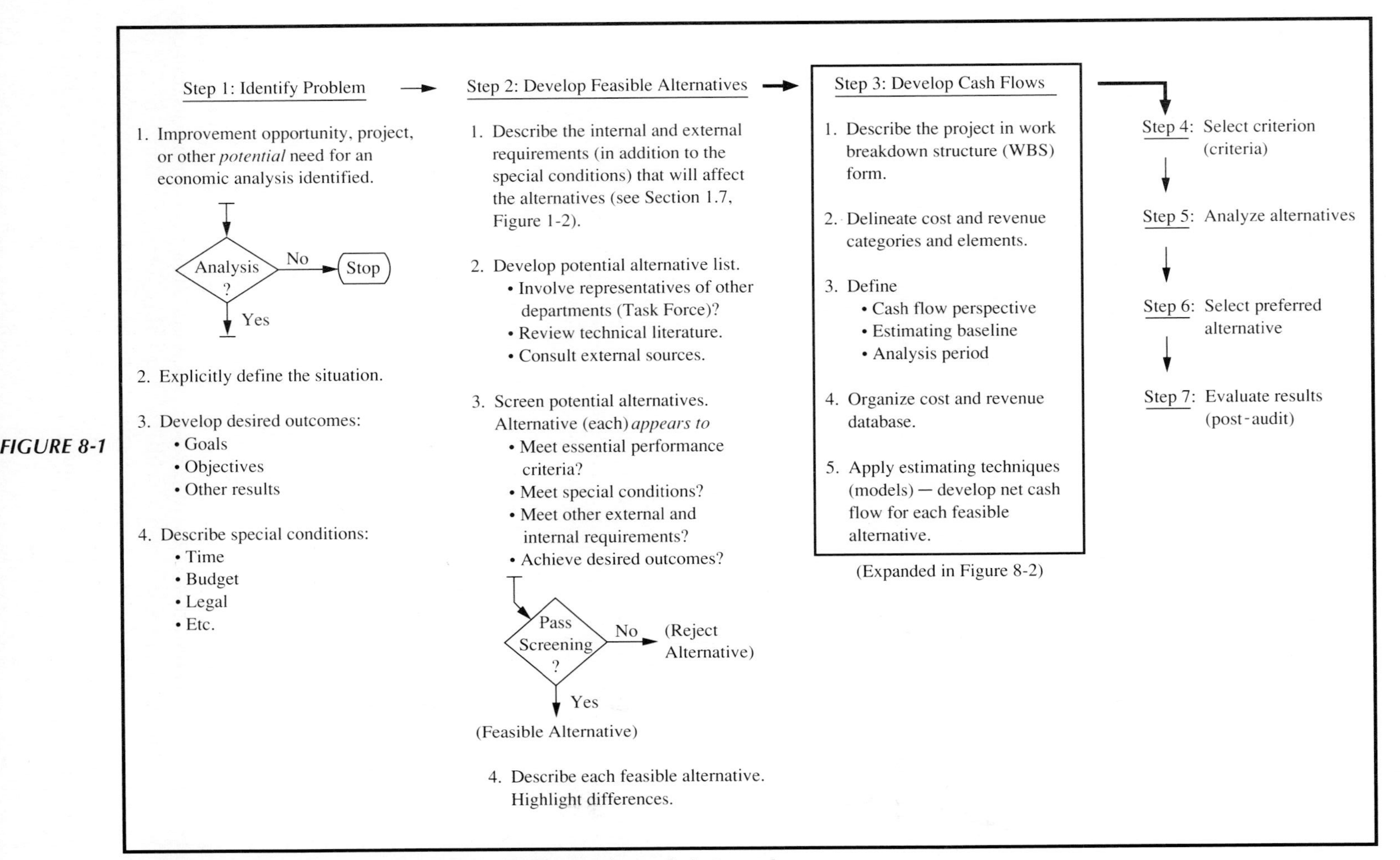

FIGURE 8-1 *The Pivotal Role of Step 3 in the Engineering Economic Analysis Procedure*

important information, such as the results to be achieved and requirements to be met that are needed in the analysis, is available from the first two steps.

The application of the concepts and methodology discussed in this chapter is an important part of engineering practice. A commercial building project is used several times as the basis for examples in Chapter 8, although any other engineering project, such as the expansion of a chemical processing plant, could have been chosen.

8.2 COMPONENTS OF AN INTEGRATED APPROACH

An integrated approach for accomplishing the activities in Step 3, and developing the net cash flows for the feasible project alternatives, is shown in Figure 8-2 (we will use the term project to refer to the undertaking that is the subject of the analysis). This integrated approach includes three basic components:

1. ***Work breakdown structure (WBS)***. This is a technique for explicitly defining, at successive levels of detail, the work elements of a project and their interrelationships (sometimes called a work element structure).
2. ***Cost and revenue structure (classification)***. The delineation of the cost and revenue categories and elements that will be estimated in developing the cash flows.
3. ***Estimating techniques (models)***. Selected mathematical models to assist with estimating the future costs and revenues during the analysis period.

These three basic components, together with integrating procedural steps, provide an organized approach for developing the net cash flows for the feasible alternatives.

As shown in Figure 8-2, the integrated approach begins with a description of the project in terms of a work breakdown structure. This project WBS is used to describe the project and each feasible alternative's unique characteristics in terms of design, labor, material requirements, and so on. Then, these variations in design, resource requirements, and other characteristics are reflected in the estimated future costs and revenues (net cash flow) for that alternative.

In order to estimate future costs and revenues for an alternative, the perspective of the cash flow must be established and an estimating baseline and analysis period defined. We discussed the perspective of a cash flow, and the concept of an estimating baseline, in Section 2.5. Cash flows are normally developed from the owner's viewpoint, and the cash flow for each alternative must be developed consistently from the same baseline. There are two procedures used for that development: (1) total cost and revenue estimating, and (2) differential estimating. Differential estimating is particularly applicable (and often used)

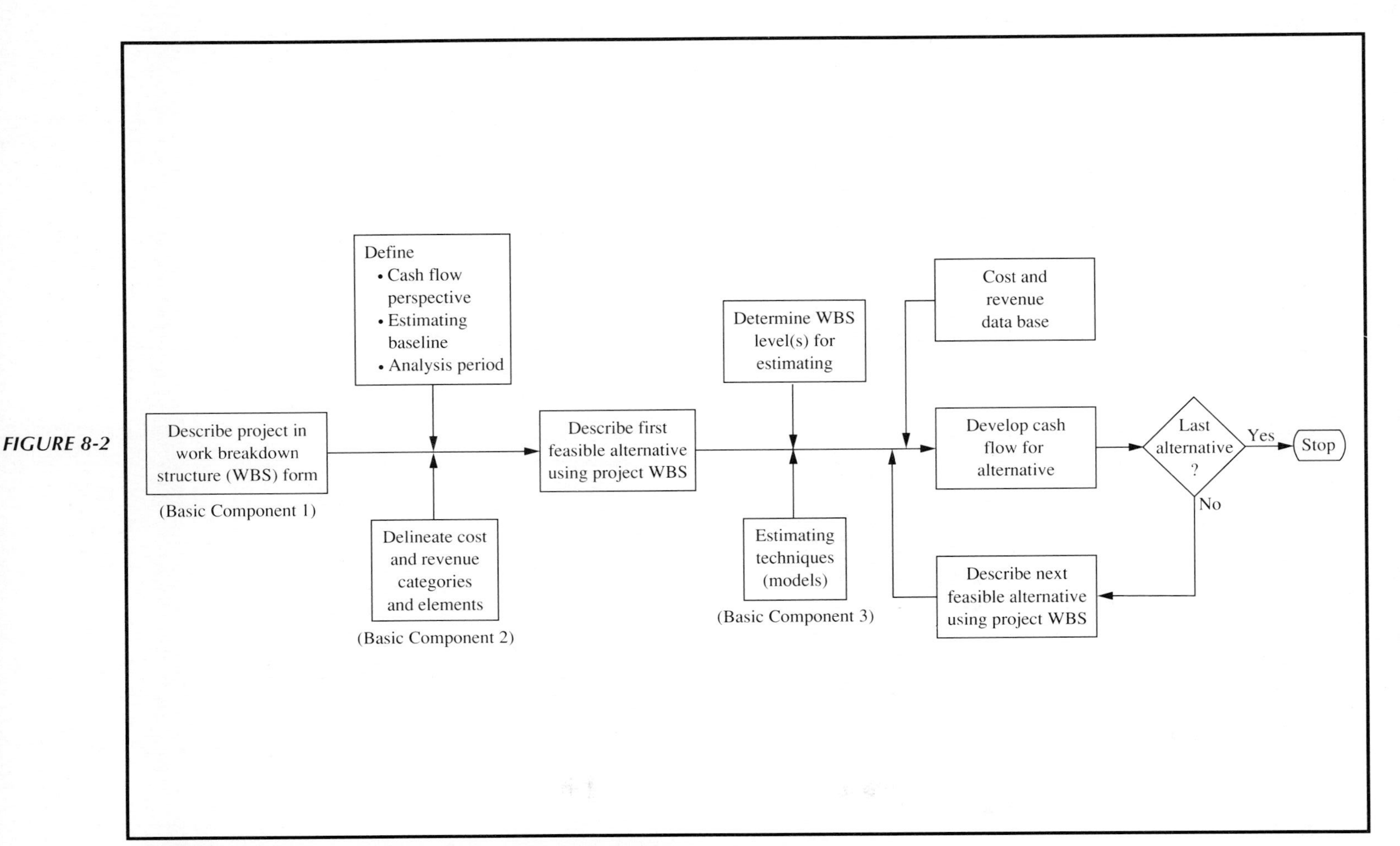

FIGURE 8-2 *Integrated Approach for Developing the Cash Flows for Feasible Alternatives*

when the project involves changes to an existing operation which serves as the baseline. However, when a new project is involved, estimating the total costs and revenues of the alternatives is necessary unless a similar operating situation is available for use as a differential estimating baseline. An important criterion in selecting the length of the analysis period is its relevance to the decision situation.

Also, before developing the net cash flow for the first alternative, other procedural steps need to be accomplished. The first is deciding what level(s) of the WBS to use for developing the cost and revenue estimates. The purpose of the study will be a primary factor in this decision. If the study is a project feasibility analysis, cost and revenue estimating will be less accurate than in the detailed economic analysis which will be used to make the final decision about a project (this is discussed further in Section 8.5.1).

Then considerable amounts of cost and revenue information from sources internal and external to the organization need to be organized, and the relevant data assembled for the study. These data, together with selected estimating techniques (models), are used to develop the future cost and revenue estimates.

At this point, the net cash flow for the first feasible alternative can be developed. Then the iterative process is continued until the net cash flows for all feasible alternatives are prepared.

8.3 THE WORK BREAKDOWN STRUCTURE (WBS) TECHNIQUE

We briefly defined a work breakdown structure (work element structure) in Section 8.2, and identified it as the first basic component in an integrated approach to developing cash flows. This technique is a basic tool in project management and is a vital aid in an engineering economy study. The WBS serves as a framework for defining all project work elements and their interrelationships, collecting and organizing information, developing relevant cost and revenue data, and integrating project management activities. If a project WBS does not exist, its development should be the first step in preparing cash flows for the feasible alternatives. It is essential in ensuring the inclusion of all work elements, eliminating duplications and overlaps between work elements, avoiding nonrelated activities, and preventing other errors that could be introduced into the study. A WBS description dictionary is often prepared for large projects to ensure that each work element in the hierarchy is uniquely defined.

Figure 8-3 shows a diagram of a typical four-level work breakdown structure. It is developed from the top (project level) down as will be demonstrated in Example 8-1. The project is divided into its major work elements (Level 2). These

FIGURE 8-3

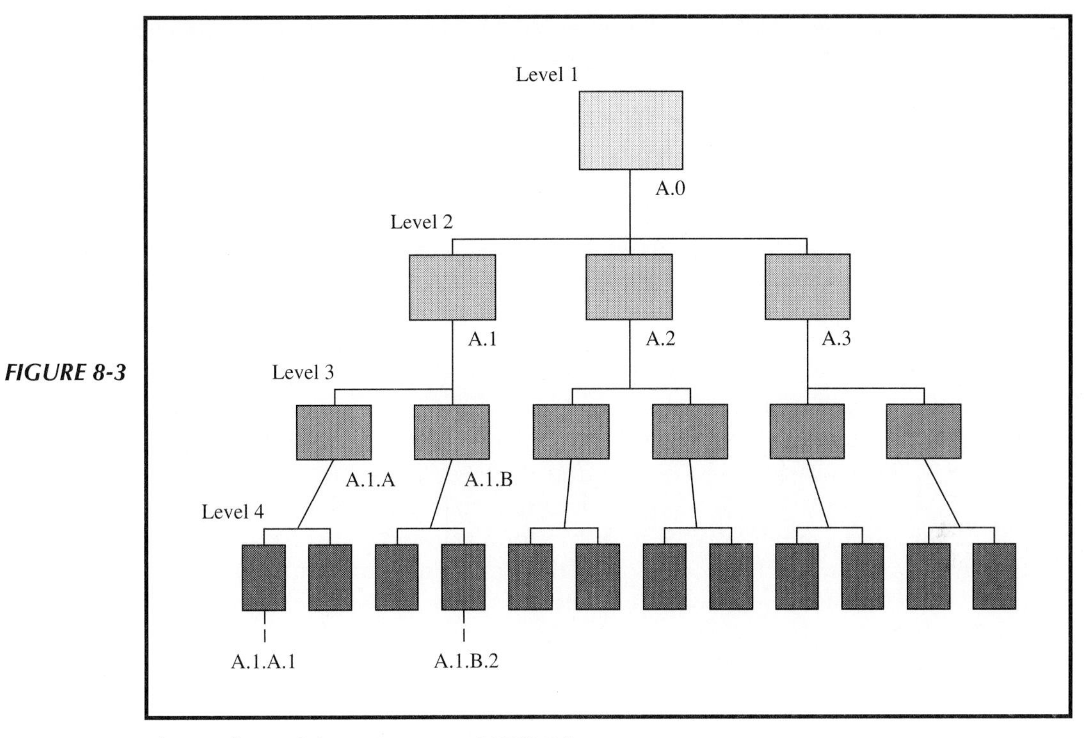

The Work Breakdown Structure (WBS) Diagram

major elements are then divided to develop Level 3. For example, an automobile (first level of the WBS) can be divided into second-level components (or work elements) such as chassis, drive train, and electrical system. Then each second-level component of the WBS can be subdivided further into third-level elements. The drive train, for example, can be subdivided into third-level components such as engine, differential, and transmission. This process is continued until the desired detail in the definition and description of the project or system is achieved.

Different numbering schemes may be used. The objectives of numbering are to indicate the interrelationships of the work elements in the hierarchy and to facilitate the manipulation and integration of data. The scheme illustrated in Figure 8-3 is an alphanumeric format. Another scheme often used is all numeric: Level 1—1.0; Level 2—1.1, 1.2, 1.3; Level 3—1.1.1, 1.1.2, 1.2.1, 1.2.2, 1.3.1, 1.3.2; and so on (i.e., similar to the organization of this book). Usually, the level is equal (except for Level 1) to the number of characters indicating the work element.

Several characteristics of a project WBS are listed below:

1. Both functional (e.g., planning) and physical (e.g., foundation) work elements are included in the WBS.

2. Some other typical functional work elements are logistical support, project management, marketing, engineering, and systems integration.
3. The physical work elements are the parts that make up a structure, product, equipment, weapon system, or similar item; they require labor, material, and other resources to produce or construct.
4. The content and resource requirements for a work element are the sum of the activities and resources of related subelements below it.
5. A project WBS usually includes recurring and nonrecurring work elements.

EXAMPLE 8-1

You have been appointed by your company to manage a project involving construction of a small commercial building with two floors of 15,000 gross square feet each. The ground floor is planned for small retail shops, and the second floor is planned as office space. Develop the first three levels of a representative WBS adequate for all project effort from the time the decision was made to proceed with the design and construction of the building until initial occupancy is completed.

Solution

There would be variation in the WBS developed by different individuals for a commercial building. However, a representative three-level WBS is shown in Figure 8-4. Level 1 is the total project. At the second level, the project is divided into seven major physical work elements and three major functional work elements. Then each of these major elements is divided into subelements as required (Level 3). The numbering scheme used in this example is all numeric. (This WBS will be used in Section 8.6 for developing an updated net cash flow estimate for the project.)

8.4 THE COST AND REVENUE STRUCTURE

The second basic component of the integrated approach for developing cash flows shown in Figure 8-2 is the cost and revenue structure. This structure is used for identifying and categorizing the costs and revenues that need to be included in the analysis. Then, detailed data are developed and organized within this structure for use with the estimating techniques of Section 8.5 to prepare the cash flow forecasts.

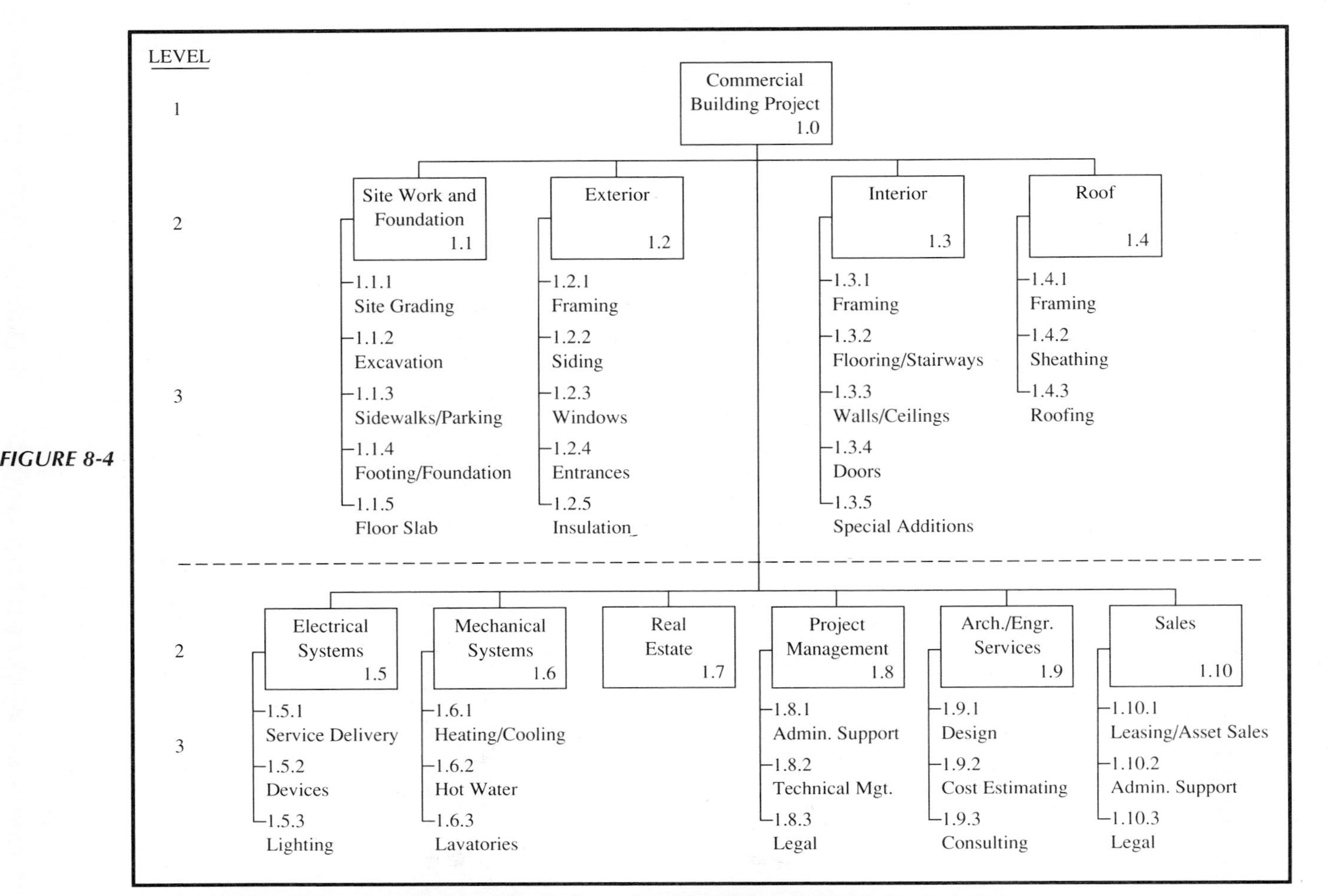

FIGURE 8-4 ***Work Breakdown Structure (Three Levels) for Commercial Building Project in Example 8-1***

8.4.1 Using the Life-Cycle Concept and the WBS

The life-cycle concept was discussed and illustrated in Section 2.2.8. The life cycle is divided into two general time periods—the acquisition phase and the operation phase—and begins with initial identification of the economic need or want (the requirement) and ends with final retirement or disposal activities. Thus, it is intended to encompass all present and future costs and revenues during this "birth to death" time horizon.

The life-cycle concept and the WBS are important aids in developing the cost and revenue structure for a project. The life cycle defines a time period, provides a perspective, and establishes a range of cost and revenue elements that need to be considered in developing cash flows. Then the WBS focuses the analyst's effort on the specific functional and physical work elements of a project, and on its related costs and revenues.

Ideally, the analysis period for a project is the life cycle of the product, structure, system, or service involved. This permits all relevant costs and revenues, both present and future, to be fully considered in decision making. Also, it makes possible the *explicit trade-off between initial costs during the acquisition phase and all future costs and revenues during the operation phase* in analyzing the feasible alternatives.

However, the accuracy of cost and revenue estimates decreases with increases in the length of the analysis period. Also, the effort required to develop cash flows increases with the length of the analysis period. Thus, a time horizon for the analysis period is selected to balance these factors and provide a sound basis for decision making.

8.4.2 Estimates Needed for a Typical Engineering Economy Study

As discussed in the previous section (8.4.1), judgment is required, based on the decision situation, to determine the analysis period, and thus how far into the future to estimate costs and revenues in an engineering economy study. This judgment should also weigh which cost and revenue elements are the most important and deserve more detailed study and which elements, even if drastically misjudged, will not produce significant changes in the estimated cash flows.

Perhaps the most serious source of errors in developing cash flows is overlooking important categories of costs and revenues. The cost and revenue structure, prepared in tabular or checklist form (as demonstrated in Example 8-2), is a good means of preventing such oversights. Technical familiarity with the project is essential in ensuring completeness of the structure, as are using the life-cycle concept and the WBS in its preparation.

Following is a brief listing of some categories of costs and revenues that are typically needed in an engineering economy study, together with some discussion of how estimates might be obtained. Some of these terms were discussed in Section 2.2.

1. *Investment (first) costs* consist of two principal categories:
   a. *Fixed capital investment,* such as for feasibility studies, design and engineering, land purchase and improvement, buildings, equipment, installation, promotional and legal fees, and startup costs.
   b. *Working capital,* such as for inventories, accounts receivable, cash for wages, materials, and other accounts payable. Working capital is a revolving fund needed to get a project started and to meet subsequent obligations. Normally, it is assumed that some or all of working capital can be recovered by the end of the life of a project.
2. *Labor costs* are a function of skill level, labor supply, and time required. Standards for the normal amount of output per labor hour have been developed for many classes of work. Standard times combined with expected wage rates provide a reasonable estimate of labor costs for repetitive jobs. Labor costs for specialized work can be predicted from bid estimates or quotes by agencies offering the service. It should be remembered that labor costs should consider fringe benefits as well as direct wages.
3. *Material costs* are dependent on the project or operating situation; that is, what is a product to Operation A may be an input to Operation B. Material costs are those associated with the physical substance(s) being worked on or transformed in the operation involved.
4. *Maintenance costs* are the ordinary costs required for the upkeep of property and minor changes required for more efficient use. Maintenance costs tend to increase with the age of an asset because more upkeep is required later in life.
5. *Property taxes and insurance* are usually expressed as an annual percentage of first cost in economic comparisons.
6. *Quality (and scrap) costs* depend upon the types of products and associated quality standards as well as upon abilities of the work force, learning time, and rework possibilities.
7. *Overhead costs* are by definition those costs that cannot be conveniently and practicably charged to particular products or services, and thus are normally prorated among the products or cost centers on some arbitrary basis. One should guard against using these arbitrary allocations in economic analyses, for the differences in overhead costs brought about by various alternatives being considered are rarely described by these rates. In general, one should consider each individual cost element included in the overhead and estimate how much, if any, each cost element is affected by each alternative.
8. *Disposal costs* are those nonrecurring costs associated with shutting down an operation, and the retirement, disposal, or sale of assets to serve the best interests of the owners. These costs are estimated based on the projected labor, material, and other costs needed to accomplish the disposal activities.
9. *Revenues* are receipts from all potential sources related to the operation. They are projected based on current market conditions, expected future

changes in the market for the products and services involved, the company's expected market share, and pricing based on competition.

10. *Salvage or market values* are typically a function of the useful life and, in the case of long lives, are relatively unimportant to the analysis result. They are usually projected from current information about the salvage and market value of similar assets.

EXAMPLE 8-2

In Example 8-1, we discussed a commercial building project which you have been appointed to manage for its design, construction, and leasing until initial occupancy is completed. A three-level WBS for the project was shown in Figure 8-4. Now develop a project cost and revenue structure. The structure is used as shown in Figure 8-2 for developing the cash flows, and for supporting project management activities. (The structure will also be used in Section 8.6 for developing an updated net cash flow estimate for the project.) The building design details will affect the initial construction costs, annual operational and maintenance costs, and subsequent market value. Also, the design details are expected to have some impact on annual leasing revenue. A 17-year time period has been selected for the analysis of alternative building designs and other economic aspects of the project.

Solution

There would be variation in the project cost and revenue structure developed by different individuals. However, the outline for a representative structure (in matrix form) is shown in Figure 8-5. Three major cost categories and a revenue category are used in this structure. The cost categories are investment (first) cost, annual operational and maintenance cost, and disposal cost (at the end of the 17-year analysis period). The revenue category includes both the annual revenues and the one-time revenue from asset sales. These major categories are further broken down into subcategories (or areas) of cost and revenue. These areas (land, personnel, material, working capital, overhead, leasing, etc.) can be further subdivided into cost and revenue elements as required. The work elements (from the WBS) form the other dimension of the matrix. Thus, a comprehensive project cost and revenue structure can be developed for organizing cost and revenue data, developing cash flows for the feasible alternatives, and supporting project management activities.

8.5 ESTIMATING TECHNIQUES (MODELS)

The third basic component of the integrated approach shown in Figure 8-2 is the estimating techniques (models). These techniques, together with the detailed

FIGURE 8-5

Work Element	A. INVESTMENT (FIRST) COST													
	Real Estate Acquisition			Design and Engineering			Construction			Working Capital		Overhead	
	Land	Personnel	Other	Personnel	Material	Other	Labor	Material	Other	Inventory	Other	Indirect	G&A
1.1 Site Work and Foundation													
1.1.1 Site Grading													
⋮													
1.10 Sales													
⋮													
1.10.3 Legal													

Work Element	B. ANNUAL OPERATIONAL & MAINTENANCE COST				
				Overhead	
	Personnel	Material	Other	Indirect	G&A
(Selected WBS Elements)					

C. DISPOSAL COST				
			Overhead	
Personnel	Material	Other	Indirect	G&A

Work Element	D. REVENUE			
	Annual Leasing			Asset Sales
	First Floor	Second Floor	Other	(at Disposal)
1.10.1 Leasing/Asset Sales				

Cost and Revenue Structure for Commercial Building Project in Example 8-2

cost and revenue data, are used to develop the net cash flow for each feasible alternative. The estimates of future costs and revenues reflected in these cash flows are a vital ingredient in any engineering economy study. Also, the purpose of estimating is to develop cash flow projections—*not to produce exact data* about the future, which is virtually impossible. Neither a preliminary estimate nor a final estimate is expected to be exact; rather, it should adequately suit the need at a reasonable cost. Preparation of precise estimates would be excessively time-consuming and expensive, and even if they were possible to obtain they would still be subject to estimation errors.

Deviations between estimated and actual outcomes result from innumerable factors. Two of the most important are human errors in making estimates and unpredictable changes in circumstances. There are usually several approaches to estimating a given quantity, and determining which approach should be used at what level of detail is a significant problem in itself.

It is useful to think of future events as resulting in part from factors that have determined these events in the past, and resulting in part from factors that are new and different. To the extent that, say, future production costs are the product of the same factors that have determined past production costs, the analyst may use explicit prediction techniques to extend past experience into forecasts of the future. However, when future production costs depend on factors that were not active in the past, the past data fail and one must rely on managerial/technical experience and judgment. Forecasting, perhaps more than any other aspect of decision making, involves the combined skill and experience of both the analyst and management.

8.5.1 Cost Estimating

Cost estimates can be classified according to detail, accuracy, and their intended use as follows:

1. ***Order of magnitude estimates:*** used in the planning and initial evaluation stage of a project.
2. ***Semidetailed or budget estimates:*** used in the preliminary or conceptual design stage of a project.
3. ***Definitive (detailed) estimates:*** used in the detailed engineering/construction stage of a project.

Order of magnitude estimates are used in selecting the feasible alternatives. They typically provide accuracy in the range of ±30 to 50% and are developed through semiformal means such as conferences, questionnaires, and generalized equations applied at Level 1 or 2 of the WBS.

Budget (semidetailed) estimates are compiled after a project has been given the "go-ahead." Their accuracy usually lies in the range of ±15%. These estimates differ in the fineness of cost breakdowns and the amount of effort spent on the estimate. Cost estimating equations applied at Levels 2 and 3 of the WBS are normally used.

Detailed estimates are used as the basis for bids, and to make detailed design decisions. Their accuracy is about ±5%. They are made from specifications, drawings, site surveys, vendor quotations, and in-house historical records, and are usually done at Level 3 and successive levels in the WBS. Often, detailed estimates are prepared with the help of computer software packages.

Thus, it is apparent that a cost estimate can vary from an instant, top-of-the-head guess to a very detailed and accurate prognostication of the future. The level of detail and accuracy of an estimate should depend on

1. Difficulty of estimating the item in question.
2. Methods or techniques employed.
3. Qualifications of the estimator(s).
4. Time and effort available and justified by the importance of the study.
5. Sensitivity of study results to the particular estimate.

As cost estimates become more detailed, the accuracy of the estimate typically improves but the cost of estimating increases dramatically. This general relationship is shown in Figure 8-6 and illustrates the idea that cost estimates should be prepared in full recognition of how accurate a particular study requires them to be.

Regardless of how estimates are made, individuals who use them should have specific recognition that the estimate will be in error to some extent. Even the use of sophisticated estimation techniques will not, in itself, eliminate error. However, it will hopefully minimize estimation errors and will at least provide better recognition of the anticipated degree of error.

FIGURE 8-6

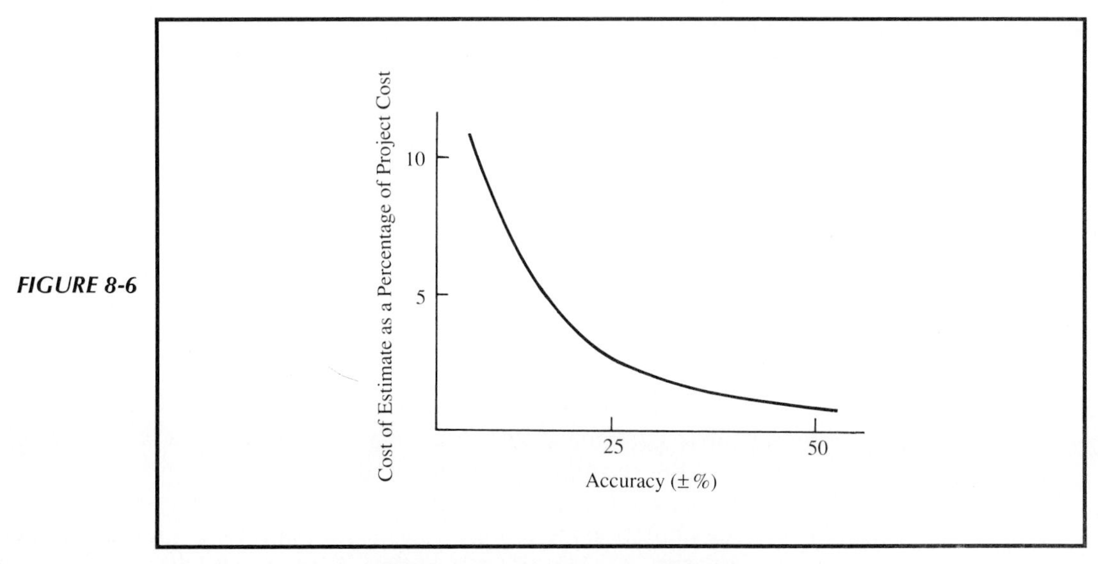

Accuracy of a Cost Estimate versus the Cost of Making It

8.5.1.1 Sources of Data for Cost Estimating

The number of information sources useful in cost estimating is too great for complete enumeration. The following four major sources of information, ordered roughly according to decreasing importance, are discussed below:

1. Accounting records.
2. Other sources within the firm.
3. Sources outside the firm.
4. Research and development.

1. *Accounting records*. It should be emphasized that although data available from the records of the accounting function are a prime source of information for economic analyses, such data are very often not suitable for direct, unadjusted use.

   A brief discussion of the accounting process and information is given in Appendix A. In its most basic sense, accounting consists of a series of procedures for keeping a detailed record of monetary transactions between established categories of assets, each of which has an accepted interpretation useful for its own purposes. The data generated by the accounting function are often inherently misleading for engineering economic analyses, not only because they are based on past results, but also because of the following limitations:

   a. The accounting system is rigidly categorized. Categories of various types of assets, liabilities, net worth, income, and expenses for a given firm may be perfectly appropriate for operating decisions and financial summaries, but rarely are they fully appropriate to the needs of economic analyses and decision making involving engineering design, project alternatives, and long term considerations.
   b. Standard accounting conventions cause misstatements of some types of financial information to be "built into" the system. These misstatements tend to be based on the philosophy that management should avoid overstating the value of its assets or understating the value of its liabilities and should therefore assess them very conservatively. This leads to such practices as (1) not changing the stated value of one's resources as they appreciate due to rising market prices, and (2) depreciating assets over a much shorter life than actually expected. As a result of such accounting practices, the analyst should always be careful about treating such resources as cheaply (or, sometimes, as expensively!) as they might be represented in the accounting records.
   c. Accounting data have illusory precision and implied authoritativeness. Although it is customary to present data to the nearest dollar or the nearest cent, the records are not nearly that accurate in general.

   In summary, accounting records are a good source of historical data, but have some limitations when used in making estimates for engineering economic analyses. Moreover, accounting records rarely contain direct statements of incremental costs or opportunity costs, both of which are essential in most economic analyses.

2. ***Other sources within the firm.*** The usual firm has a large number of people and records that may be excellent sources of estimates or information from which estimates can be made. Examples of functions within firms that keep records useful to economic analyses are sales, production, quality, purchasing, industrial engineering, and personnel. Professional colleagues, supervisors, and workers in production/service areas can provide insights or suggest sources that can be obtained readily.
3. ***Sources outside the firm.*** There are numerous sources outside the firm that provide information helpful for cost estimating. The main problem is in determining those that are most beneficial for particular needs. The following is a listing of some commonly utilized outside sources:
   a. *Published information,* including technical directories, buyer indexes, U.S. government publications, reference books, and trade journals offer a wealth of information. For instance, *Standard and Poor's Industry Surveys* gives monthly information regarding key industries. *The Statistical Abstract of the United States* is a remarkably comprehensive source of cost indexes and data. The Bureau of Labor Statistics publishes many periodicals which are good sources of labor costs, such as the *Monthly Labor Review, Employment and Earnings, Current Wage Developments, Handbook of Labor Statistics,* and the *Chartbook on Wages, Prices and Productivity.* A buyer's index of manufacturers is *Thomas' Register of American Manufacturers,* which can be used to obtain addresses of vendors to inquire for prices. An annual construction cost handbook, *Building Construction Cost Data,* is published by the R. S. Means Company, Kingston, Massachusetts. It includes standard crew sizes, unit prices, and prevailing wage rates for various regions of the country. The R. S. Means Company also publishes other volumes of cost data on mechanical and electrical work, repair and remodeling, and selected cost indexes.
   b. *Personal contacts* are excellent potential sources. Vendors, salespeople, professional acquaintances, customers, banks, government agencies, chambers of commerce, and even competitors are often willing to furnish needed information on the basis of a serious and tactful request.
4. ***Research and development (R&D).*** If the information is not published and cannot be obtained by consulting someone who knows, the only alternative may be to undertake R&D to generate it. Classic examples are developing a pilot plant and undertaking a test market program. These activities are usually expensive and may not always be successful; thus, this final step is taken only in connection with very important decisions, and when the sources mentioned above are known to be inadequate.

8.5.1.2 How Cost Estimates Are Accomplished

Cost estimates can be prepared in a number of ways, some of which can be categorized as follows:

1. A *conference* of various people who are thought to have good information or bases for estimating the quantity in question. A special version of this is the *Delphi method*, which involves cycles of questioning and feedback in which the opinions of individual participants are kept anonymous.
2. *Comparison* with similar situations or designs about which we have more information and from which we can extrapolate estimates for alternatives under consideration. The comparison method may be used to approximate the cost of a design or product that is new. This is done by taking the cost of a more complex design for a similar item as an upper bound, and the cost of a less complex item of similar design as a lower bound. The resulting approximation may not be very accurate, but the comparison method does have the virtue of setting bounds that might be useful for decision making.
3. *Quantitative techniques*, which do not always have standardized names. Some selected techniques, with the names used being generally suggestive of the approaches, are discussed in the next subsection.

8.5.1.3 Selected Cost Estimating Techniques (Models)

1. A *cost index* is a dimensionless number that indicates how costs and prices change (typically escalate) with time. Such indexes provide a convenient means of developing certain types of cost estimates. For example one simple index, $\bar{I}_n$, is given by Equation 8-1:

$$\bar{I}_n = \left(\frac{C_{n1}}{C_{k1}} + \frac{C_{n2}}{C_{k2}} + \cdots + \frac{C_{nm}}{C_{km}} + \cdots + \frac{C_{nM}}{C_{kM}} \right) \Big/ M \qquad (8\text{-}1)$$

where k = reference year (e.g., 1984)
n = year for which an index is to be determined ($n > k$)
M = total number of items in the index ($1 \leq m \leq M$)
C_{nm} = unit cost (or price) of the mth item in year n
C_{km} = unit cost (or price) of the mth item in reference year k

Notice that all M items are given equal weight in Equation 8-1.

A more popular form of Equation 8-1 is the general weighted index which places more emphasis on some items and less on others:

$$\bar{I}_n = \frac{W_1(C_{n1}/C_{k1}) + W_2(C_{n2}/C_{k2}) + \cdots + W_M(C_{nM}/C_{kM})}{W_1 + W_2 + \cdots + W_M} \qquad (8\text{-}2)$$

where $W_1, W_2, \ldots, W_M$ are the weights on items 1, 2, . . . , M. These weights can sum to any positive number, but 1.00 or 100 are often used for practical purposes.

EXAMPLE 8-3

Based on the data below, develop a weighted index for the price of a gallon of gasoline in 1989, when 1983 is the reference year having an index value of 100. The weight placed on *unleaded regular* gasoline is three times that of either leaded regular or premium because roughly three times as much is sold compared to leaded regular or premium.

	Price (Cents/Gal) in Year						
	1983	**1984**	**1985**	**1986**	**1987**	**1988**	**1989**
Regular leaded	78	86	87	91	82	90	101
Premium	85	92	95	102	91	101	112
Regular unleaded	80	88	90	96	85	95	106

Solution
In this example, k is 1983 and n is 1989. From Equation 8-2, the value of $\bar{I}_{1989}$ is

$$0.20\left(\frac{101}{78}\right) + 0.20\left(\frac{112}{85}\right) + 0.60\left(\frac{106}{80}\right) = 1.318$$

Because $\bar{I}_{1983}$ equals 100, the weighted index in 1989 would have a value of 131.8. If the index in 1994, for example, is estimated to be 184.7, it is a simple matter to determine the corresponding prices of gasoline from $\bar{I}_{1989} = 131.8$.

$$\textit{Regular leaded:}\quad 101 \text{ cents/gal}\left(\frac{184.7}{131.8}\right) = 142 \text{ cents/gal}$$

$$\textit{Premium:}\quad 112 \text{ cents/gal}\left(\frac{184.7}{131.8}\right) = 157 \text{ cents/gal}$$

$$\textit{Regular unleaded:}\quad 106 \text{ cents/gal}\left(\frac{184.7}{131.8}\right) = 149 \text{ cents/gal}$$

An estimate of the present cost of an item in year n can be obtained by multiplying the cost of the item at an earlier point in time (year k) by the ratio of the index value in year n to the index value in year k. This is sometimes referred to as the *ratio technique* of updating costs. Use of this technique allows the cost of a piece of equipment, plant, or component part to be taken from historical data with a specified base year and updated with a cost index.

EXAMPLE 8-4

A certain index for the cost of purchasing and installing utility boilers is keyed to 1974, where its baseline value was arbitrarily set at 100. Company XYZ installed a 50,000 lb/hr boiler in 1982 for $200,000 when the index had a value of 178. This same company must install another boiler of the same size in 1989. The index in 1989 is 312. What is the approximate cost of the new boiler?

Solution

Relative to 1982 when the initial boiler was installed, the cost index has risen (312/178)100 = 175%. Therefore, an approximate estimate of its cost in 1989 is $200,000(1.75) = $350,000.

Many indexes are periodically published, including the *Engineering News-Record* Construction Index which incorporates labor and material costs, and the Marshall and Stevens cost index. The *Statistical Abstract of the United States* publishes government indexes on yearly materials, labor, and construction costs. The Bureau of Labor Statistics publishes the *Producer Prices and Price Indexes* and the *Consumer Price Index Detailed Report*. Indexes of price changes are frequently used in engineering economy studies.

2. *The unit technique* involves utilizing a "per unit factor" that can be estimated effectively. Examples are

   Capital cost of plant per kilowatt of capacity.
   Fuel cost per kilowatt-hour generated.
   Capital cost per installed telephone.
   Temperature loss per 1,000 feet of steam pipe.
   Operating cost per mile.
   Maintenance cost per hour.

   Such factors, when multiplied by the appropriate unit, give a total estimate of cost or savings.

   There are limitless possibilities for breaking quantities into units that can be estimated readily. Examples are

   In different units, such as dollars per week, to convert to dollars per year.
   A proportion, rather than a number, such as % defective, to convert to number of defects.
   A number, rather than a proportion, such as number defective and number produced, to convert to % defective.
   A rate, rather than a number, such as miles per gallon, to convert to gallons consumed.
   A number, rather than a rate, such as miles and hours traveled, to convert to average speed.
   Using an adjustment factor to increase or decrease a known or estimated number, such as defectives reported, to convert to total defectives.

As a simple example, suppose that we need a preliminary estimate of the cost of a particular house. Using a unit factor of, say, \$55 per square foot and knowing that the house is approximately 2,000 square feet, we estimate its cost to be \$55 × 2,000 = \$110,000.

While the unit technique is very useful for preliminary estimating purposes, one can be dangerously misled by such average values. In general, more detailed methods can be expected to result in greater estimation accuracy.

3. *The segmenting technique* involves decomposing an uncertain quantity into parts that can be separately estimated and then added together. As an example, suppose that we desire to estimate the selling costs of product ABC in the Dakotas. The simplest possible segmenting would be to estimate selling costs separately in North Dakota and South Dakota and then to add the two together.
4. *The factor technique* is an extension of the unit method and the segmenting method in which one sums the product of several quantities or components and adds these to any components estimated directly. That is,

$$C = \sum_{d} C_d + \sum_{m} f_m U_m \tag{8-3}$$

where C = cost being estimated
C_d = cost of selected component d that is estimated directly
f_m = cost per unit of component m
U_m = number of units of component m

As a simple example, suppose that we need a slightly refined estimate of the cost of a house consisting of 2,000 square feet, two porches, and a garage. Using a unit factor of \$50 per square foot, and \$5,000 per porch and \$8,000 per garage for the two directly estimated components, we can calculate the total estimate as

$$(\$5{,}000 \times 2) + \$8{,}000 + (\$50 \times 2{,}000) = \$118{,}000$$

5. *The power-sizing technique* is a sophistication of the unit method frequently used for costing industrial plant and equipment. This method recognizes that cost varies as some power of the change in capacity or size. That is,

$$(C_A \div C_B) = (S_A \div S_B)^X \tag{8-4}$$

$$C_A = C_B(S_A/S_B)^X \tag{8-5}$$

where C_A = cost for plant A, C_B = cost for plant B (both in $ as of point in time for which estimate is desired)
S_A = size of plant A, S_B = size of plant B (both in same physical units)
X = cost-capacity factor to reflect economies of scale*

EXAMPLE 8-5

Suppose that it is desired to make a preliminary estimate of the cost of building a 600 MW fossil fuel power plant. It is known that a 200 MW plant cost $100 million in 1970 when the appropriate cost index was 400, and that cost index is now 1,200.

Solution
The power-sizing model estimate, with $X = 0.79$, is

Cost now of 200 MW plant: $100 million $\times$ (1,200 $\div$ 400) = $300 million (call it C_B)

Cost now of 600 MW plant: $C_A \div \$300 \text{ million} = (600 \div 200)^{0.79}$ (call it C_A)

$$C_A = \$300 \text{ million} \times 2.38 = \$714 \text{ million}$$

6. *Various statistical and mathematical modeling techniques* can be used for estimating or forecasting the future. Typical of the numerous mathematical modeling techniques that allow one to break down difficult problems to make more reliable estimates, but which will not be described here, are cost estimating relationships (CERs), econometric models, demographic (population characteristic) models, network models, stochastic process models, mathematical programming models, input–output tables, regression models, and exponential smoothing models. Both of the latter two models are often used to develop CERs. Also, the power-sizing model is a form of a CER.

8.5.1.4 A Manufacturing Cost-Estimating Example

Manufacturers are faced with the problem of making a product that can be sold at a competitive price so that they can make a reasonable profit. The price of their product is based on the overall cost of making the item plus a built-in profit. Some companies that make a variety of products do not have a precise idea of exactly what each product costs—to find out might be prohibitively

*May be calculated/estimated from experience. See p. 137 of W. R. Park, *Cost Engineering Analysis* (New York: John Wiley & Sons, 1973) for typical factors. For example, $X = 0.68$ for nuclear generating plants and 0.79 for fossil-fuel-generating plants.

expensive—but they still need estimates to help them make decisions about what to produce and how to price their products.

As discussed in Chapter 2, product costs are classified as direct or indirect. Direct costs are easily assignable to a specific product, while indirect costs are not easily allocated to a certain product. For instance, direct labor would be the wages of a machine operator; indirect labor would be supervision.

Manufacturing costs have a distinct relationship with production volume in that they may be fixed, variable, or step-variable. Generally, administrative costs are fixed regardless of volume, material costs vary directly with volume, and equipment cost is a step function of production level.

The general cost categories of manufacturing expense include engineering and design, development costs, tooling, manufacturing labor, materials, supervision, quality control, reliability and testing, packaging, plant overhead, general and administrative, distribution and marketing, financing, taxes, and insurance. Where do we start?

When estimating the cost of a manufactured product, we need drawings, specifications, production schedules, historical records of the company's labor cost, a bill of materials, and the process plan. The process plan tells us all operations that must be done to a product and the labor hours involved.

Engineering and design costs consist of design, analysis, and drafting, together with miscellaneous charges such as reproductions. The engineering cost may be allocated to a product on the basis of how many engineering labor hours are involved. Other major types of costs that must be estimated are

Tooling costs, which consist of repair and maintenance plus the cost of any new equipment.

Manufacturing labor costs, which are determined from standard data, historical records, or the accounting department.

Materials costs, which can be obtained from historical records, vendor quotations, and the bill of materials. Scrap allowances must be included.

Supervision is a fixed cost that is based on the salaries of supervisory personnel.

Plant overhead, which includes utilities, maintenance, and repairs. As discussed in Chapter 2, there are various methods used to allocate overhead, such as in proportion to direct labor dollars, or direct labor hours, or machine hours. If we use direct labor hours to allocate overhead, then

$$\text{overhead rate} = \frac{\text{total factory overhead}}{\text{total direct labor hours}}$$

Administrative costs are often included with the factory overhead (or burden).

The following simple example shows the general procedure for making a "per unit" product cost estimate, and illustrates the use of a typical worksheet form of the cost structure for preparing the estimate.

The worksheet in Figure 8-7 shows the determination of the cost of a throttle assembly. The 36.48 direct labor hours are multiplied by the composite labor

FIGURE 8-7

Customer: Aqua Boat Company

Model: CDX75Y — Estimator: Sam Steward

Part Name: Throttle Assembly

Part No.: 00681 — Date: July 31, 1988

Part Req'd.: 50 — Page: 1 of 1

MANUFACTURING COST	CHARGE		
	Hours	Rate	Dollars
Factory Labor	36.48	$10.54	$384.50
Planning & Liaison Labor		12%	46.14
Quality Control		11%	42.30
TOTAL LABOR			472.94
Factory Overhead		105%	496.59
General & Admin. Expense		15%	70.94
Production Material			167.17
Outside Manufacture			28.00
SUBTOTAL			1235.64
Packing Costs		5%	61.78
Premium Pay			
Total Direct Charge			1,297.42
Other Direct Charge		1%	12.97
Facility Rental			
Total Manufacturing Cost			1,310.39
Profit/Fee @ 10%			131.04
Total Selling Price			1,441.43
Quantity—(Unit)			50
Ship Set (Unit) Selling Price			$ 28.83

Manufacturing Cost Worksheet.* Source: *Adapted from T.F. McNeill and D.S. Clark, *Cost Estimating and Contract Pricing*, ***(New York: American Elsevier Publishing Co., 1966)***

rate, $10.54 per hour, to yield $384.50. Planning labor and quality control are expressed as 12% and 11% of direct labor cost, respectively. This gives a total labor cost of $472.94. Factory overhead and general and administrative expense are applied as percentages of the total labor cost. The costs of production material and parts from outside vendors are also entered on the worksheet.

Packing costs are added at a rate of 5% of all previous costs, giving a total direct charge of $1,297.42. Other direct charges are figured in to give a total manufacturing cost of $1,310.39. A 10% profit is added in giving a total selling price of $1,441.43. Since there are 50 parts in the production run, the unit selling price is $28.83.

8.5.2 Estimating Revenue

In an engineering economy study, any revenue differences between the feasible alternatives need to be carefully considered. The number of cost categories in a typical study usually far exceeds the number of revenue categories, and the oversight of a revenue source could cause a serious error in the results. Thus, the revenue impact of the feasible alternatives needs to be addressed and properly reflected in the cash flows. Also, the study assumptions regarding differences in revenue between the feasible alternatives should be explicitly stated before beginning an analysis.

The evaluation of the market and business environment for large new capital projects is, along with the related estimating of project sales, product prices, and so on, a major area of analysis. R. F. de la Mare provides a good summary discussion of economic forecasting and market analysis related to large investment projects and of the incorporation of revenue estimates into cash flows.*

Most of the selected cost estimating techniques discussed in Section 8.5.1.3 are also applicable to estimating revenue.

1. *Indexes* can be developed and applied to estimating the future sales price of a product or service. As an illustration, the solution to Example 8-3 used an index model to estimate the selling prices of a gallon of three types of gasoline in 1989. Such price estimates, developed using the index technique, can also be used to forecast sales revenue if applicable.
2. *The unit technique,* which utilizes an estimated "per unit factor" approach, can be used in a number of different revenue estimating situations. For example:

   Revenue per long-distance telephone call.
   Sales per square foot of display space.
   Revenue per ton-mile.
   Rent per square foot.
   Revenue per kilowatt-hour.

3. *The segmenting technique* involves decomposing a quantity into parts, separately estimating the parts, and then adding the estimates together. This approach can be applied to estimating revenues as well as costs. For example, estimating the annual revenue for a new toll bridge would involve

*R. F. de la Mare, *Manufacturing Systems Economics: The Life-Cycle Cost and Benefits of Industrial Assets* (London: Holt, Rinehart and Winston, 1982), pp. 123–149.

segmenting the total traffic into different traffic classes, applying a rate for each class, and adding the separate estimates together.

4. *The factor technique* extends the unit method and the segmenting method into a combination of directly estimated revenue components added to other components that were segmented into parts, separately estimated, and then summed (Equation 8-3).

EXAMPLE 8-6

The detailed design of the commercial building described in Example 8-1 affects the utilization of the gross square feet (and, thus, the net rentable space) available on each floor. Also, the size and location of the parking lot and the prime road frontage available along the property may offer some additional revenue sources. As project manager, analyze the potential revenue impacts of these considerations.

Solution

The first floor of the building has 15,000 gross square feet of retail space, and the second floor has the same amount planned for office use. Based on discussions with the project A–E contractor and the sales staff, you develop the following additional information:

1. The retail space should be designed for two different uses—60% for restaurant operation and 40% for a retail clothing store.
2. There is a high probability that all the office space on the second floor will be leased to one client.
3. An estimated 20 parking spaces can be rented on a long-term basis to two existing businesses that adjoin the property. Also, one spot along the road frontage can be leased to a sign company for erection of a billboard without impairing the primary use of the property.

Based on the above information, you estimate annual project revenue ($\hat{R}$) as follows:

$$\hat{R} = W(r_1)(12) + Y(r_2)(12) + \sum_{j=1}^{3} S_j(u_j)(d_j)$$

where W = number of parking spaces
Y = number of billboards
r_1 = rate per month per parking space = \$22
r_2 = rate per month per billboard = \$65
j = index on type of building space use
S_j = space (gross square feet) being used for purpose j
u_j = space j utilization factor (% net rentable)
d_j = rate per (rentable) square foot per year of building space used for purpose j.

Type of Space	j	d_j	u_j	S_j (ft^2)
Restaurant	1	$23	0.79	9,000
Retail Shop	2	18	0.83	6,000
Office Space	3	14	0.89	15,000

Then

$$\begin{aligned}\hat{R} &= [20(\$22)(12) + 1(\$65)(12)] + [9{,}000(0.79)(\$23) + \\ &\quad 6{,}000(0.83)(\$18) + 15{,}000(0.89)(\$14)] \\ &= \$6{,}060 + 440{,}070 = \$446{,}130\end{aligned}$$

A breakdown of the annual estimated project revenue in Example 8-6 shows that

1.4% is from miscellaneous revenue sources
98.6% is from leased building space

From a detailed design perspective, changes in annual project revenue due to changes in the building space utilization factors can be easily calculated. For example, an average 1% improvement in the ratio of rentable square feet to gross square feet would change the annual revenue as follows:

$$\begin{aligned}\Delta\hat{R} &= \sum_{j=1}^{3} S_j(u_j + 0.01)(d_j) - (\$446{,}130 - 6{,}060) \\ &= \$445{,}320 - 440{,}070 \\ &= \$5{,}250 \text{ per year}\end{aligned}$$

5. *Various statistical and mathematical modeling techniques* are also used in estimating or forecasting revenue. Of particular use are econometric models, time trend analysis (exponential models, etc.), demographic models, and regression models.

The brief discussion in this section of estimating revenue has emphasized the importance of careful analysis of project revenue in developing net cash flows for the feasible alternatives. Overlooking revenue sources, or not properly evaluating changes in revenue, can cause serious errors to show up in the cash flows and study results.

8.5.3 Estimating Cash Flows for a Typical Small Project

Before demonstrating the development of cash flows for the case study in Section 8.6, we will consider a small project typical of those often encountered in practice. To what extent does Figure 8-2 apply when the project is not large and complex? The answer is that it applies regardless of the size and complexity of the project.

When applying the integrated approach to a small project, however, several steps can be taken to reduce the level of detail to fit the specific situation.

1. *Work breakdown structure (WBS)*. The number of levels and scope of the WBS can normally be significantly reduced for a small project. Sometimes the WBS can be combined with the cost and revenue structure into a worksheet for developing the estimates (as will be demonstrated in Example 8-7). The important point is that this initial component of the integrated approach needs to be explicitly evaluated for the specific project. A WBS in the proper form and scope will facilitate the economic analysis of any project.
2. *Cost and revenue structure*. The number of cost and revenue categories and elements required can be reduced for most small projects. This second component, however, needs to be considered in detail regardless of the size of the project. For example, the number of operating and maintenance cost elements that may need to be included, even in a small project, can be quite extensive.
3. *Estimating techniques (models)*. Estimating future costs and revenues is usually reduced in complexity as the size of a project decreases. The use of techniques discussed in Sections 8.5.1 and 8.5.2, however, will still be required.

These basic components of the integrated approach apply regardless of the size of the project. Their application in small projects, however, is reduced in scope along with the cost and revenue database required.

In any engineering economy study, it is necessary to (1) define the cash flow perspective, (2) determine the estimating baseline, and (3) establish the length of the analysis (study) period. These parts of the approach do not vary with project size.

EXAMPLE 8-7

Your company is involved in the manufacture of transmission components and axles for heavy duty trucks, and is a major supplier to three truck manufacturing plants. Just-in-time inventory concepts are used at each of the three manufacturing plants. Therefore, price competitiveness, reliable delivery to meet plant production schedules, and quality of delivered product are essential to maintaining the company's position as a supplier to the three plants. Meeting such customer expectations is critical to increasing market share through the amount of components supplied to the three plants. Consequently, a project is being considered that involves replacement of some existing equipment with new automated equipment for the production of axles.

One of the feasible alternatives involves new equipment manufactured by Company A. Describe the development of its before-tax net cash flow using the integrated approach in Figure 8-2. Discuss potential sources and compilation of necessary data as appropriate (all details do not need to be given).

Solution
Some basic data related to the project are the following:

1. The equipment acquisition cost is $2,650,000 (including computer software and primary installation costs) if purchased from Company A. Other miscellaneous installation costs of $83,000 would be expensed in the first year of operation (i.e., not included in the cost basis of the equipment).
2. The analysis (study) period established by the company for this type of investment is 6 years.

The new automated equipment, if purchased from Company A, is a complete system; that is, the hardware and software do not need to be broken down further to define explicitly the system for cost and revenue estimating. Therefore, the WBS level that can be used for estimating is the total project (i.e., Level 1 of the WBS). As a result, the WBS and the cost and revenue structure can be combined in a single worksheet. Thus, in this situation a separate, detailed WBS is not required.

The cash flow perspective that should be used in this project is that of the company (owners). Since this is a project involving improvement to an existing operation, the best estimating baseline is the current operation, and the differential approach (Section 2.5) should be used. Thus, cost data from the present operation plus those obtained from the manufacturer (Company A) are the primary sources of data for estimating purposes. The estimating techniques to be used are determined by the database that is available.

A representative worksheet is shown in Figure 8-8 for summarizing the costs and revenues needed to develop the net cash flow over 6 years when equipment is purchased from Company A. The estimate of investment (first) cost is based primarily on data from the manufacturer (cost of the equipment and computer software). Internal estimates developed by the project engineering group are used for the other cost elements (installation costs, working capital, etc.).

The increased revenue estimate, based on additional market share (sales volume) as a result of the project, would be developed by the sales staff. The estimated market value for the present equipment being replaced and the new equipment at the end of 6 years could be developed using data obtained from firms involved in the resale of this type of equipment. The operational and maintenance costs would be estimated from present operating experience and expected new equipment performance data supplied by Company A.

FIGURE 8-8

A. Nonrecurring Costs and Revenues	Costs	Revenues
1. Investment (first) cost		
a. Hardware (including computer equipment)	$2,195,000	
b. Computer software	185,700	
c. Primary installation	269,300	
d. Other installation costs	83,000	
e. Working capital	28,400	
f. Project engineering and management	172,500	
Total:	$2,933,900	
2. Revenue		
a. Sale of present equipment (year 0)		$185,000
b. Sale of new equipment (year 6)		310,000

B. Recurring Annual Costs and Revenues	Costs	Revenue or Reduced Costs
1. Operational and Maintenance (O and M) Costs		
a. Direct Costs		
Labor		$201,000
Material		58,000
Other direct costs		44,600
b. Indirect Costs		
Labor/overtime		14,300
Materials and supplies		
Cost of quality (during production)		32,000
Tooling/fixtures		11,500
Maintenance	$18,600	
Utilities	4,100	
Property taxes and insurance	29,000	
Other indirect costs		5,900
2. Revenue		
Increased sales		525,000
Total:	$51,700	$892,300

Project Cost and Revenue Estimating Worksheet for Example 8-7

Based on the cost and revenue estimates shown on the worksheet, the estimated 6 year before-tax net cash flow (BTCF) for the feasible project alternative involving purchase of the new equipment from Company A is

End of Year	*BTCF (Company A)*
0	−$2,748,900
1	840,600
2	840,600
3	840,600
4	840,600
5	840,600
6	1,150,600

The cash flow amounts in year 0 and year 6 include revenue from the disposal of assets of $185,000 and $310,000, respectively, as indicated on the worksheet.

8.6 DEVELOPING CASH FLOWS (A CASE STUDY)

In the previous sections of this chapter, we discussed the work breakdown structure, cost and revenue structure, and estimating techniques related to an integrated approach for developing cash flows. In this section, we shall demonstrate developing the cash flows for a large project (the commercial building project which was the subject of Examples 8-1, 8-2, and 8-6).

Assume that as project manager of the commercial building project you want an updated semidetailed estimate (Section 8.5.1) of the project net cash flow for a time period that includes 15 years of operation after completion of construction and full tenant occupancy of the building occurs. Also, assume the present time is the beginning of 1989. The semidetailed estimate will be based on the project as defined by the WBS in Example 8-1, the cost and revenue structure in Example 8-2, and the floor space utilization assumptions and annual revenue estimates in Example 8-6. The planned schedule and assumptions related to construction, occupancy, and disposal of the building are

Year	*Schedule/Assumptions*
1(1989)	Real estate purchased; design and engineering services—75% completed; construction—30% completed.
2	Design and engineering and construction—100% completed; 4 months of full tenant occupancy.
3–17	Average occupancy: 90% first floor and 95% second floor.
17	Building is sold at the end of the year.

8.6.1 Investment (First) Cost

The investment (first) cost consists of two major components: (1) the construction costs of the office building and related facilities, and (2) other project costs incurred until initial occupancy is completed. The construction costs for this case study project are shown in Table 8-1 and described below. Then in Table 8-2, the construction cost data are combined with the other estimated cost elements to develop the total investment (first) cost for the project. This cost is then divided between 1989 (year 1 of the project) and 1990 (year 2) in accordance with the project schedule and assumptions.

TABLE 8–1 Semidetailed Estimate of Construction Costs for the Commercial Building Project

	Construction Cost Elements								
	Labor/Installation[a]			Material[a]					
WBS Element	$/Ft^2	Ft^2	Total	$/Ft^2	Ft^2	Total	Labor and Materials	G and A Overhead and Profit	Total
1.1 Site work and foundation	$2.67	18,000	$ 48,060	$2.48	18,000	$ 44,640	$ 92,700	$ 22,250	$ 114,950
1.1.3 Sidewalks/Parking	0.88	22,000	19,360	1.05	22,000	23,100	42,460	10,190	52,650
1.2 Exterior	3.64	30,000	109,200	6.38	30,000	191,400	300,600	72,140	372,740
1.3 Interior	2.72	30,000	81,600	5.50	30,000	165,000	246,600	59,180	305,780
1.4 Roof	1.19	15,000	17,850	1.61	15,000	24,150	42,000	10,080	52,080
1.5 Electrical Systems	1.73	30,000	51,900	1.71	30,000	51,300	103,200	24,770	127,970
1.6 Mechanical Systems	0.80	30,000	24,000	3.22	30,000	96,600	120,600	28,940	149,540
	Subtotals:		351,970			596,190	948,160	227,550	
					Total Construction Cost:				$1,175,710

[a] Construction overhead costs related to labor and materials are included in the dollar per gross square foot unit rates used for these cost elements (this construction overhead is similar to factory overhead in Section 8.5.1.4).

The following information applies to the construction cost estimates in Table 8-1:

1. The accuracy of a semidetailed construction cost estimate (approximately ±15%) is considered achievable in this case using the unit and segmenting techniques (Section 8.5.1.3) applied at Level 2 of the WBS (except for the element Sidewalks/Parking, which is at Level 3 of the WBS).
2. The unit cost data shown in Table 8-1 are the result of the combined experience in the geographical location of the company and the architectural-engineering firm which has provided engineering and design services in this type of construction project for 24 years.* The square footage amounts are based on the project WBS definition and available design data.
3. The general and administrative (G and A) overhead costs of the construction contractor plus profit are shown under the cost element G and A Overhead and Profit. This cost element is estimated at 24% of total labor and material costs.

The total estimated project construction cost, including sidewalks, parking, contractor overhead and profit, is $1,175,710. On the basis of 30,000 gross square feet of building space, the average cost per square foot is $39.19.

In Table 8-2, the development of the total investment (first) cost using the factor technique for the project is shown. The other cost elements involved are estimated and added to the construction cost. Then, the total for each cost element is distributed between 1989 and 1990 based on the project schedule. The following information applies to the data in Table 8-2:

1. In the construction and operation of other office buildings, the company has been averaging an initial working capital investment of 3% of initial construction material costs (Table 8-1).

$$\text{working capital} = 0.03(\$596{,}190) = \$17{,}885 \text{ (rounded)}$$

2. The real estate cost ($262,000) at this point is an actual cost. That is, the company completed the property acquisition transaction immediately prior to this updating of the estimated project cash flow.
3. The project management costs shown in the table are based on a detailed estimate of the personnel time, office space, travel, and so on during the 20 months from the beginning of the project (January 1, 1989) until full occupancy and routine operations are scheduled to be achieved August 31, 1990. The total estimated project management costs are $81,600, with 60% distributed to year 1 (12 months) and 40% to year 2 (8 months).
4. The cost of A–E services (engineering and design activities) is based on a contract rate of 8% of total construction costs:

$$\text{A–E services} = 0.08(\$1{,}175{,}710) = \$94{,}060 \text{ (rounded)}$$

*Sources of this type of data outside an organization were discussed in Section 8.5.1.1.

TABLE 8–2 Semidetailed Estimate of Total Investment (First) Cost for the Commercial Building Project

WBS Element(s)	Cost Element	Total	Distribution of Cost	
			Year 1 (1989)	Year 2 (1990)
1.1–1.6	working capital	$ 17,885		$ 17,885
1.1–1.6	construction (plus overhead)	1,175,710	$352,710	823,000
1.7	real estate	262,000	262,000	
1.8	project management	81,600	48,960	32,640
1.9	A–E services	94,060	70,545	23,515
1.10	sales (leasing)	44,615		44,615
	TOTALS	$1,675,870	$734,215	$941,655

5. The real estate management subsidiary of the company handles leasing of the available space. The estimated sales cost to achieve initial full occupancy is 10% of the first 12 months of rental income. The annual building rental income was estimated in Example 8-6 to be $446,130.

$$\text{sales (leasing) costs} = 0.10(\$446{,}130) = \$44{,}615 \text{ (rounded)}$$

Thus, the updated estimate of the total investment (first) cost for the project is $1,675,870. Of this total, $734,215 is estimated to occur in 1989 and $941,655 in 1990. On the basis of 30,000 gross square feet of building space, the average cost per square foot based on the total project investment (first) cost is $55.86.

8.6.2 Annual Operation and Maintenance Cost

A semidetailed estimate of annual operation and maintenance (O and M) costs for an office building is normally based on unit rates (dollars per gross square foot) with general and administrative (G and A) overhead costs added separately. The unit rates usually include indirect (overhead) costs associated with labor and materials. Since the company has significant experience in operating office buildings in the geographical location involved, historical data are the basis of the updated estimates.

The estimated annual O and M cost, except for the G and A overhead cost, is as follows:

Area	Ft^2	Annual Unit Cost ($\$/Ft^2$)	Total
Office Building	30,000	$2.450	$73,500
Sidewalks/Parking	20,000	0.228	4,560
			$78,060

The company G and A overhead costs associated with operating the building are based on the annual operating and maintenance costs. The G and A rate applied on this basis by the company is 19%:

$$\text{annual G and A overhead costs} = 0.19(\$78{,}060) = \$14{,}830 \text{ (rounded)}$$

and

$$\text{total annual O and M cost} = \$78{,}060 + 14{,}830 = \$92{,}890$$

The operating and maintenance cost for the initial 4 months of occupancy in 1990 would be $\frac{1}{3}$ ($92,890) = $30,963.

8.6.3 Annual Revenue and Leasing Fees

From the solution to Example 8-6, the estimated annual project revenue ($\hat{R}$) on the basis of 100% building occupancy is \$446,070. This total annual revenue is made up of \$253,170 from leasing the first floor, \$186,900 from leasing the second floor, and \$6,060 from leasing 20 parking spaces and one billboard location. Thus, the estimated revenue in 1990 for the initial 4 months of operation is \$446,130(0.33) = \$147,223.

For project years 3 through 17 (1991–2005), the estimated annual project revenue, based on an average first floor occupancy rate of 90% and a second floor rate of 95%, is

$$\begin{aligned}\hat{R} &= 0.90(\$253{,}170) + 0.95(\$186{,}900) + \$6{,}060\\ &= \$411{,}468\end{aligned}$$

The real estate management subsidiary of the company charges an 8% fee based on the annual revenue for handling all annual leasing arrangements for the project. Therefore, the annual leasing fee is

$$\text{annual fee} = 0.08(\$411{,}468) = \$32{,}917$$

8.6.4 Asset Sale Revenue and Disposal Costs

The project plan is that the office building will be sold by the company at the end of project year 17 (2005). The estimated revenue from sale of the asset is 80% of the original construction cost of the building and related facilities, plus the original cost of land and working capital. The estimated cost of selling the property is 7% of the total sales price. Thus, the estimated revenue and disposal costs associated with the sale of the asset is

$$\begin{aligned}\text{asset sale revenue} &= 0.80(\$1{,}175{,}710) + 262{,}000 + 17{,}885\\ &= \$1{,}220{,}453\end{aligned}$$

and

$$\text{sales (disposal) cost} = 0.07(\$1{,}220{,}453) = \$85{,}432$$

8.6.5 Compilation of the Updated Project Net Cash Flow

The project before-tax net cash flow (BTCF) based on the updated semidetailed analysis is shown in Table 8-3. In Columns 1–6, the annual cost and revenue cash flows previously estimated in Sections 8.6.1–8.6.4 are compiled into the project net cash flow. However, the managers of the company believe that prudent cost control in operating the office building and renegotiation of leases

TABLE 8–3 Updated Before-Tax Net Cash Flow (BTCF) for the Commercial Building Project

	(1)	(2)	(3)	(4)	(5)	(6)	(7)	(8)
End of Year k	Cost Estimates			Revenue			Revenue Growth Adjustment[a] $(1.045)^{k-2}$	
	First Cost	O & M	Sales/Fees	Annual	Asset Sale	Net BTCF		Net BTCF (Adjusted)
1 (1989)	−$734,215					−$ 734,215	1.0	−$ 734,215
2 (1990)	− 941,655	−$30,963		$ 147,223		− 825,395	1.0	− 825,395
3		− 92,890	−$32,917	411,468		285,661	1.0450	298,516
4		↓	↓	↓		↓	1.09203	311,949
5							1.14117	325,988
6							1.19252	340,658
7							1.24619	355,988
8							1.30227	372,007
9		[b]	[b]	[b]		[b]	1.36087	388,747
10							1.42211	406,241
11							1.48610	424,522
12							1.55298	443,625
13							1.62286	463,589
14							1.69589	484,450
15							1.77221	506,250
16		↓	↓	↓		↓	1.85195	529,311
17 (2005)		− 92,890	− 32,917	411,468		285,661	1.93528	552,834
17 (2005)			− 85,349		$1,220,453	1,135,104	1.0[c]	1,135,104

[a] For $3 \leq k \leq 17$

[b] The arrow indicates a uniform cash flow amount for the years indicated.

[c] Selling price, land, and working capital not affected by revenue growth.

on an annual basis will result in the project BTCF, starting in 1991, further increasing at the rate of 4.5% per year. Thus, the net BTCF (Column 6), adjusted for this additional revenue growth assumption (Column 7), is shown in Column 8.

8.6.6 After-Tax Analysis of the Updated Project Net Cash Flow

The after-tax analysis of the updated project net cash flow is shown in Table 8-4. Accomplishing engineering economy studies on an after-tax basis was discussed in Chapter 7. The following information relates to the analysis:

1. The modified ACRS method of depreciation (MACRS) is applicable to the project. This involves straight-line depreciation over a depreciable life of 31.5 years for nonresidential real property. The cost basis is the estimated construction cost of the office building and related facilities ($1,175,710). The remainder of the investment (first) cost items in Table 8-2 were not capitalized as part of the project cost basis for depreciation but charged as

TABLE 8–4 After-Tax Analysis of the Updated Net Cash Flow for the Commercial Building Project

End of Year k	(1) Net BTCF[a]	(2) Depreciation	(3) Taxable Income	(4) Income Taxes[e] (t = 0.373)	(5) After-Tax[g] Cash Flow (ATCF)
1 (1989)	−$ 734,215		−$119,505	$ 44,575	−$ 689,640
2 (1990)	−$ 825,395	$12,441[b]	34,012	− 12,686	−$ 838,081
3	298,516	37,324	261,192	− 97,425	201,091
4	311,949	↓	274,625	− 102,435	209,514
5	325,988		288,664	− 107,672	218,316
6	340,658		303,334	− 113,144	227,514
7	355,988		318,664	− 118,862	237,126
8	372,007		334,683	− 124,837	247,170
9	388,747	c	351,423	− 131,081	257,666
10	406,241		368,917	− 137,606	268,635
11	424,522		387,198	− 144,425	280,097
12	443,625		406,301	− 151,550	292,075
13	463,589		426,265	− 158,997	304,592
14	484,450		447,126	− 166,778	317,672
15	506,250		468,926	− 174,909	331,341
16	529,311		491,987	− 183,511	345,800
17 (2005)	552,834	37,324	515,510	− 192,285	360,549
17 (2005)	1,135,104		337,040[d]	− 114,594[f]	1,020,510

[a] Column 8 from Table 8–3.
[b] Four months of occupancy in 1990.
[c] The arrow indicates a uniform cash flow amount for the year indicated.
[d] Capital gain at time of sale.
[e] Column 3 entry multiplied by $-t$.
[f] Federal tax only (0.34).
[g] PW of the ATCF using i = 12% is $226,278.

a company operating expense in the year in which they occurred. Thus, the allowable annual depreciation is

$$\text{annual depreciation} = \$1{,}175{,}710/31.5 = \$37{,}324$$

A proportion of annual depreciation, depending on the month of initial occupancy or sale, applies to the first year of use (1990) and the year of sale (2005).

2. The federal corporate income tax rate for the company is 34% and the state tax rate is 5%. Therefore, the effective tax rate (t) is, from Equation 7-8,

$$\begin{aligned} t &= 0.34 + 0.05 - (0.34)(0.05) \\ &= 0.373, \text{ or } 37.3\% \end{aligned}$$

3. The capital gain is, from Equation 7-9,

$$\text{capital gain (or loss)} = \text{selling price} - \text{book value}$$

And the net selling price in 2005 of the office building and related facilities, excluding land and working capital, is

$$\begin{aligned} \text{selling price (2005)} &= (0.80)(\$1{,}175{,}710) \\ &= \$940{,}570 \text{ (rounded)} \end{aligned}$$

The book value is

$$\begin{aligned} \text{book value} &= \text{cost basis} - \text{accumulated depreciation deductions} \\ \text{book value (2005)} &= \$1{,}175{,}710 - [\$1{,}175{,}710/31.5](15.33) \\ &= \$603{,}530 \end{aligned}$$

At the time of sale at the end of 2005, $15^1/_3$ years of depreciation will have accumulated. Thus,

$$\text{capital gain} = \$940{,}570 - 603{,}530 = \$337{,}040$$

4. The state, which has a 5% corporate income tax rate, does not have a capital gains tax. Therefore, the applicable federal tax rate on capital gains is 34%.
5. The company's after-tax MARR is equal to 12%.
6. For 1989 (year 1), we estimated in Table 8-2 that \$734,215 of the project investment (first) cost will occur, of which \$119,505 (project management and A–E services) will be charged as company operating expenses and not capitalized as part of the project basis for depreciation. Also, there is no revenue from leases during 1989 and no depreciation since the office building will not be ready for use until late 1990. Therefore, taxable income (Column 3) for 1989 is, from Equation 7-6,

$$\begin{aligned} \text{taxable income} &= \text{gross income (revenue)} - \text{expenses} - \text{depreciation} \\ \text{taxable income}_{1989} &= 0 - \$119{,}505 - 0 = -\$119{,}505 \end{aligned}$$

That is, the project will contribute $119,505 to the company operating expenses, which will reduce the gross operating income of the company for tax purposes in 1989 by that amount.

7. For 1990 (year 2), we estimated in Table 8-2 that an additional $941,655 of investment (first) cost will occur, of which $100,770 will be charged as a company operating expense. Also, there is $147,223 of revenue from leases during the last 4 months of 1990, and $12,441 of allowable depreciation for the same period. Thus, the project taxable income for 1990 is

$$\text{taxable income}_{1990} = \$147{,}223 - 100{,}770 - 12{,}441$$
$$= \$34{,}012$$

Using the after-tax MARR (i = 12%), we find that the PW of the updated semidetailed estimate of the project ATCF in column 5 is $226,344. Therefore, the project meets the company's economic criterion.

Developing the net cash flow for each feasible alternative in a study is a pivotal step in the engineering economic analysis procedure (Section 2.5). An integrated approach for developing cash flows includes three major components: (1) a work breakdown structure (WBS) definition of the project, (2) a cost and revenue structure which identifies all the cost and revenue elements involved in the study, and (3) estimating techniques (models). Other considerations such as the length of the analysis period, the perspective and estimating baseline for the cash flows, and a cost and revenue database are illustrated in Figure 8-2 and discussed in the chapter.

The WBS is a powerful technique for defining all the work elements and their interrelationships for a project. This technique is a basic tool in project management, and is a vital aid in an engineering economy study. Understanding this technique and its applications are important in engineering practice.

The development of a cost and revenue structure will help to ensure that a cost element or a source of revenue is not overlooked in the analysis. Also, it assists with developing the data needed for estimating the costs and revenues involved. The life-cycle concept (Section 2.2.8) and the WBS are important aids in developing this structure for a project.

Estimating techniques (models) are used to develop the cash flows for the feasible alternatives as they are defined by the WBS. Thus, the estimating techniques form a bridge from the WBS and detailed cost and revenue data to the estimated cash flows for the feasible alternatives. Several different estimating models are discussed in Section 8.5.

In Section 8.6, a comprehensive case study is discussed which demonstrates the development of the net cash flow for a commercial building project. It includes (1) estimating the costs and revenues involved in the life cycle of an office building, (2) compiling the total project net cash flow, and (3) analyzing it with regard to the project's after-tax economic consequences. This case study illustrates the integrated approach to developing cash flows described in Figure 8-2.

8.7 PROBLEMS

8-1. Two of the basic components of an integrated approach for developing cash flows for the feasible alternatives of a project are (1) the work breakdown structure (WBS) and (2) the cost and revenue structure. (8.2)

a. Describe the concept of each component.

b. Explain the applications of each component in an engineering economy study.

8-2. Visually examine a lawnmower for home use that is (1) nonriding, (2) approximately 21 inches in cutting width, and (3) powered with a 3.5 to 5.0 horsepower air-cooled engine. Develop a WBS for this product through Level 3. (8.3)

8-3. You are planning to build a new home with approximately 2,000 to 2,500 gross square feet of living space on one floor. In addition, you are planning an attached two-car garage (with storage space) of approximately 600 gross square feet. Develop a cost and revenue structure for designing and constructing, operating (occupying) for 10 years, and then selling the home at the end of the tenth year. (8.4)

8-4. Why might a company's purchasing department be a good source of estimates for equipment required 2 years from now for a modernized production line? Which department(s) could provide good estimates of labor costs and maintenance associated with operating the production line? What information might the accounting department provide that would bear upon the incremental costs of the modernized line? (8.5)

8-5. Prepare a general index for housing construction costs in 1989 using these data: (8.5)

Type of Housing	Percent	Reference Year ($\bar{I}$ = 100)	1989
Single units	70	\$28	\$48
Duplex units	5	23 } \$/ft^2	44 } \$/ft^2
Multiple units	25	20	41

8-6. Manufacturing equipment that was purchased in 1974 for $200,000 must be replaced in 1990. What is the estimated cost of the replacement based on the following equipment cost index? (8.5)

Year	Index	Year	Index
1964	100	1987	708
1974	223	1988	779
1984	600	1989	841
1986	681	1990	972

8-7. Suppose that your brother-in-law has decided to start a company that produces synthetic lawns for lazy homeowners. He anticipates starting production in 18 months. In estimating future cash flows of the company, which of the following would be relatively easy versus relatively difficult to obtain? Also, suggest how each might be estimated with "reasonable" accuracy. (8.5)

a. Cost of land for a 10,000-square-foot building.
b. Cost of the building (cinder block construction).
c. Initial working capital.
d. Total investment (first) cost.
e. First year's labor and material costs.
f. First year's sales revenues.

8-8. In a new industrial park, telephone poles and lines must be installed. Altogether it has been estimated that 10 miles of telephone lines will be needed and each mile of line costs $14,000 (includes labor). In addition, a pole must be placed every 40 yards on the average to support the lines, and the cost of the pole and its installation is $210. What is the estimated cost of the entire job? (8.5)

8-9. If an ammonia plant that produces 500,000 pounds per year cost $2,500,000 to construct 8 years ago, what would a 1,500,000-pound-per-year plant cost now? Suppose that the construction cost index has increased an average rate of 12% per year for the past 8 years and that the cost-capacity factor (X) to reflect economy of scale is 0.65. (8.5)

8-10. Refer to the graph (top of next page) and estimate the *total* production costs for a certain chemical compound when the capacity of the plant is 2,500,000 pounds per year. (8.5)

8-11. Using the costing worksheet provided in this chapter (Figure 8-7), estimate the unit cost of manufacturing metal wire cutters in lots of 100 when these data have been obtained: (8.5)

Factory labor = 4.2 hours at $11.15/hour
Factory overhead = 150% of labor
Outside manufacture = $74.87
Production material = $26.20
Packing costs = 7% of factory labor
Profit = 12%

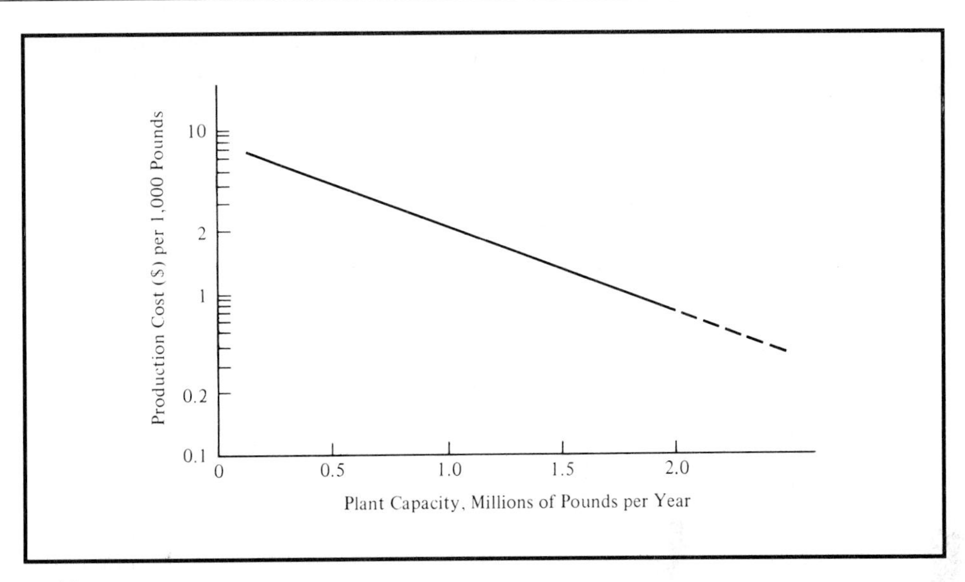

Problem 8-10

8-12. Reference problem 8-3. You have decided to build a one-floor home with 2,450 gross square feet of living space. Also, the attached two-car garage (with storage space) will have 615 gross square feet of area.

a. Develop a WBS (through Level 3) defining the work elements involved in the design and construction of the home. (8.3)

b. Develop a semidetailed estimate of your investment (first) cost associated with the project until the time of your initial occupancy of the home. (*Note:* your instructor will provide you with additional information to assist with this part of the problem.) (8.6)

8-13. Use Figure 8-2 (as a guide) and the information and results of Problem 8-12 to develop an estimated before-tax net cash flow for 10 years of ownership of the home. Assume sale (disposal) of the home at the end of 10 years. Obtain (locally) representative operating, repair, resale, and other data related to home ownership as needed to support development of the 10-year cash flow. Indicate the cost-estimating techniques used in estimating cash flows. State any assumptions you make. (8.5, 8.6)

8-14. Equation 8-5 (power-sizing technique) is sometimes referred to as an exponential cost-estimating model. The basic principle of exponential cost estimating is that in many situations increases in capacity can be achieved with less than proportionate increases in capital costs.

The basic model (Equation 8-5) can be modified to better represent a specific estimating situation. Consider the situation of an automated warehousing system for a new distribution center handling case goods (e.g., an area distribution center for a supermarket company). Equation 8-5 can be modified to improve its capability to estimate the investment cost for this project (system) by (1) segmenting the equipment and installation part of the initial cost (which can be estimated based on the exponential principle) from the other project and support cost part (engineering, purchasing, project management, etc.) of the initial cost, and (2) adjusting both parts of the initial cost with price index changes from the previous comparable system installation (in the reference year). That is, the modified form of Equation 8-5 would be

$$C_A = C_{B1}\,(S_A/S_B)^X\,(\bar{I}_{B1}) + C_{B2}\,(\bar{I}_{B2})$$

where C_A = estimated cost of new automated warehousing system
C_{B1} = equipment and installation cost of previous comparable system
C_{B2} = other project and support costs of previous comparable system
S_A = capacity of new automated warehousing system
S_B = capacity of previous comparable system
X = cost-capacity factor to reflect economies of scale
$\bar{I}_{B1}$ = cost index for equipment and installation costs
$\bar{I}_{B2}$ = cost index for other project and support costs

a. Develop the cost index value for $\bar{I}_{B1}$ and $\bar{I}_{B2}$ based on the data given below. (8.5)

Equipment and Installation Costs

Cost Element	Weight	Index (Reference Year)	Index (Current Year)
Mechanical equipment	0.41	122	201
Automation equipment	0.22	131	212
Installation hardware	0.09	118	200
Installation labor	0.28	135	184

Other Project and Support Costs

Cost Element	Weight	Index (Reference Year)	Index (Current Year)
Engineering	0.38	136	206
Project management	0.31	128	194
Purchasing	0.11	105	162
Other support	0.20	113	179

b. Develop the estimated investment (first) cost for the new automated warehousing system when the equipment and installation cost of the previous comparable system was $1,226,000; capacity of the new system is 11,000 cases of goods per 8-hour shift; capacity of the previous comparable system is 5,800 cases per 8-hour shift; the cost-capacity factor is 0.7; and other project and support costs for the previous comparable system were $234,000. (8.5)

9

DEALING WITH INFLATION AND PRICE CHANGES

When the monetary unit does not have constant value in exchange for goods and services in the market, and future price changes are expected to be significant, an erroneous choice among competing alternatives can result if price change effects are not included in an engineering economic analysis (before taxes and after taxes). The objectives of this chapter are (1) to introduce a methodology for dealing with inflation and price changes, (2) to develop and illustrate proper techniques to account for these effects in engineering economic analysis, and (3) to discuss the development of estimates of unit price changes.

The following topics are discussed in this chapter.

Terminology and basic concepts
What is the relationship between actual (current) dollars and real (constant) dollars?
Use of combined (nominal) versus real interest rates
Differential price inflation
Modeling price changes with geometric sequences of cash flows
When do we use actual dollar versus real dollar analysis?
Guidelines and strategy for application of actual and real dollar analysis
Examples of projected price changes
Case studies
Foreign exchange rates

9.1

GENERAL PRICE INFLATION

Prior to this chapter we have assumed that prices for goods and services in the marketplace are relatively unchanged over extended periods of time. Unfortunately, this is not generally a realistic assumption. *General price inflation*, which is defined here as the phenomenon of a general increase in the prices paid for goods and services bringing about a reduction in the purchasing power of the monetary unit, is a business reality that can affect the economic comparison of alternatives. The history of price changes shows that general price inflation is a much more common occurrence than general price *deflation*, which involves a general decrease in prices with an increase in the purchasing power of the monetary unit. The concepts and methodology discussed in this chapter, however, apply to any price changes.

One measure of price changes in our economy (and an estimate of general price inflation) is the Consumer Price Index (CPI). Tabulated by the U.S. Government, the CPI is a composite price index that measures price changes in housing, transportation, clothing, food, and other selected goods and services used by individuals and families. As shown in Figure 9-1, the CPI has increased from a value of 100 in base year 1967 to 340.4 in 1987. The CPI and its related annual inflation rates for this period are listed in Table 9-1. Annual general price inflation (based on these CPI rates) during this period varied from a low of 1.9% in 1986 to a high of 13.5% in 1980. The rates (in percent) shown in Table 9-1 were calculated from this expression:*

$$(\text{CPI Annual Inflation Rate})_{n+1} = \frac{(\text{CPI})_{n+1} - (\text{CPI})_n}{(\text{CPI})_n}(100)$$

For example, the CPI annual inflation rate for 1986 is

$$\begin{aligned}(\text{CPI Annual Inflation Rate})_{1986} &= \frac{(\text{CPI})_{1986} - (\text{CPI})_{1985}}{(\text{CPI})_{1985}}(100)\\ &= \frac{328.4 - 322.2}{322.2}(100) = 1.9\%\end{aligned}$$

A price index is a number that is the ratio of prices at two points in time multiplied by 100. A base year is designated, and the price for that year is used in the denominator. For example, if the price for a commodity was \$0.34 per pound in 1969 (the base year), and \$0.92 per pound in 1988, the price index for

*In this chapter only, $n(0 \le n \le N)$ is used as an index of time and $k(0 \le k \le N)$ is the base time period.

FIGURE 9-1

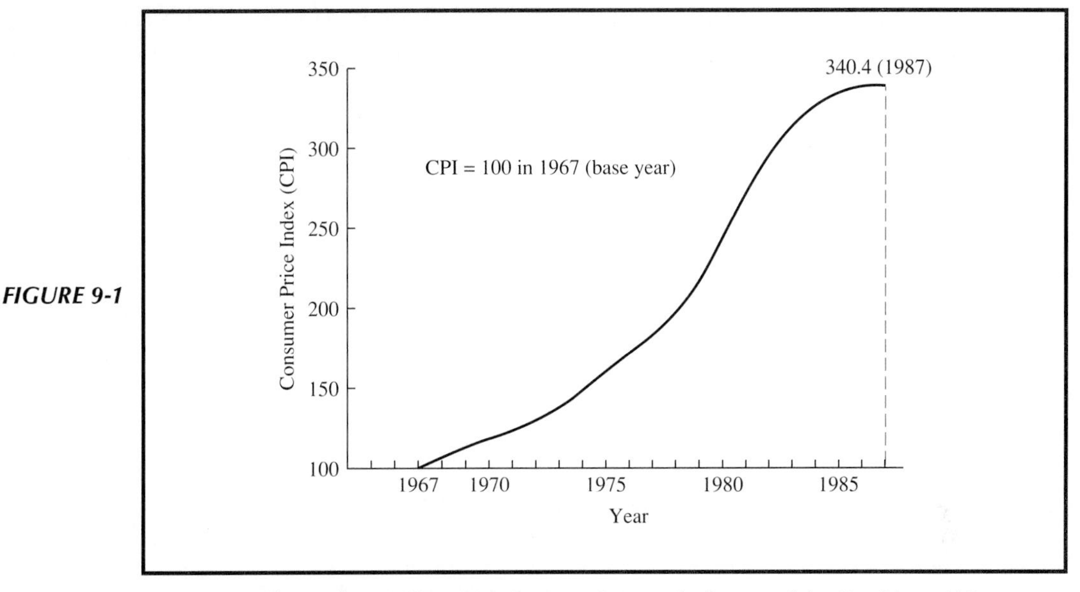

Consumer Price Index, 1967–1987.* Source:** *Economic Report of the President,* ***U.S. Government Printing Office, February, 1988.

1988 would be [(0.92)/(0.34)](100) = 270.6. This index means that the price of the commodity in 1988 is 270.6% (2.706 times) the price in 1969, the base year.

In addition to the CPI, the federal government develops a number of other price indexes from the data it gathers and analyzes. The primary agencies involved in the development of price indexes are the Department of Labor (Bureau of Labor Statistics) and the Department of Commerce (Bureau of Eco-

TABLE 9–1 Consumer Price Index (CPI) and Related Annual Inflation Rate (%), 1967–1987[a]

Year	CPI	CPI Annual Inflation Rate (%)	Year	CPI	CPI Annual Inflation Rate (%)
1967	100	2.9	1978	195.4	7.7
1968	104.2	4.2	1979	217.4	11.3
1969	109.8	5.4	1980	246.8	13.5
1970	116.3	5.9	1981	272.4	10.4
1971	121.3	4.3	1982	289.1	6.1
1972	125.3	3.3	1983	298.4	3.2
1973	133.1	6.2	1984	311.1	4.3
1974	147.7	11.0	1985	322.2	3.6
1975	161.2	9.1	1986	328.4	1.9
1976	170.5	5.8	1987	340.4	3.6
1977	181.5	6.5			

[a] *Source: Economic Report of the President,* U.S. Government Printing Office, February, 1988.

nomic Analysis). The Producer Price Index (PPI) and the Implicit Price Index for the Gross National Product (IPI-GNP) are two indexes, in addition to the CPI, that are often used to estimate general price inflation. These indexes and other available information are used by organizations to develop estimates of price changes in their cash flow estimation process. This is discussed further in Section 9.6.

9.2 TERMINOLOGY AND BASIC CONCEPTS

To facilitate the development and discussion of the methodology for including general price inflation of goods and services in engineering economy studies, we need to define and discuss some terminology and basic concepts. The dollar is used as the monetary unit in this book.

1. ***Actual dollars (A$)***: the actual number of dollars associated with a cash flow (or a non–cash flow amount such as depreciation) as of the time it occurs. For example, people typically anticipate their salaries 2 years hence in terms of actual dollars. Sometimes A$ are referred to as *current* dollars, and they include an allowance for general price inflation.
2. ***Real dollars (R$)***: dollars expressed in terms of the same purchasing power relative to a particular time. For instance, the future unit prices of a good or service that are changing rapidly are often estimated in real dollars (relative to some base year) to provide a consistent means of comparison. Often R$ are termed *constant* dollars.
3. ***General price inflation rate*** (f): a measure of the change in the purchasing power of a dollar during a specified period of time. The general price inflation rate is defined by a selected, and broadly based, index of market price changes. We define f to be the compound rate per period based on the index selected. In engineering economic analysis, the rate is projected for a future time interval and usually is expressed as an effective annual rate. Many large organizations have their own selected index that reflects the particular business environment in which they operate.
4. ***Combined (nominal) interest rate*** (i_c): the money paid for the use of capital, expressed as an effective rate (%) per interest period, that includes a market adjustment for the anticipated general price inflation rate in the economy. Thus, it represents the time value change in future cash flows that takes into account both the potential real earning power of money and the estimated general price inflation in the economy.
5. ***Real interest rate*** (i_r): the money paid for the use of capital, expressed as an effective rate (%) per interest period, that does not include a market adjustment for the anticipated general price inflation rate in the economy.

It represents the time value change in future cash flows based only on the potential real earning power of money.

6. ***Base time period*** (k): The reference or base time period used to define the purchasing power of real (constant) dollars. Often, in practice, the base time period is designated as the time of the engineering economic analysis, or reference time 0 (i.e., $k = 0$). However, k can be any designated point in time.

With an understanding of these definitions, we can delineate and illustrate some useful relationships that are important in engineering economic analysis.

9.2.1 The Relationship between Actual Dollars and Real Dollars

The relationship between actual dollars (A$) and real dollars (R$) is defined in terms of the general price inflation rate; that is, the relation is a function of f. Actual dollars as of any point in time, n, can be converted into real dollars of constant market purchasing power as of any base time period, k, by the following relation

$$(\text{R\$})_n = (\text{A\$})_n\left(\frac{1}{1+f}\right)^{n-k} = (\text{A\$})_n(P/F, f\%, n-k) \tag{9-1}$$

for a given k value. This relationship between actual dollars and real dollars applies to the unit prices or costs of fixed amounts of individual goods or services used to develop the individual cash flows. The designation for a specific type of cash flow, j, would be included as follows

$$\text{R\$}_{n,j} = \text{A\$}_{n,j}\left(\frac{1}{1+f}\right)^{n-k} = \text{A\$}_{n,j}(P/F, f\%, n-k) \tag{9-2}$$

for a given k value, where the terms $\text{R\$}_{n,j}$ and $\text{A\$}_{n,j}$ are the unit price, or cost for a fixed amount, of good or service j in time period n in real dollars and actual dollars, respectively.

EXAMPLE 9-1

Suppose that your salary is $25,000 in year 1 and will increase at 6% per year through year 4, and is expressed in A$ as follows:

Year, n	Salary (A$)
1	$25,000
2	26,500
3	28,090
4	29,775

If the general price inflation rate (f) is expected to average 8% per year, what is the R$ equivalent of these A$ salary amounts? Assume that the base time period is year 1 ($k = 1$).

Solution
By using Equation 9-2, the R$ salary equivalents are readily calculated for the base point in time, $k = 1$:

Year	Salary (R$ in Year 1)
1	$25,000($P/F$, 8%, 0) = $25,000
2	26,500(P/F, 8%, 1) = 24,537
3	28,090(P/F, 8%, 2) = 24,083
4	29,775(P/F, 8%, 3) = 23,636

In year 1 (the designated base time period for the analysis), the annual salary in actual dollars remained unchanged when converted to real dollars. This illustrates an important point. In the base time period (k), the purchasing power of an actual dollar and a real dollar is the same; that is, $R\$_{k,j} = A\$_{k,j}$. This example also illustrates the results when the actual annual rate of increase in salary (6% in this example) is less than the general price inflation rate (f). As you can see, the A$ salary cash flow shows a reasonable increase, but a decrease in the real dollar salary cash flow occurs (and thus a decrease in market purchasing power).

EXAMPLE 9-2

A recent engineering graduate has received annual salaries shown below over the past 4 years. During this time, the consumer price index (CPI) has performed as indicated. Determine the engineer's annual salaries in *year 0 dollars* ($k = 0$) if the CPI is the appropriate indicator of general price inflation for this person.

End of Year	Salary (A$)	CPI
1	$25,400	7.1%
2	27,400	5.4%
3	29,900	8.9%
4	32,600	11.2%

Solution
The engineer's salary has increased by 7.9%, 9.1%, and 9.0% in years 2, 3, and 4, respectively. These are effective annual rates. By using Equation 9-2 with each year's general price inflation taken into account separately, the R$ equivalents in year 0 dollars are calculated as follows.

End of Year	Salary (R$ in Year 0)	
1	$25,400(P/F, 7.1%, 1)	= $23,716
2	27,400(P/F, 7.1%, 1)(P/F, 5.4%, 1)	= 24,273
3	29,900(P/F, 7.1%, 1)(P/F, 5.4%, 1)(P/F, 8.9%, 1)	= 24,323
4	32,600(P/F, 7.1%, 1)(P/F, 5.4%, 1)(P/F, 8.9%, 1)(P/F, 11.2%, 1)	= 23,848

9.2.2 The Relationship between the Combined and Real Interest Rates

The basic relationship among the combined (nominal) interest rate (i_c), the real interest rate (i_r), and the general price inflation rate (f) is

$$1 + i_c = (1 + f)(1 + i_r) \tag{9-3}$$

or

$$i_c = i_r + f + i_r(f) \tag{9-4}$$

and

$$i_r = \frac{i_c - f}{1 + f} \tag{9-5}$$

Thus, the combined interest rate (Equation 9-4) is the sum of the real interest rate (i_r) and the general price inflation rate (f), plus the product of those two terms. Also, as shown in Equation 9-5, the real interest rate (i_r) can be calculated from the combined interest rate and the general price inflation rate.

EXAMPLE 9-3

An investor lends $10,000 today to be repaid in a lump sum at the end of 10 years with interest at 10% ($=i_c$) compounded annually. What is the real rate of return, assuming that general price inflation is 8%($=f$) compounded annually?

Solution

In 10 years, the investor will receive the original $10,000 plus interest that has accumulated, in actual dollars:

$$A\$_{10} = \$10{,}000(F/P, 10\%, 10) = \$25{,}937$$

The purchasing power of a dollar, however, has been reduced (or eroded) by the 8% annual general price inflation rate. That is,

$$\$1(P/F, 8\%, 10) = \$0.4632$$

Then, the $25,937 in A$ is worth only, in today's purchasing power (R$),

$$\text{R\$}_{10} = \$25{,}937 \times 0.4632 = \$12{,}014$$

The $12,014 of today's purchasing power that is returned for the use of $10,000 represents a real rate of return that may be calculated by finding the $i'_r\%$ at which

$$\$10{,}000 = \$12{,}014(P/F, i'_r\%, 10), \quad \text{or } i'_r\% = 1.85\%$$

The real interest rate can also be directly calculated from Equation 9-5 to be

$$i_r = \frac{i_c - f}{1 + f} \quad \text{or } i_r = 0.0185, \quad \text{and } i_r = 1.85\%$$

The value of i_r can be approximated by simply subtracting the general price inflation rate, f, from the $i_c = 10\%$ being charged. Thus, the real rate is approximately equal to $10\% - 8\% = 2\%$. This approximation is fairly close to the correct real interest rate at low rates of general price inflation, but becomes increasingly less accurate as this rate increases.

EXAMPLE 9-4

An investor established an individual savings account in 1981 that involves a *series* of 20 deposits (rather than the one lump-sum deposit in Example 9-3), as shown in Figure 9-2. The account is expected to compound at an average interest rate of 12% per year through the year 2001. General price inflation is expected to average 6% per year during this time.

FIGURE 9-2

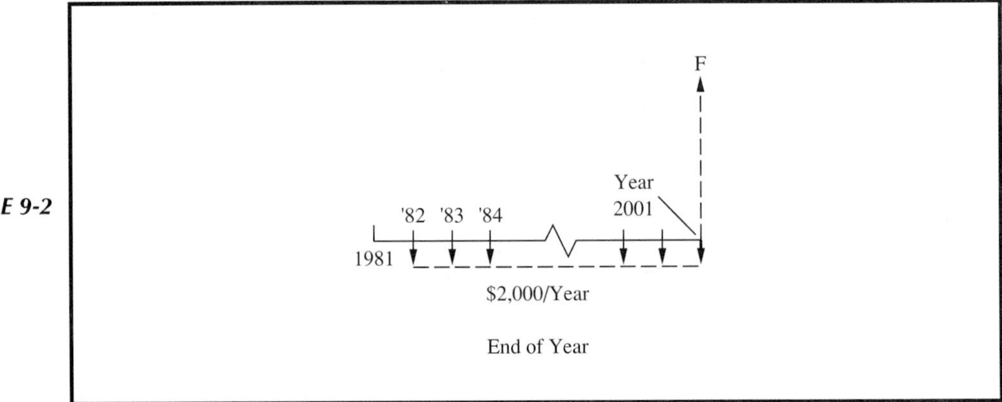

A$ Cash Flow Diagram for Example 9-4

a. What is the future worth of the savings account at the end of year 2001?
b. What is the future worth of the savings account in 1981 (base time period) spending power?

Solution

$$\begin{aligned}
&\text{(a) } F \text{ (in A\$)} = \$2{,}000(F/A, i_c = 12\%, 20) = \$144{,}105 \\
&\text{(b) } F \text{ (in R\$)} = F \text{ (in A\$)}(P/F, f\%, 20) \\
&\qquad = \$144{,}105(P/F, 6\%, 20) \\
&\qquad = \$44{,}933
\end{aligned}$$

9.2.3 What Interest Rate to Use in Engineering Economy Studies

In general, the interest rate that is appropriate for equivalence calculations in engineering economy studies depends on the type of cash flow estimates:

Method	If Cash Flows are in Terms of	Then the Interest Rate to Use Is
A	Actual dollars (A$)	Combined interest rate, i_c
B	Real dollars (R$)	Real interest rate, i_r

This table should make intuitive sense as follows. If one is estimating cash flows in terms of real dollars, the real (uninflated) interest rate is used. Similarly, if one is estimating cash flows in terms of actual (inflated) dollars, the combined (inflated) interest rate is used. Thus, one can make economic analyses using either A$ or R$ with equal validity provided that the appropriate interest rate is used for equivalence calculations.

EXAMPLE 9-5

In Example 9-1 your salary was projected to increase at the rate of 6% per year, and the general price inflation rate was expected to be 8% per year. Your resulting estimated salary for the 4 years in A$ and R$ was as follows:

End of Year, n	Salary (A$)	Salary (R$), $k=1$
1	$25,000	$25,000
2	26,500	24,537
3	28,090	24,083
4	29,775	23,636

What is the equivalent worth of the 4-year A$ and R$ salary cash flows at the end of year 1 (base year), if your personal MARR is 10% (i_c)?

Solution

(a) A$ salary cash flow:

$$P_1 = \$25{,}000 + 26{,}500(P/F, 10\%, 1) + 28{,}090(P/F, 10\%, 2) + 29{,}775(P/F, 10\%, 3)$$
$$= \$94{,}675$$

(b) R$ salary cash flow:

$$i_r = \frac{i_c - f}{1 + f} = \frac{0.10 - 0.08}{1.08} = 0.01852, \text{ or } 1.852\%$$

$$P_1 = \$25{,}000 + \$24{,}537\left(\frac{1}{1.01852}\right)^1 + 24{,}083\left(\frac{1}{1.01852}\right)^2 + 23{,}636\left(\frac{1}{1.01852}\right)^3$$

$$= \$94{,}675$$

Thus, we obtain the same equivalent worth at the end of year 1 (the base time period) for both the A$ and R$ 4-year salary cash flows when the appropriate interest rate is used for the equivalence calculations.

It is important to be consistent in using the correct interest rate for the type of analysis (A$ or R$) being done. The two mistakes made are

Interest Rate (MARR)	Type of Analysis	
	A$	R$
i_c	√ (Correct)	Mistake No. 1 Bias is against capital investment
i_r	Mistake No. 2 Bias is toward capital investment	√ (Correct)

In Mistake No. 1 the combined (nominal) interest rate (i_c), which includes an adjustment for the general price inflation rate (f), is used in equivalent worth calculations for cash flows estimated in real dollars. Since real dollars have constant purchasing power expressed in terms of the base time period (k), and do not include the effect of general price inflation, we have an inconsistency. There is a tendency to develop future cash flow estimates in terms of dollars with purchasing power at the time of the study (that is, R$ with $k = 0$), and then use the combined interest rate i_c in the analysis. The result of Mistake No. 1 is a bias against capital investment. The cash flow estimates in R$ for a project are numerically lower in value than A$ estimates with equivalent purchasing power (assuming $f > 0$). Additionally, the i_c value (which is greater than the i_r value that should be used) further reduces (understates) the present worth of the results of a proposed capital investment.

In Mistake No. 2 the cash flow estimates are in actual dollars, which include the effect of general price inflation (f), but the real interest rate (i_r) is used for equivalent worth calculations. Since the real interest rate does not include an adjustment for general price inflation, we again have an inconsistency. The effects of this mistake, opposite from those in Mistake No. 1, result in a bias toward capital investment by overstating the present worth of the future results.

9.2.4 Fixed and Responsive Annuities

Whenever future investment receipts are predetermined by contract, as in the case of a bond or a fixed annuity, these receipts do not respond to general price inflation. In cases where the future receipts are not predetermined, however, they may respond to general price inflation. The degree of response varies from case to case. To illustrate the nature of this situation, let us consider two annuities. The first annuity is fixed (unresponsive to general price inflation) and yields $2,000 per year for 10 years. The second annuity is of the same duration and yields enough future dollars to be equivalent to $2,000 per year in real purchasing power. Assuming a general price inflation rate of 6% per annum, pertinent values for the two annuities over a 10-year period are as shown in Table 9-2.

Thus, when the receipts are constant in actual dollars (unresponsive to general price inflation), their equivalent value in real dollars of investment declines over the 10-year interval to $1,117 in the final year. When receipts are fixed in

TABLE 9–2 Illustration of Fixed and Responsive Annuities with General Price Inflation Rate of 6% per Year

	Fixed Annuity		Responsive Annuity	
Year	In Actual Dollars	In Equivalent Real Dollars[a]	In Actual Dollars	In Equivalent Real Dollars[a]
1	$2,000	$1,887	$2,120	$2,000
2	2,000	1,780	2,247	2,000
3	2,000	1,679	2,382	2,000
4	2,000	1,584	2,525	2,000
5	2,000	1,495	2,676	2,000
6	2,000	1,410	2,837	2,000
7	2,000	1,330	3,007	2,000
8	2,000	1,255	3,188	2,000
9	2,000	1,184	3,379	2,000
10	2,000	1,117	3,582	2,000

[a] See Equation 9-2.

value of real dollars of investment (responsive to general price inflation), their equivalent in actual dollars rises to $3,582 by year 10.

Included in engineering economy studies are certain quantities unresponsive to general price inflation, such as depreciation, lease fees, and interest charges based on an existing contract or loan agreement. For instance, depreciation write-offs, once determined, do not increase (with present accounting practices) to keep pace with general price inflation; lease fees and interest charges typically are contractually fixed for a given period of time. Thus, it is important when doing an actual dollar analysis to recognize the quantities that are unresponsive to general price inflation, and when doing a real dollar analysis to convert these A$ quantities to R$ quantities using Equation 9-2.

9.2.5 The Impact of General Price Inflation on After-Tax Analysis

Engineering economy studies that include the effects of general price inflation present some difficulties because interest charges, depreciation deductions, and lease payments are actual dollar amounts based on past commitments. They are generally *unresponsive* to inflation. At the same time, many other types of cash flows (e.g., labor, materials, etc.) are *responsive* to inflation. Even though there are responsive and unresponsive amounts usually present in a study, the effects of general price inflation can be readily included in an engineering economic analysis. In Example 9-6, an *after-tax analysis* that includes the effects of general price inflation is presented to show the correct handling of these considerations.

EXAMPLE 9-6

The cost of newly installed production equipment at the beginning of 1989 is $180,000. It is estimated (in base year dollars, $k = 0$) that the equipment will reduce current net operating costs by $36,000 per year for 10 years, and have a $30,000 market value at the end of the tenth year. These cash flows will increase at the general price inflation rate. Due to new computer control features on the equipment, it will be necessary to contract for some maintenance support during the first 3 years. The maintenance contract will cost $2,800 per year. This equipment will be depreciated under the modified ACRS method (MACRS), and it has a recovery period (class life) of 5 years. The federal income tax rate is 34%. The state income tax rate is 6%, and state taxes are deductible from federal income taxes. The selected analysis period is 10 years; the projected general price inflation rate (f) is 8% per year; and the MARR (after taxes) is $i_c = 15\%$.

(a) Based on an actual dollar, after-tax analysis, is this capital investment justified?

(b) Develop the after-tax cash flow in real dollars.

Solution

(a) The actual dollar, after-tax economic analysis of the new production equipment is shown in Table 9-3 (Columns 1–9). The investment (first) cost,

TABLE 9–3 Example 9-6 When General Price Inflation Is 8% Per Year

End of Year (n)	(1) R$ Cash Flows	(2) A$ Adjustment $(1 + f)^{n-k}$	(3) A$ Cash Flows	(4) Contract (A$)	(5) BTCF (A$)	(6) MACRS Depreciation (A$)	(7) Taxable Income	(8) Income Taxes (t = 0.38)	(9) ATCF (A$)	(10) R$ Adjustment $[1/(1 + f)]^{n-k}$	(11) ATCF (R$)
0	−$180,000	1.0000	−$180,000		−$180,000				−$180,000	1.0000	−$180,000
1	36,000	1.0800	38,880	−$2,800	36,080	$36,000	$ 80	−$ 30	36,050	0.9259	33,379
2	36,000	1.1664	41,990	− 2,800	39,190	57,600	− 18,410	+ 6,996	46,186	0.8573	39,595
3	36,000	1.2597	45,349	− 2,800	42,549	34,560	7,989	− 3,036	39,549	0.7938	31,394
4	36,000	1.3605	48,978		48,978	20,736	28,242	− 10,732	38,246	0.7350	28,111
5	36,000	1.4693	52,895		52,895	20,736	32,159	− 12,220	40,675	0.6806	27,683
6	36,000	1.5869	57,128		57,128	10,368	46,760	− 17,769	39,359	0.6302	24,804
7	36,000	1.7138	61,697		61,697		61,697	− 23,445	38,252	0.5835	22,320
8	36,000	1.8509	66,632		66,632		66,632	− 25,320	41,312	0.5403	22,320
9	36,000	1.9990	71,964		71,964		71,964	− 27,346	44,618	0.5003	22,320
10	36,000	2.1589	77,720		77,720		77,720	− 29,534	48,186	0.4632	22,320
10	30,000	2.1589	64,767		64,767		64,767	− 24,611	40,156	0.4632	18,600

operating cost savings, and market value (in the tenth year), which were estimated in real dollars with $k = 0$ (Column 1), are adjusted to actual dollars (Columns 2 and 3) using the general price inflation rate and Equation 9-1. The maintenance contract amounts for the first 3 years (Column 4) are already in actual dollars (that is, they are unresponsive to general price inflation). The algebraic sum of Columns 3 and 4 equals the before-tax cash flow (BTCF) in actual dollars (Column 5).

In Columns 6, 7, and 8, the depreciation and income tax calculations are shown. The depreciation deductions in Column 6 are based on the MACRS method. The entries in Columns 7 and 8 are calculated as discussed in Chapter 7. The effective income tax rate (t) is equal to $0.34 + 0.06 - (0.34)(0.06) = 0.38$, or 38%. The entries in Column 8 are equal to the entries in Column 7 multiplied by $-t$. The algebraic sum of Columns 5 and 8 equals the ATCF in actual dollars (Column 9). The present worth of the actual dollar ATCF, using $i_c = 15\%$ is

$$\begin{aligned} PW(i_c) &= -\$180{,}000 + \$36{,}050(P/F, 15\%, 1) + \cdots + \$40{,}156(P/F, 15\%, 10) \\ &= \$33{,}790 \end{aligned}$$

Therefore, the project is economically justified.

(b) Next, Equation 9-1 is used to calculate the ATCF in real dollars from the entries in Column 9. The real dollar ATCF (Column 11) shows the estimated economic consequences of the new equipment in dollars that have the constant purchasing power of the base year. The actual dollar ATCF (Column 9) is in dollars that have the purchasing power of the year in which the cost or saving occurs. The comparative information provided by the ATCF in both actual dollars and real dollars helps with interpreting the results of an economic analysis. Also, as illustrated in this example, the conversion between actual dollars and real dollars can be easily done. The present worth of the real dollar ATCF (Column 11) using $i_r = (i_c - f)/(1 + f) = (0.15 - 0.08)/1.08 = 0.06481$, or 6.48%, is

$$\begin{aligned} PW(i_r) &= -\$180{,}000 + \$33{,}379(P/F, 6.48\%, 1) + \cdots + \$18{,}600(P/F, 6.48\%, 10) \\ &= \$33{,}790 \end{aligned}$$

Thus, the present worth of the real dollar ATCF is the same as the present worth calculated previously for the actual dollar ATCF.

9.2.5.1 Substituting for the General Price Inflation Rate

Now we consider the situation where the general price inflation rate is not the best estimate of future price changes for one or more of the types of cash flows in an engineering economic analysis. For these cash flows, the selected price change rates would be substituted in the calculations. For example, assume that the general price inflation rate is not considered the best estimate of price change in market value of the equipment in Example 9-6. Because of the availability of this type of used equipment, an annual change of 2% in the market value is considered the best estimate instead of the projected general price inflation rate (8%). Then the estimated market value at the end of the tenth year, using this selected price change rate instead of the general price inflation rate, is $\$30{,}000(1.02)^{10} = \$36{,}570$. This situation, which is called differential price inflation, is discussed in more detail in Section 9.3; it is discussed here only to demonstrate that it can be easily included in an actual dollar analysis.

9.2.5.2 Variation in the General Price Inflation Rate

In Example 9-6, the general price inflation rate (f) was projected to be 8% each year during the 10-year analysis period. In the case where the estimated annual rates vary during the analysis period, these varying rates would be applied successively to the costs and revenues for the years involved. (Variable general price inflation rates were illustrated in Example 9-2.) For example, assume the annual rates for Example 9-6 were estimated to vary as shown in Column 1 of Table 9-4. Then the operating cost savings and market value originally estimated in real dollars (Column 2), would be adjusted to actual dollars (Column 4) by successive application of the annual rates (as shown in Column 3). For the case where f varies during the analysis period, Equation 9-1 can be modified to show the relationship between actual dollars and real dollars. For a given $k (0 \leq k \leq N)$, the relationship is (where the symbol Π means the product)

TABLE 9–4 Variation in the General Price Inflation Rate

Year n	(1) General Price Inflation Rate (f_n)	(2) R$ Cash Flows	(3) A$ Adjustment $\left[\prod_{l=1}^{n}(1+f_l)\right]$; for $k = 0$	(4) A$ Cash Flows
0	—	−$180,000		−$180,000
1	4.0	36,000	1.04	37,440
2	5.5	36,000	$(1.04)(1.055) = 1.0972$	39,499
3	5.5	36,000	$(1.04)(1.055)^2 = 1.1576$	41,672
4	7.0	36,000	$(1.04)(1.055)^2(1.07) = 1.2386$	44,589
5	7.0	36,000	$(1.04)(1.055)^2(1.07)^2 = 1.3253$	47,710
6	7.0	36,000	$(1.04)(1.055)^2(1.07)^3 = 1.4180$	51,050
7	8.0	36,000	$(1.04)(1.055)^2(1.07)^3(1.08) = 1.5315$	55,134
8	8.0	36,000	$(1.04)(1.055)^2(1.07)^3(1.08)^2 = 1.6540$	59,544
9	8.0	36,000	$(1.04)(1.055)^2(1.07)^3(1.08)^3 = 1.7893$	64,308
10	8.0	36,000	$(1.04)(1.055)^2(1.07)^3(1.08)^4 = 1.9292$	69,440
10		30,000		57,876

$$(R\$)_n = (A\$)_n \left[\prod_{l=n+1}^{k} (1 + f_l) \right] \quad \text{when } n < k$$
$$= (A\$)_n \quad \text{when } n = k \qquad (9\text{-}6)$$
$$= (A\$)_n \left[1 / \prod_{l=k+1}^{n} (1 + f_l) \right] \quad \text{when } n > k$$

In table 9-4, Equation 9-6 was applied in Column 3. In year 0 ($n = 0$), the inverse of the second relationship was used; that is, $(A\$)_0 = (R\$)_0$. For years 1 through 10 ($1 \le n \le 10$), the inverse of the third relationship was used; that is, $(A\$)_n = (R\$)_n \prod_{l=1}^{n} (1 + f_l)$.

9.2.5.3 Calculating an Effective General Price Inflation Rate

In Table 9-4, the projected general price inflation rates varied during the 10-year analysis period. Suppose these rates are the best estimates of future price inflation in your company. However, for a study of a small capital investment project you consider the successive application of the varying annual rates in the analysis to be a refinement not justified by the situation. In this case, the analysis can be simplified by using an effective rate ($\bar{f}$) in the same way that $f = 8\%$ was used in the original solution to Example 9-6. Assume that the analysis period is 10 years for the small project. The calculation of $\bar{f}$ (based on the entries in Column 1 of Table 9-4) would be

$$\bar{f} = \left[\prod_{n=1}^{N} (1 + f_n) \right]^{1/N} - 1 \qquad (9\text{-}7)$$

$$= \left[\prod_{n=1}^{10} (1 + f_n) \right]^{1/10} - 1 = [(1.04)^1 (1.055)^2 (1.07)^3 (1.08)^4]^{0.1} - 1$$
$$= (1.9292)^{0.1} - 1$$
$$= 0.0679, \text{ or } 6.79\%$$

If this approach were applied to the original calculations in Table 9-4, the entries in Column 4 would be slightly different for years 1 through 9. However, the operating cost savings in the tenth year would be $\$36{,}000(1.0679)^{10} = \$69{,}440$, the same value calculated by successive application of the varying annual rates originally used in the table.

9.3

DIFFERENTIAL PRICE INFLATION*

In Section 9.2.5.1 we briefly discussed the situation where the general price inflation rate (f) is not considered the best estimate of future price changes for one or more types of cost and revenue cash flows in an engineering economy study. The variation between the general price inflation rate and the best estimate of future price changes for specific goods and services is called *differential price inflation*, and it is caused by factors such as technological improvements, changes in productivity, and changes in regulatory requirements. Also, a restriction in supply, an increase in demand, or a combination of both may change the value of a particular good or service relative to others. For a good or service affected by some combination of general price inflation and differential price inflation, the resulting future price changes can be represented by a *total price escalation rate*. The differential price inflation rate and the total price escalation rate are further defined as follows:

1. Differential price inflation rate (e'_j): the increment (%) of price change (in the unit price, or cost for a fixed amount), above or below the general price inflation rate, during a time period (normally annual) for good or service j.
2. Total price escalation rate (e_j): The total rate (%) of price change (in the unit price, or cost for a fixed amount) during a time period (normally annual) for good or service j. The total price escalation rate for a good or service includes the effects of both the general price inflation rate (f) and the differential price inflation rate (e'_j) on price changes.

The PPI (total for industrial commodities) is an available measure of general price inflation, like the Consumer Price Index (CPI), that may be appropriate for some applications. This index includes a number of different commodity classifications and represents the price changes that occur in large quantity purchases. The effective annual rate (%) of change in the PPI for all industrial commodities by 3-year period (during 1967–87; calculated using Equation 9-7), is shown in the first column of Table 9-5. The rates for four PPI commodity groups are also shown in Table 9-5. A comparison of these annual effective PPI rates for each 3-year period illustrates the effect of differential price inflation.

*An introductory course in engineering economy may not have time to cover the remainder of this chapter.

TABLE 9–5 Effective Annual Rate (%) of Change in Selected Price Indexes, by 3-Year Period, 1967–87

	Effective Annual Rate of Change in the PPI[a]				
3-Year Period	Total for Industrial Commodities (%)	Fuels and Related Products (%)	Lumber and Wood Products (%)	Machinery and Equipment (%)	Chemicals (%)
1967–69	2.5	1.1	7.7	3.2	0.2
1970–72	3.6	5.5	4.8	3.5	1.4
1973–75	13.3	27.4	7.0	11.0	20.3
1976–78	6.9	9.6	16.0	6.7	3.1
1979–81	13.2	29.1	2.0	10.3	13.1
1982–84	2.0	− 1.8	1.6	11.3	1.5
1985–87	− 0.3	− 9.5	1.4	1.6	1.1

[a] Source of PPI data: Economic Report of the President, U.S. Government Printing Office, February, 1988.

The basic relationship among the total price escalation rate (e_j) and the differential price inflation rate (e'_j) for a good or service, and the general price inflation rate (f), is

$$1 + e_j = (1 + f)(1 + e'_j) \tag{9-8}$$

or

$$e_j = f + e'_j + (f)(e'_j) \tag{9-9}$$

and

$$e'_j = \frac{e_j - f}{1 + f} \tag{9-10}$$

Thus, we see in Equation 9-9 that the total price escalation rate (e_j) for good or service j is the sum of the general price inflation rate and the differential price inflation rate, plus their product. As shown in Equation 9-10, the differential price inflation rate (e'_j) can be calculated from the total price escalation rate and the general price inflation rate.

In practice, the general price inflation rate (f) and the total price escalation rate (e_j) for each good and service involved are usually estimated for the analysis period. For each of these rates, different values may be used for subsets of periods within the analysis period if justified by the available data. This was illustrated in Table 9-4 for the general price inflation rate. The differential price inflation rates (e'_j) normally are not estimated directly but calculated using Equation 9-10.

In applying the total price escalation rates in an actual dollar analysis to calculate future prices, we need to keep in mind that these rates are effective rates per period and have a compounding effect when applied sequentially over time. Equation 9-11 relates the total price escalation rate (e_j) with the single payment compound amount factor (when the e_j value for a good or service does not vary by time period):

$$A\$_{n,j} = A\$_{k,j}(F/P, e_j\%, n - k) \tag{9-11}$$

for a given k, where $A\$_{n,j}$ is the projected price in actual dollars in time period n for cash flow j, and $A\$_{k,j}$ is the price in actual dollars in the base time period, k. Similarly, in a R$ analysis we would use the differential price inflation rate (e'_j) to calculate future unit prices for a good or service:

$$R\$_{n,j} = R\$_{k,j}(F/P, e'_j\%, n - k) \tag{9-12}$$

for a given k. Or we could convert projected actual dollar prices, if available, to real dollars using Equation 9-2.

EXAMPLE 9-7

Suppose that the general price inflation rate (f) is expected to average 8% per year during the next 4 years. For a certain commodity, the differential price inflation rate (e'_j) has been estimated directly and is projected to be +2%. What will the marketplace prices be for this commodity over the next 4 years if it is now selling for \$6.30 per unit ($k = 0$)?

Solution

By utilizing Equations 9-9 and 9-11, we can determine the A$ selling price of this commodity in view of its e_j estimate:

$$1 + e_j = (1.08)(1.02) \qquad \text{or} \qquad e_j = (1.08)(1.02) - 1 = 0.1016$$

End of Year	Selling Price per Unit (A$)
1	\$6.30(F/P, 10.16%, 1) = \$6.94
2	6.30(F/P, 10.16%, 2) = 7.65
3	6.30(F/P, 10.16%, 3) = 8.42
4	6.30(F/P, 10.16%, 4) = 9.28

EXAMPLE 9-8

In Example 9-7, determine what the real dollar prices will be for the commodity over the next 4 years by utilizing Equation 9-12.

Solution

End of Year	Selling Price per Unit (R$)
1	$6.30(F/P, 2%, 1) = $6.43
2	6.30(F/P, 2%, 2) = 6.55
3	6.30(F/P, 2%, 3) = 6.69
4	6.30(F/P, 2%, 4) = 6.82

9.4 MODELING PRICE CHANGES WITH GEOMETRIC CASH FLOW SEQUENCES

In Chapter 3 (Section 3.16), we discussed equivalence calculations involving projected cash flow patterns that are increasing at an effective rate, $\bar{f}$, per period. When total price escalation is included in an engineering economic analysis, projected prices of goods and services often can be modeled as increasing at a constant rate per period. Thus, the resulting end-of-period cash flow pattern is often a geometric sequence (see Section 3.16).

In Section 9.2.2, the correct interest rate used in an engineering economic analysis was shown to depend on the dollar terms of the cost and revenue cash flows; specifically, the combined interest rate (i_c) is used in an actual dollar analysis, and the real interest rate (i_r) is used in a real dollar analysis. An additional question is what $\bar{f}$ value is used for each method of analysis when cost escalation is included and a geometric sequence cash flow model is appropriate. In the following table, we see that $\bar{f}$ is equal to e_j in an A$ analysis, and equal to e'_j in a R$ analysis:

Method	Cash flows	Interest Rate (i)	Geometric Gradient ($\bar{f}$)
A	Actual Dollars (A$)	i_c	e_j
B	Real Dollars (R$)	i_r	e'_j

From this it follows that the "convenience rate" (Section 3.16) needed to calculate the present worth of a geometric cash flow sequence involving cost escalation would be

$$\textit{A\$ Analysis} \qquad\qquad \textit{R\$ Analysis}$$

$$i_{CR} = \frac{i_c - e_j}{1 + e_j} \qquad\qquad i_{CR} = \frac{i_r - e'_j}{1 + e'_j} \tag{9-13}$$

EXAMPLE 9-9

A water service public utility district is considering some new replacement pumping equipment to reduce operating costs and improve service reliability. The base time period is the present, or year 0 ($k = 0$). The estimated annual saving in year 0 dollars is $78,000. The utility district uses an 8-year analysis period for this type of equipment replacement study; the general price inflation rate is projected to be 4.6% per year; the total price escalation rate (e_j) for operating costs is projected to be 6.2% per year; the utility district is using a MARR (which includes the effect of general price inflation) of 9.5% per year; the old equipment has no net market value; and there are no taxes involved. Based on these estimates, calculate the maximum amount that could be paid for the equipment now (a) using an A$ analysis, and (b) using a R$ analysis.

Solution

(a) A$ analysis:

$$i_c = \text{MARR (as given)} = 9.5\%,\ \bar{f} = e_j = 6.2\%;\ N = 8$$

$$i_{CR} = \frac{i_c - e_j}{1 + e_j} = \frac{0.095 - 0.062}{1.062} = 0.03107, \text{ or } 3.11\%$$

From the data given, the annual savings of $78,000(1.062)^n$ for $1 \leq n \leq 8$ constitute a geometric cash flow sequence, as discussed in Section 3.16. By using Equation 3-20, we can write

$$\begin{aligned} \text{PW}(3.11\%) \text{ of savings} &= \$78{,}000(P/A, 3.11\%, 8) \\ &= \$78{,}000\left[\frac{(1.0311)^8 - 1}{0.0311(1.0311)^8}\right] = \$545{,}000 \end{aligned}$$

which is the maximum amount that ought to be paid for the equipment.

(b) R$ analysis

$$i_r = \text{MARR(Real)} = \frac{i_c - f}{1 + f} = \frac{0.095 - 0.046}{1.046} = 0.04685$$

$$\bar{f} = e'_j = \frac{e_j - f}{1 + f} = \frac{0.062 - 0.046}{1.046} = .01530$$

$$i_{CR} = \frac{i_r - e'_j}{1 + e'_j} = \frac{0.04685 - 0.0153}{1.0153} = 0.03107, \text{ or } 3.11\%$$

At this point in the R$ analysis we see that the convenience rate value is the same value calculated for the A$ analysis, and the PW of the savings would thus be the same. This should make intuitive sense because we know the present worth of a cash flow is the same in the base time period using A$ analysis or R$ analysis (Section 9.2.5, Example 9-6). Thus, the present worth of the annual savings for the new pump would be the same ($545,000) when using Method A or B.

EXAMPLE 9-10

As an individual homeowner, suppose that you are interested in purchasing a heat pump in early 1989 to replace your present heating and air-conditioning system. You calculate that on the average 24,000 kWh of electricity will be required to heat and cool your home with this heat pump. In early 1989 the cost per kWh to you is 4.5 cents, and this cost is expected to increase by 15% per year (e_j) over the next 15 years. During the same period, the general price inflation rate is expected to average 6% per year. If your present system averages 30,000 kWh of electrical consumption each year and your MARR is $10\%(=i_c)$, how much could you afford to spend now for a heat pump?

Solution

Savings per year in early 1989 dollars equal

$$\frac{6{,}000 \text{ kWh}}{\text{year}}(\$0.045/\text{kWh}) = \$270/\text{year}$$

Because these savings will escalate at 15% per year for 15 years, we need to calculate the present worth of this geometric cash flow sequence to determine how much we can afford to spend now (see also Figure 9-3):

$$i_{CR} = \frac{0.10 - 0.15}{1.15} = -0.0435, \text{ or } -4.35\%$$

$$PW = \$270(P/A, -4.35\%, 15)$$
$$= \$5{,}888$$

Thus, when the annual total price escalation rate of a commodity such as electricity exceeds the MARR, a negative convenience rate will result, as demonstrated in this example.

9.5 ACTUAL DOLLAR VERSUS REAL DOLLAR ANALYSIS

In practice, should A$ analysis or R$ analysis be utilized when total price escalation is included in an engineering economic analysis? To help answer

FIGURE 9-3

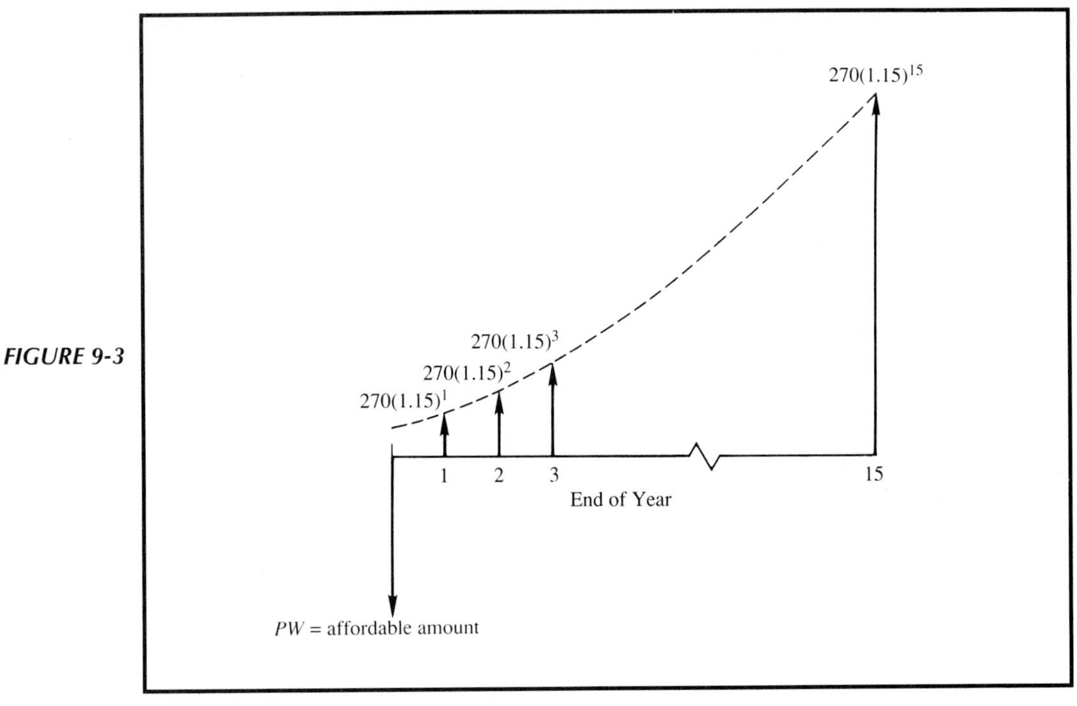

A$ Cash Flow Diagram For Example 9-10

this question, we shall discuss some guidelines that are useful in application. Four situations regarding the significance of the general price inflation and the differential price inflation components of total price escalation are as follows:

	Significance of Total Price Escalation Rate Components	
Situation	**General Price Inflation (f)**	**Differential Price Inflation (e_j)**
1	Insignificant	Insignificant
2	Significant	Insignificant
3	Insignificant	Significant
4	Significant	Significant

If Situation 1 is assumed, based on present and prospective market conditions for the goods and services involved, the analysis would be based on a projection of insignificant price escalation. Under this situation, the purchasing power of an A$ and a R$ would be the the same during all time periods (not just the base time period), and the combined (nominal) interest rate would equal the real interest rate ($i_c = i_r$). Thus, in Situation 1, an actual or a real dollar analysis would be the same. In this book, we have implicitly assumed Situation 1 prior to this chapter.

Under Situation 2, significant price escalation is assumed in the market, but the rate is identical for all goods and services. That is, total price escalation is the result of general price inflation (f) only, and no significant differential price inflation is projected. Under this assumption, price escalation can be disregarded in a before-tax analysis if no fixed (A$) annuity cash flows unresponsive to general price inflation were involved (such as lease fees, interest charges on loans, some taxes, etc.). For after-tax analysis, it would be preferable to use A$ analysis, with the general price inflation component included, since depreciation quantities (which are in actual dollars) would be involved. However, it is feasible to use R$ analysis for either before-tax or after-tax studies under this assumption if proper procedures are followed.

EXAMPLE 9-11

Consider a project requiring an investment of $20,000 which is expected to return, in terms of actual dollars, $6,000 at the end of the first year, $8,000 at the end of the second year, and $12,000 at the end of the third year. The general price inflation rate is 5% per year, and the real interest rate is 10% per year. Compare the PW of this project using before-tax A$ analysis and R$ analysis.

Solution

From Equation 9-4, the combined interest rate is 15.5%. Table 9-6 shows how the present worth of the project would be calculated to be −$1,020 using A$ analysis.

If R$ analysis is used, the cash flows should be estimated in terms of real dollars. For the data given and a general price inflation rate of 5%, if the estimator is precisely consistent, the projections in terms of real dollars (as of time of investment) should be −$20,000, $5,714, $7,256, and $10,366 for each year, respectively. Table 9-7 shows how, using a real interest rate of 10% (Equation 9-5), the net present worth of the project would be calculated to be the same (−$1,020) as when using A$ analysis.

Situation 3 involves the case where all price escalation in the analysis is the result of the differential price inflation component. The result is that the total price escalation rate for each good or service would equal the differential price

TABLE 9–6 Calculation of Present Worth with Estimates in Actual Dollars

Year, n	Outcome (A$)	$(P/F, 15.5\%, n)$	Present Worth (Rounded)
0	−$20,000	1.0000	−$20,000
1	6,000	0.8658	5,195
2	8,000	0.7496	5,997
3	12,000	0.6490	7,788
			$\sum \simeq$ −$ 1,020

TABLE 9–7 Calculation of Present Worth With Estimates in Real Dollars

Year n	Outcome ($R\$$)	(P/F, 10%, n)	Present Worth (Rounded)
0	−$20,000	1.0000	−$20,000
1	5,714	0.9091	5,195
2	7,256	0.8264	5,997
3	10,366	0.7513	7,788
			$\sum \simeq -\$ 1{,}020$

inflation rate ($e_j = e'_j$), since the general price inflation rate (f) is projected to equal 0. Under this assumption, the purchasing power of an A$ and a R$ is the same during all time periods, and the combined (nominal) interest rate equals the real interest rate ($i_c = i_r$). Thus, in Situation 3, price escalation needs to be included in an analysis because failure to do so could result in an erroneous economic decision, and an actual dollar and a real dollar analysis would be the same.

EXAMPLE 9-12

A company that routinely estimates cash flows in real dollars is attempting to evaluate a new venture. Revenues are not expected to escalate since the general price inflation rate (f) is expected to average 0% per year. The two basic cost components are labor and energy, with annual differential price inflation rates of 4% and 5%, respectively. The capital investment required for the new venture is $10,000, and the company's real MARR is 4% per year. Should the proposed venture be undertaken, based on before-tax analysis, in view of the following additional data developed by the company?

$$\text{Base time period} = \text{year } 0\ (k = 0);\ \text{Analysis period} = 5 \text{ years } (N = 5)$$

Cash Flow	*(R$) Amounts/Year*
Revenue	$5,000
Labor	1,000
Energy	1,700

Company Solution

Because the two annual differential price inflation rates were about equal, analysts in this company decided to ignore differential inflation. Their R$ analysis of present worth is the following:

$$\begin{aligned} \text{PW} &= -\$10{,}000 + (5{,}000 - 2{,}700)(P/A, 4\%, 5) \\ &= \$239.19 \end{aligned}$$

Thus, it is concluded that the new venture is marginally attractive.

TABLE 9–8 Example 9-12 with Differential Price Inflation Included

	Annual Cash Flows (R$)		
End of Year	Revenue	Labor (e'_L = 4%)	Energy (e'_E = 5%)
0	0	0	0
1	$5,000	$-\$1{,}000(1.04)^1 = -1{,}040$	$-\$1{,}700(1.05)^1 = -1{,}785$
2	5,000	$-\ 1{,}000(1.04)^2 = -1{,}082$	$-\ 1{,}700(1.05)^2 = -1{,}874$
3	5,000	$-\ 1{,}000(1.04)^3 = -1{,}125$	$-\ 1{,}700(1.05)^3 = -1{,}968$
4	5,000	$-\ 1{,}000(1.04)^4 = -1{,}170$	$-\ 1{,}700(1.05)^4 = -2{,}066$
5	5,000	$-\ 1{,}000(1.04)^5 = -1{,}217$	$-\ 1{,}700(1.05)^5 = -2{,}170$

End of Year	BTCF	
0	−$10,000	
1	2,175	
2	2,044	
3	1,907	PW (4%) = −$1,490
4	1,764	
5	1,613	

Solution Including Differential Price Inflation
The differential price inflation on revenues is 0%. However, the differential price inflation on labor (e'_L) and energy (e'_E) are

$$e'_L = \frac{0.04 - 0}{1.0} = 0.04; \quad e'_E = \frac{0.05 - 0}{1.0} = 0.05$$

The net before-tax cash flow (BTCF) is determined in Table 9-8.
With differential price inflation taken into consideration, the venture changes from a marginally attractive (PW = $239) to an unattractive (PW = −$1,490) situation. Thus, not recognizing the potential impact of differential price inflation would result in erroneous results. A particular problem that should have been recognized by the company analysts was the large difference between the revenue differential price inflation rate (0%) and the two cost rates. We should also note that the e'_j values calculated for labor and energy were the same as their e_j values since the general price inflation rate, f, is equal to 0. Also, $i_r = i_c$. That is why, under Situation 3, actual dollar and real dollar analysis are the same.

In Situation 4, both the general price inflation (f) and the differential price inflation (e'_j) components of total price escalation are estimated to be significant. Under this assumption (as in Situation 2), the purchasing power of an A$ and a R$ is not the same except in the base time period (k). Also, the combined (nominal) interest rate would not equal the real interest rate ($i_c \neq i_r$). Thus, under Situation 4, price escalation needs to be included in an analysis since failure to do so could result in an incorrect economic decision. Also, A$ analysis or R$ analysis could be used for either before-tax or after-tax studies if proper procedures are followed.

EXAMPLE 9-13

In Example 9-12, a R$ before-tax analysis was used in a company to analyze a potential venture. Now assume that $f = 6\%$ instead of 0%, the revenue price escalation rate is 11% ($e_R = 11\%$ instead of 0%), and the energy escalation rate is 9%($e_E = 9\%$ instead of 5%). All other estimates remain the same as those used in Example 9-12. With these changes the venture will be re-analyzed as a Situation 4 problem. First, another company solution will be obtained. Then, an economic analysis that includes total price escalation will be illustrated.

Company Solution

The company analysts again disregard any price escalation. Also, in this solution, they commit Mistake No. 1 (see Section 9.2.3). That is, when no price escalation is assumed they used a MARR $= i_c$ which includes the effect of general price inflation.

$$\begin{aligned} i_c &= i_r + f + (i_r)(f) \\ &= 0.04 + 0.06 + (0.04)(0.06) = 0.1024, \text{ or } 10.24\% \end{aligned}$$

$$\begin{aligned} \text{PW} &= -\$10{,}000 + (5{,}000 - 2{,}700)(P/A, 10.24\%, 5) \\ &= -\$10{,}000 + 2{,}300(3.7677) \\ &= -\$1{,}334 \end{aligned}$$

Thus, the venture does not appear economically attractive. If this company solution had used the appropriate MARR ($i_c = i_r = 4\%$) for the assumption actually made the PW of the before-tax cash flow would be the same as the company solution in Example 9-12 (PW = \$239), or marginally attractive from an economic perspective.

Solution Including Price Escalation

In this solution, an A$ analysis is used (for which a MARR $= i_c = 10.24\%$ is correct). The cost and revenue price escalation rates are

$$e_R = 11\%; \quad e_L = 4\%; \quad e_E = 9\%$$

Applying these price escalation rates to the fixed annual amounts at $k = 0$ for revenues = \$5,000, labor = \$1,000, and energy = \$1,700 over the 5-year analysis period results in the following cash flow table:

End of Year	(e_R = 11%) Revenues	(e_L = 4%) Labor	(e_E = 9%) Energy	BTCF
0	$ 0	$ 0	$ 0	−$10,000
1	5,550	−1,040	−1,853	2,657
2	6,160	−1,082	−2,020	3,058
3	7,590	−1,170	−2,400	4,020
4	8,425	−1,217	−2,616	4,592
5	9,352	−1,266	−2,851	5,235

The present worth of the before-tax cash flow (BTCF) is

$$\begin{aligned} PW &= -\$10{,}000 + \sum_{n=1}^{5} BTCF_n(P/F, 10.14\%, n) \\ &= -\$10{,}000 + \$2{,}657\left(\frac{1}{1.1014}\right)^1 + \cdots + \$5{,}235\left(\frac{1}{1.1014}\right)^5 \\ &= \$4{,}292 \end{aligned}$$

From this result, we see that the venture is economically attractive when procedures appropriate for Situation 4 are used. The same result (PW = $4,292) would be obtained from a R$ analysis using the correct procedures.

9.5.1 Summary of Guidelines

A summary of the guidelines previously discussed for Situations 1 through 4 regarding actual dollar versus real dollar analysis is shown in Table 9-9. If either the general price inflation rate or differential price inflation rate is considered significant during the analysis period, price escalation should be included in an engineering economy study (except for Situation 2, when a before-tax study is being done and no fixed cash flow annuities are involved).

The effects of inflation and price changes can be incorporated into the computer programs used to support engineering economic analyses within an organization. This way, the effects of price escalation can be routinely included. This capability is included in the BASIC computer program described in Appendix F.

9.5.2 Application Strategy

In two of the situations (1 and 3) summarized in Table 9-9, A$ and R$ analysis are the same since the general price inflation rate (f) is projected to be 0%.

TABLE 9–9 Summary of Guidelines

Situation	General Price Inflation (f)	Differential Price Inflation (e'_j)	Guidelines
1	Insignificant	Insignificant	Disregard price escalation. A$ and R$ analyses are the same since the general price inflation rate (f) is equal to 0.
2	Significant	Insignificant	May disregard price escalation in before-tax studies if fixed annuities are not involved. Otherwise, include price escalation in before-tax and after-tax studies.
3	Insignificant	Significant	Include price escalation in before-tax and after-tax studies. A$ and R$ analyses are the same since the general price inflation rate (f) is equal to 0.
4	Significant	Significant	Include price escalation in before-tax and after-tax studies.

Therefore, we do not need to decide which analysis method to use. But for the other two situations (2 and 4), which are common in professional practice, either the A$ or R$ analysis method may be used. Both methods, properly applied, result in the same present worth for a cash flow in the base time period, require the same amount of information, and do not have a practical difference in application effort.

However, there is some difference in the information content available for interpreting the economic results. The results from an A$ dollar analysis are in market purchasing power that varies with time, while the results from a R$ analysis are in constant market purchasing power defined by the base time period (k). Thus, R$ analysis provides information in terms of a constant unit of measure, while A$ analysis provides information on the actual quantities that will occur during the analysis period.

An analysis strategy that works well in professional practice is to use A$ analysis for both before-tax and after-tax studies, then, at the end of an analysis, to use Equations 9-1, 9-2, or 9-6 to provide selected cash flows (particularly the net cash flows) in R$ terms (this was demonstrated in Example 9-6). This strategy provides the additional information content that is useful with small incremental effort. In some organizations, a specific method of analysis may be specified, but, even then, selected cash flows can be easily converted to the other analysis form to assist with interpretation of the results. This strategy will be further illustrated in Section 9.7 (Example 9-15).

9.6
EXAMPLES OF PRICE CHANGE FORECASTS

In Section 9.1, we identified three general price indexes tabulated by the U.S. Government. These were the Consumer Price Index (CPI), Producers Price Index (PPI), and the Implicit Price Index for the Gross National Product (IPI-GNP). A number of other price indexes are routinely tabulated by the federal government from the data it collects and analyzes. Thus, a large amount of data is readily available about past and current price changes.

The U.S. Government does not publish forecasts of price changes. However, these forecasts may be purchased from private consulting firms engaged in the business of providing economic forecasting services.* An example of the type of forecast price change data available is given in Table 9-10. The example data shown in the table are for twelve construction material items. Similar data, which are periodically updated, are available for general price changes, as well as for most materials, goods, and services.

Many organizations use available forecasts of price changes directly in their engineering economy studies, or to assist in the preparation of their own forecasts. An example of forecast price changes prepared within an organization, using both internal and external information sources, is shown in Table 9-11.

9.7
CASE STUDIES

9.7.1 A Process Industry Case Study Including Inflation

EXAMPLE 9-14

The R Square Corporation is considering two plans, A and B, for expanding the capacity of its urea manufacturing plant. The relevant data and analyses for both plans are given below.

PLAN A

R Square Corporation's division of agricultural and chemical development operates a urea plant that was built in 1974 at a cost of $4.3 million and capacity of 300,000 lb/yr. Due to increasing use of urea-based fertilizers and

*For example, forecasts of price changes may be purchased from Data Resources, Inc. and from Wharton Econometric Associates, Inc.

TABLE 9–10 Example of Price Change Forecasts (for Materials, Goods, and Services) Available from an Economic Forecasting Service[a]

Construction Material/Component	Forecast Annual Price Change (%) from Previous Year											
	89	90	91	92	93	94	95	96	97	98	99	2000
Hardware	4.3%	3.2%	2.8%	3.8%	3.8%	3.7%	4.0%	4.2%	4.5%	4.0%	4.3%	4.2%
Plumbing Fixtures	4.3	2.8	3.7	4.1	3.7	4.3	4.2	4.5	4.4	3.9	4.0	4.6
Heating Equipment	4.8	3.9	3.7	3.9	3.6	3.9	4.1	4.2	4.4	4.1	4.1	4.0
Metal Tanks	5.3	4.6	4.4	5.4	5.5	5.3	5.9	5.8	5.8	6.0	6.7	7.0
Sheet Metal Products	2.0	1.4	2.2	3.3	4.0	3.8	4.2	4.0	4.9	5.0	4.9	4.7
Structural Metal Products	3.7	4.2	3.0	5.5	6.0	6.1	5.3	5.9	6.2	5.9	5.9	6.2
Miscellaneous Metal Products	5.4	3.5	3.7	4.9	5.2	5.6	5.7	6.0	5.8	6.1	5.9	5.9
Plastic Construction Products	1.7	3.2	2.6	4.0	4.6	4.8	4.9	5.2	5.0	5.6	5.9	5.9
Concrete Products	4.5	1.9	5.8	6.3	6.0	5.6	6.6	6.4	6.5	6.8	7.0	7.4
Asphalt Roofing	6.4	5.8	7.1	6.6	6.6	7.7	8.1	8.1	8.2	7.2	7.0	6.9
Insulation Materials	4.7	4.4	5.3	5.0	4.9	4.9	5.8	6.5	7.1	7.3	7.5	7.2
Bituminous Paving Materials	7.8	6.8	7.4	8.0	8.2	8.9	9.4	9.3	9.7	9.4	9.8	10.1

[a] *Source:* "U.S. Cost Forecasting Service Long-Term Review," Volume 13, Number 7 (Data Resources, Inc., Cost Forecasting Services, Washington, D.C.), October, 1987, Table A7. The extracted data represent only 12 of the 35 construction material and component items in Table A7.

TABLE 9–11 Example of Price Change Forecasts Developed within an Organization[a]

	Projected Annual Price Change (%) from Previous Year				
Economic Measure	**1987**	**1988**	**1989**	**1990–94**	**1995 +**
General Price Inflation					
Consumer Price Index (CPI)	3.5%	4.5%	4.5%	5.0%	5.5%
Producer Price Index (PPI)					
All Commodities	2.5	3.5	4.0	5.0	5.0
Industrial commodities	2.5	3.5	4.0	5.0	5.0
Total Price Escalation					
Petroleum Products	7.0	4.0	4.0	7.0	10.0
Labor (average)	4.1	4.3	4.3	4.3	5.5
Construction Machinery and Equipment	2.0	4.5	5.5	5.5	6.0
Concrete	1.0	3.0	4.0	5.5	6.0
Iron and Steel	1.5	3.5	4.0	4.5	5.0
Nonferrous Metals	2.5	4.0	4.0	5.5	6.0
Chemicals and Allied Products	3.0	4.0	4.0	5.5	5.5
Nonelectric Machinery	2.0	4.0	5.0	5.0	5.5
Electric Machinery	2.0	4.0	4.5	5.0	5.5
Communication Services	3.0	4.0	5.0	6.0	6.0

[a] *Source:* "Economic Outlook" (Appendix I), Tennessee Valley Authority, 1987. The terms "General Price Inflation" and "Total Price Escalation" were added by the authors.

increasing maintenance costs on the existing unit, the feasibility of constructing a new plant with 750,000 lb/yr capacity is being studied. Time does not permit an in-depth cost estimation for the new plant, so engineers decide to obtain the cost of the new plant by scaling up the cost of the old plant. The cost-capacity factor is known to be 0.65, and the construction-cost index has increased an average of 10.5% per year for the past 15 years. What cost should the engineers report to the project review committee?

Urea plant estimates:

$C_0 = \$4.3$ million: cost of plant in 1974
$S_0 = 300{,}000$ lb/yr: capacity of existing plant
$C_N = ?$: cost scaled up (but based on 1974 pricing)
$S_N = 750{,}000$ lb/yr: capacity of plant being studied
$C_{NE} =$ cost reported to committee in 1989 dollars
$X =$ cost-capacity factor $= 0.65$

Solution

By using Equation 8-5, we can determine C_N:

$$C_N = C_0\left(\frac{S_N}{S_0}\right)^X$$

$$C_N = \$4.3 \text{ million} \left(\frac{750{,}000 \text{ lb/yr}}{300{,}000 \text{ lb/yr}} \right)^{0.65}$$

$$C_N = \$7.8 \text{ million (cost basis in 1974)}$$

$$C_{NE} = C_N(F/P, 10.5\%, 15) = \$7.8 \text{ million}(1.105)^{15}$$

$$= \$34.88 \text{ million (estimated cost of new urea unit in 1989 dollars)}$$

Plan B

As an alternative plan, it is learned that a manufacturer of these units can prefabricate and install a unit on-site for a total cost of $22 million. The R Square Corporation's constructed unit has an estimated life of 15 years and the prefabricated unit has a life of 12 years. The real dollar (1989) operating costs for the company unit are $40,000 per year for years 1–10 and $30,000 per year for years 10–15. The real dollar operating costs for the prefabricated unit are $50,000 per year. The real interest rate is assumed to be $i_r = 10\%$. Which alternative should they select? What assumptions are involved?

Solution

Note: Assume negligible salvage value for both alternatives, and compare the alternatives using the annual worth method.

	Prefabricated Unit	Company Unit
Initial cost	$22,000,000	$34,880,000
Annual costs	$50,000/yr	$40,000/yr, years 1–10 $30,000/yr, years 10–15
Life	12 years	15 years

Prefabricated unit:

$$\begin{aligned} \text{AW} &= -\$50{,}000 - \$22{,}000{,}000(A/P, 10\%, 12) \\ &= -\$3{,}279{,}600 \end{aligned}$$

Company unit:

$$\begin{aligned} \text{AW} &= -\$30{,}000 - \$10{,}000(P/A, 10\%, 10)(A/P, 10\%, 15) \\ &\quad - \$34{,}880{,}000(A/P, 10\%, 15) \\ &= -\$4{,}893{,}884 \end{aligned}$$

Based on annual worth, the R Square Corporation should let the outside manufacturer construct and install the urea unit. However, numerous non-monetary considerations could shift the decision to plan A. The monetary risks associated with expanding in an uncertain and highly competitive market may well cause neither plan to be acceptable. An indefinitely long study period has been assumed in the AW analysis above.

9.7.2 A Case Study Dealing with Price Escalation in After-Tax Studies

In most engineering economy studies conducted in industry, price escalation on revenues and costs has to be considered in addition to various income tax provisions set forth by the Tax Reform Act of 1986. To illustrate this situation, a fairly comprehensive and realistic case study is presented.

EXAMPLE 9-15

A potential investment opportunity is being considered for the beginning of 1989 that requires the investment of $20,000 in added production control equipment to increase output of an assembly line. As a result of this investment, the revenue obtained from the modified assembly line is expected to increase. The following information applies to the investment opportunity.

Analysis period	10 years
Estimated useful life of the equipment	10 years
MACRS class life	5 years
Federal income tax rate	34%
State income tax rate	7.5%
Effective income tax rate (t)	39% = [0.34 + 0.075 − (0.34)(0.075)]100
Real after-tax MARR (i_r)	6%
General price inflation rate (f)	8%
Combined discount rate (i_c)	14.48% = [0.06 + 0.08 + (0.06)(0.08)]100
Increased revenue (assume that revenue escalates at the general price inflation rate of 8%)	$15,000 per year in real dollars
Market value in 10 years	10% of the investment cost in terms of real dollars

Annual Cost:

Category	Cost (R$)	Price Escalation Rate (e_j)
Material	$1,200	10%
Labor	2,500	5.5%
Energy	2,500	15%
Other costs	500	8%

Leased equipment is also required, which can be obtained for the first 5 years at a rate of $800 per year. The contract will be renegotiated at the beginning of the sixth year at an escalated value based on the general price inflation rate.

Assume the base time period (k) is the beginning of 1989 (end of year 0 in Table 9-12).

TABLE 9–12 Actual Dollar Cash Flow Analysis (with Conversion to Real Dollar ATCF) for Example 9-15

A $ After-Tax Analysis														R$ ATCF	
End-of-Year	Revenue	Initial Investment	Material	Labor	Energy	Other Cost	Leased Equipment	Total Cost	Before-Tax Cash Flow	Depreciation	Taxable Income	Cash Flow from Income Tax	A$ After-Tax Cash Flow	R$ Adjustment Factor $(1/1.08)^n$	R$ After-Tax Cash Flow
0		−$20,000						−$20,000	−$20,000				−$20,000	1.0	−$20,000
1	$16,200		−$1,320	−$2,638	−$ 2,875	−$ 540	−$ 800	− 8,173	8,028	$4,000	$ 4,028	−$1,571	6,457	0.92593	5,979
2	17,496		− 1,452	− 2,783	− 3,306	− 583	− 800	− 8,924	8,572	6,400	2,172	− 847	7,725	0.85734	6,623
3	18,896		− 1,597	− 2,936	− 3,802	− 630	− 800	− 9,765	9,131	3,840	5,291	− 2,063	7,068	0.79383	5,611
4	20,407		− 1,757	− 3,097	− 4,373	− 680	− 800	− 10,707	9,700	2,304	7,396	− 2,884	6,816	0.73503	5,010
5	22,040		− 1,933	− 3,267	− 5,028	− 735	− 800	− 11,763	10,277	2,304	7,973	− 3,109	7,168	0.68058	4,878
6	23,803		− 2,126	− 3,447	− 5,783	− 793	− 1,175	− 13,324	10,479	1,152	9,327	− 3,638	6,841	0.63017	4,311
7	25,707		− 2,338	− 3,637	− 6,650	− 857	− 1,175	− 14,657	11,050		11,050	− 4,310	6,740	0.58349	3,933
8	27,764		− 2,572	− 3,837	− 7,648	− 925	− 1,175	− 16,157	11,607		11,607	− 4,527	7,080	0.54027	3,825
9	29,985		− 2,830	− 4,048	− 8,795	− 1,000	− 1,175	− 17,846	12,139		12,139	− 4,734	7,405	0.50025	3,704
10	32,384		− 3,112	− 4,270	− 10,114	− 1,079	− 1,175	− 19,751	12,633		12,633	− 4,927	7,706	0.46319	3,569
10	4,318[a]								4,318		4,318[b]	− 1,684	2,634	0.46319	1,220
												$PW(i_c = 14.48\%) =$	$16,780		
														$PW(i_r = 6\%) =$	$16,780

[a] Estimated market value.
[b] Recovery of depreciation—taxed as ordinary income.

Perform an analysis of the present worth of this project using the appropriate income tax regulations and including the effects of total price escalation: (a) conduct an actual dollar after-tax analysis (and calculate the PW of the ATCF); (b) convert the actual dollar ATCF to a real dollar ATCF; and (c) calculate the PW of the real dollar ATCF and show that it is equivalent to the PW of the actual dollar ATCF.

Solution

(a) Actual dollar analysis: The adjustments required are described below ($k = 0$ in this case study).

1. ***Revenue***: The revenue of \$15,000 per year must be increased each year by the general price inflation rate.

$$\text{revenue in year } n = \$15{,}000(1.08)^n$$

2. ***Material, labor, energy, and other costs***: These annual costs in year n are increased each year by the appropriate total price escalation rate (e_j):

$$\begin{aligned} \text{material} &= -\$1{,}200(1.1)^n \\ \text{labor} &= -\$2{,}500(1.055)^n \\ \text{energy} &= -\$2{,}500(1.15)^n \\ \text{other cost} &= -\$500(1.08)^n \end{aligned}$$

3. ***Leased property***: The lease will be adjusted at the end of year 5 to account for 5 years of general price inflation at 8% per year:

$$\text{lease (years 6–10)} = -\$800(1.08)^5 = -\$1{,}175$$

4. ***Depreciation***: The MACRS depreciation is as follows:

Year n	Initial Cost	MACRS Recovery Percentages	MACRS Depreciation (A\$)
1	\$20,000	0.2000	\$4,000
2	20,000	0.3200	6,400
3	20,000	0.1920	3,840
4	20,000	0.1152	2,304
5	20,000	0.1152	2,304
6	20,000	0.0576	1,152

5. ***Market value***: The 10% market value is a real-dollar amount and must be increased to account for the 8% annual general price inflation rate.

$$\text{market value} = 0.1(\$20{,}000)(1.08)^{10} = \$4{,}318$$

6. ***Recaptured Depreciation***: The market value in A\$ of \$4,318 represents the recovery of depreciation and is taxed as ordinary income at the 39% rate.

The actual dollar, after-tax analysis is shown in Table 9-12. The PW of the actual dollar, after-tax cash flow (ATCF), using $i_c = 14.48\%$, is $16,780.

(b) Real dollar, after-tax cash flow: the conversion of the actual dollar ATCF to a real dollar ATCF is shown in the last two columns in Table 9-12. Equation 9-1, with $k = 0$, was used to make the conversion.

This solution method implements the strategy recommended in Section 9.5.2 of using an actual dollar analysis and then converting selected cash flows into real dollars. Reviewing the actual dollar ATCF in this case indicates an average annual positive cash flow during the analysis period of approximately $7,100 from the $20,000 investment in new equipment. However, the real dollar ATCF shows that in terms of dollars with constant purchasing power (beginning of 1989), the net positive cash flow from the investment (except for year 2) decreases from $5,979 in year 1 to $3,569 in year 10.

(c) The PW of the real dollar ATCF, using $i_r = 6\%$, is $16,780. This is the same value as the PW of the actual dollar ATCF calculated in Part (a) using $i_c = 14.48\%$.

EXAMPLE 9-15A

Repeat part (a) of Example 9-15 using the computer program in Appendix F (ATAX).

Because "revenues" and "other costs" both escalate at 8% per year and total $14,500 per year, they are entered as a net revenue. Other entries are straightforward as shown below.

```
ATAX: After Tax Investment Evaluation

Main Menu

1. Asset placed in service on, or before
   December 31, 1980.
   Applicable methods: Straight Line (SL)
                       Sum of Years Digits (SYD)
                       Declining Balance (DB)

2. Asset placed in service between January 1,
   1981 and July 31, 1986.
   Applicable method: Accelerated Cost Recovery
                      System (ACRS)

3. Asset placed in service between August 1,
   1986 to present.
   Applicable method: Modified ACRS (MACRS)
```

```
0. Exit to system.

  Enter option: 3

Menu 3: MACRS

1. Calculate depreciation of the asset.
2. Evaluate an investment.
0. Return to last menu.

  Enter option :2
```

```
Instructions

This program allows you to conduct an analysis of
an investment alternative on an after-tax basis
when inflation/escalation of revenues and costs
are involved.
Multiple replacements of an alternative are
assumed when the study period is longer than the
useful life unless a selected (coterminated)
analysis period is used. Salvage value and
estimated selling price at the end of each cycle
are specified by the user. When inflation/
escalation is considered, the reference year for
cost and revenue estimates is time k = 0.
The reference year for all cash flows, except
salvage value and selling price at disposal, is
the time at which the initial investment is made
(i.e., time 0). These are termed 'REAL DOLLARS'
in the prompts below.
```

```
*********************************************
*NOTE: Enter all percentages and rates in   *
*      DECIMAL form. For example:           *
*      10% should be entered as 0.10 or .10 *
*********************************************
Press any key to continue...

INVESTMENT

Cost of depreciable investment : 20000
Cost of non-depreciable investment : 0
Investment Tax Credit percentage (decimal) : 0.0
```

```
Often the useful life (in years) of the
investment is not identical to the study period.
Please check.

Useful life of the investment : 10
Study period : 10

Do you want to consider inflation/escalation
                                        [Y/N]?Y


REVENUES AND EXPENSES

The amount in real DOLLARS of the items below can
be CONSTANT during all, or a part, of the useful
life, or can be LUMPY
1. annual revenues
2. annual cost of utilities
3. annual cost of labor
4. annual cost of materials
5. annual lease cost

Enter C if CONSTANT/NONE else L if LUMPY

1C 2C 3C 4C 5C

Enter beginning (B) and ending (E) year in which
constant items apply, as pairs B,E:
Item 1 :1,10
Item 2 :1,10
Item 3 :1,10
Item 4 :1,10
Item 5 :1,10


REVENUES AND EXPENSES

Annual revenues in real dollars :14500
Effective annual escalation rate for revenues :
                                          0.08

Annual cost of utilities in real dollars :2500
Effective annual escalation rate for utilities :
                                          0.15
```

```
Annual cost of labor in real dollars :2500
Effective annual escalation rate for labor :
                                          0.055

Annual cost of materials in real dollars :1200
Effective annual escalation rate for materials :
                                           0.10


About LEASE COST
When lease fees are re-negotiated, the new
fee is assumed to escalate at the rate used
for replacements of capital assets.

Annual lease cost in real dollars : 800
Number of years before lease is re-negotiated :
                                               5


TAXES, INFLATION, and MARR

Incremental Income Tax Rate (decimal) : 0.39
Tax Rate on Capital Gain or Loss (decimal) :
                                          0.39
Inflation Rate for Investment Replacement
                              (decimal) : 0.08
The MARR for After-Tax Analysis (decimal) :
                                        0.1448


3.2 MACRS Depreciation

Enter MACRS Class Life of Investment
from [ 3 5 7 10 15 20 ]
The Useful Life is 10 yrs.
Class Life: 5

Please enter:
Salvage value for depreciation purposes and
selling price at time of disposal.
(in actual dollars)

In MACRS assumed salvage value equal to 0.

Selling price for year 10 : 4318.00
```

ACTUAL DOLLAR ANALYSIS

YR	UTILITY COST	LABOR COST	MATERIAL COST	LEASE COST	REVENUE	BTCF
0	0.00	0.00	0.00	0.00	-20000.00	-20000.00
1	-2875.00	-2637.50	-1320.00	-800.00	15660.00	8027.50
2	-3306.25	-2782.56	-1452.00	-800.00	16912.80	8571.99
3	-3802.19	-2935.60	-1597.20	-800.00	18265.83	9130.84
4	-4372.52	-3097.06	-1756.92	-800.00	19727.09	9700.60
5	-5028.39	-3267.40	-1932.61	-800.00	21305.26	10276.86
6	-5782.65	-3447.11	-2125.87	-1175.46	23009.68	10478.59
7	-6650.05	-3636.70	-2338.46	-1175.46	24850.46	11049.79
8	-7647.56	-3836.71	-2572.31	-1175.46	26838.49	11606.45
9	-8794.69	-4047.73	-2829.54	-1175.46	28985.58	12138.15
10	-10113.89	-4270.36	-3112.49	-1175.46	35622.42	16950.22

YR	BTCF	DEPR	TAXABLE INCOME	INCOME TAXES	ATCF	PRESENT WORTH
0	-20000.00	0.00	0.00	0.00	-20000.00	-20000.00
1	8027.50	-4000.00	4027.50	-1570.73	6456.78	5640.09
2	8571.99	-6400.00	2171.99	-847.08	7724.91	5894.33
3	9130.84	-3840.00	5290.84	-2063.43	7067.41	4710.55
4	9700.60	-2304.00	7396.60	-2884.67	6815.92	3968.32
5	10276.86	-2304.00	7972.86	-3109.41	7167.44	3645.16
6	10478.59	-1152.00	9326.59	-3637.37	6841.22	3039.18
7	11049.79	0.00	11049.79	-4309.42	6740.37	2615.63
8	11606.45	0.00	11606.45	-4526.52	7079.94	2399.90
9	12138.15	0.00	12138.15	-4733.88	7404.27	2192.38
10	16950.22	0.00	16950.22	-6610.58	10339.63	2674.29
				Present Worth:		16779.82

```
You Have a Winner
The Present Worth of the Investment is Positive.
Do you want to print this table (y/n) ? y
```

9.8 PRICE CHANGES CAUSED BY FOREIGN EXCHANGE RATES

When domestic corporations make foreign investments, the resultant cash flows that occur over time are in a different currency from U.S. dollars. Typically, foreign investments are characterized by two (or more) translations of currencies:

(1) when the initial investment is made and (2) when cash flows are returned to U.S.-based corporations. Exchange rates between currencies tend to fluctuate rather dramatically over time, so a typical question that can be anticipated is "What return (profit) did we make on our investment in the synthetic fiber plant in Thatland?" For the engineer who is designing another plant in Thatland, the question might be "What is the present worth (or internal rate of return) that our firm will obtain by constructing and operating this new plant in Thatland?"

This general situation may be conceptualized by observing that differences in exchange rates over time are analogous to price escalation rates. In this regard, we first must decide what currency to base our equivalent worth or rate of return calculations on, and then proceed to translate all cash flows into that currency using exchange rates in effect at the time of the translation. A measure of economic merit is finally calculated in view of the time value of money in the country whose currency is being used to represent all cash flows (normally U.S. dollars).

EXAMPLE 9-16

The CMOS Electronics Company is considering an investment of 50,000,000 pesos in an assembly plant located in a foreign country. Currency is expressed in pesos, and now the exchange rate is 100 pesos per 1 U.S. dollar.

The country has followed a policy of *devaluing* its currency against the dollar by 10% per year to build up its export business to the United States. This policy means that each year the number of pesos exchanged for a dollar increases by 10% ($e_j = 10\%$), so in 2 years $(1.10)^2(100) = 121$ pesos would be traded for one dollar. Labor is quite inexpensive in this country, so management of CMOS Electronics feels that the proposed plant will produce the following rather attractive after-tax cash flows in pesos.

End of Year	0	1	2	3	4	5
ATCF (millions of pesos)	−50	+20	+20	+20	+30	+30

If CMOS Electronics requires a 15% rate of return per year in U.S. dollars on its foreign investments, should this assembly plant be approved? Assume there are no unusual risks of nationalization of foreign investments in this country.

Solution

To earn a 15% rate of return in U.S. dollars, the foreign plant must earn $i_c = i_r + e_j + i_r(e_j) = 0.15 + 0.10 + 0.15(0.10) = 0.265$, which is 26.5% on its investment in pesos. Rather than work with pesos, it is easier to visualize the problem by converting pesos into dollars when evaluating this prospective investment:

Year	Pesos	Exchange Rate		Dollars
0	−50,000,000	100 pesos per $1		−500,000
1	+20,000,000	110 pesos per $1		+181,818
2	+20,000,000	121 pesos per $1		+165,289
3	+20,000,000	133.1 pesos per $1		+150,263
4	+30,000,000	146.4 pesos per $1		+204,918
5	+30,000,000	161.1 pesos per $1		+186,220
IRR:	34.6%		IRR:	22.4%
PW(26.5%):	9,165,236 pesos		PW(15%):	$91,632

The present worth of this investment at 15% annual return is now straightforward:

$$\begin{aligned} PW(15\%) &= -\$500{,}000 + \$181{,}818(P/F, 15\%, 1) + \cdots + \$186{,}220(P/F, 15\%, 5) \\ &= \$91{,}632 \end{aligned}$$

The IRR, based on dollars, is 22.4%. Therefore, the plant appears to be a good investment in economic terms.

Notice that in the table above, the IRR is 34.6% based on pesos, and the IRR based on dollars is 22.4%. The two IRRs can be reconciled by using Equation 9-5:

$$\begin{aligned} i_r(\text{IRR in \$}) &= \frac{i_c(\text{IRR in pesos}) - 0.10}{1.10} \\ &= \frac{0.346 - 0.10}{1.10} \\ &= 0.224, \text{ or } 22.4\% \end{aligned}$$

Problems that involve price changes created by exchange rates become intuitive when the analyst remembers that devaluation of foreign currency produces less expensive imports. Devaluation means that the U.S. dollar is strong relative to foreign currency. That is, fewer dollars are needed to purchase a fixed amount (barrels, tons, items) of goods and services from foreign sources or, stated differently, more units of foreign currency are required to purchase U.S. goods. This phenomenon was observed in Example 9-16.

Conversely, when exchange rates for foreign currencies drop against the U.S. dollar, the prices for imported goods and services rise. In such a situation, U.S. products are less expensive in foreign markets. For example, in 1985 the U.S. dollar was exchanged for approximately 250 Japanese yen, but in early 1988 the U.S. dollar was worth about 125 Japanese yen. As a consequence, prices on Japanese goods and services doubled in theory (but in actuality increased in the U.S. by only modest amounts). One explanation for this anomaly is that Japanese companies were willing to significantly reduce profit margins to retain their share of the U.S. market.

Inflation and price changes are an economic and business reality that can affect the comparison of alternatives and the quality of decision making in an organization. Much of this chapter has dealt with incorporating price changes into before-tax and after-tax engineering economy studies.

In this regard, it must be ascertained whether cash flows have been estimated in actual dollars or real dollars. The appropriate interest rate to use when discounting or compounding actual dollar amounts is a combined (nominal), or "marketplace," rate, while the corresponding rate to apply in real dollar analysis is the firm's real interest rate. A common error that must be avoided is estimating cash flows in real dollars and then using the combined interest rate (which includes the effect of general price inflation) in subsequent equivalency calculations.

Engineering economy studies often involve quantities that do not respond to inflation, such as depreciation, interest charges, and lease fees and other amounts established by contract. Identifying these quantities and handling them properly in an analysis are necessary to avoid erroneous economic results.

The advantages of doing an actual dollar or a real dollar analysis have been examined in the chapter. A suggested application strategy is to do actual dollar analysis and then convert the before-tax and after-tax net cash flows to real dollars to aid with the interpretation of results. Also, the use of the chapter's basic concepts in dealing with price changes caused by foreign exchange rates has been demonstrated.

9.9 PROBLEMS

9-1. Your rich aunt is going to give you end-of-year gifts of $1,000 for each of the next 10 years.

a. If general price inflation is expected to average 6% per year during the next 10 years, what is the equivalent value of these gifts at the present time? The real interest rate is considered to be 4% per year.

b. Suppose that your aunt specified that the annual gifts of $1,000 are to be increased by 6% each year to keep pace with inflation. With a real interest rate of 4%, what now is the present worth of the gifts? (9.2)

9-2. Because of general price inflation in our economy, the purchasing power of the dollar shrinks with the passage of time. If the average general price inflation rate is expected to be 8% per year into the foreseeable future, how many years will it take for the dollar's purchasing power to be one-half of what it is now? (That is, what is the future point in time when it takes 2 dollars to buy what can be purchased today for 1 dollar?) (9.2.1–9.2.3)

9-3. Which of these situations would you prefer? (9.2.1–9.2.3)

a. You invest \$2,500 in a certificate of deposit that earns an effective interest rate of 8% per year. You plan to leave the money alone for 5 years, and the general price inflation rate is expected to average 5% per year. Taxes are ignored.

b. You spend \$2,500 on an antique piece of furniture. In 5 years you believe the furniture can be sold for \$4,000. Assume that the average general price inflation rate is 5% per year. Again taxes are ignored.

9-4. Operating and maintenance costs for two alternatives have been estimated on different bases as follows.

End of Year	Alternative A—Costs Estimated in Actual (Inflated) Dollars	Alternative B—Costs Estimated in Real (Constant) Dollars with $k=0$
1	\$120,000	\$100,000
2	132,000	110,000
3	148,000	120,000
4	160,000	130,000

If the average general price inflation rate is expected to be 6% per year and money can earn a real rate of interest of 9% per year, show which alternative has the least negative present worth at time 0. (9.2.1–9.2.3)

9-5. Suppose that you deposit \$1,000 in a Swiss bank account that earns an effective interest rate of 18% per year, and you withdraw the principal after 6 years. You receive the interest each year and spend it on your favorite hobby. What is the real annual rate of return on your investment if the general price inflation rate is 10%/yr.? Be exact! (9.2.1–9.2.3)

9-6. If you buy a lathe now, it costs \$100,000. If you wait 2 years to purchase the lathe, it will cost \$135,000. Suppose you decide to purchase the lathe now, reasoning that you can earn 18% per year on your \$100,000 if you do not purchase the lathe. If the general price inflation rate (f) in the economy during the next 2 years is expected to average 12% per year, did you make the right decision? (9.2.1–9.2.3)

9-7. Your company has just issued bonds, each with a face value of \$1,000. They mature in 10 years and pay annual dividends of \$100. At present they are being sold for \$887. If the average annual general price inflation rate over the next 10 years is expected to be 6%, what is the real rate of return per year on this investment? (9.2.1–9.2.3)

9-8. A reactor vessel cost \$75,000 ten years ago. The reactor had the capacity of producing 500 pounds of product per hour. Today, it is desired to build a vessel of 1,000 pounds per hour capacity. With a general price inflation rate of 5% per year, and assuming a cost-capacity factor to reflect economies of scale, X, to be 0.75, what is the approximate future cost of the new reactor in 5 years? (8.5.1.3 and 9.2.1–9.2.3)

9-9. A man desires to have a preplanned amount in a savings account when he retires in 20 years. This amount is to be equivalent to $30,000 in today's purchasing power. If the expected average inflation rate is 7% per year and the savings account earns 5% interest, what lump sum of money should the man deposit now in his savings account? (9.2.1–9.2.3)

9-10. Determine the present worth of the following inflating cash flows that start in year 0 and continue for the next 100 years. The combined interest rate is 15% per year and the rate of general price inflation is 8% per year. (9.4)

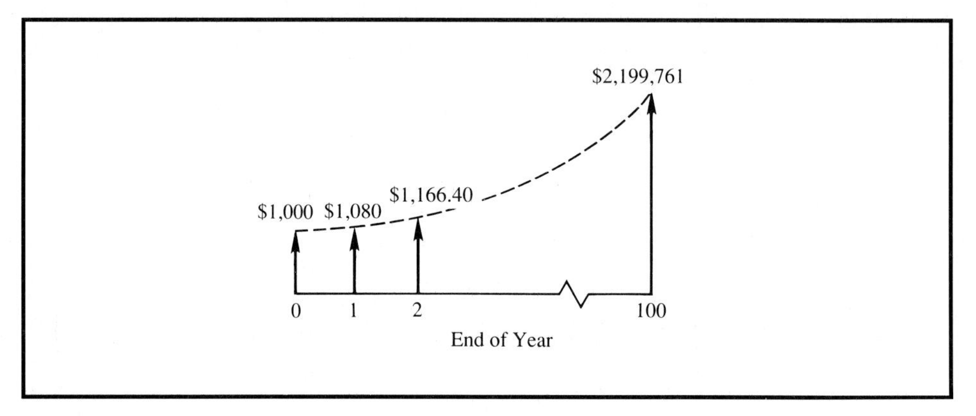

Problem 9-10

9-11. The AZROC Corporation needs to acquire a small computer system for one of its regional sales offices. The purchase price of the system has been quoted at $50,000, and the system will reduce manual office expenses by $18,000 per year in real dollars. Historically, these manual expenses have escalated at an average rate of 8% per year and this is expected to continue into the future. A maintenance agreement will also be contracted for and its cost per year in actual dollars is constant at $3,000.

What is the minimum (integer-valued) life of the system such that the new computer can be economically justified? Assume that the computer's salvage value is zero at all times. The firm's MARR is 25% and includes an adjustment for anticipated inflation in the economy. Show all calculations. (9.2.4 and 9.2.5.2)

9-12. The incremental design and installation costs of a total solar system (heating, air conditioning, hot water) in a certain Tennessee home were $14,000 in 1989. The annual savings in electricity (in 1989 dollars) have been estimated at $2,500. Assume that the life of the system is 15 years.

a. What is the internal rate of return on this investment if electricity prices do not escalate during the system's life?

b. What average annual price escalation rate on electricity would have to be experienced over the system's life to provide a rate of return of 25% for this investment? (4.6, 9.3, and 9.4)

9-13. A gas-fired heating unit is expected to meet an annual demand for thermal energy of 500 million Btu, and the unit is 80% efficient. Assume that each thousand cubic feet of natural gas, if burned at 100% efficiency, can deliver 1 million Btu. Suppose further that natural gas is now selling for $2.50 per thousand cubic feet. What is the present worth of fuel cost for this heating unit over a 12-year period if natural gas prices are expected to escalate at an average rate of 10% per year? The firm's MARR is 18%. (9.3)

9-14. A 30-year home mortgage loan of $60,000 is obtained by Mr. and Mrs. Smith at 12% nominal annual interest. Monthly payments will be made to the bank for the next 360 months in this amount:

$$\$60,000(A/P, 1\%/\text{month}, 360) = \$617.17$$

Because these payments are so high, the bank has agreed to permit the couple to make their first payment in the amount of $\$358.56(1+e)$, and payments will then escalate at this constant rate for each subsequent payment. Graphically, the situation is shown in the figure.

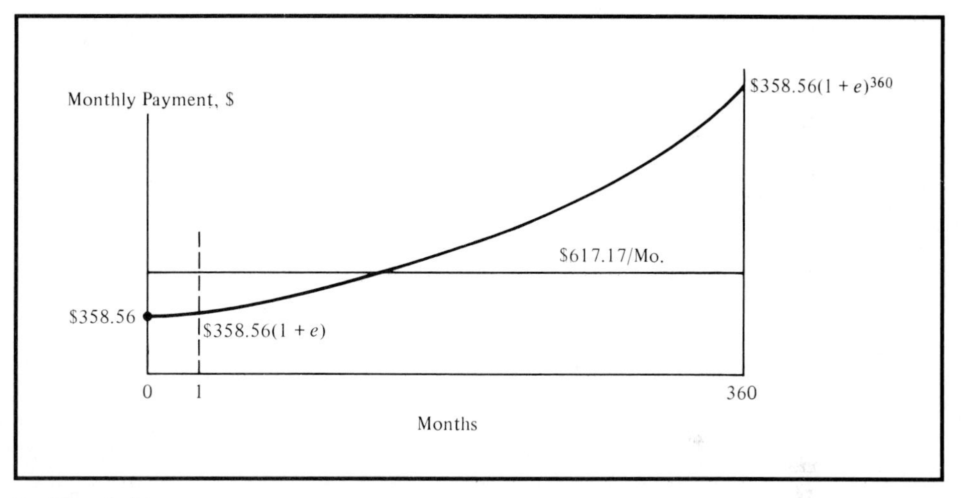

Problem 9-14

What is the constant price escalation rate that makes these two loan repayment plans equivalent (ignoring income taxes)? (9.4)

9-15. A liberal arts graduate has decided to invest 5% of her first-year's salary in a money market fund. This amounts to $1,000 at the end of the first year. She also makes payments of $120 per month on a blue BMW. She has been told that her savings should keep up with expected salary increases of 8% per year. Thus, she plans to invest an extra 8% each year starting at the end of year 2. At the end of year 1 she invests $1,000; in year 2, $1,080; in year 3, $1,166.40; and so on, up through year 10. If the general price inflation is expected to be 5% over the next 10 years and if she will earn an average of 10% per year in this account, what is the future worth of the money market fund at the end of the tenth year? (9.2)

9-16. Suppose that you have just graduated with a B.S. degree in engineering and the Omega Corporation offers you $30,000 per year as a starting salary. The company is located in New York City, and your guaranteed raises over the next 5 years will be 15% per year starting in the second year of employment. Another company offers you a position in Jonesville but says that it is willing to negotiate a starting salary with you. Guaranteed raises with the Jonesville-based company will be 8% per year over a 5-year period. What starting salary should you request of the Jonesville company if you are indifferent about where you would like to work? Ignore the effects of income taxes. Other data are the following: (9.1 and 9.2)

1. Your opportunity cost of capital (i_c) is 15% per year.
2. The average annual cost of living index for the next 5 years is 108 in New York City, and it is 100 in Jonesville (based on A$ spending power differences).
3. The general price inflation rate (f) in the economy is projected to be 8% over the next 5 years.

9-17. A small heat pump, including the duct system, now costs $2,500 to purchase and install. It has a useful life of 15 years and incurs annual maintenance of $100 per year in real (year 0) dollars over its useful life. A compressor replacement is required at the end of the eighth year at a cost of $500 in real dollars. The yearly cost of electricity for the heat pump is $680 based on prices at the beginning of the investor's time horizon. Electricity prices are projected to escalate at an annual rate of 10%. All other costs are expected to escalate at 6%, which is the projected general price inflation rate. The firm's interest rate, which includes an allowance for general price inflation, is 15%. No salvage value is expected from the heat pump at the end of 15 years.

a. What is the annual equivalent cost, expressed in actual dollars, of owning and operating this heat pump?

b. What is the annual cost in year 0 dollars of owning and operating the heat pump? (9.3 and 9.4)

9-18. Suppose that you are faced with the problem of deciding between the following two automobiles. All cost estimates are based on 15,000 driving miles per year and are expressed in year 0 dollars.

	Alternative		
	Domestic	Import	Price Escalation Rate
Negotiated purchase price (including tax)	$8,500	$9,200	—
Fuel/year	1,050	700	8%/yr
Maintenance/year	200	400	5%/yr
Insurance/year	400	440	5%/yr
Miscellaneous/year	100	125	7%/yr
Ownership period, N	5 years	5 years	—
Trade-in value at end of year N	$1,000	$2,000	6%/yr

Your personal MARR will average 12% per year over the next 5 years, and this is a "marketplace" rate of interest that includes an allowance for general price inflation ($f = 7\%$). Which automobile would you select based solely on monetary considerations? Solve using an A$ analysis. Then, for the selected alternative convert its net before-tax cash flow into real dollars. (9.3 and 9.2.5):

9-19. An electric utility company in the Northeast is trying to decide whether to switch from oil to coal at one of its generating stations. After much investigation, the problem has been reduced to these tradeoffs:

	Oil	Coal
Cost of retrofitting boilers to burn coal	—	?
Annual fuel cost (year 0 dollars)	$\$25 \times 10^6$	$\$17 \times 10^6$
Escalation rate	10%/yr	6%/yr
Life of plant	25 years	25 years

Determine the cost of retrofitting the boilers (to burn coal) that could be justified at this generating station. The utility's real annual MARR is 3% and the general price inflation rate in the economy will average 6% over the next 25 years.

a. Solve using an A$ analysis.

b. Solve using an R$ analysis. (9.3–9.5)

9-20. Referring to Table 9-10, calculate the projected effective annual price escalation rate for the period 1989–2000 for the following construction items:

a. Hardware

b. Sheet metal products

c. Insulation materials (9.2.5.3, 9.3, and 9.6)

9-21. Your company expects the market to be good over the next 5 years for a new product. Thus, additional investment in new production equipment is being considered. The estimated actual dollar before-tax cash flow is shown below for the 5-year period, and the market value of the equipment at the end of 5 years is estimated to be 20 percent of initial cost:

Year	Actual $ BTCF
0	−200,000
1	60,000
2	60,000
3	60,000
4	60,000
5	60,000
5	40,000

The effective corporate income tax rate is 40 percent; the equipment is 5-year class life MACRS property. The general price inflation rate is 10%. Based on an A$ analysis:

a. What is the after-tax internal rate of return (IRR)?

b. What is the after-tax external rate of return (ERR)? Assume the external reinvestment rate is 12%.

c. Develop the real dollar after-tax cash flow. (4.6, 4.7, and 9.2)

9-22. Because of tighter safety regulations, an improved air filtration system must be installed at a plant that produces a highly corrosive chemical compound. The investment cost of the system is $260,000 in today's (late 1989) dollars. The system has a useful life of 10 years and a class life of 5 years. MACRS depreciation is used with a zero salvage value for tax purposes. However, it is expected that the market value of the system at the end of its 10-year life will be $50,000 in today's dollars. Costs of operating and maintaining (O & M) the system, estimated in today's dollars, are expected to be $6,000 per year. Annual property tax is 4% of the investment cost and *does not inflate*. Assume that the plant has a remaining life of 20 years, and that O & M costs, replacement costs, and market value escalate at 6% per year.

If the effective income tax rate is 40%, set up a table to determine the after-tax cash flow over a 20-year period. The after-tax rate of return desired on investment capital is 12%. What is the present worth of cost of this system after income taxes have been taken into account? Develop the real dollar ATCF. (Assume the annual general price inflation rate is 4.5% over the 20-year period.) (9.2.5 and 9.3)

9-23. Referring to the case study in Section 9.7.2 (Example 9-15), solve the problem with the following changes:

1. Assume that the asset involved in the problem is one that has a MACRS class life of 7 years instead of 5 years.
2. The effective income tax rate (t) is 46% instead of 39%.
3. The real after-tax MARR (i_r) is 10% instead of 6%.
4. The price escalation rate (e_j) for labor is 9% instead of 5.5%.

9-24. Two years ago, the exchange rate for U.S. and Japanese currency was 280 yen per $1. Today the exchange rate is 140 yen per $1. If a Japanese automobile sells for 30% more in U.S. funds than it did 2 years ago, what can you conclude about the change in price per unit of a Japanese car? Assume the automobile sold 2 years ago for $10,000. (9.8)

10

DEALING WITH UNCERTAINTY

The objective of this chapter is to present and discuss several practical methods that are helpful in analyzing investment situations where risk and uncertainty exist. Several topics are presented in this chapter as we meet our stated goal:

What are risk, uncertainty, and sensitivity?
Sources of uncertainty
Common methods for dealing with uncertainty
Breakeven analysis
Sensitivity analysis
Optimistic–pessimistic estimates
Graphical sensitivity displays
Risk-adjusted MARR
Reduction in useful life
Probability functions
Monte Carlo simulation

10.1
INTRODUCTION

In previous chapters specific assumptions were stated concerning applicable revenues and costs and other quantities important to an engineering economic analysis. It was assumed that a high degree of confidence could be placed in all estimated values. That degree of confidence is sometimes called *assumed certainty*. Decisions made solely on this kind of analysis sometimes are called *decisions under certainty*. This is a rather misleading term, in that there rarely is a case in which estimated quantities can be assumed as certain.

In virtually all situations there is doubt as to the ultimate results that will be obtained from an investment. We now examine Step 4 of the seven-step procedure for conducting engineering economy studies (see Section 1.7). The motivation for dealing with risk and uncertainty is to establish the bounds of error in our estimates such that another alternative being considered may turn out to be a better choice than the one we recommended under assumed certainty. Principle 6 on page 16 further provides the basis for such an exercise.

10.2
WHAT ARE RISK, UNCERTAINTY, AND SENSITIVITY?

In this chapter we introduce the very real phenomena of *risk and uncertainty* in decision-making activities. Both are caused by lack of precise knowledge regarding future business conditions, technological developments, synergies among funded projects, and so on. We define *decisions under risk* to be those in which the analyst models the decision problem in terms of assumed possible future outcomes, or scenarios, whose probabilities of occurrence can be estimated. A *decision under uncertainty*, to the contrary, is a decision problem characterized by several unknown futures for which probabilities of occurrence cannot be estimated.

In reality the difference between risk and uncertainty is somewhat arbitrary. One contemporary school of thought posits that representative and likely future outcomes and their probabilities can always be subjectively developed.* Hence,

*R. Schlaifer, *Analysis of Decisions under Uncertainty* (New York: McGraw-Hill, 1969).

it is not unreasonable to suggest that decision making under risk is the more plausible and tractable framework for dealing with lack of perfect knowledge concerning the future. Although we may make a technical distinction between risk and uncertainty, both can cause study results to vary from predictions, and there seldom is anything significant to be gained by attempting to treat them separately. *Therefore, in the remainder of this book "risk" and "uncertainty" are used interchangeably.*

In dealing with uncertainty, it is often very helpful to determine to what degree changes in an estimate would affect an investment decision; that is, how *sensitive* a given investment is to changes in a particular factor (parameter) that is not known with certainty. If a parameter such as project life or annual revenue can be varied over a wide range without causing much effect on the investment decision, the decision under consideration is said not to be sensitive to that particular factor. Conversely, if a small change in the relative magnitude of a parameter will reverse an investment decision, the decision is highly sensitive to that parameter.

10.3 SOURCES OF UNCERTAINTY

It is useful to consider the factors that affect the uncertainty involved in an investment, so that they may be related to the economic measure of merit that must be met or exceeded in order for an investment to be justified. Similarly, they may be included as nonmonetary (irreducible) considerations in reaching a final decision.

The factors that affect uncertainty are many and varied. It would be almost impossible to list and discuss all of them. There are four major sources of uncertainty, however, which nearly always are present in economy studies.

The first factor, which is always present, is the *possible inaccuracy of the estimates used in the study*. If exact information is available regarding the items of income and expense, the resulting accuracy should be good. If, on the other hand, little factual information is available, and nearly all the values have to be estimated, the accuracy may be high or low, depending on the manner in which the estimated values are obtained. Are they sound scientific estimates or merely guesses?

The accuracy of the income figures is difficult to determine. If they are based on a considerable amount of past experience or have been determined by adequate market surveys, a fair degree of reliance may be placed on them. On the other hand, if they are merely the result of guesswork, with a considerable element of hope thrown in, they must of course be considered to contain a sizeable element of uncertainty.

A saving in existing operating expenses should involve less uncertainty. It is usually easier to determine what the saving will be since there is considerable experience and past history on which to base the estimates.

In most cases the income figures will contain more error than any other element of a study, with the possible exception of estimated operating expenses. Frequently, annual income and expenses are discovered to be the most sensitive elements in the study. There should be no large error in estimates of capital required. Uncertainty in investment capital requirements is often reflected as a *contingency* above the actual cost of plant and equipment. If we feel confident that the amount allowed for contingencies is on the high side, the resulting study is apt to be conservative.

The second key factor affecting uncertainty is the *type of business involved, in relation to the future health of the economy*. Some lines of business are notoriously less stable than others. For example, most mining enterprises are more risky than large retail food stores. However, we cannot arbitrarily say that an investment in any retail food store always involves less uncertainty than investment in mining property. Whenever capital is to be invested in an enterprise, the nature and history of the business as well as expectations of future economic conditions (e.g., interest rates) should be considered in deciding what risk is present. In this connection it becomes apparent that investment in an enterprise that is just being organized, and thus has no past history, is usually rather uncertain. This is especially true when the economy is dramatically changing because of business cycles.

A third factor affecting uncertainty is the *type of physical plant and equipment involved*. Some types of structures have rather definite economic lives and residual values. Little is known of the physical or economic lives of others, and they have almost no resale value. A good engine lathe generally can be used for many purposes in nearly any fabrication shop. Quite different would be a special type of lathe that was built to do only one unusual job. Its value would be dependent almost entirely upon the demand for the special task that it can perform. Thus, the type of physical property involved will have a direct bearing upon the accuracy of the estimated income and expenditure patterns. Where money is to be invested in specialized plant and equipment, this factor should be considered carefully.

The fourth, and very important, factor that must always be considered in evaluating uncertainty is the *length of the assumed study period*. The conditions that have been assumed in regard to income and expense must exist throughout the study period in order for us to obtain a satisfactory return on the investment. A long study period naturally decreases the probability of all the factors turning out as estimated. Therefore, a long study period, all else being equal, always increases the uncertainty in an investment.

10.4

COMMON METHODS FOR DEALING WITH UNCERTAINTY

There are numerous methods for taking uncertainty, resulting from the four major sources described above, into account. This chapter discusses and illustrates each of the following popular methods.

1. *Breakeven analysis* is commonly utilized when the selection among alternatives is heavily dependent on a single parameter, such as capacity utilization, that is uncertain. A breakeven point for the factor is determined such that two alternatives are equally desirable from an economic standpoint. It is then possible to choose between the alternatives by estimating the most likely value of the uncertain factor and comparing this estimate to the breakeven value.
2. *Sensitivity analysis* is often employed when one or more factors are subject to uncertainty. The basic questions that sensitivity analysis attempts to resolve are (a) What is the behavior of the measure of merit (e.g., present worth) to $\pm x\%$ changes in each individual factor? (b) What is the amount of change in a particular factor that will cause a reversal in preference for an alternative?
3. *Optimistic–pessimistic estimation* of parameters included in an engineering economy study has been used to establish a range of extreme values for the economic measure of merit. If the range falls *entirely* in the acceptable region (e.g., annual worth ≥ 0), the alternative is desirable. When all factors are estimated under conservative (pessimistic) conditions, an unacceptable alternative often is the result. This method directs attention to the best and worst outcomes of going ahead with an alternative and requires managerial judgment to make the final go–no go determination.
4. *Risk-adjusted minimum attractive rates of return* are sometimes utilized to deal with estimation uncertainties. This method involves the use of higher MARRs for alternatives that are classified as "highly uncertain" and lower MARRS for projects for which there appear to be fewer uncertainties.
5. *Reduction of the useful life* of an alternative is another means for attempting to include explicitly the effects of uncertainty. Here the estimated project life is reduced by a fixed percentage, for instance 50%, and each alternative is evaluated regarding its acceptability over only this reduced life span.
6. *Probability functions* for uncertain elements are estimated and directly incorporated into the analysis of alternatives. This approach involves various statistical concepts and makes use of selected descriptive measures for summarizing uncertainty in the analysis. Two procedures based on formal

probabilistic concepts are presented: (a) closed-form analysis of the measure of merit, and (b) Monte Carlo simulation. Any of the basic measures of economic merit presented in Chapter 4 could be used in connection with these two procedures.

10.5 BREAKEVEN ANALYSIS

Essentially all data employed in an economy study are uncertain simply because they represent estimates of the future. Breakeven analyses are useful when one must make a decision between alternatives that are highly sensitive to a parameter which is difficult to estimate. Through breakeven analysis, one can solve for the value of that parameter at which the conclusion is a standoff. That value is known as the *breakeven point*. (The use of breakeven points with respect to production and sales volumes was briefly discussed in Chapter 2.) If one can then estimate whether the actual outcome of that factor will be higher or lower than the breakeven point, the best alternative becomes apparent.

The following are examples of common parameters for which breakeven analyses might provide useful insights into the decision problem:

1. ***Revenue and annual cost:*** Solve for the annual revenue required to equal (break even with) annual costs. Breakeven annual costs of an alternative can also be determined in a pairwise comparison when revenues are identical for both alternatives being considered.
2. ***Rate of return:*** Solve for the rate of return at which two given alternatives are equally desirable.
3. ***Salvage value:*** Solve for future resale value that would result in indifference as to preference for an alternative.
4. ***Equipment life:*** Solve for the useful life required for an alternative to be justified.
5. ***Capacity utilization:*** Solve for the hours of utilization per year, for example, at which an alternative is justified or at which two alternatives are equally desirable.

The usual breakeven problem involving two alternatives can be most easily approached mathematically by equating the annual worths or present worths of the two alternatives expressed as a function of a parameter. In breakeven studies, project lives may or may not be equal, so care should be taken to determine whether the coterminated or repeatability assumption best fits the situation. Examples below illustrate both mathematical and graphical solutions to typical breakeven problems.

EXAMPLE 10-1

Suppose that there are two alternative electric motors that provide 100 hp output. An Alpha motor can be purchased for $1,250 and has an efficiency of 74%, an estimated life of 10 years, and estimated maintenance costs of $50 per year. A Beta motor will cost $1,600 and has an efficiency of 92%, a life of 10 years, and annual maintenance costs of $25. Annual taxes and insurance costs on either motor will be $1\frac{1}{2}\%$ of the investment. If the minimum attractive rate of return is 15%, how many hours per year would the motors have to be operated at full load for the annual costs to be equal? Assume that salvage values for both are negligible and that electricity costs $0.05 per kilowatt-hour.

Solution by Mathematics

Note: 1 hp = 0.746 kW and input = output/efficiency. If X = number of hours of operation per year, components of the equivalent annual cost, AC_α, for the Alpha motor would be as follows.

Capital recovery cost (depreciation and minimum profit):

$$\$1{,}250(A/P,\ 15\%,\ 10) = \$1{,}250(0.1993) = \$249$$

Operating cost for power:

$$(100)(0.746)(\$0.05)X/0.74 = \$5.04X$$

Maintenance cost:

$$\$50$$

Taxes and insurance:

$$\$1{,}250(0.015) = \$18.75$$

Similarly, components of the annual cost, AC_β, for the Beta motor would be as follows.

Capital recovery cost (depreciation and minimum profit):

$$\$1{,}600(A/P,\ 15\%,\ 10) = \$1{,}600(0.1993) = \$319$$

Operating cost for power:

$$(100)(0.746)(\$0.05)X/0.92 = \$4.05X$$

Maintenance cost:

$$\$25$$

Taxes and insurance:

$$\$1{,}600(0.015) = \$24$$

At the breakeven point, $AC_\alpha = AC_\beta$. Thus,

$$\begin{aligned} \$249 + \$5.04X + \$50.00 + \$18.75 &= \$319 + \$4.05X \\ &\quad + \$25.00 + \$24.00 \\ \$5.04X + \$317.75 &= \$4.05X + \$368.00 \\ \hat{X} &\simeq 51 \text{ hours/year} \end{aligned}$$

Solution by Graphics

Figure 10-1 shows a plot of the total equivalent annual costs of each motor as a function of the number of hours of operation per year. The constant annual costs (i-intercepts) are \$317.75 and \$368.00 for Alpha and Beta, respectively, and the costs that vary directly with hours of operation per year (slopes of lines) are \$5.04 and \$4.05 for Alpha and Beta, respectively. Of course, the breakeven point is the value of the independent variable at which the linear annual cost functions for the two alternatives intersect.

Hours of operation per year was the measure of business activity in Example 10-1 and was used as the variable for which a breakeven value was desired. If hours of operation had been independently estimated to be, say,

FIGURE 10-1

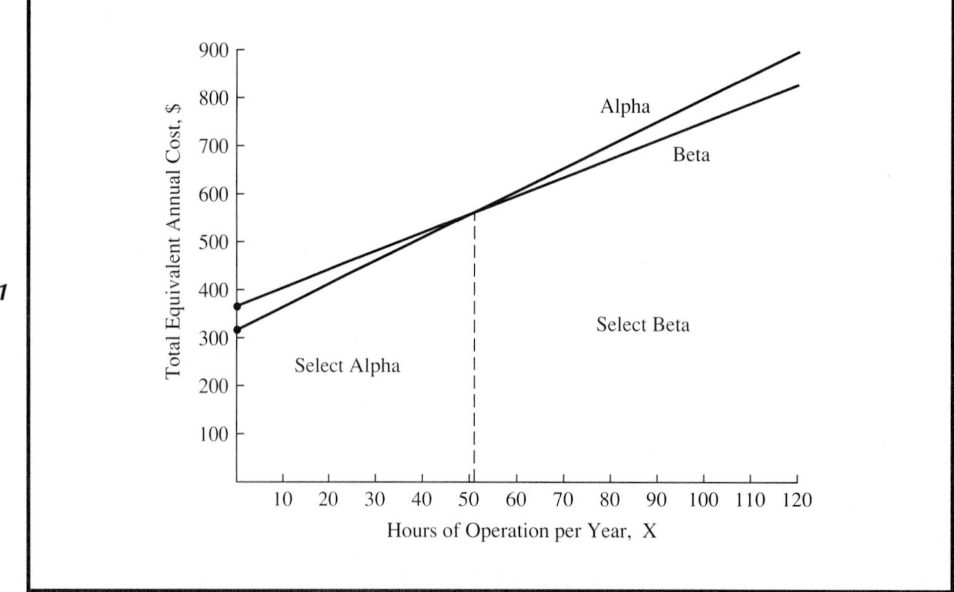

Graphical Solution of Breakeven Point for Example 10-1

200 hours per year, the choice of the Beta motor would be apparent from Figure 10-1. We next consider a simple breakeven problem in which the measure of business activity is mileage driven per year.

EXAMPLE 10-2

The Universal Postal Service is considering the possibility of putting wind deflectors on the tops of 500 of their long-haul tractors. Three types of deflectors, with the following characteristics, are being considered:

	Windshear	Blowby	Air-vantage
Investment	\$1,000	\$400	\$1,200
Drag reduction	20%	10%	25%
Maintenance/year	\$10	\$5	\$5
Life	10 years	10 years	5 years
MARR	10%/yr	10%/yr	10%/yr

If 5% in drag reduction means 2% in fuel savings per mile, how many miles do tractors have to be driven per year before the Windshear deflector is favored over the other deflectors? Over what range of miles driven per year is Air-vantage the best choice? (*Note:* Fuel cost is expected to be \$1.00/gallon and average fuel consumption is 5 miles/gallon). State any assumptions you make.

Solution

The annual operating costs of long-haul tractors equipped with the various deflectors are calculated as a function of mileage driven per year, X:

$$\text{Windshear: } (X \text{ miles/year})(0.92)(0.2 \text{ gal/mi})(\$1.00/\text{gal}) = \$0.184X/\text{yr}$$
$$\text{Blowby: } (X \text{ miles/yr})(0.96)(0.2 \text{ gal/mi})(\$1.00/\text{gal}) = \$0.192X/\text{yr}$$
$$\text{Air-vantage: } (X \text{ miles/yr})(0.90)(0.2 \text{ gal/mi})(\$1.00/\text{gal}) = \$0.180X/\text{yr}$$

Plotting equivalent annual costs (AC) of the deflectors yields the breakeven values of X shown in Figure 10-2. In summary, when $X \leq 12,831$, Blowby would be selected. If $X \geq 17,976$, the Air-vantage deflector would be chosen; otherwise, Windshear is the preferred alternative. Mathematically, the exact values can be calculated for each pair of AC equations. For example, the breakeven value between the Windshear deflector and the Blowby deflector is

$$\$1{,}000(A/P,\ 10\%,\ 10) + \$10 + \$0.184X = \$400(A/P,\ 10\%,\ 10) + \$5 + \$0.192X$$
$$172.75 + 0.184X = 70.1 + 0.192X$$
$$\hat{X} = \frac{102.65}{0.008} = 12,831 \text{ miles/year}$$

FIGURE 10-2

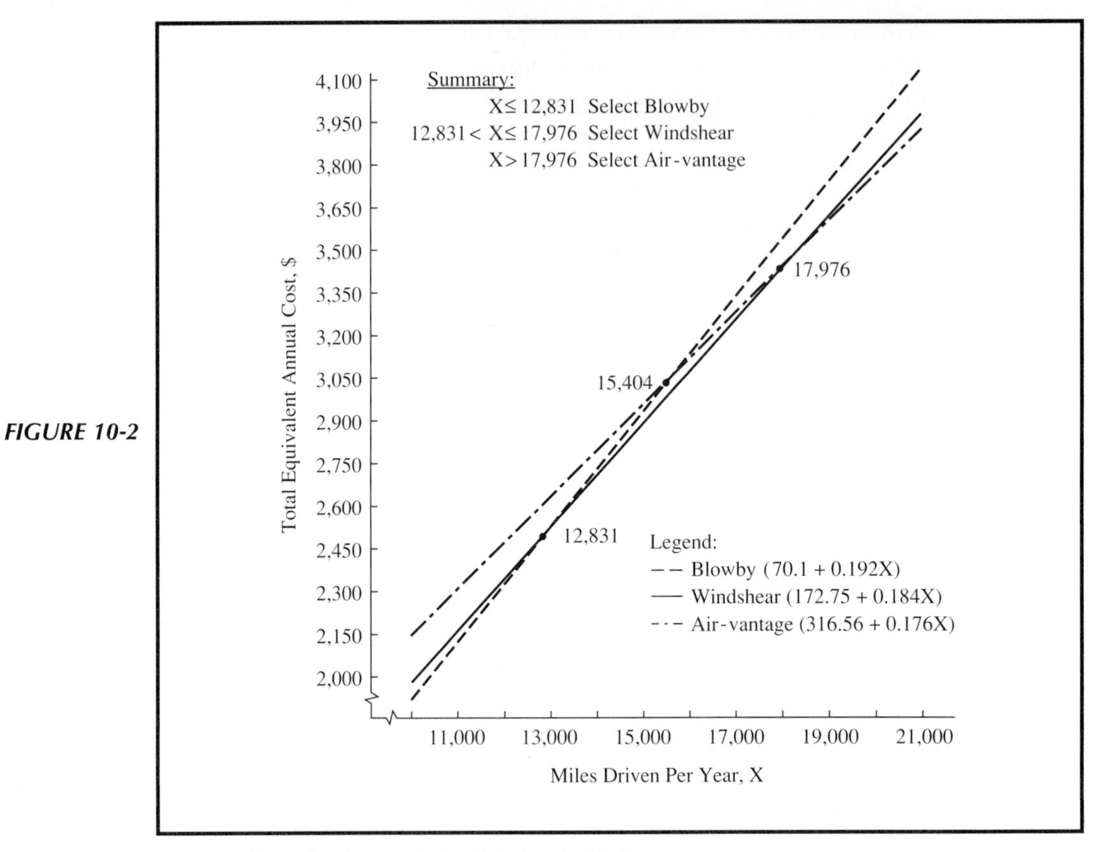

Graphical Breakeven Analysis of Example 10-2

The repeatability assumption allows the annual costs to be compared over different periods of time.

It often is helpful to know at what future date a deferred investment will be needed so that an alternative permitting deferred investment will break even with one that provides immediately for all future demands. Where only the costs of acquiring the assets by the two alternatives need to be considered, or where the annual costs through the entire life are not affected by the date of acquisition of the deferred asset, the breakeven point may be determined very easily and may be helpful in arriving at a decision between alternatives. Example 10-3 illustrates this type of breakeven study.

EXAMPLE 10-3

In the planning of a two-story municipal office building, the architect has submitted two designs. The first provides foundation and structural details so that two additional stories can be added at a later date without modifications to the original structure. This building would cost $1,400,000. The second design, without such provisions, would cost only $1,250,000. If the first plan is adopted, it is estimated that an additional two stories could be added at a later date at a cost of $850,000. If the second plan is adopted, however, considerable strengthening and reconstruction would be required, which would add $300,000 to the cost of a two-story addition. Assuming that the building is expected to be needed for 75 years, by what time would the additional two stories have to be built to make the adoption of the first design justified? (The MARR is 10%.)

Solution

The breakeven deferment period, $\hat{T}$, is determined as follows:

	Provide Now	No Provision
Present worth–cost:		
First unit	$1,400,000	$1,250,000
Second unit	$850,000($P/F$, 10%, T)	$1,150,000($P/F$, 10%, T)
Equating total present worth–costs:		
$1,400,000 + $850,000(P/F, 10%,T) = $1,250,000 + $1,150,000($P/F$, 10%, T)		

Solving, we have

$$(P/F,\ 10\%,\hat{T}) = 0.5$$

From the 10% table in Appendix C, $\hat{T} = 7$ years (approximately). Thus, if the additional space will be required in less than 7 years, it would be more economical to make immediate provision in the foundation and structural details. If the addition would not likely be needed until after 7 years, greater economy would be achieved by making no such provisions in the first structure.

There are many situations in which the relationship between the dependent and independent variables is not continuous and therefore cannot be expressed readily in mathematical terms. In other cases the relationship may be very complex so that the time required to develop a mathematical formula would be so great that it would be uneconomical to do so. In such cases a graphical solution may be used to determine the breakeven point. The following is a case where a graphical solution is used to advantage.

Products ordered from a wholesale drug company are collected in large wire-mesh baskets and taken to the shipping department for packing. Two methods of packing are used—one with heavy plastic bags and the other with cartons. Figure 10-3 shows the packing costs for the two methods as a function of the volume of the merchandise. It is apparent that the relationship of volume and cost is not a simple one and that for shipments whose volume is less than approximately 3,500 cubic inches the bag method should be used, and

FIGURE 10-3

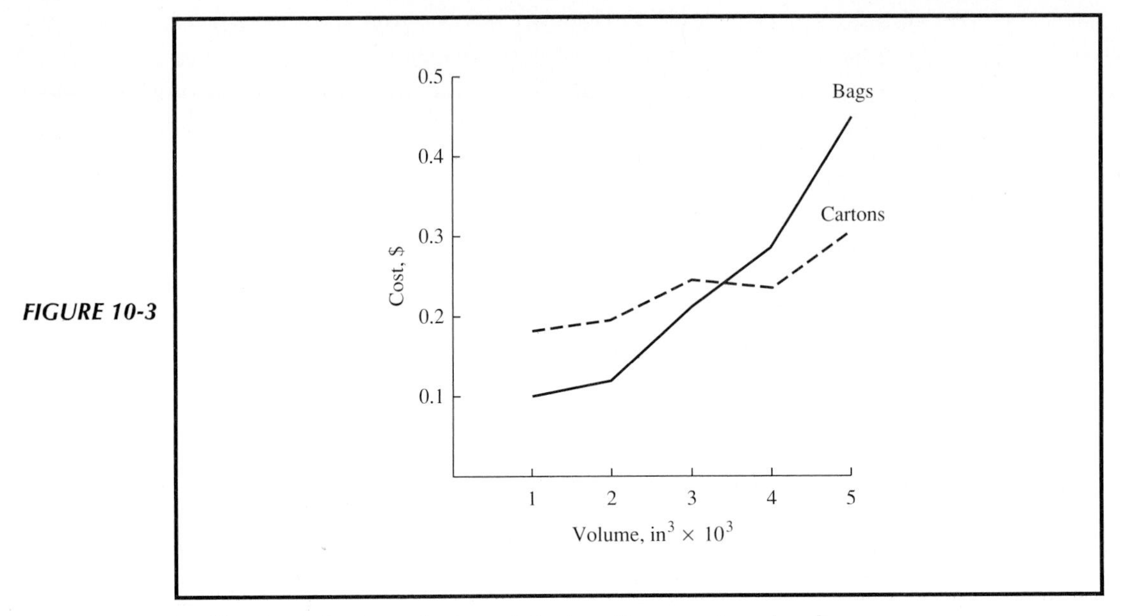

Packing Costs for Paper Sacks and Cartons Shown as Merchandise

that shipments whose volume is larger should be packed in cartons. In this instance, since the sides of the baskets are nearly vertical, the packers are able to determine which packing method to use (and the correct size of bag or carton) merely by checking the height of the merchandise in the basket against a colored vertical scale on the end of the packing table.

10.6 SENSITIVITY ANALYSIS

In a great many cases a simple breakeven analysis is not feasible because several factors may vary simultaneously as the single parameter under study is varied. In such instances, and in fact in most cases, it is helpful to determine how sensitive the situation is to the several parameters so that proper weight and consideration may be assigned to them. *Sensitivity*, in general, means the relative magnitude of change in the measure of merit (such as present worth) caused by one or more changes in estimated study parameters. Sometimes sensitivity is more specifically defined to mean the relative magnitude of the change in one or more factors that will reverse a decision among alternatives. Example 10-4 demonstrates sensitivity analysis for several parameters taken one at a time.

EXAMPLE 10-4

A machine having "most likely" cash flow estimates given below is being considered for immediate installation. Because of the new technology built into this machine, it is desired to investigate its present worth over a range of ±40% in (a) initial investment, (b) annual net cash flow, (c) salvage value, and (d) useful life.

Initial investment, I	\$11,500
Revenues/yr } A	5,000
Expenses/yr } A	2,000
Salvage value, S	1,000
Useful life, N	6 years

Draw a diagram that summarizes the sensitivity of present worth to changes in each separate parameter when the MARR = 10% per year.

Solution

The present worth of this investment for most likely estimates of parameters is

$$\begin{aligned} PW(10\%) &= -\$11{,}500 + \$3{,}000\,(P/A,\ 10\%,\ 6) + \$1{,}000\,(P/F,\ 10\%,\ 6) \\ &= +\$2{,}130 \end{aligned}$$

(a) When the initial investment varies by $\pm p\%$ the present worth is

$$\begin{aligned} PW(10\%) = &-(1 \pm p\%/100)(\$11{,}500) + \$3{,}000\,(P/A,\ 10\%,\ 6) \\ &+\$1{,}000(P/F,\ 10\%,\ 6) \end{aligned}$$

If we let $p\%$ vary in increments of 10% to ±40%, the resultant calculations of PW(10%) can be plotted as shown in Figure 10-4.

(b) The equation for present worth can be modified to reflect $\pm a\%$ changes in net annual cash flow, A:

$$\begin{aligned} PW(10\%) = &-\$11{,}500 + (1 \pm a\%/100)(\$3{,}000)\,(P/A,\ 10\%,\ 6) \\ &+\$1{,}000(P/F,\ 10\%,\ 6) \end{aligned}$$

Results are plotted in Figure 10-4 for 10% increments in A within the prescribed ±40% interval.

(c) When salvage value varies by $\pm s\%$, the present worth is

$$\begin{aligned} PW(10\%) = &-\$11{,}500 + \$3{,}000(P/A,\ 10\%,\ 6) \\ &+(1 \pm s\%/100)(\$1{,}000)(P/F,\ 10\%,\ 6) \end{aligned}$$

Results are shown in Figure 10-4 for ±40% changes in S.

FIGURE 10-4

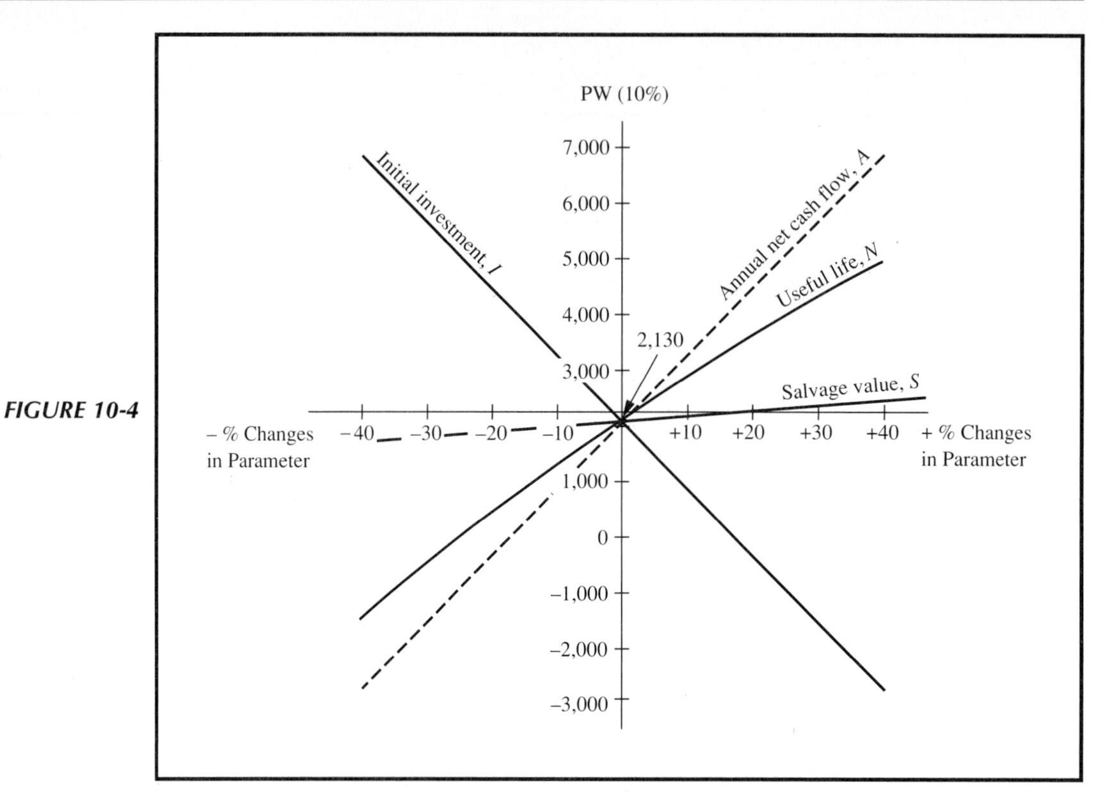

Sensitivity Analysis of Four Parameters in Example 10-4

(d) Plus and minus $n\%$ changes in useful life, as they affect PW(10%), can be represented by this equation:

$$\text{PW}(10\%) = -\$11{,}500 + \$3{,}000\,[P/A,\ 10\%,\ 6(1 \pm n\%/100)] + \$1{,}000\,[P/F,\ 10\%,\ 6(1 \pm n\%/100)]$$

When $n\%$ varies by 10% increments within the desired ±40% interval, resultant changes in PW(10%) can be calculated and plotted as shown in Figure 10-4.

10.7 ANALYZING A PROPOSED BUSINESS VENTURE

A fairly detailed illustration of the use of sensitivity analysis is provided in Example 10-5, where a new business venture is carefully scrutinized. This example extends sensitivity explorations to several parameters whose outcomes are believed to be crucial to the success of the venture. Tabular displays are utilized to summarize results of the various analyses.

EXAMPLE 10-5

A small group of investors is considering starting a premixed-concrete plant in a rapidly developing suburban area about 5 miles from a large city. The group believes that there will be a good market for premixed concrete in this area for at least the next 10 years, and that if they establish such a local plant it will be unlikely that another local plant would be established. Existing plants in the adjacent large city would, of course, continue to serve this new area. The investors believe that the plant could operate at about 75% of capacity 250 days per year since it is located in an area where the weather is mild throughout the year.

The plant will cost $100,000 and it would have a maximum capacity of 72 cubic yards of concrete per day. Its salvage value at the end of 10 years is estimated to be $20,000, which is the value of the land. To deliver the concrete, four second-hand trucks would be acquired, costing $8,000 each, having an estimated life of 5 years and a trade-in value of $500 each at the end of that time. In addition to the four truck drivers, who would be paid $50.00 per day each, four people would be required to operate the plant and office, at a cost of $175.00 per day. Annual operating and maintenance costs for the plant and office are estimated at $7,000 and for each truck at $2,250, both in view of 75% capacity utilization. Raw material costs are estimated to be $27.00 per cubic yard of concrete. Payroll taxes, vacations, and other fringe benefits would amount to 25% of the annual payroll. Annual taxes and insurance on each truck would be $500, and taxes and insurance on the plant would be $1,000 per year. The investors would not contribute any labor to the business, but a manager would be employed at an annual salary of $20,000.

Delivered, premixed concrete currently is selling for an average of $45 per cubic yard. A useful plant life of 10 years is expected, and capital invested elsewhere by these investors is earning about 15% before income taxes. It is desired to find the *annual worth* for the expected conditions above and to perform sensitivity studies for certain parameters.

Solution by AW Method

Annual revenue:

$$72 \times 250 \times \$45 \times 0.75 = \$607{,}500$$

Annual costs:

1. Capital recovery			
Plant: \$100,000(A/P, 15%, 10)			
− \$20,000(A/F, 15%, 10)	=	\$18,940	
Trucks: 4[\$8,000(A/P, 15%, 5)			
− \$500(A/F, 15%, 5)]	=	9,250	
			\$ 28,190
2. Labor:			
Plant and office: \$175 × 250	=	43,750	
Truck drivers: 4 × \$50 × 250	=	50,000	
Manager	=	20,000	
			113,750
3. Payroll taxes, fringe benefits, etc.: \$113,750 × 0.25			28,438
4. Taxes and insurance:			
Plant	=	1,000	
Trucks: \$500 × 4	=	2,000	
			3,000
5. Operations and maintenance at 75% capacity:			
Plant and office	=	7,000	
Trucks: \$2,250 × 4	=	9,000	
			16,000
6. Materials: 72 × 0.75 × 250 × \$27.00			364,500
		Total	\$553,878

The net AW for these "most likely" estimates is \$607,500 − \$553,878 = \$53,622. Apparently, the project is an attractive investment opportunity.

In Example 10-5 there are three parameters that are of great importance and that must be estimated: *capacity utilization*, the *selling price of the product*, and the *useful life of the plant*. A fourth factor—raw material costs—is important, but any significant change in this factor would probably be equally beneficial to competitors and probably would be reflected in a corresponding change in the selling price of mixed concrete. The other cost elements should be determinable with considerable accuracy. Therefore, we would like to investigate the effect of variations in the plant utilization, selling price, and useful life. Sensitivity analysis is a good method for doing this.

10.7.1 Sensitivity to Capacity Utilization

As a first step we must determine how cost factors would vary, if at all, as capacity utilization is varied. In this case, it is probable that the cost items listed under groups 1, 2, 3, and 4 in the previous tabulation would be virtually unaffected if capacity utilization should vary over a quite wide range—from 50% to 90%, for example. To meet peak demands, the same amount of plant, trucks, and personnel probably would be required. Group 5 costs for operations and maintenance would be affected somewhat. For this type of factor we must try to determine what the variation would be or make a reasonable assumption as to the probable variation. For this case it will be assumed that one half of these costs would be fixed and the other half would vary with capacity utilization by a straight-line relationship. Certain other factors, such as the cost of materials in this case, will vary in direct proportion to capacity utilization.

Using these assumptions, Table 10-1 shows how the revenue, costs, and net annual worth would change with different capacity utilizations. It will be noted that the annual worth is moderately sensitive to capacity utilization. The plant could be operated at a little less than 65% of capacity, instead of the assumed 75%, and still produce an annual worth greater than 0. Also, quite clearly, if they should be able to operate above the assumed 75% of capacity, the annual worth would be very good. This type of analysis provides those who must make the decision with a good idea as to how much leeway they have in capacity utilization and still have an acceptable venture.

10.7.2 Sensitivity to Selling Price

Examination of the sensitivity of the project to the selling price of the concrete reveals the situation shown in Table 10-2. The values in this table assume that the plant would operate at 75% of capacity; the costs would thus remain constant, with only the selling price varying. Here it will be noted that the project is quite sensitive to price. A decrease in price of 10% would drop the IRR to less than 15% (i.e., AW < 0). Since a decrease of 10% is not very large, the investors would want to make a thorough study of the price structure of concrete in the area of the proposed plant, particularly with respect to the possible effect of the increased competition that the new plant would create. If such a study reveals price instability in the market for concrete, the plant could be a risky investment.

TABLE 10–1 Annual Worth at i = 15% for Premixed-Concrete Plant for Various Capacity Utilizations (Average Selling Price Equals $45 per Cubic Yard)

	50% Capacity	65% Capacity	90% Capacity
Annual revenue	$405,000	$526,500	$729,000
Annual costs:			
Capital recovery	28,190	28,190	28,190
Labor	113,750	113,750	113,750
Payroll taxes and similar items	28,438	28,438	28,438
Taxes and insurance	3,000	3,000	3,000
Operations and maintenance[a]	13,715	15,086	17,372
Materials	243,000	315,900	+ 437,400
Total Costs	$430,093	$504,364	$628,150
AW	–$25,093	+$ 22,136	+$110,850

[a] Let x = annual operations and maintenance cost. At 75% capacity utilization, $x/2 + (x/2)(0.75) = \$16,000$, so that $x = \$18,286$ at 100% capacity utilization. Therefore, at 50% utilization, the operations and maintenance cost would be $9,143 + 0.5($9,143) = $13,715.

TABLE 10–2 Effect of Various Selling Prices on the Annual Worth for the Premixed-Concrete Plant Operating at 75% of Capacity

	Selling Price			
	$45.00	**$43.65(3%)[a]**	**$42.75(5%)[a]**	**$40.50(10%)[a]**
Annual revenue	$607,500	$589,275	$577,125	$546,750
Annual costs	553,878	553,878	553,878	553,878
AW	$ 53,622	$ 35,397	$ 23,247	−$ 7,128

[a] Percentage values shown in parentheses are reductions in price below $45.

10.7.3 Sensitivity to Useful Life

The effect of the third factor, assumed useful life of the plant, can be investigated readily. If a life of 5 years were assumed for the plant, instead of the assumed value of 10 years, the only factor in the study that would be changed would be the cost of capital recovery. If the salvage value is assumed to remain constant, the capital recovery cost over a 5-year period is

$$\$100{,}000\,(A/P,\ 15\%,\ 5) - \$20{,}000\,(A/F,\ 15\%,\ 5) = \$26{,}866 \text{ per year}$$

which is an increase of $7,926 over the initial value of $18,940. In this case the annual worth would be reduced to $45,696—a decline of 14.8%. Hence, a 50% reduction in useful life causes only a 14.8% reduction in annual worth. Clearly, the venture is rather insensitive to the assumed useful life.

With the added information supplied by the sensitivity analyses that have just been described, those who make the investment decision concerning the proposed concrete plant would be in a much better position than if they had only the initial study results, based on an assumed utilization of 75% of capacity, available to them. They would know which factors were critical and thus could seek more information about these particular items if desired.

10.8 OPTIMISTIC–PESSIMISTIC ESTIMATES

A useful method for exploring sensitivity is to estimate one or more factors in a favorable (optimistic) direction and in an unfavorable (pessimistic) direction to investigate the effect of these changes on study results. This is a simple method for including uncertainty in the analysis.

In applications of this method, the optimistic condition for a factor is often specified as a value that has one chance in twenty of being exceeded by the actual outcome. Similarly, the pessimistic condition has 19 chances out of 20 of being exceeded by the actual outcome.

As an example, consider a proposed ultrasound inspection device for which the optimistic, pessimistic, and most likely (or best) estimates are given in Table 10-3. Also shown at the end of Table 10-3 are the annual worths (AW) for all three estimation conditions. Note that the AW for optimistic conditions is highly favorable (+$73,995), while for pessimistic conditions it is quite unfavorable (−$33,100). After obtaining this information, the decision maker may be willing to make a go–no go decision on the proposed device. However, he or she should recognize that these are extreme outcomes—the optimistic AW assumes that *all* estimated factors turn out according to the optimistic estimates, and the pessimistic AW assumes that *all* estimated parameters turn out per the pessimistic estimates. It is reasonable to assume that this will not happen, but instead that different parameters may turn out to have a mixture of optimistic, most likely, and pessimistic outcomes. One good way to reflect such results is shown in Table 10-4, which shows annual worths for all combinations of estimated outcomes for three of the key factors being estimated. This could also have been done for four or more factors if they had been subject to significant variation, but the size of the matrix display would grow enormously. Even the $3 \times 3 \times 3$ matrix shown in Table 10-4 quickly becomes cumbersome because of the proliferation of numbers. There are ways around this, as we shall soon see.

10.8.1 Making Matrices Easier to Interpret

It should be recognized that the AW numbers in Table 10-4 result from estimates subject to varying degrees of uncertainty. Hence, little information of value would be lost if the numbers are rounded to the nearest thousand dollars. Further, suppose that management is most interested in the number of combinations of conditions in which the AW is, say (a) more than $50,000 and (b) less than $0. Table 10-5 shows how Table 10-4 might be changed to make it easier to interpret and use in communicating study results to management.

From Table 10-5 it is apparent that four combinations result in AW > $50,000 while nine produce AW < $0. Each combination of conditions is not necessarily equally likely. Therefore, statements such as "There are 9 chances out of 27 that we will lose money on this project" are not appropriate in this example.

10.8.2 Graphical Sensitivity Displays

An effective way of displaying and examining sensitivity is to graph the measure of merit for independent variation of all factors of interest by expressing variation for each on a common abscissa in terms of percent deviation from its

TABLE 10–3 Optimistic, Most Likely, and Pessimistic Estimates and Annual Worths for Proposed Ultrasound Device

	Estimation Condition		
	Optimistic (O)	Most Likely (M)	Pessimistic (P)
Investment	$150,000	$150,000	$150,000
Life	18 years	10 years	8 years
Salvage value	0	0	0
Annual revenues	$110,000	$ 70,000	$ 50,000
Annual expenses	$ 20,000	$ 43,000	$ 57,000
Minimum attractive rate of return	8%	8%	8%
AW	+$ 73,995	+$ 4,650	−$ 33,100

TABLE 10–4 Annual Worths ($) for All Combinations of Estimated Outcomes[a] for Annual Revenues, Annual Expenses, and Life—Proposed Ultrasound Device

	Annual Expenses								
	O			M			P		
	Life			Life			Life		
Annual Revenues	O	M	P	O	M	P	O	M	P
O	73,995	67,650	63,900	50,995	44,650	40,900	36,995	30,650	26,900
M	34,000	27,650	23,900	10,995	4,650	900	− 3,005	− 9,350	−13,100
P	14,000	7,650	3,900	−9,005	−15,350	−19,100	−23,005	−29,350	−33,100

[a] Estimates: O, optimistic; M, most likely; P, pessimistic.

TABLE 10–5 Results in Table 10–4 Made Easier to Interpret (Annual Worths in $000s)[a,b]

	Annual Expenses								
	O			M			P		
	Life			Life			Life		
Annual Revenues	O	M	P	O	M	P	O	M	P
O	(74)	(68)	(64)	(51)	45	41	37	31	27
M	34	28	24	11	5	1	<u>− 3</u>	<u>− 9</u>	<u>−13</u>
P	14	8	4	<u>− 9</u>	<u>−15</u>	<u>−19</u>	<u>−23</u>	<u>−29</u>	<u>−33</u>

[a] Estimates: O, optimistic; M, most likely, P, pessimistic.
[b] Circled entries, annual worth > $50,000 (4 out of 27 combinations); Underscored entries, annual worth < $0 (9 out of 27 combinations).

FIGURE 10-5

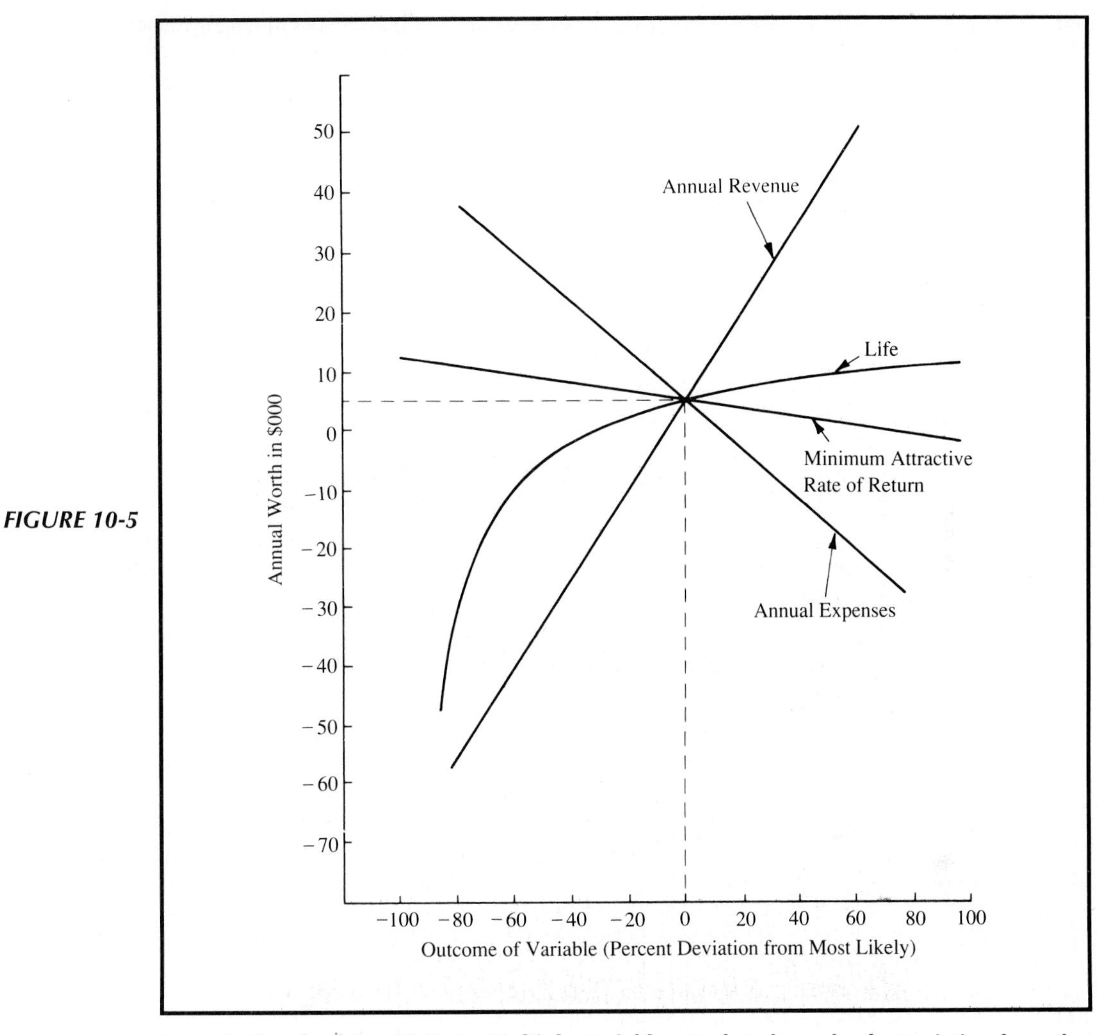

Example Graph of Sensitivity to Multiple Variables, Each Independently Deviating from the Most Likely Estimate

most likely value. This is shown in Figure 10-5 for the proposed ultrasound inspection device, for which the most likely estimates were given in Table 10-3. Figure 10-5 shows, among other things, that annual worth is relatively sensitive to changes in annual receipts, annual expenses, and reductions in project life. It also shows that annual worth is relatively insensitive to changes in the MARR and to increases in the project life.

Another type of sensitivity test that is often quite valuable is to determine the relative (or absolute) change in one or more factors that will just reverse the decision. Applied to the example in Table 10-3, this means determining the relative change in each factor that will decrease the AW by $4,650 so that it

TABLE 10–6 Sensitivity of Decision Reversal to Changes in Selected Estimates.

	Most Likely Estimate	Required Outcome[a]	Amount of Change	Change Amount as Percent of Most Likely
Investment	$150,000	$181,200	+$31,200	+20.8%
Life	10 years	7.3 years	−2.7 years	−27.0%
Salvage value	0	− 67,890	− 67,890	∞
Annual revenues	70,000	65,350	− 4,650	− 6.6%
Annual expenses	43,000	47,650	+ 4,650	+10.8%
Minimum attractive rate of return (MARR)	8%	12.5%	+4.5%	+56%

[a] To reverse decision (decrease AW to $0). Notice that reversal of AW is most sensitive to changes in annual revenues.

reaches $0. Table 10-6 shows this by using a table and bars of varying lengths to emphasize that the AW of the device is (1) most sensitive to changes in the estimated annual receipts, and (2) least sensitive to changes in the salvage value.

It is clear that even with a few factors the number of possible combinations of conditions in a sensitivity analysis can become quite large, and the task of investigating all of them might be quite time-consuming. This was made obvious in Example 10-4. Ordinarily, a sensitivity analysis involves eliminating from detailed consideration those factors for which the measure of merit is quite insensitive, and highlighting the conditions for other factors to be studied further in accordance with the degree of sensitivity of each. Thus, the number of combinations of conditions included in the analysis can perhaps be kept to a manageable size.

10.9 RISK-ADJUSTED MINIMUM ATTRACTIVE RATES OF RETURN

Uncertainty causes factors inherent to engineering economy studies, such as cash flows and project life, to become random variables in the analysis. (Simply stated, a random variable is a function that assigns a unique numerical value to each possible outcome of a probabilistic quantity.) A widely used industrial practice for explicitly including uncertainty is to increase the MARR when a project is thought to be relatively uncertain. Most likely estimates of other factors are then utilized in the study. Hence, a procedure has emerged that employs *risk-adjusted* interest rates. It should be noted, however, that many pitfalls of performing studies of financial profitability with risk-adjusted MARRs have been identified.*

In general, the preferred practice to account for uncertainty in estimates (cash flows, project life, etc.) is to deal directly with their suspected variations in terms of probability assessments rather than to manipulate the MARR as a means of reflecting the "virtually certain" versus "highly uncertain" status of a project. Intuitively, the risk-adjusted interest rate procedure makes good sense because much more certainty regarding the overall profitability of a project exists in the early years compared to, say, the last two years of its life. Increasing the MARR places emphasis on early cash flows rather than on longer-term benefits, and this would appear to compensate for time-related project uncertainties. But the question of uncertainty in cash flow amounts is not directly addressed. The following example illustrates how this method of dealing with uncertainty can lead to an illogical recommendation.

EXAMPLE 10-6

The Atlas Corporation is considering two alternatives, both affected by uncertainty to different degrees, for increasing the recovery of a precious metal from its smelting process. The following data concern capital investment requirements and estimated annual savings of both alternatives. The firm's MARR for its "risk-free" investments is 10%.

End-of-Year Cash Flow	Alternative P	Alternative Q
0	− $160,000	− $160,000
1	120,000	20,827
2	60,000	60,000
3	0	120,000
4	60,000	60,000

Because of technical considerations involved, alternative P is thought to be *more uncertain* than Q. Therefore, according to the Atlas Corporation's engineering economy handbook, the risk-adjusted MARR applied to P will be 20% and the risk-adjusted MARR for Q has been set at 17%. Which alternative should be recommended?

Solution

At the risk-free MARR of 10%, both alternatives have the same present worth of $39,659. All else being equal, alternative Q would be chosen because it is less uncertain than alternative P. Now a present worth (PW) analysis

*A. A. Robichek, and S. C. Myers, "Conceptual Problems in the Use of Risk-Adjusted Discount Rates," *Journal of Finance*, 21 (December 1966), 727–730.

is performed for the Atlas Corporation using its prescribed risk-adjusted MARRs for the two options.

$$\begin{aligned} PW_P(20\%) &= -\$160{,}000 \\ &\quad + \$120{,}000(P/F,\ 20\%,\ 1) + \$60{,}000(P/F,\ 20\%,\ 2) \\ &\quad + \$60{,}000(P/F,\ 20\%,\ 4) = \$10{,}602 \\ PW_Q(17\%) &= -\$160{,}000 + \$20{,}827(P/F,\ 17\%,\ 1) \\ &\quad + \$60{,}000(P/F,\ 17\%,\ 2) \\ &\quad + \$120{,}000(P/F,\ 17\%,\ 3) \\ &\quad + \$60{,}000(P/F,\ 17\%,\ 4) = \$8{,}575 \end{aligned}$$

Without considering uncertainty (i.e., MARR = 10%), the selection was seen to be alternative Q. But when the more uncertain alternative P is "penalized" by applying a higher risk-adjusted MARR to compute its PW, the comparison of alternatives favors alternative P! One would expect to see alternative Q recommended with this procedure. This contradictory result can be seen clearly in Figure 10-6, which demonstrates the general situation where contradictory results might be expected.

Even though the intent of the risk-adjusted MARR is to make more uncertain projects appear less economically attractive, the opposite was shown to be true in Example 10-6. Furthermore, a related shortcoming of the risk-adjusted MARR procedure is that cost-only projects are made to appear more desirable (less negative PW, for example) as the interest rate is adjusted upward to account for uncertainty. At extremely high interest rates, the alter-

FIGURE 10-6

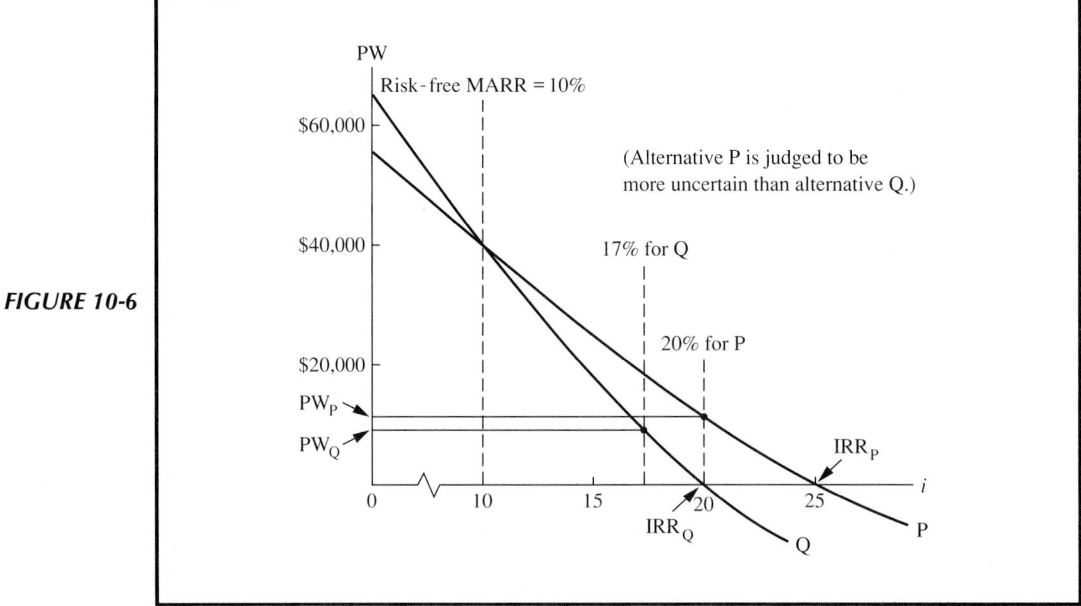

Graphical Portrayal of Risk-Adjusted Interest Rates

native having the lowest investment requirement would be favored regardless of subsequent cost cash flows. Because of such difficulties as those illustrated above, this procedure is not generally recommended as an acceptable means of dealing with uncertainty.

10.10 REDUCTION OF USEFUL LIFE

Many of the methods for dealing with uncertainty that have been discussed to this point have attempted to compensate for potential losses that could be incurred if conservative decision-making practices are not followed. Thus, dealing with uncertainty in an engineering economy study tends to lead to the adoption of conservative (pessimistic) estimates of factors so as to reduce "downside risks" of making a wrong decision.

The method considered in this section makes use of a truncated project life that is often considerably less than the estimated useful life. By dropping from consideration those revenues (savings) and costs that may occur after the reduced study period, heavy emphasis is placed on rapid recovery of investment capital in the early years of a project's life. Consequently, this method is closely related to the discounted payback technique discussed in Chapter 4, and it suffers from most of the same deficiencies that beset the payback method.

EXAMPLE 10-7

Suppose that the Atlas Corporation referred to in Example 10-6 decided not to utilize risk-adjusted interest rates as a means of recognizing uncertainty in its engineering economy studies. Instead, they have decided to truncate the study period at 75% of the most likely estimate of useful life. Hence, all cash flows past the third year would be ignored in the analysis of alternatives. By using this method, should alternative P or Q be selected when MARR = 10%?

Solution

Based on the present worth criterion, it is apparent that neither alternative would be the choice with this procedure for recognizing uncertainty.

$$\begin{aligned} PW_P(10\%) &= -\$160{,}000 + \$120{,}000(P/F, 10\%, 1) \\ &\quad + \$60{,}000(P/F, 10\%, 2) = -\$1{,}322 \\ PW_Q(10\%) &= -\$160{,}000 + \$20{,}827(P/F, 10\%, 1) \\ &\quad + \$60{,}000(P/F, 10\%, 2) \\ &\quad + \$120{,}000(P/F, 10\%, 3) = -\$1{,}322 \end{aligned}$$

EXAMPLE 10-8

A proposed new product line requires $2,000,000 in capital over a 2-year construction period. Projected receipts and expenses over this product's anticipated 8-year commercial life are shown, along with its capital requirements.

Type of Cash Flow	–1	0	1	2	3	4	5	6	7	8
						End of Year (Millions of $)				
Investment	–0.9	–1.1	0	0	0	0	0	0	0	0
Receipts	0	0	1.8	2.0	2.1	1.9	1.8	1.8	1.7	1.5
Expenses	0	0	–0.8	–0.9	–0.9	–0.9	–0.8	–0.8	–0.8	–0.7

The company's maximum payback period is 4 years (after taxes), and its after-tax MARR is 15% per year. This investment will be depreciated by the MACRS method using a 5-year class life. (MACRS rates are in Table 6-5). An effective income tax rate of 40% applies to taxable income produced by this new product.

Management of the company is quite concerned about the financial attractiveness of this venture if unforeseen circumstances (e.g., loss of market and/or technological breakthroughs) occur. They are leery of investing a large sum of capital in this product because competition is quite keen and companies that wait to enter the market may be able to purchase more cost-efficient technology. You have been given the assignment of assessing the "downside profitability" of the product when the primary concern is its staying power (life) in the marketplace. That is, determine the minimum life of the product that will produce an acceptable after-tax IRR. Draw a graph of your results and list all appropriate assumptions.

Solution

An analysis of after-tax cash flows is shown in Table 10-7 for the "most likely" product life of 8 years.

TABLE 10–7 After-Tax Analysis of Example 10-8

Year	(A) Before-Tax Cash Flow	(B) Depreciation Deduction	(C) = (A) – (B) Taxable Income	(D) = –0.4(C) Cash Flow for Income Taxes	(E) = (A) + (D) After-Tax Cash Flow
–1	–$ 900,000	—	—	—	–$ 900,000
0	– 1,100,000	—	—	—	– 1,100,000
1	1,000,000	$400,000	$600,000	–$240,000	760,000
2	1,100,000	640,000	460,000	– 184,000	916,000
3	1,200,000	384,000	816,000	– 326,400	873,600
4	1,000,000	230,400	769,600	– 307,840	692,160
5	1,000,000	230,400	769,600	– 307,840	692,160
6	1,000,000	115,200	884,800	– 353,920	646,080
7	900,000	0	900,000	– 360,000	540,000
8	800,000	0	800,000	– 320,000	480,000

FIGURE 10-7

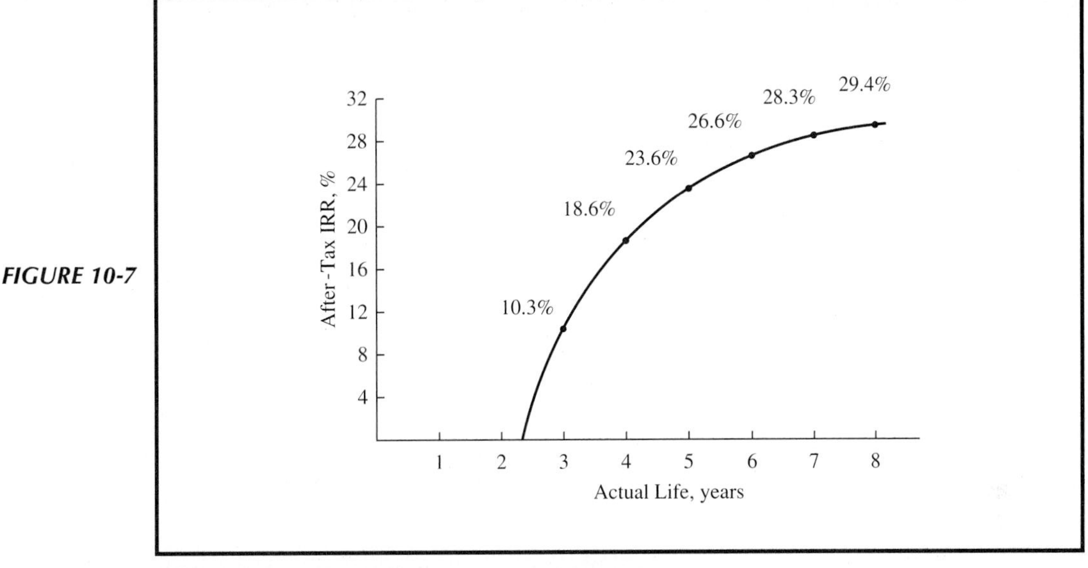

IRR for Different Product Lives in Example 10-8

It has been assumed that the residual (salvage) value of the investment is zero. Moreover, the MACRS depreciation deductions are assumed to be unaffected by the useful life of this product, and they begin in the first year of commercial operation (year 1). A plot of after-tax IRR versus actual life of the product line is shown in Figure 10-7. To make at least 15% per year after taxes on this venture, the product's life must be 4 years or more. It can be quickly determined from Table 10-7 that the *simple* payback period is 3 years. Consequently, this new product would appear to be a judicious investment as long as its actual life turns out to be 4 years or greater.

10.11 PROBABILITY FUNCTIONS—THEIR EXPECTED VALUE AND VARIANCE

Expected values and *variances* are often helpful in making decisions when uncertainty is involved. Expected values and variances are associated with the concept of probability, which is usually considered to be the long-run relative frequency with which an event occurs, or the subjectively judged likelihood that it will occur.

The expected value is the *product* of the probability that a revenue, cost, or other factor will occur and its value (numerical outcome) if it does occur, summed over all possible outcomes of the factor. Thus, if the useful life of a project is estimated to be 10 years with 0.4 probability and 15 years with 0.6 probability, the expected project life is 10(0.4) + 15(0.6) = 4 + 8 = 12 years. Factors having probabilistic outcomes are often called *random variables*.

In general, the expected value of a random variable whose discrete outcomes, x_i, occur in accordance with a known probability function can be computed as

$$E(X) = \sum_{i=1}^{n} x_i \Pr(x_i) \tag{10–1}$$

where $E(X)$ = expected value of the random variable X
$\Pr(x_i)$ = probability of outcome x_i occurring

$$\sum_{i=1}^{n} \Pr(x_i) = 1; \qquad i = 1, 2, \ldots, n \text{ outcomes}$$

The variance of a random variable is a measure of how much dispersion in possible outcomes exists about the expected value and it tends to indicate the degree of uncertainty associated with a random variable. Specifically, the variance of a discrete random variable, X, is given by the following equation:

$$V(X) = \sum_{i=1}^{n} x_i^2 \Pr(x_i) - [E(X)]^2 \tag{10–2}$$

The standard deviation of X, SD (X), is merely the square root of $V(X)$.

Expected value and variance concepts apply theoretically to long-run conditions in which it is assumed the event is going to occur repeatedly. However, application of these principles is often useful even when investments are not going to be made repeatedly over the long run.

EXAMPLE 10-9

We now apply the expected value and variance principles to the premixed-concrete plant discussed in Example 10-5. Suppose that the probabilities of attaining various capacity utilizations are

Capacity (%)	Probability
50	0.10
65	0.30
75	0.50
90	0.10

It is desired to determine the expected value and variance of *annual revenue*. Subsequently, the expected value and variance of AW for the project can be computed. By evaluating both E(AW) and V(AW) for the concrete plant, indications of the venture's average profitability and its uncertainty are obtained.

Solution for Annual Revenue:

i	Capacity (%)	(A) Probability, $Pr(x_i)$	(B) Revenue[a] x_i	(A) × (B): Expected Revenue	(C) = $(B)^2$ x_i^2	(A) × (C)
1	50	0.10	$405,000	$ 40,500	1.64×10^{11}	0.164×10^{11}
2	65	0.30	526,500	157,950	2.77×10^{11}	0.831×10^{11}
3	75	0.50	607,500	303,750	3.69×10^{11}	1.845×10^{11}
4	90	0.10	729,000	72,900	5.31×10^{11}	0.531×10^{11}
				$575,100		3.371×10^{11}

[a] From Table 10–1.

Expected value of annual revenue: $\sum$ (A × B) = $575,100
Variance of annual revenue: $\sum$ (A × C) − $(575{,}100)^2 = 6{,}400 \times 10^6$

Solution for Annual Worth:

i	Capacity (%)	(A) $Pr(x_i)$	(B) Annual Worth[a], x_i	(A) × (B): Expected AW	(C) = $(B)^2$ $(AW)^2$	(A) × (C)
1	50	0.10	−$ 25,093	−$ 2,509	0.63×10^9	0.063×10^9
2	65	0.30	22,136	6,641	0.49×10^9	0.147×10^9
3	75	0.50	53,622	26,811	2.88×10^9	1.440×10^9
4	90	0.10	100,850	10,085	10.17×10^9	1.017×10^9

[a] See Table 10-1.

Expected value of AW: $\sum$ (A × B) = $41,028
Variance of AW: $\sum$ (A × C) − $(41{,}028)^2 = 984 \times 10^6$
Standard deviation of AW: $31,370

The standard deviation of AW is less than the expected AW, and only the 50% capacity utilization situation results in a negative AW. Consequently, the investors in this undertaking may well judge the venture to be an acceptable one.

10.12

EVALUATION OF ALTERNATIVES CONSIDERING PROBABILISTIC CONSEQUENCES

There are situations, such as flood control projects, in which future losses due to natural or human-made risks can be decreased by increasing the amount of capital that is invested. Drainage ditches or dams, built to control floodwaters, may be constructed in different sizes, costing different amounts. If correctly designed and used, the larger the size, the smaller will be the resulting damage loss when a flood occurs. As we might expect, the most economical size would provide satisfactory protection against most floods, although it could be anticipated that some overloading and damage may occur at infrequent periods.

EXAMPLE 10-10

A drainage ditch in a mountain community in the West where flash floods are experienced has a capacity sufficient to carry 700 cubic feet per second. Engineering studies produce the following data regarding the probability that a given water flow in any one year will be exceeded and the cost of enlarging the ditch:

Water Flow (ft^3/sec)	Probability of a Greater Flow Occurring in Any One Year	Investment to Enlarge Ditch to Carry This Flow
700	0.20	—
1,000	0.10	$20,000
1,300	0.05	30,000
1,600	0.02	44,000
1,900	0.01	60,000

Records indicate that the average property damage amounts to $20,000 when serious overflow occurred. It is believed that this would be the average damage whenever the storm flow *was greater* than the capacity of the ditch. Reconstruction of the ditch would be financed by 40-year bonds bearing 8% interest. It is thus computed that the capital recovery cost for debt repayment (principal of the bond plus interest) would be 8.39% of the initial cost, because $(A/P, 8\%, 40) = 0.0839$. It is desired to determine the most economic ditch size (water flow capacity).

Solution

The total expected equivalent annual cost for the structure and property damage for all alternative ditch sizes would be as follows:

Water Flow (ft^3/sec)	Capital Recovery Cost	Expected Annual Property Damage[a]	Total Expected Equivalent Annual Cost
700		\$20,000(0.20) = \$4,000	\$4,000
1,000	\$20,000(0.0839) = \$1,678	20,000(0.10) = 2,000	3,678
1,300	30,000(0.0839) = 2,517	20,000(0.05) = 1,000	3,517
1,600	44,000(0.0839) = 3,692	20,000(0.02) = 400	4,092
1,900	60,000(0.0839) = 5,034	20,000(0.01) = 200	5,234

[a]These amounts are obtained by multiplying \$20,000 by the probability of greater water flow occurring.

From these calculations it may be seen that the minimum expected annual cost would be achieved by enlarging the ditch so that it would carry 1,300 cubic feet per second, with the expectation that a greater flood might occur 1 year out of 20 on the average and cause property damage of \$20,000.

It should be noted that when loss of life or limb might result such as in Example 10-10, there usually is considerable pressure to disregard pure economy and build such projects in recognition of the nonmonetary values associated with human safety.

The following example illustrates the same principles in Example 10-10 except that it applies to safety alternatives involving electrical circuits.

EXAMPLE 10-11

There are three alternatives being evaluated for the protection of electrical circuits, with the following required investments and probabilities of failure:

Alternative	Investment	Probability of Loss in Any Year
A	\$ 90,000	0.40
B	100,000	0.10
C	160,000	0.01

If a loss does occur, it will cost \$80,000 with probability 0.65, and \$120,000 with probability 0.35. The probabilities of loss in any year are independent of the probabilities associated with the resultant cost of a loss if one does occur. Each alternative has a useful life of 8 years and no salvage value at that time. The MARR is 12%, and annual maintenance cost is expected to be 10% of the capital investment. It is desired to determine which alternative is best based on expected total annual costs.

Solution

The expected cost of a loss, if it occurs, can be calculated as

$$\$80{,}000(0.65) + \$120{,}000(0.35) = \$94{,}000$$

Alternative	Capital Recovery Cost = Investment × (*A/P*, 12%, 8)	Annual Maintenance = Investment × (0.10)	Expected Annual Cost of Failure	Total Expected Equivalent Annual Cost
A	\$ 90,000(0.2013) = \$18,117	\$ 9,000	\$94,000(0.40) = \$37,600	\$64,717
B	100,000(0.2013) = 20,130	10,000	94,000(0.10) = 9,400	39,530
C	160,000(0.2013) = 32,208	16,000	94,000(0.01) = 940	49,148

Thus, alternative B is the best based on total expected annual cost, which is a long-run average cost. However, one might rationally choose alternative C so as to reduce significantly the chance of an \$80,000 or \$120,000 loss occurring in any year in return for a 24.3% increase in the total expected equivalent annual cost.

One of the major problems when expected values are to be computed is the determination of the probabilities. In most situations there is no history of previous cases for the particular venture being considered. Therefore, probabilities seldom can be based on historical data and rigorous statistical procedures. In most cases it is necessary that the analyst, or person making the decision, rely on judgment or even intuition in estimating the probabilities. This fact makes some persons hesitate to use the expected-value concept, inasmuch as they cannot see the value in applying such a technique to improve the evaluation of uncertainty when so much apparent subjectivity is present. Although this argument has merit, the fact is that economy studies always deal with future events and there must be an extensive amount of estimating. Furthermore, even if the probabilities could be based accurately on past history, there rarely is any assurance that the future will repeat the past. In such situations structured methods for assessing "subjective" probabilities are used often in practice.* Also, even if we must estimate the probabilities, the very process of doing so requires us to give some thought to the uncertainty that is inherent in all estimates going into the analysis. Such forced thinking is likely to produce better results than little or no thinking about such matters.

*For further information, see W. G. Sullivan and W. W. Claycombe, *Fundamentals of Forecasting* (Reston, Va.: Reston Publishing Company, Inc., 1977), Chap. 6.

10.13

EVALUATIONS THAT CONSIDER UNCERTAINTY, USING EXPECTED VALUES AND VARIANCES

In addition to expected values of random variables to deal with uncertainty, some situations require the variance of the variable to be also taken into account in adequately portraying the dispersion in study results. Therefore, the uncertainty associated with an alternative can be represented more realistically in terms of its variability in addition to its expected value.

The first procedure for dealing with probabilistic factors is to compute mathematically their expected values and variances. This is referred to as "closed-form" analysis. A second general procedure for treating probabilistic information is to utilize Monte Carlo simulation.

One general type of problem to which both procedures can be applied involves a cash flow that varies according to some known probability function and a project life that is certain. In this type of problem, the expected value and variance of the equivalent worth of the cash flow can be determined. A second type of problem involves cash flows that are certain and a project life with a known probability function. These first two types of problems are addressed in this section. A third type of problem involves probabilistic cash flows and probabilistic project lives. This is a more complicated type of problem and is analyzed with Monte Carlo simulation methods in a later section.

The following three examples demonstrate closed-form analysis for the first two types of problems mentioned above. In these examples, all variables are assumed to be statistically independent.

EXAMPLE 10-12

Assume that net annual benefits for a project during each year of its life are discretely estimated and have the following probabilities.

Net Annual Benefits (NAB)	Pr(NAB)
$2,000	0.40
3,000	0.50
4,000	0.10

The life is 3 years for certain and the initial investment is $7,000, with negligible salvage value. The MARR is 15%. Determine E(PW) and the probability that present worth is greater than zero [i.e., Pr (PW $\geq$ 0)].

Solution

The present worth for each value of NAB can be determined in this manner:

$$\text{PW} = -\$7{,}000 + \text{NAB}\ (P/A,\ 15\%,\ 3)$$
$$\text{PW} = -\$7{,}000 + \text{NAB}\ (2.2832)$$

If the NAB Is	Then PW Equals	Which Occurs with Probability:
$2,000	-$2,434	0.40
3,000	- 150	0.50
4,000	2,133	0.10

$$E(\text{PW}) = (0.40)\ (-\$2{,}434) + (0.50)\ (-\$150) + (0.10)\ (\$2{,}133)$$
$$= -\$835$$

Another method of working this problem is:

$$E(\text{NAB}) = \$2{,}000(0.40) + \$3{,}000(0.50) + \$4{,}000(0.10) = \$2{,}700$$
$$E(\text{PW}) = -\$7{,}000 + E(\text{NAB})(P/A,\ 15\%,\ 3)$$
$$= -\$7{,}000 + \$2{,}700\ (2.2832) = -\$835$$

The probability of each value of present worth is illustrated in Figure 10-8, which shows that Pr (PW $\geq$ 0) = 0.10.

Of perhaps more interest is the situation in which annual cash flows are random variables having estimated expected values and variances. In this

FIGURE 10-8

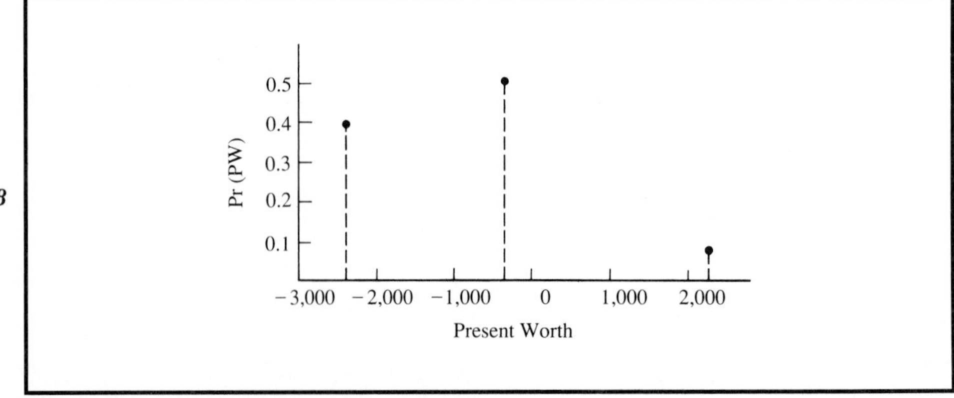

Probability of Present Worth in Example 10–12

case, the analyst frequently assumes that annual cash flows are distributed according to a normal distribution.* The example below illustrates normally distributed and independent cash flows during each year of a project that has a certain life of 3 years.

EXAMPLE 10-13

For the following cash flow estimates, find E(PW), V(PW), and Pr (IRR $\leq$ MARR). Assume that the cash flows are normally distributed and that MARR = 15%.

Year, k	Expected Value of Cash Flow, A_k	Standard Deviation (SD) of Cash Flow, A_k
0	−$7,000	0
1	3,500	$600
2	3,000	500
3	2,800	400

A graphical portrayal of these normally distributed cash flows is shown in Figure 10-9.

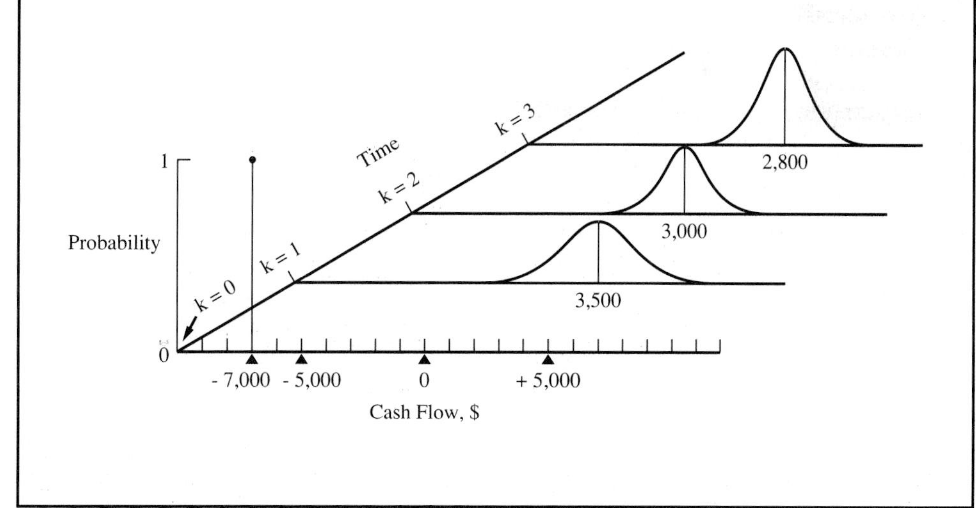

FIGURE 10-9 ***Probabilistic Cash Flows over Time***

*This frequently encountered probability function is discussed in any good statistics book, such as R. L. Scheaffer and J. T. McClare, *Statistics for Engineers* (Boston: Duxbury Press, 1982).

Solution

The expected present worth is calculated as follows:

$$E(\text{PW}) = -\$7{,}000 + \$3{,}500(P/F,\ 15\%,\ 1) + \$3{,}000(P/F,\ 15\%,\ 2) + \$2{,}800(P/F,\ 15\%,\ 3)$$
$$E(\text{PW}) = \$153$$

To determine $V(\text{PW})$, we use this relationship:

$$V(C_k A_k) = C_k^2 V(A_k)$$

where A_k is the net cash flow in year k $(0 \leq k \leq N)$ and C_k is the single-payment present worth factor $(P/F,\ 15\%,\ k)$. Note that $V(A_k) =$ (standard deviation of $A_k)^2$. Furthermore,

$$\sum_{k=1}^{N} V(C_k A_k) = \sum_{k=1}^{N} C_k^2 V(A_k)$$

Thus,

$$V(\text{PW}) = 0^2 1^2 + 600^2(P/F,\ 15\%,\ 1)^2 + 500^2(P/F,\ 15\%,\ 2)^2 + 400^2(P/F,\ 15\%,\ 3)^2 = 484{,}324$$

and

$$\text{SD(PW)} = \sqrt{V(\text{PW})} = \$696$$

where SD(PW) is the standard deviation of PW. For a decreasing PW(i) function having a unique internal rate of return, the probability that the IRR is less than the MARR is the same as the probability that PW is less than 0.

Consequently, by using the standardized normal distribution in Appendix G, we can determine the probability that PW is less than zero by utilizing the fact that the sum of independently distributed normal random variables is also normally distributed.*

*A random variable, X, is normally distributed with mean μ and standard deviation σ in accordance with the following equation:

$$f(X) = \frac{1}{\sigma\sqrt{2\pi}} \exp\left\{-\left[\frac{(X-\mu)^2}{2\sigma^2}\right]\right\}$$

The standardized normal distribution, $f(Z)$, of the variable $Z = (X - \mu)/\sigma$ has a mean of 0 and a standard deviation of 1.

$$Z = \frac{\text{PW} - E(\text{PW})}{\text{SD (PW)}} = \frac{0 - 153}{696} = -0.22$$

$$\Pr(\text{PW} \leq 0) = \Pr(Z \leq -0.22)$$

From Appendix G we find that $\Pr(Z \leq -0.22) = 0.4129$.

The approach used in Example 10-13 can be generalized with two formulas. For independent random variables, the expected value of the present worth is given by

$$E(\text{PW}) = \sum_{k=0}^{N} (1 + i)^{-k} E(A_k) \qquad (10\text{–}3)$$

where A_k is the net cash flow in year k, $E(A_k)$ is the expected value of A_k, and N is the certain project life. Similarly, the variance is given by

$$V(\text{PW}) = \sum_{k=0}^{N} (1 + i)^{-2k} V(A_k) \qquad (10\text{–}4)$$

Note that the $(1 + i)^{-2k}$ term equals either $(P/F, i, 2k)$ or $(P/F, i, k)^2$.

A different approach must be used for the second general type of problem in which the project life is a random variable. In general, the expected value of present worth can be obtained by determining the expected value of the series present worth factor and *not* by determining the expected life of the project. The following example illustrates this principle for a project having a useful life that is a random variable and uniform net annual benefits that are known with certainty.

EXAMPLE 10-14

Suppose that the probability of project life of a proposed venture is discretely distributed as follows:

Life, N	Pr(N)
2 years	0.40
3 years	0.50
4 years	0.10

The net annual benefits equal $4,000 over the life of the project, and the initial investment is $8,000 with no salvage value. Determine the expected value and variance of present worth, assuming the MARR equals 15%.

Solution

The expected value is computed as follows:

$$
\begin{aligned}
E(\text{PW}) &= -\$8{,}000 + \$4{,}000 \sum_{N=2}^{4} (P/A,\ 15\%,\ N)\ \Pr(N) \\
&= -\$8{,}000 + \$4{,}000[1.626(0.40) + 2.283(0.50) \\
&\quad + 2.855(0.10)] \\
&= \$310
\end{aligned}
$$

For a 2-year project life, the present worth is $-\$8,000 + \$4,000(1.626) = -\$1,496$ with probability 0.40. For a 3-year life, the PW is $+\$1,132$ with probability 0.50 that life will equal 3 years. Similarly, the PW is $+\$3,420$ when the project life is 4 years, with probability 0.10. The expected PW is identical to that obtained above:

$$
\begin{aligned}
E(\text{PW}) &= -\$1{,}496(0.40) + \$1{,}132(0.50) + \$3{,}420(0.10) \\
&= \$310
\end{aligned}
$$

By inspection it is apparent that $\Pr(\text{PW} \leq 0) = 0.40$.

From Equation 10-2, the variance of the present worth is determined in this manner:

$$V(\text{PW}) = E(\text{PW})^2 - [E(\text{PW})]^2$$

$$
\begin{aligned}
\text{where } E(\text{PW})^2 &= \sum_{N=2}^{4} [-\$8{,}000 + \$4{,}000(P/A,\ 15\%,\ N)]^2\ \Pr(N) \\
&= (-\$1{,}496)^2(0.40) + (\$1{,}132)^2(0.50) \\
&\quad + (\$3{,}420)^2(0.10) \\
&= 2{,}705{,}558
\end{aligned}
$$

Thus,

$$V(\text{PW}) = 2{,}705{,}558 - (310)^2 = 2{,}609{,}458$$

and

$$\text{SD}(\text{PW}) = \$1{,}615$$

The reader should realize that the examples presented in this section have been very simple and have involved straightforward cash flows. Also, all variables have been assumed to be independent. It should be apparent that additional complications in a problem formulation could result in a very complex and time-consuming solution. Monte Carlo simulation techniques are typically used in such situations.

10.14

EVALUATION OF UNCERTAINTY, USING MONTE CARLO SIMULATION*

The development of computers has resulted in the increased use of Monte Carlo simulation as an important tool for analysis of project uncertainties. For complicated problems, Monte Carlo simulation generates random outcomes for probabilistic factors so as to imitate the randomness inherent in the original problem. In this manner a solution to a rather complex problem can be inferred from the behavior of these random outcomes.

To perform a simulation analysis, the first step is to construct an analytical model that represents the actual investment opportunity. This may be as simple as developing an equation for the present worth of a proposed industrial robot in an assembly line or as complex as examining the effects of various United States–imposed trade barriers on sales of a proposed new product in international markets. The second step is development of a probability distribution from subjective or historical data for each uncertain factor in the model. Sample outcomes are randomly generated by using the probability distribution for each uncertain quantity and then utilized to determine a *trial* outcome for the model. Repeating this sampling process a large number of times leads to a frequency distribution of trial outcomes for a desired measure of merit, such as present worth or annual worth. The resulting frequency distribution can then be used to make probabilistic statements about the original problem.

To illustrate the Monte Carlo simulation procedure, suppose that the probability distribution for the useful life of a piece of machinery has been estimated as shown in Table 10-8.

The useful life can be simulated by assigning random numbers to each value such that they are proportional to the respective probabilities. (A random number is selected in a manner such that each number has an equal probability of

TABLE 10–8 Probability Distribution for Useful Life

Number of Years, N		$\Pr(N)$	
3	possible values	0.20	$\sum \Pr(N) = 1.00$
5		0.40	
7		0.25	
10		0.15	

*Adapted from W. G. Sullivan and R. Gordon Orr, "Monte Carlo Simulation Analyzes Alternatives in Uncertain Economy," *Industrial Engineering* Vol. 14, No. 11 (November 1982). Reprinted with permission from *Industrial Engineering* magazine. Copyright ©Institute of Industrial Engineers, Inc., 25 Technology Park/Atlanta, Norcross, GA.

TABLE 10–9 Assignment of Random Numbers

Number of Years, N	Random Numbers
3	00–19
5	20–59
7	60–84
10	85–99

occurrence.) Because two-digit probabilities are given in Table 10-8, random numbers can be assigned to each outcome as shown in Table 10-9. Next, a single outcome is simulated by choosing a number at random from a table of random numbers.* For example, if any random number between and including 00 and 19 is selected, the useful life is 3 years. As a further example, the random number 74 corresponds to a life of 7 years.

If the probability distribution that describes a random variable is *normal*, a slightly different approach is followed. Here the simulated outcome is based on the mean and standard deviation of the probability distribution and on a random normal deviate, which is a random number of standard deviations above or below the mean of a standardized normal distribution. An abbreviated listing of typical random normal deviates is shown in Table 10-10. For normally distributed random variables, the simulated outcome is based on Equation 10-5:

$$\text{outcome value} = \text{mean} + [\text{random normal deviate} \times \text{standard deviation}] \qquad (10\text{–}5)$$

For example, suppose that net *annual* cash flow is assumed to be normally distributed with a mean of $50,000 and a standard deviation of $10,000, as shown in the diagram on page 485.

TABLE 10–10 Random Normal Deviates (RNDs)

−1.565	0.690	−1.724	0.705	0.090
0.062	−0.072	0.778	−1.431	0.240
0.183	−1.012	−0.844	−0.227	−0.448
−0.506	2.105	0.983	0.008	0.295
1.613	−0.225	0.111	−0.642	−0.292

*The last two digits of randomly chosen telephone numbers in a telephone directory are usually quite close to being random numbers.

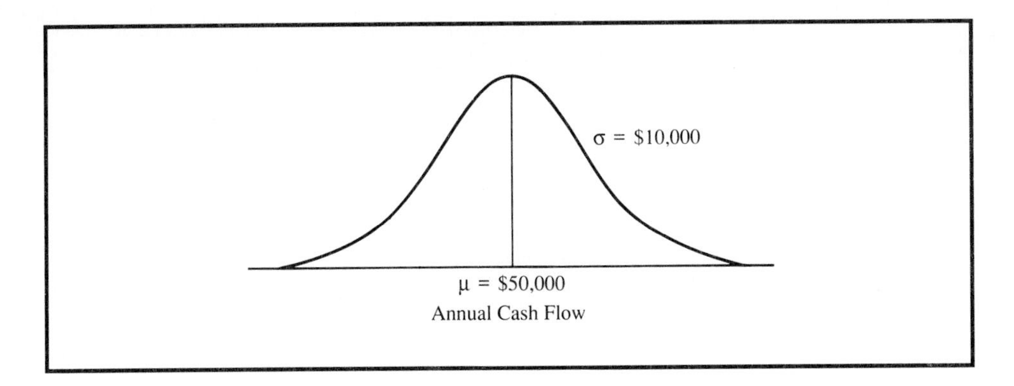

Simulated cash flows for a period of 5 years are listed in Table 10-11.

Notice that the average net annual cash flow is $248,850/5, which equals $49,770. This approximates the known mean of $50,000 with an error of 0.46%.

If the probability distribution that describes a random event is *uniform* and continuous with a minimum value of A and a maximum value of B, another procedure should be followed to determine the simulated outcome. Here the outcome can be computed with Equation 10-6:

$$\text{simulation outcome} = A + \frac{RN}{RN_m}[B - A] \qquad (10\text{–}6)$$

where RN_m is the maximum possible random number (9 if one digit is used, 99 if two, etc.) and RN is the random number actually selected. This equation should be used when the minimum outcome, A, and the maximum outcome, B, are known.

For example, suppose that the salvage value in year N is uniformly and continuously distributed between $8,000 and $12,000. A value of this random variable would be generated as follows with a random number of 74:

$$\text{simulation outcome} = \$8{,}000 + \frac{74}{99}(\$12{,}000 - \$8{,}000) = \$10{,}990$$

TABLE 10–11 Example of the Use of Random Normal Deviates

Year	Random Normal Deviate (RND)	Net Annual Cash Flow [$50,000 + RND ($10,000)]
1	0.090	$50,900
2	0.240	52,400
3	−0.448	45,520
4	0.295	52,950
5	−0.292	47,080

Proper use of these procedures, coupled with an accurate model, will result in an approximation to the actual outcome. But how many simulation trials are necessary for an *accurate* approximation of, for example, the average outcome? In general, the greater the number of trials, the more accurate that the approximation of the mean and standard deviation will be. One method to determine whether a sufficient number of trials has been conducted is to keep a running average of results. At first, this average will vary considerably from trial to trial. The amount of change between successive averages should decrease as the number of simulation trials increases. Eventually, this running (cumulative) average should level off at an accurate approximation.

EXAMPLE 10-15

This example illustrates how Monte Carlo simulation can simplify the analysis of a relatively complex problem. The estimates provided below relate to a capital investment opportunity being considered by a large manufacturer of air-conditioning equipment. Subjective probability functions have been estimated for the four independent uncertain factors as follows.

INVESTMENT

Normally distributed with a mean of −$50,000 and a standard deviation of $1,000.

USEFUL LIFE

Uniformly distributed with a minimum life of 10 years and a maximum of 14 years.

ANNUAL REVENUE

$35,000 with a probability of 0.4
$40,000 with a probability of 0.5
$45,000 with a probability of 0.1

ANNUAL EXPENSE

Normally distributed with a mean of −$30,000 and a standard deviation of $2,000.

Management of this company wishes to determine the probability that the investment will be a profitable one. The interest rate is 10%. In order to answer this question, the present worth (PW) of the venture will be simulated.

Solution

For purposes of illustrating the Monte Carlo simulation procedure, five trial outcomes are computed manually in Table 10-12. The estimate of the average present worth based on this very small sample is $19,004/5 = $3,801. For

TABLE 10–12 Monte Carlo Simulaton of PW Involving Four Independent Factors

Trial Number	Random Normal Deviate (RND_1)	Investment, I [\$50,000 + RND_1 (\$1,000)]	Three-Digit Random Numbers (RN)	Project Life, N $\left[10 + \frac{RN}{999}(14-10)\right]$	Project Life, N (Nearest Integer)
1	−1.003	\$48,997	807	13.23	13
2	−0.358	49,642	657	12.63	13
3	1.294	51,294	488	11.95	12
4	−0.019	49,981	282	11.13	11
5	0.147	50,147	504	12.02	12

Trial Number	One-Digit Random Number	Annual Revenue, R \$35,000 for 0−3; 40,000 for 4–8; 45,000 for 9	Random Normal Deviate (RND_2)	Annual Expense, E [\$30,000 + RND_2(\$2,000)]	$PW = -I + (R - E)(P/A, 10\%, N)$
1	2	\$35,000	−0.036	\$29,928	−\$12,970
2	0	35,000	0.605	31,210	− 22,724
3	4	40,000	1.470	32,940	− 3,189
4	9	45,000	1.864	33,728	+ 23,233
5	8	40,000	−1.223	27,554	+ 34,654
					+\$19,004

more accurate results, hundreds or even thousands of repetitions would be required.

The applications of Monte Carlo simulation for investigating uncertainty are many and varied. However, it must be remembered that the results can be no more accurate than the model and the probability estimates used. In all cases the procedure and rules are the same: careful study and development of the model; accurate assessment of the probabilities involved; true randomization of outcomes as required by the Monte Carlo procedure; and calculation and analysis of the results. Furthermore, a sufficiently large number of Monte Carlo trials should always be used to reduce the estimation error to an acceptable level.

10.15 PERFORMING MONTE CARLO SIMULATION WITH A COMPUTER

It is apparent from the preceding section that a Monte Carlo simulation of a complex project requiring several thousand trials could be accomplished only with the help of a computer. Indeed, there are numerous simulation programs that can be obtained from software companies and universities. To illustrate the computational features and output of a typical simulation program, Example 10-15 has been evaluated with a computer program developed by R. Gordon Orr, an industrial engineering graduate student at The University of Tennessee. The computer queries and user responses (shown in boxes) are shown in Figure 10-10. Simulation results for 3,160 trials are shown in Figure 10-11. (This number of trials was needed for the cumulative average present worth to stabilize to a variation of ±0.5%.)

The average PW is $7,759.60, which is larger than the $3,801 obtained from Table 10-11. This underscores the importance of having a sufficient number of simulation trials to ensure reasonable accuracy in Monte Carlo analyses.

The histogram in Figure 10-11 indicates that the *median present worth* of this investment is $6,700 and that the dispersion of present worth trial outcomes is considerable. The standard deviation of simulated trial outcomes is one way to measure this dispersion. Based on Figure 10-11, 59.5% of all simulation outcomes have a present worth of $0 or greater. Consequently, this project may be too risky for the cautious company to undertake because the "downside" risk of failing to realize at least a 10% return on the investment is about 4 chances out of 10. Perhaps another investment should be considered.

```
THE FOLLOWING PROGRAM USES MONTE CARLO SIMULATION
TECHNIQUES AS APPLIED TO RISK ANALYSIS PROBLEMS OF
ENGINEERING ECONOMY.

WILL YOU BE USING A REMOTE PRINTER FOR OUTPUT ? (Y OR
N) [Y]

INPUT A RANDOM NUMBER BETWEEN 1 AND 1000. [199]

MAXIMUM NUMBER OF ITERATIONS YOU WISH TO RUN ? [1000]

WHAT INTEREST RATE (PERCENT) IS TO BE USED ? [10]

THE DATA FOR EACH RANDOM VARIABLE INVOLVED MAY BE
FORMULATED AS FOLLOWS:

1.      SINGLE VALUE OR ANNUITY
2.      SINGLE VALUE WITH ARITHMETIC GRADIENT
3.      SINGLE VALUE WITH GEOMETRIC GRADIENT
4.      DISCRETE DISTRIBUTION
5.      UNIFORM DISTRIBUTION
6.      NORMAL DISTRIBUTION
7.      A SERIES OF YEARLY CASH FLOWS
8.      SALVAGE VALUE DEPENDENT ON PROJECT LIFE
9.      TRIANGULAR DISTRIBUTION

INFORMATION FOR INITIAL CASH FLOW:

    DISTRIBUTION IDENTIFICATION NUMBER = [6]

    MEAN VALUE = [-50000]

    STANDARD DEVIATION = [1000]

INFORMATION FOR YEARLY CASH FLOW:

    THIS CASH FLOW MAY CONSIST OF A NUMBER OF
    DIFFERENT ELEMENTS WHICH MAY FOLLOW DIFFERENT
    DISTRIBUTIONS.
    PLEASE INPUT THE DATA ONE ELEMENT AT A TIME AND
    YOU WILL BE PROMPTED FOR ADDITIONAL INFORMATION.

    DISTRIBUTION IDENTIFICATION NUMBER = [4]

    NUMBER OF VALUES = [3]

    INPUT VALUES IN ASCENDING ORDER:

    VALUE 1 = [35000]

                        WITH PROBABILITY [0.4]
```

FIGURE 10-10 ***Sample Monte Carlo Simulation—Computer Queries and User Responses***

```
    VALUE 2 =  [40000]
                      WITH PROBABILITY  [0.5]
    VALUE 3 =  [45000]
                      WITH PROBABILITY  [0.1]

   IS THERE ADDITIONAL ANNUAL CASH FLOW DATA? (Y OR
   N)  [Y]
   DISTRIBUTION IDENTIFICATION NUMBER =  [6]
   MEAN VALUE =  [-30000]
   STANDARD DEVIATION =  [2000]
   IS THERE ADDITIONAL ANNUAL CASH FLOW DATA? (Y OR
   N)  [N]

INFORMATION FOR SALVAGE VALUE:
   DISTRIBUTION IDENTIFICATION NUMBER =  [1]
   CASH VALUE =  [0]

INFORMATION FOR PROJECT LIFE:
   DISTRIBUTION INDENTIFICATION NUMBER =  [5]
   MINIMUM VALUE =  [10]
   MAXIMUM VALUE =  [14]

   EXPECTED VALUE OF PRESENT WORTH =            7759.60
   VARIANCE OF PRESENT WORTH =             680623960.00
   STANDARD DEVIATION OF PRESENT WORTH =       26088.77
   PROBABILITY THAT PRESENT WORTH IS GREATER THAN
   ZERO =                                         0.595

   EXPECTED VALUE OF ANNUAL WORTH =             1114.15
   VARIANCE OF ANNUAL WORTH =               14611587.00
   STANDARD DEVIATION OF ANNUAL WORTH =         3822.51
   PROBABILITY THAT ANNUAL WORTH IS GREATER THAN
   ZERO =                                         0.595
```

FIGURE 10-10

Continued

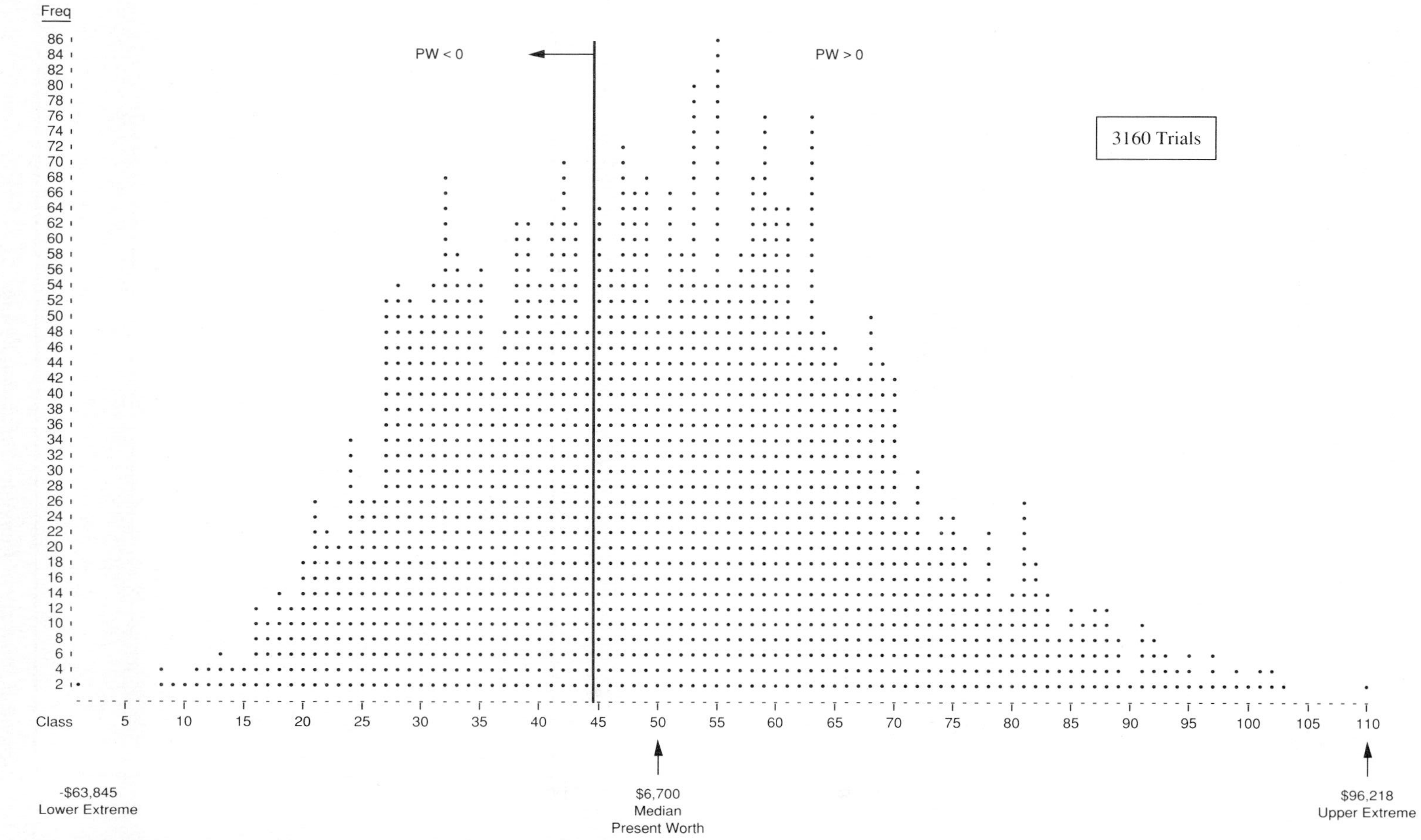

FIGURE 10-11 Histogram of Present Worth for Example 10-15

TABLE 10–13 Simulation Results for Three Mutually Exclusive Alternatives

Project	*E*(PW)	SD(PW)	*E*(PW) ÷ SD(PW)
A	$37,382	$1,999	18.70
B	49,117	2,842	17.28
C	21,816	4,784	4.56

A typical application of simulation involves the analysis of several mutually exclusive projects. In such studies, how can one compare projects that have different expected values and standard deviations of, for instance, present worth? One approach is to select the alternative that *minimizes* the probability of attaining a present worth that is less than zero. Another popular response to this question utilizes a graph of expected value (a measure of the "reward") plotted against standard deviation (an indicator of uncertainty) for each alternative. An attempt is then made to assess subjectively the trade-offs that result from choosing one alternative over another in pairwise comparisons.

To illustrate the latter concept, suppose three projects having varying degrees of uncertainty have been analyzed with Monte Carlo computer simulation, and the results shown in Table 10-13 have been obtained. These results are plotted in Figure 10-12, where it is apparent that alternative C is inferior to alternative A because of its lower E(PW) and larger standard deviation. Therefore, C offers a smaller present worth that has a greater amount of uncertainty associated with it! Unfortunately, the choice of B versus A is not so clear because the increased expected present worth of B has to be balanced against the increased uncertainty of B. This trade-off *may or may not* favor B, depending on management's attitude

FIGURE 10-12

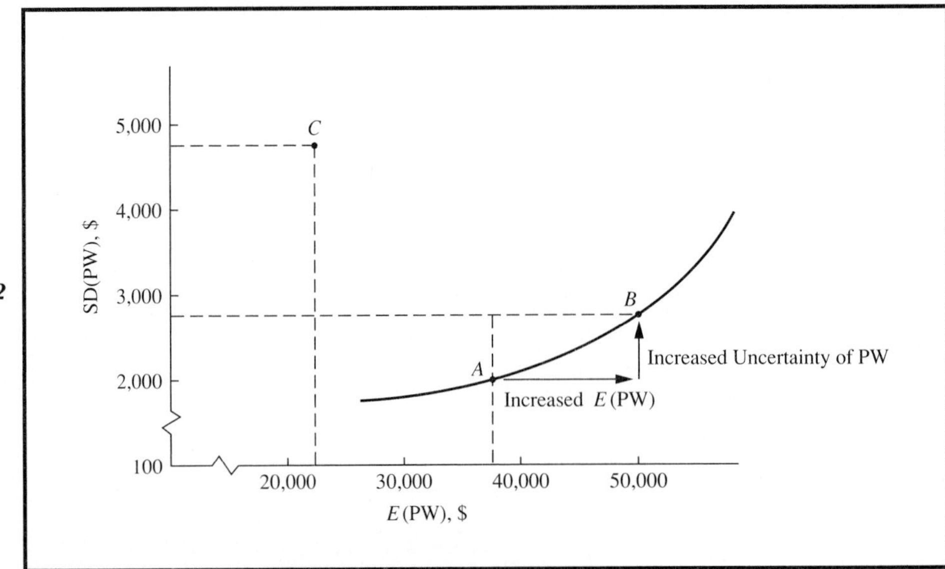

Graphical Summary of Computer Simulation Results

toward accepting the additional uncertainty associated with a larger expected reward. The comparison also presumes that alternative A is acceptable to the decision maker. One simple procedure for selecting between A and B is to rank projects based on the ratio of E(PW) to SD(PW). In this case, project A would be chosen because it has the more favorable (larger) ratio.

Engineering economy involves decision making among competing uses of scarce capital. The consequences of resultant decisions usually extend far into the future. In Chapter 10 we have explicitly dealt with the realization that the consequences (cash flows, project lives, etc.) of engineering alternatives can never be known with absolute certainty. This situation is generally referred to as *decision making under uncertainty*, and for our purposes there is no compelling reason to distinguish between risk and uncertainty.

Several of the most commonly applied and useful procedures for dealing with uncertainty in engineering economy studies have been presented in this chapter: (1) breakeven analysis, (2) sensitivity analysis, (3) optimistic–pessimistic estimates, (4) risk-adjusted MARRs, (5) reduction in useful life, and (6) probabilistic techniques such as Monte Carlo simulation. Breakeven analysis determines the value of a key parameter, such as utilization of capacity, at which the economic desirability of two alternatives is equal. This breakeven point is then compared to an independent estimate of the parameter's most likely value to assist with the selection between alternatives. Similarly, sensitivity analysis typically determines the range of values that a key parameter may have without reversing the superiority that the best alternative has over others being considered. The remaining four procedures for dealing with uncertainty are aimed at selecting the best course of action when one or more "consequences" of alternatives being evaluated lack estimation precision.

Regrettably, there is not a quick and easy answer to the question "How should uncertainty best be considered in an engineering economic evaluation?" Generally, simple procedures (e.g., breakeven analysis and sensitivity analysis) allow modest discrimination among alternatives to be made on the basis of uncertainties present, and they are relatively inexpensive to apply. Additional discrimination among alternatives is possible with more complex procedures that utilize probabilistic concepts, but their difficulty of application and expense are often prohibitive. This chapter has presented a range of techniques broad enough to enable the student/practitioner to deal effectively with the task of including uncertainty in an economic evaluation of competing designs, processes, and systems.

10.16 PROBLEMS

10-1. Explain why *risk* and *uncertainty* can be used interchangeably throughout this book. (10.2)

10-2. Why should the effects of uncertainty be considered in engineering economy studies? What are some likely sources of uncertainty in these studies? (10.3)

10-3. Describe why risk-adjusted MARRs and reduction of useful lives are *not* special cases of sensitivity analysis. (10.4)

10-4. Construct your own *nonlinear* breakeven analysis problem, develop a solution for it, and bring a one-page summary of your problem and solution to class for discussion. (10.5)

10-5. How are the optimistic and pessimistic values of parameters determined when using the optimistic–pessimistic estimation approach to dealing with uncertainty? (10.8)

10-6. Consider these two alternatives:

	Alternative 1	Alternative 2
First cost	$4,500	$6,000
Annual receipts	$1,600	$1,850
Annual expenses	$ 400	$ 500
Estimated salvage value	$ 800	$1,200
Useful life	8 years	10 years

Suppose that the salvage value of alternative 1 is known with certainty. By how much would the estimate of salvage value for alternative 2 have to vary so that the *initial* decision based on these data would be reversed? The annual minimum attractive rate of return is 15%. (10.5)

10-7

a. In Problem 10-6, determine the life of alternative 1 for which the annual worths are equal.

b. If the period of required service from either alternative 1 or 2 is exactly 7 years, which alternative should be selected in Problem 10-6? Assume that the salvage value remains unchanged. (10.5)

10-8. Hodnett County is planning to build a three-story building. It is expected that some years later three more stories will have to be added to the building. Two alternative plans have been proposed, as follows: Design A is a conventional design for a three-story building. Its estimated first cost is $750,000. It is estimated that the three additional stories will cost $1,000,000 whenever they are added. Design B has an initial first cost of $900,000. The addition of three more stories is estimated to cost $800,000 whenever they are added. The total life of the building (design A or B) will be 60 years, and its market value will be zero at that time. Maintenance and energy costs will be $5,347 per year *less* with design B for each of the 60 years. If the addition to the building with either design is made, how soon must the additional stories be constructed to justify the selection of design B? The MARR is 15% per year. (10.5)

10-9. Two electric motors are being considered to power an industrial hoist. Each is capable of providing 90 horsepower. Pertinent data for each motor are as follows:

	Motor	
	D-R	**Westhouse**
Investment	$2,500	$3,200
Electrical efficiency	0.74	0.89
Maintenance/year	$40	$60
Life	10 years	10 years
MARR	12%/year	12%/year

If the expected usage of the hoist is 500 hours per year, what would the cost of electrical energy have to be (in cents/kWh) before the D-R motor is favored over the Westhouse motor? [*Note:* 1 horsepower (hp) = 0.746 kilowatt.] (10.5)

10-10. Consider the following cash flow diagram. Plot changes in present worth to ±20% and ±40% changes in the project's life, N. Let i = 10%/yr and assume S = 0. State any other assumptions you make. (10.6)

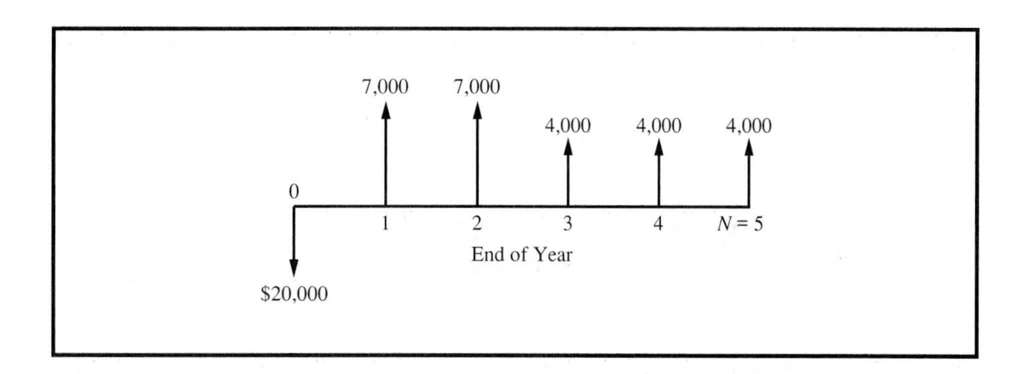

10-11. A new steam flow monitoring device must be purchased immediately by a local municipality. These "most likely" estimates have been developed by a group of engineers:

Investment	$140,000
Annual savings	$ 25,000
Useful life	12 years
Residual value (end of year 12)	$ 40,000
MARR	10%/year

Because considerable uncertainty surrounds these estimates, it is desired to evaluate the sensitivity of present worth to ±50% changes in the most likely estimates of (a) annual savings, (b) life, and (c) residual value. Graph the results and determine to which factor the decision is most sensitive. (10.6)

10-12. Suppose that for a capital investment project the optimistic–most likely–pessimistic estimates are as follows:

	Optimistic	Most Likely	Pessimistic
Investment	$80,000	$95,000	$120,000
Useful life	12 years	10 years	6 years
Salvage value	$30,000	$20,000	0
Net annual cash flow	$35,000	$30,000	$20,000
MARR	12%/yr	12%/yr	12%/yr

a. What is the annual worth for each of the three estimation conditions?

b. It is thought that the most critical elements are useful life and net annual cash inflow. Develop a table showing the annual worth for all combinations of estimates for these two factors assuming all other factors remain at their most likely values. (10.8)

10-13. A bridge is to be constructed now as part of a new road. Engineers have determined that traffic density on the new road will justify a two-lane road and bridge at the present time. Because of uncertainty regarding future use of the road, the time at which an extra two lanes will be required is currently being studied. The estimated probabilities of having to widen the bridge to four lanes at various times in the future are as follows:

Widen bridge in	Probability
3 years	0.1
4 years	0.2
5 years	0.3
6 years	0.4

The two-lane bridge will cost $200,000 and the four-lane bridge, if built at one time, will cost $350,000. The cost of widening a two-lane bridge will be an extra $200,000 plus $25,000 for every year that construction is delayed. If money can earn 12% interest, what would you recommend? (10.12)

10-14. In Problem 10-13, perform an analysis to determine how sensitive the choice of a four-lane bridge built at one time versus a four-lane bridge that is constructed in two stages is to the interest rate. Will an interest rate of 15% per year reverse the initial decision? (10.6)

10-15. In reference to Problem 10-13 suppose that instead of specifying probabilities for times at which the four-lane bridge will be required, these estimates have been made:

Pessimistic estimate	4 years
Most likely estimate	5 years
Optimistic estimate	7 years

In view of these estimates, what would you recommend? What difficulty do you have in interpreting your results? List some advantages and disadvantages of this method of preparing estimates. (10.8)

10-16. Two gasoline-fueled pumps are being considered for a certain agricultural application. Both pumps operate at a rated output of 5 hp but differ in their first cost and efficiency. Gasoline costs \$1.50 per gallon, and the MARR is 15%. Following is a summary of "most likely" data:

	Pump M	Pump N
First cost	\$1,700	\$2,220
Maintenance cost per 100 hours of operation	\$ 40	\$ 25
Fuel consumption per hour of operation	$2\frac{1}{2}$ gallons	2 gallons
Useful life	10 years	10 years
Salvage value	0	0

Because of erratic weather conditions, the most difficult parameter to estimate is hours of operation per year. Optimistic, most likely, and pessimistic estimates of annual operation are 3,000 hours, 5,000 hours, and 6,500 hours, respectively.

a. Plot annual worth of each pump as a function of hours of operation.

b. Based on results in part (a), which pump would you recommend? (10.8)

10-17. A small-town cable TV company wishes to expand its coverage into a rural area just east of town. A study indicates that 3,000 homes will be within connecting range of the proposed cable system, and the company knows the typical subscriber rate is 43% (i.e., 43% of the homes with the option to get cable TV actually do). The basic charge is \$10 per month, plus a one-time \$10 installation fee. The company must pay 3% of its gross revenue to the city government, and 20% of its gross revenue to the program sources (e.g., HBO, Showtime). The project is expected to last 10 years before requiring extensive overhaul, and there is no residual value. The company's present office, billing, and maintenance personnel can easily service 2,000 new subscribers without having to use overtime or hire new people. Now there is a choice between (1) using company people to install the new lines, taps, and distribution amplifiers, and (2) letting CABLEX, an outside company which specializes in CATV installations, do the job for a fee. The company's MARR is 15%.

a. Which alternative is more attractive if it is assumed that tornadoes will destroy the entire system in 5 years?

b. If only 35% of the homes subscribe, what action should be taken? (10.5, 10.10)

	In-House	Cablex
Materials and equipment	\$375,000	\$340,000
Labor (one-time; installation)	30,000	50,000
Maintenance	\$1,000/year	\$3,000/year

10-18. In Problem 10-6, suppose that alternative 2 is believed to be more uncertain than alternative 1. A risk-adjusted MARR of 18% will therefore be used to determine the annual worth of alternative 2. Which alternative would be recommended with this method? (10.9)

10-19. Pump M in Problem 10-16 is manufactured in a foreign country and is believed to be less reliable than pump N. To cope with this uncertainty, a risk-adjusted MARR of 20% is utilized to calculate its annual worth. When hours of operation per year total 5,000, which pump would be selected? What difficulty is encountered with this method? (10.9)

10-20. To deal with estimation uncertainties in Problem 10-6, the useful lives of alternatives 1 and 2 have been reduced to 6 years and 8 years, respectively. Does this affect the choice that should be made? If so, how? (10.10)

10-21. In a certain building project, the amount of concrete to be poured during the next week is uncertain. The foreman has estimated the following probabilities of concrete poured.

Amount (cubic yards)	Probability
1,000	0.1
1,200	0.3
1,300	0.3
1,500	0.2
2,000	0.1

Determine the expected value (amount) of concrete to be poured next week. Also compute the variance of concrete to be poured. (10.11)

10-22. Consider the following two random variables, p and q.

Price, p	$\Pr(p)$	Quantity Sold, q	$\Pr(q)$
$6	$\frac{1}{3}$	10	$\frac{1}{3}$
5	$\frac{1}{3}$	15	$\frac{1}{3}$
4	$\frac{1}{3}$	20	$\frac{1}{3}$

Assume that p and q are independent. What is the mean and variance of the probability distribution for revenue? (10.11)

10-23. Suppose that a random variable (e.g., salvage value for a piece of equipment) is normally distributed with mean = 175 and variance = 25. What is the probability that the actual salvage value is *at least* 171? (10.13)

10-24. The annual worth of project R-2 is normally distributed with a mean of $1,500 and a variance of 810,000. Determine the probability that this project's AW is less than $1,700. (10.13)

10-25. A dam is being planned for a river that is subject to frequent flooding. From past experience, the probabilities that water flow will exceed the design capacity of the dam, plus relevant cost information, are as follows:

Design of Dam	Probability of Greater Flow	Required Investment
A	0.100	$180,000
B	0.050	195,000
C	0.025	208,000
D	0.015	214,000
E	0.006	224,000

Estimated damages that occur if water flows exceed design capacity are $150,000, $160,000, $175,000, $190,000, and $210,000 for design A, B, C, D, and E, respectively. The life of the dam is expected to be 50 years, with negligible salvage value. For an interest rate of 8%, determine which design should be implemented. What nonmonetary considerations might be important to the best selection? (10.12)

10-26. A diesel generator is needed to provide auxiliary power in the event that the primary source of power is interrupted. At any given time, there is a 0.1% probability that the generator will be needed. Various generator designs are available, and more expensive generators tend to have higher reliabilities should they be called on to produce power. Estimates of reliabilities, investment costs, maintenance costs, and damages resulting from a complete power failure (i.e., the standby generator fails to operate) are given for three alternatives:

Alternative	First Cost	Operating and Maintenance Costs/Year	Reliability	Cost of Power Failure	Salvage Value
R	$200,000	$5,000	0.96	$400,000	$40,000
S	170,000	7,000	0.95	400,000	25,000
T	214,000	4,000	0.98	400,000	38,000

If the life of each generator is 10 years and MARR = 10%, which generator should be chosen? (10.12)

10-27. The owner of a ski resort is considering installing a new ski lift, which will cost $900,000. Costs, other than depreciation, for operating and maintaining the lift are estimated to be $1,500 per day when operating. The U.S. Weather Service estimates that there is a 60% probability of 80 days of skiing weather per year, 30% probability of 100 days, and 10% probability of 120 days per year. The operators of the resort estimate that during the first 80 days of adequate snow in a season, an average of 500 people will use the lift each day, at a fee of $10 each. If 20 additional days are available, the lift will be used by only 400 people per day during the extra period; and if 20 more days of skiing are available, only 300 people per day will use the lift during those days. The owners desire to recover any invested capital within 5 years, and want at least a 25% rate of return. Should the lift be installed? (10.11, 10.12)

10-28. Every time an automatic welding machine fails, the cost in idle labor and repairs is $1,000. The outage time averages 4 working hours, and the plant works 2,000 hours a year. Suppose that the machine could fail a maximum of 250 times in a given year.

The probabilities of failure during a year are assessed from a certain trade association to be no failures, 0.050; 1, 0.113; 2, 0.209; 3, 0.333; 4, 0.201; 5, 0.080; 6, 0.009; 7, 0.003; 8, 0.001; 9, 0.0008; 10, 0.00008. A standby machine with an expected life of 10 years can be procured for $10,000 with $2,000 salvage value. The annual expenses to keep it ready to run are $500. The probability of its breaking down during a standby run are nil. The minimum required rate of return is 15%. Make a recommendation regarding whether to purchase the standby machine. (10.12, 10.13)

10-29. For the following cash flow estimates, determine the $E(\text{PW})$ and $V(\text{PW})$. Also find the probability that PW will exceed $0. The cash flows are normally distributed and the MARR is 12% per year. (10.13)

End of Year	Expected Value of Cash Flow	Standard Deviation of Cash Flow
0	−$14,000	0
1	6,000	$ 800
2	4,000	400
3	4,000	400
4	8,000	1,000

10-30. The useful life of a certain machine is 5 years for certain. Its investment cost is $6,000 and its net annual savings are as follows:

Net Annual Savings	Probability
$1,000	0.3
1,800	0.5
2,500	0.2

Determine (**a**) $E(\text{PW})$, (**b**) $V(\text{PW})$, and (**c**) Pr $(\text{PW} \geq 0)$. The MARR is 10% per year. (10.13)

10-31. A proposed venture has an initial investment of $80,000, annual receipts of $30,000, and an uncertain useful life, N, as follows:

N	Probability of N
1	0.05
2	0.15
3	0.20
4	0.30
5	0.20
6	0.05
7	0.05

Determine the expected present worth of this investment when the MARR is 20% per year. (10.13)

10-32. Consider the following project, which has a minimum life of 1 year and a maximum life of 5 years.

End of Year, N	Certain Cash Flow	Probability of N
0	− $7,000	—
1	5,000	0.20
2	4,000	0.10
3	2,500	0.40
4	2,500	0.25
5	4,000	0.05

a. What is the probability that this project has a present worth greater than $1,000 if the interest rate is 20% compounded annually?
b. Determine the expected value of the project's present worth. (10.13)

10-33. Consider Problem 10-27 when, in addition to uncertainty regarding number of skiing days per year, the useful life of the venture is *also* uncertain as follows:

Life, N	Pr (N)
4	0.2
5	0.6
6	0.2

Finally, the salvage value (S) of the ski lift is a function of the venture's life:

$$S = \$10{,}000(7 - N)$$

a. Set up a table and use Monte Carlo simulation to determine five trial outcomes of the venture's annual worth. Recall that the MARR is 25% per year.
b. Based on your simulation outcomes, should the lift be installed? State any assumptions you make. (10.14)

10-34. Consider the following estimates for a new piece of manufacturing equipment:

Factor	Expected Value	Type of Probability Distributions
Investment cost	$150,000	Known with certainty
Salvage value	$ 2,000 (13 − N)	Normal, σ = $500
Annual savings	$ 70,000	Normal, σ = $4,000
Annual expenses	$ 43,000	Normal, σ = $2,000
Useful life, N	13 years	Uniform in [8, 18]
Minimum attractive rate of return (MARR)	8%/year	

a. Set up a table and simulate five trials of the equipment's present worth.
b. Compute the mean of the five trials and recommend whether the equipment should be purchased. (10.14)

10-35. Simulation results are available for two mutually exclusive alternatives. A large number of trials has been run with a computer, with the results shown in the following figure.

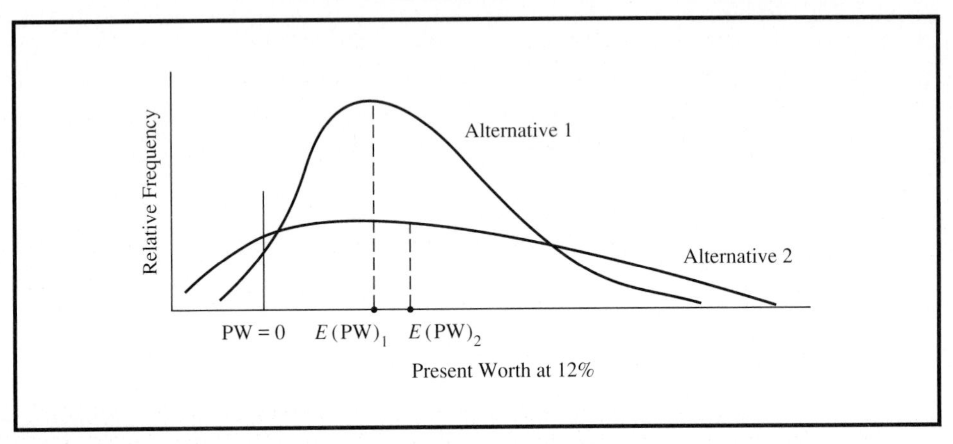

Problem 10-35

Discuss the issues that may arise when attempting to decide between these two alternatives. (10.14)

11

DEALING WITH MULTIATTRIBUTED DECISIONS

The objective of this chapter is to discuss how several relatively straightforward models can be used to evaluate alternatives in a holistic manner that encompasses the monetary and nonmonetary attributes that are integral to almost all real-life decisions. In this regard, the following topics are discussed in this chapter:

Examples of multiattributed decisions
Choice of attributes
Selection of a measurement scale
Dimensionality of the problem
Noncompensatory models
Compensatory models
A comprehensive example

11.1

INTRODUCTION

In Chapter 1 the concept of multiattributed decision making was briefly introduced, but all chapters up to Chapter 11 have dealt principally with the assessment of equivalent monetary worth of competing alternatives and proposals. As you know, *few* decisions are based strictly on dollars and cents! In this chapter our attention is directed at how diverse nonmonetary considerations (*attributes*) that arise from multiple objectives can be explicitly included in the evaluation of engineering and business ventures. By *nonmonetary* we mean that a formal market mechanism does not exist in which value, as discussed in Chapter 6, can be established for various aspects of a venture's performance such as aesthetic appeal, employee morale, and environmental enhancement.

11.2

EXAMPLES OF MULTIATTRIBUTED DECISIONS

To provide perspective and motivation for studying the subject of multiattributed decision making, two realistic examples are offered that will set the stage for other topics to follow.

A common situation encountered by a new engineering graduate is that of selecting his or her first permanent professional job. Suppose Mary Jones, a 22-year-old graduate engineer, is fortunate enough to have four acceptable job offers in writing. A choice among the four offers must be made within the next four weeks or else they become void. She is a bit perplexed concerning which offer to accept, but she decides to base her choice on these four important factors or attributes (not necessarily listed in their order of importance to her): (1) social climate of the town in which she will be working, (2) the opportunity for outdoor sports, (3) starting salary, and (4) potential for promotion and career advancement. Mary Jones next forms a table and fills it in with objective and subjective data relating to differences among the four offers. The completed table (or matrix) is shown as Table 11-1. Notice that several attributes are rated subjectively on a scale ranging from "poor" to "excellent."

It is not uncommon for monetary and nonmonetary data to be key ingredients in decision situations such as this rather elementary one. Take a minute or two to ponder which offer you would accept, given only the data in Table 11-1. Would starting salary dominate all other attributes, so that your choice would be the Apex Corporation in New York? Would you try to trade off poor social

TABLE 11–1 Job Offer Selection Problem

	Alternatives (Offers and Locations)			
Attributes	**Apex Corp., New York**	**Sycon, Inc. Los Angeles**	**Sigma Ltd. Macon, GA**	**McGraw-Wesley Flagstaff, AZ**
Social Climate	Good	Good	Fair	Poor
Weather/Sports	Poor	Excellent	Good	Very Good
Starting Salary (per annum)	$35,000	$30,000	$34,500	$31,500
Career Advancement	Fair	Good	Very Good	Excellent

climate against excellent career advancement in Flagstaff to make the McGraw-Wesley offer your top choice?

Many decision problems in industry can be reduced to matrix form similar to the foregoing job selection example. To illustrate the wide applicability of such a tabular summary of data, consider a second example involving the choice of a computer-aided design (CAD) workstation by an architectural-engineering firm. The data are summarized in Table 11-2. Three vendors and "do nothing" compose the list of feasible alternatives (choices) in this decision problem, and a total of seven attributes is judged sufficient for purposes of discriminating among the alternatives. Aside from the question of which workstation to select, other significant questions come to mind in multiattributed decision making: (1) How are the attributes chosen in the first place? (2) Who makes the subjective judgments regarding nonmonetary attributes such as "quality"? (3) What response is required—a partitioning of alternatives or a rank ordering of alternatives, for instance? Several simple, though workable and credible, models for

TABLE 11–2 CAD Workstation Selection Problem

	Alternative			
Attribute	**Vendor A**	**Vendor B**	**Vendor C**	**Reference ("Do Nothing")**
Cost of purchasing the system	$115,000	$338,950	$32,000	$0
Reduction in design time	60%	67%	50%	0
Flexibility	excellent	excellent	good	poor
Inventory control	excellent	excellent	excellent	poor
Quality	excellent	excellent	good	fair
Market share	excellent	excellent	good	fair
Machine utilization	excellent	excellent	good	poor

selecting among alternatives such as those in Tables 11-1 and 11-2 are described in this chapter.

11.3 CHOICE OF ATTRIBUTES

The choice of attributes by which to judge alternative designs, systems, products, processes, and so on is one of the most important tasks in multiattribute decision analysis. (The most important task, of course, is to identify feasible alternatives from which to select.) It has been observed that the articulation of attributes for a particular decision can, in some cases, shed enough light on the problem to make the final choice obvious to all involved!

Consider again the data in Tables 11-1 and 11-2. These general observations regarding the attributes used to discriminate among alternatives can immediately be made: (1) each attribute distinguishes at least two alternatives—in no case should identical values for an attribute apply to all alternatives; (2) each attribute captures a unique dimension or facet of the decision problem (i.e., attributes are independent and nonredundant); (3) all attributes, in a collective sense, are assumed to be sufficient for purposes of selecting the best alternative; and (4) differences in values assigned to each attribute are presumed to be meaningful in distinguishing among feasible alternatives.

In practice, selection of a set of attributes is usually the result of group consensus, and it is clearly a subjective process. The final list of attributes, both monetary and nonmonetary, is therefore heavily influenced by the decision problem at hand as well as an intuitive feel for which attributes will or will not pinpoint relevant differences among feasible alternatives. If too many attributes are chosen, the analysis will become unwieldy and difficult to manage. Too few attributes, on the other hand, will limit discrimination among alternatives. Again, judgment is required to decide what number is "too few" or "too many." If some attributes in the final list lack specificity and/or cannot be quantified, it will be necessary to subdivide them into lower-level attributes that can be measured.

To illustrate the above points, we might consider adding an attribute called "cost of operating and maintaining the system" in Table 11-2 to capture a vital dimension of the CAD system's life-cycle cost. The attribute "flexibility" should perhaps be subdivided into two other, more specific attributes such as "ability to interface with computer-aided manufacturing equipment" (such as numerically controlled machine tools) and "capability to create and analyze solid geometry representations of engineering design concepts." Finally, it would be constructive to aggregate two attributes in Table 11-2: namely, "quality" and "market share." Because there is no difference in values assigned to these two attributes across the four alternatives, they could be combined into a single attribute, possibly named "achievement of greater market share through quality improvements."

11.4 SELECTION OF A MEASUREMENT SCALE

Identifying feasible alternatives and appropriate attributes represents a large portion of the work associated with a multiattributed decision analysis. The next task is to develop metrics, or measurement scales, that permit various states of each attribute to be represented. For example, in Table 11-1 "dollars" was an obvious choice for the metric of starting salary. A subjective assessment of career advancement was made on a metric having five gradations that ranged from "poor" to "excellent." The gradations were poor, fair, good, very good, and excellent. In many problems, the metric is simply the scale upon which a physical measurement is made. For instance, anticipated noise pollution for various routings of an urban highway project might be a relevant attribute whose metric is "decibels."

11.5 DIMENSIONALITY OF THE PROBLEM

If you will once again refer to Table 11-1, you will observe that there are two basic ways to process the information presented there. *First*, you could attempt to collapse each job offer to a single metric, or dimension. For instance, all attributes could somehow be forced into their dollar equivalents *or* they could be reduced to a *utility equivalent* ranging from, say, 0 to 100. Assigning a dollar value to "good" career advancement may not be too difficult, but how about placing a dollar value on poor versus excellent social climate? Similarly, translating all job offer data to a scale of worth or utility that ranges from 0 to 100 may not be plausible to most individuals. This *first* way of dealing with the data of Table 11-1 is called single-dimensioned analysis. (The dimension corresponds to the number of metrics used to represent the attributes that discriminate among alternatives.)

Collapsing all information to a single dimension is popular in practice because many analysts believe that a complex problem can be made computationally tractable in this manner. In fact, several useful models presented later are single-dimensioned. Such models are termed *compensatory* because changes in values of a particular attribute can be offset by, or traded off against, opposing changes in another attribute.

The *second* basic way to process information in Table 11-1 is to retain the individuality of the attributes as the best alternative is being determined. That is, there is no attempt to collapse attributes to a common scale. This is referred to as full-dimensioned analysis of the multiattribute problem. For example, if r^* attributes have been chosen to characterize the alternatives under consideration, the predicted values for all r^* attributes are considered in the choice. If a metric is common to more than one attribute as in Table 11-1, we have an intermediate-dimensioned problem that is analyzed with the same models as a full-dimensioned problem would be. Several of these models are illustrated in the next section, and they are often most helpful in eliminating inferior alternatives from the analysis. We refer to these models as *noncompensatory* because trade-offs among attributes are not permissible. Thus, comparisons of alternatives must be judged on an attribute-by-attribute basis.

11.6 NONCOMPENSATORY MODELS

In this section we shall examine four noncompensatory models for making a choice where multiple attributes are present. They are (1) dominance, (2) satisficing, (3) disjunctive resolution, and (4) lexicography. In each model an attempt is made to select the best alternative in view of the full dimensionality of the problem. Example 11-1 is presented after the description of these models and will be utilized to illustrate each.

11.6.1 Dominance

Dominance is a useful screening method for eliminating inferior alternatives from the analysis. When one alternative is better than another with respect to all attributes, there is no problem in deciding between them. In this case the first alternative *dominates* the second one. By comparing each possible pair of alternatives to determine whether attribute values for one are at least as good as those for the other, it may be possible to eliminate one or more candidates from further consideration or even to select the single alternative clearly superior to all the others. Usually it will not be possible to select the best alternative based on dominance.

11.6.2 Satisficing

Satisficing, sometimes referred to as the method of feasible ranges, requires the establishment of minimum and/or maximum acceptable values (the "standard") for each attribute. Alternatives having one or more attribute values that fall outside the acceptable limits are excluded from further consideration.

The upper and lower bounds of these ranges establish two fictitious alternatives against which maximum and minimum performance expectations of feasible alternatives can be defined. By bounding the permissible values of attributes from two sides (or from one), information-processing requirements are substantially reduced. Restrictions on the domain of acceptable attribute values serve to make the evaluation problem more manageable.

Satisficing is more difficult to use than dominance because of the minimum acceptable attribute values that must be determined. Furthermore, satisficing is usually employed to evaluate feasible alternatives in more detail and to reduce the number being considered rather than to make a final choice. The satisficing principle is frequently used in practice when *satisfactory* performance on each attribute, rather than *optimal*, is good enough for decision-making purposes.

11.6.3 Disjunctive Resolution

The disjunctive method is similar to satisficing in that it relies on comparing the attributes of each alternative to a set of acceptable limits (the *standard*). The difference is that the disjunctive method evaluates each alternative on the best value achieved for any attribute. If an alternative has *just one* attribute that meets or exceeds the standard, that alternative is kept. In satisficing, *all* attributes must meet or exceed the standard in order for an alternative to be kept in the feasible set.

11.6.4 Lexicography

This model is particularly suitable for decision situations in which a single attribute is judged to be more important than all other attributes. A final choice *might* be based solely on the most acceptable value for this attribute. Comparing alternatives with respect to one attribute reduces the decision problem to a single dimension (i.e., the measurement scale of the predominant attribute). The alternative having the highest value for the most important attribute is then chosen. However, when two or more alternatives have identical values for the most important attribute, the second most important attribute must be specified and used to break the deadlock. If ties continue to occur, the analyst examines the next most important attribute until a single alternative is chosen or until all alternatives have been evaluated.

Lexicography requires that the importance of each attribute be specified to determine the order in which attributes are to be considered. If a selection is made by using one, or a few, of the attributes, lexicography does not take into account all the collected data. Lexicography does not require comparability across attributes, but it does process information in its original metric.

EXAMPLE 11-1

Mary Jones, the engineering graduate whose job offers were given in Table 11-1, has decided, based on comprehensive reasoning, to accept the Sigma position in Macon, Georgia. (Problem 11-8 will provide insight into why this choice was made.) Having moved to Macon, several other important multiattribute problems now face Mary Jones. Among them are (1) renting an apartment versus purchasing a small house, (2) what type of automobile or truck to purchase, and (3) whom to select for long-overdue dental work.

In this example we shall consider the selection of a *dentist* as a means of illustrating noncompensatory (full-dimensioned) and compensatory (single-dimensioned) models for analyzing multiattribute decision problems.

After calling many dentists in the yellow pages, Mary finds there are only four who are accepting new patients. They are Dr. Molar, Dr. Feelgood, Dr. Whoops, and Dr. Pepper. The alternatives are clear to Mary, and she has decided her objectives in selecting a dentist are to obtain high-quality dental care at a reasonable cost with minimum disruption to her schedule and little (or no) pain involved. In this regard Mary adopts these attributes to assist in gathering data and making her final choice: (1) reputation of the dentist, (2) cost per hour of dental work, (3) available office hours each week, (4) travel distance, and (5) method of anesthesia. Notice that these attributes are more or less independent in that the value of one attribute cannot be predicted by knowing the value of any other attribute.

Mary collects data by interviewing the receptionist in each dental office, talking with local townspeople, calling the Georgia Dental Association, and so on. A summary of information gathered by Mary is presented in Table 11-3.

We are now asked to determine whether a dentist can be selected by using (a) dominance, (b) satisficing, (c) disjunctive resolution and (d) lexicography.

TABLE 11–3 Summary of Information for Choice of a Dentist

	Alternatives			
Attribute	Dr. Molar	Dr. Feelgood	Dr. Whoops	Dr. Pepper
Cost ($/hr)	25	(40)	[10]	20
Method of anesthesia[a]	[Novocaine]	Acupuncture	(Hypnosis)	Laughing Gas
Driving distance (mi)	15	20	[5]	(30)
Weekly office hours	[40]	(25)	[40]	[40]
Quality of work	[Excellent]	Fair	(Poor)	Good

[ ] Best value

( ) Worst value

[a] Mary has decided that novocaine > laughing gas > acupuncture > hypnosis, where *a* > *b* means that *a* is preferred to *b*.

TABLE 11–4 Check for Dominance among Alternatives

	Paired Comparison					
Attribute	Molar vs. Feelgood	Molar vs. Whoops	Molar vs. Pepper	Feelgood vs. Whoops	Feelgood vs. Pepper	Whoops vs. Pepper
Cost	Better	Worse	Worse	Worse	Worse	Better
Anesthesia	Better	Better	Better	Better	Worse	Worse
Distance	Better	Worse	Better	Worse	Better	Better
Office hours	Better	Equal	Equal	Worse	Worse	Equal
Quality	Better	Better	Better	Better	Worse	Worse
Dominance?	Yes	No	No	No	No	No

Solution

(a) To check for dominance in Table 11-3, pairwise comparisons of each dentist's set of attributes must be inspected. There will be $4(3)/2 = 6$ pairwise comparisons necessary for the 4 dentists, and they are shown in Table 11-4. It is clear from Table 11-4 that Dr. Molar dominates Dr. Feelgood, so Dr. Feelgood will be dropped from further consideration. With dominance it is not possible to select the best dentist.

(b) To illustrate the satisficing model, acceptable limits (feasible ranges) must be established for each attribute. After considerable thought, Mary comes up with the feasible ranges given in Table 11-5.

Comparison of attribute values for each dentist against the feasible range reveals that Dr. Whoops uses a less desirable type of anesthesia (hypnosis < acupuncture) and his quality rating is also not acceptable (poor < good). Thus, Dr. Whoops joins Dr. Feelgood on Mary's list of rejects. Notice that satisficing, by itself, did not produce the best alternative.

(c) By applying the feasible ranges in Table 11-5 to the disjunctive resolution model, all dentists would be acceptable because each has at least one attribute value that meets or exceeds the minimum expectation. For instance, Dr. Whoops scores acceptably in three out of five attributes and Dr. Feelgood passes two out of five minimum expectations. Clearly, this model does not discriminate well among the four candidates.

TABLE 11–5 Feasible Ranges for Satisficing

Attribute	Minimum Acceptable Value	Maximum Acceptable Value	Unacceptable Alternative
Cost	—	$30	None (Dr. Feelgood already eliminated)
Anesthesia	Acupuncture	—	Dr. Whoops
Distance (miles)	—	30	None
Office hours	30	40	None (Dr. Feelgood already eliminated)
Quality	Good	Excellent	Dr. Whoops

TABLE 11–6 Ordinal Ranking of Dentists' Attributes

A. Results of Paired Comparisons	
Cost > Anesthesia	(Cost is more important than anesthesia)
Quality > Cost	(Quality is more important than cost)
Cost > Distance	(Cost is more important than distance)
Cost > Office Hours	(Cost is more important than office hours)
Anesthesia > Distance	(Anesthesia is more important than distance)
Anesthesia > Office Hours	(Anesthesia is more important than office hours)
Quality > Anesthesia	(Quality is more important than anesthesia)
Hours > Distance	(Hours is more important than distance)
Quality > Distance	(Quality is more important than distance)
Quality > Office Hours	(Quality is more important than office hours)
B. Attribute	*Number of times on left of > (= Ordinal ranking)*
Cost	3
Anesthesia	2
Distance	0
Office Hours	1
Quality	4

(d) Many models, including lexicography, require that all attributes first be ranked in order of importance. Perhaps the easiest way to obtain a consistent ordinal ranking is to make paired comparisons between each possible attribute combination.* This is illustrated in Table 11-6. Each attribute can be ranked according to the number of times it appears on the left-hand side of the comparison when the preferred attribute is placed on the left as shown. In this case the ranking is found to be quality > cost > anesthesia > office hours > distance.

Table 11-7 illustrates the application of lexicography to the ordinal ranking developed in Table 11-6. The final choice would be Dr. Molar because quality is the top-ranked attribute and Molar's quality rating is the best of all. If Dr. Pepper's work quality had also been rated as excellent, the choice would be made on the basis of cost. This would have resulted in the selection of Dr.

TABLE 11–7 Application of Lexicography

Attribute	Rank[a]	Alternative Ranking[b]
Cost	3	Whoops > Pepper > Molar > Goodbody
Anesthesia	2	Molar > Pepper > Goodbody > Whoops
Office Hours	1	Molar = Whoops = Pepper > Goodbody
Distance	0	Whoops > Molar > Goodbody > Pepper
Quality	4	Molar > Pepper > Goodbody > Whoops

[a] Rank of 4 most important, rank of 0 least important

[b] Selection based on the highest ranked attribute

*An ordinal ranking is simply an ordering of attributes from the most preferred to the least preferred.

Pepper. Therefore, lexicography does allow the best dentist to be chosen by Mary.

11.7 COMPENSATORY MODELS

The basic principle behind all compensatory models, which involve a single dimension, is that the values for all attributes must be converted to a common measurement scale such as *dollars* or *utiles*.* When this is done, it is possible to construct an overall dollar index or utility index for each alternative. The form of the function used to calculate the index can vary widely. For example, the converted attribute values may be added together, they may be weighted and then added, or they may be sequentially multiplied. Regardless of the functional form, the end result is that good performance in one attribute can *compensate* for poor performance in another. This allows trade-offs among attributes to be made during the process of selecting the best alternative. Because lexicography involves no trade-offs, it was classified as a full-dimensional model in Section 11.6.4.

In this section we examine three compensatory models for evaluating multi-attribute decision problems. The models are (1) nondimensional scaling, (2) the Hurwicz procedure, and (3) the additive weighting technique. Each model will be illustrated using the data of Example 11-1.

11.7.1 Nondimensional Scaling

A popular way to standardize attribute values is to convert them to nondimensional form. There are two important points to consider when doing this. First, the nondimensional values should all have a common range, such as 0 to 1, or 0 to 100. Without this constraint, the dimensionless attributes will contain implicit weighting factors. Second, all of the dimensionless attributes should follow the same trend with respect to desirability; the most preferred values should be either all small or all large. This is necessary in order to have a believable overall scale for selecting the best alternative.

Nondimensional scaling can be illustrated with the data of Example 11-1. As shown in Table 11-8, the above constraints may require that different procedures be used to *nondimensionalize* each attribute. For example, a cost-related attribute is best when it is low, but office hours are best when they are high. The goal should be to devise a nondimensionalizing procedure that rates each attribute in terms of its fractional accomplishment of the highest attainable

*A utile is a dimensionless unit of worth.

TABLE 11–8 Nondimensional Scaling for Example 11-1

Attribute	Value	Rating Procedure	Dimensionless Value
Cost	$10	(40 − Cost)/30	1.0
	20		0.67
	25		0.50
	40		0.0
Anesthesia	Hypnosis	(Relative Rank[a] −1)/3	0.0
	Acupuncture		0.33
	Laughing Gas		0.67
	Novocaine		1.0
Distance	5	(30 − Distance)/25	1.0
	15		0.60
	20		0.80
	30		0.0
Hours	25	—	0.0
	40		1.0
Quality	Poor	(Relative Rank[a] − 1)/3	0.0
	Fair		0.33
	Good		0.67
	Excellent		1.0

[a] Scale of 1 to 4, 4 being the best (from Table 11–6).

value. Table 11-3, the original table of information for Example 11-1, is restated in dimensionless terms in Table 11-9. The general procedure for converting the original data in Table 11-3 for a particular attribute to its dimensionless rating is

$$\text{Rating} = \frac{\text{worst outcome} - \text{outcome being made dimensionless}}{\text{worst outcome} - \text{best outcome}} \tag{11-1}$$

Equation 11-1 applies when large numerical values, such as dollars or driving distance, are considered to be *undesirable*. However, when large numerical

TABLE 11–9 Nondimensional Data for Example 11-1

Attribute	Dr. Molar	Dr. Feelgood	Dr. Whoops	Dr. Pepper
Cost	0.50	0.0	1.0	0.67
Method of anesthesia	1.0	0.33	0.0	0.67
Driving distance	0.60	0.80	1.0	0.0
Weekly office hours	1.0	0.0	1.0	1.0
Quality of work	1.0	0.33	0.0	0.67

values are considered *desirable* (e.g., a rating with "4" as best and "1" as worst), the relationship for converting original data to their dimensionless ratings is

$$\text{Rating} = \frac{\text{outcome being made dimensionless} - \text{worst outcome}}{\text{best outcome} - \text{worst outcome}} \tag{11-2}$$

If all the attributes in Table 11-9 are of equal importance, a score for each dentist could be found by merely summing nondimensional values in each column. The results would be Dr. Molar = 4.10, Dr. Feelgood = 1.46, Dr. Whoops = 3.00, and Dr. Pepper = 3.01. Presumably, Dr. Molar would be the best choice in this case.

11.7.2 The Hurwicz Procedure

Nondimensional attribute values may be utilized in a variety of ways. The most pessimistic approach is to assume that each alternative is only as good as its lowest performing attribute. The goal would then be to pick the alternative with the most favorable value of its worst attribute (i.e., *maxi*mum value of its *mini*mum attribute). The left-hand column of Table 11-10 illustrates this for the data of Example 11-1, and Dr. Molar would be chosen by this procedure, which is called the *maximin* rule.

At the other extreme, one could be very optimistic and choose the alternative with the most favorable value of its best attribute (i.e., *maxi*mum value of its *maxi*mum attribute). This rule, termed *maximax,* is illustrated on the right-hand side of Table 11-10. Ties with either the maximin or maximax rule can be resolved by considering the second worst or best attribute, respectively, and so on until only one alternative remains. In Table 11-10 the maximax rule leads to Dr. Pepper as the preferred choice.

The *Hurwicz procedure* provides a means of reaching an intermediate level between the pessimism of maximin and the optimism of maximax. It is based on an index of optimism, α, which is chosen to reflect the decision maker's relative attitude. For example, α could be set equal to 0 for pure pessimism and equal to 1 for pure optimism. Values between 0 and 1 would reflect intermediate

TABLE 11–10 Maximin and Maximax Rules Applied to Nondimensional Data

Alternative	Value of Worst Attribute	Value of Best Attribute	Value of Second-Best Attribute[a]
Dr. Molar	[0.50]	1.0	0.60
Dr. Feelgood	0.0	0.80	0.33
Dr. Whoops	0.0	1.0	0.0
Dr. Pepper	0.0	1.0	[0.67]

[a]When alternatives have more than one attribute with the maximum value, the second-best attribute can be selected in two different ways: 1) the maximum value can simply be repeated for alternatives where it occurs more than once, or 2) the next highest value may be selected instead. The latter method was used above.

TABLE 11–11 The Hurwicz Procedure Applied to Example 11-1

Alternative	Value of Worst Attribute (Table 11–9)	Value of Best Attribute (Table 11–9)	Weighted Sum[a]
Dr. Molar	0.50	1.0	0.75
Dr. Feelgood	0.0	0.80	0.40
Dr. Whoops	0.0	1.0	0.50
Dr. Pepper	0.0	1.0	0.50

[a]Weighted sum for each alternative = α (value of best attribute) + $(1 - \alpha)$ (value of worst attribute), where $\alpha = 0.50$.

attitudes. The optimism index is then used to weight the outcomes of maximin and maximax. The best alternative is chosen on the basis of the weighted sum. Table 11-11 illustrates the Hurwicz procedure for $\alpha = 0.5$. The value of α can be varied as shown in Figure 11-1 to analyze the sensitivity of the selection of Dr. Molar, who is judged best in Table 11-11. With the Hurwicz procedure Dr. Molar dominates all the other candidates.

An important criticism of these methods is that there is no attempt to include differential importance weightings among attributes. Up to this point attributes have been equally weighted. Comparisons are made only on the basis of best and worst values, which usually represent different attributes from one alternative to another. This leads to some exaggerated comparisons in Mary's choice of a dentist. A good example of such an extreme comparison would be the

FIGURE 11-1

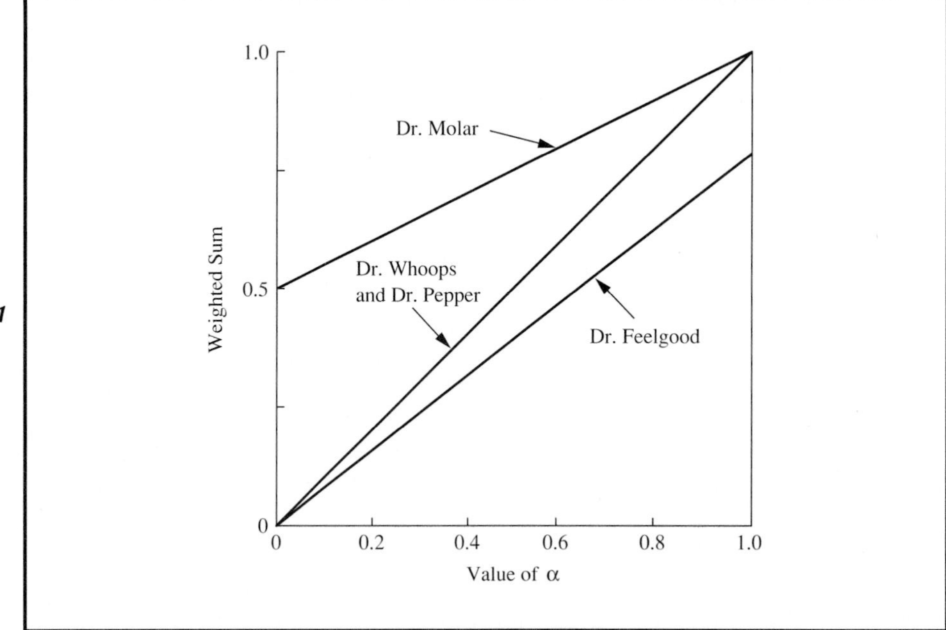

Sensitivity of Selection by the Hurwicz Procedure to Changes in α

equal ranking of Drs. Whoops and Pepper in terms of their lowest performing attributes (see Table 11-9). Dr. Pepper's poorest attribute is driving distance, while Dr. Whoops is poorest in quality of work and method of anesthesia. As a prospective patient, Mary would probably be much more concerned with the quality of work and minimizing of pain than with driving distance. Also, the Hurwicz procedure does not allow for making trade-offs among attributes.

11.7.3 The Additive Weighting Technique

Additive weighting provides for the direct use of nondimensional attributes such as those in Table 11-9 and the results of ordinal ranking as illustrated in Table 11-6. The procedure involves development of weights for attributes (based on ordinal rankings) that can be multiplied by the appropriate nondimensional attribute values to produce a *partial contribution* to the overall score for a particular alternative. When the partial contributions of all attributes are summed, the resulting set of alternative scores can be used to compare alternatives directly.

Attribute weights should be determined in two steps following the completion of ordinal ranking. *First*, relative weights are assigned to each attribute according to its ordinal ranking. The simplest procedure is to use rankings of 1,2,3 . . ., based on position with higher numbers signifying greater importance, but one might also include subjective considerations by using uneven spacing in some cases. For instance, in a case where there are four attributes, two of which are much more important than the others, the top two may be rated as 7 and 5 instead of 3 and 4. The *second* step is to normalize the relative ranking numbers. This can be done by dividing each ranking number by the sum of all the rankings. Table 11-12 summarizes these steps for Example 11-1 and demonstrates how the overall score for each alternative is determined.

Additive weighting is probably the most popular of the single-dimensional methods because it includes both the performance ratings and the importance weights of each attribute when evaluating alternatives. Furthermore, it produces recommendations that tend to agree with the intuitive feel of the decision maker concerning the best alternative. Perhaps its biggest advantage is that nondimensionalizing data and weighting attributes are separated into two distinct steps. This reduces confusion and allows for precise definition of each of these contributions. From Table 11-12 it is apparent that the additive weighting score for Dr. Molar (0.84) makes him the top choice as Mary's dentist.

11.8 A COMPREHENSIVE EXAMPLE

For many large and complex multiattribute decision problems, it is possible to apply a series of simple models from Sections 11.6 and 11.7 in order to successfully resolve the question "Which alternative is best?"

TABLE 11–12 The Additive Weighting Technique Applied to Example 11-1

Calculation of Weighting Factors			Calculation of Scores for Each Alternative[b]							
			Dr. Molar		Dr. Feelgood		Dr. Whoops		Dr. Pepper	
Attribute	Step 1: Relative Rank[a]	Step 2: Normalized Weight (A)	(B)	(A)×(B)	(B)	(A)×(B)	(B)	(A)×(B)	(B)	(A)×(B)
Cost	4	4/15 = 0.27	0.50	0.14	0.00	0.00	1.00	0.27	0.67	0.18
Anesthesia	3	3/15 = 0.20	1.00	0.20	0.33	0.07	0.00	0.00	0.67	0.13
Distance	1	1/15 = 0.07	0.60	0.04	0.80	0.06	1.00	0.07	0.00	0.00
Office Hours	2	2/15 = 0.13	1.00	0.13	0.00	0.00	1.00	0.13	1.00	0.13
Quality	5	5/15 = 0.33	1.00	0.33	0.33	0.11	0.00	0.00	0.67	0.22
Sum =	15	Sum = 1.00	Sum =	0.84		0.24		0.47		0.66

[a] Based on Table11–6: Relative Rank = Ordinal Ranking + 1. A rank of 5 is best.
[b] Data in column B are from Table 11–9.

This section presents a rather intricate problem resulting from several, often conflicting, objectives to be attained in the design of a large radiology department in a teaching hospital. The problem is attacked by using three models in succession: dominance, satisficing, and additive weighting. Fundamental objectives to be carefully considered were cost minimization, patient comfort, convenience for referring physicians (including residents), and timely and accurate diagnostic information. Four designs (alternatives) for the radiology department were proposed by a team of architects, and each design allowed important objectives to be met to varying degrees. Suitable monetary and nonmonetary attributes were chosen to distinguish among alternatives after much deliberation between hospital administrators and radiologists. For each attribute selected, a metric was specified by which the various monetary and nonmonetary considerations could be assessed. Data developed for two monetary attributes and five nonmonetary attributes are shown in Table 11-13 for each alternative. Ranges of estimates have been included to indicate the presence of uncertainty.

In reviewing the four alternatives, it can be observed that alternative IV is dominated by alternatives I and II. That is, IV is no better than I or II in each attribute. For some attributes it is *as good as* I or II, but in no instance is it better than I or II. Comparing these alternatives in a pairwise fashion results in no other instances of dominance. Consequently, with this method alternative IV would be dropped from further consideration. No further discrimination

TABLE 11–13 Illustrative Data for Evaluation of Radiology Department Designs

Attributes (and Performance Measures)	Alternatives I	II	III	IV
Monetary				
i. Construction cost (thousands of dollars)	1,000± 10	1,400± 10	1,100± 10	1,400± 10
ii. Annual operating cost (thousands of dollars)	350± 10	500± 10	450± 10	500± 15
Nonmonetary				
a. Patient travel distance (thousands of feet/year)	2,200±400	900±200	1,500±300	2,400±400
b. Physician travel distance (thousands of feet/year)	1,000±300	400±100	800±200	1,200±300
c. Information retrieval time (time to locate old records and film)	2–3 hours	15–30 minutes	5–6 hours	3–4 hours
d. Accuracy of information (excellent, very good, good, fair, poor)	E–VG	F–P	VG–G	F–P
e. Remunerative rewards (excellent, very good, good, fair, poor)	E–VG	VG–G	VG–G	VG–G

among alternatives is possible, and other methods—either separately applied or in combination with dominance—are required to arrive at the selection of the single best alternative.

The method of feasible ranges (satisficing) can be applied to the data of Table 11-13 only after upper and lower bounds for the various attributes have been established as shown in Table 11-14. Low values for some of the attributes are desirable (e.g., total construction cost, patient travel distance, and information retrieval time), while high values of certain other attributes are preferable (e.g., accuracy of information and remunerative rewards). To interpret the feasible ranges shown in Table 11-14, the *minimum* end of the range represents the worst value for an attribute that is still acceptable. On the other hand, the *maximum* end of the range represents the best possible value conceivably attainable for an attribute.

As the minimum predicted value of each attribute in Table 11-13 is compared with the least acceptable value in its feasible range (from Table 11-4), it is apparent that alternative III does not provide satisfactory performance in one attribute. The information retrieval time was predicted to be 5 to 6 hours (worse than the allowable 4 hours). Therefore, alternative II cannot be regarded as a feasible alternative and is eliminated at this point. The other alternatives are feasible with respect to the minimum performance requirements shown in Table 11-14.

The additive weighting model was next utilized to discriminate between the remaining designs: I and II. This model produces a score, or index, that is a composite measure (single-dimensioned) of the worth of important monetary and nonmonetary attributes as satisfied by each alternative. Numerous methods for constructing a score are available. The method discussed here involves these basic steps:

1. Rank attributes in order of decreasing importance.
2. Assign weights to the attributes such that their relative importance is quantified on a dimensionless scale from, for instance, 0 to 1 inclusive.
3. Scale the performance of each alternative in attaining the maximum worth attainable by each attribute.

TABLE 11–14 Feasible Ranges for Attributes

Attribute (from Table 11–13)	Minimum Acceptable (Worst Value Permissible)	Maximum Attainable (Best Realizable Value)
i	\$1,800,000	\$800,000
ii	\$ 600,000	\$300,000
a	3,000,000 ft/yr	700,000 ft/yr
b	1,800,000 ft/yr	200,000 ft/yr
c	4 hours	10 minutes
d	Poor	Excellent
e	Good	Excellent

4. Multiply the scaled performance for each attribute in step 3 by the weight of the attribute from step 2 and sum over all attributes for each alternative. The resultant score is indication of the total worth of an alternative.
5. Select the alternative that has the maximum score.

To illustrate the additive weighting technique for dealing with monetary and nonmonetary attributes, recall that alternatives III and IV of Table 11-13 have already been eliminated so that alternatives I and II remain to be evaluated. The seven attributes have been ranked in importance from high to low with these results (step 1):

Rank[a]	Attribute
7	Patient travel distance (*a*)
6	Accuracy of information (*d*)
5	Construction cost (i)
4	Information retrieval time (*c*)
3	Physician travel distance (*b*)
2	Annual operating cost (ii)
1	Remunerative rewards (*e*)

[a] To obtain ordinal rankings of attributes, refer to Section 11.5. A rank of 7 is best and a rank of 1 is worst.

Weights are next assigned to each attribute in view of its ranking, with 1.00 initially being given to patient travel distance. The second most important factor, accuracy of information, is assigned a value of 0.80 because it is about 80% as important as patient travel distance, and so on. This scheme represents a second method for establishing attribute weights (the first method is demonstrated in Table 11-12). When all attributes are considered in this manner, suppose that the following results are obtained:

Weight[a]	Attribute	Normalized Weight
1.00	Patient travel distance (*a*)	0.26
0.80	Accuracy of information (*d*)	0.21
0.65	Construction cost (i)	0.17
0.40	Information retrieval time (*c*)	0.10
0.40	Physician travel distance (*b*)	0.10
0.35	Annual operating cost (ii)	0.09
0.30	Remunerative rewards (*e*)	0.07
$\sum = 3.90$		$\sum = 1.00$

[a] For additional information refer to Chapter 8 of J. R. Canada and W. G. Sullivan, *Economic and Multi-Attribute Evaluation of Advanced Manufacturing Systems*, Englewood Cliffs, NJ: Prentice-Hall, 1989.

Normalized weights are then computed by dividing each weight above by the sum of all weights.

FIGURE 11-2

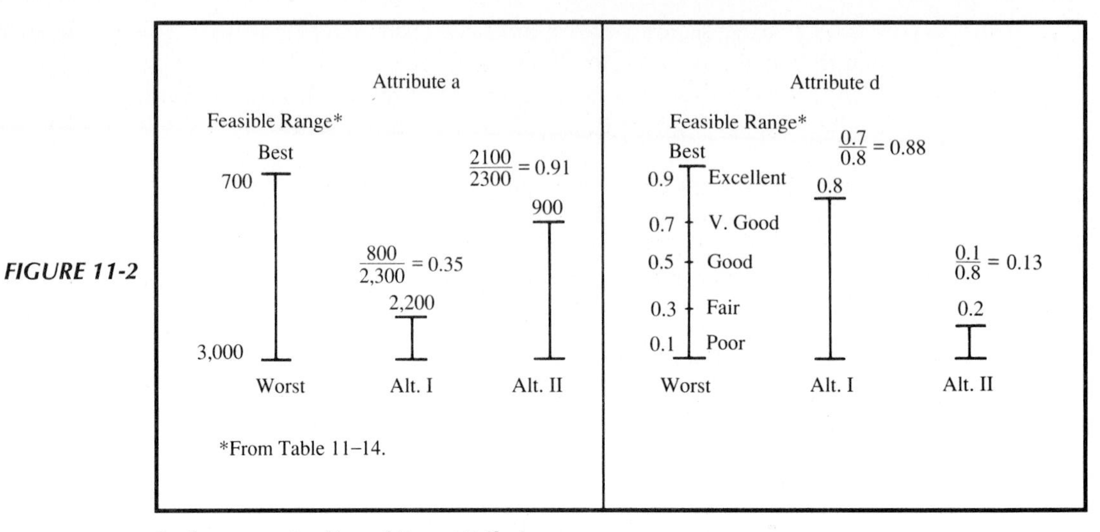

Performance Scaling of Two Attributes

The performance of each alternative is next calculated (scaled) as a simple percentage of the feasible range for each attribute presented in Table 11-14. A linear relationship between the worth of each attribute's performance and its average outcome provided in Table 11-13 is assumed here, but this relationship could be very nonlinear in other problems.*

An example of how average performance can be transformed to a (0, 1) scale in this step of the additive weighting procedure is shown in Figure 11-2 for attributes a and d. Alternative I has a value of 0.35 for attribute a and a value of 0.88 for attribute d. The corresponding values for alternative II are 0.91 and 0.13. These values are developed from Equations 11-1 and 11-2.

An overall summary of all five steps of the additive weighting procedure is presented in Table 11-15. Based on this method, alternative I would be the recommended choice because its score is higher than that of alternative II.

TABLE 11–15 Calculation of an Additive Weighted Score

Attribute	Normalized Weight	Alternative I		Alternative II	
		Rating	Product[a]	Rating	Product[a]
a	0.26	0.35	0.091	0.91	0.237
d	0.21	0.88	0.185	0.13	0.027
i	0.17	0.80	0.136	0.40	0.068
c	0.10	0.39	0.039	0.94	0.094
b	0.10	0.50	0.050	0.88	0.088
ii	0.09	0.83	0.075	0.33	0.030
e	0.07	0.75	0.053	0.25	0.018
		$\sum$ = Score =	0.629	$\sum$ = Score =	0.562

[a]Product = normalized weight × rating.

*For example, see J.R. Miller. *Professional Decision Making*, New York: Praeger, 1970.

Several methods have been described for handling multiattribute decisions. Some key points are

1. When it is desired to maximize a single criterion of choice, such as present worth, the evaluation of multiple alternatives is relatively straightforward.
2. For any decision, the objectives, available alternatives and important attributes must be clearly defined at the beginning. Construction of a decision matrix such as that in Table 11-1 helps to systematize this process.
3. Decision making can become quite complex when multiple objectives and attributes must be included in an engineering economy study.
4. Multiattribute models can be classified as either multidimensional or single dimensional. Multidimensional techniques analyze the attributes in terms of their original metrics. Single-dimensional techniques reduce attribute measurements to a common measurement scale.
5. Multidimensional, or noncompensatory models, are most useful for the initial screening of alternatives. In some instances they can be used to make a final selection, but this usually involves a high degree of subjectivity. Of the multidimensional methods discussed, dominance is probably the least selective, while satisficing is probably the most selective.
6. Single-dimensional, or compensatory models, are highly useful for making a final choice among alternatives. The additive weighting technique allows superb performance in some attributes to compensate for poor performance in others.
7. When dealing with multiattribute problems that have many attributes and alternatives to consider, it is advisable to apply a combination of several models in sequence for the purpose of reducing the selection process to a manageable activity.

11.9 PROBLEMS

11-1. Suppose that you have received your bachelor's degree, wish to earn a Master's degree, and are attempting to decide in which graduate school to enroll. Your particular age, background, undergraduate area of emphasis, monetary situation, and so on are all valid inputs for your decision. Define six attributes you would use in the selection of a graduate school and rank them in order of importance. Assign approximate weights to the attributes by using a method discussed in this chapter. Be prepared to defend your position. (11.3, 11.7)

11-2. List two advantages and two disadvantages of noncompensatory models for dealing with multiattributed decision problems. Do the same for compensatory models. (11.6, 11.7)

11-3. Discuss ways by which satisficing and the Hurwicz procedure might be used in group decision-making exercises. (11.6, 11.7)

11-4. Discuss some of the difficulties of developing nonlinear functions for nondimensional scaling of qualitative (subjective) data. (11.7)

11-5. Given the following matrix of outcomes for alternatives and attributes (with higher numbers being better), show what you can conclude using the following methods:
a. satisficing
b. dominance
c. lexicography, with rank order of attributes D > C > B > A (11.6)

	Alternative				
Attribute	**1**	**2**	**3**	**"Ideal"**	**(Minimum Acceptable)**
A	60	75	90	100	70
B	7	7	8	10	6
C	Poor	Excellent	Fair	Excellent	Good
D	7	8	8	10	6

11-6. With references to the data provided in the following table, recommend the preferred alternative by using (**a**) dominance, (**b**) satisficing, (**c**) disjunctive resolution, and (**d**) lexicography. (11.6)

	Alternative				
Attribute	**Vendor I**	**Vendor II**	**Vendor III**	**Retain Existing System**	**Worst Acceptable Value**
A. Reduction in throughput time	75%	60%	84%	—	50%
B. Flexibility	Good	Excellent	Good	Poor	Good
C. Reliability	Excellent	Good	Very Good	—	Good
D. Quality	Good	Excellent	Excellent	Fair	Good
E. Cost of system (PW of life-cycle cost)	$270,000	$310,000	$214,000	$0	$350,000

Pairwise Comparisons:
1. A < B
2. A = C
3. A < D
4. A < E
5. B < C
6. B > D
7. B < E
8. C < D
9. C < E
10. D < E

11-7. Three large industrial centrifuge designs are being considered for a new chemical plant.

a. By using the following data, recommend a preferred design with each method that was discussed in this chapter for dealing with nonmonetary attributes.

b. How would you modify your analysis if two or more of the attributes were found to be *dependent* (e.g., maintenance and product quality)? (11.6, 11.7)

		Design			
Attribute	**Weight**	A	B	C	**Feasible Range**
Initial cost	0.25	$140,000	$180,000	$100,000	$80,000–$180,000
Maintenance	0.10	Good	Excellent	Fair	Fair–excellent
Safety	0.15	Not known	Good	Excellent	Good–excellent
Reliability	0.20	98%	99%	94%	94–99%
Product quality	0.30	Good	Excellent	Good	Fair–excellent

11-8. Mary Jones utilized the additive weighting technique to select a job with Sigma Ltd. in Macon, Georgia. The importance weights she placed on the four attributes in Table 11-1 were social climate = 1.00, starting salary = 0.50, career advancement = 0.33, and weather/sports = 0.25. Nondimensional values given to her ratings in Table 11-1 were excellent = 1.00, very good = 0.70, good = 0.40, fair = 0.25, and poor = 0.10.

a. Normalize Mary's importance weights.

b. Develop nondimensional values for the starting salary attribute.

c. Use results of (a) and (b) in a decision matrix to see if Mary's choice was indeed consistent with results obtained from the additive weighting technique. (11.7)

11-9. Two highway alignments have been proposed for access to a new manufacturing plant. Based on the data and information given below, a comparison can be made of the two alignments. The better one must be chosen as the proposed highway alignment that will be the connector between an interstate highway and the proposed site. The following attributes and data are to be taken into consideration:

1. Grades, curve, and terrain
2. Rivers, creeks, lakes, and sinkholes
3. Roadway crossings
4. Route length
5. Churches, cemetaries, and residential areas
6. Noise, air, and water pollution

Attributes	**Alignment A**	**Alignment B**
Monetary:		
Land	$ 4,044,662	$ 4,390,000
Bridges	10,134,000	8,701,000
Pavement	4,112,500	4,462,500
Grade and drainage	7,050,000	7,650,000
Erosion control	470,000	510,000
Clearing and replanting	188,000	204,000
Total	$25,999,162	$25,917,500

Miscellaneous:	Alignment A	Alignment B
Route length	4.7 miles	5.1 miles
Maintenance	Moderate (6)	High (3)
Noise pollution	Very good (6)	Good (5)
Cost savings (on gas)	Excellent	Poor
Access ability to another major roadway	U.S. Highway 41	None
Impact on wildlife	Little	Little
Relocation of residences	2	3
Road condition	Flat	Hilly

Use a combination of models presented in this chapter to recommend a highway alignment. Show all work. (11.6, 11.7)

11-10. Use all the multiattributed models of this chapter to decide which of the following job offers you should accept. Attempt to develop the data where question marks appear for your particular situation in view of the attributes shown. (11.6, 11.7)

	Job Offer				
Attribute	1	2	3	Weight	Feasible Range
Location	Phoenix	Buffalo	Raleigh	?	—
Annual salary	\$36,000	\$37,500	\$33,000	?	?
Proximity to relatives	?	?	?	?	?
Quality of leisure time	?	?	?	?	?
Promotion potential	Fair	Excellent	Excellent	?	?
Commuting time/day	1 hour	$1\frac{1}{2}$ hours	$\frac{1}{2}$ hour	?	?
Fringe benefits	Excellent	Very good	Good	?	?
Type of work	Factory	Hospital	Government	?	?

11-11.

a. Use the additive weighting technique to make a selection of one of the three automobiles for which data are given below. State your assumptions regarding miles driven each year, life of the automobile (how long *you* would keep it), market (resale) value at end of life, interest cost, price of fuel, cost of annual maintenance, and other subjectively based determinations. (11.7)

b. Use the data developed in part (a) and the Hurwicz procedure with $\alpha = 0.70$ to select which automobile you should buy. Do your answers in parts (a) and (b) agree? Explain why they should (or should not) agree. (11.7)

Attribute	Alternative		
	Domestic 1	Domestic 2	Foreign
Price	$8,400	$10,000	$9,300
Gas mileage	25 mpg	30 mpg	35 mpg
Type of fuel	Gasoline	Gasoline	Diesel
Comfort	Very good	Excellent	Excellent
Aesthetic appeal	5 out of 10	7 out of 10	9 out of 10
Passengers	4	6	4
Ease of servicing	Excellent	Very good	Good
Performance on road	Fair	Very good	Very good
Stereo system	Poor	Good	Excellent
Ease of cleaning upholstery	Excellent	Very good	Poor
Luggage space	Very good	Excellent	Poor

11-12. You have volunteered to serve as a judge in a Midwestern contest to select Sunshine, the most wholesome pig in the world. Your assessments of the four finalists for each of the attributes used to distinguish among semi-finalists are as follows:

Attribute	Contestant			
	I	II	III	IV
Facial quality	Cute, but plump	Sad eyes, great snout	Big lips, small ears	A real killer!
Poise[a]	10	8	8	3
Body tone[a]	5	10	7	8
Weight (lb)	400	325	300	380
Coloring	Brown	Spotted, black and white	Gray	Brown and white
Disposition	Friendly	Tranquil	Easily excited	Sour

[a] Scaled from 1 to 10, with 10 being the highest rating possible.

a. Use dominance, feasible ranges, lexicography, and additive weighting to select your winner. Develop your own feasible ranges and weights for the attributes. (11.6, 11.7)

b. If there were two other judges, discuss how the final selection of this year's "Sunshine" might be made. (11.7)

11-13. You have decided to buy a new luxury automobile and are willing to spend a maximum of $20,000 from your savings account. (Money not spent will remain in the account, where it earns effective interest of 12% per year.) The choice has been narrowed to three cars having these attribute values:

Attribute	Alternative			Feasible Range
	Domestic 1	Domestic 2	Foreign	
Negotiated price	$18,400	$20,000	$19,300	$0–$20,000
Gas mileage (average)	25 mpg	30 mpg	35 mpg	20–50 mpg
Type of fuel	Gasoline	Gasoline	Diesel	Gasoline or diesel
Comfort	Very good	Excellent	Fair	Fair–excellent
Aesthetic appeal	4 out of 10	8 out of 10	9 out of 10	4–10
Number of passengers	6	6	4	2–6
Ease of servicing	Excellent	Very good	Good	Fair–excellent
Performance on road	Fair	Excellent	Very good	Fair–excellent

Use four methods for dealing with nonmonetary attributes (dominance, feasible ranges, lexicography, and additive weighting) and determine whether a selection can be made with each. You will need to develop additional data that reflects your preferences. (11.6, 11.7)

11-14. The additive weighting model is a decision tool that aggregates information from different, independent criteria to arrive at an overall score for each course of action being evaluated. The alternative with the highest score is preferred. The general form of the model is

$$V_j = \sum_{i=1}^{n} w_i x_{ij}$$

where V_j = the score for the jth alternative.
w_i = the weight assigned to the ith decision attribute ($1 \leq i \leq n$).
x_{ij} = the rating assigned to the ith criterion, which reflects the performance of alternative j relative to maximum attainment of the attribute.

Consider the table on page 529, in view of the above definitions, and determine the value of each "?" shown. (11.7)

i	w_i	Rank	Decision Factor		Alternative j (1) Keep Existing Machine Tool	(2) Purchase a New Machine Tool
1	1.0	1	Annual cost of ownership (capital recovery cost)	Rank	?	?
				x_{ij}	1.0	0.7
2	?	4	Flexibility in types of jobs scheduled	Rank	2	1
				x_{ij}	0.8	1.0
3	0.8	2	Ease of training and operation	Rank	1	2
				x_{ij}	?	0.5
4	0.7	?	Time savings per part produced	Rank	2	1
				x_{ij}	0.7	1.0
				V_j	2.69	2.30
				V_j(Normalized)	1.00	?

12

DEALING WITH REPLACEMENT PROBLEMS

Replacement decisions are critically important to an operating organization. Thus, the objectives of this chapter are to discuss the considerations involved in replacement studies and to address the key question of whether an asset should be kept one or more years or immediately replaced with the best available challenger. To achieve these objectives, the following topics are included in Chapter 12:

Reasons for the replacement of assets
The effect of income taxes
Factors that must be considered in replacement studies
What is the investment value of existing assets?
The economic life of the challenger(s)
The economic life of a defender
Replacement versus augmentation
When should abandonment of an asset be considered?
A replacement case history

12.1 INTRODUCTION

Business firms and individuals must frequently decide whether existing assets should be continued in service or whether available new assets will better and more economically meet current and future needs. These decisions are being made with increasing frequency as the pressures of worldwide competition intensify and technology produces more rapid changes. Thus, the *replacement problem*, as it commonly is called, requires careful engineering economy studies in order to arrive at sound decisions and maintain a competitive position.

Unfortunately, replacement problems sometimes are accompanied by unpleasant financial facts. Often it turns out that earlier decisions, particularly concerning the anticipated useful life of existing assets, were not as good as might be desired, especially when hindsight can be applied. Frequently, new and apparently superior assets appear on the scene. Sometimes replacement situations arise because there are completely unanticipated, and even unreasonable, changes in market conditions. Consequently, there is a tendency to regard replacement with some apprehension, whereas in fact it often represents economic opportunity.

Economic studies of replacement alternatives are performed in the same manner as economic studies of any two or more alternatives—the only difference is that one alternative is to keep an *existing (old) asset*, which can be descriptively called the *defender*. The one or more alternative *replacement (new) assets* are then called *challengers*.

12.2 REASONS FOR REPLACEMENT

It should be recognized that not all existing assets replaced are physically scrapped or even disposed of by the owner. Frequently, new assets are acquired to perform the services of existing assets, with the existing assets not retired but merely transferred to some other use—often an "inferior" use, such as standby service. Of course, many assets are sold by their present owners and are used by one or more other owners before reaching the scrap heap.

The four major reasons for replacement are as follows:

1. ***Physical impairment:*** The existing asset is worn out, owing to normal use or accident, and no longer will render its intended function unless extensive repairs are made.

2. ***Inadequacy:*** The existing asset does not have sufficient capacity to fill the current and expected demands. Here, clearly, the requirements have changed from those that were anticipated at the time the asset was acquired. This condition does not necessarily imply physical impairment, but to meet the new demands the asset must be either supplemented or replaced.
3. ***Obsolescence:*** This may be of two types: (a) functional, or (b) economic. Both types result in loss of profits. In the case of functional obsolescence, there has been a decrease in the demand for the output of the asset; thus, a loss in revenue follows. For example, the market may wish a higher quality product than it previously demanded. Economic obsolescence is the result of the existence of a new asset that will produce at lower cost than can be obtained with the old asset.
4. ***Rental or lease possibilities:*** This is a variation of obsolescence, except that the replacement asset does not necessarily have to be different, in any respect, from the existing asset. The possible economic advantage is due to advantageous financial factors that sometimes may accrue from leasing. These usually involve income tax considerations.

For purposes of replacement studies, the following is a distinction between economic life and other types of lives for typical assets.

Economic life is the period of time extending from date of installation to date of retirement (by demotion or disposal) from the primary intended service. The need for retirement is signaled in an engineering economy study when the equivalent cost of a new asset (challenger) is less than the equivalent cost of keeping the present asset (defender) for an additional period of time.

Ownership life is the period between date of acquisition and date of disposal by a specific owner. A given asset may have several different lives for a given owner. For example, a car may serve as the primary family car for several years and then serve only for local job commuting for several more years.

Physical life is the period between original acquisition and final disposal of an asset over its succession of owners. For example, the car above may have several owners over its existence.

Depreciable life is the period over which depreciation deductions are scheduled to be made.

12.3 INCOME TAXES IN REPLACEMENT STUDIES

The replacement of assets often results in capital gains or losses, or gains or losses from the sale of land or *depreciable property*, as was discussed in Chapter

7. Consequently, to obtain an accurate economic analysis in such cases the studies must be made on an *after-tax basis*. In specific cases the situation may be complicated by whether the organization has had other such gains or losses within the tax year, inasmuch as such losses must be offset against any gains before the full tax-credit benefits apply. However, because it often is difficult to determine the existence of such gains or losses throughout an organization without undue loss of time, many companies follow the practice of considering each case separately and *applying the full tax benefits or penalties* to the gains or losses associated with each study, unless the amounts involved are very large.

It is evident that the existence of a taxable gain or loss, in connection with replacement, can have a considerable effect on the results. A prospective gain from the disposal of assets can be reduced by as much as 40% or 50%, depending on the effective income tax rate that is being used in a particular problem. Consequently, one's normal propensity for disposal or retention of existing assets can be influenced considerably by income tax effects.

The correct consideration of the income tax aspects of retirement and replacement situations may well result in conclusions exactly opposite from the conclusions one may reach by intuition or even by conducting a before-tax study. For example, intuition often favors disposal of an asset at an apparent profit. However, a taxable gain reduces the net after-tax cash flow realized from such a disposal. Further, although intuition often runs counter to disposal of an asset at an apparent loss, a loss that is deductible from taxable income increases the net after-tax cash flow realized from such a disposal.

12.4 FACTORS THAT MUST BE CONSIDERED IN REPLACEMENT STUDIES

Following is a list of frequently encountered factors that must be considered in replacement studies.

1. Recognition and acceptance of past "error."
2. The possible existence of a sunk cost.
3. Remaining life of the old asset (defender).
4. Economic life for the proposed replacement asset (challenger).
5. Method of handling unamortized values.
6. Possible capital gains or losses.

Once a proper viewpoint has been established with respect to these items, little difficulty is experienced in making replacement studies and in arriving at sound decisions.

When an asset's book value is greater than its current market value, the difference frequently has been designated as a past "error." Such "errors" also arise when capacity is inadequate, maintenance costs are higher than anticipated, and so forth. This designation is unfortunate because in most cases these differences are not the result of errors but of honest inability to foresee future conditions. The distinction is important in establishing a proper perspective in replacement analyses.

Any unamortized values (i.e., unallocated depreciation of an asset's initial cost) arising from an existing asset under consideration for replacement are strictly the result of *past* decisions—the initial decision to invest in that asset and decisions as to the method and number of years to be used for depreciation purposes. Such losses are *sunk costs* and thus have no relevance to the replacement decisions that must be made (*except to the extent that they result in a tax saving*).

Acceptance of these facts may be made easier by posing a hypothetical question: "What will be the costs of my competitor whose similar property is completely depreciated and who therefore has no past 'errors' to consider?" In other words, we must decide whether we wish to live in the *past*, with its errors and discrepancies, or to be in a sound competitive position in the *future*. A common reaction is, "I can't afford to take the loss in value of the existing asset that will result if the replacement is made." The fact is that the loss already has occurred, whether or not it could be afforded, and it exists whether or not the replacement is made.

It must be remembered, of course, that the existence of a sunk cost may mean that there will be a resulting income tax saving that should be considered. (This is illustrated in the after-tax solution to Example 12-2.)

12.5 INVESTMENT VALUE OF EXISTING ASSETS FOR REPLACEMENT STUDIES

Recognition of the nonrelevance of book values and sunk costs leads to the proper viewpoint to utilize in placing value on existing assets for replacement study purposes. Clearly, the value of an existing asset at any time is determined by either (1) what amount of money will be tied up if it is kept rather than disposed of, or (2) the present worth of any income that can be obtained in the future through its profitable use. For replacement study purposes the second measure of value is not relevant; we do not yet know whether the asset should be retained—a requirement for it to earn future profits. Instead, we wish to determine whether it should be retained or replaced. Consequently, it is certain that the *minimum* present value upon which capital recovery costs can be based is the amount of capital that will be tied up in the asset if it is retained, but that could be recovered immediately if it were replaced.

Thus the *present realizable* market value (modified by any income tax effects) is the correct investment amount to be assigned to an existing asset in replacement studies. A good way to reason that this is true is to make use of the *opportunity cost* or *opportunity foregone principle*. That is, if it should be decided to keep the existing asset, one is giving up the opportunity to obtain the net realizable market value at that time. Thus this represents the *opportunity cost* of keeping the defender.

There is one addendum to the rationale stated above: If any new investment expenditure (such as for overhaul) is needed to upgrade the existing asset so that it will be competitive in level of service with the challenger, the extra amount should be added to the present realizable market value to determine the total investment in the existing asset for replacement study purposes. In summary, we have established that

> Total investment in the defender asset = Present realizable market value of the defender + Expenditures required to upgrade the defender to competitive status with the challenger

Application of the foregoing principle does not rule out the possibility that the true value of an existing asset, as an operating property, may turn out to be greater than the investment value assigned to it for engineering economy study purposes. If this happens, then one can say that, on the basis of the investment value used for replacement study purposes, the asset turned out to be more profitable than initially anticipated.

EXAMPLE 12-1

The investment cost of a machine purchased 5 years ago was $20,000. It has been depreciated by the straight-line method at $2,000 per year and its current book value is $10,000. The market value of the machine, if sold now, is $5,000, and it would cost $2,000 to overhaul the machine to make it serviceable for another 5 years. What are the total investment in the defender and its unamortized value?

Solution

The investment in an existing asset is its present realizable market value plus any required expenditures to make it serviceable (and comparable) relative to new machines that may be available. Therefore, the investment cost of keeping the present machine is $5,000 + $2,000 = $7,000. If this machine were sold for $5,000, the unamortized value would be $10,000 − $5,000 = $5,000.

12.6
METHODS OF HANDLING LOSSES DUE TO UNAMORTIZED VALUES

Although unamortized values are sunk costs and have no place in before-tax replacement studies, they do cause considerable concern on the part of business owners and managers. Therefore, an understanding of how they are handled may provide some peace of mind to those making engineering economy studies. Several methods are used for dealing with unamortized values. Some are sound; others are incorrect.

One incorrect method is, in effect, a pretense that no unamortized loss has occurred. This is accomplished by using book value as the investment cost for the old equipment in the replacement study. People who do this usually justify their action by saying that if the old equipment is kept, they can continue to deduct depreciation on it until the entire amount of invested capital is recovered. Actually, such wishful thinking could be realized only if one had no competition. If a competitive situation exists, one's competitor may obtain the more economical new equipment and be able to sell at lower prices.

A second incorrect method of handling unamortized values involves adding them to the cost of the challenger and then computing future depreciation costs with this total if the replacement is made. Users of this procedure believe that they can thus require the challenger to repay the unamortized value "that the replacement has caused." This line of reasoning is fallacious on two counts. First, the unamortized value is a fact, regardless of whether the replacement is made or not. Second, in a competitive environment, it probably will not be possible to make the challenger pay the sunk cost. (One's selling price usually is dictated by competition and not by one's desire to recover a sunk cost.)

The *correct* handling of unamortized values requires that they be recognized for what they are—losses of capital. One of the most common procedures is to charge them directly to the profit and loss account (see Appendix A). This considers them a loss in the current operating period and thus deducts them from current earnings. This is a satisfactory procedure except for the fact that unusually large losses will have a serious effect upon the stated accounting profits of the current period.

A more satisfactory method is to provide a surplus account against which such losses may be charged. Thus, a certain sum is set aside each accounting period to build up and maintain this surplus. When any losses occur from unamortized values, they are charged against the surplus account and current profits are not affected seriously. The periodic contribution to the surplus account is in the nature of an insurance premium to prevent profits from being affected by losses from unamortized values.

12.7

A TYPICAL REPLACEMENT PROBLEM

The following is a typical replacement situation used to illustrate a number of factors that must be considered in replacement studies. It is first solved on a before-tax basis. Then an after-tax analysis is performed to include MACRS depreciation for the new (challenger) asset, and straight-line depreciation for the defender. Mixtures of depreciation methods will exist throughout the 1990s.

EXAMPLE 12-2

In early 1989 an operating manager of a large chemical plant became concerned about the performance of an important pump in a cascading process. He presented his chief engineer with the following information and requested that a replacement analysis be performed.

Five years ago the plant purchased pump A, including a driving motor, for $17,000. At the time of purchase it was estimated that it would have an ADR guideline period (i.e., useful life) of 10 years from Table 6-2 and a salvage value at the end of that time of 10% of the initial cost. Depreciation has been charged in the accounting records on this basis by use of the *straight-line* method. Considerable difficulty has been experienced with the pump; annual replacement of the impeller and bearings has been required at a cost of $1,750. Normal annual operating and maintenance costs have been $3,250. Annual taxes and insurance are 2% of the first cost. It appears that the pump will continue to operate for another 9 years, the foreseeable demand period, if the present maintenance and repair practice is continued. It is estimated that if it is continued in service its ultimate market, or residual, value will be about $200.

An alternative to keeping the existing pump in service is to immediately sell it and to purchase a new and different type of pump (pump B) for $16,000. A cash market value of $750 could be obtained for the existing pump. A 9-year guideline period (useful life) would be assigned to the new pump and an estimated market value would be 20% of the initial cost at the end of year 9. Thus, the new pump has a MACRS class life of 5 years (see Table 6-4). Operating and maintenance costs for the new pump are estimated at $3,000 per year. Annual taxes and insurance would total 2% of first cost.

This company is in a 40% effective income tax bracket on ordinary income and pays 40% on long-term capital gains. It has a MARR of 10% on its capital before taxes and 6% after taxes. The data for Example 12-2 are summarized in Table 12-1.

In a before-tax analysis of the defender and challenger, care must be taken to identify correctly the investment amount in the existing pump. From the

TABLE 12–1 Summary of Information for Example 12-2

Existing pump A (defender)		
Investment when purchased 5 years ago		\$17,000
Estimated useful life when originally purchased		10 years
Estimated salvage value at the end of 10 years		\$ 1,700
Annual depreciation (\$17,000 – \$1,700)/10		\$ 1,530
Annual expenses:		
Replacement of impeller and bearings	\$1,750	
Operating and maintenance	3,250	
Taxes and insurance: \$17,000 × 2%	340	
		\$ 5,340
Present market value		\$ 750
Market value at the end of 9 additional years		\$ 200
Current book value [\$17,000 – 5(\$1,530)]		\$ 9,350
Book value at the end of 9 additional years		\$ 0
Replacement pump B (challenger)		
Investment		\$16,000
Estimated useful life (ADR guideline period)		9 years
MACRS class life		5 years
Estimated market value at the end of 9 years		\$ 3,200
Annual expenses:		
Operating and maintenance	\$3,000	
Taxes and insurance: \$16,000 × 2%	320	
		\$ 3,320
Effective income tax rate = 40%		
MARR (before taxes) = 10%		
and MARR (after taxes) = 6%		

problem statement, this would be the current market value of \$750. Notice that the investment amount of pump A ignores the original purchase price of \$17,000 and the present book value of

$$\$17,000 - 5\left(\frac{\$17,000 - \$1,700}{10}\right) = \$9,350$$

By using the principles discussed thus far in this chapter, a before-tax analysis of *equivalent annual worth* (AW) of pump A and pump B can be made.

Solution of Example 12-2 by the AW Method (before Taxes)

Study Period = 9 Years	Keep Old Pump A	Replacement Pump B
Annual costs:		
Expenses	–\$5,340	–\$3,320
Capital recovery		
–(\$750 – \$200)(*A/P*, 10%, 9) – \$200(0.10)	– 115	
–(\$16,000 – \$3,200)(*A/P*, 10%, 9) – \$3,200(0.10)		– 2,542
Total AW	–\$5,455	–\$5,862

Since \$5,862 > \$5,455, the replacement pump apparently is not justified and the defender should be kept at least one more year. We could also make the analysis using other methods (e.g., PW) and the indicated choice would be the same.

It now must be acknowledged that a before-tax analysis often is not a valid basis for a replacement decision because of the effect of depreciation and any significant capital gain or capital loss (unamortized value) on income taxes. An after-tax analysis is a valid basis for such decisions.

Solution of Example 12-2 by the AW Method (after Taxes)

Tables 12-2 and 12-3 show computations of the after-tax cash flows for the existing and replacement pumps of Example 12-2. Note in Table 12-2 that the amounts for the defender alternative on the year 0 line have reversed signs in parentheses. This is done because, as a computational convenience, the amounts first were determined by assuming the sale and subsequent replacement of the existing pump A. *The reversed signs in parentheses indicate the effect that would result from keeping pump A.* Also shown in Table 12-3 are the after-tax investment in the replacement pump B and its after-tax market value.

After the one-time and annual effects have been determined in Tables 12-2 and 12-3, the next step in an after-tax replacement study involves the equivalence calculations, which basically are no different from any other study of alternatives, except, of course, that an after-tax minimum attractive rate of return is used. The following is the after-tax AW analysis for Example 12-2.

$$
\begin{aligned}
\text{AW of defender (pump A)} &= -\$4{,}190\,(A/P,\ 6\%,\ 9) - \$2{,}592 \\
&\quad -\$612\,(F/A,\ 6\%,\ 4)(A/F,\ 6\%,\ 9) \\
&\quad +\$120\,(A/F,\ 6\%,\ 9) \\
&= -\$3{,}432
\end{aligned}
$$

TABLE 12–2 After-Tax Cash Flow Computations for the Defender (Existing Pump A) in Example 12-2

Year, k	Before-Tax Cash Flow	Straight-Line Depreciation	Taxable Income	Cash Flow from Income Tax (40%)	After-Tax Cash Flow
0	(−)\$ 750		(+)−\$8,600[a]	(−)+\$3,440	(−)\$4,190
1–5	− 5,340	\$1,530	− 6,870	+ 2,748	− 2,592
6–9	− 5,340	0	− 5,340	+ 2,136	− 3,204
9	\$ 200		+ 200	− 80	+ 120

[a] If sold for \$750, there would be an \$8,600 long-term capital loss on disposal (\$9,350 − \$750). If \$8,600 capital gains elsewhere were offset by this loss, there would be no tax on the capital gain, saving 40% × \$8,600 = \$3,440. If pump A is not sold, this loss is not realized and the resulting income taxes on long-term capital gains would be \$3,440 higher than if the pump had been sold.

TABLE 12–3 After-Tax Cash Flow Computations for the Challenger (Replacement Pump B) in Example 12-2

		Depreciation							
Year, k	Before-Tax Cash Flow	Investment	×	1989 MACRS Percentages	=	Amount	Taxable Income	Cash Flow from Income Tax (40%)	After-Tax Cash Flow
0	−$16,000								−$16,000
1	− 3,320	$16,000	×	20.00%	=	$3,200	−$6,520	+$2,608	− 712
2	− 3,320	16,000	×	32.00%	=	5,120	− 8,440	+ 3,376	+ 56
3	− 3,320	16,000	×	19.20%	=	3,072	− 6,392	+ 2,557	− 763
4	− 3,320	16,000	×	11.52%	=	1,843	− 5,163	+ 2,065	− 1,255
5	− 3,320	16,000	×	11.52%	=	1,843	− 5,163	+ 2,065	− 1,255
6	− 3,320	16,000	×	5.76%	=	922	− 4,242	+ 1,697	− 1,623
7–9	− 3,320					0	− 3,320	+ 1,328	− 1,992
9	+ 3,200						+ 3,200[a]	− 1,280	+ 1,920

[a] If sold for $3,200, there is a $3,200 recovery of depreciation taxable at 40%.

$$\begin{aligned}\text{AW of challenger (pump B)} &= -\$16{,}000\,(A/P,\ 6\%,\ 9)\\ &\quad -[\$712\,(P/F,\ 6\%,\ 1) - \$56\,(P/F,\ 6\%,\ 2)\\ &\quad -\$763\,(P/F,\ 6\%,\ 3)\\ &\quad - \cdots - \$1{,}992\,(P/F,\ 6\%,\ 9)]\,(A/P,\ 6\%,\ 9)\\ &\quad +\$1{,}920\,(A/F,\ 6\%,\ 9)\\ &= -\$3{,}375\end{aligned}$$

Based strictly on monetary outcomes, the challenger is the better choice. Because the AWs of both pumps are essentially identical, other considerations, such as the improved reliability of the new pump, may cause pump B to be chosen regardless of economic considerations. The after-tax annual costs of both alternatives are considerably less than their before-tax annual costs.

Also notice that an after-tax analysis may *reverse* the results of the before-tax analysis for the same problem. Identical recommendations should not necessarily always be expected, particularly for analyses in which one or more alternatives involve large unamortized values that produce substantial savings in income taxes.

We will now describe how to determine economic lives of depreciable assets and how economic analyses should be performed in the frequently encountered situation when the economic life of the defender differs from that for the challenger. For simplicity several of the examples illustrating these principles do not explicitly include income taxes. However, the principles used in determining economic lives apply on an after-tax basis whenever income taxes are applicable.

12.8

DETERMINING THE ECONOMIC LIFE OF A NEW ASSET (CHALLENGER)

In all engineering economy studies up to this point, we have made the analyses with assumed "useful lives" for all alternatives under consideration, including challengers in replacement studies. For any new asset, an *economic life* can be computed if various operating and maintenance costs are known (or can be estimated) in addition to its year-by-year salvage values. The economic life minimizes the equivalent cost of owning and operating an asset, and it is often shorter than useful and/or physical life. *Determination of an economic life for the challenger(s) and defender is important in a replacement analysis because of the fundamental principle that new and existing assets should be compared over their economic lives*. One reason this is often not done in practice lies in the difficulty of obtaining accurate future cost and salvage value estimates for an alternative.

Determination of economic lives through formal methods can be accomplished with before-tax or after-tax cash flow estimates. Example 12-3 illustrates one such method for determining the life, N^*, of a new asset at which before-tax equivalent annual costs of ownership and operation are minimized. This life is called the economic life.

EXAMPLE 12-3

A new forklift truck will require an investment of $20,000 and is expected to have year-end salvage values and expenses as shown in columns 2 and 5, respectively, of Table 12-4. If the before-tax MARR is 10%, how long should the asset be retained in service?

Solution

The solution to this problem is obtained by completing columns 3, 4, 6, and 7 of Table 12-4. In the solution the customary year-end occurrence of all cash flows is assumed. The actual depreciation for any year is the difference between the beginning and year-end salvage values. In this example, depreciation is not computed according to any formal method but rather results from expected forces in the marketplace. The opportunity cost of capital in year k is 10% of the capital unrecovered after cumulative depreciation at the beginning of each year has been subtracted from the original investment. The values in column 7 are the equivalent annual costs that would be incurred each year if the asset were retained in service for the number of years shown, and then replaced at the end of the year. The minimum annual cost occurs at the end of year N^*.

From the values shown in column 7, it is apparent that the asset will have minimum annual cost if it is kept in service only 3 years (i.e., $N^* = 3$).

The computational approach in the example above, as shown in Table 12-4, was to determine the total annual cost for each year (sometimes called *marginal*, or *year-by-year cost*) and then to convert these into an equivalent annual cost (AC). The AC for any life can also be calculated using the more familiar capital recovery formulas in Chapter 4. For example, for a life of 2 years, the AC can be calculated as follows:

$$(\$20{,}000 - \$11{,}250(A/P,\ 10\%,\ 2) + \$11{,}250(10\%) \\ + [\$2{,}000(P/F,\ 10\%,\ 1) \\ + \$3{,}000(P/F,\ 10\%,\ 2)](A/P,\ 10\%,\ 2) = \$8{,}643$$

which checks with the corresponding row in column 7 of Table 12-4.

TABLE 12–4 Determination of the Economic Life (N^*) of a New Asset (Example 12-3)

	Cost of service for kth Year					Equivalent Annual Cost (AC) if Retired at End of Year k
(1) Year, k	(2) Salvage Value at Year End	(3) Actual Depreciation During Year	(4) Cost of Capital at 10%	(5) Annual Expenses	(6) [=(3)+(4)+(5)] Total Cost for Year, TC_k	(7) $AC = \left[\sum_{k=1}^{5} TC_k(P/F, 10\%, k)\right](A/P, 10\%, k)$
0	$20,000	—	—	—	—	—
1	15,000	$5,000	$2,000	$ 2,000	$ 9,000	$9,000
2	11,250	3,750	1,500	3,000	8,250	8,643
3	8,500	2,750	1,130	4,620	8,500	8,600 ← minimum ($=N^*$)
4	6,500	2,000	850	8,000	10,850	9,082
5	4,750	1,750	650	12,000	14,400	9,965

12.9

THE ECONOMIC LIFE OF A DEFENDER

In replacement analyses, we must also determine the economic life that is most favorable to the defender. When a major outlay for defender alteration or overhaul is needed, the life that will yield the least equivalent annual cost is likely to be the period that will elapse before the next major alteration or overhaul will be needed. Alternatively, when there is no defender market value now or later (and no outlay for alteration or overhaul) and when defender operating expenses are expected to increase annually, the remaining life that will yield the least equivalent annual cost will be 1 year (or possibly less).

When market values are greater than zero and expected to decline from year to year, it may be necessary to calculate the "apparent" remaining economic life. This is done in the same manner as in Example 12-3 for a new asset. The only difference is that the present realizable market value of the defender is considered to be its investment value.

Regardless of how the "apparent" economic remaining life for the defender is determined, a decision to keep the defender does not mean that it should be kept only for this period of time. Indeed, the defender should be kept longer than the "apparent" economic life as long as its *marginal cost* (total cost for an additional year of service) is less than the minimum AC for the best challenger.

EXAMPLE 12-4

Suppose that it is desired to determine how much longer an old forklift truck should remain in service before it is replaced by the new truck (challenger) for which data were given in Example 12-3 and Table 12-4. The old truck (defender) is 2 years old, originally cost $13,000, and has a present realizable market value of $5,000. If kept, its salvage value and operating expenses are expected to be as follows:

Year from Now	Salvage Value	Operating Expenses
1	$4,000	$5,500
2	3,000	6,600
3	2,000	7,800
4	1,000	8,800

Determine the most economical period to keep the old truck before replacing it (if at all) with the present challenger of Example 12-3.

TABLE 12–5 Determination of the Economic Life of an Old Asset (Example 12-4)

(1) Year from Now	(2) Depreciation for Year	(3) Cost of Capital at 10%[a]	(4) Operating Expenses	(5) Total Cost for Year (Marginal Cost)	(6) Equivalent Annual Cost (AC)
1	$1,000	$500	$5,500	$ 7,000	$7,000
2	1,000	400	6,600	8,000	7,475
3	1,000	300	7,800	9,100	7,966
4	1,000	200	8,800	10,000	8,406

[a] Based on a present realizable market value of $5,000.

Solution

Table 12-5 shows the calculation of total cost for each year (marginal cost) and AC for each year for the defender using the same format as that used in Table 12-4. Note that the minimum AC of $7,000 corresponds to keeping the old truck for 1 more year. However, the marginal cost of keeping the truck for the second year is $8,000, which is still less than the minimum AC for the challenger (i.e., $8,600, from Example 12-3). The marginal cost for keeping the defender the third year and beyond is greater than the $8,600 minimum AC for the challenger. Based on currently available data shown above, it would be most economical to keep the defender for 2 more years and then to replace it with the challenger. This situation is portrayed graphically in Figure 12-1.

Example 12-4 assumes that there is only one challenger alternative available. In this situation if the old asset (defender) is retained beyond the point where its marginal (year-by-year) costs exceed the minimum AC for the challenger, the difference in costs continues to grow and replacement becomes more urgent, as illustrated to the right of the intersection in Figure 12-1.

Figure 12-2 illustrates the effect of improved new challengers in the future. If an improved challenger X becomes available before replacement with the new asset of Figure 12-1, then a new replacement study probably should take place to consider the improved challenger. If there is a possibility of a further-improved challenger Y as of, say, 4 years later, it may be better still to postpone replacement until that challenger becomes available. Although retention of the old asset beyond its breakeven point with the best available challenger has a cost that may well grow with time, this cost of waiting can, in some instances, be worthwhile if it permits purchase of an improved asset having economies that offset the cost of waiting. Of course, a decision to postpone a replacement may "buy time and information" also. Because technological change tends to be sudden and dramatic rather than uniform and gradual, new challengers with significantly improved features can arise sporadically and can change replacement plans substantially.

When replacement is not signaled by the engineering economy study, more information may become available before the next "challenge" to the defender,

FIGURE 12-1

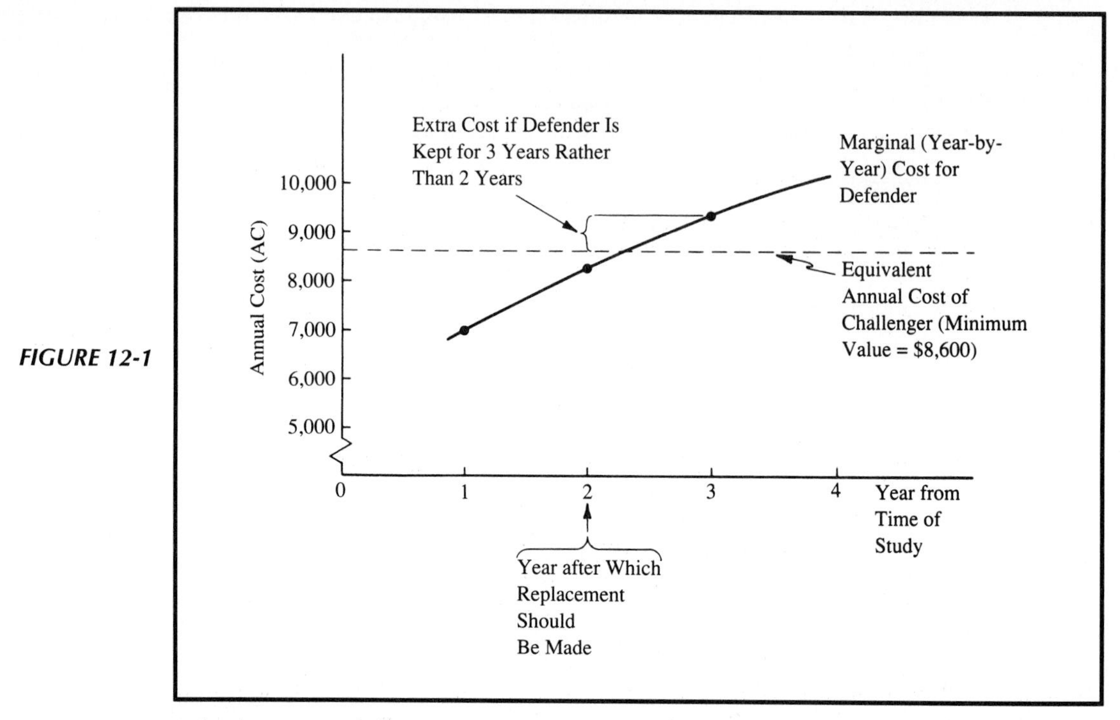

Defender versus Challenger Fork Trucks. (Based on Examples 12-3 and 12-4)

FIGURE 12-2

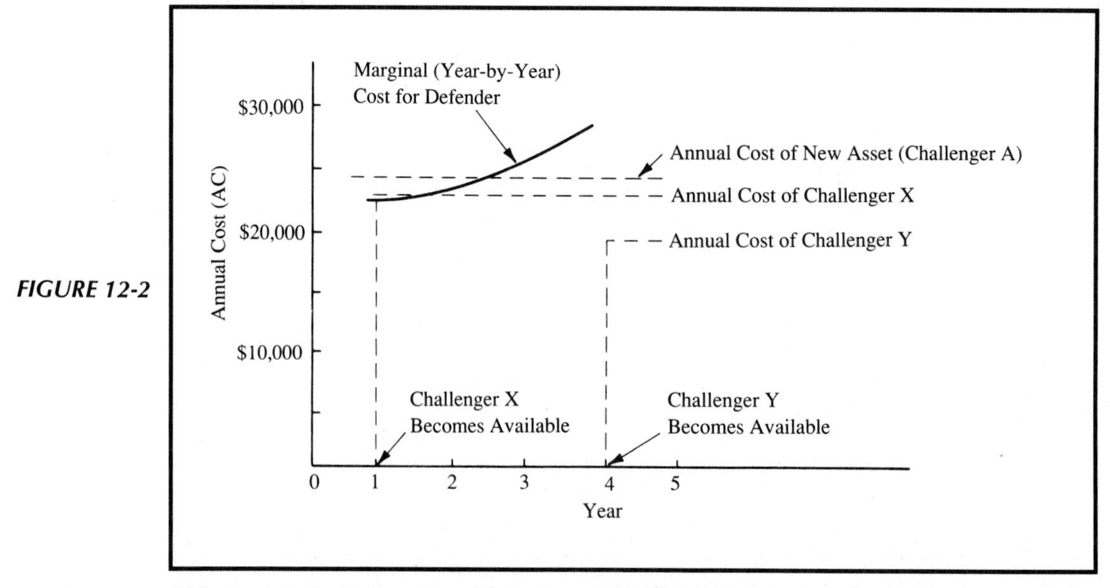

Old versus New Asset Costs with Improved Challengers Becoming Available in the Future

and hence the next comparison should include that additional information. *Postponement* generally should mean a postponement of the decision "when to replace," not the decision to postpone replacement until a specified future date.

12.10 COMPARISONS IN WHICH DEFENDER'S ECONOMIC LIFE DIFFERS FROM THAT OF THE CHALLENGER

When the lives of the replacement alternatives are equal, as in Example 12-2, economic comparisons of alternatives can be made by any of the correct analysis methods without complication. However, when the expected economic lives of the defender and challenger(s) differ, this difference should be taken into account, and the analysis may be complicated.

Chapter 5 described the two assumptions that are commonly used for economic comparisons of alternatives having different lives: (1) *repeatability* and (2) *cotermination*. The repeatability assumption involves two main stipulations:

1. The period of needed service for which the alternatives are being compared is either indefinitely long or a length of time equal to a common multiple of the lives of the alternatives.
2. What is estimated to happen in the first life span will happen in all succeeding life spans, if any, for each alternative.

For replacement analyses, the first condition above may be acceptable, but normally the second is not reasonable for the defender. The defender is typically an older piece of equipment with a modest current realizable market value (selling price). An identical replacement, even if it could be found, probably would have an installed cost far in excess of the current market value of the defender.

Failure to meet the second stipulation can be circumvented if the period of needed service is assumed to be indefinitely long and if we recognize that the analysis is really to determine if *now* is the time to replace the defender. When the defender is replaced, it will be by the challenger—the best available replacement.

Example 12-4 involving the defender versus challenger fork trucks made use of the *repeatability* assumption. That is, it was assumed that the particular challenger analyzed in Table 12-4 would have a minimum AC regardless of when it replaces the defender. Figure 12-3 shows time diagrams of the cost consequences of keeping the defender for 2 more years versus adopting

FIGURE 12-3

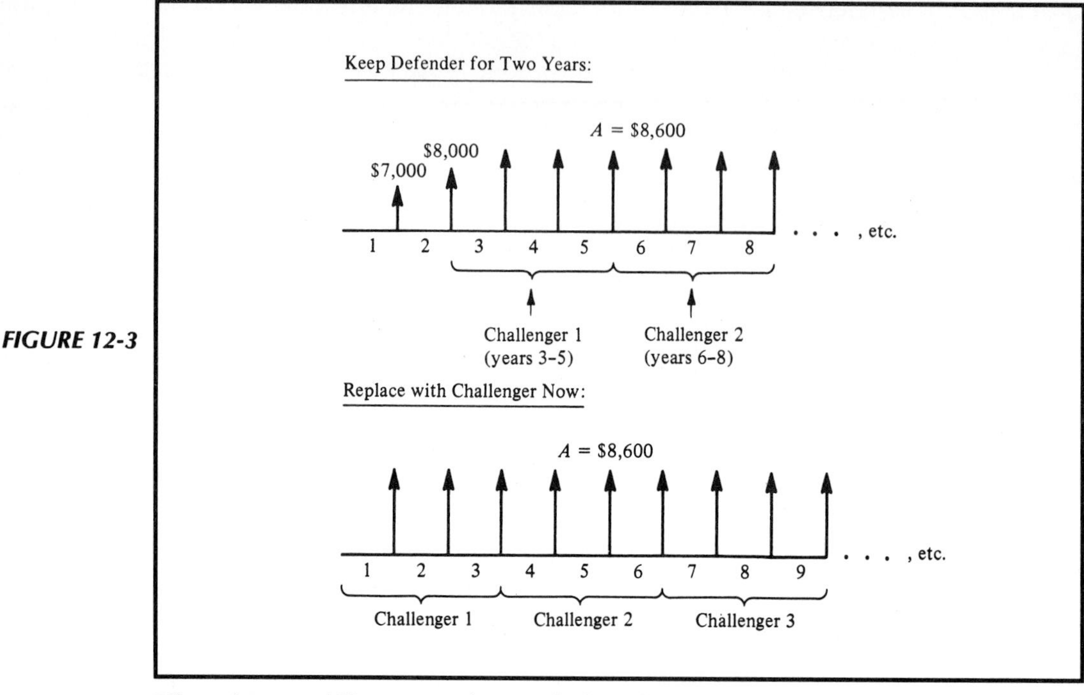

Effect of Repeatability Assumption Applied to Alternatives in Example 12-4

the challenger now, with the challenger costs to be repeated into the indefinite future. Recall that the economic life of the challenger is 3 years. *It can be seen in Figure 12-3 that the only difference between the alternatives is in years 1 and 2.*

The *repeatability* assumption, applied to replacement problems, simplifies their economic comparison by permitting conservative comparison of the alternatives by any correct analysis method. Whenever the *repeatability* assumption is not applicable, the *coterminated* assumption may be used and involves using a finite study period for all alternatives.

As described in Chapter 5, use of the *coterminated* assumption requires detailing what, and when, cash flows are expected to occur for each alternative and then determining which is most economical, using any of the correct economic analysis methods. When the effects of inflation are to be considered in replacement studies, it is recommended that the coterminated assumption be used.

EXAMPLE 12-5

Suppose that we are faced with the same replacement problem as in Example 12-3 except that the period of needed service (or the firm's planning horizon) is (a) 3 years, or (b) 4 years. In each case, which alternative should be selected?

Solution

(a) For a planning horizon of 3 years, one might intuitively think that either the defender should be kept for 3 years or it should be replaced immediately by the challenger to serve the next 3 years. From Table 12-5 the AC for the defender for 3 years is $7,966, and from Table 12-4 the AC for the challenger for 3 years is $8,600. Thus, following this reasoning, the defender would be kept for 3 years. However, this is not quite right. Focusing on the "Total Cost for Year" columns, one can see that the defender has the lowest cost in the first 2 years, but in the third year its cost is $9,100, whereas the cost of the first year of service for the challenger is only $9,000. Hence it would be economical to replace the defender after the second year. This conclusion can be confirmed by enumerating all replacement possibilities and their respective costs and then computing the AC for each, as will be done for the 4-year planning horizon in part (b).

(b) For a planning horizon of 4 years, the alternatives and their respective costs for each year and the AC of each are given in Table 12-6. Thus, the most economical alternative is to keep the defender for 2 years and the challenger for 2 years. The decision to keep the defender for 2 years happens to be the same as when the repeatability assumption was used, but of course this would not necessarily be true in general.

12.11 REPLACEMENT VERSUS AUGMENTATION

Conditions frequently arise wherein an existing asset is not of sufficient capacity to meet increased demands that have developed. In such cases there usually are two alternatives open: (1) to augment the existing asset, or (2) to replace it with a new asset of sufficient capacity to meet the new demands.

TABLE 12–6 Determination of Economic Life of Defender for a Planning Horizon of 4 Years

Keep Defender for	Keep Challenger for	Total (Marginal) Costs for Each Year					AC at 10% for 4 Years
		1	2	3	4		
0 years	4 years	$9,000	$8,250	$8,500	$10,850		$9,082
1	3	7,000	9,000	8,250	8,500		8,301
2	2	7,000	8,000	9,000	8,250	(min)→	8,005
3	1	7,000	8,000	9,100	9,000		8,190
4	0	7,000	8,000	9,100	10,000		8,406

Example 12-6 provides a situation involving augmentation versus replacement and includes the consideration of income taxes. The period of required service is *coterminated* at 5 years, and general price inflation during this time must be considered in the analysis. Furthermore, both alternatives qualify for MACRS depreciation.

EXAMPLE 12-6

In 1984 (5 years ago) a construction company purchased an air compressor for $2,000. Its useful life was estimated to be 10 years, but its ACRS class life was 5 years. Annual out-of-pocket expenses for operations, maintenance, taxes, and insurance are running at approximately $1,050.

The demands for compressed air have increased to the point where either an additional small compressor, costing $1,000, will have to be added, or the old compressor will have to be replaced with a larger compressor, which will cost $4,000. It is estimated that the new, small compressor would require annual expenses of $300, and that the annual expenses for the larger, more modern compressor would be $500. The new small (augmenting) unit would have a 5-year MACRS class life and a $550 estimated market value at the end of year 5. Depreciation charges for the large unit also will be based on a 5-year MACRS class life, and it is expected that the actual market value of the large unit at the end of the 5-year study period (or period of needed service) would be $2,000. If the existing compressor is replaced, it will bring a present market value of $200. If the old compressor is used for another 5 years, it is believed that its market value at that time will be negligible.

All cost and market value estimates above are expressed in terms of 1989 dollars. The company earns about 20% on its capital before income taxes and 10% after taxes, being in a 50% effective income tax bracket. General price inflation will affect all operating costs and market values at an average rate of 6% per year. Should the existing compressor be augmented or replaced, based on an after-tax analysis of equivalent annual costs and a 5-year study period?

Solution

Table 12-7 shows the after-tax cash flow computations for Example 12-6. This example includes numerous realistic conditions that are present in today's economic analyses: augmentation of an existing asset versus its complete replacement, inflation of cash flows, MACRS depreciation, and an opportunity cost of keeping an existing asset. A brief description of some of the main points in Table 12-7 follows.

If the existing compressor is sold for $200, this is a gain of capital (depreciation recovery) taxed at the ordinary income rate of 50% since the book value after 5 years with ACRS depreciation is zero. If the old compressor is *kept*, the opportunity cost is $200 and the income taxes avoided would be $200(0.50) = $100. Before-tax cash flows are inflated by 6% per year as shown,

TABLE 12–7 After-Tax Cash Flow Analysis for Augmentation Versus Replacement Alternatives in Example 12-6

Alternative	Year, k	Before-Tax Cash Flow	MACRS Depreciation	Taxable Income	Income Taxes (at 50%)	After-Tax Cash Flow	
Augment Old Compressor	0	(−) \$200[a] − \$1,000		(−) + \$ 200[b]	(+) − \$ 100[c]	−\$1,100	
	1	− (1,050 + 300)$(1.06)^1$	\$ 200[d]	− 1,631	816	− 615	
	2	− (1,050 + 300)$(1.06)^2$	320	− 1,837	918	− 598	
	3	− (1,050 + 300)$(1.06)^3$	192	− 1,800	900	− 708	AW(10%) =
	4	− (1,050 + 300)$(1.06)^4$	115	− 1,820	910	− 794	−\$935
	5	− (1,050 + 300)$(1.06)^5$	115	− 1,922	961	− 846	
	5	550$(1.06)^5$ − 58		678[e]	− 339	339	
Purchase New Compressor	0	−\$4,000				−\$4,000	
	1	− 500$(1.06)^1$	\$ 800	− 1,330	665	135	
	2	− 500$(1.06)^2$	1,280	− 1,842	921	359	
	3	− 500$(1.06)^3$	768	− 1,364	682	86	AW(10%) =
	4	− 500$(1.06)^4$	461	− 1,092	546	− 85	−\$1,126
	5	− 500$(1.06)^5$	461	− 1,130	565	− 104	
	5	2,000$(1.06)^5$ − 230		2,446	− 1,223	1,223	

[a] Market value of existing asset, which becomes part of the investment cost.
[b] Recaptured depreciation for existing asset (reduces taxable income when asset is kept).
[c] Income tax avoided (saved) when existing asset is retained.
[d] Existing compressor is fully depreciated.
[e] Inflated market value (A\$) less book value of \$58 at end of year 5 (depreciation recapture).

and this includes the estimated market value of the small unit. Because a zero salvage value for the MACRS method of depreciation does not occur until the end of year 6, the entire amount of a compressor's inflated market value less its book value at the end of year 5 is subject to ordinary income taxation at 50%. The after-tax equivalent annual cost at 10% of this alternative is $935.

If the new, larger compressor is purchased, the MACRS depreciation method would result in a book value of $230 at the end of year 5, producing an actual dollar depreciation recovery amounting to $2,446. The after-tax cash flows that result from this alternative have an equivalent annual cost at 10% of $1,126.

Since $935 < $1,126, augmenting the existing compressor would be the recommended choice.

12.12 RETIREMENT WITHOUT REPLACEMENT (ABANDONMENT)

Consider a project for which the period of required service is finite and which has *positive* net cash flows following an initial capital investment at the end of year 0. Salvage values, or abandonment values, are estimated for the end of each remaining year in the project's life. In view of an opportunity cost of capital of $i\%$ per year, should the project be undertaken? What is the best year to abandon the project, given that we have decided to implement it?

For this type of problem, the following assumptions are applicable:

1. Once a capital investment has been made, the firm desires to postpone the decision to abandon a project as long as its annual worth is positive.
2. The existing project will be terminated at the best abandonment time and will not be replaced by the firm.

The abandonment problem is similar to determining the economic life of an asset for which all cash flows are negative. In abandonment problems, however, annual benefits (positive cash flows) are present.

EXAMPLE 12-7

A baling machine for salvaged paper is being considered by the XYZ Company. After-tax annual net receipts have been estimated for the machine in addition to the after-tax abandonment value at the end of each year.

	End of Year:							
	0	1	2	3	4	5	6	7
Net After-Tax receipts	−\$50,000[a]	\$10,000	\$15,000	\$18,000	\$13,000	\$ 9,000	\$ 6,000	\$ 5,000
Abandonment Value	\$50,000	40,000	32,000	25,000	21,000	18,000	17,000	15,000

[a] Investment cost.

The firm's after-tax cost of capital is 12%. What is the optimal abandonment time if the firm has already decided to acquire the machine and use it no longer than 7 years?

Solution

The annual worths that result from deciding now to keep the machine exactly 1, 2, 3, 4, 5, 6, and 7 years are shown below. As you can see, annual worth is maximized by retaining the machine 6 years, so this is the best replacement interval.

Keep 1 year:

$$\begin{aligned} \text{AW}(12\%) &= -\$50{,}000(A/P,\ 12\%,\ 1) + (\$40{,}000 + 10{,}000)(A/F,\ 12\%,\ 1) \\ &= -\$6{,}000 \end{aligned}$$

Keep 2 years:

$$\begin{aligned} \text{AW}(12\%) &= [-\$50{,}000 + \$10{,}000(P/F,\ 12\%,\ 1) \\ &\quad + (\$15{,}000 + 32{,}000)(P/F,\ 12\%,\ 2)](A/P,\ 12\%,\ 2) \\ &= -\$2{,}132 \end{aligned}$$

In the same manner, the net AW for years 3–7 can be computed as follows:

Keep 3 years: AW(12%) = \$622
Keep 4 years: AW(12%) = \$1,747
Keep 5 years: AW(12%) = \$2,012
Keep 6 years: AW(12%) = \$2,121
Keep 7 years: AW(12%) = \$2,006

In some cases it may be decided that although an existing asset is to be retired from its current use, it will not be replaced or removed from all service. Although it may not be able to compete economically in its existing usage, it may be desirable and even economical to retain it as a standby unit or for some different use. The cost to retain the asset under such conditions may be quite low, owing to its relatively low present realizable resale value and perhaps low costs for operation and maintenance. There often are income tax considerations that also bear on the true cost of retaining the asset. This type of situation is illustrated in the following example.

EXAMPLE 12-8

What would be the cost of retaining, as a standby unit for 5 years, the existing compressor discussed in Example 12-6? Assume that the before-tax annual cost for operation and maintenance (O & M) as a standby would be \$155 (in early 1989 dollars). General price inflation of 6% per year should be considered in the analysis. The first cash flow for O & M occurs at the end of 1989.

Solution

First, it is evident that if the compressor is retained, the company will forgo an after-tax recovery of \$100, which would have resulted from its disposal. Thus, the annual capital recovery cost is \$100(A/P, 10%, 5) = \$26.38. Because the depreciation charges for the existing asset are \$0 per year, the \$155 for operation and maintenance results in an equivalent annual cost of \$155 (P/A, 3.77%, 5) (A/P, 10%, 5) = \$183 when f = 6% per year (refer to Section 3.16). Hence, the equivalent annual reduction in taxable income is \$183. At the 50% rate, income taxes would be reduced by \$91.50 and the net effect would be an annual after-tax cost of \$91.50.

Thus, the total equivalent annual cost of keeping the old compressor on standby service for 5 years is \$26.38 + \$91.50 = \$117.88.

12.13 DETERMINATION OF VALUE BY REPLACEMENT THEORY

As was pointed out in Chapter 6, it frequently is necessary to determine (estimate or assess) the value of old assets, often a considerable number of years after their acquisition. During the years since the assets were acquired, price levels will have changed and technological progress will have occurred. As a result, the valuation process is difficult.

If proper cost data are available, a reasonably accurate value of old assets can be obtained by applying replacement theory. As has been pointed out in this chapter, an asset has economic value only if it can be operated profitably in competition with the most economically efficient asset available. The economic value of an existing asset is the maximum amount on which depreciation and interest can be charged while permitting it to compete on an even basis with the most efficient new asset. Such a comparison is illustrated in Example 12-9.

EXAMPLE 12-9

An existing asset can be utilized for 5 more years and will experience annual expenses of \$17,000. A potential new asset would require an investment of \$50,000 and have annual expenses of \$14,000 over a 5-year life. If either alternative is expected to have zero salvage value at the end of 5 years and capital should earn 20% before taxes, what is the current value of the old asset at which it is equally as economical as the new asset? Inflation can be ignored.

Solution

Letting V = value of the existing asset, we obtain

	Existing Asset	New Asset
Capital recovery amount		
$V(A/P, 20\%, 5)$	$0.3344V	
$50,000(A/P, 20%, 5)		$16,719
Annual expenses	17,000	14,000
Total Annual Costs	$17,000 + $0.3344V	$30,719

Equating the total equivalent annual costs and solving, we find that V = \$41,026. This is a good measure of the present economic value of the old asset if the new asset is the best alternative available.

Because the alternatives have the same economic lives, it would also be easy to obtain the same answer by equating present worths of costs over the 5-year period. Thus,

$$V + \$17{,}000(P/A, 20\%, 5) = \$50{,}000 + \$14{,}000(P/A, 20\%, 5)$$
$$V = \$41{,}026$$

12.14 A REPLACEMENT PROBLEM CASE STUDY (CIRCA 1984)

The Operations Engineering Group of the Power Engineering Corporation has been studying the possibility of replacing a large boiler in one of their petroleum refining processes. A new boiler with roughly the same size and reliability can be purchased from a well-known U.S. manufacturer. It has been determined to be the best challenger currently available. The existing boiler (defender) was purchased in 1979 (5 years before) and has experienced what the firm's engineers feel to be excessive operation and maintenance (O&M) expenses. Cost projections by the Operations Engineering Group (in actual, or inflation-adjusted, dollars) are summarized below, along with other relevant information and assumptions.

Defender: Original cost was $200,000; ADR guideline life (see Table 6-2) for straight-line depreciation was 10 years; original estimate of salvage value for computing depreciation was $22,000; O&M expenses for the next year (sixth year of life for the defender) are $80,000 and will increase by approximately $10,000 per year thereafter; present estimated net market value is $20,000; modification cost now to upgrade the defender to make it comparable in performance to the challenger is $100,000 (no investment tax credit can be claimed); estimated market value of the modified defender in 5 years is negligible; the modification costs are fully depreciable over 5 years by using the straight-line method.

Challenger: Initial cost (including purchase price, freight, installation, and related expenses) is $400,000; ACRS guideline period is 5 years; salvage value for tax purposes is $0; estimated net market value at the end of 5 years is $165,000; annual O&M expenses are $48,000.

Assumptions: Five-year service requirement (planning horizon); depreciable amount of the modified defender equals its *book value* at the end of year 5 less original salvage value, *plus modification costs*; an investment credit of 8% can be claimed on new assets; effective tax rate on ordinary income is 50%; capital gains (losses) create a tax liability (savings) which is taxed at 28%; the firm's after-tax MARR is 18%; sufficient capital is available to implement the challenger (if it is selected).*

Assignment: You have been requested to compare the two alternatives on an after-tax basis in view of the information given above. Use a tabular setup and show all calculations.

Solution

The determination of the after-tax cash flow for the *defender* is summarized in Table 12-8. The year 0 row has two separate entries—the first showing the opportunity cost of keeping the defender. The current book value of the defender is $200,000 − 5($17,800) = $111,000, and the associated *capital loss* would be $91,000 if the defender is sold. Because we are considering the possibility of retaining the defender, this entry would be reversed in sign to indicate an increase in taxable income relative to the firm's position if the defender had been sold. Similarly, the income tax saving of 28% × $91,000 that results *if* the defender had been sold gives rise to an "opportunity tax liability" of $25,480 when the defender is kept. (In a large firm the absence of a tax savings of $25,480 due to keeping the defender will create an additional tax liability for the year). A $22,000 loss on disposal at the end of year 5

*In 1984 an investment tax credit of 8% could be claimed for new assets used in the production process. A 28% rate applied to taxable capital gains (or losses when used to offset gains).

TABLE 12–8 After-Tax Analysis of Existing Boiler (Defender)

End of Year	Before-Tax Cash Flow	Depreciation[a]	Taxable Income	Cash Flow for Taxes	After-Tax Cash Flow
0	(−)$ 20,000		(+)$ 91,000	(−) $25,480	(−)$ 45,480
0	− 100,000				− 100,000
1	− 80,000	$37,800	− 117,800	58,900	− 21,100
2	− 90,000	37,800	− 127,800	63,900	− 26,100
3	− 100,000	37,800	− 137,800	68,900	− 31,100
4	− 110,000	37,800	− 147,800	73,900	− 36,100
5	− 120,000	37,800	− 157,800	78,900	− 41,100
5	0		− 22,000	6,160	+ 6,160

[a] Depreciation $= \frac{\$200{,}000-\$22{,}000}{10} + \frac{\$100{,}000}{5} = \$37{,}800/\text{yr}.$

would be claimed because the market value of the modified defender is $0 while its book value is $22,000. A tax credit of 28% × $22,000 = $6,160 will result. From Table 12-8, the present worth of the after-tax cash flow for the defender is computed at a MARR of 18% and found to be −$234,927.

The computation of after-tax cash flow for the *challenger* is provided in Table 12-9. An investment tax credit of 8% × $400,000 = $32,000 is taken in year 0. Depreciation is determined with 1984 ACRS percentages. Because the book value with the ACRS method is zero at the end of year 5, the estimated market value of $165,000 at that time is fully taxable as "recovery of previously charged depreciation." The income tax owed is 50% × $165,000 = $82,500. At a MARR of 18%, the present worth of the challenger is −$284,383.

From the foregoing analysis, you would probably recommend keeping the defender and modifying the existing boiler. Because present worths differ by less than $50,000, nonmonetary considerations could easily swing the decision in favor of the challenger. A list of these considerations might include operating safety, reliability of the system, and improved control over the boiler effluents.

TABLE 12–9 After-Tax Analysis of a New Boiler (Challenger)

End of Year	Before-Tax Cash Flow	Depreciation[a]	Taxable Income	Cash Flow for Taxes	After-Tax Cash Flow
0	−$400,000			$32,000	− $368,000
1	− 48,000	$60,000	−$108,000	54,000	6,000
2	− 48,000	88,000	− 136,000	68,000	20,000
3	− 48,000	84,000	− 132,000	66,000	18,000
4	− 48,000	84,000	− 132,000	66,000	18,000
5	− 48,000	84,000	− 132,000	66,000	18,000
5	165,000		165,000	− 82,500	82,500

[a] Use 1984 recovery percentages: 15% in year 1, 22% in year 2, 21% in year 3, 21% in year 4, and 21% in year 5.

There are six factors that must be considered to varying degrees in replacement decisions. The *first* is the possibility that if replacement is deferred, assets available in the future may be improvements over those presently available. When technology is changing rapidly, we hesitate to acquire an asset that very soon may be obsolete economically. Consequently, in making replacement decisions, a knowledge of the state and trend of technological development in the area involved is important.

The *second* factor relates to probable variations in the future market value of an asset. If study shows that the most likely disposal date is a considerable number of years distant, moderate changes in the assumed disposal value are not likely to make any substantial change in the analysis. On the other hand, if only a few years are involved, prospective changes in the market value can have a considerable effect on the year-to-year economy.

Where the trade-in or disposal value is changing rapidly, we must be certain that a sufficient number of years is considered in determining the economy of an existing asset. For example, on the basis of keeping the old asset 1 more year, it might appear to be more economical to make the replacement now. If, however, a 3- or 4-year retention period is used, the old asset might be more economical than a new one. Thus, in general, slow rates of decrease in market value, with increasing operation and/or maintenance costs, tend to favor early replacement, whereas rapid rates of decrease in market value favor later replacement.

A *third* factor is the probability that the enterprise will grow in the immediate future. If it is evident that such growth will require increased output unobtainable from the existing asset, it is obvious that replacement or augmentation will have to be made in the near future. The resulting problem is basically to determine whether replacement or augmentation is cheaper, and at what time either should occur. The only complicating factor is the possibility that the desired equipment may not be available when needed if replacement is deferred, or that the increased demand may develop earlier than anticipated. To avoid these possibilities, companies sometimes prefer to make the replacement somewhat earlier than necessary and then attempt to develop the market.

A *fourth* factor is somewhat counter to the previous one. Replacement frequently results in excess capacity, since new facilities and equipment usually are more efficient than those they replace. If there is no actual use for such excess capacity, it has no value and thus should not be given any consideration. On the other hand, if the excess capacity can be used for some function not rendered by the old asset, this should be considered in the replacement study.

The possibility that the purchase price of the replacement asset will change in the future is a *fifth* factor that must be considered. Although it is difficult to predict what price changes will occur in the future, the long history of inflation that has occurred causes many to believe that the trend is apt to continue. Such a condition tends to favor earlier replacement. However, this practice should not be followed blindly, particularly where the need for the service is expected to continue for many years. Two additional facts must also be considered. First, the earlier the replacement, the sooner the new asset will have to be replaced—at higher first cost. Second, it is likely that technological improvements will occur, and by deferring replacement it is possible that a more efficient asset may be obtained. Therefore, before too much emphasis is given to probable price increases, other possible future changes also should be considered.

Budgetary and personnel considerations constitute a *sixth* factor that frequently affects replacement decisions. Most companies do not have unlimited funds, and many projects are usually competing for the funds that they do have. In such cases replacement studies supply information on the basis of which decisions can be made regarding the timing of replacements as funds become available.

12.15 PROBLEMS

12-1. A small plant has four space heaters that were purchased 3 years ago for $1,300 each. They will last 10 more years, and will have no salvage value at that time. An expansion to the plant can be heated by two additional space heaters that will last 10 years, have no salvage value, and cost $1,800 each. An alternative method of heating is to install a central heating system that will cost $6,400, last 10 years, and have a salvage value of $4,000. If central heating is installed, the present four space heaters can be sold for $200 each. Compare the equivalent annual cost of heating with space heaters and with central heating. Assume that fuel costs will be the same for either system and that the minimum attractive rate of return (before taxes) for the company is 15%. What would you recommend? (12.5–12.7)

12-2. One year ago a machine was purchased at a cost of $2,000 to be useful for 6 years. However, the machine has failed to perform properly and has cost $500 per year for repairs, adjustments, and shutdowns. A new machine is available to accomplish the functions desired and has an initial cost of $3,500. Its maintenance costs are expected to be $50 per year during its service life of 5 years. The approximate market value of the presently owned machine has been estimated to be roughly $1,200. If the operating costs (other than maintenance) for both machines are equal, show whether it is economical to purchase the new machine. Perform a before-tax study using an interest rate of 12%, and assume that salvage values will be negligible. (12.5–12.7)

12-3. Suppose that you have an old car, which is a real gas guzzler. It is 10 years old and could be sold to a local dealer for $400 cash. The annual maintenance costs will average $800 per year into the foreseeable future, and the car averages only 10 miles per gallon. Gasoline costs $1.50 per gallon and your driving averages 15,000 miles per year. You now have an opportunity to replace the old car with a better one that costs $8,000. If you buy it, you will pay cash. Because of a 2-year warranty, the maintenance costs are expected to be negligible. This car averages 30 miles per gallon. Use the incremental internal rate of return method and specify which alternative you should select. Utilize a 2-year comparison period and assume your new car can be sold for $5,000 at end of year 2. Ignore the affects of income taxes and let your MARR be 15%. State any other assumptions you make. (12.5–12.7).

12-4. A pipeline contractor is considering the purchase of certain pieces of earthmoving equipment to reduce his costs. His alternatives are as follows.

Plan A: Retain a backhoe already in use. The contractor still owes $10,000 on this machine but can sell it on the open market for $15,000. This machine will last another 6 years, with maintenance, insurance, and labor costs totaling $28,000 per year.

Plan B: Purchase a trenching machine and a highlift to do the trenching and backfill work and sell the current backhoe on the open market. The new machines will cost $55,000, will last 6 years with a salvage value of $15,000, and will reduce annual maintenance, insurance, and labor costs to $20,000.

If money is worth 8% before taxes to the contractor, which plan should he follow? Use the present worth procedure to evaluate both plans on a before-tax basis. (12.5–12.7)

12-5. A 3-year-old asset that was originally purchased for $4,500 is being considered for replacement. The new asset under consideration would cost $6,000. The engineering department has made the following estimates of the operating and maintenance costs of the two alternatives.

Year	Old Asset	New Asset
1	$2,000	$ 500
2	2,400	1,500
3	—	2,500
4	—	3,500
5	—	4,500

The dealer has agreed to place a $2,000 trade-in value on the old asset if the new one is purchased now. It is estimated that the residual value for either of the assets will be zero at any time in the future. If the before-tax rate of return is 12%, make an equivalent annual cost analysis of this situation and recommend which course of action should be taken now. Use the repeatability assumption and give a short written explanation of your answer. (12.5–12.7)

12-6. You have a machine that cost $30,000 2 years ago—it has a present market value of $5,000. Operating costs total $2,000 per year so long as the machine is in use. In 2 more years it will no longer be useful and you can sell it for $500 scrap value. You are considering replacing this machine with a new model incorporating the latest technology—this new model will cost you $20,000 now. It has an annual operating cost of $1,000 with a useful life of 8 years and negligible salvage value. If you delay the purchase of the new model, the cost will be $24,000 because of the installation difficulties that do not exist now. When your present machine is retired, you have no hopes of getting another one like it, since it was a very limited model. If money is worth 10% before taxes, what should you do? (12.5–12.7)

12-7. A diesel engine was installed 10 years ago at a cost of $50,000. It has a present realizable market value of $14,000. If kept, it can be expected to last 5 more years, have operating expenses of $14,000 per year, and have a salvage value of $8,000 at the end of the 5 years. This engine can be replaced with an improved version costing $65,000 and having an expected life of 20 years. This improved version will have estimated annual operating expenses of $9,000 and an ultimate salvage value of $13,000. It is thought that an engine will be needed indefinitely and that the results of the economy study would not be affected by the consideration of income taxes. Using a before-tax MARR of 15%, make an annual-cost analysis to determine whether to keep or replace the old engine. (12.5–12.7)

12-8.

a. The replacement of a boring machine is being considered by the Reardorn Furniture Company. The new, improved machine will cost $30,000 installed, and will have an estimated useful life of 12 years and a $2,000 salvage value at that time. It is estimated that annual operating and maintenance costs will average $16,000 per year. The present machine has a book value of $6,000 and a present market value of $4,000. Data for the present machine for the next 3 years are as follows:

Year	Salvage Value at End of Year	Book Value at End of Year	Operating and Maintenance Costs During the Year
1	$3,000	$4,500	$20,000
2	2,500	3,000	25,000
3	2,000	1,500	30,000

Using a before-tax interest rate of 15%, make an *annual-cost comparison* to determine whether it is economical to make the replacement now.

b. If the operating and maintenance costs for the present machine had been estimated to be $15,000, $18,000, and $23,000 in years 1, 2, and 3, respectively, what replacement strategy should be recommended? (12.7)

12-9. A diesel engine was installed 5 years ago at a cost of $80,000. Its present market value is $45,000. If the present engine is kept, it can be expected to last 5 more years, have operating expenses of $12,000 per year, and have a market value of $20,000 at that time. This engine is depreciated by the straight-line method using a 10-year life and a salvage value of $10,000 for tax purposes. The old engine can be replaced at the present time with a rebuilt diesel engine that will cost $60,000, have operating expenses of $9,000 per year, and have a life of 5 years. The salvage value for tax purposes is $15,000 and straight-line depreciation will be used. Its market value at the end of 5 years is expected to be zero. The effective income tax rate is 40%. Depreciation recapture is taxed at 40%. If the desired after-tax MARR is 12%, which alternative would you recommend? (12.7, 12.14)

12-10. Suppose that it is desired to make an after-tax analysis for the situation posed in Problem 12-7. The existing engine is being depreciated by the straight-line method over 15 years; an estimated salvage value of $8,000 is used for depreciation purposes. Assume that if the replacement is made, the improved version will be depreciated by the MACRS method, using a 5-year class life and a zero salvage value. Also, assume that annual expenses will affect taxes at 40%, and that any capital gain or capital loss will affect taxes at 40%. Use the equivalent annual cost method to determine if the replacement is justified by earning an after-tax MARR of 10% or more. (12.7, 12.10)

12-11. Use the present worth method to select the better alternative shown below. Let the minimum attractive rate of return equal 10% after taxes and the effective income tax rate be 40%. State any assumptions you make. Capital gains (losses) are taxed at 40%. The firm is known to be profitable in its overall operation. (12.7, 12.14)

Alternative A: Retain an already owned machine in service for 8 more years.
Alternative B: Sell old machine and rent a new one for 8 years.

Data	Alternative A	Alternative B
Labor per year	$300,000	$250,000
Material cost/year	$250,000	$100,000
Insurance and property taxes/year	4% of first cost	None
Maintenance/year	$8,000	None
Rental cost/year	None	$100,000

Alternative A:
Cost of old machine 5 years ago = $500,000
Book value now = $350,000
Depreciation with straight-line method, 15-year life = $30,000/year
Estimated salvage at end of useful life = $50,000
Present market value = $150,000

12-12. You have a machine that was purchased 4 years ago and was set up on a 5-year straight-line depreciation schedule with no salvage value. The original cost was $150,000, and the machine can last for 10 years or more. A new machine is now available at a cost of only $100,000. It can be depreciated over 6 years with the MACRS method, and it has no salvage value. Annual operating cost of the new machine is only $5,000 while the operating cost of the present one is $20,000/year. The new machine has a useful life greater than 10 years. You find that $40,000 is the best price you can get if you sell the present machine now. Your best projection for the future is that you will need the service provided by either of the two machines for the next 5 years. The market value of the old machine is estimated at $2,000 in 5 years, but the market value of the new machine is estimated at $5,000 in 5 years. If the after-tax MARR is 10%, should you sell the old machine and purchase the new one? You do not need both. Assume that the company is in a 40% income tax bracket. (12.7, 12.14)

12-13. Robert Roe has just purchased a 4-year-old used car, paying $3,000 for it. A friend has suggested that he should determine in advance how long he should keep the car so as to ensure the greatest overall economy. Robert has decided that, because of style changes, he would not want to keep the car longer than 4 years, and he has estimated the out-of-pocket costs and trade-in values for years 1 through 4 as follows:

	Year 1	Year 2	Year 3	Year 4
Operating costs	$ 950	$1,050	$1,100	$1,550
Trade-in value at end of year	2,250	1,800	1,450	1,160

If Robert's capital is worth 12%, at the end of which year should he dispose of the car? (12.8, 12.9)

12-14. Determine the economic service life of a machine that has a first cost of $5,000 and estimated operating costs and year-end salvage values as follows. Assume that the interest rate is (a) 0%; (b) 10%. (12.8, 12.9)

Year, k	Operating Cost for Year, c_k	Salvage Value, S
1	$ 800	$4,000
2	900	3,500
3	1,100	3,000
4	1,100	2,000
5	1,300	1,500
6	1,400	1,000

12-15. Consider a piece of equipment that initially cost $8,000 and has these estimated annual operating and maintenance costs and salvage values:

End of Year	Operating and Maintenance Costs for Year	Salvage Value
1	$3,000	$4,700
2	3,000	3,200
3	3,500	2,200
4	4,000	1,450
5	4,500	950
6	5,250	600
7	6,250	300
8	7,750	0

If the cost of money is 7%, determine the most economical time to replace this equipment. (12.8, 12.9)

12-16. A motorcycle can be purchased for $2,500 when new. There follows a schedule of annual operating expenses for each year and trade-in values at the end of each year. Assume that these amounts would be repeated for future replacements, and that the motorcycle will not be kept more than 3 years. If capital is worth 10% before taxes, determine at which year's end the cycle should be replaced so that equivalent annual costs will be minimized. (12.8, 12.9)

	Year 1	Year 2	Year 3
Operating expenses for year	$ 850	$ 950	$1,025
Trade-in value at end of year	1,700	1,400	1,000

12-17. The company for which you are working must improve some of its facilities to meet increasing sales. The existing facilities, which have been fully depreciated, can be used for another 5 years provided that some new materials-handling equipment, costing $25,000, is added. If this procedure is followed, neither the old nor the new equipment would have any salvage value at the end of 5 years, when it all would have to be replaced. An alternative is to dismantle the existing facilities and build entirely new ones at a cost of $600,000. The new installation would have a depreciable life of 20 years, and it would result in a reduction of at least $109,000 per year in out-of-pocket expenses, as compared with the first alternative. The company has an effective income tax rate of 40% and has a 20% MARR before taxes and a 10% MARR after taxes. It uses straight-line depreciation for accounting and tax purposes and finds that its book values agree quite well with actual salvage values. Coterminate the analysis period at 5 years. (a) Make a before-tax analysis and recommendation, using the IRR procedure. (b) Make an after-tax analysis using the IRR procedure. Would your recommendation be the same as in part (a)? Explain your answer. (12.7, 12.11)

12-18. A steel highway bridge must either be reinforced or replaced. Reinforcement would cost $22,000 and would make the bridge adequate for an additional 5 years of service. If the bridge is torn down, the scrap value of the steel would exceed the demolition cost by $14,000. If it is reinforced, it is estimated that its net salvage value would be increased by $16,000 at the time it is retired from service. A new prestressed concrete bridge would cost $140,000 and would meet the foreseeable requirements of the next 40 years. Such a bridge would have no scrap or salvage value. It is estimated that the annual maintenance costs of the reinforced bridge would exceed those of the concrete bridge by $3,200. Assume that money costs the state 10% and that the state pays no federal income taxes. What would you recommend? (12.5–12.7)

12-19. Four years ago the Attaboy Lawn Mower Company purchased a piece of equipment for their assembly line. Because of increasing maintenance costs for this equipment, a new piece of machinery is being considered. The cost characteristics of the defender (present equipment) and the challenger are as follows:

Defender	Challenger
Original cost = $9,000	Purchase cost = $13,000
Maintenance = $300 in year 1, increasing by 10% per year thereafter.	Maintenance = $100 in year 1, increasing by 10% per year thereafter.
Original estimated salvage value for tax purposes = 0	Market value = $3,000 at the end of year 5.
Original estimated life = 9 yrs.	MACRS class life = 5 years

Suppose that a $3,200 market value is available now for the defender. Perform an after-tax analysis, using an after-tax MARR of 10% to determine which alternative to select. The effective income tax rate is 40% on ordinary income, and capital gains (losses) are taxed at 40%. Straight-line depreciation is applied to the defender, while the challenger qualifies for MACRS depreciation. (12.7, 12.14)

12-20. A company is considering replacing a turret lathe with a single-spindle screw machine. The turret lathe was purchased 6 years previously at a cost of $80,000 and depreciation has been figured on a 10-year, straight-line basis, using zero salvage value. It can now be sold for $15,000, and if retained would operate satisfactorily for 4 more years and have zero salvage value. The screw machine is estimated to have a useful life of 10 years. MACRS depreciation would be used with a class life of 5 years. It would require only 50% attendance of an operator who receives $7.50 per hour. The machines would have equal capacities and would be operated 8 hours per day, 250 days per year. Maintenance on the turret lathe has been $3,000 per year; for the screw machine it is estimated to be $1,500 per year. Taxes and insurance on each machine would be 2% of the first cost annually. If capital is worth 10% to the company, after taxes, and the company has a 40% income tax rate, what is the maximum price it can afford to pay for the screw machine? Capital gains (losses) are taxed at 40%. Use the repeatability assumption. (12.10, 12.14)

12-21. Ten years ago a corporation built a warehouse at a cost of $400,000 in an area that since has developed into a major retail location. At the time the warehouse was constructed it was estimated to have a depreciable life of 20 years, with no salvage value, and straight-line depreciation has been used. The corporation now finds it would be more convenient to have its warehouse in a less congested location, and can sell the old warehouse for $250,000. A new warehouse, in the desired new location, would cost $500,000, have a MACRS class life of 10 years with no salvage value, and there would be an annual saving of $4,000 per year in the operation and maintenance costs. Taxes and insurance on the old warehouse have been 5% of the first cost per year, while for the new warehouse they are estimated to be only 3% per year. The corporation has a 40% income tax rate and a 40% long-term capital gains tax rate. Capital is worth not less than 12% after taxes. What would you recommend on the basis of an after-tax IRR analysis? (12.7, 12.14)

12-22. Five years ago an airline installed a baggage conveyor system in a terminal, knowing that within a few years it would move to a new section of the terminal and that this equipment would then have to be moved. The original cost of the installation was $120,000, and through accelerated depreciation methods the company has been able to write off the entire cost. It now finds that it will cost $40,000 to move and reinstall the conveyor. This cost would be recovered over the next 5 years, which the airline believes is a good estimate of the remaining useful life of the system if moved. It finds that it can purchase a somewhat more efficient conveyor system for an installed cost of $120,000, and this system would result in an estimated reduction in annual operating and maintenance costs of $6,000 in year 0 dollars during its estimated 5-year MACRS class life. Annual operating and maintenance costs are expected to inflate by 6% per year. In both alternatives, MACRS depreciation is used. A small airline company, which will occupy the present space, has offered to buy the old conveyor for $90,000.

Annual property taxes and insurance on the present equipment have been $1,500, but it is estimated that they would increase to $1,800 if the equipment is moved and reinstalled. For the new system it is estimated that these would be about $2,750 per year. All other costs would be about equal for the two alternatives. The company is in the 40% income tax bracket. It wishes to obtain at least 10%, after taxes, on any invested capital. What would you recommend? (12.11)

12-23. It has been decided to replace an existing machine process with a newer, more productive process which costs $80,000 and has an estimated market value of $20,000 at the end of its service life of 10 years. Installation charges for the new process will amount to $3,000—this is not added to first cost but will be an expensed item during the first year of operation. MACRS depreciation will be utilized, and the class life is 5 years. The new process will reduce direct costs (labor, insurance, maintenance, rework, etc.) by $10,000 in the first year, and this amount is expected to grow by 5% each year thereafter during the 10-year life. It is also known that the book value of the *old* machine process is $10,000 but that its fair market value is $14,000. Capital gains are taxed as ordinary income, and the effective income tax rate is 40%. Determine the prospective *after-tax* rate of return for the incremental cash flow associated with the *new* machine process if it is believed that the existing process will perform satisfactorily for 10 more years. (12.14)

12-24. A large firm desires to automate one of its high-volume production lines. Thus, it proposes to substitute capital for labor. Formulate this general replacement situation as a multiattributed decision problem (refer to Chapter 11) and discuss how the firm might go about justifying the investment. (12.2–12.4)

13

CAPITAL FINANCING AND ALLOCATION

For ease of presentation and discussion, we have divided this chapter into two main areas: (1) the long-term *sourcing* of capital for a firm (capital financing) and (2) the *expenditure* of capital through development, selection, and implementation of specific projects (capital allocation). Our aim is to impart to the student an understanding of these basic components of the capital budgeting process so that the important role of the engineer in this complex and strategic function will be made clear. The following subjects are discussed in this chapter:

Capital financing: borrowed funds versus equity
Financing with borrowed capital
Financing with common stock
Other sources of equity capital
The after-tax weighted average cost of capital
The minimum attractive rate of return
Leasing as a source of capital
Capital allocation among independent projects
Allocating capital using risk catergories
Capital allocation: policy and procedures

13.1 INTRODUCTION

Capital financing and allocation, which were briefly discussed in Section 1.6, are challenging and difficult functions that take place at the highest levels of management, such as the office of the chief executive officer or the Board of Directors. However, information produced at lower levels in the organizational hierarchy directly affects those investment proposals that are ultimately considered in the overall capital budget. For example, many alternatives of design and specification will usually have been considered before a major engineering project is considered in top management's capital budget, and the related decisions are an inherent part of the project package (portfolio) that is recommended to upper management. Appropriate procedures for evaluation of these alternatives should be available and uniformly applied to ensure that their merits are properly considered at all levels of the capital appropriation process.

Capital financing and allocation are normally simultaneous decision processes regarding *how much* and *where* resources will be obtained and expended for future use, particularly in the production of future goods and services. The scope of these activities encompasses

1. How the money is acquired and from what sources.
2. How individual capital project alternatives (and combinations of alternatives) are identified and evaluated.
3. How minimum requirements of acceptability are set.
4. How final project selections are made.
5. How postmortem reviews are conducted.

13.2 CAPITAL FINANCING

As has been discussed in previous chapters, capital plays a very important role in engineering and business projects. Although the economy-study analyst seldom engages in obtaining the capital for projects, the method by which the capital is to be obtained, and whether it is equity or borrowed capital, may be of great importance, inasmuch as the costs of obtaining capital and the restrictions that may be imposed upon its use can be quite different. Thus, just as in the case of other ingredients to be used in an engineering project, whether capital is available, how it must be obtained, and the costs of obtaining it are of significant importance in an engineering economy study. Many well-engineered projects have failed because of improper or too costly financing.

Most engineering economy studies are concerned with the *total* capital used without regard for source; this method, in effect, evaluates the *project* rather than the interests of any particular group of capital suppliers. The illustrations and problems in this book normally evaluate the *project* because in most analyses the choice between alternatives can be made independently of sources of funds to be used. Hence, up to this point, the firm's overall "pool of capital" has been regarded as the source of investment funds, and its "cost" in Chapter 7, for example, was the after-tax MARR. However, the sources of capital may be important in investment analyses in which different sources can be used for different alternatives (e.g., leasing versus borrowing versus equity).

13.2.1 Differences Between Sources of Borrowed Capital and Equity

In previous chapters various differences in the uses of equity and borrowed capital have been mentioned. These can be summarized as follows:

1. Equity capital is supplied and used by its owners in the expectation that a *profit* will be earned. There is no assurance that a profit will, in fact, be gained or that the invested capital will be recovered. Similarly, there are no limitations placed on the use of the funds except those imposed by the owners themselves. There is no *explicit* cost for the use of such capital, in the ordinary sense of a tax-deductible cost.
2. When borrowed funds are used, interest must be paid to the suppliers of the capital, and the debt must be repaid at a specified time. The suppliers of debt capital do not share in the profits resulting from the use of the capital; the interest they receive, of course, comes out of the firm's revenues. In many instances the borrower pledges some type of security to ensure that the money will be repaid. Sometimes the terms of the loan may place limitations on the uses to which the funds may be put, and in some cases restrictions may be put on further borrowing. Interest paid for the use of borrowed funds is a tax-deductible expense for the firm.

The sources of capital to a company affect its overall financial performance. Where equity capital is available, it is possible to make engineering economic evaluations without assigning the cost of the capital directly as an expense. It is not necessary to worry about interest charges to be met whether business is good or bad. At the other extreme is the case where capital cannot be obtained at any reasonable cost. If capital is required but cannot be obtained, the project cannot be undertaken. The fact that capital cannot be obtained does not necessarily mean that the project itself is not feasible and sound. In most engineering economy studies we wish to determine the economic feasibility of the project, without regard to financing problems. In effect, we assume that adequate equity financing is available, or that it is desirable to separate the financing problems from the inherent economic merits of the project. This viewpoint has considerable merit; in most instances and in most companies if one project is not approved, some other available one will be undertaken with whatever funds are available. Further, even if part or all of the required funds are not available, a knowledge of the economic merits of a project may make it clear that further investigation should be carried out to determine whether the financing can be done to economic advantage. Thus the financing study becomes, in effect, a separate economic study.

13.2.2 Types of Business Organizations

A. Individual Ownership Several types of business organizations have been devised and are employed for obtaining and using capital in business ventures. The simplest and oldest form is that of individual ownership. An individual uses his or her own capital to establish a business and is the sole owner. Thus, the individual owner controls the enterprise and is entitled to whatever benefits and profits that accrue, and must assume any losses that occur.

This type of organization has serious limitations relating to continuity and growth. Obviously, the organization ceases upon the death of the owner. Because life is uncertain, the life of the organization is uncertain. Equity funds for expansion must, in general, come from the after-tax profits of the enterprise. They thus are seriously limited. At the same time, it is difficult for such an organization to obtain borrowed capital, particularly in large amounts and for an extended period of time, owing to the uncertainty of the life of the organization. Thus, except for very small businesses, the individual-ownership form of organization is not a satisfactory one for financing and operating most enterprises in our modern economy.

B. The Partnership One obvious solution to the limited amount of capital that ordinarily can be raised through individual ownership is for two or more persons to become partners and pool their resources so that the required capital will be obtained. This method was used to a great extent in the United States during the nineteenth century.

The partnership has a number of advantages. It is bound by few legal requirements as to its accounts, procedures, tax forms, and other items of operation. Dissolution of the partnership may take place at any time by mere agreement of the partners with practically no consideration of outside persons. It provides an easy method whereby two persons of differing talents may enter into business, each carrying those burdens that he or she can best handle; this is often the case where one partner is a technician and the other a salesperson.

The partnership, however, has four serious disadvantages. First, the amount of capital that can be accumulated is definitely limited. Second, the life of the partnership is determined by the lives of the individual partners. When *any* partner dies, the partnership automatically ends. Third, there may be serious disagreement among the individual partners. Fourth, each member of the partnership is liable for all the debts of the partnership. This particular disadvantage is one of the most serious.

C. The Corporation The corporation is a form of organization that was originated to avoid many of the disadvantages of the individual and partnership forms of ownership of business enterprises. A corporation is a fictitious being, recognized by law, that can engage in almost any type of business transaction in which a real person could occupy himself or herself. It operates under a charter that is granted by a state and is endowed by this charter with certain rights and privileges, such as perpetual life without regard to any change in the person of its owners, the stockholders. In payment for these privileges and the enjoyment of legal entity, the corporation is subject to certain restrictions. It is limited in its field of action by the provisions of its charter. In order to enter new fields of enterprise, it must apply for a revision of its charter or obtain a new one. Special taxes are also assessed against it.

The capital of a corporation is acquired through the sale of stock. The purchasers of the stock are part owners, usually called stockholders, of the corporation and its assets. In this manner the ownership may be spread throughout the entire world, and as a result enormous sums of capital can feasibly be accumulated. With few exceptions, the stockholders of a corporation, although they are the owners and are entitled to share in the profits, are not liable for the debts of the corporation. They are thus never compelled to suffer any loss beyond the value of their stock. Because the life of a corporation is continuous or indefinite, long-term investments can be made and the future faced with some degree of certainty. This makes debt capital (particularly long-term) easier to obtain, and generally at a lower interest cost, for corporations than for individual and partnership types of business organizations.

The widespread ownership that is possible in corporations usually results in the responsibility for operation being delegated by the owners to a group of hired managers. Frequently, the management may own very little or no stock in the corporation. At the same time, individual stockholders may exercise little or no significant influence in the running of the corporation and may be interested only in the annual dividends they receive. As a result, the management sometimes tends to make decisions on the basis of what is best for themselves rather than what is best for the stockholders.

One factor that does not favor the corporation form of business organization is that, except in limited circumstances, the profits of a corporation are subject to double taxation. That is, after the corporation income tax is paid, any remaining profits that are distributed as dividends to stockholders are again taxed as income to those stockholders. Thus, if the corporation's income tax rate is 40%, and all remaining profits are distributed to stockholders paying an average of 25% income taxes, \$1.00 earned by the corporation becomes $\$1.00(1 - 0.40) = \$0.60$ distributed to the stockholder, which becomes $\$0.60(1 - 0.25) = \$0.45$ net after taxes. Hence, the actual total income tax rate is $(\$1.00 - \$0.45)/\$1.00 = 0.55 = 55\%$.

Stock certificates are issued as evidence of stock ownership. The value of stock commonly is measured in three ways. Market value is the price it will bring if sold on the market. Book value is determined by dividing the net equity (net worth) of the corporation by the number of shares outstanding, assuming that all the stock is of one type. A third measure is the price–earnings ratio, which is the stock's price divided by the annual after-tax net income (earnings).

There are a number of types of stock, but two are of primary importance. These are *common stock,* which represents ordinary ownership without special guarantees of return on investment; and *preferred stock,* which has certain privileges and restrictions not available for common stock. For instance, dividends on common stock are *not* paid until the fixed percentage return on preferred stock has been paid.

13.3 FINANCING WITH DEBT CAPITAL

There are many situations where the use of borrowed capital is preferable to the use of equity capital. Expansion through the use of equity capital requires either the existing owners to supply more capital or the sale of additional stock to others, which results in decreasing the percentage ownership of the existing stockholders. If the additional capital is needed for a fairly definite period of time and there is considerable assurance that the existing or future cash flow can readily pay the costs and provide for the repayment of borrowed capital, it may be to the advantage of the existing owners to obtain the needed capital by borrowing. It is common for 5 to 30% of the total capital of private, competitive corporations to be from debt sources.

If additional debt capital is needed only for a short period of time, usually less than 5 years and more frequently less than 2 years, it may be borrowed from a bank or other lending agency by the signing of a short-term note. Such a note is merely a promise to repay the amount borrowed, with interest, at a fixed future date or dates. The lending agency may require something of tangible value as security for the loan, or at least it will make certain the financial position of the borrowing organization is such that there is minimal risk involved.

If capital is obtained through short-term borrowing and the corporation needs the capital for a long time, it must refinance the loan every 2 years or so. Obviously, this impedes long-range planning and investment in projects that have long lives and that may ultimately be quite profitable but may not produce much cash flow during the first few years. Under such conditions there is considerable uncertainty as to whether the money for repayment of a short-term loan will be available when required, or whether refinancing will be available at reasonable cost if this is needed. Because long commitments of capital are required in most projects, corporations usually resort to bond issues for obtaining long-term debt capital.

13.3.1 Long-Term Bonds

A *bond* is essentially a long-term note given to the lender by the borrower, stipulating the terms of repayment and other conditions. In return for the money loaned, the corporation promises to repay the loan and interest upon it at a specified rate. In addition, the corporation may give a deed to certain of its assets that becomes effective if it defaults in the payment of interest or principal as promised. Through these provisions the bondholder has a more stable and secure investment than does the holder of common or preferred stock. Because the bond merely represents corporate indebtedness, the bondholder has no voice in the affairs of the business, at least for as long as his interest is paid, and of course he is not entitled to any share of the profits.

Bonds usually are issued in units of from $100 to $1,000 each, which is known as the *face value,* or *par value,* of the bond. This is to be repaid the lender at the end of a specified period of time. When the face value has been repaid, the bond is said to have been *retired,* or *redeemed.* The interest rate quoted on the bond is called the *bond rate,* and the periodic interest payment due is computed as the face value times the bond interest rate per period.

A description of what happens during the normal life cycle of a bond can be illustrated by the diagram and three-step explanation of Figure 13-1.

FIGURE 13-1

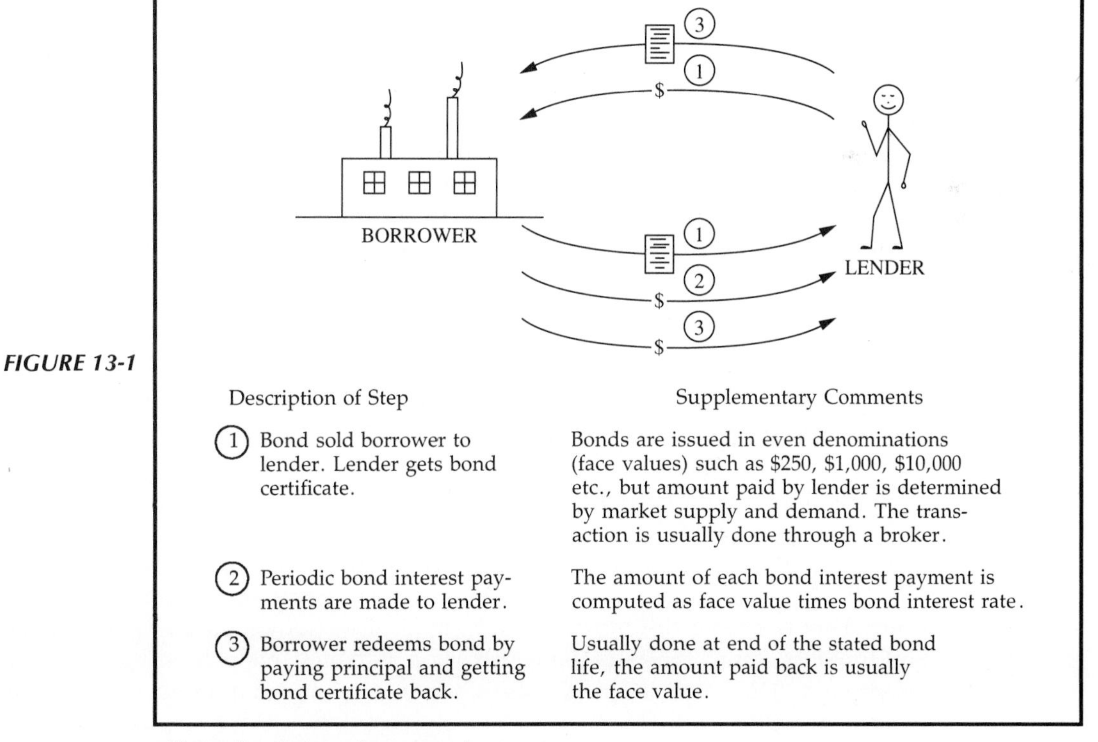

Life Cycle of Financing a Bond

Interest may be paid in either of two ways. The name of the owner of *registered bonds* is recorded on the record books of the corporation, and interest payments are sent to the owner as they become due without any action on his or her part. *Coupon bonds* have a coupon attached to the bond for each interest payment that will come due during the life of the bond. When an interest payment is due, the holder clips the corresponding coupon from the bond and can convert it into cash, usually at any bank. A registered bond thus requires no action on the part of the holder, but it is not as easily transferred to new ownership as is the coupon bond.

13.3.2 Bond Retirement

Because stock represents ownership, it is unnecessary for a corporation to make absolute provision for payments (dividends) to the stockholders. If profits remain after the operating expenses are paid, part or all of these can be divided among the stockholders. Bonds, on the other hand, represent debt, and the interest upon them is a cost of doing business. In addition to this periodic cost, the corporation must look forward to the day when the bonds become due and the principal must be repaid to the bondholders. Provision may be made for repaying the principal by two different methods.

If the business has prospered and general market conditions are good when the bonds come due, the corporation may be able to sell a new issue of bonds and use the proceeds to pay off the holders of the old issue. If conditions are right, the new issue may bear a lower rate of interest than the original bonds. If this is the case, the corporation maintains the desired capital at a decreased cost. On the other hand, if business conditions are bad when the time for refinancing arrives, the bond market may not be favorable, and it may be impossible to sell a new bond issue, or possible only at an increased interest rate. This could be a serious handicap to the corporation. In addition, the bondholders wish to have assurance that provision is being made so that there will be no doubt concerning the availability of funds with which their bonds will be retired.

When it is desired to repay long-term loans and thus reduce corporate indebtedness, a systematic program frequently is adopted for repayment of a bond issue when it becomes due. Some such provision, planned in advance, gives assurance to the bondholders and makes the bonds more attractive to the investing public; it may also allow the bonds to be issued at a lower rate of interest.

In many cases the corporation periodically sets aside definite sums which, with the interest they earn, will accumulate to the amount needed to retire the bonds at the time they are due. Because it is convenient to have these periodic deposits equal in amount, the retirement procedure becomes a sinking fund. This is one of the most common uses of a sinking fund. By its use the bondholders know that adequate provision is being made to safeguard their investment. The corporation knows in advance what the annual cost for bond retirement will be.

If a bond issue of \$100,000 in 10-year bonds, in \$1,000 units, paying 10% nominal interest in semiannual payments, must be retired by the use of a sinking fund that earns 8% compounded semiannually, the semiannual cost for retirement will be as follows:

$$\begin{aligned} A &= F(A/F, i\%, N) \\ F &= \$100{,}000 \\ i &= 8\%/2 = 4\% \text{ per period} \\ N &= 2 \times 10 = 20 \text{ periods} \end{aligned}$$

Thus,

$$A = \$100{,}000(0.0336) = \$3{,}358$$

In addition, the semiannual interest on the bonds must be paid. This would be calculated as follows:

$$\begin{aligned} \text{interest} &= \$100{,}000 \times \frac{0.10}{2} = \$5{,}000 \\ \text{total semiannual cost} &= \$3{,}358 + \$5{,}000 = \$8{,}358 \\ \text{annual cost} &= \$8{,}358 \times 2 = \$16{,}716 \end{aligned}$$

The total cost for interest and retirement of the entire bond issue over 20 periods (10 years) will be

$$\$8{,}358 \times 20 = \$167{,}160$$

13.3.3 Bond Value

A bond provides an excellent example of commercial value being the present worth of the future net cash flows that are expected to be received through ownership of a property, as discussed in Chapters 4 and 6. Thus, the value of a bond, at any time, is the present worth of future cash receipts. For a bond, let

Z = face, or par, value
C = redemption or disposal price (usually equal to Z)
r = bond rate (nominal interest) per interest period
N = number of periods before redemption
i = bond *yield* rate per period
V_N = value (price) of the bond N periods prior to redemption

The owner of a bond is paid two types of payments by the borrower. The first consists of the series of periodic interest payments he or she will receive until the bond is retired. There will be N such payments, each amounting to rZ. These constitute an annuity of N payments. In addition, when the bond is retired or sold, the bondholder will receive a single payment in amount equal to C. The present worth of the bond is the sum of present worths of these two types of payments at the bond's yield rate:

$$V_N = C(P/F, i\%, N) + rZ(P/A, i\%, N) \qquad (13\text{–}1)$$

EXAMPLE 13-1

Find the current price (present worth) of a 10-year 6% bond, interest payable semiannually, that is redeemable at par value, if bought to yield 10% per year. The face value of the bond is $1,000.

$$N = 10 \times 2 = 20 \text{ periods}$$
$$R = 6\%/2 = 3\% \text{ per period}$$
$$i = 10\%/2 = 5\% \text{ per period}$$
$$C = Z = \$1,000$$

Solution

Using Equation 13-1, we obtain

$$V_N = \$1,000\,(P/F, 5\%, 20) + \$1,000\,(0.03)\,(P/A, 5\%, 20)$$
$$= \$376.89 + \$373.87 = \$750.76$$

EXAMPLE 13-2

A bond with a face value of $5,000 pays interest of 8% annually. This bond will be redeemed at par value at the end of its 20-year life, and the first interest payment is due one year from now.

(a) How much should be paid now for this bond in order to receive a yield of 10% per year on the investment?

(b) If this bond is purchased now for $4,600, what yield would the buyer receive?

Solution

(a) By using Equation 13-1, the value of V_N can be determined:

$$V_N = \$5,000\,(P/F, 10\%, 20) + \$5,000\,(0.08)\,(P/A, 10\%, 20)$$
$$= \$743.00 + \$3,405.44 = \$4,148.44$$

(b) Here we are given $V_N = \$4,600$ and we must find the value of $i\%$ in Equation 13-1:

$$\$4,600 = \$5,000\,(P/F, i\%.\ 20) + \$5,000\,(0.08)\,(P/A, i\%, 20)$$

To solve for $i\%$, we can resort to an iterative trial and error procedure (e.g., try 8.5%, 9.0%, etc.), or we can use the computer program in Appendix E to determine that $i\% = 8.9\%$.

13.4

FINANCING WITH COMMON STOCK

The issuance of common stock is an important source of new capital utilized to finance new, modernized, and/or expanded operations of a corporation. As shown earlier in Figure 7-1, other sources of equity capital include preferred stock, retained earnings, and depreciation reserves.

Establishing value for a share of common stock is not so straightforward as placing value on a bond. Valuation of common stock is actually a controversial subject because of numerous assumptions regarding the future of dividend growth rates, future stock prices, perceived riskiness of the investment, projected after-tax earnings, and so on.* Because the value of common stock must be a measure of the earnings that will be received through ownership of the stock, it is dependent upon several factors. However, these can probably all be summed up under two headings—dividends and market price. The market price will be affected by dividends as well as by general economic conditions, future prospects of the corporation, the money market, and the preferences of the investing public. In addition, it may be drastically altered by speculation, which may bring about a market price that is not a true measure of the actual worth of the stock. Likewise, the mere payment of a large dividend is not proof that the stock has great value. If the corporation runs a consistently profitable business and is well managed, the holders of common stock will likely realize a good return on their investment. If this is not the case, they may not only fail to get any profit but may also lose part or all of their capital. Because common stocks represent ownership of business enterprises, their value is affected by changes in the general level of market prices. Thus, during times of inflation their value rises.

In this section, we present a very simple approach for the valuation of common stock and the estimation of per share rate of return expected by the investor. The approach is termed the *dividend valuation model*. Other approaches, such as the earnings model and the investment-opportunities model, are discussed in any good finance textbook.

The owner of a share of common stock in a corporation is entitled to receive cash dividends declared by the company as well as the price of the stock at the time it is sold. If we let after-tax value of cash dividends received during year k equal D'_k, the current value of a share of common stock in the dividend valuation model can be approximated by the present worth of future cash receipts during an N-year ownership period:

*See, for example, Franco Modigliani and Merton H. Miller, "The Cost of Capital, Corporation Finance, and the Theory of Investment," *American Economic Review*, June 1958, pp. 261–297; and D. Durand, "The Cost of Capital in an Imperfect Market: a Reply to Modigliani and Miller," *American Economic Review*, September 1959, pp. 639–655.

$$P_0 \simeq \frac{D'_1}{(1+e_a)} + \frac{D'_2}{(1+e_a)^2} + \cdots + \frac{D'_N}{(1+e_a)^N} + \frac{P_N}{(1+e_a)^N} \quad (13\text{–}2)$$

where e_a = rate of return per year (%) required by common stockholders (This is the after-tax cost of equity to the corporation.)
P_0 = current value of a share of common stock
P_N = selling price of a share of common stock at the end of N years

The value of e_a must be sufficient to compensate the shareholder for his or her time value of money and for the risk that is believed to be associated with the investment. How one estimates the value of P_N is a further complication in determining P_0.

The dividend valuation model incorporates conservative assumptions that dividends are *constant* over the indefinitely long life of a corporation and that $P_0 = P_N$. In this case, the current price of a share of common stock equals the present worth of an infinite series of dividend receipts that remain constant in amount:

$$P_0 = D'\,(P/A,\, e_a,\, \infty) = \frac{D'}{e_a} \quad (13\text{–}3)$$

Thus, if the current selling price of a share of common stock is known and the annual dividend for the past year is also known, the return to equity (common stock) is conservatively estimated to be

$$e_a = \frac{D'}{P_0} \quad (13\text{–}4)$$

When the future price of the security is assumed to grow at a rate of g (expressed as a decimal) each year, the cost of equity can be approximated by Equation 13-5.

$$e_a = \frac{D'}{P_0} + g \quad (13\text{–}5)$$

Suppose that a share of common stock is priced at \$100 and a dividend of \$8 is currently paid annually. The expected annual growth in price is 4% per year. If an investor is willing to purchase this security based on the assumption that dividends remain constant and the price grows at 4% annually, the expected return is about \$8/\$100 + 0.04 = 0.12, or 12% per year. A second, less risky security being considered may sell for \$100 and pay a dividend of \$10 annually, with g = 0. In this case, e_a = 10%. If the investor is indifferent between the two securities, an additional expected return of 2% is required to compensate for the extra risk associated with the first investment.

The determination of the cost of all types of equity is difficult in practice. For purposes of this book, the opportunity cost principle and Equations 13-4 and 13-

5 provide a basic, though oversimplified, point of departure for approximating this quantity.

EXAMPLE 13-3

The CMX Corporation is expected to generate perpetual after-tax net earnings of $1,600,000 per year with its existing assets. This firm produces a stable product and has been in business for 75 years. Furthermore, it is 100% equity financed, with 200,000 shares of common stock outstanding, and has a long-standing policy of declaring an annual dividend that is 50% of its after-tax earnings. The remaining 50% of earnings is retained for cash reserves, equipment replacement, and so on.

(a) If investors require a 15% return on their investment, how much would they be willing to pay for a share of CMX common stock if dividends remain constant?

(b) An optimistic investor who owns 1,000 shares of CMX stock believes that its net earnings will grow at a rate of 5% per year in the future. What is the rate of return on CMX stock expected by this investor?

Solution

(a) From Equation 13-3 the current selling price of a share of CMX common stock should be [$1,600,000 (0.5)/200,000 shares] / 0.15 = $26.67.

(b) The return to equity based on Equation 13-5 would be approximately ($4.00/$26.67) + 0.05 = 0.20, or 20% per year.

13.5 FINANCING WITH PREFERRED STOCK

Preferred stock also represents ownership, but the owner has certain additional privileges and restrictions not assigned to the holder of common stock. Preferred stockholders are guaranteed a definite dividend on their stock, usually a percentage of its par value, before the holders of the common stock may receive any return. In case of dissolution of the corporation, the assets must be used to satisfy the claims of the preferred stockholders before those of the holders of common stock. Preferred stockholders usually, but not always, have voting rights. Occasionally they are granted certain privileges, such as the election of special representatives on the board of directors, if their preferred dividends are not paid for a specified period.

Because the dividend rate is fixed, preferred stock is a more conservative investment than common stock and has many of the features of long-term bonds. For this reason, the market value of such stock is less likely to fluctuate. Moreover, the value of preferred stock is determined similarly to bond valuation discussed in Section 13.3.3 since *fixed* dividends (typically 6–8% per year) are paid on this particular type of equity.

13.6 FINANCING THROUGH RETAINED EARNINGS

Another important source of internal capital for expansion of existing enterprises is retained profits that are reinvested in the business instead of being paid to the owners. Although this method of financing is used by most companies, there are three factors that tend to limit its use.

Probably the greatest deterrent is the fact that the owners (the stockholders in the case of a corporation) usually expect and demand that they receive some profits from their investment. Therefore, it *usually* is necessary for a large portion (maybe 50% or more) of the profits to be paid to the owners in the form of dividends. This is essential to ensure the continued availability of equity (ownership) capital when it is needed. However, it usually is possible to obtain part of the needed capital for expansion by retaining a portion of the profits. Although such a retention of profits reduces the immediate amount of the dividends per share of stock, it increases the book value of the stock and should also result in greater future dividends and/or market resale value for the stock. Many investors prefer to have some of the profits retained and reinvested so as to help in increasing the value of their stock.

The second, also serious, limitation on the use of retained profits is the fact that income taxes must be paid upon them twice and deducted from them before they may be used. With federal taxes taking 28% of the individual's taxable income and 34% of the taxable income of most corporations, this severely limits the amount available after dividends and personal income taxes have been paid.

A third, and less important, deterrent is the fact that as profits are retained and used, the total annual profits and the profits per share of stock should increase. This gives the impression that a company is able to pay higher wage rates. Such an implication frequently is used by unions in wage negotiations without acknowledgment that the larger profits are due to the investment of increased amounts of capital by each shareholder through retained profits. This disadvantage may be quite easily avoided by issuing stock dividends in lieu of the retained profits, thereby maintaining a fairly constant rate of profit per share.

The "cost" of this type of capital (retained earnings) is normally assumed to be identical to the rate of return expected by common stockholders. The reason is that retained earnings are reinvested within the corporation, and their opportunity cost to the owners of the firm (common stockholders) should be at least e_a, or else these funds ought to be distributed as dividends.

13.7

DEPRECIATION FUNDS AS A SOURCE OF CAPITAL

As was explained in Chapter 6, funds that are set aside out of revenue as an allowance for depreciation are usually retained and used in a business. These funds are available for reinvestment and must be utilized to the best advantage. Thus, they are an important internal source of capital for financing new projects, as shown in Figure 7-1.

Because one of the purposes of depreciation accounting is to provide for replacement of a property when required, we might conclude that depreciation funds provide only for such replacement and never for new equipment of a different type. This, however, is only partially true. In a great many instances when a particular property or piece of equipment has lost all further economic value, the function for which it was originally purchased no longer exists. Under such conditions, we do not wish to replace it with a similar piece of equipment. Instead, other needs have developed, and different equipment or property is needed. The accumulated depreciation funds may thus be used to meet the new needs.

In other instances a piece of equipment may continue to be used after its original value has been recovered through normal depreciation procedures. Here again the accumulated funds are available for other use until the original equipment must be replaced. If depreciation procedures are used that provide for the recovery of a large portion of the initial cost during the first few years of life, there usually will be excess funds available before the equipment must be replaced.

In effect, the depreciation funds provide a revolving investment fund that may be used to the best possible advantage. The funds are thus an important source of capital for financing new ventures within an existing enterprise. Obviously, depreciation funds must be managed so that required capital is available for replacing essential equipment when the time for replacement arrives.

13.8

THE AFTER-TAX WEIGHTED AVERAGE COST OF CAPITAL

Most businesses do not operate entirely on either equity capital or borrowed capital. Instead, in most instances good practice calls for much of the capital to be obtained from equity sources, a smaller portion being borrowed. To obtain the desired amount of equity capital, a sufficient rate of dividends must be maintained to make the investment attractive to investors in view of the risks involved. Because of the absence of security, the rate that must be paid for equity capital is virtually always greater than must be paid to obtain borrowed capital.

One factor that is usually important to decisions regarding acquisition of borrowed versus equity capital is the probable effect of income taxes on the cost of that capital to the firm. Any interest paid for the use of money borrowed by a firm is deductible from income (profits) reported for federal and state income tax purposes. Hence, taxes are saved when borrowed funds are used to finance a project. The costs of various types of equity funds are customarily stated as after-tax amounts (or rates).

For example, suppose that \$10,000 is borrowed by a certain firm at an annual interest rate of 8%, and the effective income tax rate is 40%. The annual interest is

$$P \times i = \$10{,}000 \times 8\% = \$800$$

This \$800 reduces the net profits or income on which taxes must be paid and hence results in *saving* income taxes in the amount of \$800 × 40% = \$320. Therefore, the net after-tax cost of the interest is \$800 − \$320 = \$480, and the after-tax interest rate is \$480/\$10,000 = 4.8%. In other words, if there are other profits so that \$320 of the interest cost can be charged against positive taxable income, the \$10,000 of debt capital need produce only \$480 of added income rather than \$800 to satisfy the demands of the suppliers of the capital. Without other taxable profits, the maximum the \$10,000 would have to earn, in order to satisfy the interest demands, would be \$800.

On the other hand, dividends paid by a firm for the use of equity capital are not deductible from the income or profits of the firm, on which income taxes must be paid. Hence, if the firm plans to pay 8% dividends after taxes in order to satisfy the stockholders who supply \$10,000 of equity capital, the investment would have to earn \$1,333 [= \$10,000 × 8% / (1.0 − 0.4)] before taxes. At a 40% income tax rate, \$1,333 × 0.40 = \$533 would be paid in income taxes, leaving \$800 to satisfy the stockholders. Consequently, in this case the before-tax cost of equity *to the firm* is 13.3%, and the after-tax rate is 8%. Recall that the before-tax cost of borrowed capital was 8% and its after-tax cost was 4.8%. A meaningful comparison, *with income taxes being considered*, shows that the borrowed capital at 4.8% (under most conditions) is much less costly for financing a project than equity capital at 8% from the viewpoint of this firm.

When both types of capital are used for the general financing of the enterprise, the *bare* cost of the capital is not that paid for equity or for borrowed capital. Instead, it is some intermediate rate, depending upon the proportions of each type of capital used. For example, suppose (for the case above involving \$10,000) that capital might be borrowed at 8% or obtained from stockholders at

TABLE 13–1 Example of Average Cost of Capital for Various Percentages of Equity and Borrowed Capital, Both Before and After Income Taxes

	Borrowed Capital				
	0%	20%	40%	60%	100%
1. Total capital	\$10,000	\$10,000	\$10,000	\$10,000	\$10,000
2. Borrowed capital [(1) × % borrowed]	0	2,000	4,000	6,000	10,000
3. Before-tax interest cost/yr [(2) × 8%]	0	160	320	480	800
4. After-tax interest cost/yr [(3) × (1−0.40)]	0	96	192	288	480
5. Equity capital [(1) −(2)]	10,000	8,000	6,000	4,000	0
6. Dividends/yr [(5) × 8%]	800	640	480	320	0
7. Before-tax earnings required to pay dividends [(6)/(1 − 0.40)]	1,333	1,067	800	533	0
8. Total cost before taxes [(3) + (7)]	1,333	1,227	1,120	1,013	800
9. Total cost after taxes [(4) + (6)]	800	736	672	608	480
10. Average rate before taxes [(8)/(1)]	13.3%	12.3%	11.2%	10.1%	8.0%
11. Average rate after taxes [(9)/(1)]	8.00%	7.36%	6.72%	6.08%	4.80%

8%. The effective income tax rate is 40% and various mixes of debt and equity capital are being considered. Table 13-1 shows the average cost of several combinations of debt and equity capital both before and after considering the effect of income taxes.

In general, the after-tax weighted average cost of capital, i_a, can be determined by

$$i_a = (1 - t)(i_b)(B/V) + e_a(C/V) + e_a^* (P/V) \tag{13–6}$$

where t = effective income tax rate for the firm
i_b = cost of borrowed capital, before taxes
B/V = total capitalization (V) divided by total long-term borrowed capital (B)
e_a = after-tax cost of common stock to the corporation
C/V = fraction of total capitalization (V) made up of total investments by common stockholders (C)
e_a^* = after-tax cost of preferred stock
P/V = fraction of total capitalization (V) composed of total investments by preferred stockholders (P)

EXAMPLE 13-4

A corporation that pays an effective income tax rate of 40% has put together the following table of financial data.

	Capitalization	
Source	Fraction of Total Capitalization	Cost of Source (Annual Percentage)
Long-term bonds	0.38	8.4%
Preferred stock	0.13	8.8%
Common stock	0.49	16.5%

What is this firm's after-tax weighted average cost of capital?

Solution

With Equation 13-6, we can quickly calculate the value of i_a:

$$i_a = (1 - 0.40)(8.40\%)(0.38) + (8.8\%)(0.13) + 16.5\%(0.49)$$
$$= 11.14\%$$

In some engineering economy studies it is clear that only borrowed capital or only equity capital will be employed. In such cases the corresponding cost of capital is sometimes used in interest and equivalence calculations. In most studies, however, all money is assumed to come from a common capital investment pool, and considerations of financing are taken up separately from the study of project profitability. The following section discusses several approaches that firms utilize to determine a minimum attractive rate of return, which is then used as the interest rate in equivalence calculations and the economic criterion for project feasibility.

13.9 DETERMINING THE MINIMUM ATTRACTIVE RATE OF RETURN

The minimum attractive rate of return (MARR) is usually a policy issue resolved by the top management of an organization in view of numerous considerations. Among these considerations are

1. The amount of money available for investment, and the source and cost of these funds (i.e., equity funds or borrowed funds).

2. The number of good projects available for investment and their purpose (i.e., whether they sustain present operations and are *essential*, or expand on present operations and are *elective*).
3. The amount of perceived risk that is associated with investment opportunities available to the firm, and the projected cost of administering projects over short planning horizons versus long planning horizons.
4. The type of organization involved (i.e., government, public utility, or competitive industry).

In theory the MARR should be chosen to maximize the economic well-being of an organization, subject to the types of considerations listed above. How an individual firm accomplishes this in practice is far from clear-cut and is frequently open to criticism. One popular approach to establishing a MARR involves the *opportunity cost* viewpoint described in Chapter 2, and it results from the phenomenon of *capital rationing*.*

Rationing of capital is necessary when the amount of available capital is insufficient to sponsor all worthy investment opportunities. A simple example of capital rationing is given in Figure 13-2, where the cumulative investment

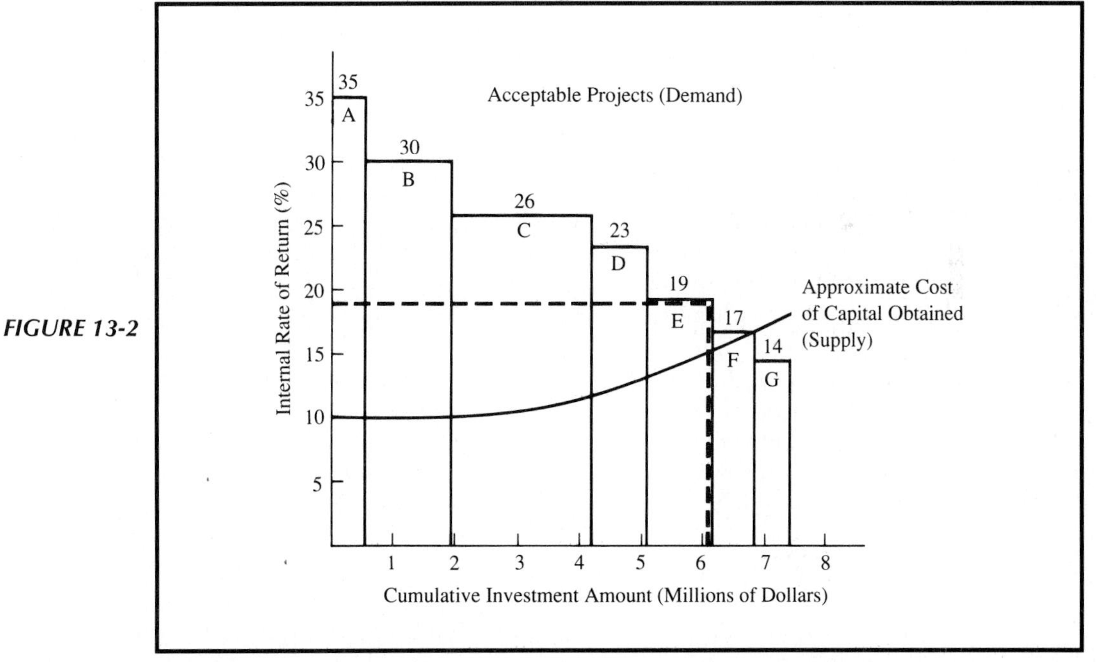

FIGURE 13-2

Determination of the MARR Based on the Opportunity Cost Viewpoint

*G. Donaldson. "Strategic Hurdle Rates for Capital Investment," *Harvard Business Review*, March–April 1972.

requirements of seven independent projects are plotted against the prospective internal rate of return of each. Figure 13-2 shows a limit of $6 million on available capital. In view of this limitation the last funded project would be E, with a prospective IRR, or *hurdle rate*, of 19%. In this case the minimum attractive rate of return by the opportunity cost principle would be 19%. By *not* investing $1 million in project E, the firm would presumably be forgoing the chance to realize a 19% return. As the amount of investment capital and opportunities available change over time, the firm's MARR will also change.

Superimposed on Figure 13-2 is the approximate cost of obtaining the $6 million to illustrate that project E is acceptable only as long as its IRR exceeds the cost of raising the last $1 million. As shown in Figure 13-2, the cost of capital will tend to increase gradually as larger sums of money are acquired through increased borrowing and/or new issuances of common stock (equity).† One last observation in connection with Figure 13-2 is that the perceived risk associated with financing and undertaking the seven projects has been determined by top management to be acceptable.

Representative MARRs for government agencies, public utilities, and private industries in 1989 are indicated in Figure 13-3. Even though it is not possible to provide consistent trends in MARRs throughout selected industries with a high degree of confidence, typical MARRs utilized among organizations and several underlying causes for their differences are depicted on the four scales shown in Figure 13-3.

For municipalities and government agencies, the MARRs are usually based on the before-tax cost of borrowed money, which ranges from about 8 to 10%. There is little risk to the investor and almost all capital is obtained through borrowing.

In the case of regulated public utilities (electricity, telephone, gas, water, etc.), the after-tax weighted average cost of capital is frequently used to establish a MARR. From Figure 13-3 it can be seen that these MARRs lie in the neighborhood of 11 to 14% and that borrowed funds constitute 40 to 60% of a utility's capitalization. Because of the stable nature of these companies, a large fraction of funds acquired is customarily borrowed and the associated risk borne by the equity investor in the utility industry is relatively low.

Private, competitive industries frequently employ the opportunity cost viewpoint toward choosing an after-tax MARR. (Engineering economy studies in competitive industry as well as in public utilities are nearly always made on an after-tax basis.) These MARRs typically cover a rather broad range of 15 to 20% or higher, principally due to the small amount of borrowed capital in the firm's capitalization which typically ranges from around 35% to none at all! As more equity capital is utilized by a firm, it is customary to experience higher after-tax return and/or capital growth expectations by owners of the equity capital.

†This happens because the investor in bonds might require a higher return as compensation for greater risk caused by a smaller proportion of equity capital available to absorb capital and operating losses. Similarly, the equity investor might expect a higher return because this capital is more speculative in nature (there is a greater likelihood of reduced earnings in economic downturns).

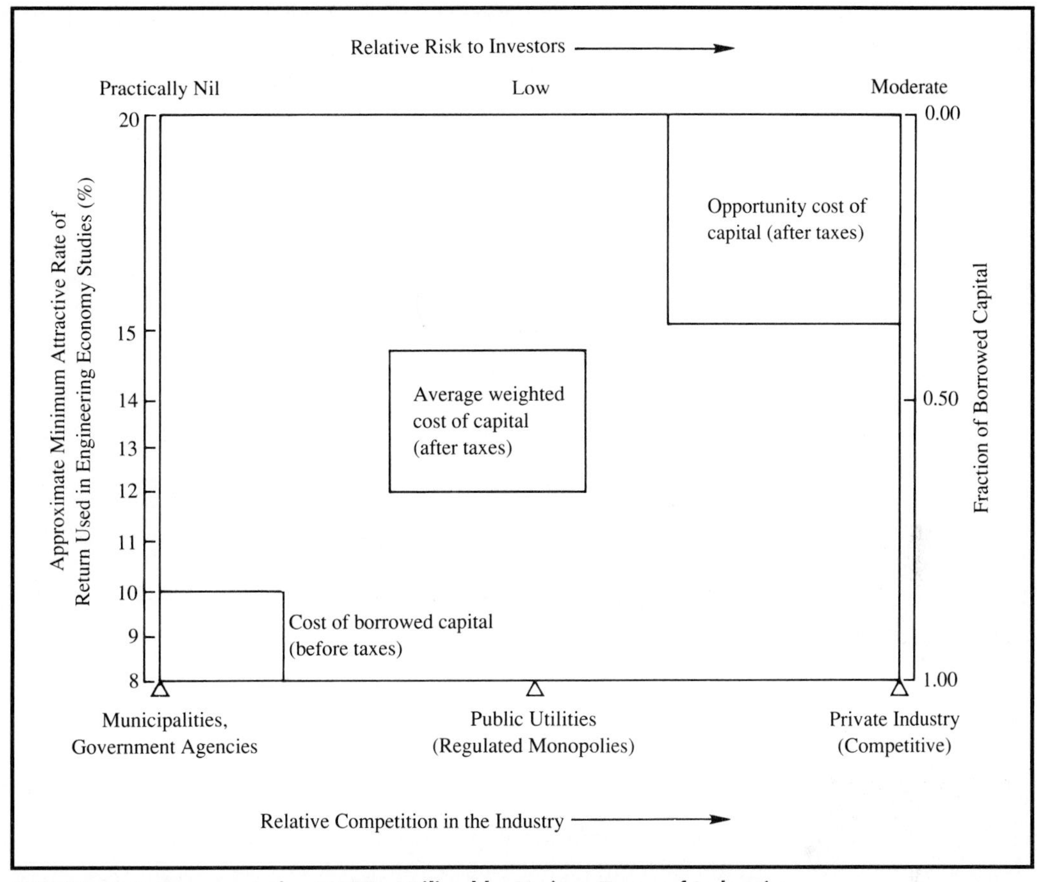

FIGURE 13-3 Representative MARRs Utilized by Various Types of Industries

EXAMPLE 13-5

Consider the following schedule which shows prospective internal rates of return for a company's portfolio of capital investment projects (this is the *demand* for capital):

Expected IRR	*Investment Requirements (Thousands of Dollars)*	*Cumulative Investment*
40% and over	$ 2,200	$ 2,200
30–39.9%	3,400	5,600
20–29.9%	6,800	12,400
10–19.9%	14,200	26,600
Below 10%	22,800	49,400

If the supply of capital obtained from internal and external sources has a cost of 15% for the first $5,000,000 invested and then increases 1% for every

FIGURE 13-4

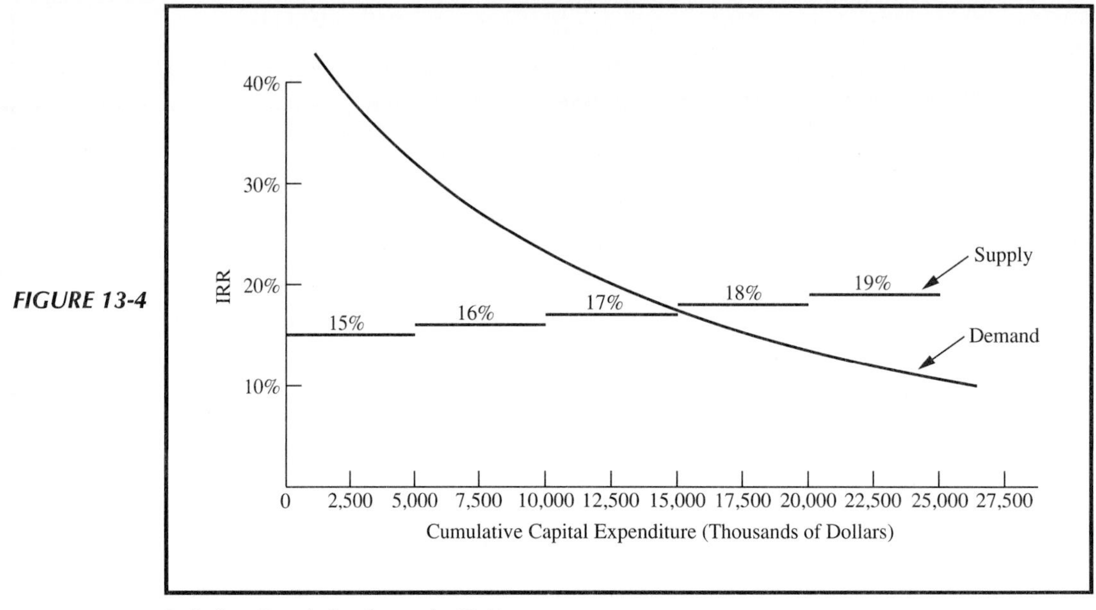

Solution Graph for Example 13-5

$5,000,000 thereafter, what is this company's MARR when using an opportunity cost viewpoint?

Solution

Cumulative capital demand versus supply can be plotted against prospective IRR as shown in Figure 13-4. The point of intersection is approximately 17%, which represents a realistic estimate of this company's MARR using the opportunity cost viewpoint.

13.10 LEASING AS A SOURCE OF CAPITAL

As mentioned at the beginning of this chapter, management must address several fundamental issues when obtaining and spending money for long-term investments: (1) How much money is needed for capital expenditures?, (2) How much money can be made available for this purpose? and (3) What priorities should be used for allocating funds to projects? Normally, the first two of these issues are answered separately, but in the case of leasing they are resolved simultaneously. As a result, leases are unique in the way their economic advantages and disadvantages are analyzed. The purpose of this section is to demonstrate correct evaluation procedures for simple "lease versus purchase" situations.

The decision to lease or purchase an asset represents a situation where the source of capital may affect which alternative is eventually chosen. Leasing is a source of capital generally regarded as a long-term liability similar to a mortgage, whereas the purchase of an asset typically uses funds from the firm's overall pool of capital (much of which is equity). Before we consider examples of lease–buy problems, some background information about leases is provided below.

Rent is the amount that one has to pay for the use of property not owned. A lease is a type of contract by which an individual or a corporation acquires the use of an asset for a specified period of time at a specified rent. A large number of leasing methods have been utilized in real estate financing for hundreds of years, but only within the last 35 years has leasing of industrial equipment become popular. Most frequently, automobiles, trucks, buildings, or various types of equipment are involved. Such alternatives may arise either in connection with a consideration of the purchase of completely new assets or in connection with the possible replacement of existing assets. In most cases it is essential to consider the income tax implications so that we can accurately compare the alternatives of purchasing versus leasing or retention versus leasing.

For corporations, the rent paid on leased assets used in their trade or business is generally deductible as a business expense. To make lease payments deductible as rent, the contract must be for a true leasing arrangement instead of a conditional sales agreement. In a true lease the corporation using the property (lessee) does not acquire ownership in or title to the asset, whereas a conditional sales contract will transfer to the lessee an equity interest or title to the asset being leased. Hence the test of whether or not lease payments qualify as business expenses lies in distinguishing between a "true" lease and a conditional sale.*

For our purposes, it is assumed that a true lease exists and that an asset may be acquired through *leasing* or *purchasing*. Before the decision is made regarding leasing or purchasing, however, the services of the asset in question must have been shown to be economically justifiable. That is, from an investment standpoint the equipment will "pay its own way," and the decision regarding how best to finance it constitutes the "lease versus purchase decision."

The reasons companies decide to lease rather than purchase are not as simple and clear as might be expected. We quite naturally would think that a decision in favor of leasing would be based on the fact that it was more economical, yet this very often is not the basis for the decision. One study involving companies that had leased fleets of automobiles and trucks revealed that in the majority of cases, no detailed comparison of the costs of leasing versus purchasing had been made. The major reasons given by these companies were, in order of frequency, as follows:

*For further information, see *Tax Guide for Small Business*, U.S. Internal Revenue Service Publication 334, published annually.

1. Leasing freed needed working capital, or enabled needed equipment to be acquired without the company going into debt.
2. Leasing was thought to have income tax advantages which made it less expensive than purchasing.
3. Leasing reduced maintenance and administrative problems; for example, it eliminated the necessity for a company to be a fleet operator, leaving this to the specialist.
4. Leasing provided flexibility when technological change is inherent to the asset being acquired.

Inasmuch as leasing is only one of several ways of obtaining working capital, a decision to lease should consider the cost of obtaining capital by all possible methods. In the study cited, it was found that very few of the companies had made such a comparison.

A number of studies have shown that there is no real income tax advantage in leasing. This is particularly true since accelerated methods (e.g., ACRS) have been permitted for depreciation. Assuming a given purchase price, the firm offering a lease contract (lessor) can charge no more for depreciation than can the owner of assets. If assets are leased, the annual lease payments are deducted in computing income taxes; if the assets are purchased, annual depreciation is deducted. Most companies have now come to realize that leasing does not offer major tax advantages, but the myth still persists in some quarters.

There may or may not be savings in maintenance costs through leasing. This will depend on the actual circumstances, which should be carefully evaluated in each case. There is no doubt that leasing usually does simplify maintenance problems, and this may be an important factor. Also, many indirect costs, which frequently are difficult to determine, will usually be associated with ownership.

These same studies concluded that a true advantage of leasing lies in allowing a firm to obtain modern equipment subject to rapid technological change. Further, leasing for this purpose typically provides an effective hedge against obsolescence and inflation.

The following two examples illustrate correct methods of handling a lease versus purchase study on an after-tax basis; they utilize the tabular format presented in Chapter 7 (Figure 7-3).

EXAMPLE 13-6

A company is considering building a small office complex at a cost of $100,000 on land that would cost $20,000. It is estimated that the land could be sold at its cost at the end of 50 years, and that the building, although fully depreciated on a straight-line basis over the 50 years, probably would have a salvage value of $20,000. Annual expenses for maintenance, property taxes, insurance, and so on, are estimated at $5,250. An alternative is to lease a building for $16,500 per year for a 50-year period. The company ordinarily obtains about 20% on its capital before taxes and, being in a 40% income tax bracket, about 12% after taxes. What should it do?

TABLE 13–2 After-Tax Cash Flows for Example 13-6

Year	(A) Before-Tax Cash Flow	(B) Depreciation	(C) = (A) − (B) Taxable Income	D = −(C) × 40% Cash Flow for Income Taxes	(E) = (A) + (D) After-Tax Cash Flow
Purchase Building					
0	−$120,000				−$120,000
1–50	− 5,250	$2,000	−$ 7,250	+$2,900	− 2,350
50	+ 40,000		+ 20,000[a]	− 8,000	+ 32,000
Lease Building					
1–50	−$ 16,500		−$16,500	+$6,600	−$ 9,900

[a]Capital gain on building.

Solution (Using IRR Method)
Table 13-2 will facilitate analysis of the two alternatives. The IRR on the extra capital required to purchase rather than to lease the building is the $i'\%$ at which

$$-\$120{,}000 + (-\$2{,}350 + \$9{,}990)(P/A,\ i'\%,\ 50) + \$32{,}000(P/F,\ i'\%,\ 50) = 0$$

The $i'\%$ can be found to be approximately 6%. Since $6\% < 12\%$, it would be better to lease the building. Of course, any of the other theoretically correct methods would result in the same recommendation.

EXAMPLE 13-7

An industrial forklift truck can be purchased for $30,000 or leased for a fixed amount of $9,200 per year payable at the *beginning* of each year. The lease contract provides that maintenance expenses are borne by the lessor. Regardless of whether the truck is purchased or leased, the study period is 6 years. If purchased, annual maintenance expenses are expected to be $1,000 in year 0 purchasing power and they will inflate at 5% per year over the study period. The market value of the truck is expected to be negligible after 6 years of normal use. Depreciation is determined with the MACRS method using a 5-year class life (deductions occur over 6 years). The effective income tax rate is 40% and the after-tax MARR, which includes an allowance for general price inflation, is 15%.

Use the annual worth method and determine whether the forklift truck should be purchased or leased. This firm is profitable in its overall business activity.

Solution
The effects of general price inflation and income taxes on the after-tax cash flows of both alternatives are shown in Table 13-3. The purchase alternative

TABLE 13–3 After-Tax Cash Flows for Example 13-7

Year	(A) Before-Tax Cash Flow	(B) Depreciation[a]	(C) = (A) − (B) Taxable Income	D = −(C) × 40% Cash Flow for Income Taxes	(E) = (A) + (D) After-Tax Cash Flow
Purchase Truck (A $ Study)[b]					
0	−$30,000				−$30,000
1	− 1,050	$6,000	−$ 7,050	$2,820	1,770
2	− 1,102	9,600	− 10,702	4,281	3,179
3	− 1,158	5,760	− 6,918	2,767	1,609
4	− 1,216	3,456	− 4,672	1,869	653
5	− 1,276	3,456	− 4,732	1,893	617
6	− 1,340	1,728	− 3,068	1,227	− 113
Lease Truck (A $ Study)[c]					
0	−$ 9,200		−$ 9,200	$3,680	−$ 5,520
1–5	− 9,200		− 9,200	3,680	− 5,520
6	0	0	0	0	0

[a] MACRS rates are given in Table 6-5.
[b] The AW at MARR = 15% is −$6,439.
[c] The AW at MARR = 15% is −$7,167.

is less costly than the lease alternative (−$6,439 > −$7,167) and would probably be selected. If capital is not readily available, the firm may elect to lease the forklift truck because the difference in annual equivalent worths is small. Furthermore, if estimates of maintenance expenses and general price inflation are believed to be inaccurate, the firm would tend to favor leasing as a hedge against an uncertain technological future.

Rather than make use of tabular procedures illustrated in Examples 13-6 and 13-7, models may be developed that yield the same equivalent worths (e.g., present worths) for the lease and purchase alternatives. They are summarized below.

13.10.1 Cost of the Lease Alternative

The after-tax cost of a lease during year k is given by

$$c_k = L_k(1 - t)$$

where c_k = after-tax lease cost during year k
L_k = before-tax lease cost during year k
t = effective income tax rate

If i, the interest rate the firm expects from the use of its money, is known and fixed, the present worth of the after-tax cost of the lease during its life of N years is given by

$$PW_{\text{Lease}}(i\%) = \sum_{k=1}^{N} \frac{L_k(1-t)}{(1+i)^k} \tag{13–7}$$

It should be noted that annual maintenance costs are not included in Equation 13-7 since they are assumed to be borne by the equipment supplier and are included in the annual lease cost L_k. Furthermore, end-of-year cash flows are assumed.

13.10.2 Cost of the Purchase Alternative

The after-tax cost of equipment when it is purchased is a function of the expected annual costs during the life of the equipment as well as the purchase price and expected salvage value. The present worth of the after-tax cost of purchased equipment is given by

$$\text{PW}_{\text{Buy}}(i\%) = I - \frac{S}{(1+i)^N} + \sum_{k=1}^{N} \frac{M_k(1-t) - d_k(t)}{(1+i)^k} \tag{13–8}$$

where I = initial purchase price of equipment
S = expected salvage value at end of year N
i = interest rate per year
N = life of equipment in years
M_k = maintenance cost during year k
t = effective income tax rate
d_k = depreciation during year k

It should be noted that the salvage value and depreciation credit terms of Equation 13-8 are negative since they reduce costs. Again, end-of-year cash flow convention is assumed.

13.11 CAPITAL ALLOCATION

To this point we have been discussing capital financing topics that deal with (1) how a company obtains capital (from what sources) and (2) how much capital, and at what cost, the company must invest to maintain a successful business enterprise in the years ahead.

The remainder of this chapter examines the capital expenditure decision-making process, also referred to as capital allocation. This process involves the planning, evaluation, and management of capital projects. In fact, much of this book has dealt with concepts and techniques required to make correct capital expenditure decisions involving engineering projects. Now our task is to place them in the broader context of upper management's responsibility for proper planning, measurement, and control of the firm's overall portfolio of capital investments.

Consequently, the rest of Chapter 13 serves as a bridge from the practice of engineering economy to upper management's use of information produced by economy studies as funds are allocated within a capital expenditure budget. Such a budget applies to a portfolio of recommended projects drawn from the *entire* organization. Because of limited capital in practically any budget period, some worthy projects may not be funded immediately and will have to wait for consideration until a future time.

We now provide examples of typical capital allocation problems. Next, general corporate policy and procedure for making capital expenditure decisions are discussed. Finally, a typical set of corporate capital allocation guidelines is presented to illustrate how the capital budgeting process actually works in the industrial world.

13.12 ALLOCATING CAPITAL AMONG INDEPENDENT PROJECTS

Companies are constantly presented with independent opportunities in which they can invest capital across the organization. These opportunities usually represent a collection of the best mutually exclusive alternatives for improving operations in all areas of the company (e.g., manufacturing, research and development, etc.). In most cases the amount of available capital is limited, and additional capital can be obtained only at increasing incremental cost, as shown in Figure 13-2. Thus, companies have a problem of budgeting, or allocating, available capital to numerous possible uses.

One popular approach to capital budgeting utilizes the present worth (PW) criterion and was previously discussed in Chapter 5. If project risks are about equal, the procedure is to compute the PW for each investment opportunity and then determine the combination of projects that maximizes PW subject to various constraints on the availability of capital. The following example provides a brief review of this procedure.

EXAMPLE 13-8

Consider these five independent projects and determine the best allocation of capital among them if no more than $300,000 is available to invest.

Independent Project	Initial Capital Outlay	PW
A	−$100,000	$25,000
B	− 125,000	30,000
C	− 150,000	35,000
D	− 75,000	40,000

Solution

All possible combinations of these projects taken two, three, and four at a time are shown in Table 13-4, together with the total PW and initial capital outlay of each. After eliminating those combinations that violate the $300,000 funds constraint, the proper selection of projects would be ABD and the maximum present worth is $95,000. The process of enumerating combinations of projects having nearly identical risks is best accomplished with a computer when large numbers of projects are being evaluated.

Methods for determining to which possible projects available funds should be allocated seem to require the exercise of judgment in most realistic capital budgeting problems. Example 13-9 illustrates such a problem and possible methods of solution.

EXAMPLE 13-9

Assume that a firm has five investment opportunities (projects) available, which require the indicated amounts of capital and which have economic

TABLE 13–4 Project Combinations for Example 13-8

Combinations	Total PW ($\times 10^3$)	Total Capital Outlay ($\times 10^3$)
AB	$ 55	$225
AC	60	250
AD	65	175
BC	65	275
BD	70	200
CD	75	225
ABC	90	375
ACD	100	325
BCD	105	350
ABD	95	300* **Best**
ABCD	130	450

TABLE 13–5 Prospective Projects for a Firm[a]

Project	Investment	Life (years)	Rate of Return (%)
A	$40,000	5	7
B	15,000	5	10
C	20,000	10	8
D	25,000	15	6
E	10,000	4	5

[a] Here we assume that the indicated rates of return for these projects can be repeated indefinitely by subsequent "replacements."

lives and prospective after-tax internal rates of return as shown in Table 13-5. Further, assume that the five ventures are independent of each other—investment in one does not prevent investment in any other, and none is dependent upon the undertaking of another.

Now let it be assumed that the company has unlimited funds available, or at least sufficient funds to finance all these projects, and that capital funds cost the company 6% after taxes. For these conditions the company probably would decide to undertake all projects that offered a return of at least 6%, and thus projects A, B, C, and D would be financed. However, such a conclusion would assume that the risks associated with each project are reasonable in the light of the prospective internal rate of return or are no greater than those encountered in the normal projects of the company.

Unfortunately, in most cases the amount of capital is limited, either by absolute amount or increasing cost. If the total of capital funds available is $60,000, the decision becomes more difficult. Here it would be helpful to list the projects in order of decreasing profitability in Table 13-6 (omitting the undesirable project E). Here it is clear that a complication exists. We quite naturally would wish to undertake those ventures that have the greatest profit potential. However, if projects B and C are undertaken, there will not be sufficient capital for financing project A, which offers the next greatest rate of return. Projects B, C, and D could be undertaken and would provide an annual return of $4,600 (= $15,000 × 10% + $20,000 × 8% + $25,000 × 6%). If project A were undertaken, together with either B or C, the total annual

TABLE 13–6 Prospective Projects of Table 13-5 Ordered by Internal Rate of Return

Project	Investment	Life (years)	Rate of Return (%)
B	$15,000	5	10
C	20,000	10	8
A	40,000	5	7
D	25,000	15	6

return would not exceed $4,600.* A further complicating factor is the fact that project D involves a longer life than the others. It is thus apparent that we might not always decide to adopt the alternative that offers the greatest profit potential.

The problem of allocating limited capital becomes even more complex when the risks associated with the various available projects are not the same. As mentioned in Chapter 4, many companies calculate the *payback period* as a preliminary index of a given project's risk. More reliable measures of risk are obtained, when warranted, with techniques described in Chapter 10. Assume that the risks associated with project B are determined to be higher than the average risk associated with projects undertaken by the firm, and that those associated with project C are lower than average. The company thus might rank the projects according to their overall desirability as in Table 13-7. Under these conditions the company might decide to finance projects C and A, thus avoiding one project with a higher-than-average risk and another having the lowest prospective return and longest life of the group.

13.13 ALLOCATING CAPITAL USING RISK CATEGORIES

Another means of allocating capital among independent projects while explicitly considering risk is to place projects into two or more risk categories and to determine in advance what approximate proportion of the available capital will be invested in each category. Then one can provisionally pick which projects earn the highest returns within the capital available for each risk category. After this is done, if one judges that a certain provisionally rejected project with a certain risk and rate of return is more desirable than some other project with a relatively higher risk and higher rate of return (or relatively lower risk and lower rate of return), then a trade-off can be made. This is possible as long as the total capital required and the resulting amount of capital invested in each

TABLE 13–7 Prospective Projects of Table 13-6 Ordered According to Overall Desirability

Project	Investment	Life (years)	Rate of Return (%)	Risk Rating
C	$20,000	10	8	Lower
A	40,000	5	7	Average
B	15,000	5	10	Higher
D	25,000	15	6	Average

*This assumes that the leftover capital could earn no more than 6% per year.

risk category is reasonably in balance relative to the objectives of the firm and the risks and returns involved.

EXAMPLE 13-10

Table 13-8 illustrates twelve projects (F through R) categorized according to high, medium, or low risk. (The consideration of these projects is completely separate from the previous consideration of projects A through E in Example 13-9.) Their prospective after-tax internal rates of return are also indicated in Table 13-8.

Suppose that the firm has $15,000,000 of capital and management desires to invest approximately 1/3, or $5,000,000, in each risk category as long as the return seems commensurate with the risk. Thus, the firm would conditionally accept projects F, G, and H in the high-risk category and projects K and L in the medium-risk category. In the low-risk category the firm could conditionally accept either projects N and P or project O. Suppose it is judged that low-risk project N having an internal rate of return of 21% on an investment of $1,500,000 is preferred to the provisionally accepted medium-risk project L earning 24% on the same investment amount. Hence, project O is the apparent remaining provisional choice in the low-risk category. Other trade-offs between risk categories can now be considered for any combination of projects, keeping in mind the $15,000,000 total capital constraint. Suppose that after consideration of many combinations, the only remaining trade-off that is judged wise is to accept medium-risk project R, earning 22%, rather than high-risk project H, earning 23%.

Summarizing the results of these judgmental decisions, we find that the final acceptances are projects F and G, requiring $4,000,000 in the high-risk category; projects K and R, requiring $4,500,000 in the medium-risk category; and projects N and O, requiring $6,500,000 in the low-risk category. Thus, the final allocation of capital among the risk categories is a bit more conservative than the initially planned one-third split.

TABLE 13–8 Prospective Projects within Three Risk Categories

High Risk			Medium Risk			Low Risk		
Project	Invest-ment	Rate of Return (%)	Project	Invest-ment	Rate of Return (%)	Project	Invest-ment	Rate of Return (%)
F	$2,000,000	30	K	$3,500,000	25	N	$1,500,000	21
G	2,000,000	28	L	1,500,000	24	O	5,000,000	19
H	1,000,000	23	R	1,000,000	22	P	3,500,000	17
J	1,500,000	16				Q	6,500,000	14
M	2,500,000	15						

13.14

TYPICAL CORPORATE CAPITAL ALLOCATION POLICY AND PROCEDURES

There is always the very real possibility that a student of engineering economy has been immersed in such a maze of details by this point that he or she has lost sight of the "enterprise context" in which various types of calculations are performed to evaluate proposed capital expenditures. Therefore, our objective in the remainder of this chapter is to focus on how the results of an engineering economic analysis are utilized in the corporate capital allocation process. We accomplish this first by briefly describing the successive steps that a typical medium- or large-sized company follows in implementing its capital allocation policy and evaluation procedure. Second, we present illustrative forms that are used to assist with capital investment decisions in a large corporation.

The student should pay close attention to how measures of financial merit, such as present worth and internal rate of return, are utilized in these capital appropriation requests. Furthermore, the guidelines that are presented later deal explicitly with other topics covered in this book such as sensitivity analysis, intangible (nonmonetary) considerations, cash flow estimation (including inflation), and use of computers to perform incremental cash flow analyses. Also observe that in the capital allocation process it is extremely important to *document key assumptions* regarding sales forecasts, actions of competitors, patterns in future production costs, and so on for each proposed venture.

Typical corporate capital allocation policy and procedure consist of several sequential steps:

1. Preliminary planning and screening
2. Annual capital expenditure budget
3. Cost of enterprise capital
4. Capital expenditure policies and evaluation procedures
5. Project implementation and post-audit review
6. Communication

13.14.1 Preliminary Planning and Screening

A considerable amount of planning must be accomplished before capital expenditure decisions can be made. The main purpose of capital expenditure planning is to make sure that long-term goals of the organization can be attained. These long-term goals and strategic plans directly tie profit plans to capital budgets. While budget periods normally range from 3 to 10 years, most large- and medium-sized firms use a 5-year period and small firms use a 3- to 5-year period.

A list of proposed capital project expenditures is prepared by plant and functional managers and forwarded to top management. As these project proposals go up the ranks of the organizational hierarchy, some are eliminated while others are added. To assist management in completing the capital budgeting process, the proposed projects must be classified in some manner. No matter what size the firm is, the two most common methods of classifying proposed projects are by operating division (project type and purpose) and by project size in dollars.

Once the proposed projects are classified, it is necessary to rank them within the portfolio according to various selection criteria. The profitability of invested capital and adherence to the long-term strategy and goals of the business are often ranked as the top two criteria. Two methods frequently used by companies to measure economic merit in the planning stages of a project are the payback period and IRR methods. Projects with long payback periods and/or low IRRs are dropped from further consideration unless there are extenuating circumstances to retain them in the investment portfolio (e.g., projects that must be funded to ensure compliance with legal requirements).

13.14.2 Annual Capital Expenditure Budget

A normal procedure for developing an annual capital expenditure budget within a firm is for division and functional level managers to develop the list of proposed projects, which is then submitted to higher level management for review and approval. Large projects are normally listed individually while smaller projects may be combined into summary groups or categories. As might be expected, especially in large companies, top-level management and the board of directors will usually approve the overall capital budget while division and functional-level managers decide on the allocation of capital to most of the individual projects.

Certainly, every year a company will have some projects that can be called *noneconomic*. A noneconomic project is defined as a project that requires capital investment but provides little or no monetary return. Most companies separate economic and noneconomic projects when requesting funding, and some firms further divide noneconomic projects into various categories such as sustaining, regulatory and environmental, safety and health, and administrative.

For various reasons, not all profitable projects are accepted. A project may be turned down at two stages of the capital budgeting process, the first being at the planning and selection stage and the second at the implementation stage. While productivity of capital is important, the two major reasons for rejecting a proposed project at either stage are incompatibility with company goals and objectives, and unavailability of capital.

13.14.3 Cost of Enterprise Capital

In the long-range planning of a business enterprise, a company decides how big it wants to be, how fast it wants to grow, how much capital it needs, and how it will acquire the capital funds needed. As discussed in Section 13.8, the acquisition of these funds from either internal or external sources determines the cost of capital. The most common approach to determining the cost of capital is the after-tax weighted average. Companies using an after-tax cost of capital normally use the current market rate as a basis of cost for each source, and the weighting is based on either the planned debt/equity ratio or the book value of securities.

Some firms use the acquisition cost of capital as the minimum acceptable return for capital expenditure planning while others use this as a starting point in developing the MARR for each division. The latter is more prominent in medium-size firms, although most firms tend to use one companywide rate. A revision of a firm's cost of capital is usually made annually.

13.14.4 Capital Expenditure Policies and Procedures

This activity can be subdivided into two broad parts: (1) management approval levels for projects of different sizes, and (2) management control over specific capital expenditures.

Three typical plans for delegating management responsibility for project approvals are summarized as follows.

1. Whenever proposed projects are clearly "good" in terms of economic desirability according to operating division analysis, the division is given approval power as long as control can be maintained over the total amount invested by each division and as long as the division analyses are considered reliable.
2. Whenever projects represent the execution of policies already established by headquarters, such as routine replacements, the division is given the power to commit funds within the limits of appropriate controls.
3. Whenever a project requires a total commitment of more than a certain amount, this request is sent to higher levels within the organization. This is often coupled with a budget limitation regarding the maximum total investment that a division may undertake in a budget period.

To illustrate the idea of a larger investment requiring higher administrative approval, the limitations for a particular firm might be as follows:

If the total investment is . . .		Then Approval Is Required Through
More Than	**But Less Than or Equal**	
$ 5,000	$ 100,000	Plant manager
100,000	1,000,000	Division vice-president
1,000,000	2,500,000	President
2,500,000	—	Board of directors

The purpose of these policies is to streamline the capital expenditure planning and control process by delegating authority to various management levels to approve projects that can be handled effectively at these levels. This permits top management to concentrate on the most significant capital demands.

Establishing capital expenditure policies is a major responsibility of top management, but responsibility for developing sound economic selection criteria tends to vary by organization. However, regardless of which group develops these criteria, they are applied when a project is proposed and again when it is ready for implementation. Upon initial proposal or later during serious funding consideration, it is customary practice to develop cash flow estimates for a particular project. This information gives some indication of the project size, profitability, and its overall impact on the firm, and it also plays a key role in the development of long-range cash flow forecasts for the company.

13.14.5 Project Implementation and Post-Audit Review

The implementation time for a project may be short or it may be very long, and responsibility for project implementation customarily rests with division management and the project sponsor. Approximately 2 to 6 months before implementation, an appropriations request (AR) must be submitted and approved. During the time that a project is being implemented, a periodic progress report is usually submitted to the proper levels of management. This is used to ensure that the project is on schedule and that management is aware of any problems that may have arisen. Often a project will have a cost overrun due to the difficulty in estimating future cash flows. Most companies allow a 10% overrun without requiring a new AR to be submitted.

In most firms, it is the responsibility of division management to conduct a post-audit review after a project has attained operational status (see Step 7 of the engineering economic analysis procedure discussed in Section 2.5). This review is usually seen as a constructive learning experience that includes a review of project operations and financial performance. The primary objectives of the post-audit appraisal are (1) to determine whether project objectives are achieved, (2) to discover degree of conformance to plan and ascertain where variance occurred, (3) to encourage more careful estimates in the original proposal, and (4) to learn from the results, to identify problems, and to promote better estimates in the future. The post-audit appraisal varies from 3 months to 2 years after start-up but is normally done after 1 year of operation.

13.14.6 Communication

If project proposals are to be transmitted from one organizational unit to another for review and approval, there must be effective means for communication that may range from standard forms to personal appearances. In communicating proposed projects to higher levels in the management structure, it is desirable to use a format that is as standardized as possible to help ensure uniformity and completeness of information and evaluation. In general, the technical and marketing aspects of each proposed project should be completely described in the manner most appropriate to each individual case. However, financial summaries of all proposals should be standardized so that they may be consistently and fairly evaluated. An example of industrial guidelines used in practice is given in Section 13.15.

13.15 CAPITAL ALLOCATION PRACTICES IN CORPORATION A

The corporate guidelines and forms illustrated in this section are well designed analysis and control procedures. However, they should not be accepted as the best available or as necessarily appropriate for use in companies searching for "cookbook" procedures. Clearly, an organization's capital allocation guidelines should be designed and written around its own set of business goals and objectives and long-term strategy. The materials for Corporation A typify fundamental calculations required to develop and analyze cash flows associated with capital investment proposals. It is our intent to demonstrate that previous topics covered in this book are integral components of the process used by top management to execute their responsibility for judiciously allocating capital in building a company's future.

Corporation A is a Fortune 100 company that is known for its thorough and careful evaluation of proposed capital investment opportunities. The capital allocation procedure summarized below requires information concerning estimated project cash flow profiles and nonmonetary considerations, sensitivity analyses of key parameters, inflation adjustments, business history of the product involved, and so forth. Be sure to observe the sequence of steps comprising this company's procedure and the timing of various approvals required to move a capital investment proposal forward to final acceptance and funding. This company requires that a plant appropriation request (PAR) be prepared for capital expenditures that exceed \$50,000. Who makes the final approvals for PARs then depends on the amount of capital requested. For example, when \$20,000,000 or more is being sought, the Board of Directors must grant final approval.

Company Procedure For Initiation, Preparation, Review and Approval of Plant Appropriation Requests*

I. Application of the Procedure

Responsibilities for initiation, preparation, review, and approval, as well as the routine to be followed in processing a Plant Appropriation Request (PAR) requiring Corporate Executive Office (CEO) or Board of Directors (BOD) approval, are specified in this Company Procedure. Principal topics covered in this Procedure include

- Approvals in Conjunction with Acquisition/Disposition Proposals
- Notification of Initiation (NOI)
- Amount of PAR and Scope of Project
- PAR Reviews
- Approval, Implementation, and Follow-up

II. Transactions Requiring a PAR

Any (1) facility investment, (2) purchase or sale of land, (3) sale or lease of Company-owned facilities, (4) plant relocation, (5) sublease of facilities to external parties (unless such transactions are normally part of the component's business), and/or (6) lease commitment requiring CEO or BOD approval must be supported by an approved PAR before any commitment is made. The calculations to determine approval amounts are discussed in the PAR Instructions provided in Section VI.

III. Purpose of the Notification of Initiation

The use of the Notification of Initiation (NOI) is mandatory for PARs requiring CEO or BOD approval. The principal purpose of the NOI is for the use of the initiating manager in securing the concurrence of the approving authority prior to proceeding with the preparation and review of a PAR. The NOI establishes the time schedule for review and approval and specifies the Functional Reviews required by the approving authority.

The initiating manager provides basic information pertaining to a project as follows:

- A summary description of the proposed project including a preliminary evaluation of the compatibility with strategic plan.

*Courtesy of the General Electric Company.

- *Estimated* expenditures (investment and expense).
- Proposed target dates for endorsement or approval by the approving authorities.
- Recommendation of Functional Reviews and critical issues to be considered.

The initiating manager will then forward the NOI through channels to the final approving authority for completion of

- Approval or modification of the initiating manager's proposed schedule and Functional Reviews.
- Specifications for additional Functional Reviews and/or additional instructions.

The primary purpose of the preliminary evaluation of compatibility with strategic plan is to confirm that the strategic plan, of which the proposed investment is a part, remains current and viable. It is not intended as a detailed strategic review of the PAR or as a detailed evaluation of the merits of the project. In the case of projects not included in an approved strategic plan, the scope of the preliminary evaluation of compatibility will usually be broadened to a more detailed critique of the supporting assumptions and documentation of the consistency of the proposed investment with the approved strategic plan.

IV. Timing of NOI

An NOI should be prepared after the preliminary analyses have been completed and immediately upon the decision to seek funds, but *prior* to the preparation of the PAR.

Return of the signed NOI by the responsible Executive Officer, through channels, constitutes approval for the initiating manager to proceed. The initiating manager will then provide copies of the approved NOI to the designated reviewers.

V. Amount of PAR and Scope of Project

In defining the total amount to be requested and the scope of a project, all purchases, costs, activities, and manpower essential to plan and complete the undertaking must be documented, identifying all items of expense. The project scope should be such that the proposal can be considered independently without regard to other plant investment needs. Also, the evaluation of the financial benefits must be based solely on the benefits achieved by the project itself.

VI. What is Required in a Plant Appropriation Request

All PARs should include the following:

1. Executive Summary—should be limited to two pages, and in most instances should cover the items outlined below in roughly the order

suggested. This is not an all-inclusive list, however, and the approach should be modified as the facts warrant.

a. Narrative Description of Project—including the following where applicable:
   - Component requesting, product line involved
   - Category—for example: cost reduction, capacity, pollution control, etc.
   - Project description and location
   - Amount, amounts previously requested, timing
   - How project will be implemented (for example: purchase equipment, lease facility, rearrange, etc.)
   - Why project is required
   - Expected results
   - Project phases
   - Consistency with strategic plan and budget
   - Degree of flexibility, for example:
     —If lease, minimum payment commitment
     —If purchase, possible alternate uses of proposed equipment
b. Financial Benefits—brief description of the basis for determining project financial benefits (for example: incremental sales, cost improvements, etc.). Identify key assumptions on which benefits are based and sensitivity of benefits to changes in these assumptions. State the discounted funds flow rate of return (DCRR), and the payback period. Comment on upside and downside potential. If financial benefits cannot be quantified, state other benefits on which approval should be based.
c. Alternatives—other alternatives considered and reason for selection of recommended approach, including relative financial benefits.
d. Risks—summary and assessment of key risks to project success. Quantify impact on DCRR.
e. Conclusion—brief summary of financial and nonfinancial reasons for approval.

2. Exhibits A through C
   a. Exhibit A—contains the basic financial data required for PARs.
   b. Exhibit B, Incremental Funds Flow Worksheet—facilitates calculation of the funds flow on reported and inflation adjusted bases.
   c. Exhibit C. Key Assumptions, Evaluation of Alternatives, and Sensitivity Analysis—the main purpose of the exhibit is to facilitate the communication, understanding, and appraisal of information for a proposed project with regard to
      - Alternative courses of action considered.
      - Principal advantages and disadvantages of this request and the alternatives considered, as well as reasons for rejecting the alternatives.
      - Key assumptions basic to achieving project objectives.
      - Key assumptions upon which individual alternatives are based.
      - Evaluation of risk associated with this request.

PLANT APPROPRIATION REQUEST (PAR) No. ____________

1. .... Sector
.... Group
.... Division
.... Department
.... Product Line
.... Project Location

APPROVAL REQUIRED
....
(Dollar amounts in thousands)

2. SUMMARY DESCRIPTION OF PROPOSED PROJECT

3. PROJECT EXPENDITURES

	This request	Previously approved	Future requests	Total project
Basis for approval				
Related expense......................	______	______	______	______
Total	======	======	======	======

4. KEY FINANCIAL MEASUREMENTS

	Reported	Inflation Adjusted
DCRR*	______%	______%
Payback period (years)	______	______

Chart of cumulative funds flow

—— Reported
---- Inflation Adjusted

Amount

0

19 19 19 19 19 19 19 19 19 19

*Discounted funds flow rate of return (DCRR) is synonymous with IRR.

EXHIBIT A

5. BUSINESS HISTORY AND FORECAST OF . . . . DEPARTMENT/OPERATION

	REPORTED						INFLATION ADJUSTED–c)			
	Market	Sales		Net Income				Net Income		
	Position –a)	Amt.	Price Index –b)	Amt.	ROS*	ROI*	Sales	Amt.	ROS	ROI
Year										
a. Last five years										
19										
19										
19										
19										
19										
b. Forecast with proposed project:										
Current year										
19										
Next five years										
19										
19										
19										
19										
19										

6. BUSINESS HISTORY AND FORECAST OF . . . .
. . . . .PRODUCT LINE

a. Last five years										
19										
19										
19										
19										
19										
b. Forecast with proposed project:										
Current year										
19										
Next five years										
19										
19										
19										
19										
19										
c. Increment resulting from project:										
Current year										
19										
Next five years										
19										
19										
19										
19										
19										

(a –Basis –
(b –19 = 100; Basis –
(c –Base year –

Federal Income tax rate used –

*ROS is "return on sales"; ROI is "return on investment."

EXHIBIT A Continued

7. SUMMARY OF PROJECT EXPENDITURES

	This request	Previously approved	Future requests	Total project
Investment expenditures				
Associated deferred charges				
Lease commitments not capitalized plus lease related expenses	____	____	____	____
Sub-total - Basis for approval				
Patterns and tooling				
All other related expense				
Grand total				
As a memo:				
All other starting costs				____
Trade-in value of surplus equipment				____

8. CATEGORY(S):

	Total	Investment	Expense
Category (Prime)			
Category (Other)	____	____	____
Total, this request	====	====	====

9. ESTIMATED GAIN (LOSS) IN NET INCOME TO OTHER G.E. COMPONENTS:

2 calendar years following project completion	
19__	19__

10. FACILITY TO BE REPLACED

First cost . . . .
Year purchased . . . .
Book value . . . .

Description of facility and proposed disposition

11. STARTING DATE . . . .
month/year

COMPLETION DATE . . . .
month/year

12. UTILIZATION ANTICIPATED IN FIRST YEAR AFTER PROJECT COMPLETION - 19. . . .
. . . . %
Basis . . . .

13. NUMBER OF EMPLOYEES
Location

	Before	After	Before	After
Manufacturing				
All other	____	____	____	____
Total	====	====	====	====

EXHIBIT A Continued

14. PERFORMANCE ON CLOSED APPROPRIATIONS

Appropriations past 3 years – a)

	Total expenditures	Project incremental net income 1st year	Project incremental net income 2nd year
Forecast – b)			
Actual			
VF%* – c)			

Board appropriations past 5 years

	Total expenditures	Project incremental net income 1st year	Project incremental net income 2nd year
Forecast – b)			
Actual			
VF%* – c)			

(a – CEO approval and above
(b – In appropriation requests
(c – Variance from forecast

15. PRINCIPAL COMPETITORS

Name	Estimated rank or market position: This year	Estimated rank or market position: Last year

16. APPROPRIATION ENDORSED AND SUPPORTING FINANCIAL DATA CERTIFIED BY

Manager – Finance

Appropriation endorsed or approved by

Manager – Marketing

Manager – Employee Relations

Manager – Engineering

Manager – Manufacturing

Legal Counsel

Manager – Strategic/Operational Planning

Department General Manager — Date

Division General Manager — Date

Group Executive — Date

Sector Executive — Date

Exhibit A
Page 4 of 4

*VF% is the percentage variance from the original forecast.

EXHIBIT A Continued

INCREMENTAL FUNDS FLOW WORKSHEET

(in thousands)	19_	19_	19_	19_	19_	19_	19_	19_	19_	19_
REPORTED FUNDS FLOW										
Incremental increase/decrease in net income resulting from the project based on:										
Cost reductions										
Incremental volume										
Changes in average selling price										
Investment credit										
Total –a)										
Provision for incremental depreciation expense										
Increased/decreased incremental investment in:										
Plant and equipment –b)										
Inventories										
Receivables										
All other										
Reported annual funds flow										
Reported cumulative funds flow										
INFLATION ADJUSTED FUNDS FLOW										
Reported annual funds flow										
CPI (1st year of investment = 100.0)										
Inflation adjusted annual funds flow										
Inflation adjusted cumulative funds flow										

Residual investment value at the end of funds flow period: Reported_______ Inflation adjusted_______

(a– Must agree with incremental net income as shown in Section 6 (c) of Exhibit A.

(b– Gross expenditures only, exclusive of provision for depreciation expense.

EXHIBIT B

KEY ASSUMPTIONS
EVALUATION OF ALTERNATIVES
AND
SENSITIVITY ANALYSIS

1. Brief description of major alternatives

   This Request –

   Alternative I –

   Alternative II –

   Alternative III –

2. Principal advantages/disadvantages and reasons for rejecting alternatives

   This Request –

   Alternative I –

   Alternative II –

   Alternative III –

3. Key assumptions basic to achieving project objectives upon which This Request is based

   a.

   b.

   c.

   d.

   e.

EXHIBIT C

KEY ASSUMPTIONS
EVALUATION OF ALTERNATIVES
AND
SENSITIVITY ANALYSIS

4. Key assumptions upon which individual alternatives are based

Alternative I a.

b.

c.

Alternative II a.

b.

c.

Alternative III a.

b.

c.

5. Sensitivity Analysis – Evaluate the project downside risk by identifying "worst case" changes in key variables. The project's upside potential, if relevant, may also be identified here, or in the text of the appropriation.

Variable (Selling Prices, Market Share, Material Costs, etc.)	Assumption Reflected in Project Funds Flow	Management Assessment of "Worst Case" Result	Downside DCRR	
			Reported	Inflation adjusted

EXHIBIT C Continued

VII. PAR Reviews

PAR reviews are classified into three main categories: Financial, Strategic, and Functional.

A. Financial Review

A Financial Review is mandatory in addition to any other accounting or treasury participation in the PAR Review. The purpose of the Financial Review is to provide a financial evaluation of the attractiveness of the proposal and to verify that the financial information is complete, properly presented and meets the requirement for a detailed financial expression of the proposed investment.

B. Strategic Review

A Strategic Review is mandatory. The purpose is to determine that the proposal is compatible with the most recently approved Strategic Plan and with the Sector mission, strategy, objectives and goals. The Strategic Review will be performed by Corporate Strategy Review. After receipt of the Financial and any Functional Review letters, the Strategic Reviewer will prepare a Strategic Review Report. The Strategic Review Report will include an assessment of the purpose/need for the proposed project and its compatibility with the approved Strategic Plan, an evaluation of the principal benefits, issues and risks associated with the project, and conclusions and recommendations.

C. Functional Reviews

The purpose of any Functional Review approved by an Executive Officer is to ensure the effectiveness and practicability of the proposed investment (as described in the PAR) to produce intended results within the estimated project cost. Functional Reviews may be designated to cover such areas/considerations as:

- Employee Relations
- Engineering
- International
- Legal
- Manufacturing
- Marketing
- Research and Development
- Treasury
- Real Estate and Construction Operation
- Other

D. Endorsement Prior to Final Approval

When this work is completed, the Strategic Reviewer will assemble and forward to the cognizant Sector Executive the PAR review package (Strategic Review

Report and Review letters) to obtain the endorsement of the Sector Executive for transmittal of the PAR package to those Senior Vice Presidents—Corporate Components designated by the CEO for review and comment. The position of the designated Senior Vice Presidents will be documented in a brief letter addressed to the CEO with copies to the Manager—Corporate Strategy Review.

VIII. Notification of Final Approval

For PARs that require CEO approval, the documents provided by Corporate Strategy Review will enable the CEO to return a signed copy which authorizes the project. Corporate Investment Strategy Review will notify the initiating manager and others who need to know upon receipt of the signed approval document.

For PARs that require BOD approval, the Secretary of the BOD will notify Corporate Finance Staff of the action taken by the BOD. Corporate Finance Staff, in turn, will notify the initiating manager and others will need to know.

IX. Implementation and Follow-up

A. Change in Scope or Results, Requirements for Additional Funds

During the implementation of an approved PAR, it is the responsibility of the component to inform the approving authority in writing of any major change in the scope of the project, such as site location, reduced profit forecast, or cost overrun. In these instances, further commitment of funds must receive the same careful planning and correlation of investment opportunities with the Sector's and Strategic Business Unit's Strategic Plans as the original proposal.

B. Follow-Up Reporting

The following two reports must be prepared periodically and submitted to Corporate Finance Staff.

1. Status reports on projects covered by open appropriations approved by the CEO or BOD. These reports must reflect and explain deviations from the approved PARs with respect to estimates of timing, amount of expenditure, and benefits realized.
2. Closed appropriation reports on projects approved by the CEO or BOD after the appropriations are closed and results can be appraised. These reports must summarize and explain variances between estimated and actual expenditures and benefits.

This chapter has provided a broad overview of capital financing (sourcing) and capital allocation (expenditure). Our discussion of capital financing has dealt with where companies get money to continue to grow and prosper, and how much it costs to obtain this capital. In this regard, differences between borrowed capital and owner's (equity) capital were made clear to the student. We further noted that the after-tax weighted average cost of capital is often used by a company to establish policy concerning what the MARR should be. Leasing, as a source of capital, was also briefly described and two lease versus purchase examples were analyzed.

Our treatment of capital allocation among independent investment opportunities has been built on two important observations. First, the primary concern in capital expenditure activity is to ensure the survival of the company by implementing ideas to maximize future shareholder wealth, which is equivalent to maximization of shareholder present worth. Second, engineering economic analysis plays a vital role in deciding which project alternative from a mutually exclusive set is recommended for funding approval and included in a company's overall capital investment portfolio. We concluded our examination of capital allocation by providing a typical set of corporate operating guidelines for making investment decisions. We hope these guidelines have confirmed for students the widespread usage of engineering economy principles in U.S. corporations.

3.16 PROBLEMS

13-1. Describe how an organization's capital financing activities affect the practice of engineering economy. (13.2)

13-2. Why do most engineering economic analyses normally assume that an "investor's pool of capital" is being used to sponsor a capital expenditure instead of a specific source of capital (e.g., equity versus borrowed funds)? (13.2)

13-3. List five possible sources of funds to a corporation for sponsoring large, capital-intensive projects. (13.3–13.7)

13-4. Briefly describe the six basic steps associated with a company's capital allocation procedure. (13.14)

13-5. Contrast the limits on liability for debts of a partnership versus those of a corporation. (13.2)

13-6.

a. What is equity capital, and explain how it is different from debt capital.

b. Why do bondholders, on the average, receive a lower return than do holders of common stock in the same corporation?

c. Why would a fast-growing company possibly *not* desire to use large amounts of debt capital in its operation? (13.2)

13-7.

a. List at least four characteristics of a corporation.

b. Give an example of how the profits of a corporation are "double taxed" and discuss how this may retard investment in a firm's stock. (13.2)

13-8. A corporation sold an issue of 20-year bonds, having a total face value of \$5,000,000, for \$4,750,000. The bonds bear interest at 10%, payable semiannually. The company wishes to establish a sinking fund for retiring the bond issue and will make semiannual deposits that will earn 8%, compounded semiannually. Compute the annual cost for interest and redemption of these bonds. (13.3)

13-9. A 20-year bond with a face value of \$5,000 is offered for sale at \$3,800. The rate of interest on the bond is 7%, paid semiannually. This bond is now 10 years old (i.e., the owner has received 20 semiannual interest payments). If the bond is purchased for \$3,800, what effective rate of interest would be realized on this investment opportunity? (13.3)

13-10.

a. A company has issued 10-year bonds, with a face value of \$1,000,000, in \$1,000 units. Interest at 8% is paid quarterly. If an investor desires to earn 10% nominal interest on \$10,000 worth of these bonds, what would the selling price have to be?

b. If the company plans to redeem these bonds in total at the end of 10 years and establishes a sinking fund that earns 8%, compounded semiannually, for this purpose, what is the *annual* cost of interest and redemption? (13.3)

13-11. You bought a \$1,000 bond that paid interest at the rate of 7%, payable semiannually, and held it for 10 years. You then sold it at a price that resulted in a yield of 10% nominal on your capital. What was the selling price? (13.3)

13-12. The Yog Company, a privately owned business, has an opportunity to buy the Small Company, from which it purchases a large amount of its raw materials. The purchase price is to be \$500,000. It is estimated that the Yog Company can realize a saving of \$75,000 annually from operating the Small Company instead of having to purchase its raw materials from it. The anticipated profit of \$75,000 is exclusive of any financial expense that might be involved in the transaction but includes provision for writing off the investment over a 20-year period. The Yog Company does not have \$500,000 available that it can use to buy the Small Company. However, it can sell a \$500,000 bond issue and use the proceeds for this purpose. The bonds will be issued as 20-year, 10% bonds. Should the Yog Company issue the bonds and buy the Small Company? (13.3)

13-13. The Yog Manufacturing Company's common stock is presently selling for \$32 per share and annual dividends have been constant at \$2.40 per share. If an investor believes that the price of a share of common stock will grow at 5% per year into the foreseeable future, what is the approximate cost of equity to Yog? What assumptions did you make? (13.4)

13-14. Reconsider the situation of problem 13-12 when $500,000 in *new* common stock is issued rather than a bond issue. The cost of equity is that obtained in Problem 13-13. Should Yog purchase the Small Company with equity funds? What difficulties could this pose to present stockholders? (13.4)

13-15. During the first 5 years of its life, a small corporation, which has a capitalization of $2,000,000 represented by 2,000 shares of common stock, has paid no dividends, in order to finance its expansion out of retained profits. It now needs $1,000,000 in additional capital to finance and stock two new warehouses. Discuss briefly the advantages and disadvantages of obtaining the required capital from (**a**) selling additional common stock, (**b**) borrowing on a 5-year bank loan at 10% interest, and (**c**) issuing 8%, 10-year bonds which would contain a provision that the corporation could not incur further indebtedness until the bond issue was retired. (13.3, 13.4)

13-16. Consider a small manufacturing firm in which there are 1,000 shares of common stock divided amoung three owners as follows:

Owner 1: 250 shares
Owner 2: 375 shares
Owner 3: 375 shares

Each share has been "valued" at $500, based on what the owners know they can get for their stock. The firm needs to obtain an additional $125,000 to purchase some new equipment, repay short-term bank notes, and substantially increase raw materials inventories. Their current after-tax return on equity is 12% and the effective income tax rate is 32%. Two plans have been developed for raising cash.

Plan I: Owners 2 and 3 agree to each make 125 shares of their stock available at $500/share.
Plan II: Instead of selling stock (i.e., ownership in the firm), the three owners believe they can sell 8-year mortgage bonds, bearing 10% interest each year, to raise the needed $125,000. The bonds would be sold in $1,000 units and redeemed in full at the end of 8 years. A 6% sinking fund will be used to guarantee repayment of the loan.

By spending the $125,000 as indicated above, the owners are certain they can increase revenues by $40,000 per year (on the average), exclusive of financing costs. Assume that depreciation associated with the purchase of new equipment is negligible in this situation. Which plan would you recommend for acquiring the needed $125,000? Show all calculations, and state any assumptions that you feel are required. (13.4)

13-17. Determine the before-tax and after-tax weighted average cost of capital for a firm that has this capital structure:

Amount	Source of Capital	Cost of Capital (%)	Financing Cost per Year
$ 3 million	Short-term bank loans	10	$0.30 million
$ 7 million	Mortgage bonds	7	$0.49 million
$ 4 million	Preferred stock	8	$0.32 million
$11 million	Common stock and retained earnings	13	$1.43 million

Assume that the firm's effective income tax rate is 40% and that a 13% rate of return to purchasers of common stock represents a satisfactory opportunity cost of equity capital. (13.8)

13-18. The after-tax weighted average cost of capital (K_a') to a large utility company is defined as follows:

$$K_a' = (1 - t)(\lambda)i_b + (1 - \lambda)e_a$$

where t = effective income tax rate
λ = fraction of total capitalization in borrowed funds
i_b = marketplace before-tax cost of borrowed funds
e_a = after-tax cost of equity funds

If the *real* (inflation-free) annual returns on borrowed funds and equity capital are roughly constant at 2% and 5%, respectively, what is the value of K_a' when inflation averages 7% per year? Let $\lambda = 0.50$ and $t = 0.40$. (13.8)

13-19. Refer to Example 13-7. If annual maintenance expenses can range from \$800 to \$1,300 per year and inflation can vary from 3% to 8% per year, determine whether the forklift truck should be purchased or leased for each combination of extreme values. (13.10)

Annual Maintenance	Annual Inflation Rate (%)	Recommendation
\$ 800	3	?
800	8	?
1,300	3	?
1,300	8	?

13-20. An existing piece of equipment has been performing poorly and needs replacing. More modern equipment can be *purchased* for cash using retained earnings (equity funds), or it can be *leased* from a reputable firm. If purchased, the equipment will cost \$20,000 and have a depreciable life of 5 years with no salvage value. Straight-line depreciation is used by the firm. Because of improved operating characteristics of the equipment, raw materials savings of \$5,000 per year are expected to result relative to continued use of the present equipment. However, labor costs for the new equipment will most likely increase by \$2,000 per year and maintenance will go up by \$1,000 per year. To lease the new equipment requires a refundable deposit of \$2,000, and the end-of-year leasing fee is \$6,000. Annual materials savings and extra labor costs will be the same when purchasing or leasing the equipment, but the company will provide maintenance for its equipment as part of the leasing fee. The desired after-tax rate of return (IRR) is 15% and the effective income tax rate is 50%. If purchased, it is believed that the equipment can be sold at the end of 5 years for \$1,500 even though \$0 was used in calculating depreciation. An investment credit of 10% can be used to offset income taxes at the time of the purchase (year 0). Determine whether the company should buy or lease the new equipment, assuming that it has been decided to replace the present equipment. (13.10)

13-21. Determine the more economical means of acquiring a business machine if you may either (1) purchase the machine for $5,000 with a probable resale value of $1,000 at the end of 5 years, or (2) lease the machine at an annual rate of $900/year for 5 years with an initial deposit of $500 refundable upon returning the machine in good condition. If you own the machine, you will depreciate it for tax purposes at an annual rate of $800. All lease charges are deductible for income tax purposes. As owner or lessee you will pay all expenses associated with the operation of the machine.

a. Compare these alternatives by use of the annual worth method. The after-tax minimum attractive rate of return is 10% and the effective income tax rate is 50%.

b. How high could the annual leasing fees be such that leasing remains the more desirable alternative? (13.10)

13-22. A company has $290,000 for investment in new projects during the coming year. Projects currently being considered are as follows:

Project	Capital Required	Life (years)	Estimated Annual Rate of Profit (%)	Risk
A	$ 60,000	5	10	Low
B	100,000	3	20	Average
C	150,000	5	8	Average
D	110,000	8	15	High

The company follows a general policy of not committing capital for a longer period than 8 years, and it prefers 5 years or less. Uncommitted capital would remain temporarily in a bank, where it would earn at least 8%. Which projects do you recommend? (13.12, 13.13)

13-23. A company has $20,000 to invest in elective projects and has taken the position that at least half of these funds should be spent on medium- and low-risk projects. Based on the investments shown in the following table, categorized by risk, what subset of these independent capital investment opportunities should be recommended? (13.12, 13.13)

Project	Investment	Prospective Annual Return After Taxes (%)
High Risk	*Amount ($$10^3$)*	
M	4,000	32
N	2,000	28
O	6,000	22
P	5,500	17
Medium Risk		
S	8,000	22
T	7,500	20
U	10,000	16
V	5,000	12
W	6,200	10

Low Risk		
C	12,000	16
D	8,000	14
E	10,000	13
F	5,000	8

13-24. A company has $100,000 available for investment in new projects. Four projects are available as follows:

Project	Required Capital	Life, Years	Expected Rate of Return (%)
A	$30,000	5	10
B	50,000	8	15
C	20,000	10	8
D	10,000	4	20

Projects A, B, and D are rated about equal and normal in risk. Project C is considered to involve somewhat lower risk. Any capital uninvested will remain in a bank where it will earn 6% interest and be available for any later use that may arise. Which projects would you recommend the company undertake? (13.12)

13-25.

a. List some of the factors that a high-technology electronics firm might consider in establishing a minimum attractive rate of return for its investments.

b. What other objectives, in addition to maximizing the future worth (or present worth) of the firm, would a high-technology company consider in its capital investment decision making? (13.9)

13-26. A small corporation having a capitalization of $200,000, represented by 2,000 shares of common stock, has been in operation for 5 years. During this time it has paid no dividends, in order to be able to finance its growth through retained profits. It now needs $100,000 in additional capital to finance expansion. It is considering three methods of obtaining the capital: (a) attempting to issue $100,000 in new common stock; (b) borrowing from a bank at 8% interest; and (c) selling 5-year bonds bearing interest at 7%, with the restriction that no further indebtedness can be incurred during the life of the bond issue. Discuss briefly the advantages and disadvantages of each method of financing. (13.3, 13.4)

13-27. The Holiday Car Company has offered to finance your new $12,000 automobile as follows:

Purchase price = $12,000 (includes all taxes, etc.)
Down payment = $4,000
Amount to be financed = $8,000

You tell the credit manager that you think the 15% finance charge is quite high and that you have $12,000 cash with which to purchase the automobile. "Nonsense," she says, "Why invest in a depreciating asset (the car) when you can finance it through us and invest your $12,000 in a bond (a nondepreciating asset) that pays 12%?"

She turns to her computer, enters all this information and comes back with the following "proof" that her claim is correct. (*Note:* BOY is an abbreviation for "beginning of year.")

	Your Loan Payments (at 15%)			Bonds at 12%	
Year	Principal at BOY	Principal Repayment	Interest	Principal at BOY	Interest
1	$8,000	$1,600[a]	$1,200	$ 8,000	$ 960.0
2	6,400	1,600	960	8,960	1,075.2
3	4,800	1,600	720	10,035.2	1,204.3
4	3,200	1,600	480	11,239.4	1,348.6
5	1,600	1,600	240	12,588.2	1,510.6
		Total	$3,600	Total	$6,098.7[b]

[a] Her company's analysis assumes a uniform repayment each year of the loan principal.

[b] $\$8,000(1.12)^5 - \$8,000 = \$14,098.73 - \$8,000 = \$6,098.73$.

She states that the total interest cost on your car loan from her company is $3,600 (annual payments are assumed to simplify the comparison), while the earned interest on a bond paying 12% per year accumulates to $6,098.73 at the end of 5 years. "Therefore, you come out ahead by financing your car rather than paying cash for it!"

Comment on the credit manager's analysis, and *quantitatively* show why you think it is correct or incorrect. Should the comparison be based solely on interest earned or paid? Should the $8,000 principal that is common to both plans be considered? What are the *incremental* cash flows between the two financing alternatives? (13.3, 13.10)

14

EVALUATING PUBLIC PROJECTS WITH THE BENEFIT-COST RATIO METHOD

The objectives of this chapter are (1) to describe many of the unique characteristics of public projects, and (2) to learn how to use the benefit–cost (B/C) ratio method for selecting the best alternative from a set of mutually exclusive or independent projects.

The following subjects are covered in this chapter:

Differences between public and private projects
Self-liquidating projects
Multiple-purpose projects
Difficulties inherent in engineering economy studies of public works
What interest rate should be used for public projects?
The benefit–cost ratio method
Evaluating independent projects by B/C ratios
Added benefits or reduced costs?
Comparison of mutually exclusive alternatives
After-the-fact justifications
Distributional considerations
Omitting qualitative information

14.1 INTRODUCTION

Public projects are those authorized, financed, and operated by governmental agencies—federal, state, or local. Such public works are numerous, may be of any size, and frequently are much larger than private ventures. Because they require the expenditure of capital, such projects are subject to the principles of engineering economy with respect to their design, acquisition, and operation. However, because they are public projects, a number of important special factors exist which are not ordinarily found in privately financed and operated businesses.

14.2 DIFFERENCES BETWEEN PUBLIC AND PRIVATE PROJECTS

As a consequence of the differences listed in Table 14-1, it often is not possible to make economy studies and investment decisions for public works projects in exactly the same manner as for privately owned projects. Different decision criteria must be used, and this creates problems for the public (which pays the bill), for those who must make the decisions, and for those who must manage public works projects.

14.3 SELF-LIQUIDATING PROJECTS

The term *self-liquidating project* is often applied to a governmental project that is expected to earn direct revenue sufficient to repay the cost in a specified period of time. Such projects usually provide utility services, such as water, electric power, sewage disposal, or irrigation water. In addition, toll bridges, tunnels, and highways are built and operated in this manner.

TABLE 14–1 Some Basic Differences between Privately Owned and Publicly Owned Projects

	Private	Public
1. Purpose	Provide goods or services at a profit Provide jobs Promote technology Improve living standards	Protect health Protect lives and property Provide services (at no profit) Provide jobs
2. Sources of capital	Private investors and lenders	Taxation Private lenders
3. Method of financing	Individual ownership Partnerships Corporations	Direct payment from taxes Loans without interest Loans at low interest Self-liquidating bonds Indirect subsidies Guarantee of private loans
4. Multiple purposes	Moderate	Common, such as electrical power, flood control irrigation, recreation, and education
5. Life of individual projects	Usually relatively short (5 to 20 years)	Usually relatively long (20 to 60 years)
6. Relationship of suppliers of capital to project	Direct	Indirect, or none
7. Conflict of purposes	Moderate	Quite common (dam for flood control and environment preservation)
8. Conflict of interests	Moderate	Very common (between agencies)
9. Effect of politics	Little to moderate	Frequent factors; short-term tenure of decision makers Pressure groups Financial and residential restrictions, etc.
10. Measurement of efficiency	Rate of return on capital	Very difficult; no direct comparison with private projects

Self-liquidating projects are not expected to earn profits or pay income taxes. Neither do they pay property taxes, although in some cases they do make *in lieu* payments in place of the property or franchise taxes that would have been paid to the cities, counties, or states involved had the project been built and operated by private ownership. For example, the U.S. government agreed to pay the states of Arizona and Nevada $300,000 each annually for 50 years in lieu of taxes that might have accrued if Hoover Dam had been privately constructed and operated. In most cases the in lieu payments are considerably less than the actual property or franchise taxes would be. Furthermore, once the in lieu payments are agreed upon, usually at the time the project is originated, they virtually never are changed thereafter. Such is not the situation in the case of property taxes.

Whether self-liquidating projects should or should not pay the same taxes as privately owned projects is not within the scope of this text. Such payments undoubtedly would make it much easier to compare the economy of similar publicly and privately operated activities. However, the fact that public projects do not have to earn profits or pay certain taxes must be given proper consideration in making engineering economy studies of such projects and in making decisions regarding them.

14.4 MULTIPLE-PURPOSE PROJECTS

Many governmental and private projects have more than one purpose or function. A governmental project, for example, may be intended for flood control, irrigation, and generation of electric power. A private project may utilize by-product gases in a petroleum refinery to generate steam and electricity for refinery use and to cogenerate electricity for public sale. Such projects commonly are called *multiple-purpose projects*. By designing and building them to serve more than one purpose, greater overall economy can be achieved. This is very important in projects involving very large sums of capital and utilization of natural resources, such as rivers. It is not uncommon for a public project to have four or five purposes. This usually is desirable, but, at the same time, it creates economy and managerial problems because of overlapping utilization of facilities and, sometimes, conflict of interest between the several purposes and agencies involved. An understanding of the problems involved in such situations is essential to anyone who wishes to make economy studies of such projects or to understand the cost data and political issues arising from them.

A number of basic problems may arise in connection with multipurpose public works. These may be illustrated by the simple example of a dam, shown in Figure 14-1, which is to be built in a semiarid portion of the United States, primarily to provide control against serious floods. It is at once apparent that, if the flow of the water impounded behind the dam could be regulated and diverted onto the adjoining land to provide irrigation water, the value of the land would be increased tremendously. This would result in an increase in the nation's resources, and it thus appears desirable to expand the project into one having two purposes: flood control and irrigation.

FIGURE 14-1

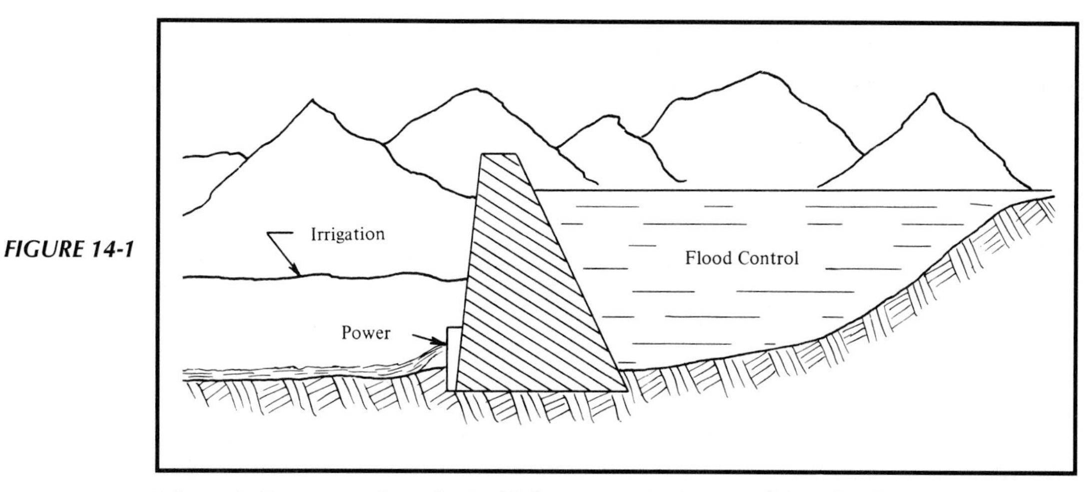

Schematic Representation of a Multiple-Purpose Project Involving Flood Control, Irrigation, and Power

The existence of a dam with a high water level on one side and a much lower level on the other side, at least during a good portion of the year, immediately suggests that some of the nation's resources will be wasted unless some of the water is permitted to run through turbines to generate electric power that can be sold to customers in the adjacent territory. Thus, the project is expanded to have a third purpose—the generation of power.

In the semiarid surroundings the creation of a large lake, such as would exist behind the dam, would provide valuable recreation facilities for hunting, fishing, boating, camping, and so on. This gives a fourth purpose to the project. All of these purposes have desirable economic and social worth, and what started as a single-purpose project has ended by having four purposes. Not to develop the project to fill all four functions probably would mean that valuable national resources would be wasted. On the other hand, however, the potential loss of land (e.g., for mining minerals and retaining ecological balance, etc.) must be factored into the analysis as disbenefits.

If the project is built for four purposes, the important fact that one dam will serve all of them will result in at least three basic problems. The *first* is the allocation of the cost of the dam to the four purposes that it will serve. Assume, for example, that the dam will cost $10,000,000. What portion of this amount should be assigned to flood control? What amounts to irrigation, power generation, and recreation? If the identical dam will serve all four purposes, deciding how much of its cost should be assigned to any one of the purposes obviously might present considerable difficulties. One extreme would be to decide that, since the dam was required for flood control, its entire cost should be assigned to this purpose. If this were done, the cost of the electric power and the irrigation water would obviously be far less than if a considerable part or all of the $10,000,000 had been assigned to these functions.

Another extreme would be to assign each purpose with an amount equal to the total cost of the dam. Such a procedure would have two effects. First, it would undoubtedly make some of the purposes so costly—for example, recreation and irrigation—that they could not be undertaken. Second, by using this procedure, the cost of the dam might be returned to the government several times. Although many taxpayers would appreciate such a profitable occurrence, it is hardly feasible.

The *second* basic problem in multiple-purpose projects is the matter of conflict of interest among the several purposes. This may be illustrated by the matter of the water level maintained behind the dam. For flood control it would be best to have the reservoir empty most of the year to provide maximum storage capacity during the season when floods might occur. Such a condition would be most unsatisfactory for power generation. For this purpose it would be desirable to maintain as high a level as possible behind the dam during most of the year. For recreational purposes also a fairly constant level would be desirable. The requirements for irrigation probably would be somewhere between those for flood control and for power generation. Furthermore, while the lake created might be ardently approved by boating and recreation enthusiasts, it might be opposed with equal vigor by ecology-minded groups. Thus, some very definite conflicts of interest usually arise in connection with multiple-purpose works. As a result, compromise decisions must be made.

A *third* problem connected with multiple-purpose public works is the matter of politics. Since the various purposes are likely to be desired or opposed by various segments of the public and by various interest groups that will be affected, it is inevitable that such projects frequently become political issues. This often has an effect upon cost allocations and thus on the overall economy of these projects.

The net result of these three factors is that the cost allocations made in multiple-purpose public projects tend to be arbitrary. As a consequence, production and selling costs of the services provided also are arbitrary. Because of this, they cannot be used as valid "yardsticks" with which similar private projects can be compared to determine the relative efficiencies of public and private ownership.

14.5 DIFFICULTIES INHERENT IN ENGINEERING ECONOMY STUDIES OF PUBLIC WORKS

In view of the various problems that have been cited, engineers and other individuals sometimes raise the question as to whether engineering economy studies of governmental projects should be attempted. It must be admitted that such economy studies frequently cannot be made in as complete and satisfactory a manner as in the case of studies of privately financed projects. However, decisions regarding the investment and use of capital in public projects must be made—by the public, by elected or appointed officials and, very importantly, by managers, who usually cannot use the same measures of operating efficiency that are available to managers of private ventures. The alternative to basing such decisions on the best possible economy studies is to base them on hunch and expediency. It is therefore essential that economy studies be made in the best possible manner, but with a sound understanding of the nature of such activities and of all the background, conditions, and limitations connected with them.

It must be recognized that engineering economy studies of public works can be made from several viewpoints, each applicable for certain conditions. Many studies are made from the point of view of the governmental body involved. In comparing alternative structures or methods for accomplishing the same objective, such a point of view is satisfactory. The point of view of the citizens in a restricted area is used in other studies, such as local self-liquidating or service projects where those who must pay the costs are the direct beneficiaries of the project. In other cases the entire nation may be affected, so the economy study should reflect a very broad point of view. Such studies are apt to involve many "irreducible" costs and benefits that cannot be assigned to specific persons or groups. In each case as many factors as possible should be evaluated in monetary terms, because this usually will be most helpful in making decisions regarding the expenditure of public funds. However, it must also be recognized that nonmonetary factors often exist and are frequently of great importance in final decisions.

There are a number of difficulties inherent in public works that must be considered in making engineering economy studies and economic decisions regarding them. Some of these are as follows:

1. There is no profit standard to be used as a measure of financial effectiveness. Most public works are intended to be nonprofit.
2. There is no easy-to-quantify monetary measure of many of the benefits provided by a public project.
3. There frequently is little or no direct connection between the project and the public, who are the owners.
4. Whenever public funds are used, there is apt to be political influence. This can have very serious effects on the economy of projects, from conception through operation, and knowledge of its actual or possible existence is an important factor to be considered.
5. The usual profit motive as a stimulus to effective operation is absent. This may have a marked effect on the effectiveness of a project, from conception through operation. This does not mean that all public works are inefficient or that many managers and employees of them are not trying to do, and are actually doing, an effective job. But the fact remains that the direct stimuli present in privately owned companies are lacking and that this may have a considerable effect on project economics.
6. Public works usually are much more circumscribed by legal restrictions than are private companies. Their ability to obtain capital usually is restricted. Frequently, their area of operations is restricted as, for example,

where a municipally owned power company cannot sell power outside the city limits.

7. In many cases decisions concerning public works, particularly their conception and authorization, are made by elected officials whose tenure of office is very uncertain. *As a result, immediate costs and benefits may be stressed, to the detriment of long-range economy.*

14.6 WHAT INTEREST RATE SHOULD BE USED FOR PUBLIC PROJECTS?

The interest rate should play the same formal role in the evaluation of public sector investment projects that it does in the private sector. The rationale for its use is somewhat different. In the private sector, it is used because it leads directly to the private sector goal of profit maximization or cost minimization. In the public sector, its basic function is similar, in that it should lead to a maximization of social benefits, provided that these have been appropriately measured. Choice of an interest rate will lead to determination of how available funds may be allocated best among competing projects.

Three main considerations may bear on what interest rate to use in governmental economy studies:

1. The interest rate on borrowed capital.
2. The opportunity cost of capital to the governmental agency.
3. The opportunity cost of capital to the taxpayers.

In general, it is appropriate to use the borrowing rate only for cases in which money is borrowed specifically for the project(s) under investigation and use of that money will not cause other worthy projects to be forgone.

The opportunity cost (interest rate) encompasses the annual rate of profit (or other social or personal benefit) to the constituency served by the government agency or the composite of taxpayers who will eventually pay for the government project. If projects are chosen so that the return rate on all accepted projects is higher than that on any of the rejected projects, the interest rate used in the economic analysis is that associated with the best opportunity forgone. If this is done for all projects and investment capital available within a government agency, the result is a *governmental opportunity cost*. If, on the other hand, one considers the best opportunities available to the taxpayers if the money were not obtained through taxes for use by the governmental agency, the result is a *taxpayers' opportunity cost*.

Theory suggests that in usual governmental economic analyses the interest rate should be the largest of the three listed above. Generally, the taxpayers' opportunity cost is the highest of the three. As an indicative example, a federal government directive* in 1972 specified that an interest rate of 10% should be used in engineering economy studies for a wide-range of federal projects. This 10%, it can be argued, is at least a rough approximation of the average return taxpayers could be obtaining from the use of that money. In any case, it is greater than typical rates of interest that the federal government has been paying for the use of borrowed money.

14.7 THE BENEFIT–COST (B/C) RATIO METHOD

Basically, an engineering economy study of a public project is no different from any other economy study. Any of the methods of Chapter 4 could be used for such a study. From a practical viewpoint, however, because profit almost never is involved and most projects have multiple benefits, some of which cannot be measured precisely in terms of dollars, many economy studies of public projects are made by (1) comparing the equivalent worth of costs or (2) determining the ratio of the equivalent worth of benefits to the equivalent worth of costs at an effective interest rate of $i\%$ per period.

The benefit–cost (B/C) ratio can be defined as the ratio of the equivalent worth of benefits to the equivalent worth of costs. The equivalent worth utilized is customarily present worth (PW) or annual worth (AW), but it can also be future worth (FW). The B/C ratio is also referred to as the savings–investment ratio (SIR) by some governmental agencies.

Two commonly used formulations of the B/C ratio (here expressed in terms of AW) are as follows:

1. Conventional B/C ratio:

$$\text{B/C} = \frac{\text{AW (benefits of the proposed project)}}{\text{AW (total costs of the proposed project)}} = \frac{\text{B}}{\text{CR} + (\text{O\&M})} \tag{14–1}$$

where AW(•) = annual worth of (•)

B = annual worth of benefits of the proposed project

CR = capital recovery cost (i.e., the equivalent annual cost of the initial investment, I, including an allowance for salvage value, if any).

O&M = equivalent annual operating and maintenance costs of the proposed project

*Office of Management and Budget, "Discount Rates to Be Used in Evaluating Time Distributed Costs and Benefits," *Circular No. A-94* (rev.), March 27, 1972. The 10% was an inflation-free interest rate, which means that the *combined* interest rate (i_c) would be higher, as explained in Chapter 9.

2. Modified B/C ratio:

$$B/C = \frac{B - (O \& M)}{CR} \tag{14–2}$$

The numerator of the modified B/C ratio expresses the equivalent worth of the benefits minus the equivalent annual operating and maintenance costs; the denominator includes only the annual equivalent investment (i.e., capital recovery) costs. A project is acceptable when the B/C ratios as defined in Equations 14-1 and 14-2 are greater than or equal to 1.0.

Equations 14-1 and 14-2 can be rewritten in terms of PW as follows:

1. Conventional B/C ratio:

$$B/C = \frac{\text{PW (benefits of the proposed project)}}{\text{PW (total costs of the proposed project)}} = \frac{PW(B)}{I + PW\ (O\&M)} \tag{14–3}$$

where $PW\ (\bullet) =$ present worth of $(\bullet)$
$PW(B) =$ present worth of benefits of the proposed project
$I =$ initial investment of the proposed project
$PW(O\&M) =$ present worth of operating and maintenance costs

2. Modified B/C ratio:

$$B/C = \frac{PW(B) - PW(O\&M)}{I} \tag{14–4}$$

A salvage value (if any) is normally treated as a reduction to the present equivalent investment in the denominator of Equations 14-3 and 14-4.

Both B/C methods give consistent answers regarding whether the ratio is greater than or equal to 1.0, and both yield the same recommended choice when comparing *mutually exclusive* investment alternatives. As discussed in Section 14.10, an incremental procedure is required when comparing mutually exclusive alternatives with the B/C ratio method.

When *independent* projects are being evaluated, it is possible that the conventional B/C ratio and the modified B/C ratio will produce project rankings (i.e., the best project, the next best, and so on) that are not identical. This is illustrated in Example 14.4. In this situation the B/C ratio method may prove to be an inconsistent method for allocating capital among available investment opportunities. In general, independent projects are deemed worthwhile investments if their B/C ratios are $\geq$ 1.0.

Another subject of interest concerns the classification of certain types of items (e.g., salvage value) as "added benefits" rather than as "reduced costs." This and similar judgments involving various types of costs affect the B/C ratio, however it is defined. In Section 14.9, we demonstrate that the final recommendation of projects is not altered by such classification issues.

EXAMPLE 14-1

Determine the B/C ratio for a project with the following data.

First cost	\$20,000
Project life	5 years
Salvage value	\$ 4,000
Annual benefits	\$10,000
Annual O&M costs	\$ 4,400
Interest rate	8%

Solution

A cash flow diagram, showing benefits for this project as upward arrows and costs as downward arrows, is provided in Figure 14-2. The conventional B/C ratio and modified B/C ratio, based on AW, are computed as follows:

$$\begin{aligned}\text{CR} &= (\$20{,}000 - \$4{,}000)(A/P, 8\%, 5) + \$4{,}000(0.08)\\ &= \$4{,}327\end{aligned}$$

$$\begin{aligned}\text{Conventional B/C ratio} &= \frac{\text{B}}{\text{CR} + (\text{O\&M})} = \frac{\$10{,}000}{\$4{,}327 + \$4{,}400}\\ &= 1.146 > 1.0\end{aligned}$$

$$\begin{aligned}\text{Modified B/C ratio} &= \frac{\text{B} - (\text{O\&M})}{\text{CR}} = \frac{\$10{,}000 - \$4{,}400}{\$4{,}327}\\ &= 1.294 > 1.0\end{aligned}$$

Since B/C > 1.0 (by either ratio), this investment opportunity is worthwhile.

FIGURE 14-2

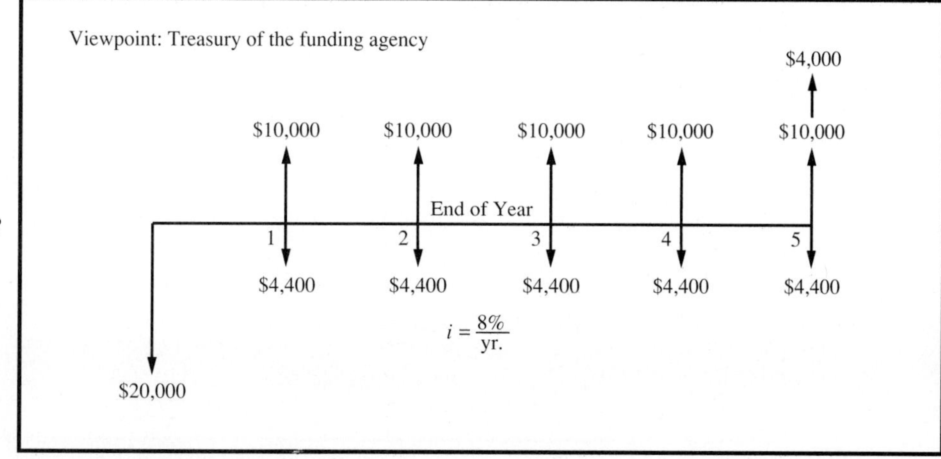

Cash Flow Diagram for Example 14-1

14.8

EVALUATING INDEPENDENT PROJECTS BY B/C RATIOS

If we are interested in evaluating independent projects, we can plot the B/C ratios for each of the projects on a graph as shown in Figure 14-3. (See Sections 5.6 and 5.7 for a review of independent versus mutually exclusive alternatives.) When B − (O & M) = CR, as for project 3, the benefit–cost ratio is exactly one, corresponding to an equivalent worth (e.g., PW) equal to zero for the project. If the B/C ratio is greater than 1, the equivalent worth is greater than 0, as with project 2. If the B/C ratio is less than 1, as with project 1, the equivalent worth is less than 0. The best project among a group of *independent* projects is the project having the highest B/C ratio. In Figure 14-3 project 2 is the best project, and the second best project is project 3 even though it is marginally acceptable (B/C = 1.0).

EXAMPLE 14-2

It is desired to compare the two projects in Example 5-10 by the B/C ratio method, with annual revenue to be considered as the annual benefit. Assume that the projects are independent investment opportunities rather than mutually exclusive alternatives. The minimum attractive rate of return is 10%, and

FIGURE 14-3

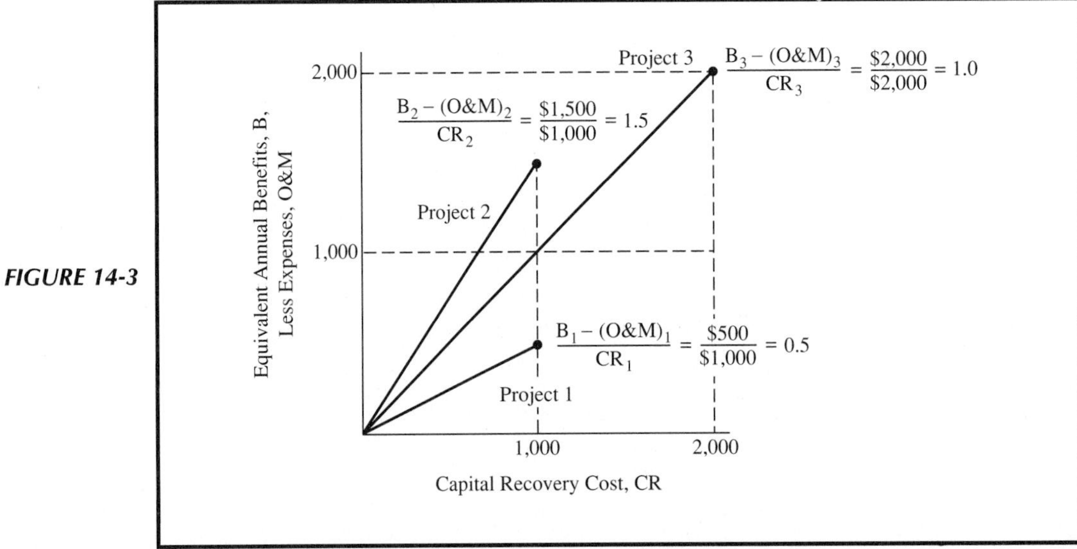

Graph of Modified Benefit-Cost Ratios for Independent Projects

a restatement of relevant data is shown in the following table. Which project(s) should be chosen?

	Project	
	I	II
Investment	$3,500	$5,000
Annual benefits to user	1,900	2,500
Annual O&M expenses to supplier	645	1,383
Useful life	4 years	8 years
Net salvage value	0	0

Solution

By using Equation 14-2, we obtain these results:

	Project	
	I	II
Annual benefits minus O&M expenses:		
$1,900−$645	$1,255	
$2,500−$1,383		$1,117
CR costs:		
$3,500($A/P$, 10%, 4)	1,104	
$5,000($A/P$, 10%, 8)		937
Modified B/C ratio:		
$1,255/$1,104	1.14	
$1,117/$937		1.19

Thus, project II is the better of the two, but both are satisfactory because the B/C ratios are > 1.0.

14.9 ADDED BENEFITS VERSUS REDUCED COSTS FOR INDEPENDENT PROJECTS

It is often necessary for the analyst to classify certain cash flows as either added benefits or reduced costs. How critical is the proper assignment of a cash flow item? Is the outcome of an analysis affected by classifying a reduced cost as a benefit? *An arbitrary decision as to the classification of a benefit or a cost has no effect on the acceptability of a project.* The following demonstrates how a cash flow can be classified as either a reduced cost or an added benefit without changing the suitability of the project.

Consider the case where b and c represent the present worths of benefits and costs, respectively, of a project. Let m ($m > 0$) represent the present worth of a cash flow not included in b or c. We further stipulate that $c - m > 0$.

If m is classified as an additional benefit, then

$$\text{Present Worth} = (b + m) - c = b - c + m$$

If m is a reduced cost, then as a second case we have

$$\text{Present Worth} = b - (c - m) = b - c + m$$

For the first case, the conventional benefit–cost ratio is

$$\text{B/C} = \frac{b + m}{c}$$

and for the second case, it is

$$\text{B/C} = \frac{b}{c - m}$$

Although the ratios are different, they can be compared to the measure of acceptability, namely, B/C $\geq$ 1:

$$\frac{b + m}{c} \geq 1 \text{ which means } b + m > c, \text{ and}$$

$$\frac{b}{c - m} \geq 1 \text{ which means } b > c - m, \text{ or } b + m > c$$

Therefore, regardless of the classification of a cash flow item as an additional benefit or a reduced cost, the present worth of the project will be the same and the *acceptability* of the project based on the B/C ratio will be unaffected.

EXAMPLE 14-3

When the present worth method is used in a benefit–cost analysis, the comparison of investment-type projects having *different lives* assumes implicitly that cash flows from shorter-lived projects are reinvested at i% (the MARR) over the life of the longer-lived project.*

*For further reading, see Chapter 8 of Lynn E. Bussey, *The Economic Analysis of Industrial Projects* (Englewood Cliffs, NJ: Prentice-Hall, 1978).

TABLE 14–2 Benefits, Costs, and Salvage Values for Five Independent Projects

Project	Life, Years	Initial Investment	Annual Cost	Annual Benefit	Salvage Value
1	5	$200,000	$10,000	$75,000	$20,000
2	6	175,000	24,900	71,000	12,000
3	8	95,000	28,000	38,000	4,000
4	4	80,000	22,000	36,000	2,000
5	10	150,000	12,950	52,000	9,000

Data for five independent projects are given in Table 14-2, and the MARR is 12% per year. Use the present worth method to compute a conventional B/C ratio for each project, and determine which projects (if any) should be selected, if (**a**) salvage value is considered a benefit, and (**b**) salvage value is considered a reduced cost.

Solution

(a) If the salvage value is classified as an additional benefit, Table 14-3 shows the conventional B/C ratio for the five projects. For example, the PW of project 1 in Table 14-3 is [−$200,000 − $10,000 (*P/A*, 12%, 5)] + [$75,000 (*P/A*, 12%, 5) + $20,000 (*P/F*, 12%, 5)] = $45,659. The conventional B/C ratio for project 1 is [$75,000 (*P/A*, 12%, 5) + $20,000 (*P/F*, 12%, 5)] / [$200,000 + $10,000 (*P/A*, 12%, 5)] = 1.193.

(b) When the salvage value is treated as a reduced cost, the conventional B/C ratios are different but do not change the acceptability of a project, as shown in Table 14-4. (The same set of recommended projects is obtained when using the modified B/C ratio method.)

14.10 COMPARISON OF MUTUALLY EXCLUSIVE ALTERNATIVES

The incremental benefit–cost ratio method is used when several *mutually exclusive* alternatives are being considered. In this situation the objective is to find

TABLE 14–3 Salvage Value Classified as an Additional Benefit

Project	PW	Conventional B/C Ratio	Acceptable?
1	$45,659	1.193	Yes
2	20,615	1.074	Yes
3	− 43,708	0.813	No
4	− 32,427	0.788	No
5	73,538	1.330	Yes

TABLE 14–4 Salvage Value Classified as a Reduced Cost

Project	PW	Conventional B/C Ratio	Acceptable?
1	$45,659	1.203	Yes
2	20,615	1.076	Yes
3	− 43,708	0.812	No
4	− 32,427	0.786	No
5	73,538	1.334	Yes

the alternative with the maximum profit potential, which is not necessarily the alternative having the largest overall B/C ratio. Here we define "profit potential" to be synonymous with the equivalent (present, annual, future) worth of an alternative.

When comparing mutually exclusive alternatives, *they are first ranked in order of increasing total equivalent worth of costs*. (Rank-ordering can be based on equivalent annual, present, or future worth of costs.) Starting with the lowest equivalent cost alternative (often the do-nothing alternative), this alternative is compared with the next higher equivalent cost alternative. The difference (Δ) in their respective equivalent benefits and costs is used to determine the incremental benefit–cost ratio, $\Delta B/\Delta C$. If the value of this incremental ratio is greater than or equal to one, then the higher equivalent cost alternative is chosen. That alternative is then compared to the next higher equivalent cost alternative, and so on until the last alternative has been evaluated. When making a comparison, if the value of the $\Delta B/\Delta C$ ratio is less than one, the lower equivalent cost alternative is selected. As long as the incremental ratio is greater than or equal to one, the higher equivalent cost alternative should be chosen, since it represents greater profit potential (i.e., it maximizes PW, AW, and FW at the MARR).

The equations for determining the incremental benefit–cost ratio for two mutually exclusive alternatives, D and E, are

Conventional incremental B/C ratio for the cash flow increment D → E:

$$\Delta B/\Delta C = \frac{B_D \rightarrow B_E}{[CR_D + (O\&M)_D] \rightarrow [CR_E + (O\&M)_E]} \tag{14–5}$$

Modified incremental B/C ratio for the cash flow increment D → E:

$$\Delta B/\Delta C = \frac{[B_D - (O\&M)_D] \rightarrow [B_E - (O\&M)_E]}{CR_D \rightarrow CR_E} \tag{14–6}$$

In Equations 14-5 and 14-6, alternative E has the larger-valued total equivalent annual cost (although present or future equivalents could have been used). Thus, alternative E is preferred when $\Delta B/\Delta C \geq 1$. Otherwise alternative D is chosen. The following examples illustrate the incremental analysis procedure.

EXAMPLE 14-4

Consider two major mutually exclusive transportation alternatives, D and E, for which equivalent annual benefits to the public (recipients of government services) are \$22.5 million and \$35 million, respectively. The equivalent annual capital recovery (CR) costs and equivalent annual operating and maintenance (O&M) costs to the government are estimated as follows.

	Alternatives D	E
CR	\$10 million	\$10 million
O&M	5 million	15 million
Total Equivalent Annual Cost	\$15 million	\$25 million

(a) Calculate the modified B/C ratio for each alternative.
(b) Compute the conventional B/C ratio for each alternative.
(c) Which alternative would you recommend and why?

Solution

(a) The modified B/C ratio for alternative D is (\$22.5 − \$5.0)/\$10 = 1.75, and it is (\$35.0 − \$15.0)/\$10 = 2.0 for alternative E.

(b) The conventional B/C ratio for alternative D is \$22.50/(\$10 + \$5) = 1.5, and for alternative E it is \$35.0/(\$10 + \$15) = 1.4.

(c) A proper choice is *not* necessarily made by maximizing the B/C ratio with either the conventional or the modified B/C ratio. Instead, *we must examine the incremental benefits and incremental costs* to discover which alternative is better.

Modified Incremental B/C Ratio

Examine the cash flow increment D → E, where alternative D was found to be acceptable in (a) and E has the larger-valued equivalent annual cost:

$$\frac{(\$35 - \$15) - (\$22.5 - \$5)}{\$10 - \$10} = \text{not defined}$$

The modified ratio cannot be used in this case because of division by zero.

Conventional Incremental B/C Ratio

Examine D → E, where D was determined to be acceptable in (b):

$$\frac{\$35 - \$22.5}{(\$10 + \$15) - (\$5 + \$10)} = 1.25$$

Hence, the cash flow increment has a B/C ratio > 1 and E should be selected. This is confirmed by computing the AW of E (\$35 − \$25 = \$10) and comparing it with the AW of D (\$22.5 − \$15 = \$7.5). Because $AW_E > AW_D$, alternative E is the better choice.

As a point of interest, if the alternatives in Example 14-4 had been *independent* rather than mutually exclusive, the conventional ratio in part (b) would rank D better than E (1.5 > 1.4). However, the modified ratio would have ranked E better than D (2.0 > 1.75). Hence, the two B/C ratio methods can produce different rank orderings of *independent* investment alternatives.

EXAMPLE 14-5

In dealing with a certain problem, six mutually exclusive alternatives are available. The equivalent annual benefits and costs and the conventional B/C ratios are as shown in Table 14-5. Because all the B/C ratios exceed 1, it is not at once apparent which alternative should be selected. We might be tempted to select alternative B because it has the highest B/C ratio and its annual costs are much less than those of the other alternatives. On the other hand, if a project is justified when its B/C ratio exceeds 1, then we might be tempted to select alternative F because it produces the largest annual benefits. Actually, neither of these decisions would be correct. Which is the best alternative?

Solution

If the data are computed and arranged according to increasing equivalent annual cost as shown in Table 14-5, the proper decision becomes clear. In accordance with the principles set forth in Chapter 5, alternative D is the base alternative because it involves the minimum expenditure of capital. The incremental benefits and incremental costs, *compared with the acceptable alternative having the next lower annual costs*, are computed. Then the *incremental* B/C ratios are computed, as shown in the next-to-last column of Table 14-6. It will be noted that, as in the case of computing incremental rates of return, whenever an incremental B/C ratio is less than 1.0, that alternative is eliminated from further consideration, and each successive alternative is compared with the last, previous alternative having an acceptable B/C ratio.

TABLE 14–5 Equivalent Annual Benefits, Costs, and Benefit–Cost Ratios for Alternatives in Example 14-5

Alternative	Annual Benefits (\$000s)	Annual Costs (\$000s)	Conventional B/C Ratio
D	900	600	1.50
B	1,600	800	2.00
A	1,660	850	1.95
C	1,200	900	1.33
E	1,825	1,000	1.83
F	1,875	1,100	1.70

TABLE 14–6 Incremental Benefit–Cost Ratios for Alternatives in Example 14-5

Increment Considered	Incremental Benefit ($000s)	Incremental Cost ($000s)	Incremental B/C Ratio	Justified
Do Nothing → D	900	600	1.5	Yes
D → B	700	200	3.5	Yes
B → A	60	50	1.2	Yes
A → C	−460	50	Negative	No
A → E	165	150	1.1	Yes
E → F	50	100	0.5	No

Thus, in Table 14-6, alternatives C and F are not acceptable. Based on the incremental B/C ratios, and assuming capital is not limited, it is clear that alternative E should be selected because it provides maximum benefits while having a B/C ratio greater than 1.0 on each increment of cost.

EXAMPLE 14-6

Suppose that a certain government agency is considering the same set of three *independent* projects described earlier in Example 5-16. The cost ceiling for sponsoring these public projects is $200 million, and a 10% interest rate is required in such undertakings. By using a benefit–cost ratio, which project(s) should be initiated? A restatement of relevant data is as follows:

Project	Present Worth of Total Costs	Present Worth of Benefits	Useful Life, N
A	$93 million	$98.9 million	15 years
B	55 million	58.4 million	10 years
C	71 million	98.0 million	30 years

Solution

From three independent projects, a total of $2^3 = 8$ *mutually exclusive* combinations can be developed. By using the procedure described in Chapter 5 (see Example 5-14), Table 14-7, showing mutually exclusive combinations of the three independent projects, is obtained. Alternative 7 is eliminated because its PW of costs exceeds the $200 million available.

An *incremental analysis* must be performed to make the correct choice. The comparisons are as follows in order of increasing PW of total costs:

Comparison of Combinations	ΔB/ΔC	Decision
8 (do nothing) → 2	1.0634	Select combination 2
2 → 3	2.4750	Select combination 3
3 → 1	0.0409	Keep combination 3
3 → 6	1.0618	Select combination 6
6 → 4	0.0409	Keep combination 6
6 → 5	1.0658	Select combination 5

TABLE 14–7 Benefit–Cost Ratios for Combinations in Example 14-6

	Project					
Mutually Exclusive Combination	A	B	C	Present Worth of Total Costs	Present Worth of Benefits	Conventional B/C Ratio
1	1	0	0	93.0	98.9	1.0634
2	0	1	0	55.0	58.4	1.0618
3	0	0	1	71.0	98.0	1.3803
4	1	1	0	148.0	157.3	1.0628
5	1	0	1	164.0	196.9	1.2000
6	0	1	1	126.0	156.4	1.2413
7 (Eliminate)	1	1	1	219.0	255.3	1.1658
8	0	0	0	0	0	—

As a result of an incremental analysis, the decision would be to select combination 5 (projects A and C). Notice that this choice agrees with that of Example 5-16. It is also important to observe that the correct alternative *was not chosen* with the total B/C ratio computed for each. Based on maximizing the B/C ratio, alternative 3 would have been incorrectly chosen. The B/C ratio suffers from the same general drawbacks that characterize rate of return methods when used to select among mutually exclusive alternatives. Therefore, an incremental analysis is required when utilizing the B/C ratio method (and rate of return methods) to evaluate correctly mutually exclusive investment alternatives.

14.11 AFTER-THE-FACT JUSTIFICATIONS*

Many critics charge that benefit–cost analysis is a method of using numbers to justify or reject a project, depending on the views of the group paying for the analysis. In this regard, a congressional subcommittee concluded that "the most significant factor in evaluating a benefit–cost study is the name of the sponsor. Benefit–cost studies generally are formulated after basic positions on an issue are taken by the respective parties. The results of competing studies predictably reflect the respective positions of the parties on the issue." (p. 55).

An illustration of the misuse of information from benefit–cost studies is when the Bureau of Reclamation's analysis in 1967 supported the proposed Nebraska Mid-State project. The purpose of the project was to divert water from the

*James T. Campen, *Benefit, Cost, and Beyond: The Political Economy of Benefit-Cost Analysis*, Cambridge, MA: Ballinger Publishing Company, 1986. All quotes in Sections 14.11–14.13 are taken from this source.

Platte River in order to irrigate crop land. The Bureau of Reclamation computed a B/C ratio of 1.24 for the project. This ratio was based on several inappropriate assumptions, including the following four (p. 53):

1. An artificially low discount rate of 3.125% was used.
2. A project life of 100 years was used, rather than the generally more accepted and supportable assumption of 50 years.
3. The wildlife and fish benefits were counted as positive when in actuality the project would have resulted in the Platte River running dry over half of the 30-year time period that was on record (1931–60). This would destroy in-stream fishery, eliminate waterfowl habitats on 150 miles of the river, and reduce the prospects for survival of several endangered species (including the whooping crane, the sandhill crane, and the bald eagle).
4. Increased farm output was valued at support prices that would require a substantial federal subsidy.

Campen states that "the common core of these criticisms is not so much the fact that benefit–cost analysis is used to justify particular positions but that it is being used in this way at the same time that it is presented as a scientific unbiased method of analysis" (pp. 52–53). In order for an analysis to be fair and reliable, it must be performed by an unbiased group, or by a party that includes people from each of the interest groups involved. For instance when the Nebraska Mid-State project was re-evaluated by an impartial third party, the benefit-cost ratio was estimated to be 0.23. Unfortunately, analyses are generally performed by people who have already formed a strong opinion on the subject for which the analysis is to be used.

14.12 DISTRIBUTIONAL CONSIDERATIONS

Another shortcoming of benefit–cost analysis is that benefits and costs can cancel each other, showing no regard for who gains or loses from the project under consideration. Two reasons for being critical of this lack of distributional equity in benefit-cost studies are (1) public policy generally "operates to reduce economic inequality by improving the well-being of disadvantaged groups," and (2) there is little concern for the equality or inequality of people who are in the same general economic circumstances (p. 56).

Public policy is generally viewed as one method for reducing the economic inequalities for the poor, residents of underdeveloped regions, and racial minorities. Of course there are many cases where it is not possible to make distributional judgments, but there are others where it is clear. Conceivably, a project that would have adverse consequences on the less fortunate would be readily acceptable with a benefit–cost ratio $\geq$ 1 (based on total discounted costs), only because positive benefits exist for those who possibly have more influence and are better off economically.

A lack of understanding of the distributional consequences of a project may produce inequalities for individuals who are in approximately the same economic circumstances. The relevance of this inequality can be shown in the following example:

> Consider a proposal to raise property taxes by 50 percent on all properties with odd-numbered street addresses and simultaneously to lower property taxes by 50 percent on all properties with even-numbered street addresses. A conventional benefit–cost analysis of this proposal would conclude that its net benefits were approximately zero, and an analysis of its impact on the overall level of inequality in the size distribution of income would also show no significant effects. Nevertheless, such a proposal would be generally, and rightly, condemned as consisting of an arbitrary and unfair redistribution of income (p. 56).

A more realistic example is a project that would cause a chemical plant to stop the production of a particular substance because of an undesirable by-product that pollutes the water supply of a town that is 50 miles downstream. The plant may have to lay off several workers, which would be a loss, while the town downstream would have a cleaner water supply, and possibly, certain individuals would have reductions in medical bills. A benefit–cost ratio would not show the distributional consequences; it would only show the net monetary effect.

14.13 OMITTING QUALITATIVE INFORMATION

We can now appreciate that many problems exist when using unreliable monetary values for irreducible (nonmonetary) considerations in a benefit–cost analysis; but what happens when these unreliable values are completely omitted? When only easily quantifiable information is used in the analysis, the importance of irreducibles is totally neglected, and there is favoritism toward projects that have mostly monetary benefits and that are less capital intensive. Busy decision makers want a single number to summarize the acceptance or rejection of a project. No matter how heavily the irreducibles are emphasized in the discussion of a project, most managers will go to the bottom line to get a single number and will use it to make a decision. A 1980 Congressional committee concluded that "whenever some quantification is done—no matter how speculative or limited—the number tends to get into the public domain and the qualifications tend to get forgotten. . . . The number is the thing" (p. 68).

From the discussion and examples of public projects presented in this chapter, it is apparent that because of the methods of financing, the absence of tax and profit requirements, and political and social factors, the same criteria frequently cannot be applied to such works as are used in evaluating privately financed projects. Neither should public projects be used as yardsticks with which to compare private projects. Nevertheless, whenever possible, public works should be justified on an economic basis to ensure that the public obtains the maximum return from the tax money that is spent. Whether an engineer is working on such projects, is called upon to serve as a consultant, or only fills the role of a taxpayer, he or she is bound by professional ethics to do the utmost to see that these projects are carried out in the best possible manner within the limitations of the legislation enacted for their authorization.

The benefit–cost (B/C) ratio has remained a popular method for evaluating the financial performance of public projects. Both the conventional and modified B/C ratio methods have been explained and illustrated in Chapter 14. A final note of caution: The best project among a *mutually exclusive* set of projects is not necessarily the one that maximizes the conventional or the modified B/C ratio. We have seen in this chapter that an incremental analysis approach to evaluating benefits and costs is necessary to ensure the correct choice. We have also shown that the project having the largest B/C ratio among *independent* investment opportunities is the best project.

14.14 PROBLEMS

14-1.

- **a.** What is a self-liquidating project?
- **b.** What is a multiple-purpose project?
- **c.** Why is it difficult to make a logical assignment of costs in a multiple-purpose project?
- **d.** Why is it difficult to assess the efficiency of operation in most multiple-purpose projects? (14.3, 14.4)

14-2. Describe how engineering economy studies that involve public projects differ from those that are conducted by private organizations. (14.5)

14-3. Discuss some of the considerations that go into determining an interest rate to be used in evaluating alternatives in the public sector. (14.6)

14-4. Select some public works project that currently is being proposed or carried out in your area. Obtain the economic analysis for the project. Determine the following:

a. Are there any intangibles included in the benefits?
b. If there are intangibles, would the project be economically justified if the intangible benefits were omitted?
c. Would most people in the affected area agree on the values assigned to intangible benefits? (14.5)

14-5. A state Resources Development Department has proposed building a dam and hydroelectric project that will remedy a flood situation on a mountain river, generate power, provide water for irrigation and domestic use, and provide certain recreational facilities for boating and fishing. The construction costs would be

Dam	$40,000,000
Access roads	2,000,000
Power plant	4,000,000
Transmission lines	1,500,000
Fish ladders and elevators	800,000
Irrigation and water canals	3,000,000

It is proposed to finance the project by issuing 8%, tax-exempt, 40-year bonds.

It is estimated that the annual operation and maintenance costs will be $1,250,000 for the power generating and distributing facilities and $750,000 for all other portions of the project. In addition, the state will pay $400,000 annually to the county where the project is located in lieu of property taxes. Estimates of the annual benefits and revenues are as follows:

Flood control	$ 900,000
Sale of power	2,700,000
Sale of water	1,600,000
Recreation benefits	1,800,000
Income from sports concessions	200,000

a. Determine the B/C ratio for the project, using the estimated values of the benefits as stated.
b. Would the elimination of the benefits that you consider to be intangible leave the project justified economically? (14.7)

14-6. Five independent projects are available for funding by a certain public agency. The following tabulation shows the equivalent annual benefits and costs for each.

Project	Annual Benefits	Annual Costs
A	$1,800,000	$2,000,000
B	5,600,000	4,200,000
C	8,400,000	6,800,000
D	2,600,000	2,800,000
E	6,600,000	5,400,000

a. Assume that the projects are of the type for which the benefits can be determined with considerable certainty and that the agency is willing to invest money as long as the B/C ratio is at least 1. Which alternatives should be selected as best?
b. What is the rank ordering of projects, from best to worst?
c. If the projects involved intangible benefits which required considerable judgment in assigning their values, would this affect your recommendation? (14.8)

14-7. In developing a publicly owned, commercial waterfront area, three possible independent plans are being considered. Their costs and estimated benefits are as follows:

	Present Worth ($000s)	
Plan	**Costs**	**Benefits**
A	123,000	139,000
B	135,000	150,000
C	99,000	114,000

a. Which plan(s) should be adopted, if any, if the controlling board wishes to invest any amount required provided that the B/C ratio on the required investment is at least 1.0? (14.8)

b. Suppose that 10% of the costs of each plan are reclassified as "disbenefits." What percentage change in the B/C ratio of each plan results from the reclassification? (14.9)

c. Comment on why the rank orderings in (a) are unaffected by the change in (b). (14.9)

14-8. Consider the two types of equipment below and determine which is the better choice if a firm desires to invest as long as the benefit–cost ratio is greater than or equal to 1.0. The firm's MARR is 10% per year. Ignore income taxes. Assume repeatability and show all work. (14.10)

	Equipment Type	
	RS-422	**RS-511**
Initial Investment	$500	$1,750
Useful Life (years)	6	12
Market (salvage) Value	125	375
Annual Benefits	238	388
Annual O&M Costs	108	113

14-9. Consider the following mutually exclusive alternatives:

Alternative	Equivalent Annual Cost of Project	Expected Annual Flood Damage	Annual Benefits
I. No flood control	0	$100,000	0
II. Construct levees	$ 30,000	80,000	$112,000
III. Build small dam	100,000	5,000	110,000

Which alternative would be chosen according to these decision criteria:

a. Maximum benefit?

b. Minimum cost?

c. Maximum benefits minus costs?

d. Largest investment having an incremental B/C ratio larger than 1.0?

e. Largest B/C ratio?

Which project *should* be chosen? (14.10)

14-10.* A river that passes through private lands is formed from four branches of water that flow from a national forest. Some flooding occurs each year, and a major flood generally occurs every few years. If small earthen dams are placed on each of the four branches, the chances of major flooding would be practically eliminated. Construction of one or more dams would reduce the amount of flooding by varying degrees.

Other potential benefits from the dams are the reduction of damages to fire and logging roads in the forest, the value of the dammed water for protection against fires, and the increased recreational use. The following summary contains the estimated benefits and costs associated with building one or more dams.

		Costs		Benefits		
Option	Dam Sites	Construction	Annual Maintenance	Annual Flood	Annual Fire	Annual Recreation
A	1	$3,120,000	$ 52,000	$520,000	$ 52,000	$ 78,000
B	1&2	3,900,000	91,000	630,000	104,000	78,000
C	1,2,&3	7,020,000	130,000	728,000	156,000	156,000
D	1,2,3,&4	9,100,000	156,000	780,000	182,000	182,000

The equation used to calculate B/C ratios is

$$\text{B/C ratio} = \frac{\text{annual flood and fire savings} + \text{recreational benefits}}{\text{equivalent annual construction costs} + \text{maintenance}}$$

Benefits and costs are to be compared using the annual worth method with an interest rate of 8%, and useful lives of the dams are 100 years. (14.10)

a. Which of the four options would you recommend? Show why.

b. If fire benefits are reclassified as reduced costs, would the choice in (a) be affected? Show your work.

14-11. A state-sponsored Forest Management Bureau is evaluating alternative routes for a new road into a formerly inaccessible region. Three mutually exclusive plans for routing the road provide different benefits, as indicated in the following table.

Route	Construction Cost	Annual Maintenance Cost	Annual Savings in Fire Damage	Annual Recreational Benefit	Annual Timber Access Benefit
A	$185,000	$2,000	$ 5,000	$3,000	$ 500
B	220,000	3,000	7,000	6,500	1,500
C	290,000	4,000	12,000	6,000	2,800

*Fashioned after a problem from James L. Riggs, *Engineering Economics*, New York: McGraw-Hill, 1977, pp. 432–434.

The roads are assumed to have an economic life of 50 years, and the interest rate is 8%. Which route should be selected according to the B/C ratio method? (14.10)

14-12. An area on the Colorado River is subject to periodic flood damage which occurs, on the average, every 2 years and results in $2,000,000 loss. It has been proposed that the river channel should be straightened and deepened, at a cost of $2,500,000, to reduce the probable damage to not over $1,600,000 for each occurrence during a period of 20 years before it would have to again be deepened. This procedure would also involve annual expenditures of $80,000 for minimal maintenance. One legislator in the area has proposed that a better solution would be to construct a flood-control dam at a cost of $8,500,000, which would last indefinitely with annual maintenance costs of not over $50,000. He estimates that this project would reduce the probable annual flood damage to not over $450,000. In addition, this solution would provide a substantial amount of irrigation water that would produce an annual revenue of $175,000 and recreational facilities, which he estimates would be worth at least $45,000 per year to the adjacent populace. A second legislator believes that the dam should be built and that the river channel also should be straightened and deepened, noting that the total cost of $11,000,000 would reduce the probable annual flood loss to not over $350,000, while providing the same irrigation and recreational benefits. If the state's capital is worth 10%, determine the B/C ratios and the increment B/C ratio. Recommend which alternative should be adopted. (14.10)

14-13. Ten years ago the port of Secoma built a new pier containing a large amount of steel work, at a cost of $300,000, estimating that it would have a life of 50 years. The annual maintenance cost, much of it for painting and repair caused by the environment, has turned out to be unexpectedly high, averaging $27,000. The port manager has proposed to the port commission that this pier be replaced immediately with a reinforced concrete pier at an initial cost of $600,000. He assures them that this pier will have a life of at least 50 years, with annual maintenance costs of not over $2,000. He presents the following figures as justification for the replacement, having determined that the net salvage value of the existing pier is $40,000.

Annual Cost of Present Pier		Annual Cost of Proposed Pier	
Depreciation: $300,000/50	$ 6,000	Depreciation: $600,000/50	$12,000
Maintenance cost	27,000	Maintenance cost	2,000
Total	$33,000	Total	$14,000

He has stated that since the port earns a net profit of over $3,000,000 per year, the project could be financed out of annual earnings and there would thus be no interest cost, and an annual saving of $19,000 would be obtained by making the replacement.

a. Comment on the port manager's analysis.

b. Make your own analysis and recommendation regarding the proposal. (14.10)

A

ACCOUNTING AND ITS RELATIONSHIP TO ENGINEERING ECONOMY

A.1 INTRODUCTION

Engineering economy studies are made for the purpose of determining whether capital should be invested in a project or whether it should be utilized differently than it presently is being used. Engineering economy studies always deal, at least for one of the alternatives being considered, with something that currently is not being done. Economy studies thus provide information upon which investment and managerial decisions about future operations can be based. Thus, the engineer doing an economic analysis might be termed an *alternatives fortune-teller*.

After a decision to invest capital in a project has been made and the capital has been invested, those who supply and manage the capital want to know the financial results. Therefore, procedures are established so that financial events relating to the investment can be recorded and summarized and financial productivity determined. At the same time, through the use of proper financial information, controls can be established and utilized to aid in guiding the venture toward the desired financial goals. Financial accounting and cost accounting are the procedures that provide these necessary services in a business organization. Accounting studies thus are concerned with *past* and *current* financial events. Thus, the accountant might be termed a *financial historian*.

The accountant is somewhat like a data recorder in a scientific experiment. Such a recorder reads the pertinent gauges and meters and records all the essential data during the course of an experiment. From these it is possible to determine the results of the experiment and to prepare a report. Similarly, the accountant records all significant financial events connected with an investment, and from these data he or she can determine what the results have been and can prepare financial reports. By taking cognizance of what is happening during the course of an experiment and making suitable corrections, thereby gaining more information and better results from the experiment, engineers and managers must rely on accounting reports to make corrective decisions in order to improve the current and future financial performance of the business.

Accounting is generally a source of much of the past financial data needed in making estimates of future financial conditions. Accounting is also a prime source of data for *postmortem*, or after-the-fact, analyses that might be made regarding how well an investment project has turned out compared to the results that were predicted in the engineering economy study.

A proper understanding of the origins and meaning of accounting data is needed in order to properly use or not use those data in making projections into the future and in comparing actual versus predicted results.

A.2 ACCOUNTING FUNDAMENTALS

Accounting is often referred to as the language of business. Engineers should make serious efforts to learn about a firm's accounting practice so that they can better communicate with top management. This section contains an extremely brief and simplified exposition of the elements of financial accounting in recording and summarizing transactions affecting the finances of the enterprise. These fundamentals apply to any entity (such as an individual, corporation, governmental unit, etc.), called here a *firm*.

All accounting is based on the *fundamental accounting equation*, which is

$$\text{assets} = \text{liabilities} + \text{owners' equity} \tag{A–1}$$

where *assets* are those things of monetary value that the firm possesses, *liabilities* are those things of monetary value that the firm owes, and *owners' equity* is the worth of what the firm owes to its stockholders (also referred to as *equities, net worth,* etc.). For example, typical accounts in each term of Equation A-1 are as follows:

Asset Accounts	=	**Liability Accounts**	+	**Owners' Equity Accounts**
Cash		Short-term debt		
Receivables		Payables		Capital stock
Inventories		Long-term debt		
Equipment				Retained earnings
Buildings				(income retained in
Land				the firm)

The fundamental accounting equation defines the format of the *balance sheet,* which is one of the two most common accounting statements, and which shows the financial position of the firm at any given point in time.

Another important, and rather obvious, accounting relationship is

$$\text{revenues} - \text{expenses} = \text{profit (or loss)} \tag{A-2}$$

This relationship defines the format of the *income statement* (also commonly known as a *profit-and-loss statement*), which summarizes the revenue and expense results of operations *over a period of time.* Equation A-1 can be expanded to take account of profit defined in Equation A-2:

$$\begin{aligned}\text{assets} = &\text{liabilities} + (\text{beginning owners' equity} \\ &+ \text{revenue} - \text{expenses})\end{aligned} \tag{A-3}$$

Profit is the increase in money value (not to be confused with cash) that results from a firm's operations and is available for distribution to stockholders. It therefore represents the return on owners' invested capital.

A useful analogy is that a balance sheet is like a "snapshot" of the firm at an instant in time, while an income statement is a summarized "moving picture" of the firm over an interval of time. It is also useful to note that a revenue serves to increase owners' interests in a firm, while an expense serves to decrease the owners' equity amount for a firm.

To illustrate the workings of accounts in reflecting the decisions and actions of a firm, suppose that an individual decides to undertake an investment opportunity and the following sequence of events occurs over a period of 1 year:

1. Organize XYZ firm and invest \$3,000 cash as capital.
2. Purchase equipment for a total cost of \$2,000 by paying cash.
3. Borrow \$1,500 through note to bank.
4. Manufacture year's supply of inventory through the following:
   (a) Pay \$1,200 cash for labor.
   (b) Incur \$400 account payable for material.
   (c) Recognize the partial loss in value (depreciation) of the equipment amounting to \$500.

5. Sell on credit all goods produced for year, 1,000 units at $3 each. Recognize that the accounting cost of these goods is $2,100, resulting in an increase in equity (through profits) of $900.
6. Collect $2,200 of account receivable.
7. Pay $300 of account payable and $1,000 of bank note.

A simplified version of the accounting entries recording the same information in a format that reflects the effects on the fundamental accounting equation (with a "+" denoting an increase and a "−" denoting a decrease) is shown in Table A-1. A summary of results is shown in Figure A-1.

It should be noted that the profit for a period serves to increase the value of the owners' equity in the firm by that amount. Also, it is significant that the net cash flow from operation of $700 (= $2,200 − $1,200 − $300) is not the same as profit. This was recognized in transaction 4(c), in which capital consumption (depreciation) for equipment of $500 was declared. Depreciation serves to convert part of an asset into an expense which is then reflected in a firm's profits, as seen in Equation A-2. Thus, the profit was $900, or $200 more than the net cash flow. For purposes of financial accounting, revenue is recognized when it is earned and expenses are recognized when they are incurred.

One important and potentially misleading indicator of after-the-fact financial performance that can be obtained from Figure A-1 is "annual rate of return." If the invested capital is taken to be the owners' (equity) investment, the annual rate of return at the end of this particular year is $900/$3,900 = 23%.

FIGURE A-1 Balance Sheet and Income Statement Resulting from Transactions Shown in Table A-1

XYZ Firm
Balance Sheet
As of Dec. 31, 19xx

Assets		**Liabilities and Owners' Equity**	
Cash	$2,200	Bank note	$ 500
Accounts receivable	800	Accounts payable	100
Equipment	1,500	Equity	3,900
Total	$4,500	Total	$4,500

XYZ Firm
Income Statement
for Year Ending Dec. 31, 19xx

			Cash Flow
Operating revenues (Sales)		$3,000	$ 2,200
Operating costs (Inventory depleted)			
Labor	$1,200		−1,200
Material	400		− 300
Depreciation	500		0
		$2,100	
Net income (Profits)		$ 900	$ 700

TABLE A–1 Accounting Effects of Transactions—XYZ Firm

	Account	Transaction 1	2	3	4	5	6	7	Balances at End of Year	
Assets	Cash	+ $3,000	− $2,000	+ $1,500	− $1,200		+ 2,200	− 1,300	+ $2,200	$4,500
	Account receivable					+ $3,000	− 2,200		+ 800	
	Inventory				+ 2,100	− 2,100			0	
	Equipment		+ 2,000		− 500				+ 1,500	
equals										
Liabilities	Account payable				+ 400			− 300	+ 100	$4,500
	Bank note			+ 1,500				− 1,000	+ 500	
plus										
Owners' equity	Equity	+ 3,000				+ 900			+ 3,900	

Financial statements are usually most meaningful if figures are shown for 2 or more years (or other reporting periods such as quarters or months), or for two or more individuals or firms. Such comparative figures can be used to reflect trends or financial indications that are useful in enabling investors and management to determine the effectiveness of investments *after* they have been made.

A.3 COST ACCOUNTING

Cost accounting, or management accounting, is a phase of accounting that is of particular importance in engineering economic analysis because it is concerned principally with decision making and control in a firm. Consequently, cost accounting is the source of much of the cost data needed in making engineering economy studies. Modern cost accounting may satisfy any or all of the following objectives:

1. Determination of the actual cost of products or services.
2. Provision of a rational basis for pricing goods or services.
3. Provision of a means for controlling expenditures.
4. Provision of information on which operating decisions may be based and by means of which operating decisions may be evaluated.

Although the basic objectives of cost accounting are simple, the exact determination of costs usually is not. As a result, some of the procedures used are arbitrary devices that make it possible to obtain reasonably accurate answers for most cases but which may contain a considerable percentage of error in other cases, particularly with respect to the actual cash flow involved.

A.4 THE ELEMENTS OF COST

One of the first problems in cost accounting is that of determining the elements of cost that arise in the production of an article or the rendering of a service. A study of how these costs occur gives an indication of the accounting procedure that must be established to give satisfactory cost information. Also, an understanding of the procedure that is used to account for these costs makes it possible to use them more intelligently.

From an engineering and managerial viewpoint in manufacturing enterprises, it is common to consider the general elements of cost to be *direct materials*, *direct labor*, and *overhead*. Such terms as *burden* and *indirect costs* are often used synonymously with overhead, and overhead costs are often divided into several subcategories.

Ordinarily, the materials that can be conveniently and economically charged directly to the cost of the product are called direct materials. Several guiding principles are used when we decide whether a material is classified as a direct material. In general, direct materials should be readily measurable, be of the same quantity in identical products, and be used in economically significant amounts. Those materials that do not meet these criteria are classified as *indirect materials* and are a part of the charges for overhead. For example, the exact amount of glue used in making a chair would be difficult to determine. Still more difficult would be the measurement of the exact amount of coal that was used to produce the steam that generated the electricity that was used to heat the glue. Some reasonable line must be drawn beyond which no attempt is made to measure directly the material that is used for each unit of production.

Labor costs also are ordinarily divided into *direct* and *indirect* categories. Direct labor costs are those which can be conveniently and easily charged to the product or service in question. Other labor costs, such as for supervisors and material handlers, are charged as indirect labor and are thus included as part of overhead costs. It is often imperative to know what is included in direct labor and direct material cost data before attempting to use them in engineering economy studies.

In addition to indirect materials and indirect labor there are numerous other cost items that must be incurred in the production of products or the rendering of services. Property taxes must be paid; accounting and personnel departments must be maintained; buildings and equipment must be purchased and maintained; supervision must be provided. It is essential that these necessary *overhead* costs be allocated to each unit produced in proper proportion to the benefits received. Proper allocation of these overhead costs is not easy, and some factual, yet reasonably simple, method of allocation must be used.

As might be expected where solutions attempt to meet conflicting requirements, such as exist in overhead-cost allocation, the resulting procedures are empirical approximations that are accurate in some cases and less accurate in others.

There are many methods of allocating overhead costs among the products or services produced. The most commonly used methods involve allocation in proportion to direct labor cost, or direct labor hours, or direct materials cost, or sum of direct labor and direct materials cost, or machine hours. In each of these methods, it is necessary to know what the total of the overhead costs has been if postmortem costs are being determined, or to estimate what the total overhead costs will be if predicted or standard costs are being determined. In either case, *the total overhead costs will be associated with a certain level of production*. This is an important condition that should always be remembered when we are dealing with unit-cost data. They can be correct only for the conditions for which they were determined.

To illustrate one method of allocation of overhead costs, consider the method that assumes overhead is incurred in direct proportion to the cost of direct labor used. With this method the overhead rate (overhead per dollar of direct labor) and the overhead cost per unit would be

$$\text{overhead rate} = \frac{\text{total overhead in dollars for period}}{\text{direct labor in dollars for period}}$$

$$\text{overhead cost/unit} = \text{overhead rate} \times \text{direct labor cost/unit} \qquad \text{(A–4)}$$

Suppose that for a future period (say quarter) the total overhead cost is expected to be \$100,000 and the total direct labor cost is expected to be \$50,000. From this, the overhead rate = \$100,000/\$50,000 = \$2 per dollar of direct labor cost. Suppose further that for a given unit of production (or job) the direct labor cost is expected to be \$60. From Equation A-4, the overhead cost for the unit of production would be 60 × \$2 = \$120.

This method obviously is simple and easy to apply. In many cases it gives quite satisfactory results. However, in many other instances, it gives only very approximate results because some items of overhead, such as depreciation and taxes, have very little relationship to labor costs. Quite different total costs may be obtained for the same product when different procedures are used for the allocation of overhead costs. The magnitude of the difference will depend on the extent to which each method produces or fails to produce results that realistically capture the facts. The choice of an overhead allocation method should be based on what seems most reasonable under the circumstances. Several new approaches to cost management are discussed in Section A.7.

A.5 COST ACCOUNTING EXAMPLE

This relatively simple example involves a job order system in which costs are assigned to work by job number. Schematically, this process is illustrated in the following diagram.

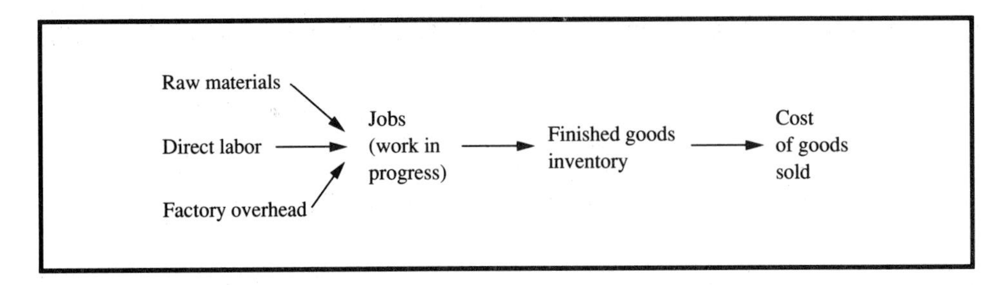

Costs are assigned to jobs in this manner:

1. Raw materials attach to jobs via material requisitions.
2. Direct labor attaches to jobs via direct labor tickets.
3. Overhead cannot be attached to jobs directly, but must have an allocation procedure that relates it to one of the resource factors, such as direct labor, which is already accumulated by the job.

Consider how an order for 100 tennis rackets accumulates costs at the Bowling Sporting Good Company:

Job #161	100 tennis rackets
Labor Rate	$7 per hour
Leather	50 yards at $2 per yard
Gut	300 yards at $0.50 per yard
Graphite	180 pounds at $3 per pound
Labor hours for the job	200 hours
Total annual factory overhead costs	$600,000
Total annual direct labor hours	200,000 hours

The three major costs are now attached to the job. Direct labor and material costs are straightforward:

Job #161		
Direct labor:	200 × 7 =	$1,400
Direct material:	leather: 50 × $2 =	100
	gut: 300 × $0.5 =	150
	graphite: 180 × $3 =	540
Direct (prime) costs		$2,190

Notice that this is not the total cost. We must somehow find a way to attach (allocate) factory costs that cannot be directly identified to the job but are nevertheless involved in producing the 100 rackets. Costs such as the power to run the graphite molding machine, the depreciation on this machine, the depreciation of the factory building, and the plant manager's salary constitute overhead for this company. These overhead costs are part of the cost structure of the 100 rackets but cannot be directly traced to the job. For instance, do we really know how much machine obsolescence is attributable to the 100 rackets? Probably not. Therefore, we must allocate these overhead costs to the 100 rackets using the overhead rate determined as follows:

$$\text{Overhead rate} = \frac{\$600{,}000}{200{,}000} = \$3 \text{ per direct labor hour}$$

This means that \$600 (\$3 × 200) of the total annual overhead cost of \$600,000 would be allocated to Job #161. Thus, the total cost of Job #161 would be

Direct labor	\$1,400
Direct materials	790
Factory overhead	600
	\$2,790

The cost of manufacturing each racket is thus \$27.90.

A.6 THE USE OF ACCOUNTING COSTS IN ENGINEERING ECONOMY STUDIES

When we recognize that accounting costs are linked to a definite set of conditions, and that they are the result of certain arbitrary decisions concerning the allocation of overhead costs, it is apparent that they should not be used without modification in cases where the conditions are different from those for which they were determined. Engineering economy studies invariably deal with situations that now are *not* being done. Thus, ordinary accounting costs normally cannot be used without modification in these economy studies. However, if we understand how the accounting costs were determined, we should be able to break them down into their component elements, and then we often find that these cost elements will supply much of the cost information that is needed for an engineering economy study. Thus, an understanding of the basic objectives and procedures of cost accounting will enable the engineer doing an economic analysis to make best use of available cost information and to avoid needless work and serious mistakes.

One should not assume that the figures contained in accounting reports are absolutely correct and indicative, even though they have been prepared with the utmost care by highly professional accountants. This is because accounting procedures often must include certain assumptions that are based on subjective judgment or the current tax laws. For example, the years of life on which depreciation expense for a particular asset is based has to be determined or assumed and the estimate may turn out to have caused unrealistic depreciation expenses and book values in accounting reports. Also, there are many accepted practices in accounting that may provide unrealistic information for management control purposes. For example, the net book value of an asset is generally declared in the balance sheet at the original price (cost basis) minus any accumulated depreciation, even though it may be recognized that the true value of the asset at a particular time is far above or below this reported book value. Today's accepted practices are giving way to a new generation of cost accounting procedures, as described in the next section.

A.7 BRINGING COST MANAGEMENT UP TO DATE[1]

Today's increasingly rapid technological advancement is changing the basis of competition. It is more important than ever that companies transform their manufacturing facilities into modern, automated plants capable of producing superior products at competitive cost and with reliable production schedules. Many firms have recognized the importance of manufacturing as a competitive weapon and are beginning to incorporate advanced manufacturing technologies such as just-in-time (JIT), robotics, computer-aided design, and flexible manufacturing systems into their facilities. While the impact of technological change has been felt in the manufacturing process, it has not yet been addressed by cost accounting.

One consequence of the new technologies is a significant restructuring of manufacturing cost patterns—the direct labor and inventory components of product cost are decreasing while depreciation, engineering, and data processing costs are increasing. In fact, the direct labor component of product cost now accounts for only 8–12% of manufacturing cost in many industries. The substitution of technology for direct labor is illustrated in Figure A-2.

Technology is not, of course, a new phenomenon. Since the dawn of the Industrial Revolution, manufacturing has developed increasingly sophisticated manufacturing processes (hardware). What is different today is that technology is a significant component of manufacturing cost. In many industries, it exceeds the traditional prime cost of direct labor. A second factor is the growing role of software. Software facilitates the transfer of manufacturing knowledge from the worker to the technology. From a cost accounting perspective, this is a critical change, because the cost behavior patterns of automated manufacturing will be radically different than those of traditional manufacturing. The laborer and the manufacturing process are not longer synchronous. The technology controls the pace of manufacturing; the laborer assists and monitors. Simply stated, the problem is that cost management practices have not kept pace with the new manufacturing environment.

Manufacturing companies are experiencing increasing problems with determining product cost, cost justifying advanced manufacturing technologies, and monitoring the efficiency and effectiveness of automated plants. A survey of US manufacturers by Computer Aided Manufacturing-International (CAM-I) of Arlington, TX, and the National Association of Accountants (NAA) of Montvale, NJ, confirmed that the most common cost accounting practice for determining the overhead component of product cost is to utilize labor-based burden rates. Because product cost techniques do not mirror the manufacturing process, they obscure accurate product cost.

Today's reported product cost is highly distorted by high overhead rates inflated by many costs that should be traceable to the product, rather than arbitrarily allocated. Inaccurate product cost information often results in incorrect management decisions.

Assume a company uses a traditional cost accounting system that applies overhead based on direct labor dollars (Figure A-3). The product cost is computed as $550 with a sales price of $660, resulting in a reported net profit of $110 per unit.

The manufacture of the product involves a significant number of automated processes, however. Automated production requires a heavy investment in depreciation, software,

[1] Source: James A. Brimson, "Bringing Cost Management Up to Date, *Manufacturing Engineering*, June 1988, pp. 49–51. Reproduced with permission of the Society of Manufacturing Engineers, Dearborn, MI.

FIGURE A-2

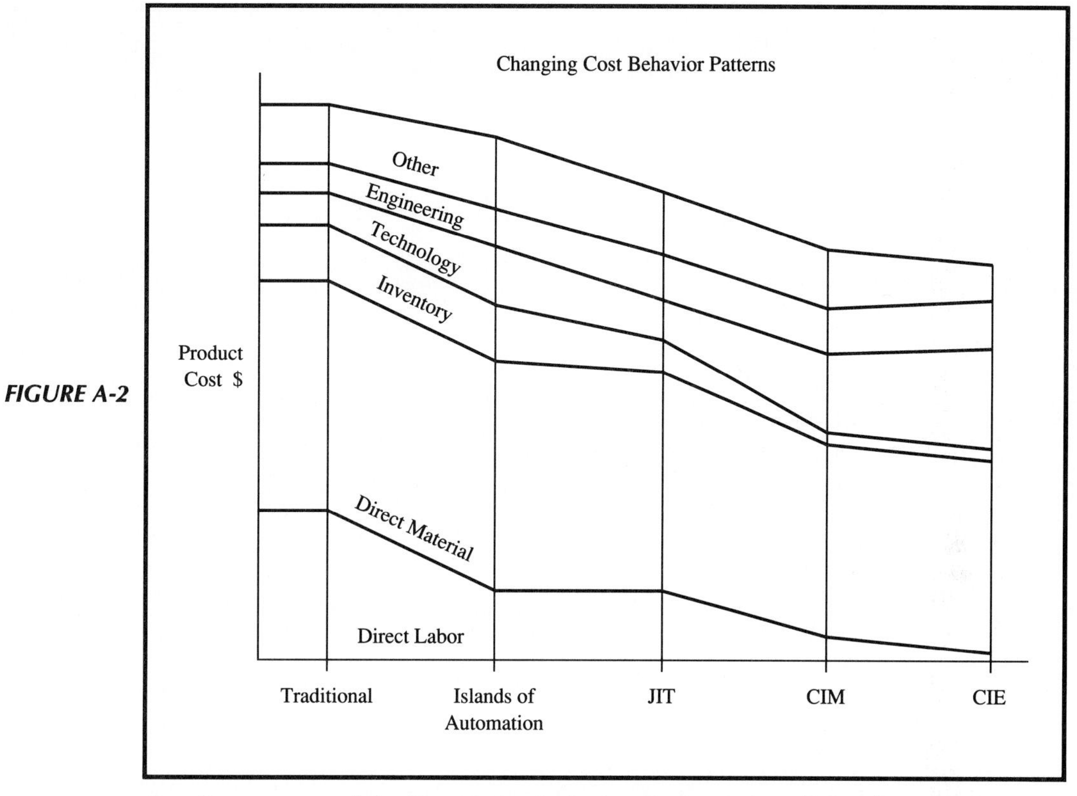

One Consequence of the New Technologies is a Restructuring of Manufacturing Cost Patterns

FIGURE A-3

Traditional Cost Accounting

Sales price		$660
Direct labor	$ 50	
Direct material	300	
Overhead	200	
Total production cost		550
Net profit		$110

A Traditional Cost Accounting System Applies Overhead Based on Direct Labor Dollars

FIGURE A-4

New Cost Accounting			
Sales			$660
Cost			
Directly traceable			
Direct labor	$ 50		
Direct material	300		
Technology	200		
Scrap and rework	50		
Imputed cost			
Raw materials inventory	20		
WIP inventory	60		
Other direct cost	90		
		770	
Nontraceable overhead		155	
Total cost			925
Net loss			($265)

New Cost Management Techniques, which Consider the Ramifications of Automation, Produce Different Financial Results

and maintenance support. An analysis of these costs and others traced to the product results in a completely different financial outcome (Figure A-4). The new product cost after the analysis was $925. With a sales price of $660, the company was losing $265 on each unit produced.

A primary reason for this distortion is that the overhead rate is inflated by potentially traceable direct costs. The inflated overhead cost is then allocated to the products using a direct labor base. For this allocation method to be correct, there must be a direct relation between labor and technology. In fact, this relation is diametric rather than complementary. In other words, the reported product cost is based on a choice of accounting methods that do not mirror the manufacturing process.

Existing cost accounting systems and cost management practices do not adequately support the objectives of advanced manufacturing for the following reasons:

- They do not adequately trace costs to products and processes.
- They do not isolate the cost of unnecessary activities. These create waste rather than customer-perceived value.
- They do not penalize overproduction. In fact, the philosophy of overhead absorption based on production volumes encourages companies to keep utilization in excess of demand.
- They do not adequately identify and report the cost of quality deficiencies in products or processes.

- They focus on controlling the production process while significant costs determined at the design and development phases of a product's life cycle are inadequately identified.
- They employ performance measurements that often conflict with strategic manufacturing objectives.
- They cannot adequately evaluate the importance of nonfinancial measures such as quality, throughput, and flexibility.
- They do not support the justification of new investments in advanced manufacturing technology and fail to monitor the benefits obtained.

The time for change is long overdue, but—until recently—new cost management concepts did not exist. While some good ideas had been suggested, much work still had to be done before an adequate conceptual foundation was developed. This was the genesis of CAM-I's Cost Management System (CMS) project.

In early 1986, CAM-I formed a consortium of industrial organizations, professional accounting firms, universities, and government agencies for the purpose of defining the role of cost management in today's manufacturing environment. The goal of the coalition was to provide an international forum where cost management experts could share ideas and experiences, and to provide a mechanism for consolidating the knowledge of practices that had proven successful in an automated environment. Through this interaction, the group established unified approaches to cost management and encouraged implementation of new ideas. The consortium identified three phases for the CMS project, each to last approximately one year:

Phase I (1986)—Conceptual Design: a collection of cost management principles applicable to a modern manufacturing environment

Phase II (1987)—Systems Design: further refinement of cost management principles using a systems design methodology

Phase III (1988)—Implementation: proof of concepts and development of implementation strategies.

Key concepts developed in Phase I of the CMS project that differentiate a cost management system from traditional cost accounting systems include the following:

Activity accounting. A company accomplishes its goals and objectives through its selection of activities. Activities, in turn, consume a firm's resources (labor, machine, travel, and so forth). To control cost and improve product costing, a firm must control and monitor its activities. For example, a customer order will trigger a series of activities ranging from marketing through purchasing, production control, manufacturing, and shipping. All of these activities are required to make a product and deliver it to the customer. These activities can, and should, be assigned directly to the product.

Cost drivers. A cost driver is an event that affects the cost/performance of a group of related activities. To control cost, it is important to understand how these activities relate. By controlling the cost driver, unnecessary costs can be eliminated.

Direct traceability. Direct traceability involves attributing costs to those products or processes that consume the resources. Many costs that are today included in overhead could be cost effectively traced to products. Direct traceability would provide more accurate product cost.

Non-value-added costs. Companies should distinguish between non-value-added and value-added costs. Recent experiences of manufacturers have reinforced the idea that successful companies establish programs for continual improvement and elimination of waste. In a manufacturing process there are many activities that add no value as perceived by the consumer. Examples of these non-value-added activities are scrap and rework, inventory holding costs, and machine down-time. These activities add unnecessary cost to the product. Identifying the cost of non-value-added activities provides visibility for their reduction and elimination.

Technology accounting. Traditional cost accounting practices identify touch labor and material as the prime direct costs. Technology costs, which are defined as plant and equipment and information systems, are buried in the third cost element—overhead. Technology is currently addressed through depreciation methodologies less comprehensive than what is needed for companies to remain competitive in today's technology-dominant environment. The objective of technology accounting is to charge technology costs to products directly.

Impact of time on cost. Companies should charge the imputed cost of holding assets to the product cost. Holding assets represents an important non-value-added cost. These assets must be financed through internal cash or external debt and equity. Traditionally, the cost associated with holding assets has been either buried in overhead or ignored on the financial reports. This cost can be calculated as an imputed cost for management reporting purposes.

Life cycle management. Studies have shown that 70–90% of a product's cost is determined prior to the production stage, but accounting systems focus almost exclusively on physical production costs. Companies should monitor the profitability (revenue and costs) of products and manufacturing processes throughout their life cycles. Today, companies report "period" performance on a monthly and yearly basis. The long-term performance of products and manufacturing processes is not typically monitored. Life cycle reporting can be an important step in focusing management's attention on long-term profitability.

Integrated performance measurement. As companies implement modern manufacturing techniques such as JIT and Total Quality Control (TQC), the importance of physical measures (lead time, quality statistics, flexibility, and so forth) increases. Today, each of these measures is independently calculated, and there is no objective criterion for ranking relative importance. It should be the role of cost management systems to translate physical measures into financial terms, and vice versa.

Integrated investment management. Traditional methods of investment justification are inadequate for justifying investment in advanced manufacturing technologies. Using traditional accounting data, benefits have been elusive and cost estimates often unrealistic. Cost management systems must develop more effective tools for evaluating the impact of new technology. Additionally, the output of the investment decision represents plans and budgets that should be linked to the control systems to help ensure that the benefits are achieved.

Target costing. Target cost is a market-based cost that is necessary to capture a predetermined market share. It is typically less than manufacturing standard cost. To achieve target cost, a company's cost management system must identify cost-reduction opportunities.

During 1987, the CMS project focused on a cost management systems design. Based on the key concepts of the CMS conceptual design, the Phase II goal was to provide the general design and architecture of a cost management system for an advanced manufacturing company. The expected results of the transition from labor-intensive manufacturing to automated production were addressed. Sponsoring companies will use this design as the basis for creating a cost management system to meet their own unique requirements.

Additionally, several concepts from the design were expanded through the development of research papers. Recognized academicians from the US, Europe, and Japan addressed cost management topics such as target costing, the impact of flexible manufacturing on cost accounting, and cost drivers and their causal relationships.

The objective of CMS Phase III is to prove the concepts developed by sponsor companies. A study of these implementations will allow the concepts to be validated and modified as necessary. This approach allows all sponsors to benefit from the validation process without incurring the risk of implementation. In addition, the experiences and tools developed will provide more cost-effective future implementations.

During 1988, the CMS research agenda will continue to fill knowledge voids discovered during Phases I and II. Such issues as how to measure and evaluate manufacturing flexibility, how to effectively manage capacity, and how to portray the technological health of a manufacturing facility will be addressed.

Today's competitive environment dictates that accounting must move from a passive after-the-fact reporting system to a tool that proactively helps companies manage their business. Manufacturing excellence increasingly depends on understanding economic cause and effect. Current cost accounting systems do not reflect these factors. Companies that implement new cost management concepts will have a valuable tool to assist management in formulating decisions.

B

GLOSSARY OF COMMONLY USED SYMBOLS AND TERMINOLOGY

ECONOMIC ANALYSIS METHODS AND COSTS

AC	Annual cost
AW	Annual worth
AW-C	Annual worth–cost (same as AC)
CR	Capital recovery cost (annual cost of depreciation plus interest on investment)
E_k	Expenses for kth compounding period
ERR	External rate of return
FW	Future worth
IRR	Internal rate of return
MARR	Minimum attractive rate of return
PW	Present worth
PW-C	Present worth–cost
R_k	Receipts (revenues) for kth compounding period
S_N	Salvage value during compounding period N

COMMONLY USED SYMBOLS

ϵ	Reinvestment rate per interest period (as used for ERR method)
e_j'	Differential price inflation rate per interest period above or below the general price inflation rate for good or service j
e_j	Total price escalation rate per interest period for good or service j
f	General price inflation rate per interest period
I	Investment cost of an asset (or activity)
i	Effective interest rate per interest period
i_c	Effective combined interest rate per interest period
i_r	Effective real interest rate per interest period
i'	Interest or rate of return (to be determined)
k	Index for compounding periods
r	Nominal interest rate per year
$\underline{r}$	Nominal interest rate per year, compounded continuously
N	Number of compounding periods
P	Present sum of money (present worth); the equivalent worth of one or more cash flows at a relative point in time called the present
F	Future sum of money (future worth); the equivalent worth of one or more cash flows at a relative point in time called the future
A	End-of-period cash flows (or equivalent end-of-period values) in a uniform series continuing for a specified number of periods
G	Uniform period-by-period increase or decrease in cash flows or amounts (the arithmetic gradient)
$\bar{A}$	Amount of money (or equivalent value) flowing continuously and uniformly during each and every period continuing for a specific number of periods

Functional Forms (Symbols) for Compound Interest Factors

Functional Format (Symbol)	Name of Factor
All Cash Flows Discrete: End-of-Period Compounding	
$(F/P, i\%, N)$	Compound amount factor (single payment)
$(P/F, i\%, N)$	Present worth factor (single payment)
$(A/F, i\%, N)$	Sinking fund factor
$(A/P, i\%, N)$	Capital recovery factor
$(F/A, i\%, N)$	Compound amount factor (uniform series)
$(P/A, i\%, N)$	Present worth factor (uniform series)
$(A/G, i\%, N)$	Arithmetic gradient conversion factor (to uniform series)
$(P/G, i\%, N)$	Arithmetic gradient conversion factor (to present value)
All Cash Flows Discrete: Continuous Compounding	
$(F/P, \underline{r}\%, N)$	Continuous compounding: compound amount factor (single payment)
$(P/F, \underline{r}\%, N)$	Continuous compounding: present worth factor (single payment)
$(A/F, \underline{r}\%, N)$	Continuous compounding: sinking fund factor
$(A/P, \underline{r}\%, N)$	Continuous compounding: capital recovery factor
$(F/A, \underline{r}\%, N)$	Continuous compounding: compound amount factor (uniform series)
$(P/A, \underline{r}\%, N)$	Continuous compounding: present worth factor (uniform series)
Continuous, Uniform Cash Flows: Continuous Compounding (Payments During a Continuous Series of Periods)	
$(\bar{A}/F, \underline{r}\%, N)$	Continuous compounding: sinking fund factor (continuous uniform payments)
$(\bar{A}/P, \underline{r}\%, N)$	Continuous compounding: capital recovery factor (continuous, uniform payments)
$(F/\bar{A}, \underline{r}\%, N)$	Continuous compounding: compound amount factor (continuous, uniform payments)
$(P/\bar{A}, \underline{r}\%, N)$	Continuous compounding: present worth factor (continuous, uniform payments)

Source: American National Standard Publication ANSI Z94.5–1972, published by the American Society of Mechanical Engineers, New York, except that use of $\underline{r}$ is added for continuous compounding.

TECHNICAL TERMS USED IN ENGINEERING ECONOMY*

AMORTIZATION—(a)(1) As applied to a capital asset, the distribution of the initial cost by periodic charges to operations as in depreciation. Most property applies with indefinite life; (2) the reduction of a debt by either periodic or irregular payments; (b) a plan to pay off a financial obligation according to some prearranged program.

*Adapted from *American National Standard Publication ANSI Z94.5–1972,* published by the American Society of Mechanical Engineers, New York.

ANNUAL EQUIVALENT—(a) In *Time Value of Money* (qv), a uniform annual amount for a prescribed number of years that is equivalent in value to the present worth of any sequence of financial events for a given interest rate; (b) one of a sequence of equal end-of-year payments which would have the same financial effect when interest is considered as another payment or sequence of payments which are not necessarily equal in amount or equally spaced in time.

ANNUITY—(a) An amount of money payable to a beneficiary at regular intervals for a prescribed period of time out of a fund reserved for that purpose; (b) a series of equal payments occurring at equal periods of time.

ANNUITY FUND—A fund that is reserved for payment of annuities. The present worth of funds required to support future annuity payments.

ASSET—An accounting term for capital owned by a company.

BOOK VALUE—The recorded current value of an asset. First cost less accumulated depreciation, amortization, or depletion; (b) original cost of an asset less the accumulated depreciation; (c) the worth of a property as shown on the accounting records of a company. It is ordinarily taken to mean the original cost of the property less the amounts that have been charged as depreciation expense.

BREAKEVEN CHART—A graphic representation of the relation between total income and total costs for various levels of production and sales indicating areas of profit and loss.

BREAKEVEN POINT(S)—(a)(1) In business operations, the rate of operations, output, or sales at which income is sufficient to equal operating cost, or operating cost plus additional obligations that may be specified; (2) the operating condition, such as output, at which two alternatives are equal in economy. (b) The percentage of capacity operation of a manufacturing plant at which income will just cover expenses.

CAPACITY FACTOR—(a) The ratio of average load to maximum capacity; (b) the ratio between average load and the total capacity of the apparatus, which is the optimum load; (c) the ratio of the average actual use to the available capacity.

CAPITAL—(a) The financial resources involved in establishing and sustaining an enterprise or project (see *Investment* and *Working Capital*); (b) a term describing wealth that may be utilized to economic advantage. The form that this wealth takes may be as cash, land, equipment, patents, raw materials, finished product, etc.

CAPITAL RECOVERY—(a) Charging periodically to operations amounts that will ultimately equal the amount of capital expenditure (see *Amortization*, *Depletion*, and *Depreciation*); (b) the replacement of the original cost of an asset plus interest; (c) the process of regaining the net investment in a project by means of revenue in excess of the costs from the project. (Usually implies amortization of principal plus interest on the diminishing unrecovered balance.)

CAPITAL RECOVERY FACTOR—A factor used to calculate the sum of money required at the end of each of a series of periods to regain the net investment of a project plus the compounded interest on the unrecovered balance.

CAPITALIZED COST—(a) The present worth of a uniform series of periodic costs that continue for an indefinitely long time (hypothetically infinite). Not to be confused with a capitalized expenditure; (b) the value at the purchase date of the first life of the asset of all expenditures to be made in reference to this asset over an infinite period of time. This cost can also be regarded as the sum of capital which, if invested in a fund earning a stipulated interest rate, will be sufficient to provide for all payments required to maintain the asset in perpetual service.

CASH FLOW—(a) The flowback of profit plus depreciation from a given project; (b) the actual dollars passing into and out of the treasury of a financial venture.

COMMON COSTS—Costs that cannot be identified with a given output of products, operations, or services.

COMPOUND AMOUNT—The future worth of a sum invested (or loaned) at compound interest.

COMBINED INTEREST RATE (i_c)—The cost of capital, expressed as an effective rate (percent) per interest period, including a market adjustment for the anticipated general price inflation rate in the economy. Thus, it represents the time value change in future cash flows and takes into account both the potential real earning power of money and the estimated general price inflation in the economy.

COMPOUND AMOUNT FACTOR—(a) The function of interest rate and time that determines the compound amount from a stated initial sum; (b) a factor which when multiplied by the single sum or uniform series of payments will give the future worth at compound interest of such single sum or series.

COMPOUND INTEREST—(a) The type of interest that is periodically added to the amount of investment (or loan) so that subsequent interest is based on the cumulative amount; (b) the interest charges under the condition that interest is charged on any previous interest earned in any time period, as well as on the principal.

COMPOUNDING, CONTINUOUS—(a) A compound interest situation in which the compounding period is zero and the number of periods infinitely great. A mathematical concept that is practical for dealing with frequent compounding and small interest rates; (b) a mathematical procedure for evaluating compound interest factors based on a continuous interest function rather than discrete interest periods.

COMPOUNDING PERIOD—The time interval between dates at which interest is paid and added to the amount of an investment or loan. Designates frequency of compounding.

DECISION MAKING—A program of action undertaken as a result of established policy to influence the final decision.

DECISIONS UNDER CERTAINTY—Simple decisions that assume complete information and no uncertainty connected with the analysis of the decisions.

DECISIONS UNDER RISK—A decision problem in which the analyst elects to consider several possible futures, the probabilities of which can be estimated.

DECISIONS UNDER UNCERTAINTY—A decision for which the analyst elects to consider several possible futures, the probabilities of which *cannot* be estimated.

DECLINING BALANCE DEPRECIATION—Also known as *percent on diminishing value*. A method of computing depreciation in which the annual charge is a fixed percentage of the depreciated book value at the beginning of the year to which the depreciation applies.

DEMAND FACTOR—(a) The ratio of the maximum instantaneous production rate to the production rate for which the equipment was designed; (b) the ratio between the maximum power demand and the total connected load of the system.

DEPLETION—(a) A form of capital recovery applicable to extractive property (e.g., mines). Can be a unit-of-output basis the same as straight-line depreciation related to original or current appraisal of extent and value of deposit (known as *cost depletion*). Can also be a percentage of income received from extractions (known as *percentage depletion*); (b) a lessening of the value of an asset due to a decrease in the quantity available. It is similar to depreciation except that it refers to such natural resources as coal, oil, and timber in forests.

DEPRECIATED BOOK VALUE—The first cost of the capitalized asset minus the accumulation of annual depreciation cost charges.

DEPRECIATION—(a)(1) Decline in value of a capitalized asset; (2) a form of capital recovery applicable to a property with two or more years' life span, in which an appropriate portion of the asset's value is periodically charged to current operations; (b) the loss of value because of obsolescence or due to attrition. In accounting, depreciation is the allocation of this loss of value according to some plan.

DIFFERENTIAL PRICE INFLATION RATE (e_j')—The percentage increment of price change (in the unit price, or cost for a fixed amount), above or below the general price inflation rate, during a time period for good or service j. (The subscript j is used to label different goods and services.)

DISCOUNTED CASH FLOW—(a) The present worth of a sequence in time of sums of money when the sequence is considered as a flow of cash into and/or out of an economic unit; (b) an investment analysis that compares the present worth of projected receipts and disbursements occurring at designated future times in order to estimate the rate of return from the investment or project.

EARNING VALUE—The present worth of an income producer's probable future net earnings, as prognosticated on the basis of recent and present expense and earnings and the business outlook.

ECONOMIC GOOD—Anything that is useful, transferable, and not abundant.

ECONOMIC MODEL—A mathematical expression or tabular relation that expresses the interaction of technical and economic variables applying to a specific problem.

ECONOMIC RETURN—The profit derived from a project or business enterprise without consideration of obligations to financial contributors and claims of others based on profit.

ECONOMY—The cost or profit situation regarding a practical enterprise or project, as in *economy study, engineering economy, engineering economic analysis.*

EFFECTIVE INTEREST—The true value of interest rate computed by equations for compound interest for a 1-year period.

ENDOWMENT—A fund established for the support of some project or succession of donations or financial obligations.

ENGINEERING ECONOMY—(a) The application of engineering or mathematical analysis and synthesis to economic decisions; (b) a body of knowledge and techniques concerned with the evaluation of the worth of commodities and services relative to their cost; (c) the economic analysis of engineering alternatives.

ESTIMATE—The true magnitude as closely as it can be determined by the exercise of sound judgment based on approximate computations; not to be confused with offhand approximations that are little better than outright guesses.

EXPECTED RETURN—The profit anticipated from a venture.

EXPECTED YIELD—The ratio expected return/investment, usually expressed as a percentage on an annual basis.

EXPENSE—Synonymous with cost for purposes of this book.

FIRST COST—The initial cost of a capitalized property, including transportation, installation, preparation for service, and other related initial expenditures.

FUTURE WORTH—(a) The equivalent value at a designated future date based on *time value of money*; (b) the monetary sum, at a given future time, which is equivalent to one or more sums at given earlier times when interest is compounded at a given rate.

GENERAL PRICE INFLATION RATE (f)—A general measure of the change in the purchasing power of a dollar during a specified period of time. The general price inflation rate is defined by a selected, and broadly based, index of market price changes. We define f to be the compound general price inflation rate per period based on the index selected or developed. In engineering economic analysis, the rate is projected for a future time interval and usually expressed as an effective annual rate. Many large organizations have their own index of general price inflation that reflects the particular business environment in which they operate.

GOING-CONCERN VALUE—The difference between the value of a property as it stands possessed of its going elements and the value of the property alone as it would stand at completion of construction as a bare or inert assembly of physical parts.

INCREMENTAL COST—The additional cost that will be incurred as the result of increasing the output one more unit. Conversely, it can be defined as the cost that will not be incurred if the output is reduced one unit. More technically, it is the variation in output resulting from a unit change in input. It is known as the marginal cost.

INCREMENTAL REVENUE—The additional revenue resulting from the sale of one more item.

IN-PLACE VALUE—A value of a physical property—market value plus costs of transportation to site and installation.

INTANGIBLES—(a) In engineering economy studies, conditions or economy factors that cannot be readily evaluated in *quantitative* terms as in money; (b) in accounting, the assets that cannot be reliably evaluated (e.g., *Aesthetics*).

INTEREST —(a)(1) Financial share in a project or enterprise; (2) periodic compensation for the lending of money; (3) in economy study, synonymous with *required return*, expected profit, or *charge* for the use of capital. (b) The cost for the use of capital. Sometimes referred to as the *Time Value of Money*.

INTEREST RATE—The ratio of the interest payment to the principal for a given unit of time; usually expressed as a percentage of the principal.

INTEREST RATE, EFFECTIVE—An interest rate for a stated period (per year unless otherwise specified) that is the equivalent of a smaller rate of interest that is more frequently compounded.

INTEREST RATE, NOMINAL—The customary type of interest rate designation on an annual basis without consideration of compounding periods. The usual basis for computing periodic interest payments.

INVESTMENT—(a) As applied to an enterprise as a whole, the cost (or present value) of all the properties and funds necessary to establish and maintain the enterprise as a going concern. The *Capital* tied up in the enterprise or project. (b) Any expenditure that has substantial and enduring value (at least two years' anticipated life) and is therefore capitalized.

INVESTOR'S METHOD—(see *Discounted Cash Flow*)

IRREDUCIBLES—(a) A term that may be used for the class of intangible conditions or economy factors that can only be *qualitatively* appraised (e.g., ethical considerations); (b) matters that cannot readily be reduced to estimated money receipts and expenses.

LIFE—(a) Economic: that period of time after which a machine or facility should be discarded or replaced because of its excessive costs or reduced profitability. The economic impairment may be absolute or relative. (b) Physical: that period of time after which a machine or facility can no longer be repaired in order to perform its design function properly. (c) Service: the period of time that a machine or facility will satisfactorily perform its function without major overhaul.

LIFE CYCLE—The time span that begins with the identification of the economic need or want (the requirement) for a product, structure, system, or service, and ends with retirement and disposal activities.

LIFE CYCLE COST—A summation of all the costs, both recurring and nonrecurring, related to a product, structure, system, or service during its life span (cycle).

LOAD FACTOR—(a) A ratio that applies to physical plant or equipment: average load/maximum demand, usually expressed as a percentage. Equivalent to percent of capacity operation if facilities just accommodate the maximum demand; (b) the ratio of average load to maximum load.

MARKET VALUE—(see *Salvage Value*)

MARGINAL ANALYSIS—An economic concept concerned with those elements of costs and revenue associated directly with a specific course of action, normally using available current costs and revenue as a base, and usually independent of traditional accounting allocation procedures.

MARGINAL COST—(a) The cost of one additional unit of production, activity, or service; (b) the rate of change of cost with production or output.

MATHESON FORMULA—A title for the formula used for *Declining Balance Depreciation* (qv).

NOMINAL INTEREST—The number employed to describe the annual percentage rate on a loan.

OBSOLESCENCE—(a) The condition of being out of date. A loss of value occasioned by new developments that place the older property at a competitive disadvantage. A factor in depreciation; (b) a decrease in the value of an asset brought about by the development of new and more economical methods, processes, and/or machinery; (c) the loss of usefulness or worth of a product or facility as a result of the appearance of better and/or more economical products, methods, or facilities.

PAYOFF PERIOD—(a) Regarding an investment, the number of years (or months) required for the related profit or savings in operating cost to equal the amount of said investment; (b) the period of time at which a machine, facility, or other investment has produced sufficient net revenue to recover its investment costs.

PERPETUAL ENDOWMENT—An endowment with hypothetically infinite life. (see *Capitalized Cost* and *Endowment*.)

PRESENT WORTH—(a) The equivalent value at the present, based on *Time Value of Money*; (b)(1) the monetary sum equivalent to a future sum(s) when interest is compounded at a given rate; (2) the discounted value of future sums.

PRESENT WORTH FACTOR—(a) A mathematical expression also known as the present value of an annuity of one; (b) one of a set of mathematical formulas used to facilitate calculation of present worth in economic analyses involving compound interest.

PROFITABILITY INDEX—The rate of return in an economy study or investment decision when calculated by the *Discounted Cash Flow Method* or the *Investor's Method* (qv).

RATE OF RETURN—(a) The interest rate at which the present worth of the cash flows on a project is zero; (b) the interest rate earned by an investment.

REAL INTEREST RATE (i_r)—The cost of capital, expressed as an effective rate (percent) per interest period, not including a market adjustment for the anticipated general price inflation rate in the economy. It represents the time value change in future cash flows based only on the potential real earning power of money.

REPLACEMENT POLICY—A set of decision rules (usually optimal) for the replacement of facilities that wear out, deteriorate, or fail over a period of time. Replacement models are generally concerned with weighing the increasing operating costs (and possibly decreasing revenues) associated with aging equipment against the net proceeds from alternative equipment.

REPLACEMENT STUDY—An economic analysis involving the comparison of an existing facility and a facility proposed to supplant the existing facility.

REQUIRED RETURN—The *minimum* return or profit necessary to justify an investment. Often termed *interest*, *expected* return or profit, or *charge for the use of capital*. It is the minimum acceptable percentage, no more and no less.

REQUIRED YIELD—The ratio of required return over amount of investment, usually expressed as a percentage on an annual basis.

RETIREMENT OF DEBT—The termination of a debt obligation by appropriate settlement with lender—understood to be in full amount unless partial settlement is specified.

SALVAGE VALUE—(a) The cost recovered or which could be recovered from a used property when removed, sold, or scrapped. A factor in appraisal of property value and in computing depreciation; (b) the market value of a machine or facility at any point in time. Normally, an estimate of an asset's net market value at the end of its estimated life.

SENSITIVITY—The relative magnitude of the change in one or more elements of an engineering economy problem that will reverse a decision among alternatives.

SIMPLE INTEREST—(a) Interest that is not compounded—is not added to the income-producing investment or loan; (b) the interest charges under the condition that interest at any time is charged only on the principal.

SINKING FUND—(a) A fund accumulated by periodic deposits and reserved exclusively for a specific purpose, such as retirement of a debt or replacement of a property; (b) a fund created by making periodic deposits (usually equal) at compound interest in order to accumulate a given sum at a given future time for some specific purpose.

SINKING FUND DEPRECIATION—(a) A method of computing depreciation in which the periodic amount is presumed to be deposited in a *sinking fund* that earns interest at a specified rate. Sinking fund may be real but is usually hypothetical; (b) a method of depreciation where a fixed sum of money is regularly deposited at compound interest in a real or imaginary fund in order to accumulate an amount equal to the total depreciation of an asset at the end of the asset's estimated life. The sinking fund depreciation in any year equals the sinking fund deposit plus interest in the sinking fund balance.

SINKING FUND FACTOR—(a) The function of interest rate and time that determines the cumulative amount of a sinking fund resulting from specified periodic deposits. Future worth per unit of uniform amounts; (b) the mathematical formulas used to facilitate sinking fund calculations.

STRAIGHT-LINE DEPRECIATION—Method of depreciation whereby the amount to be recovered (written off) is spread uniformly over the estimated life of the asset in terms of time periods or units of output.

STUDY PERIOD—In economy study, the length of time that is presumed to be covered in the schedule of events and appraisal of results. Often the anticipated life of the project under consideration, but a shorter time may be more appropriate for decision making.

SUM-OF-DIGITS METHOD—Also known as sum-of-the-years'-digits method. A method of computing depreciation in which the amount for any year is based on the ratio of (years of remaining life) to $(1 + 2 + 3 + \cdots + N)$, the total anticipated life.

SUNK COST—(a) The unrecovered balance of an investment. It is a cost, already paid, that is not relevant to the decision concerning the future that is being made. Capital already invested that for some reason cannot be retrieved; (b) a past cost that has no relevance with respect to future receipts and disbursements of a facility undergoing an engineering economy study. This concept implies that since a past outlay is the same regardless of the alternative selected, it should not influence the choice between alternatives.

TANGIBLES—Things that can be *quantitatively* measured or valued, such as items of cost and physical assets.

TIME VALUE OF MONEY—(a) The cumulative effect of elapsed time on the money value of an event, based on the earning power of equivalent invested funds (see *Future Worth* and *Present Worth*); (b) the expected interest rate that capital should or will earn.

TOTAL PRICE ESCALATION RATE (e_j)—The total rate (%) of price change (in the unit price, or cost for a fixed amount) during a time period for good or service j. The total price escalation rate for a good or service includes the effects of both the general price inflation rate (f) and the differential price inflation rate (e_j') on price changes. The rate is projected for a future time interval and usually expressed as an effective annual rate.

TRACEABLE COSTS—Costs which can be identified with a given product, operation, or service.

UNRECOVERED INVESTMENT BALANCE—The investment amount outstanding in a venture, allowing for profits generated and including the time value of money.

VALUATION OR APPRAISAL—The art of estimating the fair-exchange value of specific properties.

WORKING CAPITAL—(a) That portion of investment represented by *current assets* (assets that are not capitalized) less the *current liabilities*. The capital necessary to sustain operations; (b) those funds that are required to make the enterprise or project a going concern.

YIELD—The ratio of return or profit over the associated investment, expressed as a percentage or decimal, usually on an annual basis.

C

INTEREST AND ANNUITY TABLES FOR DISCRETE COMPOUNDING

(For various values of i from $\frac{1}{4}\%$ to 50%)

$i =$ effective interest rate per period (usually 1 year)
$N =$ number of compounding periods

$$(F/P, i\%, N) = (1 + i)^N$$

$$(P/F, i\%, N) = \frac{1}{(1 + i)^N}$$

$$(F/A, i\%, N) = \frac{(1 + i)^N - 1}{i}$$

$$(P/A, i\%, N) = \frac{(1 + i)^N - 1}{i(1 + i)^N}$$

$$(A/F, i\%, N) = \frac{i}{(1 + i)^N - 1}$$

$$(A/P, i\%, N) = \frac{i(1 + i)^N}{(1 + i)^N - 1}$$

$$(P/G, i\%, N) = \frac{1}{i}\left[\frac{(1 + i)^N - 1}{i(1 + i)^N} - \frac{N}{(1 + i)^N}\right]$$

$$(A/G, i\%, N) = \frac{1}{i} - \frac{N}{(1 + i)^N - 1}$$

Interest tables for selected integer-valued interest rates have been included in Appendixes C and D. The following computer program, written in BASIC, can be used on most personal computers (IBM or compatible) to obtain interest tables for discrete cash flows and discrete or continuous compounding of integer or fractional interest rates expressed in decimal form. On the $5\frac{1}{4}$" diskette that may be obtained from your instructor, the program is called up by typing ITABLE and pressing the Return (Enter) key.

The IBM Personal Computer DOS
Version 3.20 (C)Copyright International Business Machines Corp 1981, 1986
(C)Copyright Microsoft 1981, 1986

```
10 REM PROGRAM ITABLE.BAS
20 COLOR 10,0:CLS : KEY OFF
30 PRINT "               ITABLE: Interest tables"
40 PRINT "               -----------------------"
50 PRINT "This program computes interest tables for discrete"
60 PRINT "compounding and continuous compounding. End-of-Period"
70 PRINT "cash flow convention is assumed. Enter I per period as"
80 PRINT "a decimal for discrete compounding and R per period as"
90 PRINT "a decimal for continuous compounding. N is the number"
100 PRINT "of compounding periods."
110 PRINT
120 PRINT "Press any key to continue.."
130 ANS$ = INKEY$: IF ANS$ = "" THEN 130
140 CLS
150 PRINT
160 PRINT "       Type of Compounding"
170 PRINT "       -------------------"
180 PRINT
190 PRINT "    1. Discrete Compounding"
200 PRINT "    2. Continuous Compounding"
210 PRINT "    0. Exit"
```

```
220 PRINT
230 INPUT "       Select option : ", OPT$
240 IF LEN(OPT$) <> 1 OR INSTR("012", OPT$) = 0 THEN 140
250 IF OPT$ = "1" THEN 350
260 IF OPT$ = "2" THEN 290
270 IF OPT$ = "0" THEN END
280 GOTO 140
290 PRINT
300 INPUT "Enter the nominal interest rate as a decimal :", I
310 IF I <= 0 OR I > 1 THEN BEEP: PRINT " (0 < Rate ≤ 1)": GOTO 290
320 I = EXP(I) - 1
330 T1$ = "Continuous Compounding -- Nominal Rate (R) = "
340 GOTO 390
350 PRINT
360 INPUT "Enter the effective interest rate as a decimal :", I
370 IF I <= 0 OR I > 1 THEN BEEP: PRINT " (0 < Rate ≤ 1)": GOTO 350
380 T1$ = "Discrete Compounding -- Effective Rate (I) = "
390 NROW = CSRLIN: IF NROW > 21 THEN CLS : NROW = CSRLIN
400 PRINT
410 PRINT "Enter starting and ending values for tabulation"
420 LOCATE NROW + 2, 1: PRINT "Start:      End:"
430 LOCATE NROW + 2, 7: INPUT "", Q
440 LOCATE NROW + 2, 17: INPUT "", R
450 IF Q < 1 THEN Q = 1
460 GOSUB 570
470 PRINT
480 INPUT "Do you want to print this table [Y/N]?", ANS$
490 IF LEN(ANS$) <> 1 OR INSTR("yYnN", ANS$) = 0 THEN 470
500 IF ANS$ = "Y" OR ANS$ = "y" THEN GOSUB 770
510 GOSUB 950
520 PRINT
530 INPUT "Do you want to print this table [Y/N]?", ANS$
540 IF LEN(ANS$) <> 1 OR INSTR("yYnN", ANS$) = 0 THEN 530
550 IF ANS$ = "Y" OR ANS$ = "y" THEN GOSUB 1160
560 GOTO 140
570 CLS : PRINT : PRINT T1$; 100 * I; "%": PRINT
580 PRINT "  N       F/P        P/F        P/A        A/P"
590 PRINT " ---     ------     ------     ------     ------"
600 NLIN = 5
610 FOR N = Q TO R
620 NLIN = NLIN + 1
630 A = (1 + I) ^ N
640 C = 1 / A
650 D = (A - 1) / (I * A)
660 E = 1 / D
670 IF NLIN <= 21 THEN GOTO 740
680 PRINT "Press any key to continue..."
690 ANS$ = INKEY$: IF ANS$ = "" THEN 690
700 NLIN = 5
710 CLS : PRINT : PRINT T1$; 100 * I; "%": PRINT
720 PRINT "  N       F/P        P/F        P/A        A/P"
730 PRINT " ---     ------     ------     ------     ------"
740 PRINT USING "###  #####.#### #####.#### #####.#### #####.####"; N, A, C, D, E
750 NEXT N
760 RETURN
770 LPRINT CHR$(12): LPRINT : LPRINT T1$; 100 * I; "%": LPRINT
780 LPRINT "  N       F/P        P/F        P/A        A/P"
790 LPRINT " ---     ------     ------     ------     ------"
800 NLIN = 5
810 FOR N = Q TO R
820 NLIN = NLIN + 1
830 A = (1 + I) ^ N
840 C = 1 / A
850 D = (A - 1) / (I * A)
```

```
860 E = 1 / D
870 IF NLIN <= 55 THEN 920
880 NLIN = 5: LPRINT CHR$(12)
890 LPRINT : LPRINT T1$; 100 * I; "%": LPRINT
900 LPRINT "  N        F/P         P/F         P/A         A/P"
910 LPRINT " ---      ------      ------      ------      ------"
920 LPRINT USING "###  #####.#### #####.#### #####.#### #####.####"; N, A, C, D, E
930 NEXT N
940 RETURN
950 CLS : PRINT : PRINT T1$; 100 * I; "%": PRINT
960 PRINT "  N        A/F         F/A         P/G         A/G"
970 PRINT " ---      ------      ------      ------      ------"
980 NLIN = 5
990 FOR N = Q TO R
1000 NLIN = NLIN + 1
1010 A = (1 + I) ^ N
1020 F = I / (A - 1)
1030 G = 1 / F
1040 P = ((A - 1) / (I * A) - N / A) / I
1050 T = 1 / I - N / (A - 1)
1060 IF NLIN <= 21 THEN GOTO 1130
1070 PRINT "Press any key to continue..."
1080 ANS$ = INKEY$: IF ANS$ = "" THEN 1080
1090 NLIN = 5
1100 CLS : PRINT : PRINT T1$; 100 * I; "%": PRINT
1110 PRINT "  N        A/F         F/A         P/G         A/G"
1120 PRINT " ---      ------      ------      ------      ------"
1130 PRINT USING "###  #####.#### #####.#### #####.#### #####.####"; N, F, G, P, T
1140 NEXT N
1150 RETURN
1160 LPRINT CHR$(12): LPRINT : LPRINT T1$; 100 * I; "%": LPRINT
1170 LPRINT "  N        A/F         F/A         P/G         A/G"
1180 LPRINT " ---      ------      ------      ------      ------"
1190 NLIN = 5
1200 FOR N = Q TO R
1210 NLIN = NLIN + 1
1220 A = (1 + I) ^ N
1230 F = I / (A - 1)
1240 G = 1 / F
1250 P = ((A - 1) / (I * A) - N / A) / I
1260 T = 1 / I - N / (A - 1)
1270 IF NLIN <= 55 THEN 1320
1280 NLIN = 5: LPRINT CHR$(12)
1290 LPRINT : LPRINT T1$; 100 * I; "%": LPRINT
1300 LPRINT "  N        A/F         F/A         P/G         A/G"
1310 LPRINT " ---      ------      ------      ------      ------"
1320 LPRINT USING "###  #####.#### #####.#### #####.#### #####.####"; N, F, G, P, T
1330 NEXT N
1340 RETURN
```

EXAMPLES

```
THIS PROGRAM COMPUTES INTEREST TABLES FOR DISCRETE
(DIS) COMPOUNDING AND CONTINUOUS (CON) COMPOUNDING.
END-OF-PERIOD CASH FLOW CONVENTION IS ASSUMED. ENTER
'I' PER PERIOD AS A DECIMAL FOR DISCRETE COMPOUNDING
AND 'R' PER PERIOD AS A DECIMAL FOR CONTINUOUS
COMPOUNDING. 'N' IS THE NUMBER OF COMPOUNDING PERIODS.
```

A

```
ENTER VALUE OF INTEREST RATE AS A DECIMAL? 0.1
DISCRETE (DIS) OR CONTINUOUS (CON) COMPOUNDING? dis

ENTER STARTING AND ENDING VALUES OF N
FOR TABULATIONS. (SEPARATE BY COMMA)? 1,10
```

THE EFFECTIVE RATE (I) = 10%

N	F/P	P/F	P/A	A/P
1	1.1000	0.9091	0.9091	1.1000
2	1.2100	0.8264	1.7355	0.5762
3	1.3310	0.7513	2.4869	0.4021
4	1.4641	0.6830	3.1699	0.3155
5	1.6105	0.6209	3.7908	0.2638
6	1.7716	0.5645	4.3553	0.2296
7	1.9487	0.5132	4.8684	0.2054
8	2.1436	0.4665	5.3349	0.1874
9	2.3579	0.4241	5.7590	0.1736
10	2.5937	0.3855	6.1446	0.1627

THE EFFECTIVE RATE (I) = 10%

N	A/F	F/A	P/G	A/G
1	1.0000	1.0000	0.0000	0.0000
2	0.4762	2.1000	0.8264	0.4762
3	0.3021	3.3100	2.3291	0.9366
4	0.2155	4.6410	4.3781	1.3812
5	0.1638	6.1051	6.8618	1.8101
6	0.1296	7.7156	9.6842	2.2236
7	0.1054	9.4872	12.7631	2.6216
8	0.0874	11.4395	16.0287	3.0045
9	0.0736	13.5795	19.4215	3.3724
10	0.0627	15.9374	22.8914	3.7255

```
DO YOU WANT TO PRINT THIS TABLE (Y/N)? Y
DO YOU WANT TO RUN FOR ANOTHER INTEREST RATE
                              VALUE (Y/N)? Y
```

B

```
ENTER VALUE OF INTEREST RATE AS A DECIMAL? 0.5
DISCRETE (DIS) OR CONTINUOUS (CON) COMPOUNDING? dis

ENTER STARTING AND ENDING VALUES OF N FOR TABULATIONS
(SEPARATE BY COMMA)? 10,25
```

THE EFFECTIVE RATE (I) = 50%

N	F/P	P/F	P/A	A/P
10	57.6650	0.0173	1.9653	0.5088
11	86.4976	0.0116	1.9769	0.5058
12	129.7463	0.0077	1.9846	0.5039
13	194.6195	0.0051	1.9897	0.5026
14	291.9293	0.0034	1.9931	0.5017
15	437.8939	0.0023	1.9954	0.5011
16	656.8408	0.0015	1.9970	0.5008
17	985.2612	0.0010	1.9980	0.5005
18	1477.8920	0.0007	1.9986	0.5003
19	2216.8380	0.0005	1.9991	0.5002
20	3325.2570	0.0003	1.9994	0.5002
21	4987.8850	0.0002	1.9996	0.5001
22	7481.8280	0.0001	1.9997	0.5001
23	11222.7400	0.0001	1.9998	0.5000
24	16834.1100	0.0001	1.9999	0.5000
25	25251.1700	0.0000	1.9999	0.5000

THE EFFECTIVE RATE (I) = 50%

N	A/F	F/A	P/G	A/G
10	0.0088	113.3301	3.5838	1.8235
11	0.0058	170.9951	3.6994	1.8713
12	0.0039	257.4927	3.7842	1.9068
13	0.0026	387.2390	3.8459	1.9329
14	0.0017	581.8585	3.8904	1.9519
15	0.0011	873.7878	3.9224	1.9657
16	0.0008	1311.6820	3.9452	1.9756
17	0.0005	1968.5230	3.9614	1.9827
18	0.0003	2953.7840	3.9729	1.9878
19	0.0002	4431.6760	3.9811	1.9914
20	0.0002	6648.5130	3.9868	1.9940
21	0.0001	9973.7690	3.9908	1.9958
22	0.0001	14961.6600	3.9936	1.9971
23	0.0000	22443.4800	3.9955	1.9980
24	0.0000	33666.2200	3.9969	1.9986
25	0.0000	50500.3400	3.9979	1.9990

```
DO YOU WANT TO PRINT THIS TABLE (Y/N)? Y
DO YOU WANT TO RUN FOR ANOTHER INTEREST RATE
                                VALUE (Y/N)? Y
```

C

```
ENTER VALUE OF INTEREST RATE AS A DECIMAL ? 0.006667
DISCRETE (DIS) OR CONTINUOUS (CON) COMPOUNDING? dis

ENTER STARTING AND ENDING VALUES OF N
FOR TABULATIONS (SEPARATE BY COMMA)? 10,15
```

THE EFFECTIVE RATE (I) = 0.6667%

N	F/P	P/F	P/A	A/P
10	1.0687	0.9357	9.6429	0.1037
11	1.0758	0.9295	10.5724	0.0946
12	1.0830	0.9234	11.4958	0.0870
13	1.0902	0.9172	12.4130	0.0806
14	1.0975	0.9112	13.3242	0.0751
15	1.1048	0.9051	14.2293	0.0703

THE EFFECTIVE RATE (I) = 0.6667%

N	A/F	F/A	P/G	A/G
10	0.0970	10.3054	42.8664	4.4454
11	0.0879	11.3741	52.1631	4.9339
12	0.0803	12.4500	62.3204	5.4212
13	0.0739	13.5330	73.3249	5.9071
14	0.0684	14.6232	85.1728	6.3923
15	0.0636	15.7207	97.8417	6.8761

TABLE C–1 Discrete Compounding; $i = 1/4\%$

	Single Payment		Uniform Series				
	Compound Amount Factor	Present Worth Factor	Compound Amount Factor	Present Worth Factor	Sinking Fund Factor	Capital Recovery Factor	
N	To find *F* Given *P* *F/P*	To find *P* Given *F* *P/F*	To find *F* Given *A* *F/A*	To find *P* Given *A* *P/A*	To find *A* Given *F* *A/F*	To find *A* Given *P* *A/P*	*N*
1	1.0025	0.9975	1.0000	0.9975	1.0000	1.0025	1
2	1.0050	0.9950	2.0025	1.9925	0.4994	0.5019	2
3	1.0075	0.9925	3.0075	2.9851	0.3325	0.3350	3
4	1.0100	0.9901	4.0150	3.9751	0.2491	0.2516	4
5	1.0126	0.9876	5.0250	4.9627	0.1990	0.2015	5
6	1.0151	0.9851	6.0376	5.9478	0.1656	0.1681	6
7	1.0176	0.9827	7.0527	6.9305	0.1418	0.1443	7
8	1.0202	0.9802	8.0703	7.9107	0.1239	0.1264	8
9	1.0227	0.9778	9.0905	8.8885	0.1100	0.1125	9
10	1.0253	0.9753	10.1132	9.8638	0.0989	0.1014	10
11	1.0278	0.9729	11.1385	10.8367	0.0898	0.0923	11
12	1.0304	0.9705	12.1663	11.8072	0.0822	0.0847	12
13	1.0330	0.9681	13.1967	12.7753	0.0758	0.0783	13
14	1.0356	0.9656	14.2297	13.7409	0.0703	0.0728	14
15	1.0382	0.9632	15.2653	14.7041	0.0655	0.0680	15
16	1.0408	0.9608	16.3035	15.6650	0.0613	0.0638	16
17	1.0434	0.9584	17.3442	16.6234	0.0577	0.0602	17
18	1.0460	0.9561	18.3876	17.5795	0.0544	0.0569	18
19	1.0486	0.9537	19.4335	18.5331	0.0515	0.0540	19
20	1.0512	0.9513	20.4821	19.4844	0.0488	0.0513	20
21	1.0538	0.9489	21.5333	20.4333	0.0464	0.0489	21
22	1.0565	0.9466	22.5872	21.3799	0.0443	0.0468	22
23	1.0591	0.9442	23.6436	22.3241	0.0423	0.0448	23
24	1.0618	0.9418	24.7027	23.2659	0.0405	0.0430	24
25	1.0644	0.9395	25.7645	24.2054	0.0388	0.0413	25
26	1.0671	0.9371	26.8289	25.1425	0.0373	0.0398	26
27	1.0697	0.9348	27.8959	26.0773	0.0358	0.0383	27
28	1.0724	0.9325	28.9657	27.0098	0.0345	0.0370	28
29	1.0751	0.9302	30.0381	27.9399	0.0333	0.0358	29
30	1.0778	0.9278	31.1132	28.8678	0.0321	0.0346	30
35	1.0913	0.9163	36.5291	33.4723	0.0274	0.0299	35
40	1.1050	0.9050	42.0130	38.0197	0.0238	0.0263	40
45	1.1189	0.8937	47.5659	42.5107	0.0210	0.0235	45
50	1.1330	0.8826	53.1884	46.9460	0.0188	0.0213	50
55	1.1472	0.8717	58.8817	51.3262	0.0170	0.0195	55
60	1.1616	0.8609	64.6464	55.6521	0.0155	0.0180	60
65	1.1762	0.8502	70.4836	59.9244	0.0142	0.0167	65
70	1.1910	0.8396	76.3941	64.1436	0.0131	0.0156	70
75	1.2059	0.8292	82.3788	68.3105	0.0121	0.0146	75
80	1.2211	0.8189	88.4388	72.4257	0.0113	0.0138	80
85	1.2364	0.8088	94.5748	76.490	0.0106	0.0131	85
90	1.2520	0.7987	100.788	80.504	0.0099	0.0124	90
95	1.2677	0.7888	107.079	84.467	0.0093	0.0118	95
100	1.2836	0.7790	113.45	88.382	0.0088	0.0113	100

TABLE C–2 Discrete Compounding; $i = 1/2\%$

	Single Payment		Uniform Series				
	Compound Amount Factor	Present Worth Factor	Compound Amount Factor	Present Worth Factor	Sinking Fund Factor	Capital Recovery Factor	
N	To find *F* Given *P* *F/P*	To find *P* Given *F* *P/F*	To find *F* Given *A* *F/A*	To find *P* Given *A* *P/A*	To find *A* Given *F* *A/F*	To find *A* Given *P* *A/P*	*N*
1	1.0050	0.9950	1.0000	0.9950	1.0000	1.0050	1
2	1.0100	0.9901	2.0050	1.9851	0.4988	0.5038	2
3	1.0151	0.9851	3.0150	2.9702	0.3317	0.3367	3
4	1.0202	0.9802	4.0301	3.9505	0.2481	0.2531	4
5	1.0253	0.9754	5.0502	4.9259	0.1980	0.2030	5
6	1.0304	0.9705	6.0755	5.8964	0.1646	0.1696	6
7	1.0355	0.9657	7.1059	6.8621	0.1407	0.1457	7
8	1.0407	0.9609	8.1414	7.8229	0.1228	0.1278	8
9	1.0459	0.9561	9.1821	8.7790	0.1089	0.1139	9
10	1.0511	0.9513	10.2280	9.7304	0.0978	0.1028	10
11	1.0564	0.9466	11.2791	10.6770	0.0887	0.0937	11
12	1.0617	0.9419	12.3355	11.6189	0.0811	0.0861	12
13	1.0670	0.9372	13.3972	12.5561	0.0746	0.0796	13
14	1.0723	0.9326	14.4642	13.4887	0.0691	0.0741	14
15	1.0777	0.9279	15.5365	14.4166	0.0644	0.0694	15
16	1.0831	0.9233	16.6142	15.3399	0.0602	0.0652	16
17	1.0885	0.9187	17.6973	16.2586	0.0565	0.0615	17
18	1.0939	0.9141	18.7857	17.1727	0.0532	0.0582	18
19	1.0994	0.9096	19.8797	18.0823	0.0503	0.0553	19
20	1.1049	0.9051	20.9791	18.9874	0.0477	0.0527	20
21	1.1104	0.9006	22.0839	19.8879	0.0453	0.0503	21
22	1.1160	0.8961	23.1944	20.7840	0.0431	0.0481	22
23	1.1216	0.8916	24.3103	21.6756	0.0411	0.0461	23
24	1.1272	0.8872	25.4319	22.5628	0.0393	0.0443	24
25	1.1328	0.8828	26.5590	23.4456	0.0377	0.0427	25
26	1.1385	0.8784	27.6918	24.3240	0.0361	0.0411	26
27	1.1442	0.8740	28.8303	25.1980	0.0347	0.0397	27
28	1.1499	0.8697	29.9744	26.0676	0.0334	0.0384	28
29	1.1556	0.8653	31.1243	26.9330	0.0321	0.0371	29
30	1.1614	0.8610	32.2799	27.7940	0.0310	0.0360	30
35	1.1907	0.8398	38.1453	32.0353	0.0262	0.0312	35
40	1.2208	0.8191	44.1587	36.1721	0.0226	0.0276	40
45	1.2516	0.7990	50.3240	40.2071	0.0199	0.0249	45
50	1.2832	0.7793	56.6450	44.1427	0.0177	0.0227	50
55	1.3156	0.7601	63.1256	47.9813	0.0158	0.0208	55
60	1.3488	0.7414	69.7698	51.7254	0.0143	0.0193	60
65	1.3829	0.7231	76.5818	55.3773	0.0131	0.0181	65
70	1.4178	0.7053	83.5658	58.9393	0.0120	0.0170	70
75	1.4536	0.6879	90.7262	62.4135	0.0110	0.0160	75
80	1.4903	0.6710	98.0674	65.8022	0.0102	0.0152	80
85	1.5280	0.6545	105.594	69.107	0.0095	0.0145	85
90	1.5666	0.6383	113.311	72.331	0.0088	0.0138	90
95	1.6061	0.6226	121.22	75.475	0.0082	0.0132	95
100	1.6467	0.6073	129.33	78.542	0.0077	0.0127	100
∞				200.0		0.0050	∞

TABLE C–3 Discrete Compounding; $i = 3/4\%$

	Single Payment		Uniform Series				
	Compound Amount Factor	Present Worth Factor	Compound Amount Factor	Present Worth Factor	Sinking Fund Factor	Capital Recovery Factor	
N	To find *F* Given *P* *F/P*	To find *P* Given *F* *P/F*	To find *F* Given *A* *F/A*	To find *P* Given *A* *P/A*	To find *A* Given *F* *A/F*	To find *A* Given *P* *A/P*	*N*
1	1.0075	0.9926	1.0000	0.9926	1.0000	1.0075	1
2	1.0151	0.9852	2.0075	1.9777	0.4981	0.5056	2
3	1.0227	0.9778	3.0226	2.9556	0.3308	0.3383	3
4	1.0303	0.9706	4.0452	3.9261	0.2472	0.2547	4
5	1.0381	0.9633	5.0756	4.8894	0.1970	0.2045	5
6	1.0459	0.9562	6.1136	5.8456	0.1636	0.1711	6
7	1.0537	0.9490	7.1595	6.7946	0.1397	0.1472	7
8	1.0616	0.9420	8.2132	7.7366	0.1218	0.1293	8
9	1.0696	0.9350	9.2748	8.6716	0.1078	0.1153	9
10	1.0776	0.9280	10.3443	9.5996	0.0967	0.1042	10
11	1.0857	0.9211	11.4219	10.5207	0.0876	0.0951	11
12	1.0938	0.9142	12.5076	11.4349	0.0800	0.0875	12
13	1.1020	0.9074	13.6014	12.3423	0.0735	0.0810	13
14	1.1103	0.9007	14.7034	13.2430	0.0680	0.0755	14
15	1.1186	0.8940	15.8136	14.1370	0.0632	0.0707	15
16	1.1270	0.8873	16.9322	15.0243	0.0591	0.0666	16
17	1.1354	0.8807	18.0592	15.9050	0.0554	0.0629	17
18	1.1440	0.8742	19.1947	16.7791	0.0521	0.0596	18
19	1.1525	0.8676	20.3386	17.6468	0.0492	0.0567	19
20	1.1612	0.8612	21.4912	18.5080	0.0465	0.0540	20
21	1.1699	0.8548	22.6523	19.3628	0.0441	0.0516	21
22	1.1787	0.8484	23.8222	20.2112	0.0420	0.0495	22
23	1.1875	0.8421	25.0009	21.0533	0.0400	0.0475	23
24	1.1964	0.8358	26.1884	21.8891	0.0382	0.0457	24
25	1.2054	0.8296	27.3848	22.7187	0.0365	0.0440	25
26	1.2144	0.8234	28.5902	23.5421	0.0350	0.0425	26
27	1.2235	0.8173	29.8046	24.3594	0.0336	0.0411	27
28	1.2327	0.8112	31.0282	25.1707	0.0322	0.0397	28
29	1.2420	0.8052	32.2609	25.9758	0.0310	0.0385	29
30	1.2513	0.7992	33.5028	26.7750	0.0298	0.0373	30
35	1.2989	0.7699	39.8537	30.6826	0.0251	0.0326	35
40	1.3483	0.7416	46.4464	34.4469	0.0215	0.0290	40
45	1.3997	0.7145	53.2900	38.0731	0.0188	0.0263	45
50	1.4530	0.6883	60.3941	41.5664	0.0166	0.0241	50
55	1.5083	0.6630	67.7686	44.9315	0.0148	0.0223	55
60	1.5657	0.6387	75.4239	48.1733	0.0133	0.0208	60
65	1.6253	0.6153	83.3706	51.2962	0.0120	0.0195	65
70	1.6871	0.5927	91.6198	54.3045	0.0109	0.0184	70
75	1.7514	0.5710	100.183	57.2026	0.0100	0.0175	75
80	1.8180	0.5500	109.072	59.9943	0.0092	0.0167	80
85	1.8872	0.5299	118.300	62.6837	0.0085	0.0160	85
90	1.9591	0.5104	127.879	65.275	0.0078	0.0153	90
95	2.0337	0.4917	137.822	67.770	0.0073	0.0148	95
100	2.1111	0.4737	148.14	70.175	0.0068	0.0143	100

TABLE C–4 Discrete Compounding; $i = 1\%$

	Single Payment		Uniform Series				
	Compound Amount Factor	Present Worth Factor	Compound Amount Factor	Present Worth Factor	Sinking Fund Factor	Capital Recovery Factor	
N	To find *F* Given *P* *F/P*	To find *P* Given *F* *P/F*	To find *F* Given *A* *F/A*	To find *P* Given *A* *P/A*	To find *A* Given *F* *A/F*	To find *A* Given *P* *A/P*	*N*
1	1.0100	0.9901	1.0000	0.9901	1.0000	1.0100	1
2	1.0201	0.9803	2.0100	1.9704	0.4975	0.5075	2
3	1.0303	0.9706	3.0301	2.9410	0.3300	0.3400	3
4	1.0406	0.9610	4.0604	3.9020	0.2463	0.2563	4
5	1.0510	0.9515	5.1010	4.8534	0.1960	0.2060	5
6	1.0615	0.9420	6.1520	5.7955	0.1625	0.1725	6
7	1.0721	0.9327	7.2135	6.7282	0.1386	0.1486	7
8	1.0829	0.9235	8.2857	7.6517	0.1207	0.1307	8
9	1.0937	0.9143	9.3685	8.5660	0.1067	0.1167	9
10	1.1046	0.9053	10.4622	9.4713	0.0956	0.1056	10
11	1.1157	0.8963	11.5668	10.3676	0.0865	0.0965	11
12	1.1268	0.8874	12.6825	11.2551	0.0788	0.0888	12
13	1.1381	0.8787	13.8093	12.1337	0.0724	0.0824	13
14	1.1495	0.8700	14.9474	13.0037	0.0669	0.0769	14
15	1.1610	0.8613	16.0969	13.8650	0.0621	0.0721	15
16	1.1726	0.8528	17.2578	14.7178	0.0579	0.0679	16
17	1.1843	0.8444	18.4304	15.5622	0.0543	0.0643	17
18	1.1961	0.8360	19.6147	16.3982	0.0510	0.0610	18
19	1.2081	0.8277	20.8109	17.2260	0.0481	0.0581	19
20	1.2202	0.8195	22.0190	18.0455	0.0454	0.0554	20
21	1.2324	0.8114	23.2391	18.8570	0.0430	0.0530	21
22	1.2447	0.8034	24.4715	19.6603	0.0409	0.0509	22
23	1.2572	0.7954	25.7162	20.4558	0.0389	0.0489	23
24	1.2697	0.7876	26.9734	21.2434	0.0371	0.0471	24
25	1.2824	0.7798	28.2431	22.0231	0.0354	0.0454	25
26	1.2953	0.7720	29.5256	22.7952	0.0339	0.0439	26
27	1.3082	0.7644	30.8208	23.5596	0.0324	0.0424	27
28	1.3213	0.7568	32.1290	24.3164	0.0311	0.0411	28
29	1.3345	0.7493	33.4503	25.0657	0.0299	0.0399	29
30	1.3478	0.7419	34.7848	25.8077	0.0287	0.0387	30
35	1.4166	0.7059	41.6602	29.4085	0.0240	0.0340	35
40	1.4889	0.6717	48.8863	32.8346	0.0205	0.0305	40
45	1.5648	0.6391	56.4809	36.0945	0.0177	0.0277	45
50	1.6446	0.6080	64.4630	39.1961	0.0155	0.0255	50
55	1.7285	0.5785	72.8523	42.1471	0.0137	0.0237	55
60	1.8167	0.5505	81.6695	44.9550	0.0122	0.0222	60
65	1.9094	0.5237	90.9364	47.6265	0.0110	0.0210	65
70	2.0068	0.4983	100.676	50.1684	0.0099	0.0199	70
75	2.1091	0.4741	110.912	52.5870	0.0090	0.0190	75
80	2.2167	0.4511	121.671	54.8881	0.0082	0.0182	80
85	2.3298	0.4292	132.979	57.0776	0.0075	0.0175	85
90	2.4486	0.4084	144.86	59.161	0.0069	0.0169	90
95	2.5735	0.3886	157.35	61.143	0.0064	0.0164	95
100	2.7048	0.3697	170.48	63.029	0.0059	0.0159	100
∞				100.000		0.0100	∞

TABLE C–5 Discrete Compounding; $i = 1^1/2\%$

	Single Payment		Uniform Series				
	Compound Amount Factor	Present Worth Factor	Compound Amount Factor	Present Worth Factor	Sinking Fund Factor	Capital Recovery Factor	
N	To find *F* Given *P* *F/P*	To find *P* Given *F* *P/F*	To find *F* Given *A* *F/A*	To find *P* Given *A* *P/A*	To find *A* Given *F* *A/F*	To find *A* Given *P* *A/P*	*N*
1	1.0150	0.9852	1.0000	0.9852	1.0000	1.0150	1
2	1.0302	0.9707	2.0150	1.9559	0.4963	0.5113	2
3	1.0457	0.9563	3.0452	2.9122	0.3284	0.3434	3
4	1.0614	0.9422	4.0909	3.8544	0.2444	0.2594	4
5	1.0773	0.9283	5.1523	4.7826	0.1941	0.2091	5
6	1.0934	0.9145	6.2295	5.6972	0.1605	0.1755	6
7	1.1098	0.9010	7.3230	6.5982	0.1366	0.1516	7
8	1.1265	0.8877	8.4328	7.4859	0.1186	0.1336	8
9	1.1434	0.8746	9.5593	8.3605	0.1046	0.1196	9
10	1.1605	0.8617	10.7027	9.2222	0.0934	0.1084	10
11	1.1779	0.8489	11.8632	10.0711	0.0843	0.0993	11
12	1.1956	0.8364	13.0412	10.9075	0.0767	0.0917	12
13	1.2136	0.8240	14.2368	11.7315	0.0702	0.0852	13
14	1.2318	0.8118	15.4504	12.5434	0.0647	0.0797	14
15	1.2502	0.7999	16.6821	13.3432	0.0599	0.0749	15
16	1.2690	0.7880	17.9323	14.1312	0.0558	0.0708	16
17	1.2880	0.7764	19.2013	14.9076	0.0521	0.0671	17
18	1.3073	0.7649	20.4893	15.6725	0.0488	0.0638	18
19	1.3270	0.7536	21.7967	16.4261	0.0459	0.0609	19
20	1.3469	0.7425	23.1236	17.1686	0.0432	0.0582	20
21	1.3671	0.7315	24.4705	17.9001	0.0409	0.0559	21
22	1.3876	0.7207	25.8375	18.6208	0.0387	0.0537	22
23	1.4084	0.7100	27.2251	19.3308	0.0367	0.0517	23
24	1.4295	0.6995	28.6335	20.0304	0.0349	0.0499	24
25	1.4509	0.6892	30.0630	20.7196	0.0333	0.0483	25
26	1.4727	0.6790	31.5139	21.3986	0.0317	0.0467	26
27	1.4948	0.6690	32.9866	22.0676	0.0303	0.0453	27
28	1.5172	0.6591	34.4814	22.7267	0.0290	0.0440	28
29	1.5400	0.6494	35.9986	23.3761	0.0278	0.0428	29
30	1.5631	0.6398	37.5386	24.0158	0.0266	0.0416	30
35	1.6839	0.5939	45.5920	27.0756	0.0219	0.0369	35
40	1.8140	0.5513	54.2678	29.9158	0.0184	0.0334	40
45	1.9542	0.5117	63.6141	32.5523	0.0157	0.0307	45
50	2.1052	0.4750	73.6827	34.9997	0.0136	0.0286	50
55	2.2679	0.4409	84.5294	37.2714	0.0118	0.0268	55
60	2.4432	0.4093	96.2145	39.3802	0.0104	0.0254	60
65	2.6320	0.3799	108.803	41.3378	0.0092	0.0242	65
70	2.8355	0.3527	122.364	43.1548	0.0082	0.0232	70
75	3.0546	0.3274	136.97	44.8416	0.0073	0.0223	75
80	3.2907	0.3039	152.71	46.4073	0.0065	0.0215	80
85	3.5450	0.2821	169.66	47.8607	0.0059	0.0209	85
90	3.8189	0.2619	187.93	49.2098	0.0053	0.0203	90
95	4.1141	0.2431	207.61	50.4622	0.0048	0.0198	95
100	4.4320	0.2256	228.80	51.6247	0.0044	0.0194	100
∞				66.667		0.0150	∞

TABLE C–6 Discrete Compounding; $i = 2\%$

	Single Payment		Uniform Series				
	Compound Amount Factor	Present Worth Factor	Compound Amount Factor	Present Worth Factor	Sinking Fund Factor	Capital Recovery Factor	
	To find *F* Given *P*	To find *P* Given *F*	To find *F* Given *A*	To find *P* Given *A*	To find *A* Given *F*	To find *A* Given *P*	
N	*F/P*	*P/F*	*F/A*	*P/A*	*A/F*	*A/P*	*N*
1	1.0200	0.9804	1.0000	0.9804	1.0000	1.0200	1
2	1.0404	0.9612	2.0200	1.9416	0.4950	0.5150	2
3	1.0612	0.9423	3.0604	2.8839	0.3268	0.3468	3
4	1.0824	0.9238	4.1216	3.8077	0.2426	0.2626	4
5	1.1041	0.9057	5.2040	4.7135	0.1922	0.2122	5
6	1.1262	0.8880	6.3081	5.6014	0.1585	0.1785	6
7	1.1487	0.8706	7.4343	6.4720	0.1345	0.1545	7
8	1.1717	0.8535	8.5830	7.3255	0.1165	0.1365	8
9	1.1951	0.8368	9.7546	8.1622	0.1025	0.1225	9
10	1.2190	0.8203	10.9497	8.9826	0.0913	0.1113	10
11	1.2434	0.8043	12.1687	9.7868	0.0822	0.1022	11
12	1.2682	0.7885	13.4121	10.5753	0.0746	0.0946	12
13	1.2936	0.7730	14.6803	11.3484	0.0681	0.0881	13
14	1.3195	0.7579	15.9739	12.1062	0.0626	0.0826	14
15	1.3459	0.7430	17.2934	12.8493	0.0578	0.0778	15
16	1.3728	0.7284	18.6393	13.5777	0.0537	0.0737	16
17	1.4002	0.7142	20.0121	14.2919	0.0500	0.0700	17
18	1.4282	0.7002	21.4123	14.9920	0.0467	0.0667	18
19	1.4568	0.6864	22.8405	15.6785	0.0438	0.0638	19
20	1.4859	0.6730	24.2974	16.3514	0.0412	0.0612	20
21	1.5157	0.6598	25.7833	17.0112	0.0388	0.0588	21
22	1.5460	0.6468	27.2990	17.6580	0.0366	0.0566	22
23	1.5769	0.6342	28.8449	18.2922	0.0347	0.0547	23
24	1.6084	0.6217	30.4218	18.9139	0.0329	0.0529	24
25	1.6406	0.6095	32.0303	19.5234	0.0312	0.0512	25
26	1.6734	0.5976	33.6709	20.1210	0.0297	0.0497	26
27	1.7069	0.5859	35.3443	20.7069	0.0283	0.0483	27
28	1.7410	0.5744	37.0512	21.2813	0.0270	0.0470	28
29	1.7758	0.5631	38.7922	21.8444	0.0258	0.0458	29
30	1.8114	0.5521	40.5681	22.3964	0.0246	0.0446	30
35	1.9999	0.5000	49.9944	24.9986	0.0200	0.0400	35
40	2.2080	0.4529	60.4019	27.3555	0.0166	0.0366	40
45	2.4379	0.4102	71.8927	29.4902	0.0139	0.0339	45
50	2.6916	0.3715	84.5793	31.4236	0.0118	0.0318	50
55	2.9717	0.3365	98.5864	33.1748	0.0101	0.0301	55
60	3.2810	0.3048	114.051	34.7609	0.0088	0.0288	60
65	3.6225	0.2761	131.126	36.1975	0.0076	0.0276	65
70	3.9996	0.2500	149.978	37.4986	0.0067	0.0267	70
75	4.4158	0.2265	170.792	38.6771	0.0059	0.0259	75
80	4.8754	0.2051	193.772	39.7445	0.0052	0.0252	80
85	5.3829	0.1858	219.144	40.7113	0.0046	0.0246	85
90	5.9431	0.1683	247.16	41.5869	0.0040	0.0240	90
95	6.5617	0.1524	278.08	42.3800	0.0036	0.0236	95
100	7.2446	0.1380	312.23	43.0983	0.0032	0.0232	100
∞				50.0000		0.0200	∞

TABLE C–7 Discrete Compounding; $i = 3\%$

	Single Payment		Uniform Series				
	Compound Amount Factor	Present Worth Factor	Compound Amount Factor	Present Worth Factor	Sinking Fund Factor	Capital Recovery Factor	
N	To find *F* Given *P* *F/P*	To find *P* Given *F* *P/F*	To find *F* Given *A* *F/A*	To find *P* Given *A* *P/A*	To find *A* Given *F* *A/F*	To find *A* Given *P* *A/P*	*N*
1	1.0300	0.9709	1.0000	0.9709	1.0000	1.0300	1
2	1.0609	0.9426	2.0300	1.9135	0.4926	0.5226	2
3	1.0927	0.9151	3.0909	2.8286	0.3235	0.3535	3
4	1.1255	0.8885	4.1836	3.7171	0.2390	0.2690	4
5	1.1593	0.8626	5.3091	4.5797	0.1884	0.2184	5
6	1.1941	0.8375	6.4684	5.4172	0.1546	0.1846	6
7	1.2299	0.8131	7.6625	6.2303	0.1305	0.1605	7
8	1.2668	0.7894	8.8923	7.0197	0.1125	0.1425	8
9	1.3048	0.7664	10.1591	7.7861	0.0984	0.1284	9
10	1.3439	0.7441	11.4639	8.5302	0.0872	0.1172	10
11	1.3842	0.7224	12.8078	9.2526	0.0781	0.1081	11
12	1.4258	0.7014	14.1920	9.9540	0.0705	0.1005	12
13	1.4685	0.6810	15.6178	10.6349	0.0640	0.0940	13
14	1.5126	0.6611	17.0863	11.2961	0.0585	0.0885	14
15	1.5580	0.6419	18.5989	11.9379	0.0538	0.0838	15
16	1.6047	0.6232	20.1569	12.5611	0.0496	0.0796	16
17	1.6528	0.6050	21.7616	13.1661	0.0460	0.0760	17
18	1.7024	0.5874	23.4144	13.7535	0.0427	0.0727	18
19	1.7535	0.5703	25.1168	14.3238	0.0398	0.0698	19
20	1.8061	0.5537	26.8703	14.8775	0.0372	0.0672	20
21	1.8603	0.5375	28.6765	15.4150	0.0349	0.0649	21
22	1.9161	0.5219	30.5367	15.9369	0.0327	0.0627	22
23	1.9736	0.5067	32.4528	16.4436	0.0308	0.0608	23
24	2.0328	0.4919	34.4264	16.9355	0.0290	0.0590	24
25	2.0938	0.4776	36.4592	17.4131	0.0274	0.0574	25
26	2.1566	0.4637	38.5530	17.8768	0.0259	0.0559	26
27	2.2213	0.4502	40.7096	18.3270	0.0246	0.0546	27
28	2.2879	0.4371	42.9309	18.7641	0.0233	0.0533	28
29	2.3566	0.4243	45.2188	19.1884	0.0221	0.0521	29
30	2.4273	0.4120	47.5754	19.6004	0.0210	0.0510	30
35	2.8139	0.3554	60.4620	21.4872	0.0165	0.0465	35
40	3.2620	0.3066	75.4012	23.1148	0.0133	0.0433	40
45	3.7816	0.2644	92.7197	24.5187	0.0108	0.0408	45
50	4.3839	0.2281	112.797	25.7298	0.0089	0.0389	50
55	5.0821	0.1968	136.071	26.7744	0.0073	0.0373	55
60	5.8916	0.1697	163.053	27.6756	0.0061	0.0361	60
65	6.8300	0.1464	194.332	28.4529	0.0051	0.0351	65
70	7.9178	0.1263	230.594	29.1234	0.0043	0.0343	70
75	9.1789	0.1089	272.630	29.7018	0.0037	0.0337	75
80	10.6409	0.0940	321.362	30.2008	0.0031	0.0331	80
85	12.3357	0.0811	377.856	30.6311	0.0026	0.0326	85
90	14.3004	0.0699	443.35	31.0024	0.0023	0.0323	90
95	16.5781	0.0603	519.27	31.3227	0.0019	0.0319	95
100	19.2186	0.0520	607.29	31.5989	0.0016	0.0316	100
∞				33.3333		0.0300	∞

TABLE C–8 Discrete Compounding; $i = 4\%$

	Single Payment		Uniform Series				
	Compound Amount Factor	Present Worth Factor	Compound Amount Factor	Present Worth Factor	Sinking Fund Factor	Capital Recovery Factor	
	To find *F* Given *P*	To find *P* Given *F*	To find *F* Given *A*	To find *P* Given *A*	To find *A* Given *F*	To find *A* Given *P*	
N	*F/P*	*P/F*	*F/A*	*P/A*	*A/F*	*A/P*	*N*
1	1.0400	0.9615	1.0000	0.9615	1.0000	1.0400	1
2	1.0816	0.9246	2.0400	1.8861	0.4902	0.5302	2
3	1.1249	0.8890	3.1216	2.7751	0.3203	0.3603	3
4	1.1699	0.8548	4.2465	3.6299	0.2355	0.2755	4
5	1.2167	0.8219	5.4163	4.4518	0.1846	0.2246	5
6	1.2653	0.7903	6.6330	5.2421	0.1508	0.1908	6
7	1.3159	0.7599	7.8983	6.0021	0.1266	0.1666	7
8	1.3686	0.7307	9.2142	6.7327	0.1085	0.1485	8
9	1.4233	0.7026	10.5828	7.4353	0.0945	0.1345	9
10	1.4802	0.6756	12.0061	8.1109	0.0833	0.1233	10
11	1.5395	0.6496	13.4863	8.7605	0.0741	0.1141	11
12	1.6010	0.6246	15.0258	9.3851	0.0666	0.1066	12
13	1.6651	0.6006	16.6268	9.9856	0.0601	0.1001	13
14	1.7317	0.5775	18.2919	10.5631	0.0547	0.0947	14
15	1.8009	0.5553	20.0236	11.1184	0.0499	0.0899	15
16	1.8730	0.5339	21.8245	11.6523	0.0458	0.0858	16
17	1.9479	0.5134	23.6975	12.1657	0.0422	0.0822	17
18	2.0258	0.4936	25.6454	12.6593	0.0390	0.0790	18
19	2.1068	0.4746	27.6712	13.1339	0.0361	0.0761	19
20	2.1911	0.4564	29.7781	13.5903	0.0336	0.0736	20
21	2.2788	0.4388	31.9692	14.0292	0.0313	0.0713	21
22	2.3699	0.4220	34.2480	14.4511	0.0292	0.0692	22
23	2.4647	0.4057	36.6179	14.8568	0.0273	0.0673	23
24	2.5633	0.3901	39.0826	15.2470	0.0256	0.0656	24
25	2.6658	0.3751	41.6459	15.6221	0.0240	0.0640	25
26	2.7725	0.3607	44.3117	15.9828	0.0226	0.0626	26
27	2.8834	0.3468	47.0842	16.3296	0.0212	0.0612	27
28	2.9987	0.3335	49.9676	16.6631	0.0200	0.0600	28
29	3.1187	0.3207	52.9663	16.9837	0.0189	0.0589	29
30	3.2434	0.3083	56.0849	17.2920	0.0178	0.0578	30
35	3.9461	0.2534	73.6522	18.6646	0.0136	0.0536	35
40	4.8010	0.2083	95.0255	19.7928	0.0105	0.0505	40
45	5.8412	0.1712	121.029	20.7200	0.0083	0.0483	45
50	7.1067	0.1407	152.667	21.4822	0.0066	0.0466	50
55	8.6464	0.1157	191.159	22.1086	0.0052	0.0452	55
60	10.5196	0.0951	237.991	22.6235	0.0042	0.0442	60
65	12.7987	0.0781	294.968	23.0467	0.0034	0.0434	65
70	15.5716	0.0642	364.290	23.3945	0.0027	0.0427	70
75	18.9452	0.0528	448.631	23.6804	0.0022	0.0422	75
80	23.0498	0.0434	551.245	23.9154	0.0018	0.0418	80
85	28.0436	0.0357	676.090	24.1085	0.0015	0.0415	85
90	34.1193	0.0293	827.98	24.2673	0.0012	0.0412	90
95	41.5113	0.0241	1012.78	24.3978	0.0010	0.0410	95
100	50.5049	0.0198	1237.62	24.5050	0.0008	0.0408	100

TABLE C–9 Discrete Compounding; i = 5%

	Single Payment		Uniform Series				
	Compound Amount Factor	Present Worth Factor	Compound Amount Factor	Present Worth Factor	Sinking Fund Factor	Capital Recovery Factor	
N	To find *F* Given *P* *F/P*	To find *P* Given *F* *P/F*	To find *F* Given *A* *F/A*	To find *P* Given *A* *P/A*	To find *A* Given *F* *A/F*	To find *A* Given *P* *A/P*	*N*
1	1.0500	0.9524	1.0000	0.9524	1.0000	1.0500	1
2	1.1025	0.9070	2.0500	1.8594	0.4878	0.5378	2
3	1.1576	0.8638	3.1525	2.7232	0.3172	0.3672	3
4	1.2155	0.8227	4.3101	3.5460	0.2320	0.2820	4
5	1.2763	0.7835	5.5256	4.3295	0.1810	0.2310	5
6	1.3401	0.7462	6.8019	5.0757	0.1470	0.1970	6
7	1.4071	0.7107	8.1420	5.7864	0.1228	0.1728	7
8	1.4775	0.6768	9.5491	6.4632	0.1047	0.1547	8
9	1.5513	0.6446	11.0266	7.1078	0.0907	0.1407	9
10	1.6289	0.6139	12.5779	7.7217	0.0795	0.1295	10
11	1.7103	0.5847	14.2068	8.3064	0.0704	0.1204	11
12	1.7959	0.5568	15.9171	8.8633	0.0628	0.1128	12
13	1.8856	0.5303	17.7130	9.3936	0.0565	0.1065	13
14	1.9799	0.5051	19.5986	9.8986	0.0510	0.1010	14
15	2.0789	0.4810	21.5786	10.3797	0.0463	0.0963	15
16	2.1829	0.4581	23.6575	10.8378	0.0423	0.0923	16
17	2.2920	0.4363	25.8404	11.2741	0.0387	0.0887	17
18	2.4066	0.4155	28.1324	11.6896	0.0355	0.0855	18
19	2.5269	0.3957	30.5390	12.0853	0.0327	0.0827	19
20	2.6533	0.3769	33.0659	12.4622	0.0302	0.0802	20
21	2.7860	0.3589	35.7192	12.8212	0.0280	0.0780	21
22	2.9253	0.3418	38.5052	13.1630	0.0260	0.0760	22
23	3.0715	0.3256	41.4305	13.4886	0.0241	0.0741	23
24	3.2251	0.3101	44.5020	13.7986	0.0225	0.0725	24
25	3.3864	0.2953	47.7271	14.0939	0.0210	0.0710	25
26	3.5557	0.2812	51.1134	14.3752	0.0196	0.0696	26
27	3.7335	0.2678	54.6691	14.6430	0.0183	0.0683	27
28	3.9201	0.2551	58.4026	14.8981	0.0171	0.0671	28
29	4.1161	0.2429	62.3227	15.1411	0.0160	0.0660	29
30	4.3219	0.2314	66.4388	15.3725	0.0151	0.0651	30
35	5.5160	0.1813	90.3203	16.3742	0.0111	0.0611	35
40	7.0400	0.1420	120.800	17.1591	0.0083	0.0583	40
45	8.9850	0.1113	159.700	17.7741	0.0063	0.0563	45
50	11.4674	0.0872	209.348	18.2559	0.0048	0.0548	50
55	14.6356	0.0683	272.713	18.6335	0.0037	0.0537	55
60	18.6792	0.0535	353.584	18.9293	0.0028	0.0528	60
65	23.8399	0.0419	456.798	19.1611	0.0022	0.0522	65
70	30.4264	0.0329	588.528	19.3427	0.0017	0.0517	70
75	38.8327	0.0258	756.653	19.4850	0.0013	0.0513	75
80	49.5614	0.0202	971.228	19.5965	0.0010	0.0510	80
85	63.2543	0.0158	1245.09	19.6838	0.0008	0.0508	85
90	80.7303	0.0124	1594.61	19.7523	0.0006	0.0506	90
95	103.035	0.0097	2040.69	19.8059	0.0005	0.0505	95
100	131.501	0.0076	2610.02	19.8479	0.0004	0.0504	100
∞				20.0000		0.0500	∞

TABLE C–10 Discrete Compounding; i = 6%

	Single Payment		Uniform Series				
	Compound Amount Factor	Present Worth Factor	Compound Amount Factor	Present Worth Factor	Sinking Fund Factor	Capital Recovery Factor	
N	To find *F* Given *P* *F/P*	To find *P* Given *F* *P/F*	To find *F* Given *A* *F/A*	To find *P* Given *A* *P/A*	To find *A* Given *F* *A/F*	To find *A* Given *P* *A/P*	*N*
1	1.0600	0.9434	1.0000	0.9434	1.0000	1.0600	1
2	1.1236	0.8900	2.0600	1.8334	0.4854	0.5454	2
3	1.1910	0.8396	3.1836	2.6730	0.3141	0.3741	3
4	1.2625	0.7921	4.3746	3.4651	0.2286	0.2886	4
5	1.3382	0.7473	5.6371	4.2124	0.1774	0.2374	5
6	1.4185	0.7050	6.9753	4.9173	0.1434	0.2034	6
7	1.5036	0.6651	8.3938	5.5824	0.1191	0.1791	7
8	1.5938	0.6274	9.8975	6.2098	0.1010	0.1610	8
9	1.6895	0.5919	11.4913	6.8017	0.0870	0.1470	9
10	1.7908	0.5584	13.1808	7.3601	0.0759	0.1359	10
11	1.8983	0.5268	14.9716	7.8869	0.0668	0.1268	11
12	2.0122	0.4970	16.8699	8.3838	0.0593	0.1193	12
13	2.1329	0.4688	18.8821	8.8527	0.0530	0.1130	13
14	2.2609	0.4423	21.0151	9.2950	0.0476	0.1076	14
15	2.3966	0.4173	23.2760	9.7122	0.0430	0.1030	15
16	2.5404	0.3936	25.6725	10.1059	0.0390	0.0990	16
17	2.6928	0.3714	28.2129	10.4773	0.0354	0.0954	17
18	2.8543	0.3503	30.9056	10.8276	0.0324	0.0924	18
19	3.0256	0.3305	33.7600	11.1581	0.0296	0.0896	19
20	3.2071	0.3118	36.7856	11.4699	0.0272	0.0872	20
21	3.3996	0.2942	39.9927	11.7641	0.0250	0.0850	21
22	3.6035	0.2775	43.3923	12.0416	0.0230	0.0830	22
23	3.8197	0.2618	46.9958	12.3034	0.0213	0.0813	23
24	4.0489	0.2470	50.8155	12.5504	0.0197	0.0797	24
25	4.2919	0.2330	54.8645	12.7834	0.0182	0.0782	25
26	4.5494	0.2198	59.1563	13.0032	0.0169	0.0769	26
27	4.8223	0.2074	63.7057	13.2105	0.0157	0.0757	27
28	5.1117	0.1956	68.5281	13.4062	0.0146	0.0746	28
29	5.4184	0.1846	73.6397	13.5907	0.0136	0.0736	29
30	5.7435	0.1741	79.0581	13.7648	0.0126	0.0726	30
35	7.6861	0.1301	111.435	14.4982	0.0090	0.0690	35
40	10.2857	0.0972	154.762	15.0463	0.0065	0.0665	40
45	13.7646	0.0727	212.743	15.4558	0.0047	0.0647	45
50	18.4201	0.0543	290.336	15.7619	0.0034	0.0634	50
55	24.6503	0.0406	394.172	15.9905	0.0025	0.0625	55
60	32.9876	0.0303	533.128	16.1614	0.0019	0.0619	60
65	44.1449	0.0227	719.082	16.2891	0.0014	0.0614	65
70	59.0758	0.0169	967.931	16.3845	0.0010	0.0610	70
75	79.0568	0.0126	1300.95	16.4558	0.0008	0.0608	75
80	105.796	0.0095	1746.60	16.5091	0.0006	0.0606	80
85	141.579	0.0071	2342.98	16.5489	0.0004	0.0604	85
90	189.464	0.0053	3141.07	16.5787	0.0003	0.0603	90
95	253.546	0.0039	4209.10	16.6009	0.0002	0.0602	95
100	339.301	0.0029	5638.36	16.6175	0.0002	0.0602	100

TABLE C–11 Discrete Compounding; $i = 7\%$

	Single Payment		Uniform Series				
	Compound Amount Factor	Present Worth Factor	Compound Amount Factor	Present Worth Factor	Sinking Fund Factor	Capital Recovery Factor	
N	To find *F* Given *P* *F/P*	To find *P* Given *F* *P/F*	To find *F* Given *A* *F/A*	To find *P* Given *A* *P/A*	To find *A* Given *F* *A/F*	To find *A* Given *P* *A/P*	*N*
1	1.0700	0.9346	1.0000	0.9346	1.0000	1.0700	1
2	1.1449	0.8734	2.0700	1.8080	0.4831	0.5531	2
3	1.2250	0.8163	3.2149	2.6243	0.3111	0.3811	3
4	1.3108	0.7629	4.4399	3.3872	0.2252	0.2952	4
5	1.4026	0.7130	5.7507	4.1002	0.1739	0.2439	5
6	1.5007	0.6663	7.1533	4.7665	0.1398	0.2098	6
7	1.6058	0.6227	8.6540	5.3893	0.1156	0.1856	7
8	1.7182	0.5820	10.2598	5.9713	0.0975	0.1675	8
9	1.8385	0.5439	11.9780	6.5152	0.0835	0.1535	9
10	1.9672	0.5083	13.8164	7.0236	0.0724	0.1424	10
11	2.1049	0.4751	15.7836	7.4987	0.0634	0.1334	11
12	2.2522	0.4440	17.8884	7.9427	0.0559	0.1259	12
13	2.4098	0.4150	20.1406	8.3576	0.0497	0.1197	13
14	2.5785	0.3878	22.5505	8.7455	0.0443	0.1143	14
15	2.7590	0.3624	25.1290	9.1079	0.0398	0.1098	15
16	2.9522	0.3387	27.8880	9.4466	0.0359	0.1059	16
17	3.1588	0.3166	30.8402	9.7632	0.0324	0.1024	17
18	3.3799	0.2959	33.9990	10.0591	0.0294	0.0994	18
19	3.6165	0.2765	37.3790	10.3356	0.0268	0.0968	19
20	3.8697	0.2584	40.9955	10.5940	0.0244	0.0944	20
21	4.1406	0.2415	44.8652	10.8355	0.0223	0.0923	21
22	4.4304	0.2257	49.0057	11.0612	0.0204	0.0904	22
23	4.7405	0.2109	53.4361	11.2722	0.0187	0.0887	23
24	5.0724	0.1971	58.1766	11.4693	0.0172	0.0872	24
25	5.4274	0.1842	63.2490	11.6536	0.0158	0.0858	25
26	5.8074	0.1722	68.6764	11.8258	0.0146	0.0846	26
27	6.2139	0.1609	74.4838	11.9867	0.0134	0.0834	27
28	6.6488	0.1504	80.6977	12.1371	0.0124	0.0824	28
29	7.1143	0.1406	87.3465	12.2777	0.0114	0.0814	29
30	7.6123	0.1314	94.4607	12.4090	0.0106	0.0806	30
35	10.6766	0.0937	138.237	12.9477	0.0072	0.0772	35
40	14.9744	0.0668	199.635	13.3317	0.0050	0.0750	40
45	21.0024	0.0476	285.749	13.6055	0.0035	0.0735	45
50	29.4570	0.0339	406.529	13.8007	0.0025	0.0725	50
55	41.3150	0.0242	575.928	13.9399	0.0017	0.0717	55
60	57.9464	0.0173	813.520	14.0392	0.0012	0.0712	60
65	81.2728	0.0123	1146.75	14.1099	0.0009	0.0709	65
70	113.989	0.0088	1614.13	14.1604	0.0006	0.0706	70
75	159.876	0.0063	2269.66	14.1964	0.0004	0.0704	75
80	224.234	0.0045	3189.06	14.2220	0.0003	0.0703	80
85	314.500	0.0032	4478.57	14.2403	0.0002	0.0702	85
90	441.102	0.0023	6287.18	14.2533	0.0002	0.0702	90
95	618.669	0.0016	8823.85	14.2626	0.0001	0.0701	95
100	867.715	0.0012	12381.7	14.2693	[a]	0.0701	100

[a] Less than 0.0001.

TABLE C–12 Discrete Compounding; $i = 8\%$

	Single Payment		Uniform Series				
	Compound Amount Factor	Present Worth Factor	Compound Amount Factor	Present Worth Factor	Sinking Fund Factor	Capital Recovery Factor	
N	To find *F* Given *P* *F/P*	To find *P* Given *F* *P/F*	To find *F* Given *A* *F/A*	To find *P* Given *A* *P/A*	To find *A* Given *F* *A/F*	To find *A* Given *P* *A/P*	*N*
1	1.0800	0.9259	1.0000	0.9259	1.0000	1.0800	1
2	1.1664	0.8573	2.0800	1.7833	0.4808	0.5608	2
3	1.2597	0.7938	3.2464	2.5771	0.3080	0.3880	3
4	1.3605	0.7350	4.5061	3.3121	0.2219	0.3019	4
5	1.4693	0.6806	5.8666	3.9927	0.1705	0.2505	5
6	1.5869	0.6302	7.3359	4.6229	0.1363	0.2163	6
7	1.7138	0.5835	8.9228	5.2064	0.1121	0.1921	7
8	1.8509	0.5403	10.6366	5.7466	0.0940	0.1740	8
9	1.9990	0.5002	12.4876	6.2469	0.0801	0.1601	9
10	2.1589	0.4632	14.4866	6.7101	0.0690	0.1490	10
11	2.3316	0.4289	16.6455	7.1390	0.0601	0.1401	11
12	2.5182	0.3971	18.9771	7.5361	0.0527	0.1327	12
13	2.7196	0.3677	21.4953	7.9038	0.0465	0.1265	13
14	2.9372	0.3405	24.2149	8.2442	0.0413	0.1213	14
15	3.1722	0.3152	27.1521	8.5595	0.0368	0.1168	15
16	3.4259	0.2919	30.3243	8.8514	0.0330	0.1130	16
17	3.7000	0.2703	33.7502	9.1216	0.0296	0.1096	17
18	3.9960	0.2502	37.4502	9.3719	0.0267	0.1067	18
19	4.3157	0.2317	41.4463	9.6036	0.0241	0.1041	19
20	4.6610	0.2145	45.7620	9.8181	0.0219	0.1019	20
21	5.0338	0.1987	50.4229	10.0168	0.0198	0.0998	21
22	5.4365	0.1839	55.4567	10.2007	0.0180	0.0980	22
23	5.8715	0.1703	60.8933	10.3711	0.0164	0.0964	23
24	6.3412	0.1577	66.7647	10.5288	0.0150	0.0950	24
25	6.8485	0.1460	73.1059	10.6748	0.0137	0.0937	25
26	7.3964	0.1352	79.9544	10.8100	0.0125	0.0925	26
27	7.9881	0.1252	87.3507	10.9352	0.0114	0.0914	27
28	8.6271	0.1159	95.3388	11.0511	0.0105	0.0905	28
29	9.3173	0.1073	103.966	11.1584	0.0096	0.0896	29
30	10.0627	0.0994	113.283	11.2578	0.0088	0.0888	30
35	14.7853	0.0676	172.317	11.6546	0.0058	0.0858	35
40	21.7245	0.0460	259.056	11.9246	0.0039	0.0839	40
45	31.9204	0.0313	386.506	12.1084	0.0026	0.0826	45
50	46.9016	0.0213	573.770	12.2335	0.0017	0.0817	50
55	68.9138	0.0145	848.923	12.3186	0.0012	0.0812	55
60	101.257	0.0099	1253.21	12.3766	0.0008	0.0808	60
65	148.780	0.0067	1847.25	12.4160	0.0005	0.0805	65
70	218.606	0.0046	2720.08	12.4428	0.0004	0.0804	70
75	321.204	0.0031	4002.55	12.4611	0.0002	0.0802	75
80	471.955	0.0021	5886.93	12.4735	0.0002	0.0802	80
85	693.456	0.0014	8655.71	12.4820	0.0001	0.0801	85
90	1018.92	0.0010	12723.9	12.4877	[a]	0.0801	90
95	1497.12	0.0007	18071.5	12.4917	[a]	0.0801	95
100	2199.76	0.0005	27484.5	12.4943	[a]	0.0800	100
∞				12.5000		0.0800	∞

[a] Less than 0.0001.

TABLE C–13 Discrete Compounding; $i = 9\%$

	Single Payment		Uniform Series				
	Compound Amount Factor	Present Worth Factor	Compound Amount Factor	Present Worth Factor	Sinking Fund Factor	Capital Recovery Factor	
	To find *F* Given *P*	To find *P* Given *F*	To find *F* Given *A*	To find *P* Given *A*	To find *A* Given *F*	To find *A* Given *P*	
N	*F/P*	*P/F*	*F/A*	*P/A*	*A/F*	*A/P*	*N*
1	1.0900	0.9174	1.000	0.917	1.00000	1.09000	1
2	1.1881	0.8417	2.090	1.759	0.47847	0.56847	2
3	1.2950	0.7722	3.278	2.531	0.30505	0.39505	3
4	1.4116	0.7084	4.573	3.240	0.21867	0.30867	4
5	1.5386	0.6499	5.985	3.890	0.16709	0.25709	5
6	1.6771	0.5963	7.523	4.486	0.13292	0.22292	6
7	1.8280	0.5470	9.200	5.033	0.10869	0.19869	7
8	1.9926	0.5019	11.028	5.535	0.09067	0.18067	8
9	2.1719	0.4604	13.021	5.995	0.07680	0.16680	9
10	2.3674	0.4224	15.193	6.418	0.06582	0.15582	10
11	2.5804	0.3875	17.560	6.805	0.05695	0.14695	11
12	2.8127	0.3555	20.141	7.161	0.04965	0.13965	12
13	3.0658	0.3262	22.953	7.487	0.04357	0.13357	13
14	3.3417	0.2992	26.019	7.786	0.03843	0.12843	14
15	3.6425	0.2745	29.361	8.061	0.03406	0.12406	15
16	3.9703	0.2519	33.003	8.313	0.03030	0.12030	16
17	4.3276	0.2311	36.974	8.544	0.02705	0.11705	17
18	4.7171	0.2120	41.301	8.756	0.02421	0.11421	18
19	5.1417	0.1945	46.018	8.950	0.02173	0.11173	19
20	5.6044	0.1784	51.160	9.129	0.01955	0.10955	20
21	6.1088	0.1637	56.765	9.292	0.01762	0.10762	21
22	6.6586	0.1502	62.873	9.442	0.01590	0.10590	22
23	7.2579	0.1378	69.532	9.580	0.01438	0.10438	23
24	7.9111	0.1264	76.790	9.707	0.01302	0.10302	24
25	8.6231	0.1160	84.701	9.823	0.01181	0.10181	25
26	9.3992	0.1064	93.324	9.929	0.01072	0.10072	26
27	10.2451	0.0976	102.723	10.027	0.00973	0.09973	27
28	11.1671	0.0895	112.968	10.116	0.00885	0.09885	28
29	12.1722	0.0822	124.135	10.198	0.00806	0.09806	29
30	13.2677	0.0753	136.308	10.274	0.00734	0.09734	30
35	20.4140	0.0490	215.711	10.567	0.00464	0.09464	35
40	31.4094	0.0318	337.882	10.757	0.00296	0.09296	40
45	48.3273	0.0207	525.859	10.881	0.00190	0.09190	45
50	74.3575	0.0134	815.084	10.962	0.00123	0.09123	50
55	114.4083	0.0087	1260.092	11.014	0.00079	0.09079	55
60	176.0313	0.0057	1944.792	11.048	0.00051	0.09051	60
65	270.8460	0.0037	2998.288	11.070	0.00033	0.09033	65
70	416.7301	0.0024	4619.223	11.084	0.00022	0.09022	70
75	641.1909	0.0016	7113.232	11.094	0.00014	0.09014	75
80	986.5517	0.0010	10950.574	11.100	0.00009	0.09009	80
85	1517.9320	0.0007	16854.800	11.104	0.00006	0.09006	85
90	2235.5266	0.0004	25939.184	11.106	0.00004	0.09004	90
95	3593.4971	0.0003	39916.635	11.108	0.00003	0.09003	95
100	5529.0408	0.0002	61422.675	11.109	0.00002	0.09002	100

TABLE C–14 Discrete Compounding; $i = 10\%$

	Single Payment		Uniform Series				
	Compound Amount Factor	Present Worth Factor	Compound Amount Factor	Present Worth Factor	Sinking Fund Factor	Capital Recovery Factor	
	To find *F* Given *P*	To find *P* Given *F*	To find *F* Given *A*	To find *P* Given *A*	To find *A* Given *F*	To find *A* Given *P*	
N	*F/P*	*P/F*	*F/A*	*P/A*	*A/F*	*A/P*	*N*
1	1.1000	0.9091	1.0000	0.9091	1.0000	1.1000	1
2	1.2100	0.8264	2.1000	1.7355	0.4762	0.5762	2
3	1.3310	0.7513	3.3100	2.4869	0.3021	0.4021	3
4	1.4641	0.6830	4.6410	3.1699	0.2155	0.3155	4
5	1.6105	0.6209	6.1051	3.7908	0.1638	0.2638	5
6	1.7716	0.5645	7.7156	4.3553	0.1296	0.2296	6
7	1.9487	0.5132	9.4872	4.8684	0.1054	0.2054	7
8	2.1436	0.4665	11.4359	5.3349	0.0874	0.1874	8
9	2.3579	0.4241	13.5795	5.7590	0.0736	0.1736	9
10	2.5937	0.3855	15.9374	6.1446	0.0627	0.1627	10
11	2.8531	0.3505	18.5312	6.4951	0.0540	0.1540	11
12	3.1384	0.3186	21.3843	6.8137	0.0468	0.1468	12
13	3.4523	0.2897	24.5227	7.1034	0.0408	0.1408	13
14	3.7975	0.2633	27.9750	7.3667	0.0357	0.1357	14
15	4.1772	0.2394	31.7725	7.6061	0.0315	0.1315	15
16	4.5950	0.2176	35.9497	7.8237	0.0278	0.1278	16
17	5.0545	0.1978	40.5447	8.0216	0.0247	0.1247	17
18	5.5599	0.1799	45.5992	8.2014	0.0219	0.1219	18
19	6.1159	0.1635	51.1591	8.3649	0.0195	0.1195	19
20	6.7275	0.1486	57.2750	8.5136	0.0175	0.1175	20
21	7.4002	0.1351	64.0025	8.6487	0.0156	0.1156	21
22	8.1403	0.1228	71.4027	8.7715	0.0140	0.1140	22
23	8.9543	0.1117	79.5430	8.8832	0.0126	0.1126	23
24	9.8497	0.1015	88.4973	8.9847	0.0113	0.1113	24
25	10.8347	0.0923	98.3470	9.0770	0.0102	0.1102	25
26	11.9182	0.0839	109.182	9.1609	0.0092	0.1092	26
27	13.1100	0.0763	121.100	9.2372	0.0083	0.1083	27
28	14.4210	0.0693	134.210	9.3066	0.0075	0.1075	28
29	15.8631	0.0630	148.631	9.3696	0.0067	0.1067	29
30	17.4494	0.0573	164.494	9.4269	0.0061	0.1061	30
35	28.1024	0.0356	271.024	9.6442	0.0037	0.1037	35
40	45.2592	0.0221	442.592	9.7791	0.0023	0.1023	40
45	72.8904	0.0137	718.905	9.8628	0.0014	0.1014	45
50	117.391	0.0085	1163.91	9.9148	0.0009	0.1009	50
55	189.059	0.0053	1880.59	9.9471	0.0005	0.1005	55
60	304.481	0.0033	3034.81	9.9672	0.0003	0.1003	60
65	490.370	0.0020	4893.71	9.9796	0.0002	0.1002	65
70	789.746	0.0013	7887.47	9.9873	0.0001	0.1001	70
75	1271.89	0.0008	12708.9	9.9921	[a]	0.1001	75
80	2048.40	0.0005	20474.0	9.9951	[a]	0.1000	80
85	3298.97	0.0003	32979.7	9.9970	[a]	0.1000	85
90	5313.02	0.0002	53120.2	9.9981	[a]	0.1000	90
95	8556.67	0.0001	85556.7	9.9988	[a]	0.1000	95
100	13780.6	[a]	137796	9.9993	[a]	0.1000	100
∞				10.0000		0.1000	∞

[a] Less than 0.0001.

TABLE C–15 Discrete Compounding; $i = 11\%$

	Single Payment		Uniform Series				
	Compound Amount Factor	Present Worth Factor	Compound Amount Factor	Present Worth Factor	Sinking Fund Factor	Capital Recovery Factor	
N	To find *F* Given *P* *F/P*	To find *P* Given *F* *P/F*	To find *F* Given *A* *F/A*	To find *P* Given *A* *P/A*	To find *A* Given *F* *A/F*	To find *A* Given *P* *A/P*	*N*
1	1.1100	0.9009	1.000	0.901	1.00000	1.11000	1
2	1.2321	0.8116	2.110	1.713	0.47393	0.58393	2
3	1.3676	0.7312	3.342	2.444	0.29921	0.40921	4
4	1.5181	0.6587	4.710	3.102	0.21233	0.32233	4
5	1.6851	0.5935	6.228	3.696	0.16057	0.27057	5
6	1.8704	0.5346	7.913	4.231	0.12638	0.23638	6
7	2.0762	0.4817	9.783	4.712	0.10222	0.21222	7
8	2.3045	0.4339	11.859	5.146	0.08432	0.19432	8
9	2.5581	0.3909	14.164	5.537	0.07060	0.18060	9
10	2.8394	0.3522	16.722	5.889	0.05980	0.16980	10
11	3.1518	0.3173	19.561	6.207	0.05112	0.16112	11
12	3.4984	0.2858	22.713	6.492	0.04403	0.15403	12
13	3.8833	0.2575	26.212	6.750	0.03815	0.14815	13
14	4.3104	0.2320	30.095	6.982	0.03323	0.14323	14
15	4.7846	0.2090	34.405	7.191	0.02907	0.13907	15
16	5.3109	0.1883	39.190	7.379	0.02552	0.13552	16
17	5.8951	0.1696	44.501	7.549	0.02247	0.13247	17
18	6.5436	0.1528	50.396	7.702	0.01984	0.12984	18
19	7.2633	0.1377	56.939	7.839	0.01756	0.12756	19
20	8.0623	0.1240	64.203	7.963	0.01558	0.12558	20
21	8.9492	0.1117	72.265	8.075	0.01384	0.12384	21
22	9.9336	0.1007	81.214	8.176	0.01231	0.12231	22
23	11.0263	0.0907	91.148	8.266	0.01097	0.12097	23
24	12.2392	0.0817	102.174	8.348	0.00979	0.11979	24
25	13.5855	0.0736	114.413	8.422	0.00874	0.11874	25
26	15.0799	0.0663	127.999	8.488	0.00781	0.11781	26
27	16.7386	0.0597	143.079	8.548	0.00699	0.11699	27
28	18.5799	0.0538	159.817	8.602	0.00626	0.11626	28
29	20.6237	0.0485	178.397	8.650	0.00561	0.11561	29
30	22.8923	0.0437	199.021	8.694	0.00502	0.11502	30
31	25.4104	0.0394	221.913	8.733	0.00451	0.11451	31
32	28.2056	0.0355	247.324	8.769	0.00404	0.11404	32
33	31.3082	0.0319	275.529	8.801	0.00363	0.11363	33
34	34.7521	0.0288	306.837	8.829	0.00326	0.11326	34
35	38.5749	0.0259	341.590	8.855	0.00293	0.1129	35
40	65.0009	0.0154	581.826	8.951	0.00172	0.1117	40
45	109.5302	0.0091	986.639	9.008	0.00101	0.1110	45
50	184.5648	0.0054	1688.771	9.042	0.00060	0.1106	50
55	311.0025	0.0032	2818.204	9.062	0.0004	0.1104	55
60	524.0573	0.0019	4755.066	9.074	0.0002	0.1102	60
70	1488.019	0.0007	13518.356	9.085	0.0001	0.1101	70
80	4225.113	0.0002	38401.027	9.089	a	0.1100	80
∞				9.091	a	0.1100	∞

[a] Less than 0.0001.

TABLE C–16 Discrete Compounding; $i = 12\%$

	Single Payment		Uniform Series				
	Compound Amount Factor	Present Worth Factor	Compound Amount Factor	Present Worth Factor	Sinking Fund Factor	Capital Recovery Factor	
N	To find F Given P F/P	To find P Given F P/F	To find F Given A F/A	To find P Given A P/A	To find A Given F A/F	To find A Given P A/P	N
1	1.1200	0.8929	1.0000	0.8929	1.0000	1.1200	1
2	1.2544	0.7972	2.1200	1.6901	0.4717	0.5917	2
3	1.4049	0.7118	3.3744	2.4018	0.2963	0.4163	3
4	1.5735	0.6355	4.7793	3.0373	0.2092	0.3292	4
5	1.7623	0.5674	6.3528	3.6048	0.1574	0.2774	5
6	1.9738	0.5066	8.1152	4.1114	0.1232	0.2432	6
7	2.2107	0.4523	10.0890	4.5638	0.0991	0.2191	7
8	2.4760	0.4039	12.2997	4.9676	0.0813	0.2013	8
9	2.7731	0.3606	14.7757	5.3282	0.0677	0.1877	9
10	3.1058	0.3220	17.5487	5.6502	0.0570	0.1770	10
11	3.4785	0.2875	20.6546	5.9377	0.0484	0.1684	11
12	3.8960	0.2567	24.1331	6.1944	0.0414	0.1614	12
13	4.3635	0.2292	28.0291	6.4235	0.0357	0.1557	13
14	4.8871	0.2046	32.3926	6.6282	0.0309	0.1509	14
15	5.4736	0.1827	37.2797	6.8109	0.0268	0.1468	15
16	6.1304	0.1631	42.7533	6.9740	0.0234	0.1434	16
17	6.8660	0.1456	48.8837	7.1196	0.0205	0.1405	17
18	7.6900	0.1300	55.7497	7.2497	0.0179	0.1379	18
19	8.6128	0.1161	63.4397	7.3658	0.0158	0.1358	19
20	9.6463	0.1037	72.0524	7.4694	0.0139	0.1339	20
21	10.8038	0.0926	81.6987	7.5620	0.0122	0.1322	21
22	12.1003	0.0826	92.5026	7.6446	0.0108	0.1308	22
23	13.5523	0.0738	104.603	7.7184	0.0096	0.1296	23
24	15.1786	0.0659	118.155	7.7843	0.0085	0.1285	24
25	17.0001	0.0588	133.334	7.8431	0.0075	0.1275	25
26	19.0401	0.0525	150.334	7.8957	0.0067	0.1267	26
27	21.3249	0.0469	169.374	7.9426	0.0059	0.1259	27
28	23.8839	0.0419	190.699	7.9844	0.0052	0.1252	28
29	26.7499	0.0374	214.583	8.0218	0.0047	0.1247	29
30	29.9599	0.0334	241.333	8.0552	0.0041	0.1241	30
35	52.7996	0.0189	431.663	8.1755	0.0023	0.1223	35
40	93.0509	0.0107	767.091	8.2438	0.0013	0.1213	40
45	163.988	0.0061	1358.23	8.2825	0.0007	0.1207	45
50	289.002	0.0035	2400.02	8.3045	0.0004	0.1204	50
55	509.320	0.0020	4236.00	8.3170	0.0002	0.1202	55
60	897.596	0.0011	7471.63	8.3240	0.0001	0.1201	60
65	1581.87	0.0006	13173.9	8.3281	[a]	0.1201	65
70	2787.80	0.0004	23223.3	8.3303	[a]	0.1200	70
75	4913.05	0.0002	40933.8	8.3316	[a]	0.1200	75
80	8658.47	0.0001	72145.6	8.3324	[a]	0.1200	80
∞				8.333		0.1200	∞

[a] Less than 0.0001.

TABLE C–17 Discrete Compounding; $i = 15\%$

	Single Payment		Uniform Series				
	Compound Amount Factor	Present Worth Factor	Compound Amount Factor	Present Worth Factor	Sinking Fund Factor	Capital Recovery Factor	
N	To find *F* Given *P* *F/P*	To find *P* Given *F* *P/F*	To find *F* Given *A* *F/A*	To find *P* Given *A* *P/A*	To find *A* Given *F* *A/F*	To find *A* Given *P* *A/P*	*N*
1	1.1500	0.8696	1.0000	0.8696	1.0000	1.1500	1
2	1.3225	0.7561	2.1500	1.6257	0.4651	0.6151	2
3	1.5209	0.6575	3.4725	2.2832	0.2880	0.4380	3
4	1.7490	0.5718	4.9934	2.8550	0.2003	0.3503	4
5	2.0114	0.4972	6.7424	3.3522	0.1483	0.2983	5
6	2.3131	0.4323	8.7537	3.7845	0.1142	0.2642	6
7	2.6600	0.3759	11.0668	4.1604	0.0904	0.2404	7
8	3.0590	0.3269	13.7268	4.4873	0.0729	0.2229	8
9	3.5179	0.2843	16.7858	4.7716	0.0596	0.2096	9
10	4.0456	0.2472	20.3037	5.0188	0.0493	0.1993	10
11	4.6524	0.2149	24.3493	5.2337	0.0411	0.1911	11
12	5.3502	0.1869	29.0017	5.4206	0.0345	0.1845	12
13	6.1528	0.1625	34.3519	5.5831	0.0291	0.1791	13
14	7.0757	0.1413	40.5047	5.7245	0.0247	0.1747	14
15	8.1371	0.1229	47.5804	5.8474	0.0210	0.1710	15
16	9.3576	0.1069	55.7175	5.9542	0.0179	0.1679	16
17	10.7613	0.0929	65.0751	6.0472	0.0154	0.1654	17
18	12.3755	0.0808	75.8363	6.1280	0.0132	0.1632	18
19	14.2318	0.0703	88.2118	6.1982	0.0113	0.1613	19
20	16.3665	0.0611	102.444	6.2593	0.0098	0.1598	20
21	18.8215	0.0531	118.810	6.3125	0.0084	0.1584	21
22	21.6447	0.0462	137.632	6.3587	0.0073	0.1573	22
23	24.8915	0.0402	159.276	6.3988	0.0063	0.1563	23
24	28.6252	0.0349	184.168	6.4338	0.0054	0.1554	24
25	32.9189	0.0304	212.793	6.4641	0.0047	0.1547	25
26	37.8568	0.0264	245.712	6.4906	0.0041	0.1541	26
27	43.5353	0.0230	283.569	6.5135	0.0035	0.1535	27
28	50.0656	0.0200	327.104	6.5335	0.0031	0.1531	28
29	57.5754	0.0174	377.170	6.5509	0.0027	0.1527	29
30	66.2118	0.0151	434.745	6.5660	0.0023	0.1523	30
35	133.176	0.0075	881.170	6.6166	0.0011	0.1511	35
40	267.863	0.0037	1779.09	6.6418	0.0006	0.1506	40
45	538.769	0.0019	3585.13	6.6543	0.0003	0.1503	45
50	1083.66	0.0009	7217.71	6.6605	0.0001	0.1501	50
55	2179.62	0.0005	14524.1	6.6636	[a]	0.1501	55
60	4384.00	0.0002	29220.0	6.6651	[a]	0.1500	60
65	8817.78	0.0001	58778.5	6.6659	[a]	0.1500	65
70	17735.7	[a]	118231	6.6663	[a]	0.1500	70
75	35672.8	[a]	237812	6.6665	[a]	0.1500	75
80	71750.8	[a]	478332	6.6666	[a]	0.1500	80
∞				6.667		0.1500	∞

[a] Less than 0.0001.

TABLE C–18 Discrete Compounding; $i = 18\%$

	Single Payment		Uniform Series				
	Compound Amount Factor	Present Worth Factor	Compound Amount Factor	Present Worth Factor	Sinking Fund Factor	Capital Recovery Factor	
N	To find *F* Given *P* *F/P*	To find *P* Given *F* *P/F*	To find *F* Given *A* *F/A*	To find *P* Given *A* *P/A*	To find *A* Given *F* *A/F*	To find *A* Given *P* *A/P*	*N*
1	1.1800	0.8475	1.000	0.847	1.00000	1.18000	1
2	1.3924	0.7182	2.180	1.566	0.45872	0.63872	2
3	1.6430	0.6086	3.572	2.174	0.27992	0.45992	3
4	1.9388	0.5158	5.215	2.690	0.19174	0.37174	4
5	2.2878	0.4371	7.154	3.127	0.13978	0.31978	5
6	2.6996	0.3704	9.442	3.498	0.10591	0.28591	6
7	3.1855	0.3139	12.142	3.812	0.08236	0.26236	7
8	3.7589	0.2660	15.327	4.078	0.06524	0.24524	8
9	4.4355	0.2255	19.086	4.303	0.05239	0.23239	9
10	5.2338	0.1911	23.521	4.494	0.04251	0.22251	10
11	6.1759	0.1619	28.755	4.656	0.03478	0.21478	11
12	7.2876	0.1372	34.931	4.793	0.02863	0.20863	12
13	8.5994	0.1163	42.219	4.910	0.02369	0.20369	13
14	10.1472	0.0985	50.818	5.008	0.01968	0.19968	14
15	11.9737	0.0835	60.965	5.092	0.01640	0.19640	15
16	14.1290	0.0708	72.939	5.162	0.01371	0.19371	16
17	16.6722	0.0600	87.068	5.222	0.01149	0.19149	17
18	19.6733	0.0508	103.740	5.273	0.00964	0.18964	18
19	23.2144	0.0431	123.414	5.316	0.00810	0.18810	19
20	27.3930	0.0365	146.628	5.353	0.00682	0.18682	20
21	32.3238	0.0309	174.021	5.384	0.00575	0.18575	21
22	38.1421	0.0262	206.345	5.410	0.00485	0.18485	22
23	45.0076	0.0222	244.487	5.432	0.00409	0.18409	23
24	53.1090	0.0188	289.494	5.451	0.00345	0.18345	24
25	62.6686	0.0160	342.603	5.467	0.00292	0.18292	25
26	73.9490	0.0135	405.272	5.480	0.00247	0.18247	26
27	87.2598	0.0115	479.221	5.492	0.00209	0.18209	27
28	102.9665	0.0097	566.481	5.502	0.00177	0.18177	28
29	121.5005	0.0082	669.447	5.510	0.00149	0.18149	29
30	143.3706	0.0070	790.948	5.517	0.00126	0.18126	30
31	169.1774	0.0059	934.319	5.523	0.00107	0.18107	31
32	199.6293	0.0050	1103.496	5.528	0.00091	0.18091	32
33	235.5625	0.0042	1303.125	5.532	0.00077	0.18077	33
34	277.9638	0.0036	1538.688	5.536	0.00065	0.18065	34
35	327.9973	0.0030	1816.652	5.539	0.00055	0.18055	35
40	750.3783	0.0013	4163.213	5.548	0.00024	0.18024	40
45	1716.6839	0.0006	9531.577	5.552	0.00010	0.18010	45
50	3927.3569	0.0003	21813.094	5.554	a	0.18005	50
∞				5.556	a	0.18000	∞

[a] Less than 0.0001.

TABLE C–19 Discrete Compounding; i = 20%

	Single Payment		Uniform Series				
	Compound Amount Factor	Present Worth Factor	Compound Amount Factor	Present Worth Factor	Sinking Fund Factor	Capital Recovery Factor	
N	To find *F* Given *P* *F/P*	To find *P* Given *F* *P/F*	To find *F* Given *A* *F/A*	To find *P* Given *A* *P/A*	To find *A* Given *F* *A/F*	To find *A* Given *P* *A/P*	*N*
1	1.2000	0.8333	1.0000	0.8333	1.0000	1.2000	1
2	1.4400	0.6944	2.2000	1.5278	0.4545	0.6545	2
3	1.7280	0.5787	3.6400	2.1065	0.2747	0.4747	3
4	2.0736	0.4823	5.3680	2.5887	0.1863	0.3863	4
5	2.4883	0.4019	7.4416	2.9906	0.1344	0.3344	5
6	2.9860	0.3349	9.9299	3.3255	0.1007	0.3007	6
7	3.5832	0.2791	12.9159	3.6046	0.0774	0.2774	7
8	4.2998	0.2326	16.4991	3.8372	0.0606	0.2606	8
9	5.1598	0.1938	20.7989	4.0310	0.0481	0.2481	9
10	6.1917	0.1615	25.9587	4.1925	0.0385	0.2385	10
11	7.4301	0.1346	32.1504	4.3271	0.0311	0.2311	11
12	8.9161	0.1122	39.5805	4.4392	0.0253	0.2253	12
13	10.6993	0.0935	48.4966	4.5327	0.0206	0.2206	13
14	12.8392	0.0779	59.1959	4.6106	0.0169	0.2169	14
15	15.4070	0.0649	72.0351	4.6755	0.0139	0.2139	15
16	18.4884	0.0541	87.4421	4.7296	0.0114	0.2114	16
17	22.1861	0.0451	105.931	4.7746	0.0094	0.2094	17
18	26.6233	0.0376	128.117	4.8122	0.0078	0.2078	18
19	31.9480	0.0313	154.740	4.8435	0.0065	0.2065	19
20	38.3376	0.0261	186.688	4.8696	0.0054	0.2054	20
21	46.0051	0.0217	225.026	4.8913	0.0044	0.2044	21
22	55.2061	0.0181	271.031	4.9094	0.0037	0.2037	22
23	66.2474	0.0151	326.237	4.9245	0.0031	0.2031	23
24	79.4968	0.0126	392.484	4.9371	0.0025	0.2025	24
25	95.3962	0.0105	471.981	4.9476	0.0021	0.2021	25
26	114.475	0.0087	567.377	4.9563	0.0018	0.2018	26
27	137.371	0.0073	681.853	4.9636	0.0015	0.2015	27
28	164.845	0.0061	819.223	4.9697	0.0012	0.2012	28
29	197.814	0.0051	984.068	4.9747	0.0010	0.2010	29
30	237.376	0.0042	1181.88	4.9789	0.0008	0.2008	30
35	590.668	0.0017	2948.34	4.9915	0.0003	0.2003	35
40	1469.77	0.0007	7343.85	4.9966	0.0001	0.2001	40
45	3657.26	0.0003	18281.3	4.9986	[a]	0.2001	45
50	9100.43	0.0001	45497.2	4.9995	[a]	0.2000	50
55	22644.8	[a]	113219	4.9998	[a]	0.2000	55
60	56347.5	[a]	281732	4.9999	[a]	0.2000	60
∞				5.0000	[a]	0.2000	∞

[a] Less than 0.0001.

TABLE C–20 Discrete Compounding; $i = 25\%$

	Single Payment		Uniform Series				
	Compound Amount Factor	Present Worth Factor	Compound Amount Factor	Present Worth Factor	Sinking Fund Factor	Capital Recovery Factor	
N	To find *F* Given *P* *F/P*	To find *P* Given *F* *P/F*	To find *F* Given *A* *F/A*	To find *P* Given *A* *P/A*	To find *A* Given *F* *A/F*	To find *A* Given *P* *A/P*	*N*
1	1.2500	0.8000	1.0000	0.8000	1.0000	1.2500	1
2	1.5625	0.6400	2.2500	1.4400	0.4444	0.6944	2
3	1.9531	0.5120	3.8125	1.9520	0.2623	0.5123	3
4	2.4414	0.4096	5.7656	2.3616	0.1734	0.4234	4
5	3.0518	0.3277	8.2070	2.6893	0.1218	0.3718	5
6	3.8147	0.2621	11.2588	2.9514	0.0888	0.3388	6
7	4.7684	0.2097	15.0735	3.1611	0.0663	0.3163	7
8	5.9605	0.1678	19.8419	3.3289	0.0504	0.3004	8
9	7.4506	0.1342	25.8023	3.4631	0.0388	0.2888	9
10	9.3132	0.1074	33.2529	3.5705	0.0301	0.2801	10
11	11.6415	0.0859	42.5661	3.6564	0.0235	0.2735	11
12	14.5519	0.0687	54.2077	3.7251	0.0184	0.2684	12
13	18.1899	0.0550	68.7596	3.7801	0.0145	0.2645	13
14	22.7374	0.0440	86.9495	3.8241	0.0115	0.2615	14
15	28.4217	0.0352	109.687	3.8593	0.0091	0.2591	15
16	35.5271	0.0281	138.109	3.8874	0.0072	0.2572	16
17	44.4089	0.0225	173.636	3.9099	0.0058	0.2558	17
18	55.5112	0.0180	218.045	3.9279	0.0046	0.2546	18
19	69.3889	0.0144	273.556	3.9424	0.0037	0.2537	19
20	86.7362	0.0115	342.945	3.9539	0.0029	0.2529	20
21	108.420	0.0092	429.681	3.9631	0.0023	0.2523	21
22	135.525	0.0074	538.101	3.9705	0.0019	0.2519	22
23	169.407	0.0059	673.626	3.9764	0.0015	0.2515	23
24	211.758	0.0047	843.033	3.9811	0.0012	0.2512	24
25	264.698	0.0038	1054.79	3.9849	0.0009	0.2509	25
26	330.872	0.0030	1319.49	3.9879	0.0008	0.2508	26
27	413.590	0.0024	1650.36	3.9903	0.0006	0.2506	27
28	516.988	0.0019	2063.95	3.9923	0.0005	0.2505	28
29	646.235	0.0015	2580.94	3.9938	0.0004	0.2504	29
30	807.794	0.0012	3227.17	3.9950	0.0003	0.2503	30
35	2465.19	0.0004	9856.76	3.9984	0.0001	0.2501	35
40	7523.16	0.0001	30088.7	3.9995	[a]	0.2500	40
45	22958.9	[a]	91831.5	3.9998	[a]	0.2500	45
50	70064.9	[a]	280256	3.9999	[a]	0.2500	50
∞				4.0000		0.2500	∞

[a] Less than 0.0001.

TABLE C–21 Discrete Compounding; $i = 30\%$

	Single Payment		Uniform Series				
	Compound Amount Factor	Present Worth Factor	Compound Amount Factor	Present Worth Factor	Sinking Fund Factor	Capital Recovery Factor	
	To find *F* Given *P*	To find *P* Given *F*	To find *F* Given *A*	To find *P* Given *A*	To find *A* Given *F*	To find *A* Given *P*	
N	*F/P*	*P/F*	*F/A*	*P/A*	*A/F*	*A/P*	*N*
1	1.3000	0.7692	1.000	0.769	1.0000	1.3000	1
2	1.6900	0.5917	2.300	1.361	0.4348	0.7348	2
3	2.1970	0.4552	3.990	1.816	0.2506	0.5506	3
4	2.8561	0.3501	6.187	2.166	0.1616	0.4616	4
5	3.7129	0.2693	9.043	2.436	0.1106	0.4106	5
6	4.8268	0.2072	12.756	2.643	0.0784	0.3784	6
7	6.2749	0.1594	17.583	2.802	0.0569	0.3569	7
8	8.1573	0.1226	23.858	2.925	0.0419	0.3419	8
9	10.604	0.0943	32.015	3.019	0.0312	0.3312	9
10	13.786	0.0725	42.619	3.092	0.0235	0.3235	10
11	17.922	0.0558	56.405	3.147	0.0177	0.3177	11
12	23.298	0.0429	74.327	3.190	0.0135	0.3135	12
13	30.287	0.0330	97.625	3.223	0.0102	0.3102	13
14	39.374	0.0254	127.91	3.249	0.0078	0.3078	14
15	51.186	0.0195	167.29	3.268	0.0060	0.3060	15
16	66.542	0.0150	218.47	3.283	0.0046	0.3046	16
17	86.504	0.0116	285.01	3.295	0.0035	0.3035	17
18	112.46	0.0089	371.52	3.304	0.0027	0.3027	18
19	146.19	0.0068	483.97	3.311	0.0021	0.3021	19
20	190.05	0.0053	630.16	3.316	0.0016	0.3016	20
21	247.06	0.0040	820.21	3.320	0.0012	0.3012	21
22	321.18	0.0031	1067.3	3.323	0.0009	0.3009	22
23	417.54	0.0024	1388.5	3.325	0.0007	0.3007	23
24	542.80	0.0018	1806.0	3.327	0.0005	0.3005	24
25	705.64	0.0014	2348.8	3.329	0.0004	0.3004	25
26	917.33	0.0011	3054.4	3.330	0.0003	3.3003	26
27	1192.5	0.0008	3971.8	3.331	0.0003	0.3003	27
28	1550.3	0.0006	5164.3	3.331	0.0002	0.3002	28
29	2015.4	0.0005	6714.6	3.332	0.0002	0.3002	29
30	2620.0	0.0004	8730.0	3.332	0.0001	0.3001	30
31	3406.0	0.0003	11350.	3.332	[a]	0.3001	31
32	4427.8	0.0002	14756.	3.333	[a]	0.3001	32
33	5756.1	0.0002	19184.	3.333	[a]	0.3001	33
34	7483.0	0.0001	24940.	3.333	[a]	0.3000	34
35	9727.8	0.0001	32423.	3.333	[a]	0.3000	35
∞				3.333	[a]	0.3000	∞

[a] Less than 0.0001.

TABLE C–22 Discrete Compounding; $i = 40\%$

	Single Payment		Uniform Series				
	Compound Amount Factor	Present Worth Factor	Compound Amount Factor	Present Worth Factor	Sinking Fund Factor	Capital Recovery Factor	
N	To find *F* Given *P* *F/P*	To find *P* Given *F* *P/F*	To find *F* Given *A* *F/A*	To find *P* Given *A* *P/A*	To find *A* Given *F* *A/F*	To find *A* Given *P* *A/P*	*N*
1	1.4000	0.7143	1.000	0.714	1.000	1.4000	1
2	1.9600	0.5102	2.400	1.224	0.4167	0.8167	2
3	2.7440	0.3644	4.360	1.589	0.2294	0.6294	3
4	3.8416	0.2603	7.104	1.849	0.1408	0.5408	4
5	5.3782	0.1859	10.946	2.035	0.0934	0.4914	5
6	7.5295	0.1328	16.324	2.168	0.0613	0.4613	6
7	10.541	0.0949	23.853	2.263	0.0419	0.4419	7
8	14.758	0.0678	34.395	2.331	0.0291	0.4291	8
9	20.661	0.0484	49.153	2.379	0.0203	0.4203	9
10	28.925	0.0346	69.814	2.414	0.0143	0.4143	10
11	40.496	0.0247	98.739	2.438	0.0101	0.4101	11
12	56.694	0.0176	139.23	2.456	0.0072	0.4072	12
13	79.371	0.0126	195.93	2.469	0.0051	0.4051	13
14	111.12	0.0090	275.30	2.478	0.0036	0.4036	14
15	155.57	0.0064	386.42	2.484	0.0026	0.4026	15
16	217.80	0.0046	541.99	2.489	0.0018	0.4019	16
17	304.91	0.0033	759.78	2.492	0.0013	0.4013	17
18	426.88	0.0023	1064.7	2.494	0.0009	0.4009	18
19	597.63	0.0017	1491.6	2.496	0.0007	0.4007	19
20	836.68	0.0012	2089.2	2.497	0.0005	0.4005	20
21	1171.4	0.0009	2925.9	2.498	0.0003	0.4003	21
22	1639.9	0.0006	4097.2	2.498	0.0002	0.4002	22
23	2295.9	0.0004	5737.1	2.499	0.0002	0.4002	23
24	3214.2	0.0003	8033.0	2.499	0.0001	0.4001	24
25	4499.9	0.0002	11247.	2.499	[a]	0.4001	25
26	6299.8	0.0002	15747.	2.500	[a]	0.4001	26
27	8819.8	0.0001	22047.	2.500	[a]	0.4000	27
28	12348.	0.0001	30867.	2.500	[a]	0.4000	28
29	17287.	0.0001	43214.	2.500	[a]	0.4000	29
30	24201.	[a]	60501.	2.500	[a]	0.4000	30
∞				2.500		0.4000	∞

[a] Less than 0.0001.

TABLE C–23 Discrete Compounding; $i = 50\%$

	Single Payment		Uniform Series				
	Compound Amount Factor	Present Worth Factor	Compound Amount Factor	Present Worth Factor	Sinking Fund Factor	Capital Recovery Factor	
N	To find *F* Given *P* *F/P*	To find *P* Given *F* *P/F*	To find *F* Given *A* *F/A*	To find *P* Given *A* *P/A*	To find *A* Given *F* *A/F*	To find *A* Given *P* *A/P*	*N*
1	1.5000	0.6667	1.000	0.667	1.0000	1.5000	1
2	2.2500	0.4444	2.500	1.111	0.4000	0.9000	2
3	3.3750	0.2963	4.750	1.407	0.2101	0.7105	3
4	5.0625	0.1975	8.125	1.605	0.1231	0.6231	4
5	7.5938	0.1317	13.188	1.737	0.0758	0.5758	5
6	11.391	0.0878	20.781	1.824	0.0481	0.5481	6
7	17.086	0.0585	32.172	1.883	0.0311	0.5311	7
8	25.629	0.0390	49.258	1.922	0.0203	0.5203	8
9	38.443	0.0260	74.887	1.948	0.0134	0.5134	9
10	57.665	0.0173	113.33	1.965	0.0088	0.5088	10
11	86.498	0.0116	171.00	1.977	0.0059	0.5059	11
12	129.75	0.0077	257.49	1.985	0.0039	0.5039	12
13	194.62	0.0051	387.24	1.990	0.0026	0.5026	13
14	291.93	0.0034	581.86	1.993	0.0017	0.5017	14
15	437.89	0.0023	873.79	1.995	0.0011	0.5011	15
16	656.84	0.0015	1311.7	1.997	0.0008	0.5008	16
17	985.26	0.0010	1968.5	1.998	0.0005	0.5005	17
18	1477.9	0.0007	2953.8	1.999	0.0003	0.5003	18
19	2216.8	0.0005	4431.7	1.999	0.0002	0.5002	19
20	3325.3	0.0003	6648.5	1.999	0.0002	0.5002	20
21	4987.9	0.0002	9973.8	2.000	0.0001	0.5001	21
22	7481.8	0.0001	14962.	2.000	[a]	0.5001	22
23	11223.	0.0001	22443.	2.000	[a]	0.5000	23
24	16834.	0.0001	33666.	2.000	[a]	0.5000	24
25	25251.	[a]	50500.	2.000	[a]	0.5000	25
∞				2.000		0.5000	∞

[a] Less than 0.0001.

TABLE C–24 Gradient to Present Worth Conversion Factor for Discrete Compounding (to Find *P*, Given *G*)

$$(P/G, i\%, N) = \frac{1}{i}\left[\frac{(1+i)^N - 1}{i(1+i)^N} - \frac{N}{(1+i)^N}\right]$$

N	1%	2%	5%	8%	10%	12%	15%	20%	25%	30%	50%	N
1	0.00	0.00	0.00	0.00	0.00	0.00	0.00	0.00	0.00	0.00	0.00	1
2	0.98	0.96	0.91	0.86	0.83	0.80	0.76	0.69	0.64	0.59	0.44	2
3	2.92	2.85	2.63	2.45	2.33	2.22	2.07	1.85	1.66	1.50	1.04	3
4	5.80	5.62	5.10	4.65	4.38	4.13	3.79	3.30	2.89	2.55	1.63	4
5	9.61	9.24	8.24	7.37	6.86	6.40	5.78	4.91	4.20	3.63	2.16	5
6	14.32	13.68	11.97	10.52	9.68	8.93	7.94	6.58	5.51	4.67	2.60	6
7	19.92	18.90	16.23	14.02	12.76	11.64	10.19	8.26	6.77	5.62	2.95	7
8	26.38	24.88	20.97	17.81	16.03	14.47	12.48	9.88	7.95	6.48	3.22	8
9	33.69	31.57	26.13	21.81	19.42	17.36	14.75	11.43	9.02	7.23	3.43	9
10	41.84	38.95	31.65	25.98	22.89	20.25	16.98	12.89	9.99	7.89	3.58	10
11	50.80	47.00	37.50	30.27	26.40	23.13	19.13	14.23	10.85	8.45	3.70	11
12	60.57	55.67	43.62	34.63	29.90	25.95	21.18	15.47	11.60	8.92	3.78	12
15	94.48	85.20	63.29	47.89	40.15	33.92	26.69	18.51	13.33	9.92	3.92	15
20	165.46	144.60	98.49	69.09	55.41	44.97	33.58	21.74	14.89	10.70	3.99	20
25	252.89	214.26	134.23	87.80	67.70	53.10	38.03	23.43	15.56	10.98	4.00	25
30	355.00	291.72	168.62	103.46	77.08	58.78	40.75	24.26	15.83	11.07	—	30
35	470.15	374.88	200.58	116.09	83.99	62.61	42.36	24.66	15.94	11.10	—	35
40	596.85	461.99	229.55	126.04	88.95	65.12	43.28	24.85	15.98	11.11	—	40
45	733.70	551.56	255.31	133.73	92.45	66.73	43.81	24.93	15.99	—	—	45
50	879.41	642.36	277.91	139.59	94.89	67.76	44.10	24.97	16.00	—	—	50
60	1192.80	823.70	314.34	147.30	97.70	68.81	44.34	24.99	—	—	—	60
70	1528.64	999.83	340.84	151.53	98.99	69.21	44.42	—	—	—	—	70
80	1879.87	1166.79	359.65	153.80	99.56	69.36	44.47	—	—	—	—	80
90	2240.55	1322.17	372.75	154.99	99.81	—	—	—	—	—	—	90
100	2605.76	1464.75	381.75	155.61	99.92	—	—	—	—	—	—	100

TABLE C–25 Gradient to Uniform Series Conversion Factor for Discrete Compounding (to Find A, Given G)

$$(A/G,\ i\%,\ N) = \frac{1}{i} - \frac{N}{(1+i)^N - 1}$$

N	1%	2%	5%	8%	10%	12%	15%	20%	25%	30%	50%	N
1	0.0001	0.0000	0.00	0.0000	0.0000	0.0000	0.0000	0.0000	0.0000	0.00	0.00	1
2	0.4974	0.4950	0.49	0.4808	0.4762	0.4717	0.4651	0.4545	0.4444	0.43	0.40	2
3	0.9932	0.9868	0.97	0.9487	0.9366	0.9246	0.9071	0.8791	0.8525	0.83	0.74	3
4	1.4874	1.4752	1.44	1.4040	1.3812	1.3589	1.3263	1.2742	1.2249	1.18	1.02	4
5	1.9799	1.9604	1.90	1.8465	1.8101	1.7746	1.7228	1.6405	1.5631	1.49	1.24	5
6	2.4708	2.4422	2.36	2.2763	2.2236	2.1720	2.0972	1.9788	1.8683	1.77	1.42	6
7	2.9600	2.9208	2.81	2.6937	2.6216	2.5515	2.4498	2.2902	2.1424	2.01	1.56	7
8	3.4476	3.3961	3.24	3.0985	3.0045	2.9131	2.7813	2.5756	2.3872	2.22	1.68	8
9	3.9335	3.8680	3.68	3.4910	3.3724	3.2574	3.0922	2.8364	2.6048	2.40	1.76	9
10	4.4177	4.3367	4.10	3.8713	3.7255	3.5847	3.3832	3.0739	2.7971	2.55	1.82	10
11	4.9003	4.8021	4.51	4.2395	4.0641	3.8953	3.6549	3.2893	2.9663	2.68	1.87	11
12	5.3813	5.2642	4.92	4.5957	4.3884	4.1897	3.9082	3.4841	3.1145	2.80	1.91	12
15	6.8141	6.6309	6.10	5.5945	5.2789	4.9803	4.5650	3.9588	3.4530	3.03	1.97	15
20	9.1692	8.8433	7.90	7.0369	6.5081	6.0202	5.3651	4.4643	3.7667	3.23	1.99	20
25	11.4829	10.9744	9.52	8.2254	7.4580	6.7708	5.8834	4.7352	3.9052	3.30	2.00	25
30	13.7555	13.0251	10.97	9.1897	8.1762	7.2974	6.2066	4.8731	3.9628	3.32	—	30
35	15.9869	14.9961	12.25	9.9611	8.7086	7.6577	6.4019	4.9406	3.9858	3.33	—	35
40	18.1774	16.8885	13.38	10.5699	9.0962	7.8988	6.5168	4.9728	3.9947	3.33	—	40
45	20.3271	18.7033	14.36	11.0447	9.3740	8.0572	6.5830	4.9877	3.9980	—	—	45
50	22.4362	20.4420	15.22	11.4107	9.5704	8.1597	6.6205	4.9945	3.9993	—	—	50
60	26.5331	23.6961	16.61	11.9015	9.8023	8.2664	6.6530	4.9989	—	—	—	60
70	30.4701	26.6632	17.62	12.1783	9.9113	8.3082	6.6627	—	—	—	—	70
80	34.2490	29.3572	18.35	12.3301	9.9609	8.3241	6.6656	—	—	—	—	80
90	37.8723	31.7929	18.87	12.4116	9.9831	—	—	—	—	—	—	90
100	41.3424	33.9863	19.23	12.4545	9.9927	—	—	—	—	—	—	100

D

INTEREST AND ANNUITY TABLES FOR CONTINUOUS COMPOUNDING

(For various values of r from 1% to 25%)

$r =$ nominal interest rate per period, compounded continuously
$N =$ number of compounding periods

$$(F/P, r\%, N) = e^{rN}$$

$$(P/F, r\%, N) = e^{-rN} = \frac{1}{e^{rN}}$$

$$(F/A, r\%, N) = \frac{e^{rN} - 1}{e^{r} - 1}$$

$$(P/A, r\%, N) = \frac{e^{rN} - 1}{e^{rN}(e^{r} - 1)}$$

$$(F/\bar{A}, r\%, N) = \frac{e^{rN} - 1}{r}$$

$$(P/\bar{A}, r\%, N) = \frac{e^{rN} - 1}{re^{rN}}$$

The computer program called ITABLE that is listed in Appendix C can be utilized to produce interest tables for discrete cash lows and *continuous compounding* of integer- or fractional-valued interest rates expressed in decimal form. On the diskette provided by your instructor, the program is used by typing ITABLE and pressing the Return (Enter) key.

EXAMPLES

```
THIS PROGRAM COMPUTES INTEREST TABLES FOR DISCRETE
(DIS) COMPOUNDING AND CONTINUOUS (CON) COMPOUNDING.
END-OF-PERIOD CASH FLOW CONVENTION IS ASSUMED. ENTER
'I' PER PERIOD AS A DECIMAL FOR DISCRETE COMPOUNDING
AND 'R' PER PERIOD AS A DECIMAL FOR CONTINUOUS
COMPOUNDING. 'N' IS THE NUMBER OF COMPOUNDING PERIODS.
```

A

```
ENTER VALUE OF INTEREST RATE AS A DECIMAL ? 0.05
DISCRETE (DIS) OR CONTINUOUS (CON) COMPOUNDING? con
ENTER STARTING AND ENDING VALUES OF N
FOR TABULATIONS. (SEPARATE BY COMMA)? 1,10
```

THE NOMINAL RATE (R) = 5.0%

N	F/P	P/F	P/A	A/P
1	1.0513	0.9512	0.9512	1.0513
2	1.1052	0.9048	1.8561	0.5388
3	1.1618	0.8607	2.7168	0.3681
4	1.2214	0.8187	3.5355	0.2828
5	1.2840	0.7788	4.3143	0.2318
6	1.3499	0.7408	5.0551	0.1978
7	1.4191	0.7047	5.7598	0.1736

N	F/P	P/F	P/A	A/P
8	1.4918	0.6703	6.4301	0.1555
9	1.5683	0.6376	7.0678	0.1415
10	1.6487	0.6065	7.6743	0.1303

THE NOMINAL RATE (R) = 5.0%

N	A/F	F/A	P/G	A/G
1	1.0000	1.0000	0.0000	0.0000
2	0.4875	2.0513	0.9048	0.4875
3	0.3168	3.1564	2.6262	0.9667
4	0.2316	4.3183	5.0824	1.4375
5	0.1805	5.5397	8.1976	1.9001
6	0.1465	6.8237	11.9017	2.3544
7	0.1223	8.1736	16.1298	2.8004
8	0.1042	9.5926	20.8221	3.2382
9	0.0902	11.0845	25.9231	3.6678
10	0.0790	12.6528	31.3819	4.0892

DO YOU WANT TO PRINT THIS TABLE (Y/N)? **Y**
DO YOU WANT TO RUN FOR ANOTHER INTEREST RATE
VALUE (Y/N)?**Y**

B

ENTER VALUE OF INTEREST RATE AS A DECIMAL? **0.123**
DISCRETE (DIS) OR CONTINUOUS (CON) COMPOUNDING? **CON**

ENTER STARTING AND ENDING VALUES OF N
FOR TABULATIONS (SEPARATE BY COMMA)? **5,15**

THE NOMINAL RATE (R) = 12.3%

N	F/P	P/F	P/A	A/P
5	1.8497	0.5406	3.5097	0.2849
6	2.0917	0.4781	3.9877	0.2508
7	2.3655	0.4227	4.4105	0.2267
8	2.6751	0.3738	4.7843	0.2090
9	3.0253	0.3305	5.1148	0.1955
10	3.4212	0.2923	5.4071	0.1849
11	3.8690	0.2585	5.6656	0.1765
12	4.3754	0.2286	5.8941	0.1697
13	4.9481	0.2021	6.0962	0.1640
14	5.5957	0.1787	6.2749	0.1594
15	6.3281	0.1580	6.4330	0.1554

THE NOMINAL RATE (R) = 12.3%

N	A/F	F/A	P/G	A/G
5	0.1540	6.4917	6.1615	1.7556
6	0.1199	8.3413	8.5519	2.1446
7	0.0958	10.4331	11.0883	2.1541
8	0.0781	12.7986	13.7050	2.8646
9	0.0646	15.4737	16.3494	3.1965
10	0.0541	18.4990	18.9800	3.5102
11	0.0456	21.9202	21.5647	3.8063
12	0.0388	25.7892	24.0787	4.0852
13	0.0332	30.1646	26.5039	4.3476
14	0.0285	35.1127	28.8271	4.5940
15	0.0246	40.7084	31.0395	4.8251

TABLE D–1 Continuous Compounding; $r = 1\%$

	Discrete Flows				Continuous Flows		
	Single Payment		Uniform Series		Uniform Series		
	Compound Amount Factor	Present Worth Factor	Compound Amount Factor	Present Worth Factor	Compound Amount Factor	Present Worth Factor	
	To find *F* Given *P*	To find *P* Given *F*	To find *F* Given *A*	To find *P* Given *A*	To find *F* Given $\bar{A}$	To find *P* Given $\bar{A}$	
N	*F/P*	*P/F*	*F/A*	*P/A*	$F/\bar{A}$	$P/\bar{A}$	*N*
1	1.0101	0.9900	1.0000	0.9900	1.0050	0.9950	1
2	1.0202	0.9802	2.0101	1.9703	2.0201	1.9801	2
3	1.0305	0.9704	3.0303	2.9407	3.0455	2.9554	3
4	1.0408	0.9608	4.0607	3.9015	4.0811	3.9211	4
5	1.0513	0.9512	5.1015	4.8527	5.1271	4.8771	5
6	1.0618	0.9418	6.1528	5.7945	6.1837	5.8235	6
7	1.0725	0.9324	7.2146	6.7269	7.2508	6.7606	7
8	1.0833	0.9231	8.2871	7.6500	8.3287	7.6884	8
9	1.0942	0.9139	9.3704	8.5639	9.4174	8.6069	9
10	1.1052	0.9048	10.4646	9.4688	10.5171	9.5163	10
11	1.1163	0.8958	11.5698	10.3646	11.6278	10.4166	11
12	1.1275	0.8869	12.6860	11.2515	12.7497	11.3080	12
13	1.1388	0.8781	13.8135	12.1296	13.8828	12.1905	13
14	1.1503	0.8694	14.9524	12.9990	15.0274	13.0642	14
15	1.1618	0.8607	16.1026	13.8597	16.1834	13.9292	15
16	1.1735	0.8521	17.2645	14.7118	17.3511	14.7856	16
17	1.1853	0.8437	18.4380	15.5555	18.5305	15.6335	17
18	1.1972	0.8353	19.6233	16.3908	19.7217	16.4730	18
19	1.2092	0.8270	20.8205	17.2177	20.9250	17.3041	19
20	1.2214	0.8187	22.0298	18.0365	22.1403	18.1269	20
21	1.2337	0.8106	23.2512	18.8470	23.3678	18.9416	21
22	1.2461	0.8025	24.4849	19.6496	24.6077	19.7481	22
23	1.2586	0.7945	25.7309	20.4441	25.8600	20.5466	23
24	1.2712	0.7866	26.9895	21.2307	27.1249	21.3372	24
25	1.2840	0.7788	28.2608	22.0095	28.4025	22.1199	25
26	1.2969	0.7711	29.5448	22.7806	29.6930	22.8948	26
27	1.3100	0.7634	30.8417	23.5439	30.9964	23.6621	27
28	1.3231	0.7558	32.1517	24.2997	32.3130	24.4216	28
29	1.3364	0.7483	33.4748	25.0480	33.6427	25.1736	29
30	1.3499	0.7408	34.8113	25.7888	34.9859	25.9182	30
35	1.4191	0.7047	41.6976	29.3838	41.9068	29.5312	35
40	1.4918	0.6703	48.9370	32.8034	49.1825	32.9680	40
45	1.5683	0.6376	56.5476	36.0563	56.8312	36.2372	45
50	1.6487	0.6065	64.5483	39.1505	64.8721	39.3469	50
55	1.7333	0.5769	72.9593	42.0939	73.3253	42.3050	55
60	1.8221	0.5488	81.8015	44.8936	82.2119	45.1188	60
65	1.9155	0.5220	91.0971	47.5569	91.5541	47.7954	65
70	2.0138	0.4966	100.869	50.0902	101.375	50.3415	70
75	2.1170	0.4724	111.143	52.5000	111.700	52.7633	75
80	2.2255	0.4493	121.942	54.7923	122.554	55.0671	80
85	2.3396	0.4274	133.296	56.9727	133.965	57.2585	85
90	2.4596	0.4066	145.232	59.0468	145.960	59.3430	90
95	2.5857	0.3867	157.780	61.0198	158.571	61.3259	95
100	2.7183	0.3679	170.971	62.8965	171.828	63.2121	100

TABLE D–2 Continuous Compounding; $r = 2\%$

	Discrete Flows				Continuous Flows		
	Single Payment		Uniform Series		Uniform Series		
	Compound Amount Factor	Present Worth Factor	Compound Amount Factor	Present Worth Factor	Compound Amount Factor	Present Worth Factor	
N	To find *F* Given *P* *F/P*	To find *P* Given *F* *P/F*	To find *F* Given *A* *F/A*	To find *P* Given *A* *P/A*	To find *F* Given $\bar{A}$ $F/\bar{A}$	To find *P* Given $\bar{A}$ $P/\bar{A}$	*N*
1	1.0202	0.9802	1.0000	0.9802	1.0101	0.9901	1
2	1.0408	0.9608	2.0202	1.9410	2.0405	1.9605	2
3	1.0618	0.9418	3.0610	2.8828	3.0918	2.9118	3
4	1.0833	0.9231	4.1228	3.8059	4.1644	3.8442	4
5	1.1052	0.9048	5.2061	4.7107	5.2585	4.7581	5
6	1.1275	0.8869	6.3113	5.5976	6.3748	5.6540	6
7	1.1503	0.8694	7.4388	6.4670	7.5137	6.5321	7
8	1.1735	0.8521	8.5891	7.3191	8.6755	7.3928	8
9	1.1972	0.8353	9.7626	8.1544	9.8609	8.2365	9
10	1.2214	0.8187	10.9598	8.9731	11.0701	9.0635	10
11	1.2461	0.8025	12.1812	9.7756	12.3038	9.8741	11
12	1.2712	0.7866	13.4273	10.5623	13.5625	10.6686	12
13	1.2969	0.7711	14.6985	11.3333	14.8465	11.4474	13
14	1.3231	0.7558	15.9955	12.0891	16.1565	12.2108	14
15	1.3499	0.7408	17.3186	12.8299	17.4929	12.9591	15
16	1.3771	0.7261	18.6685	13.5561	18.8564	13.6925	16
17	1.4049	0.7118	20.0456	14.2678	20.2474	14.4115	17
18	1.4333	0.6977	21.4505	14.9655	21.6665	15.1162	18
19	1.4623	0.6839	22.8839	15.6494	23.1142	15.8069	19
20	1.4918	0.6703	24.3461	16.3197	24.5912	16.4840	20
21	1.5220	0.6570	25.8380	16.9768	26.0981	17.1477	21
22	1.5527	0.6440	27.3599	17.6208	27.6354	17.7982	22
23	1.5841	0.6313	28.9126	18.2521	29.2037	18.4358	23
24	1.6161	0.6188	30.4967	18.8709	30.8037	19.0608	24
25	1.6487	0.6065	32.1128	19.4774	32.4361	19.6735	25
26	1.6820	0.5945	33.7615	20.0719	34.1014	20.2740	26
27	1.7160	0.5827	35.4435	20.6547	35.8003	20.8626	27
28	1.7507	0.5712	37.1595	21.2259	37.5336	21.4395	28
29	1.7860	0.5599	38.9102	21.7858	39.3019	22.0051	29
30	1.8221	0.5488	40.6962	22.3346	41.1059	22.5594	30
35	2.0138	0.4966	50.1824	24.9199	50.6876	25.1707	35
40	2.2255	0.4493	60.6663	27.2591	61.2770	27.5336	40
45	2.4596	0.4066	72.2528	29.3758	72.9802	29.6715	45
50	2.7183	0.3679	85.0578	31.2910	85.9141	31.6060	50
55	3.0042	0.3329	99.2096	33.0240	100.208	33.3564	55
60	3.3201	0.3012	114.850	34.5921	116.006	34.9403	60
65	3.6693	0.2725	132.135	36.0109	133.465	36.3734	65
70	4.0552	0.2466	151.238	37.2947	152.760	37.6702	70
75	4.4817	0.2231	172.349	38.4564	174.084	38.8435	75
80	4.9530	0.2019	195.682	39.5075	197.652	39.9052	80
85	5.4739	0.1827	221.468	40.4585	223.697	40.8658	85
90	6.0496	0.1653	249.966	41.3191	252.482	41.7351	90
95	6.6859	0.1496	281.461	42.0978	284.295	42.5216	95
100	7.3891	0.1353	316.269	42.8023	319.453	43.2332	100

TABLE D–3 Continuous Compounding; $r = 3\%$

	Discrete Flows				Continuous Flows		
	Single Payment		Uniform Series		Uniform Series		
	Compound Amount Factor	Present Worth Factor	Compound Amount Factor	Present Worth Factor	Compound Amount Factor	Present Worth Factor	
N	To find *F* Given *P* *F/P*	To find *P* Given *F* *P/F*	To find *F* Given *A* *F/A*	To find *P* Given *A* *P/A*	To find *F* Given $\bar{A}$ $F/\bar{A}$	To find *P* Given $\bar{A}$ $P/\bar{A}$	*N*
1	1.0305	0.9704	1.0000	0.9704	1.0152	0.9851	1
2	1.0618	0.9418	2.0305	1.9122	2.0612	1.9412	2
3	1.0942	0.9139	3.0923	2.8261	3.1391	2.8690	3
4	1.1275	0.8869	4.1865	3.7131	4.2499	3.7693	4
5	1.1618	0.8607	5.3140	4.5738	5.3945	4.6431	5
6	1.1972	0.8353	6.4758	5.4090	6.5739	5.4910	6
7	1.2337	0.8106	7.6730	6.2196	7.7893	6.3139	7
8	1.2712	0.7866	8.9067	7.0063	9.0416	7.1124	8
9	1.3100	0.7634	10.1779	7.7696	10.3321	7.8874	9
10	1.3499	0.7408	11.4879	8.5105	11.6620	8.6394	10
11	1.3910	0.7189	12.8378	9.2294	13.0323	9.3692	11
12	1.4333	0.6977	14.2287	9.9271	14.4443	10.0775	12
13	1.4770	0.6771	15.6621	10.6041	15.8994	10.7648	13
14	1.5220	0.6570	17.1390	11.2612	17.3987	11.4318	14
15	1.5683	0.6376	18.6610	11.8988	18.9437	12.0791	15
16	1.6161	0.6188	20.2293	12.5176	20.5358	12.7072	16
17	1.6653	0.6005	21.8454	13.1181	22.1764	13.3168	17
18	1.7160	0.5827	23.5107	13.7008	23.8669	13.9084	18
19	1.7683	0.5665	25.2267	14.2663	25.6089	14.4825	19
20	1.8221	0.5488	26.9950	14.8151	27.4040	15.0396	20
21	1.8776	0.5326	28.8171	15.3477	29.2537	15.5803	21
22	1.9348	0.5169	30.6947	15.8646	31.1597	16.1050	22
23	1.9937	0.5016	32.6295	16.3662	33.1239	16.6141	23
24	2.0544	0.4868	34.6232	16.8529	35.1478	17.1083	24
25	2.1170	0.4724	36.6776	17.3253	37.2333	17.5878	25
26	2.1815	0.4584	38.7946	17.7837	39.3824	18.0531	26
27	2.2479	0.4449	40.9761	18.2285	41.5969	18.5047	27
28	2.3164	0.4317	43.2240	18.6603	43.8789	18.9430	28
29	2.3869	0.4190	45.5404	19.0792	46.2304	19.3683	29
30	2.4596	0.4066	47.9273	19.4858	48.6534	19.7810	30
35	2.8577	0.3499	60.9975	21.3453	61.9217	21.6687	35
40	3.3201	0.3012	76.1830	22.9459	77.3372	23.2935	40
45	3.8574	0.2592	93.8260	24.3235	95.2475	24.6920	45
50	4.4817	0.2231	114.324	25.5092	116.056	25.8957	50
55	5.2070	0.1920	138.140	26.5297	140.233	26.9317	55
60	6.0496	0.1653	165.809	27.4081	168.322	27.8234	60
65	7.0287	0.1423	197.957	28.1641	200.956	28.5909	65
70	8.1662	0.1225	235.307	28.8149	238.872	29.2515	70
75	9.4877	0.1054	278.702	29.3750	282.924	29.8200	75
80	11.0232	0.0907	329.119	29.8570	334.106	30.3094	80
85	12.8071	0.0781	387.696	30.2720	393.570	30.7306	85
90	14.8797	0.0672	455.753	30.6291	462.658	31.0931	90
95	17.2878	0.0578	534.823	30.9365	542.926	31.4052	95
100	20.0855	0.0498	626.690	31.2010	636.185	31.6738	100

TABLE D–4 Continuous Compounding; $r = 5\%$

	Discrete Flows				Continuous Flows		
	Single Payment		Uniform Series		Uniform Series		
	Compound Amount Factor	Present Worth Factor	Compound Amount Factor	Present Worth Factor	Compound Amount Factor	Present Worth Factor	
N	To find *F* Given *P* *F/P*	To find *P* Given *F* *P/F*	To find *F* Given *A* *F/A*	To find *P* Given *A* *P/A*	To find *F* Given $\bar{A}$ $F/\bar{A}$	To find *P* Given $\bar{A}$ $P/\bar{A}$	N
1	1.0513	0.9512	1.0000	0.9512	1.0254	0.9754	1
2	1.1052	0.9048	2.0513	1.8561	2.1034	1.9033	2
3	1.1618	0.8607	3.1564	2.7168	3.2367	2.7858	3
4	1.2214	0.8187	4.3183	3.5355	4.4281	3.6254	4
5	1.2840	0.7788	5.5397	4.3143	5.6805	4.4240	5
6	1.3499	0.7408	6.8237	5.0551	6.9972	5.1836	6
7	1.4191	0.7047	8.1736	5.7598	8.3814	5.9062	7
8	1.4918	0.6703	9.5926	6.4301	9.8365	6.5936	8
9	1.5683	0.6376	11.0845	7.0678	11.3662	7.2474	9
10	1.6487	0.6065	12.6528	7.6743	12.9744	7.8694	10
11	1.7333	0.5769	14.3015	8.2512	14.6651	8.4610	11
12	1.8221	0.5488	16.0347	8.8001	16.4424	9.0238	12
13	1.9155	0.5220	17.8569	9.3221	18.3108	9.5591	13
14	2.0138	0.4966	19.7724	9.8187	20.2751	10.0683	14
15	2.1170	0.4724	21.7862	10.2911	22.3400	10.5527	15
16	2.2255	0.4493	23.9032	10.7404	24.5108	11.0134	16
17	2.3396	0.4274	26.1287	11.1678	26.7929	11.4517	17
18	2.4596	0.4066	28.4683	11.5744	29.1921	11.8686	18
19	2.5857	0.3867	30.9279	11.9611	31.7142	12.2652	19
20	2.7183	0.3679	33.5137	12.3290	34.3656	12.6424	20
21	2.8577	0.3499	36.2319	12.6789	37.1530	13.0012	21
22	3.0042	0.3329	39.0896	13.0118	40.0833	13.3426	22
23	3.1582	0.3166	42.0938	13.3284	43.1639	13.6673	23
24	3.3201	0.3012	45.2519	13.6296	46.4023	13.9761	24
25	3.4903	0.2865	48.5721	13.9161	49.8069	14.2699	25
26	3.6693	0.2725	52.0624	14.1887	53.3859	14.5494	26
27	3.8574	0.2592	55.7317	14.4479	57.1485	14.8152	27
28	4.0552	0.2466	59.5891	14.6945	61.1040	15.0681	28
29	4.2631	0.2346	63.6443	14.9291	65.2623	15.3086	29
30	4.4817	0.2231	67.9074	15.1522	69.6338	15.5374	30
35	5.7546	0.1738	92.7346	16.1149	95.0921	16.5245	35
40	7.3891	0.1353	124.613	16.8646	127.781	17.2933	40
45	9.4877	0.1054	165.546	17.4484	169.755	17.8920	45
50	12.1825	0.0821	218.105	17.9032	223.650	18.3583	50
55	15.6426	0.0639	285.592	18.2573	292.853	18.7214	55
60	20.0855	0.0498	372.247	18.5331	381.711	19.0043	60
65	25.7903	0.0388	483.515	18.7479	495.807	19.2245	65
70	33.1155	0.0302	626.385	18.9152	642.309	19.3961	70
75	42.5211	0.0235	809.834	19.0455	830.422	19.5296	75
80	54.5981	0.0183	1045.39	19.1469	1071.963	19.6337	80
85	70.1054	0.0143	1347.84	19.2260	1382.108	19.7147	85
90	90.0171	0.0111	1736.20	19.2875	1780.342	19.7778	90
95	115.584	0.0087	2234.87	19.3354	2291.686	19.8270	95
100	148.413	0.0067	2875.17	19.3727	2948.263	19.8652	100

TABLE D–5 Continuous Compounding; $r = 8\%$

	Discrete Flows				Continuous Flows		
	Single Payment		Uniform Series		Uniform Series		
	Compound Amount Factor	Present Worth Factor	Compound Amount Factor	Present Worth Factor	Compound Amount Factor	Present Worth Factor	
N	To find *F* Given *P* *F/P*	To find *P* Given *F* *P/F*	To find *F* Given *A* *F/A*	To find *P* Given *A* *P/A*	To find *F* Given $\bar{A}$ $F/\bar{A}$	To find *P* Given $\bar{A}$ $P/\bar{A}$	*N*
1	1.0833	0.9231	1.0000	0.9231	1.0411	0.9610	1
2	1.1735	0.8521	2.0833	1.7753	2.1689	1.8482	2
3	1.2712	0.7866	3.2568	2.5619	3.3906	2.6672	3
4	1.3771	0.7261	4.5280	3.2880	4.7141	3.4231	4
5	1.4918	0.6703	5.9052	3.9584	6.1478	4.1210	5
6	1.6161	0.6188	7.3970	4.5771	7.7009	4.7652	6
7	1.7507	0.5712	9.0131	5.1483	9.3834	5.3599	7
8	1.8965	0.5273	10.7637	5.6756	11.2060	5.9088	8
9	2.0544	0.4868	12.6602	6.1624	13.1804	6.4156	9
10	2.2255	0.4493	14.7147	6.6117	15.3193	6.8834	10
11	2.4109	0.4148	16.9402	7.0265	17.6362	7.3152	11
12	2.6117	0.3829	19.3511	7.4094	20.1462	7.7138	12
13	2.8292	0.3535	21.9628	7.7629	22.8652	8.0818	13
14	3.0649	0.3263	24.7920	8.0891	25.8107	8.4215	14
15	3.3201	0.3012	27.8569	8.3903	29.0015	8.7351	15
16	3.5966	0.2780	31.1770	8.6684	32.4580	9.0245	16
17	3.8962	0.2567	34.7736	8.9250	36.2024	9.2917	17
18	4.2207	0.2369	38.6698	9.1620	40.2587	9.5384	18
19	4.5722	0.2187	42.8905	9.3807	44.6528	9.7661	19
20	4.9530	0.2019	47.4627	9.5826	49.4129	9.9763	20
21	5.3656	0.1864	52.4158	9.7689	54.5694	10.1703	21
22	5.8124	0.1720	57.7813	9.9410	60.1555	10.3494	22
23	6.2965	0.1588	63.5938	10.0998	66.2067	10.5148	23
24	6.8120	0.1466	69.8903	10.2464	72.7620	10.6674	24
25	7.3891	0.1353	76.7113	10.3817	79.8632	10.8083	25
26	8.0045	0.1249	84.1003	10.5067	87.5559	10.9384	26
27	8.6711	0.1153	92.1048	10.6220	95.8892	11.0584	27
28	9.3933	0.1065	100.776	10.7285	104.917	11.1693	28
29	10.1757	0.0983	110.169	10.8267	114.696	11.2716	29
30	11.0232	0.0907	120.345	10.9174	125.290	11.3660	30
35	16.4446	0.0608	185.439	11.2765	193.058	11.7399	35
40	24.5325	0.0408	282.547	11.5172	294.157	11.9905	40
45	36.5982	0.0273	427.416	11.6786	444.978	12.1585	45
50	54.5982	0.0183	643.535	11.7868	669.977	12.2711	50
55	81.4509	0.0123	965.947	11.8593	1005.64	12.3465	55
60	121.510	0.0082	1446.93	11.9079	1506.38	12.3971	60
65	181.272	0.0055	2164.47	11.9404	2253.40	12.4310	65
70	270.426	0.0037	3234.91	11.9623	3367.83	12.4538	70
75	403.429	0.0025	4831.83	11.9769	5030.36	12.4690	75
80	601.845	0.0017	7214.15	11.9867	7510.56	12.4792	80
85	897.847	0.0011	10768.1	11.9933	11210.6	12.4861	85
90	1339.43	0.0007	16070.1	11.9977	16730.4	12.4907	90
95	1998.20	0.0005	23979.7	12.0007	24964.9	12.4937	95
100	2980.96	0.0003	35779.3	12.0026	37249.5	12.4958	100

TABLE D–6 Continuous Compounding; $r = 10\%$

	Discrete Flows				Continuous Flows		
	Single Payment		Uniform Series		Uniform Series		
	Compound Amount Factor	Present Worth Factor	Compound Amount Factor	Present Worth Factor	Compound Amount Factor	Present Worth Factor	
N	To find *F* Given *P* *F/P*	To find *P* Given *F* *P/F*	To find *F* Given *A* *F/A*	To find *P* Given *A* *P/A*	To find *F* Given $\bar{A}$ $F/\bar{A}$	To find *P* Given $\bar{A}$ $P/\bar{A}$	*N*
1	1.1052	0.9048	1.0000	0.9048	1.0517	0.9516	1
2	1.2214	0.8187	2.1052	1.7236	2.2140	1.8127	2
3	1.3499	0.7408	3.3266	2.4644	3.4986	2.5918	3
4	1.4918	0.6703	4.6764	3.1347	4.9182	3.2968	4
5	1.6487	0.6065	6.1683	3.7412	6.4872	3.9347	5
6	1.8221	0.5488	7.8170	4.2900	8.2212	4.5119	6
7	2.0138	0.4966	9.6391	4.7866	10.1375	5.0341	7
8	2.2255	0.4493	11.6528	5.2360	12.2554	5.5067	8
9	2.4596	0.4066	13.8784	5.6425	14.5960	5.9343	9
10	2.7183	0.3679	16.3380	6.0104	17.1828	6.3212	10
11	3.0042	0.3329	19.0563	6.3433	20.0417	6.6713	11
12	3.3201	0.3012	22.0604	6.6445	23.2012	6.9881	12
13	3.6693	0.2725	25.3806	6.9170	26.6930	7.2747	13
14	4.0552	0.2466	29.0499	7.1636	30.5520	7.5340	14
15	4.4817	0.2231	33.1051	7.3867	34.8169	7.7687	15
16	4.9530	0.2019	37.5867	7.5886	39.5303	7.9810	16
17	5.4739	0.1827	42.5398	7.7713	44.7395	8.1732	17
18	6.0496	0.1653	48.0137	7.9366	50.4965	8.3470	18
19	6.6859	0.1496	54.0634	8.0862	56.8589	8.5043	19
20	7.3891	0.1353	60.7493	8.2215	63.8906	8.6466	20
21	8.1662	0.1225	68.1383	8.3440	71.6617	8.7754	21
22	9.0250	0.1108	76.3045	8.4548	80.2501	8.8920	22
23	9.9742	0.1003	85.3295	8.5550	89.7418	8.9974	23
24	11.0232	0.0907	95.3037	8.6458	100.232	9.0928	24
25	12.1825	0.0821	106.327	8.7278	111.825	9.1791	25
26	13.4637	0.0743	118.509	8.8021	124.637	9.2573	26
27	14.8797	0.0672	131.973	8.8693	138.797	9.3279	27
28	16.4446	0.0608	146.853	8.9301	154.446	9.3919	28
29	18.1741	0.0550	163.298	8.9852	171.741	9.4498	29
30	20.0855	0.0498	181.472	9.0349	190.855	9.5021	30
35	33.1155	0.0302	305.364	9.2212	321.154	9.6980	35
40	54.5981	0.0183	509.629	9.3342	535.982	9.8168	40
45	90.0171	0.0111	846.404	9.4027	890.171	9.8889	45
50	148.413	0.0067	1401.65	9.4443	1474.13	9.9326	50
55	244.692	0.0041	2317.10	9.4695	2436.92	9.9591	55
60	403.429	0.0025	3826.43	9.4848	4024.29	9.9752	60
65	665.142	0.0015	6314.88	9.4940	6641.42	9.9850	65
70	1096.63	0.0009	10417.6	9.4997	10956.3	9.9909	70
75	1808.04	0.0006	17182.0	9.5031	18070.7	9.9945	75
80	2980.96	0.0003	28334.4	9.5051	29799.6	9.9966	80
85	4914.77	0.0002	46721.7	9.5064	49137.7	9.9980	85
90	8103.08	0.0001	77037.3	9.5072	81020.8	9.9988	90
95	13359.7	[a]	127019	9.5076	133587	9.9993	95
100	22026.5	[a]	209425	9.5079	220255	9.9995	100

[a] Less than 0.0001.

TABLE D–7 Continuous Compounding; r = 12%

	Discrete Flows				Continuous Flows		
	Single Payment		Uniform Series		Uniform Series		
	Compound Amount Factor	Present Worth Factor	Compound Amount Factor	Present Worth Factor	Compound Amount Factor	Present Worth Factor	
N	To find *F* Given *P* *F/P*	To find *P* Given *F* *P/F*	To find *F* Given *A* *F/A*	To find *P* Given *A* *P/A*	To find *F* Given *A* $F/\bar{A}$	To find *P* Given *A* $P/\bar{A}$	*N*
1	1.1275	0.8869	1.0000	0.8869	1.0625	0.9423	1
2	1.2712	0.7866	2.1275	1.6735	2.2604	1.7781	2
3	1.4333	0.6977	3.3987	2.3712	3.6111	2.5194	3
4	1.6161	0.6188	4.8321	2.9900	5.1340	3.1768	4
5	1.8221	0.5488	6.4481	3.5388	6.8510	3.7599	5
6	2.0544	0.4868	8.2703	4.0256	8.7869	4.2771	6
7	2.3164	0.4317	10.3247	4.4573	10.9679	4.7357	7
8	2.6117	0.3829	12.6411	4.8402	13.4308	5.1426	8
9	2.9447	0.3396	15.2528	5.1798	16.2057	5.5034	9
10	3.3201	0.3012	18.1974	5.4810	19.3343	5.8234	10
11	3.7434	0.2671	21.5176	5.7481	22.8618	6.1072	11
12	4.2207	0.2369	25.2610	5.9850	26.8391	6.3589	12
13	4.7588	0.2101	29.4817	6.1952	31.3235	6.5822	13
14	5.3656	0.1864	34.2405	6.3815	36.3796	6.7802	14
15	6.0496	0.1653	39.6061	6.5468	42.0804	6.9558	15
16	6.8210	0.1466	45.6557	6.6934	48.5080	7.1116	16
17	7.6906	0.1300	52.4767	6.8235	55.7551	7.2498	17
18	8.6711	0.1153	60.1673	6.9388	63.9261	7.3723	18
19	9.7767	0.1023	68.8384	7.0411	73.1390	7.4810	19
20	11.0232	0.0907	78.6151	7.1318	83.5265	7.5774	20
21	12.4286	0.0805	89.6383	7.2123	95.2383	7.6628	21
22	14.0132	0.0714	102.067	7.2836	108.443	7.7387	22
23	15.7998	0.0633	116.080	7.3469	123.332	7.8059	23
24	17.8143	0.0561	131.880	7.4030	140.119	7.8655	24
25	20.0855	0.0498	149.694	7.4528	159.046	7.9184	25
26	22.6464	0.0442	169.780	7.4970	180.386	7.9654	26
27	25.5337	0.0392	192.426	7.5362	204.448	8.0070	27
28	28.7892	0.0347	217.960	7.5709	231.577	8.0439	28
29	32.4597	0.0308	246.749	7.6017	262.164	8.0766	29
30	36.5982	0.0273	279.209	7.6290	296.652	8.1056	30
35	66.6863	0.0150	515.200	7.7257	547.386	8.2084	35
40	121.510	0.0082	945.203	7.7788	1004.25	8.2648	40
45	221.406	0.0045	1728.72	7.8079	1836.72	8.2957	45
50	403.429	0.0025	3156.38	7.8239	3353.57	8.3127	50
55	735.095	0.0014	5757.75	7.8327	6117.46	8.3220	55
60	1339.43	0.0007	10497.8	7.8375	11153.6	8.3271	60
65	2440.60	0.0004	19134.6	7.8401	20330.0	8.3299	65
70	4447.07	0.0002	34872.0	7.8416	37050.6	8.3315	70
75	8103.08	0.0001	63547.3	7.8424	67517.4	8.3323	75
80	4764.8	[a]	115797	7.8428	123032	8.3328	80

[a] Less than 0.0001.

TABLE D–8 Continuous Compounding; $r = 15\%$

	Discrete Flows				Continuous Flows		
	Single Payment		Uniform Series		Uniform Series		
	Compound Amount Factor	Present Worth Factor	Compound Amount Factor	Present Worth Factor	Compound Amount Factor	Present Worth Factor	
N	To find *F* Given *P* *F/P*	To find *P* Given *F* *P/F*	To find *F* Given *A* *F/A*	To find *P* Given *A* *P/A*	To find *F* Given $\bar{A}$ $F/\bar{A}$	To find *P* Given $\bar{A}$ $P/\bar{A}$	*N*
1	1.1618	0.8607	1.0000	0.8607	1.0789	0.9286	1
2	1.3499	0.7408	2.1618	1.6015	2.3324	1.7279	2
3	1.5683	0.6376	3.5117	2.2392	3.7887	2.4158	3
4	1.8221	0.5488	5.0800	2.7880	5.4808	3.0079	4
5	2.1170	0.4724	6.9021	3.2603	7.4467	3.5176	5
6	2.4596	0.4066	9.0191	3.6669	9.7307	3.9562	6
7	2.8577	0.3499	11.4787	4.0168	12.3843	4.3337	7
8	3.3201	0.3012	14.3364	4.3180	15.4674	4.6587	8
9	3.8574	0.2592	17.6565	4.5773	19.0495	4.9384	9
10	4.4817	0.2231	21.5139	4.8004	23.2113	5.1791	10
11	5.2070	0.1920	25.9956	4.9925	28.0465	5.3863	11
12	6.0496	0.1653	31.2026	5.1578	33.6643	5.5647	12
13	7.0287	0.1423	37.2522	5.3000	40.1913	5.7182	13
14	8.1662	0.1225	44.2809	5.4225	47.7745	5.8503	14
15	9.4877	0.1054	52.4471	5.5279	56.5849	5.9640	15
16	11.0232	0.0907	61.9348	5.6186	66.8212	6.0619	16
17	12.8071	0.0781	72.9580	5.6967	78.7140	6.1461	17
18	14.8797	0.0672	85.7651	5.7639	92.5315	6.2186	18
19	17.2878	0.0578	100.645	5.8217	108.585	6.2810	19
20	20.0855	0.0498	117.933	5.8715	127.237	6.3348	20
21	23.3361	0.0429	138.018	5.9144	148.907	6.3810	21
22	27.1126	0.0369	161.354	5.9513	174.084	6.4208	22
23	31.5004	0.0317	188.467	5.9830	203.336	6.4550	23
24	36.5982	0.0273	219.967	6.0103	237.322	6.4845	24
25	42.4024	0.0235	256.565	6.0338	276.807	6.5099	25
26	49.5211	0.0202	299.087	6.0541	322.683	6.5317	26
27	57.3975	0.0174	348.489	6.0715	375.983	6.5505	27
28	66.6863	0.0150	405.886	6.0865	437.909	6.5667	28
29	77.4785	0.0129	472.573	6.0994	509.856	6.5806	29
30	90.0171	0.0111	550.051	6.1105	593.448	6.5926	30
35	190.566	0.0052	1171.36	6.1467	1263.78	6.6317	35
40	403.429	0.0025	2486.67	6.1638	2682.86	6.6501	40
45	854.059	0.0012	5271.19	6.1719	5687.06	6.6589	45
50	1808.04	0.0006	11166.0	6.1757	12046.9	6.6630	50
55	3827.63	0.0003	23645.3	6.1775	25510.8	6.6649	55
60	8103.08	0.0001	50064.1	6.1784	54013.9	6.6658	60
65	17154.2	[a]	105993	6.1788	114355	6.6663	65
70	36315.5	[a]	224393	6.1790	242097	6.6665	70
75	76879.9	[a]	475047	6.1791	512526	6.6666	75
80	162755	[a]	1005680	6.1791	1085030	6.6666	80

[a] Less than 0.0001.

TABLE D–9 Continuous Compounding; $r = 20\%$

	Discrete Flows				Continuous Flows		
	Single Payment		Uniform Series		Uniform Series		
	Compound Amount Factor	Present Worth Factor	Compound Amount Factor	Present Worth Factor	Compound Amount Factor	Present Worth Factor	
N	To find *F* Given *P* *F/P*	To find *P* Given *F* *P/F*	To find *F* Given *A* *F/A*	To find *P* Given *A* *P/A*	To find *F* Given $\bar{A}$ $F/\bar{A}$	To find *P* Given $\bar{A}$ $P/\bar{A}$	*N*
1	1.2214	0.8187	1.0000	0.8187	1.1070	0.9063	1
2	1.4918	0.6703	2.2214	1.4891	2.4591	1.6484	2
3	1.8221	0.5488	3.7132	2.0379	4.1106	2.2559	3
4	2.2255	0.4493	5.5353	2.4872	6.1277	2.7534	4
5	2.7183	0.3679	7.7609	2.8551	8.5914	3.1606	5
6	3.3201	0.3012	10.4792	3.1563	11.6006	3.4940	6
7	4.0552	0.2466	13.7993	3.4029	15.2760	3.7670	7
8	4.9530	0.2019	17.8545	3.6048	19.7652	3.9905	8
9	6.0496	0.1653	22.8075	3.7701	25.2482	4.1735	9
10	7.3891	0.1353	28.8572	3.9054	31.9453	4.3233	10
11	9.0250	0.1108	36.2462	4.0162	40.1251	4.4460	11
12	11.0232	0.0907	45.2712	4.1069	50.1159	4.5464	12
13	13.4637	0.0743	56.2944	4.1812	62.3187	4.6286	13
14	16.4446	0.0608	69.7581	4.2420	77.2232	4.6959	14
15	20.0855	0.0498	86.2028	4.2918	95.4277	4.7511	15
16	24.5325	0.0408	106.288	4.3325	117.633	4.7962	16
17	29.9641	0.0334	130.821	4.3659	144.820	4.8331	17
18	36.5982	0.0273	160.785	4.3932	177.991	4.8634	18
19	44.7012	0.0224	197.383	4.4156	218.506	4.8881	19
20	54.5981	0.0183	242.084	4.4339	267.991	4.9084	20
21	66.6863	0.0150	296.682	4.4489	328.432	4.9250	21
22	81.4509	0.0123	363.369	4.4612	402.254	4.9386	22
23	99.4843	0.0101	444.820	4.4713	492.422	4.9497	23
24	121.510	0.0082	544.304	4.4795	602.552	4.9589	24
25	148.413	0.0067	665.814	4.4862	737.066	4.9663	25
26	181.272	0.0055	814.227	4.4917	901.361	4.9724	26
27	221.406	0.0045	995.500	4.4963	1102.03	4.9774	27
28	270.426	0.0037	1216.91	4.5000	1347.13	4.9815	28
29	330.299	0.0030	1487.33	4.5030	1646.50	4.9849	29
30	403.429	0.0025	1817.63	4.5055	2012.14	4.9876	30
35	1096.63	0.0009	4948.60	4.5125	5478.17	4.9954	35
40	2980.96	0.0003	13459.4	4.5151	14899.8	4.9983	40
45	8103.08	0.0001	36594.3	4.5161	40510.4	4.9994	45
50	22026.5	[a]	99481.4	4.5165	110127	4.9998	50
55	59874.1	[a]	270426	4.5166	299366	4.9999	55
60	162755	[a]	735103	4.5166	813769	5.0000	60

[a] Less than 0.0001.

TABLE D–10 Continuous Compounding; $r = 25\%$

	Discrete Flows				Continuous Flows		
	Single Payment		Uniform Series		Uniform Series		
Discrete Flows	Compound Amount Factor	Present Worth Factor	Compound Amount Factor	Present Worth Factor	Compound Amount Factor Continuous Flows	Present Worth Factor	
N	To find F Given P F/P	To find P Given F P/F	To find F Given A F/A	To find P Given A P/A	To find F Given $\bar{A}$ $F/\bar{A}$	To find P Given $\bar{A}$ $P/\bar{A}$	N
1	1.2840	0.7788	1.0000	0.7788	1.1361	0.8848	1
2	1.6487	0.6065	2.2840	1.3853	2.5949	1.5739	2
3	2.1170	0.4724	3.9327	1.8577	4.4680	2.1105	3
4	2.7183	0.3679	6.0497	2.2256	6.8731	2.5285	4
5	3.4903	0.2865	8.7680	2.5121	9.9614	2.8540	5
6	4.4817	0.2231	12.2584	2.7352	13.9268	3.1075	6
7	5.7546	0.1738	16.7401	2.9090	19.0184	3.3049	7
8	7.3891	0.1353	22.4947	3.0443	25.5562	3.4587	8
9	9.4877	0.1054	29.8837	3.1497	33.9509	3.5784	9
10	12.1825	0.0821	39.3715	3.2318	44.7300	3.6717	10
11	15.6426	0.0639	51.5539	3.2957	58.5705	3.7443	11
12	20.0855	0.0498	67.1966	3.3455	76.3421	3.8009	12
13	25.7903	0.0388	87.2821	3.3843	99.1614	3.8449	13
14	33.1155	0.0302	113.073	3.4145	128.462	3.8792	14
15	42.5211	0.0235	146.188	3.4380	166.084	3.9059	15
16	54.5982	0.0183	188.709	3.4563	214.393	3.9267	16
17	70.1054	0.0143	243.307	3.4706	276.422	3.9429	17
18	90.0171	0.0111	313.413	3.4817	356.068	3.9556	18
19	115.584	0.0087	403.430	3.4904	458.337	3.9654	19
20	148.413	0.0067	519.014	3.4971	589.653	3.9730	20
21	190.566	0.0052	667.427	3.5023	758.265	3.9790	21
22	244.692	0.0041	857.993	3.5064	974.768	3.9837	22
23	314.191	0.0032	1102.69	3.5096	1252.76	3.9873	23
24	403.429	0.0025	1416.88	3.5121	1609.72	3.9901	24
25	518.013	0.0019	1820.30	3.5140	2068.05	3.9923	25
26	665.142	0.0015	2338.31	3.5155	2656.57	3.9940	26
27	854.059	0.0012	3003.46	3.5167	3412.23	3.9953	27
28	1096.63	0.0009	3857.52	3.5176	4382.53	3.9964	28
29	1408.10	0.0007	4954.15	3.5183	5628.42	3.9972	29
30	1808.04	0.0006	6362.26	3.5189	7228.17	3.9978	30
35	6310.69	0.0002	22215.2	3.5203	25238.8	3.9994	35
40	22026.5	[a]	77547.5	3.5207	88101.9	3.9998	40
45	76879.9	[a]	270676	3.5208	307516	3.9999	45
50	268337	[a]	944762	3.5208	1073350	4.0000	50

E

COMPUTER PROGRAM FOR BEFORE-TAX ENGINEERING ECONOMY STUDIES

E.1 GENERAL DESCRIPTION

Each instructor who adopts this book for classroom use will receive a 5-1/4" floppy diskette that contains computer programs in Appendixes C, E, and F. All programs are written in BASIC and run under MS-DOS, version 3.2, on an IBM or IBM-compatible personal computer. A minimum of 256K RAM is recommended to run the largest program. Students are encouraged to obtain a copy of their instructor's diskette for use in their engineering economy course.

This appendix presents and demonstrates a computer program named BTAX that has wide applicability to many problems and homework exercises in this book. The program is written as a conversational computer program that offers these capabilities:

1. A choice of any one or all of seven methods for computing a measure of merit in an engineering economy study. These methods are present worth, future worth, annual worth, internal rate of return, external rate of return, simple payback period, and discounted payback period.
2. A choice of discrete or continuous compounding of a periodic rate of interest (e.g., an effective annual or nominal interest rate).
3. The ability to conduct sensitivity studies and correct data entry errors with a minimum of effort.
4. The calculation of various measures of merit at fractional positive or negative rates of interest that sometimes arise in inflation-related problems.
5. The ability to deal with negative time periods that occur when capital investment costs are spread over several years. Measures of merit are computed relative to the end of year 0, which is usually defined as the start of commercial operation of a venture.

To facilitate the use of this computer program for conducting before-tax engineering economy studies, a program listing of BTAX and several examples of the program's application are presented in this Appendix. The user may call BTAX by simply typing "BTAX" and pressing the Return (Enter) key. Instructions are built into the program so that its operation is self-explanatory.

E.2 EXAMPLE PROBLEMS

Several examples from Chapters 3 and 4 are presented to illustrate the versatility of the engineering economy computer program called BTAX. It should be noted that after-tax cash flows of Chapters 7 and 12 could also be reduced to a desired measure of merit with this program. Users of this program are encouraged to first backup the diskette using the instructions given in Appendix F.

Example 4-2 with Three Different Cash Flow Values

```
HOW MANY DIFFERENT CASH FLOW VALUES? 3
ENTER CASH FLOW, FIRST PERIOD, LAST PERIOD
?-25000,0,0
?8000,1,4
?13000,5,5

YEAR        CASH FLOW
0           -25,000.00
1             8,000.00
2             8,000.00
3             8,000.00
4             8,000.00
5            13,000.00
```

```
WANT TO MAKE ANY CORRECTIONS OR ADDITIONS (Y,N) ?N
WHAT WOULD YOU LIKE TO DO (TYPE PW,FW,AW,IRR,ERR,SPP,
DPP,TABLE,? OR END)?PW ENTER INTEREST RATE AS
DECIMAL, (FOR EXAMPLE ENTER 10% AS .1) ?0.2
DO YOU WANT CONTINUOUS (CON) OR DISCRETE (DIS)
COMPOUNDING ?DIS
==========> PW = 934.285
```

Example 4-2 with Cash Flow Values for Each Individual Year

```
HOW MANY DIFFERENT CASH FLOW VALUES?6
ENTER CASH FLOW, FIRST PERIOD, LAST PERIOD
?-25000,0,0
?8000,1,1
?8000,2,2
?8000,3,3
?8000,4,4
?13000,5,5

YEAR      CASH FLOW
0         -25,000.00
1           8,000.00
2           8,000.00
3           8,000.00
4           8,000.00
5          13,000.00

WANT TO MAKE ANY CORRECTIONS OR ADDITIONS (Y,N)?N
WHAT WOULD YOU LIKE TO DO (TYPE PW,FW,AW,IRR,ERR,SPP,
DPP,TABLE,? OR END)?PW
ENTER INTEREST RATE AS DECIMAL, (FOR EXAMPLE ENTER
10% AS .1)?.2
DO YOU WANT CONTINUOUS (CON) OR DISCRETE (DIS)
COMPOUNDING ?DIS
==========> PW = 934.285
```

Example 4-2 With Continuous Compounding

```
WANT ANOTHER RUN? (Y,N) ?Y
TYPE NEW,OLD OR CHANGE (TYPE '?' FOR
EXPLANATION) ?OLD
WHAT WOULD YOU LIKE TO DO (TYPE PW, FW, AW, IRR, ERR,
SPP, DPP, TABLE,? OR END) ?PW
ENTER INTEREST RATE AS DECIMAL, (FOR EXAMPLE
ENTER 10% AS .1) ?0.2
DO YOU WANT CONTINUOUS (CON) OR DISCRETE (DIS)
COMPOUNDING ?CON
=========> PW = -320.036
```

Examples 4-5, 4-7, and 4-9 (in Order)

```
WANT ANOTHER RUN ? (Y/N) ?Y
TYPE NEW,OLD OR CHANGE (TYPE '?' FOR EXPLANATION) ??
TO ENTER COMPLETE NEW CASH FLOW TYPE ========>
NEW
TO KEEP OLD CASH FLOWS WITH NO CHANGE TYPE ==>
OLD
TO MAKE SOME CHANGES ON OLD CASH FLOWS TYPE =>
CHANGE
?OLD
WHAT WOULD YOU LIKE TO DO (TYPE PW,FW,AW,IRR,ERR,SPP,
DPP,TABLE,? OR END)
?AW
ENTER INTEREST RATE AS DECIMAL, (FOR EXAMPLE
ENTER 10% AS .1)
?.2
DO YOU WANT CONTINUOUS (CON) OR DISCRETE (DIS)
COMPOUNDING ?DIS
==========> AW = 312.406
```

```
WANT ANOTHER RUN ? (Y/N) ?Y
TYPE NEW,OLD OR CHANGE (TYPE '?' FOR EXPLANATION) ?OLD
WHAT WOULD YOU LIKE TO DO (TYPE PW,FW,AW,IRR,ERR,SPP,
DPP,TABLE,? OR END)
?FW
ENTER INTEREST RATE AS DECIMAL, (FOR EXAMPLE
ENTER 10% AS .1)
?.2
DO YOU WANT CONTINUOUS (CON) OR DISCRETE (DIS)
COMPOUNDING ?DIS
==========> FW = 2324.8
```

```
WANT ANOTHER RUN ? (Y,N) ?Y
TYPE NEW,OLD OR CHANGE (TYPE '?' FOR EXPLANATION) ?OLD
WHAT WOULD YOU LIKE TO DO (TYPE PW,FW,AW,IRR,ERR,SPP,
DPP,TABLE,? OR END)
?IRR
DO YOU WANT DISCRETE (DIS) OR CONTINUOUS (CON)
INTEREST RATE
?DIS
=========> IRR IS BETWEEN 21.577 AND 21.578 %
```

Example 3-18 to Illustrate a Negative Interest Rate

```
WANT ANOTHER RUN ? (Y,N) ?Y
TYPE NEW,OLD OR CHANGE (TYPE '?' FOR EXPLANATION) ?
EXAMPLE
PLEASE RETYPE
?NEW
```

```
HOW MANY DIFFERENT CASH FLOW VALUES
?1
ENTER CASH FLOW, FIRST PERIOD, LAST PERIOD
?600,1,15

YEAR      CASH FLOW
1         600.00
2         600.00
3         600.00
.         .
.         .
.         .
14        600.00
15        600.00

WANT TO MAKE ANY CORRECTIONS OR ADDITIONS (Y,N)
?N
WHAT WOULD YOU LIKE TO DO (TYPE PW,FW,AW,IRR,ERR,
SPP,DPP,TABLE,? OR END)
?PW
ENTER INTEREST RATE AS DECIMAL, (FOR EXAMPLE
ENTER 10% AS .1)
?-0.0175
DO YOU WANT CONTINUOUS (CON) OR DISCRETE (DIS)
COMPOUNDING ?DIS
==========> PW = 10395.5
```

Example 4-17: Industrial Case Study

```
WANT ANOTHER RUN ? (Y,N) ?Y
TYPE NEW,OLD OR CHANGE (TYPE '?' FOR EXPLANATION) ?NEW
HOW MANY DIFFERENT CASH FLOW VALUES
?17
ENTER CASH FLOW, FIRST PERIOD, LAST PERIOD
?-40000,0,0
?-522100,1,1
?-777000,2,2
?-46400,3,3
?373600,4,4
?385800,5,5
?265800,6,6
?385800,7,7
?385800,8,8
?385800,9,9
?385800,10,10
?265800,11,11
?385800,12,12
?385800,13,13
?385800,14,14
?385800,15,15
?471200,16,16
```

```
YEAR        CASH FLOW
0           - 40,000.00
1           -522,100.00
2           -777,000.00
3           - 46,400.00
4             373,600.00
5             385,800.00
6             265,800.00
7             385,800.00
8             385,800.00
9             385,800.00
10            385,800.00
11            265,800.00
12            385,800.00
13            385,800.00
14            385,800.00
15            385,000.00
16            471,200.00
```

```
WANT TO MAKE ANY CORRECTIONS OR ADDITIONS (Y,N)
?N
WHAT WOULD YOU LIKE TO DO (TYPE PW,FW,AW,IRR,ERR,
SPP,DPP,TABLE,? OR END)
?IRR
DO YOU WANT DISCRETE (DIS) OR CONTINUOUS (CON)
INTEREST RATE
?DIS
==========> IRR IS BETWEEN 18.342 AND 18.343 %
```

```
WANT ANOTHER RUN ? (Y,N) ?Y
TYPE NEW,OLD OR CHANGE (TYPE '?' FOR EXPLANATION)
?OLD
WHAT WOULD YOU LIKE TO DO (TYPE PW,FW,AW,IRR,ERR,
SPP,DPP,TABLE,? OR END)
?PW
ENTER INTEREST RATE AS DECIMAL, (FOR EXAMPLE
ENTER 10% AS .1)
?.2
DO YOU WANT CONTINUOUS (CON) OR DISCRETE (DIS)
COMPOUNDING ?DIS
========> PW = -87137.3
```

Example 4-17 to Illustrate Sensitivity Analysis: What Is the Effect of Not Recovering Working Capital of $85,400 at the End of the Project?

```
WANT ANOTHER RUN ? (Y,N) ?Y
TYPE NEW,OLD OR CHANGE (TYPE '?' FOR EXPLANATION)
```

```
?CHANGE
WOULD YOU LIKE TO SEE THE OLD CASH FLOW TABLE
(YES OR NO) ?NO
ENTER YEAR (FOR CORRECTION OR ADDITION) ?16
ENTER CASH FLOW FOR ABOVE YEAR ? 385800
WANT TO MAKE ANY CORRECTIONS OR ADDITIONS (Y,N)
?N
WHAT WOULD YOU LIKE TO DO (TYPE
PW,FW,AW,IRR,ERR,SPP,DPP,TABLE,? OR END)
?IRR
DO YOU WANT DISCRETE (DIS) OR CONTINUOUS (CON)
INTEREST RATE
?DIS
=========> IRR IS BETWEEN 18.24 AND 18.241%
```

```
WANT ANOTHER RUN ? (Y,N) ?Y
TYPE NEW,OLD OR CHANGE (TYPE '?' FOR EXPLANATION)
?OLD
WHAT WOULD YOU LIKE TO DO (TYPE PW,FW,AW,IRR,ERR,
SPP,DPP,TABLE,? OR END)
?PW
ENTER INTEREST RATE AS DECIMAL, (FOR EXAMPLE
ENTER 10% AS .1)
?.2
DO YOU WANT CONTINUOUS (CON) OR DISCRETE (DIS)
COMPOUNDING ?DIS
==========> PW = -91756.4
```

F

COMPUTER PROGRAM FOR CALCULATING DEPRECIATION AND PERFORMING AFTER-TAX ENGINEERING ECONOMY STUDIES

F.1 GENERAL DESCRIPTION

This BASIC program (ATAX) is a computational tool for analyzing the present worth of after-tax problems of Chapters 7 through 12. It is designed for cash flow studies that include inflation/escalation or that ignore the effects of such price changes. Furthermore, the program is written to compute depreciation by methods discussed in Chapter 6. ATAX contains files of ACRS and MACRS percentages that may be updated if necessary. ATAX is an interactive, menu-driven program suitable for personal computers and, as such, contains instructions to the user regarding its required data inputs and key assumptions. Copies of the program can be obtained from the instructor's master diskette; the program is suitable for IBM (or IBM-compatible) personal computers and utilizes the MS-DOS 3.2 operating system.

All cash flows, except salvage values, are referenced in terms of purchasing power to the time of the initial investment (i.e., time 0). They may be entered as annuities or nonuniform amounts. Salvage values are entered as actual dollar amounts at the time of an asset's disposal. The program selects the appropriate method of depreciation based upon when the asset was placed in service (with a choice of three methods for assets placed in service on or before December 31, 1980). The ACRS and MACRS percentages have been included in the program so that the user need only supply the class life of the asset.

Appendix F takes the user through several examples while explaining, in detail, the steps that must be followed to execute the program correctly.

F.2 PROCEDURE FOR LOADING AND USING THE PROGRAM

Before using the Engineering Economy diskette, make a backup copy of the disk.

For A Computer with One or Two Floppy Disk Drives

- Boot the computer with a disk containing the Disk Operating System (DOS) in Drive A.
- With the DOS disk still in Drive A, type

```
FORMAT B: <RETURN>
```

- The computer will prompt you to place a blank disk in Drive B. Once you have done this, press the <RETURN> key again.
- When the formatting process is complete, the computer will ask if you want to format another disk. Response by typing

```
NO <RETURN>
```

- With the DOS disk still in Drive A, type

```
DISKCOPY A: B: <RETURN>
```

- The computer will then prompt you to remove the DOS disk from Drive A and insert the original disk in its place as well as placing the blank, formatted disk into Drive B. Once you have done this, press <RETURN>.

- Once the copying process is complete, store your original disk in a safe place and use the backup disk as your regular "working" disk.

*For a Computer System with a Hard Disk Drive**

- Boot the computer.

To Copy the Program onto a Hard Disk

- To copy the program onto your hard disk, first make a separate directory (here, the directory is labeled EE) by typing

```
MD C:\EE
```

- Place the backup copy of the program disk in Drive A and type

```
COPY A:*.* C:\EE
```

All the programs on the Engineering Economy diskette have been written in BASIC. If you wish to run under BASIC, there exists ATAX.BAS. However, ATAX.EXE runs much faster and it does not require BASIC. Therefore, its use is recommended for all purposes.

ATAX uses the data file ACRS.DAT, which contains the ACRS and modified ACRS percentages. This file should be in the same directory as ATAX.

To run the program, log into your directory with CD C:\EE which contains the programs and enter

```
ATAX <CR>
```

The main menu is displayed and you are now ready to use the program.

F.3 CALCULATING DEPRECIATION

An overview of the three sets of depreciation calculations and two types of after-tax PW evaluations in ATAX is provided in Figure F-1.

The ATAX program leads the user through a sequence of menus, beginning with the Main Menu (see Figure F-2). This menu asks the user to specify when the asset under consideration was placed in service. To illustrate the many features of ATAX, we consider a series of sample problems.

PROBLEM 1

An asset is placed in service before December 31, 1980. It has a useful life of 10 years. The initial cost of the investment is $6,400 and the salvage value is $500. Calculate the depreciation for this asset.

*If the user desires to backup the Engineering Economy diskette, follow the instructions for the two-drive system above.

FIGURE F-1

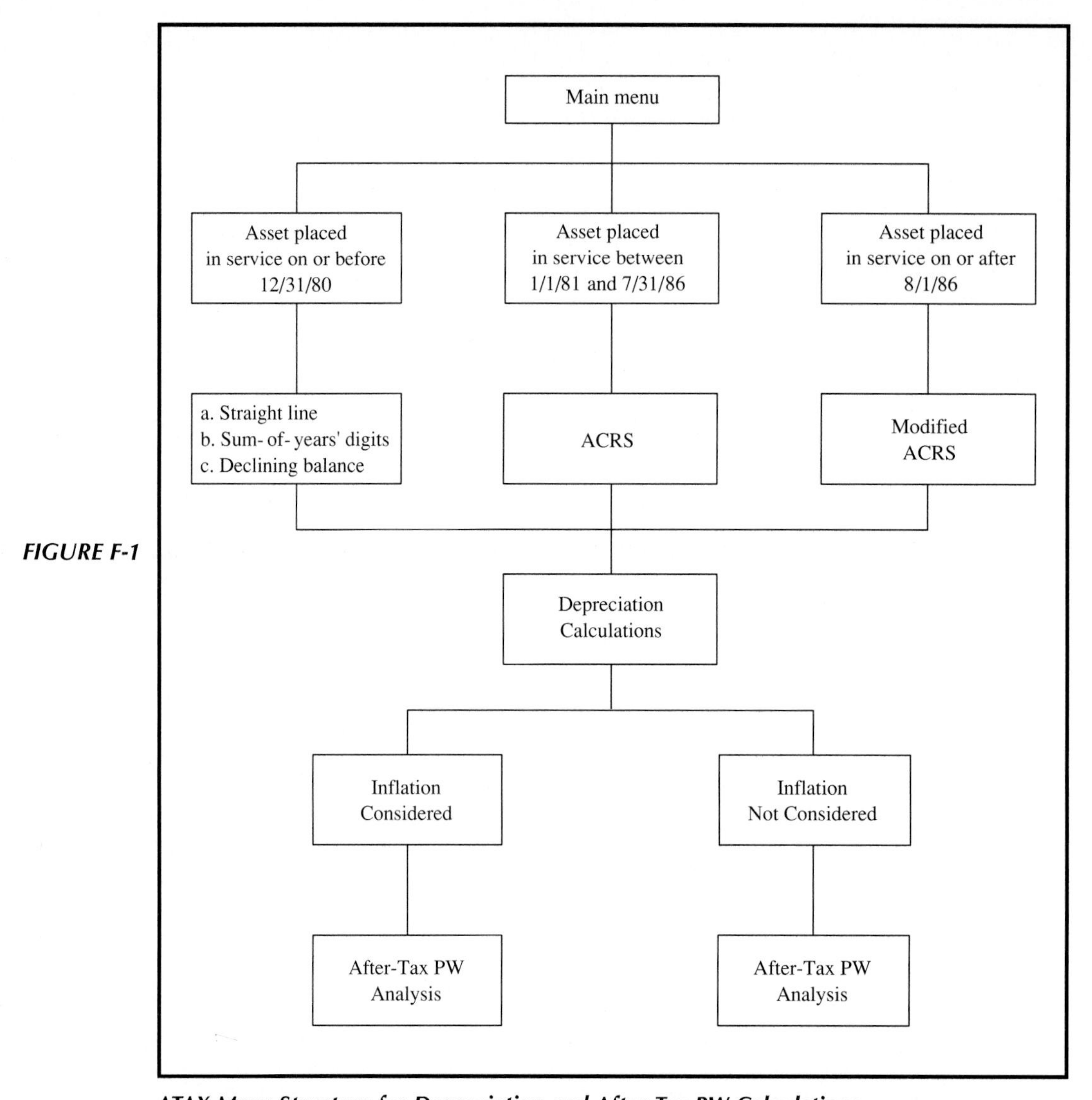

ATAX Menu Structure for Depreciation and After-Tax PW Calculations

Solution

1. Beginning with the Main Menu (Figure F-2), select option 1 since the asset was placed in service prior to December 31, 1980.
2. Menu 1 (Figure F-3) will come up next and ask whether you want to calculate the depreciation, evaluate an investment, or return to the previous menu. Since we wish to calculate the depreciation, we select option 1.
3. The next menu, Menu 1.1 (Figure F-4), gives the user a choice of three methods of depreciation that may be used for assets placed in service on or before December 31, 1980, and a chance to return to the previous menu. Let us begin by looking at option 1, for the straight-line method.

FIGURE F-2

```
ATAX: After Tax Investment Evaluation
Main Menu
________________________________________________________
1. Asset placed in service on or before December 31,
   1980.
   Applicable methods: Straight Line          (SL)
                       Sum of Years Digits    (SYD)
                       Declining Balance      (DB)
2. Asset placed in service between January 1, 1981
   and July 31, 1986.
   Applicable method:  Accelerated Cost Recovery
                       System                 (ACRS)
3. Asset placed in service between August 1, 1986
   to present.
   Applicable method:  Modified ACRS          (MACRS)
0. Exit to system.
   Enter option:       1
```

Main Menu for ATAX

FIGURE F-3

```
Menu 1: SL, SYD, DB
___________________
1. Calculate depreciation of the asset.
2. Evaluate an Investment.
0. Return to last menu.
   Enter option :1
```

Menu for Assets in Service on or before December 31, 1980

FIGURE F-4

```
Menu 1.1: Depreciation methods
______________________________
1. Straight Line        (SL).
2. Sum of Years Digits  (SYD).
3. Declining Balance    (DB).
0. Return to last menu.
   Enter option: 1
```

Depreciation Options for Assets Placed in Service on or before December 31, 1980

FIGURE F-5

```
1.1.1 Straight Line Method

Enter the useful life of the asset in years (N): 10
Enter the investment (I): 6400
Enter salvage value at the end of year 10 (S): 500

You entered:
N = 10
I = 6400
S = 500
Is this correct [Y/N]? Y
```

Straight-Line Depreciation Data Entry for Problem 1

4. The straight-line method menu (Figure F-5) asks us to enter the life of the asset in years (N), the initial investment (I), and the salvage value (S) at the end of the asset's life. We enter the following:

   10 (for N)
   6400 (for I)
   500 (for S)

   The program will then tell us what we entered (after we press <RETURN>) and ask if the listed information is correct. If we say **n** (for no), the program will repeat the questions.
5. The next screen (Figure F-6) provides a table showing the end of year (EOY) book value, depreciation, and cumulative depreciation for each year in the life of the asset. It also asks if we wish to print the table.

FIGURE F-6

Straight Line Depreciation

Yr	EOY Book Value	Depreciation	Cumulative
1	5810.00	590.00	590.00
2	5220.00	590.00	1180.00
3	4630.00	590.00	1770.00
4	4040.00	590.00	2360.00
5	3450.00	590.00	2950.00
6	2860.00	590.00	3540.00
7	2270.00	590.00	4130.00
8	1680.00	590.00	4720.00
9	1090.00	590.00	5310.00
10	550.00	590.00	5900.00

```
I = 6400   N = 10   S = 500

Do you want to print the table [Y/N]? Y
```

Annual Depreciation, Book Value, and Cumulative Depreciation for Problem 1

FIGURE F-7

```
3. MACRS Method

Enter the Useful Life of the Asset in Years (N): 10
Enter the Investment (I): 6400
Enter the Class Life from
[ 3 5 7 10 15 20 ]
Class: 7
You entered:
  N    = 10
  I    = 6400
Class  = 7
Is this correct [Y/N]? Y
```

MACRS Data Entries for Problem 2

PROBLEM 2

We now assume that the asset in Problem 1 is placed in service after August 1, 1986. Calculate the depreciation for the asset.

Solution

1. This time we selection option 3 from the Main Menu (Figure F-2).
2. On Menu 3, selection option 1.
3. The MACRS Method Menu (Figure F-7) asks for the same information as does the menu for the ACRS method except that you will enter the MACRS class life instead of the ACRS class life.
4. The table for MACRS depreciation is shown in Figure F-8. Notice that for the MACRS method it takes 8 years to fully depreciate the 7-year asset due to the half-year convention.

FIGURE F-8

MACRS Depreciation

Yr	EOY Book Value	Depreciation	MACRS Rate	Cumulative
1	5485.44	914.56	0.1429	914.56
2	3918.08	1567.36	0.2449	2481.92
3	2798.72	1119.36	0.1749	3601.28
4	1999.36	799.36	0.1249	4400.64
5	1427.84	571.52	0.0893	4972.16
6	856.32	571.52	0.0893	5543.68
7	285.44	570.88	0.0892	6114.56
8	0.00	285.44	0.0446	6400.00
9	0.00	0.00	0.0000	6400.00
10	0.00	0.00	0.0000	6400.00

I = 6400 N = 10 MACRS CLASS = 7

Do you want to print the table [Y/N]? Y

MACRS Depreciation Results for Problem 2

Note. The ACRS and MACRS percentages reside on a file called ACRS.DAT. This ASCII file can be edited and modified as the rates change in the future. Its contents are as follows:

- Rows 1 and 2 contain the number of ACRS and MACRS class lives, respectively.
- The next several rows are for the ACRS method. Each group starts with the ACRS class life followed by percentages (in percentage form).
- The remaining rows are for the MACRS method. Each group starts with the MACRS class life followed by percentages (in decimal form).

F.4 EVALUATING AN INVESTMENT

Now that we have looked at various methods for calculating the depreciation of an asset, we examine how the program may be used to evaluate an investment by using the present worth method.

To illustrate this, we consider a fairly complex problem that would be difficult to solve manually. We shall see just how easy it is to solve such a problem using ATAX.

PROBLEM 3

An asset is placed in service in 1980 and is to be depreciated using straight-line depreciation. The following information applies to this asset:

Initial Cost	= $20,000
Investment Tax Credit	= 10%
Useful Life	= 10 years
Study Period	= 10 years
Annual Revenues (R$)	= $15,000 with an annual increase (escalation) of 8%
Annual Cost of Utilities (R$)	= $2,400 with an annual escalation rate of 10%/ year
Annual Labor Costs (R$)	= $3,000 with an annual escalation rate of 5%/ year
Annual Cost of Materials (R$)	= $2,000 with an annual escalation rate of 9%/ year
Annual Leasing Costs (A$)	= $900
Lease Period	= 5 years

Note: All costs apply to years 1–10.

Incremental Income Tax Rate	= 50%
Tax Rate on Capital Gain/Loss	= 28%
Inflation Rate (for replacement of investment)	= 6%
After-Tax MARR	= 12%
Salvage Value for Depreciation Purposes (at end of year 10, A$)	= $4,000
Market Value (at end of year 10, A$)	= $0

Evaluate the investment using the present worth method.

Solution

1. On the Main Menu (Figure F-2) we select option 1 (since the investment was placed in service prior to December 31, 1980).
2. On Menu 1 (Figure F-3) we select option 2 because we want to evaluate the investment.

FIGURE F-9

```
Instructions

This program allows you to conduct an analysis of an
investment alternative on an after-tax basis when
inflation/escalation of revenues and costs are
involved. Multiple replacements of an alternative
are assumed when the study period is longer than the
useful life. Conversely the study period can be
shorter than the useful life. Salvage value and
estimated selling price at the end of each cycle are
specified by the user. When inflation/escalation is
considered, the reference year for cost and revenue
estimates is time k=0. The reference year for all
cash flows, except salvage value and selling price
at disposal, is the time at which the initial
investment is made (i.e., time 0). These are termed
'REAL DOLLARS' in the prompts below.

    NOTE: Enter all percentages and rates in
          DECIMAL form. For example:
          10% should be entered as 0.10 or .10

Press any key to continue
```

Description of How to Evaluate an Investment

3. We are now asked if we want a description of the program. If we say **y** (yes), we get a screen as shown in Figure F-9.
4. The next screen consists of the Investment Menu (Figure F-10). On this menu, we are asked to enter the initial investment cost that will be depreciated, the nondepreciable investment cost (e.g., land), the investment tax credit percentage,

FIGURE F-10

```
INVESTMENT

Cost of depreciable investment : 20000
Cost of non-depreciable investment : 0
Investment Tax Credit percentage (as a decimal) : 0.10

Often the Useful Life of the investment is not
identical to the Study Period. Please check.

Useful Life of the investment : 10
Study Period : 10

Do you want to consider inflation/escalation [Y/N]? Y
```

Capital Investment–Related Inputs for Problem 3

the useful life, and the study period (the number of years over which we analyze the investment). We are also asked whether inflation/escalation is to be considered. In problem 3 our response is **y**. The data entries for Problem 3 are shown in Figure F-10. *Note*: All percentages MUST be entered in their decimal form.

5. The Revenues and Expenses Menu (Figure F-11) is next. On this menu, we enter all costs, revenues, escalation rates, and the length of the lease. After we have entered all of these, the program asks if the costs entered apply to the entire useful life of the investment. In Problem 3 they are constant during the full 10 years.
6. The next screen is the Taxes and Inflation Menu (Figure F-12). This is where we enter the incremental income tax rate, the capital gain/loss tax rate, the inflation rate, and the after-tax MARR.
7. Next, Menu 1.2 (Figure F-13) gives us a choice of three methods of depreciation since our asset was placed in service on or before December 31, 1980. The problem statement called for the straight-line method, which is option 1.
8. The next screen gives us the Salvage Value Menu (Figure F-14), which asks us to enter the salvage value (in A$) for the last year in the asset's life. ATAX then asks if we want to see all the input data. If we respond with **y** (yes), we will see a listing of our inputs, as shown on Figures F-15, F-16, and F-17. After showing all our inputs, the program will ask if we want to print the input data. Answering **y** (yes) will give a printout of the same lists.
9. The Actual Dollar Analysis screen (Figure F-18) appears next and lists all costs and revenues that occur in each year and shows the resulting before-tax cash flows (BTCF). When we press <RETURN>, the program asks if we want to print this table.
10. The next screen (Figure F-19) gives us the table we are used to seeing when performing after-tax studies. It lists, for each year in the study period, the before-tax cash flows, the depreciation, the taxable income, the income taxes, the after-tax cash flows, and the present worth (for each year). The individual present worths are summed to give the total present worth that we wanted. Based upon the sign of this value (positive or negative), the program tells us if we have a winner or a loser. As you can see, we have a winner. As before, the program asks if we wish to print the table (when we press <RETURN>).

```
REVENUES AND EXPENSES

The amount in real DOLLARS of the items below can be
CONSTANT during all, or a part, of the Useful Life,
or can be LUMPY
        1. annual revenues
        2. annual cost of utilities
        3. annual cost of labor
        4. annual cost of materials
        5. annual lease cost

Enter C if CONSTANT/NONE else L if LUMPY

1C 2C 3C 4C 5C

Enter beginning (B) and ending (E) year in which
constant items apply, as pairs B,E:
Item 1 :1, 10
Item 2 :1, 10
Item 3 :1, 10
Item 4 :1, 10
Item 5 :1, 10

REVENUES AND EXPENSES

Annual revenues in real dollars :15000
Effective annual escalation rate for revenues :0.08

Annual cost of utilities in real dollars :2400
Effective annual escalation rate for utilities :0.10

Annual cost of labor in real dollars :3000
Effective annual escalation rate for labor :0.05

Annual cost of materials in real dollars :2000
Effective annual escalation rate for materials :0.09

About LEASE COST

When lease fees are re-negotiated, the new
fee is assumed to escalate at the rate used
for replacements of capital assets.

Annual lease cost in real dollars :900
Number of years before lease is re-negotiated :5
```

FIGURE F-11

Revenue and Expense Inputs for Problem 3

FIGURE F-12

```
TAXES AND INFLATION

Incremental Income Tax Rate (as a decimal) : .50
Tax Rate on Capital Gain or Loss (decimal) : .28
Inflation Rate for Investment Replacement (in
decimal) : .06
The M.A.R.R. for After-Tax Analysis (as a decimal) :.12
```

Remaining Data Entries for Problem 3

FIGURE F-13

```
1.2 Depreciation methods

1. Straight Line       (SL).
2. Sum of Years Digits (SYD).
3. Declining Balance   (DB).

   Enter option: 1
```

Depreciation Choices for Problem 3

FIGURE F-14

```
Please enter:
 Salvage Value for depreciation purposes and
 Selling Price at time of disposal.
        (in actual dollars)

Salvage value for year 10 : 4000
Selling price for year 10 : 0

Do you want to see input data [Y/N]? Y
Do you want to print input data [Y/N]? Y
```

Salvage Value (for Purposes of Calculating Depreciation) and Actual Selling Price Entries for Problem 3

FIGURE F-15

```
DEPRECIATION

Depreciation method . . . . . . . . .           :SL
Period of study in years . . . . . . .          : 10
Useful life of investment in years  . .         : 10
Tax life of investment in years . . . .         : 10
First cost of investment . . . . . . .          : 20000
Non-depreciable investment  . . . . .           : 0
Recovery year non-depreciable investment        : 0
Salvage Value in actual dollars yr. 10          :   4000.00
Selling Price in actual dollars yr. 10          :      0.00

Press any key to continue . . .
```

Listing of Depreciation Data for Problem 3

FIGURE F-16

```
Yr REVENUES UTILITIES LABOR    MATERIALS LEASE
 1 15000.00 2400.00   3000.00 2000.00    900.00
 2 15000.00 2400.00   3000.00 2000.00    900.00
 3 15000.00 2400.00   3000.00 2000.00    900.00
 4 15000.00 2400.00   3000.00 2000.00    900.00
 5 15000.00 2400.00   3000.00 2000.00    900.00
 6 15000.00 2400.00   3000.00 2000.00    900.00
 7 15000.00 2400.00   3000.00 2000.00    900.00
 8 15000.00 2400.00   3000.00 2000.00    900.00
 9 15000.00 2400.00   3000.00 2000.00    900.00
10 15000.00 2400.00   3000.00 2000.00    900.00
Press any key to continue. . .


Effective annual escalation rate for revenues   : .08
Effecitve annual escalation rate for utilities  : .1
Effective annual escalation rate for labor  .   : .05
Effective annual escalation rate for materials  : .09
Life of the lease in years  . . . . . . . . .   : 5

Press any key to continue. . .
```

Listing of Revenue and Expense Data for Problem 3

TAX, INFLATION, AND INTEREST RATES

Investment tax credit percentage . . .	: .1
Incremental income tax rate	: .5
Tax rate on capital gain or loss . . .	: .28
General inflation rate	: .06
After-Tax, combined discount rate, MARR	: .12

Press any key to continue. . .

FIGURE F-17 *Listing of Remaining Data Inputs to Problem 3*

ACTUAL DOLLAR ANALYSIS

YR	UTILITY COST	LABOR COST	MATERIAL COST	LEASE COST	REVENUE	BTCF
0	0.00	0.00	0.00	0.00	-20000.00	-20000.00
1	-2640.00	-3150.00	-2180.00	-900.00	16200.00	7330.00
2	-2904.00	-3307.50	-2376.20	-900.00	17496.00	8008.30
3	-3194.40	-3472.88	-2590.06	-900.00	18895.68	8738.35
4	-3513.84	-3646.52	-2823.16	-900.00	20407.33	9523.81
5	-3865.22	-3828.84	-3077.25	-900.00	22039.92	10368.61
6	-4251.75	-4020.29	-3354.20	-1204.40	23803.12	10972.48
7	-4676.92	-4221.30	-3656.08	-1204.40	25707.36	11948.66
8	-5144.61	-4432.37	-3985.13	-1204.40	27763.95	12997.45
9	-5659.07	-4653.98	-4343.79	-1204.40	29985.07	14123.82
10	-6224.98	-4886.68	-4734.73	-1204.40	32383.88	15333.08

Press any key to continue. . .

FIGURE F-18 Actual Dollar Analysis (Before Taxes) for Problem 3

YR	BTCF	DEPR	TAXABLE INCOME	INCOME TAXES	ATCF	PRESENT WORTH
0	-20000.00	0.00	0.00	2000.00	-18000.00	-18000.00
1	7330.00	-1600.00	5730.00	-2865.00	4465.00	3986.61
2	8008.30	-1600.00	6408.30	-3204.15	4804.15	3829.84
3	8738.35	-1600.00	7138.35	-3569.17	5169.17	3679.32
4	9523.81	-1600.00	7923.81	-3961.91	5561.91	3534.69
5	10368.61	-1600.00	8768.61	-4384.30	5984.30	3395.65
6	10972.48	-1600.00	9372.48	-4686.24	6286.24	3184.80
7	11948.66	-1600.00	10348.66	-5174.33	6774.33	3064.36
8	12997.45	-1600.00	11397.45	-5698.72	7298.72	2947.83
9	14123.82	-1600.00	12523.82	-6261.91	7861.91	2835.08
10	15333.08	-1600.00	9733.08	-5746.54	9586.54	3086.61
					Total Present Worth:	15544.80

You Have a Winner
The Total Present Worth of the Investment is Positive.
Do you want to print this table [Y/N]? Y

FIGURE F-19 After-Tax Analysis of Problem 3 and Calculation of Present Worth

G

STANDARDIZED NORMAL DISTRIBUTION FUNCTION

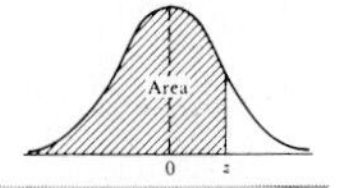

Areas Under the Normal Curve[a]

z	0.00	0.01	0.02	0.03	0.04	0.05	0.06	0.07	0.08	0.09
−3.4	0.0003	0.0003	0.0003	0.0003	0.0003	0.0003	0.0003	0.0003	0.0003	0.0002
−3.3	0.0005	0.0005	0.0005	0.0004	0.0004	0.0004	0.0004	0.0004	0.0004	0.0003
−3.2	0.0007	0.0007	0.0006	0.0006	0.0006	0.0006	0.0006	0.0005	0.0005	0.0005
−3.1	0.0010	0.0009	0.0009	0.0009	0.0008	0.0008	0.0008	0.0007	0.0007	0.0007
−3.0	0.0013	0.0013	0.0013	0.0012	0.0012	0.0011	0.0011	0.0011	0.0010	0.0010
−2.9	0.0019	0.0018	0.0017	0.0017	0.0016	0.0016	0.0015	0.0015	0.0014	0.0014
−2.8	0.0026	0.0025	0.0024	0.0023	0.0023	0.0022	0.0021	0.0021	0.0020	0.0019
−2.7	0.0035	0.0034	0.0033	0.0032	0.0031	0.0030	0.0029	0.0028	0.0027	0.0026
−2.6	0.0047	0.0045	0.0044	0.0043	0.0041	0.0040	0.0039	0.0038	0.0037	0.0036
−2.5	0.0062	0.0060	0.0059	0.0057	0.0055	0.0054	0.0052	0.0051	0.0049	0.0048
−2.4	0.0082	0.0080	0.0078	0.0075	0.0073	0.0071	0.0069	0.0068	0.0066	0.0064
−2.3	0.0107	0.0104	0.0103	0.0099	0.0096	0.0094	0.0091	0.0089	0.0087	0.0084
−2.2	0.0139	0.0136	0.0132	0.0129	0.0125	0.0122	0.0119	0.0116	0.0113	0.0110
−2.1	0.0179	0.0174	0.0170	0.0166	0.0162	0.0158	0.0154	0.0150	0.0146	0.0143
−2.0	0.0228	0.0222	0.0217	0.0212	0.0207	0.0202	0.0197	0.0192	0.0118	0.0183
−1.9	0.0287	0.0281	0.0274	0.0268	0.0262	0.0256	0.0250	0.0244	0.0239	0.0233
−1.8	0.0359	0.0352	0.0344	0.0336	0.0329	0.0322	0.0314	0.0307	0.0301	0.0294
−1.7	0.0446	0.0436	0.0427	0.0418	0.0409	0.0401	0.0392	0.0384	0.0375	0.0367
−1.6	0.0548	0.0537	0.0526	0.0516	0.0505	0.0495	0.0485	0.0475	0.0465	0.0455
−1.5	0.0668	0.0655	0.0643	0.0630	0.0618	0.0606	0.0594	0.0582	0.0571	0.0559
−1.4	0.0808	0.0793	0.0778	0.0764	0.0749	0.0735	0.0722	0.0708	0.0694	0.0681
−1.3	0.0968	0.0951	0.0934	0.0918	0.0901	0.0885	0.0869	0.0853	0.0838	0.0823
−1.2	0.1151	0.1131	0.1112	0.1093	0.1075	0.1056	0.1038	0.1020	0.1003	0.0985
−1.1	0.1357	0.1335	0.1314	0.1292	0.1271	0.1251	0.1230	0.1210	0.1190	0.1170
−1.0	0.1587	0.1562	0.1539	0.1515	0.1492	0.1469	0.1446	0.1423	0.1401	0.1379
−0.9	0.1841	0.1841	0.1788	0.1762	0.1736	0.1711	0.1685	0.1660	0.1635	0.1611
−0.8	0.2119	0.2090	0.2061	0.2033	0.2005	0.1977	0.1949	0.1922	0.1894	0.1867
−0.7	0.2420	0.2389	0.2358	0.2327	0.2296	0.2266	0.2236	0.2206	0.2177	0.2148
−0.6	0.2743	0.2709	0.2676	0.2643	0.2611	0.2578	0.2546	0.2514	0.2483	0.2451
−0.5	0.3085	0.3050	0.3015	0.2981	0.2946	0.2912	0.2877	0.2843	0.2810	0.2776
−0.4	0.3446	0.3409	0.3372	0.3336	0.3300	0.3264	0.3228	0.3192	0.3156	0.3121
−0.3	0.3821	0.3783	0.3745	0.3707	0.3669	0.3632	0.3594	0.3557	0.3520	0.3483
−0.2	0.4207	0.4168	0.4129	0.4090	0.4052	0.4013	0.3974	0.3936	0.3897	0.3859
−0.1	0.4602	0.4562	0.4522	0.4483	0.4443	0.4404	0.4364	0.4325	0.4286	0.4247
−0.0	0.5000	0.4960	0.4920	0.4880	0.4840	0.4801	0.4761	0.4721	0.4681	0.4641
0.0	0.5000	0.5040	0.5080	0.5120	0.5160	0.5199	0.5239	0.5279	0.5319	0.5359
0.1	0.5398	0.5438	0.5478	0.5517	0.5557	0.5596	0.5636	0.5675	0.5714	0.5753
0.2	0.5793	0.5832	0.5871	0.5910	0.5948	0.5987	0.6026	0.6064	0.6103	0.6141
0.3	0.6179	0.6217	0.6255	0.6293	0.6331	0.6368	0.6406	0.6443	0.6480	0.6517
0.4	0.6554	0.6591	0.6628	0.6664	0.6700	0.6736	0.6772	0.6808	0.6844	0.6879
0.5	0.6915	0.6950	0.6985	0.7019	0.7054	0.7088	0.7123	0.7157	0.7190	0.7224
0.6	0.7257	0.7291	0.7324	0.7357	0.7389	0.7422	0.7454	0.7486	0.7517	0.7549
0.7	0.7580	0.7611	0.7642	0.7673	0.7704	0.7734	0.7764	0.7794	0.7823	0.7852
0.8	0.7881	0.7910	0.7939	0.7967	0.7995	0.8023	0.8051	0.8078	0.8106	0.8133
0.9	0.8159	0.8186	0.8212	0.8238	0.8264	0.8289	0.8315	0.8340	0.8365	0.8389

Areas Under the Normal Curve[a] (Continued)

z	0.00	0.01	0.02	0.03	0.04	0.05	0.06	0.07	0.08	0.09
1.0	0.8413	0.8438	0.8461	0.8485	0.8508	0.8531	0.8554	0.8577	0.8599	0.8621
1.1	0.8643	0.8665	0.8686	0.8708	0.8729	0.8749	0.8770	0.8790	0.8810	0.8830
1.2	0.8849	0.8869	0.8888	0.8907	0.8925	0.8944	0.8962	0.8980	0.8997	0.9015
1.3	0.9032	0.9049	0.9066	0.9082	0.9099	0.9115	0.9131	0.9147	0.9162	0.9177
1.4	0.9192	0.9207	0.9222	0.9236	0.9251	0.9265	0.9278	0.9292	0.9306	0.9319
1.5	0.9332	0.9345	0.9357	0.9370	0.9382	0.9394	0.9406	0.9418	0.9429	0.9441
1.6	0.9452	0.9463	0.9474	0.9484	0.9495	0.9505	0.9515	0.9525	0.9535	0.9545
1.7	0.9554	0.9564	0.9573	0.9582	0.9591	0.9599	0.9608	0.9616	0.9625	0.9633
1.8	0.9641	0.9649	0.9656	0.9664	0.9671	0.9678	0.9686	0.9693	0.9699	0.9706
1.9	0.9713	0.9719	0.9726	0.9732	0.9738	0.9744	0.9750	0.9756	0.9761	0.9767
2.0	0.9772	0.9778	0.9783	0.9788	0.9793	0.9798	0.9803	0.9808	0.9812	0.9817
2.1	0.9821	0.9826	0.9830	0.9834	0.9838	0.9842	0.9846	0.9850	0.9854	0.9857
2.2	0.9861	0.9864	0.9868	0.9871	0.9875	0.9878	0.9881	0.9884	0.9887	0.9890
2.3	0.9893	0.9896	0.9898	0.9901	0.9904	0.9906	0.9909	0.9911	0.9913	0.9916
2.4	0.9918	0.9920	0.9922	0.9925	0.9927	0.9929	0.9931	0.9932	0.9934	0.9936
2.5	0.9938	0.9940	0.9941	0.9943	0.9945	0.9946	0.9948	0.9949	0.9951	0.9952
2.6	0.9953	0.9955	0.9956	0.9957	0.9959	0.9960	0.9961	0.9962	0.9963	0.9964
2.7	0.9965	0.9966	0.9967	0.9968	0.9969	0.9970	0.9971	0.9972	0.9973	0.9974
2.8	0.9974	0.9975	0.9976	0.9977	0.9977	0.9978	0.9979	0.9979	0.9980	0.9981
2.9	0.9981	0.9982	0.9982	0.9983	0.9984	0.9984	0.9985	0.9985	0.9986	0.9986
3.0	0.9987	0.9987	0.9987	0.9988	0.9988	0.9989	0.9989	0.9989	0.9990	0.9990
3.1	0.9990	0.9991	0.9991	0.9991	0.9992	0.9992	0.9992	0.9992	0.9993	0.9993
3.2	0.9993	0.9993	0.9994	0.9994	0.9994	0.9994	0.9994	0.9995	0.9995	0.9995
3.3	0.9995	0.9995	0.9995	0.9996	0.9996	0.9996	0.9996	0.9996	0.9996	0.9997
3.4	0.9997	0.9997	0.9997	0.9997	0.9997	0.9997	0.9997	0.9997	0.9997	0.9998

[a] From Ronald E. Walpole and Raymond H. Myers, *Probability and Statistics for Engineers and Scientists*, 2nd ed., New York: Macmillan, 1978, p. 513.

H

SELECTED REFERENCES

American Telephone and Telegraph Company, Engineering Department. *Engineering Economy*, 3rd ed. New York: American Telephone and Telegraph Co., 1977.

Arthur Andersen & Co. *Tax Reform 1986: Analysis and Planning*, Subject File AA3010, Item 27, 1986.

Au, T., and T. P. Au. *Engineering Economics for Capital Investment Analysis*. Boston: Allyn and Bacon, 1983.

Bandy, Dale D. et. al., eds. *1989 Federal Tax Course*. Prentice-Hall, Englewood Cliffs, NJ, 1988.

Barish, N. N., and S. Kaplan. *Economic Analysis for Engineering and Managerial Decision Making*. New York: McGraw-Hill Book Company, 1978.

Bierman, H., Jr., and S. Smidt. *The Capital Budgeting Decision*, 6th ed. New York: Macmillan, 1984.

Blank, L. T., and A. J. Tarquin. *Engineering Economy*, 2nd ed. New York: McGraw-Hill, 1983.

Bussey, Lynn E. *The Economic Analysis of Industrial Projects*. Englewood Cliffs, NJ: Prentice-Hall, 1978.

Campen, J. T. *Benefit, Cost, and Beyond*. Cambridge, MA: Ballinger Publishing Company, 1986.

Canada, J. R., and J. A. White. *Capital Investment Decision Analysis for Management and Engineering*. Englewood Cliffs, NJ: Prentice-Hall, 1980.

Canada, J. R., and W. G. Sullivan. *Economic and Multiattribute Analysis of Advanced Manufacturing Systems*. Englewood Cliffs, NJ: Prentice-Hall, 1989.

Chemical Engineering. Published biweekly by McGraw-Hill, New York.

Cochrane, J. L., and M. Zeleny. *Multiple Criteria Decision Making*. Columbia, SC: University of South Carolina, 1973.

Collier, C. A., and W. B. Ledbetter. *Engineering Cost Analysis*. 2nd ed. New York: Harper & Row, 1987.

Commerce Clearing House, Inc. *Explanation of Tax Reform Act of 1986*. 4025 W. Peterson Ave., Chicago, 1987.

de la Mare, R. F. *Manufacturing Systems Economics*. London: Cassell Educational, 1982.

Engineering Economist, The. A quarterly journal jointly published by the Engineering Economy Division of the American Society for Engineering Education and the Institute of Industrial Engineers. Published by IIE, Norcross, GA.

Engineering News-Record. Published monthly by McGraw-Hill, New York.

English, J. M., ed. *Cost Effectiveness: Economic Evaluation of Engineering Systems*. New York: John Wiley & Sons, 1968.

Fabrycky, W. J., and G. J. Thuesen. *Economic Decision Analysis*. 3rd ed. Englewood Cliffs, NJ: Prentice-Hall, 1984.

Fleischer, G. A. *Risk and Uncertainty: Non-deterministic Decision Making in Engineering Economy*, Norcross, GA: Institute of Industrial Engineers, Publication EE-75-1, 1975.

Fleischer, G. A. *Engineering Economy: Capital Allocation Theory*. Monterey, California: Brooks/Cole Engineering Division of Wadsworth, Inc., 1984.

Grant, E. L., W. G. Ireson, and R. S. Leavenworth. *Principles of Engineering Economy*, 7th ed. New York: John Wiley & Sons, 1982.

Goicoechea, A., D. R. Hansen, and L. Duckstein. *Multiobjective Decision Analysis with Engineering and Business Applications*. New York: John Wiley and Sons, 1982.

Happel, J., and D. Jordan. *Chemical Process Economics*, 2nd ed. New York: Marcel Dekker, 1975.

Harvard Business Review. Published bimonthly by the Harvard University Press, Boston.

Hull, J. C. *The Evaluation of Risk in Business Investment*. New York: Pergamon Press, 1980.

Industrial Engineering. A monthly magazine published by the Institute of Industrial Engineers, Norcross, GA.

Internal Revenue Service Publication 534. *Depreciation*. U.S. Government Printing Office, revised periodically (December 1987).

Jelen, F. C., and J. H. Black. *Cost and Optimization Engineering*, 2nd ed. New York: McGraw-Hill, 1983.

Jeynes, P. H. *Profitability and Economic Choice*. Ames, IA: Iowa State University Press, 1968.

Jones, B. W. *Inflation in Engineering Economic Analysis*. New York: John Wiley & Sons, 1982.

Kenney, R. L., and H. Raiffa. *Decisions with Multiple Objectives: Preferences and Value Tradeoffs*. New York: John Wiley & Sons, 1976.

Kleinfeld, Ira H. *Engineering and Managerial Economics*. New York: Holt, Rinehart & Winston, 1986.

Lasser, J. K. *Your Income Tax*. New York: Simon & Schuster (see latest edition).

Machinery and Allied Products Institute. *MAPI Replacement Manual*. Washington, DC: Machinery and Allied Products Institute, 1950.

Mallik, A. K. *Engineering Economy with Computer Applications*. Mahomet, IL: Engineering Technology, 1979.

Mao, James. *Quantitative Analysis of Financial Decisions*. New York: Macmillan, 1969.

Matthews, Lawrence M. *Estimating Manufacturing Costs: A Practical Guide for Managers and Estimators*. New York: McGraw-Hill, 1983.

Mayer, R. R. *Capital Expenditure Analysis for Managers and Engineers*. Prospect Heights, IL: Waveland Press, 1978.

Merrett, A. J., and A. Sykes. *The Finance and Analysis of Capital Projects*. New York: John Wiley & Sons, 1963.

Miller, J. R. *Professional Decision Making*. New York: Praeger, 1970.

Mishan, E. J. *Cost–Benefit Analysis*. New York: Praeger Publishers, 1976.

Morris, W. T. *Engineering Economic Analysis*. Reston, VA: Reston Publishing, 1976.

Morris, William T. *Decision Analysis*. Columbus, OH: Grid, 1977.

Newnan, Donald G. *Engineering Economic Analysis*, 3rd ed. San Jose, CA: Engineering Press, 1988.

Oakford, R. V. *Capital Budgeting: A Quantitative Evaluation of Investment Alternatives*. New York: John Wiley & Sons, 1970.

Ostwald, P. F. *Cost Estimating for Engineering and Management*, 2nd ed. Englewood Cliffs, NJ: Prentice-Hall, 1984.

Park, William R., and D. E. Jackson. *Cost Engineering Analysis: A Guide to Economic Evaluation of Engineering Projects*, 2nd ed. New York: John Wiley & Sons, 1984.

Peters, M. S., and K. D. Timmerhaus. *Plant Design and Economics for Chemical Engineers*, 2nd ed. New York: McGraw-Hill, 1968.

Porter, M. E. *Competitive Strategy: Techniques for Analyzing Industries and Competitors*. New York: The Free Press, 1980.

Reisman, A. *Managerial and Engineering Economics*. Boston: Allyn and Bacon, 1971.

Riggs, J. L., and T. M. West. *Engineering Economics*, 3rd ed. New York: McGraw-Hill, 1987.

Rose, L. M. *Engineering Investment Decisions: Planning Under Uncertainty*. Amsterdam: Elsevier, 1976.

Schlaifer, R. *Probability and Statistics for Business Decisions*. New York: McGraw-Hill, 1959.

Smith, G. W. *Engineering Economy: The Analysis of Capital Expenditures*, 4th ed. Ames, IA: The Iowa State University Press, 1987.

Stermole, F. J., and J. M. Stermole. *Economic Evaluation and Investment Decision Methods*, 6th ed. Golden, CO: Investment Evaluations Corp., 1987.

Stevens, G. T. *Economic and Financial Analysis of Capital Investment*. New York: John Wiley & Sons, 1979.

Stewart, R. D. *Cost Estimating*. New York: John Wiley & Sons, 1982.

Stewart, Rodney D., and Wyskida, Richard M., eds. *Cost Estimators' Reference Manual*. New York: John Wiley & Sons, 1987.

Sullivan, W. G., and W. W. Claycombe. *Fundamentals of Forecasting*. Reston, VA: Reston Publishing, 1977.

Taylor, G. A. *Managerial and Engineering Economy*, 3rd ed. New York: Van Nostrand Reinhold, 1980.

Terborgh, George. *Business Investment Management*. Washington, DC: Machinery and Allied Products Institute, 1967.

Thuesen, G. J., and W. J. Fabrycky. *Engineering Economy*, 7th ed. Englewood Cliffs, NJ: Prentice-Hall, 1989.

VanHorne, J. C. *Financial Management and Policy*, 5th ed. Englewood Cliffs, NJ: Prentice-Hall, 1980.

Weingartner, H. M. *Mathematical Programming and the Analysis of Capital Budgeting Problems*. Englewood Cliffs, NJ: Prentice-Hall, 1975.

Wellington, A. M. *The Economic Theory of Railway Location*. New York: John Wiley & Sons, 1887.

White, J. A., M. H. Agee, and K. E. Case. *Principles of Engineering Economic Analysis*, 2nd ed. New York: John Wiley & Sons, 1984.

Woods, D. R. *Financial Decision Making in the Process Industry*. Englewood Cliffs, NJ: Prentice-Hall, 1975.

Zeleny, M. *Multiple Criteria Decision Making*. New York: McGraw-Hill, 1982.

I

ANSWERS TO SELECTED EVEN-NUMBERED PROBLEMS

2-6. **a.** $X' = \$8{,}400{,}000$ per year. **b.** $X' = \$7{,}392{,}000$ per year (12%lower). **c.** $X' = \$8{,}400{,}000$ per year (no change).

2-8. C_u is a minimum at $12,000,000 per year.

2-10. Yes (Charter Service = $8,600; Commercial Airlines = $8,800).

2-12. Aluminum =$6.51 per piece; steel =$7.73 per piece

3-2. No, this financing plan does not involve simple interest.

3-4. Additional interest due to the compounding effect is $312.12.

3-6. $F = \$3{,}714$.

3-8. Total interest paid in this problem is $4,200; in problem 3-7 the total interest is $4,531.74.

3-10. The cash flow diagram is

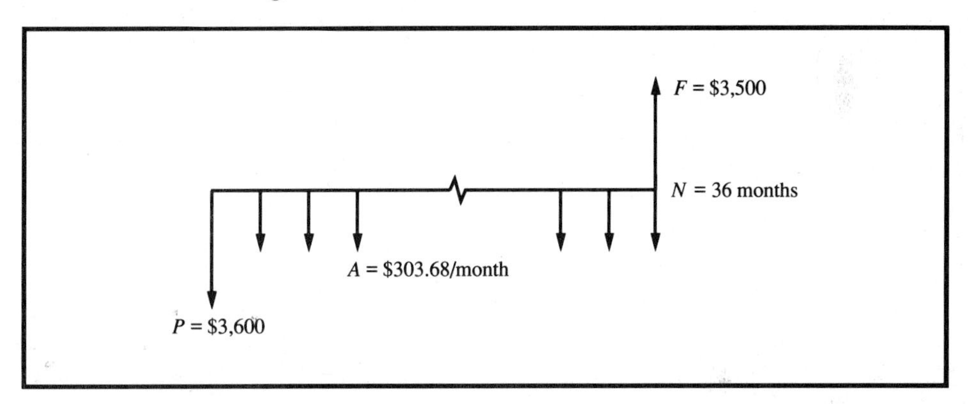

3-12. $P = \$17{,}943.60$.

3-14. $A = \$4{,}417$.

3-16. $F = \$14{,}490$. (choice E).

3-18. $N = 49$ years (to nearest year).

3-20. $A = \$681.59$ over a 50-year period.

3-22. $F = \$9{,}409.10$.

3-24. $A = \$361$.

3-26. $F = \$3{,}500$.

3-28. Repairs should be made now.

3-30. **a.** $P = \$1{,}288.75$. **b.** $A = \$264.71$. **c.** $F = \$2{,}511.37$.

3-32. $A = \$189.63$ (12 payments).

3-34. How much remains? $26,745.18

3-36. This is a good investment (P = \$7,521).

3-38. Install the insulation ($P > 0$).

3-40. G = \$3,944.00.

3-42. B = \$11,046.77.

3-44. P = \$4,684.65.

3-46. You can afford to spend as much as \$268,307.

3-48. **a.** 10.25% **b.** 10.38% **c.** 10.52% **d.** 10.51%

3-50. i = 14%.

3-52. The interest rate is about 26%.

3-54. N = 30 months.

3-56. 89 months (to nearest month)

3-58. **a.** \$7,302.40 **b.** \$7,505.60 **c.** \$7,464.10

3-60. **a.** \$169.60 = monthly payment (effective annual rate = 6.17%). **b.** i = 1.51% per quarter.

3-62. A = \$558.62.

3-64. F = \$8,651.60.

3-66. **a.** \$543.68 **b.** \$7,408.91 **c.** \$3,668.30 **d.** \$2,054.65

3-68. F = \$8,441.40.

3-70. A = \$1,117.27.

3-72. **a.** \$345,205 **b.** \$64,624

3-74. N = 5 years.

3-76. The difference in present equivalent worths is \$634.50.

4-2. PW = −\$171,453; do not invest in the new product line.

4-4. **a.** PW = −\$13,424. **b.** CR = \$1,828.

4-6. AW = −\$4,429.

4-8. He paid \$5,668 (PW on September 1).

4-10. ERR = 13.7%.

4-14. **a.** F = \$12,094. **b.** P = \$451.24.

4-16. Effective interest rate = 9.3% per year.

4-18. The effective interest rate per year is 41%.

4-20. Nominal interest rate is 8.28% per year (0.69% per month); effective interest is 8.6% per year.

4-22. P = \$25,058.

4-24. IRR = 21.4%.

4-26. a. FW = −\$1,128. **b.** Discounted payback period is greater than 5 years.

4-28. AW = \$35,648; establish the new company.

4-30. The company can afford to spend as much as \$10,993.

4-32. a. When demand is 4,000 units per year, the part should be manufactured. **b.** The required minimum demand to justify purchasing the equipment for in-house manufacture is 1,185 units per year.

4-34. a. Multiple interest rates do exist (at 0.5% and 28.8%) **b.** Cash inflow < cash outflow (by inspection); the project should not be pursued.

5-2. a. Select alternative 3. **b.** Analysis of the IRR of incremental cash flows shows that alternative 3 is again the choice.

5-4. a. Recommend alternative C. **b.** With the ERR method, alternative C is selected.

5-6. Alternative A is the choice.

5-8. a. Recommend the Big Mack truck. **b.** Select Wiltsbilt.

5-10. Choose process S.

5-12. Analysis of the IRR of incremental cash flows shows that equipment C should be chosen.

5-14. Select alternative N.

5-16. a. Choose the apartment house. **b.** The apartment house is again selected.

5-18. Correct application of the IRR method (incremental analysis) agrees with the PW method and leads to the selection of alternative III. The payback method is a measure of liquidity rather than profitability and should not be compared against IRR and PW results.

5-20. Select plan A.

5-22. Capitalized worth (CW) of Circle D is −\$24,084; CW of Qwik Sump is −\$22,324. There is a 15.6% difference between the PW and CW of the Circle D pump (using PW as the base).

5-24. a. There are 16 combinations, including "do nothing," to consider. **b.** All but 3 combinations are feasible.

5-26. Select crane B with the optional auger attachment.

5-28. Choose the combination consisting of A1 and C1.

5-30. Proposals A1, B1, and C1 should be approved. The remaining \$200,000 of investment capital will be spent elsewhere in the firm where the MARR can be earned.

6-18. See Table 6-7 for analogous derivation of 5-year MACRS rates.

6-20. a. Total investment (unadjusted basis) = \$17,200. **b.** \$16,200 − \$9,720 = \$6,480.

6-22. **a.** d_3 = \$3,428.57; BV_5 = \$42,857.14 (N = 14) **b.** d_3 = \$5,485.71; BV_5 = \$32,571.43 (N = 14) **c.** d_3 = \$6,297.38; BV_5 = \$27,759.86 (N = 14) **d.** d_3 = \$10,494; BV_5 = \$13,392

6-24. **a.** 5 years **b.** \$31,500 **c.** \$94,500

6-26. **a.** The difference is \$1,660 ($BV - MV$). **b.** $BV - MV$ = \$481.22. **c.** $BV - MV$ = −\$1,500.

6-28. The capital recovery accounts are \$9,600 (EOY 1), \$12,480 (EOY 2), \$7,488 (EOY 3), \$4,492.80 (EOY 4), \$4,078.08 (EOY 5), and \$1,935.36 (EOY 6). The equivalent annual capital recovery amount at i = 12% is \$8,322.

6-30. **a.** d_1 = \$24,000(0.2)(0.875) assuming a 10-year class life, d_2 = \$3,960, etc. **b.** d_1 = \$24,000(0.10) = \$2,400; d_2 = \$24,000(0.18) = \$4,320, etc., using rates from Table 6-5.

6-32. In year 1 the value of the depletion unit is \$0.40/Mcf; in year 2 the value of the depletion unit is \$0.37/Mcf.

6-34. **a.** \$1,500 **b.** \$4,000

7-2. **a.** \$18,850 **b.** \$71,150 **c.** \$130,000

7-4. 39.94%

7-6. **a.** Bob Brown had the greater after-tax income (\$25,468). **b.** 15%

7-8. **a.** She owes \$6,360. **b.** 11% interest (taxable)

7-10. After taxes, he broke even (IRR = 0%).

7-12. Consider "used"→"new" cash flow: The present worth at 12% on this increment is \$2,439. Thus, the new equipment should be recommended.

7-14. **a.** PW = −\$770; do not purchase. **b.** PW = −\$811.

7-16. The equipment must produce \$31,200 (before taxes).

7-18. ERR = 15.8% (assuming t = 0.39).

7-20. Annual operating expenses could increase by as much as \$1,774 (depreciation calculated over 5 years with one-half year convention).

7-22. **a.** 25% **b.** d_1 = \$12,861; d_2 = \$22,041; etc. **c.** \$10,000 **d.** ATCF_1 = \$14,144; etc. **e.** Reject the project (PW = −\$25,080).

7-24. Net savings would have to be approximately \$0.90 per pallet to justify the purchase.

7-26. **a.** IRR = 25%. **b.** ERR = 12.2%.

7-28. PW at 15% is \$899.

7-30. PW at 18% is −\$28,521.

7-32. **a.** ATCF_1 = \$6,700,000. **b.** PW(12%) = −\$2,143,660.

8-6. C_{90} = \$871,750.

8-8. C = \$232,400.

8-10. C =$1,000.

8-14. a. $\bar{I}_{B1}$ = 1.56 (156%); $\bar{I}_{B2}$ = 1.53 (153%). **b.** C_A = $3,351,600

9-2. 9 years.

9-4. PW(A) =$388,475; PW(B) =$369,080.

9-6. No (F_2 = $139,240).

9-8. C_a = $262,220.

9-10. PW = $15,403.

9-12. a. IRR = 16.06% **b.** e_j = 7.7%

9-14. e = 0.5%.

9-16. $31,370.

9-18. PW (Domestic) = −$15,371; PW(Import) = −$14,845.

9-20. a. $\bar{f}$ = 3.9% **b.** $\bar{f}$ = 3.69% **c.** $\bar{f}$ = 5.88%

9-22. PW = −$357,807.

9-24. Average decrease of 17.5% per year (in Yen).

10-6. The initial decision is to select alternative 1. The salvage value of alternative 2 would have to be $2,051 (an increase of $851) before this alternative is preferred.

10-8. The additional stories (floors) must be needed in 4 years or less to justify design B.

10-10. Present worth is most sensitive to changes in annual savings and least sensitive to changes in residual value.

10-12. a. AW = $23,330 (optimistic), AW = $14,325 (most likely), and AW = −$9,184 (pessimistic). **b.** There are 9 entries in the table. When both elements are at their optimistic level, for example, the AW is $20,495.

10-14. An interest rate of 15% will not reverse the initial recommendation to build a four-lane bridge now; when *i* exceeds 17%, the decision is to build a two-lane bridge now.

10-16. Except for below 500 (approx.) hours of annual operation, pump N is less expensive and would be selected based on the three estimates given.

10-18. Select alternative 1.

10-20. Assuming salvage values remain constant, the initial choice is not affected.

10-22. The expected value (mean) is $75; the variance is 577.8.

10-24. 0.587

10-26. Select alternative T.

10-28. Purchase the standby machine.

10-30. a. $444.36 **b.** 4.024×10^6 **c.** 0.7

10-32. **a.** 0.7 **b.** $841.90

10-34. In most simulations, the equipment should be recommended.

11-2. Noncompensatory models: advantage = quick and easy to use; disadvantage = may not lead to a single choice. Compensatory models: advantage = complex problems are broken into manageable pieces; disadvantage = use of a single measurement scale may not be credible.

11-4. The major difficulty lies in justifying the objective choice of some nonlinear scoring rationale.

11-6. **a.** No alternatives can be eliminated. **b.** Remove "retain existing system" from further consideration. **c.** No additional alternatives can be eliminated. **d.** Select vendor III.

11-8. **a.** Weight on "social climate" = 0.48. **b.** Nondimensional value of "starting salary" for Sigma Ltd. is 0.9. **c.** Sigma is confirmed as Mary's best choice.

11-10. Accept offer 3 with satisficing; accept offer 2 with lexicography; accept offer 3 with nondimensional scaling and additive weighting.

11-12. Lexicography selects contestant II; additive weighting recommends contestant I.

11-14. $W_2 = 0.5$; $V_2 = 0.86$ (normalized). Select alternative 1.

12-2. AW of present machine = −$833; it should be retained.

12-4. Select plan B (PW = −$138,007).

12-6. Choose to acquire the challenger (PW = −$25,335).

12-8. **a.** The new machine should be purchased immediately. **b.** Keep present machine 2 more years, then replace it with the new machine.

12-10. Annual worths are almost identical (−$10,530); nonmonetary considerations would probably favor selection of the challenger.

12-12. No, keep the defender for at least 1 more year.

12-14. **a.** 3 years **b.** 3 years.

12-16. Keep the motorcycle for 3 years.

12-18. Reinforce the existing bridge now and build a new concrete bridge in 5 years.

12-20. The maximum price is $79,790.

12-22. Recommend purchasing the challenger (PW = −$72,853).

13-8. Annual cost of interest and redemption is $605,000.

13-10. **a.** $8,744.56 **b.** $1,472,000

13-12. Yes, Yog should issue bonds and buy the company.

13-14. Annual cost of equity = $62,500.

13-16. Recommend plan I.

13-18. $K_a' = 8.92\%$.

13-20. Lease the equipment (PW $= -\$6{,}034$).

13-22. Recommend B, D, and A.

13-24. Select D, B, and A.

14-6. a. Projects B, C, and E are acceptable. **b.** B, C, and E **c.** Probably project B would be selected.

14-8. Select RS 511.

14-10. a. Choose option B. **b.** Again, B would be selected.

14-12. Improving the channel should be recommended.

INDEX